T0134402

Frontiers in Functional Equations and Analytic Inequalities

George A. Anastassiou • John Michael Rassias
Editors

Frontiers in Functional Equations and Analytic Inequalities

 Springer

Editors
George A. Anastassiou
Department of Mathematical Sciences
University of Memphis
Memphis, TN, USA

John Michael Rassias
National and Kapodistrian
University of Athens
Pedagogical Department E. E.,
Section of Mathematics and Informatics
Athens, Greece

ISBN 978-3-030-28952-2   ISBN 978-3-030-28950-8   (eBook)
https://doi.org/10.1007/978-3-030-28950-8

Mathematics Subject Classification: 26A33, 26D07, 26D10, 26D15, 26D20, 39Bxx, 30D05

This Springer imprint is published by the registered company Springer Nature Switzerland AG.
The registered company address is: Gewerbestrasse 11, 6330 Cham, Switzerland

*To all the Great Mathematicians from Euclid
(325 BC–265 BC) until Stanislaw Ulam
(1909 AD–1984 AD)*

Preface

This edited volume, *Frontiers in Functional Equations and Analytic Inequalities,* investigates functional equations and analytic inequalities in the broad sense and is written by authors who are in these mathematical fields for the past 50 years. It starts with fundamental notions of functional equations and analytic inequalities and several methods of solving functional equations and analytic inequalities in various spaces. It follows the progress of functional equations and analytic inequalities in the last five decades via excellent research chapters dedicated to the approximation and stability results of different types of functional equations and functional inequalities due to Abel, Cauchy, Euler, Lagrange, Jensen, Schröder, d'Alembert, Poisson, Drygas, Golab-Schinzel, etc., as well as very interesting new research results in analytic inequalities related to Cauchy-Schwarz, Hölder, Minkowski, Heisenberg, Landau, Ostrowski, Poincare, Sobolev, Hilbert, Hardy, Littlewood, and Polya inequalities. Clearly, error estimates are expressed in terms of inequalities. This book is a forum for exchanging ideas among eminent mathematicians from many parts of the world, as a tribute to the *Frontiers of Functional Equations and Analytic Inequalities.* It is intended to boost the cooperation among mathematicians working on a broad variety of pure and applied mathematical areas. This transatlantic collection of mathematical ideas and methods comprises a wide area of applications in which equations, inequalities, and computational techniques pertinent to their solutions play a core role, resulting in tremendous influence on our everyday life, because new tools have been developed and revolutionary research results have been achieved, bringing scientists of exact sciences even closer, by fostering the emergence of new approaches, techniques, and perspectives in functional equations, analytic inequalities, etc. Notice analytic inequalities is the backbone of pure and applied mathematics. Based on them, we find all kinds of important estimates, necessary to derive important conclusions in most of mathematics. Opial inequalities of all kinds including fractional are used to establish uniqueness of initial value and boundary problems in ordinary and partial differential equations. Also derive upper bounds of solutions. Ostrowski inequalities are employed in quadrature problems in connection to numerical analysis. Gruss inequalities are applied in probability in covariance problems. Also, Ostrowski inequalities help find estimates in expected value in probability and statistics. Furthermore, analytic inequalities have a great interest on their own merit, they present very surprising results, and thus they deserve continuous research. At AMS-MathSciNet, searching for "inequalities," we found over 30,676 articles and books!!!

This project belongs to *constructive* and *computational mathematics*: that is, the solution to the problem is given via computations and constructively. It is presented as a visually completely described theoretical object, i.e., you see it and you know it. It has nothing to do with existential mathematics where the solution is proved to exist, but you do not know it! All in all, this book makes a great pillar of modern mathematics based on classical and contemporary methods.

The functional equations part contains the following interesting topics:

- Hyperstability of a linear functional equation on restricted domains: F. Skof solved the Hyers-Ulam problem for additive mappings on a restricted domain. Also, S.M. Jung investigated the Hyers-Ulam stability for additive and quadratic mappings on restricted domains, and John M. Rassias investigated the Hyers-Ulam stability of mixed-type mappings on restricted domains. Hyers-Ulam's stability results to a three-point boundary value problem of nonlinear fractional-order differential equations are established. Furthermore in this part, different types of Ulam stability concepts for the aforesaid problem of nonlinear FDEs are introduced. Finally, the whole analysis is verified by some adequate examples.
- Topological degree theory and Ulam's stability analysis of a boundary value problem of fractional differential equations: Here, the existence and uniqueness of positive solution to a class of nonlinear fractional-order differential equations with boundary conditions are studied. By using fixed-point theorems on contraction mapping together with topological degree theory, some sufficient conditions in order to obtain the existence and uniqueness of positive solution for the considered problem are investigated.
- On a variant of μ-Wilson's functional equation with an endomorphism: The main goal of this topic is to find the solutions (f, g) of the generalized variant of μ-d'Alembert's functional equation and μ-Wilson's functional equation.
- On the additivity of maps preserving triple Jordan product $A^*B + \lambda B^*A$ on *-algebras.
- General solution and Hyers-Ulam stability of duotrigintic functional equation in multi-Banach spaces: In this part, the general form of a new duotrigintic functional equation is introduced. Then, the general solution and the generalized Hyers-Ulam stability of such functional equations in multi-Banach spaces are investigated.
- Stabilities of MIQD and MIQA functional equations via fixed-point technique: The stabilities of multiplicative inverse quadratic difference· and multiplicative inverse quadratic adjoint functional equations in the setting of non-Archimedean fields via the fixed-point method are investigated.
- Hyers-Ulam stability of first-order differential equation via integral inequality: In this part, the nonlinear integral inequality of Gollwitzer type is derived.

- Stability of n-dimensional functional equations in Banach spaces and fuzzy normed spaces: The general solution of a new generalized additive functional equation is investigated, and its generalized Hyers-Ulam stability in Banach spaces and stability in fuzzy normed spaces are discussed by using two different fundamental methods.
- Measure zero stability problem for the Drygas functional equation with complex involution: The Hyers-Ulam stability theorem for the famous σ-Drygas functional equation $f(x + y) + f(x + \sigma(y)) = 2f(x) + f(y) + f\sigma((y))$, where σ is an involution, is established.
- Fourier transforms and Ulam stabilities of linear differential equations: The purpose of this part is to study the Hyers-Ulam stability and the generalized Hyers-Ulam stability of the general linear differential equations of first-order and second-order with constant coefficients using the classical Fourier transform method.
- A class of functional equations of type d'Alembert on monoids: The solutions of the functional equation $f (x y) - f(\sigma(y)x) = g(x)h(y)$ are obtained, where σ is an involutive automorphism.
- Hyers-Ulam stability of a discrete diamond-alpha derivative equation: Here, the Hyers-Ulam stability (HUS) of a certain first-order linear constant coefficient discrete diamond-alpha derivative equation is established. In particular, for each parameter value, it is determined whether the equation has HUS, and if so, whether there exists a minimum HUS constant.
- Hyers-Ulam stability for a first-order linear proportional nabla difference operator: The Hyers-Ulam stability (HUS) of a certain first-order proportional nabla difference equation with a sign-alternating coefficient is established.
- Solution of generalized Jensen and quadratic functional equation.
- On some functional equations with applications in networks: Functional equations appear in many applications. They provide a powerful tool for narrowing the models used to describe many phenomena. On the other hand, a boundary value problem theory to investigate the solution of a special functional equation is discussed.
- Approximate solutions of an (AQQ) additive-quadratic-quartic functional equation: In this paper, the authors prove some stability and hyperstability results for an interesting new mixed type (AQQ), additive-quadratic-quartic functional equation.

The inequalities part of the book deals with:

- Quantitative complex Korovkin theory via inequalities
- Ostrowski-type inequalities involving sublinear integrals with applications to Choquet and Shilkret integrals
- Inequalities for special strong differential superordinations using a generalized Salagean operator and Ruscheweyh derivative
- Conformable fractional Landau, Hilbert-Pachpatte, Ostrowski, Opial, Poincare, and Sobolev inequalities
- Inequalities for h-quasiconvex functions

- Local fractional inequalities of Opial, Hilbert-Pachpatte, Ostrowski, comparison of means, Poincare, Sobolev, Landau, and Polya-Ostrowski
- Hermite-Hadamard-type integral inequalities for twice differentiable generalized convex mappings
- Hardy-type inequalities using conformable calculus
- Inequalities for symmetrized or anti-symmetrized inner products of complex-valued functions
- Generalized finite Hilbert transform and inequalities
- Hermite-Hadamard inequalities for composite convex functions
- Error estimates for approximate solutions of delay Volterra integral equations
- Harmonic and trace inequalities over Lipschitz domains
- Dirichlet beta function via generalized Mathieu series
- About Levinson inequality
- Integral norm inequalities for operators on differential forms
- Hadamard integral inequality for harmonically convex functions
- Norm inequalities for singular integrals related to operators and Dirac-Harmonic equations
- Inequalities for analytic functions induced by a fractional integral operator

This volume's results are expected to find applications in many areas of pure and applied mathematics, especially in ordinary and partial differential equations and fractional differential equations. As such, this book is suitable for researchers, graduate students, and related seminars, also to be in all science and engineering libraries.

The preparation of volume took place during 2018–2019 in Memphis, Tennessee, USA, and Athens, Greece.

Memphis, TN, USA   George A. Anastassiou
Athens, Greece   John Michael Rassias
May 1, 2019

Contents

**Norm Inequalities for Singular Integrals Related to Operators
and Dirac-Harmonic Equations** 713
Ravi P. Agarwal, Shusen Ding, and Yuming Xing

**Inequalities for Analytic Functions Defined by a Fractional Integral
Operator** 731
Alina Alb Lupaş

Part I
Introduction

Complex Korovkin Theory via Inequalities: A Quantitative Approach

George A. Anastassiou

Abstract Let K be a compact convex subset of $\mathbb{C}$ and $C(K, \mathbb{C})$ be the space of continuous functions from K into $\mathbb{C}$. We consider bounded linear operators from $C(K, \mathbb{C})$ into itself. We assume that these are bounded by companion real positive linear operators. We study quantitatively the rate of convergence of the approximation and high order approximation of these complex operators to the unit operators. Our results are inequalities of Korovkin type involving the complex modulus of continuity of the engaged function or its derivatives and basic test functions.

2010 Mathematics Subject Classification 41A17, 41A25, 41A36

1 Introduction

The study of the convergence of positive linear operators became more intensive and attractive when P. Korovkin (1953) proved his famous theorem (see [6], p. 14).

Korovkin's First Theorem Let $[a, b]$ be a compact interval in $\mathbb{R}$ and $(L_n)_{n \in \mathbb{N}}$ be a sequence of positive linear operators L_n mapping $C([a, b])$ into itself. Assume that $(L_n f)$ converges uniformly to f for the three test functions $f = 1, x, x^2$. Then $(L_n f)$ converges uniformly to f on $[a, b]$ for all functions of $f \in C([a, b])$.

So a lot of authors since then have worked on the theoretical aspects of the above convergence. But R. A. Mamedov (1959) (see [7]) was the first to put Korovkin's theorem in a quantitative scheme.

Mamedov's Theorem Let $\{L_n\}_{n \in \mathbb{N}}$ be a sequence of positive linear operators in the space $C([a, b])$, for which $L_n 1 = 1$, $L_n(t, x) = x + \alpha_n(x)$, $L_n\left(t^2, x\right) = x^2 + \beta_n(x)$. Then it holds

G. A. Anastassiou (✉)
Department of Mathematical Sciences, University of Memphis, Memphis, TN, USA
e-mail: ganastss@memphis.edu

© Springer Nature Switzerland AG 2019
G. A. Anastassiou and J. M. Rassias (eds.), *Frontiers in Functional Equations and Analytic Inequalities*, https://doi.org/10.1007/978-3-030-28950-8_1

$$\|L_n (f, x) - f (x)\|_\infty \leq 3\omega_1 \left(f, \sqrt{d_n} \right), \tag{1}$$

where ω_1 is the first modulus of continuity and $d_n = \left\| \beta_n (x) - 2x\alpha_n (x) \right\|_\infty$.

An improvement of the last result was the following.

Shisha and Mond's Theorem (1968, see [9]) Let $[a, b] \subset \mathbb{R}$ be a compact interval. Let $\{L_n\}_{n \in \mathbb{N}}$ be a sequence of positive linear operators acting on $C ([a, b])$. For $n = 1, 2, \ldots$, suppose $L_n (1)$ is bounded. Let $f \in C ([a, b])$. Then for $n = 1, 2, \ldots$, it holds

$$\|L_n f - f\|_\infty \leq \|f\|_\infty \cdot \|L_n 1 - 1\|_\infty + \|L_n (1) + 1\|_\infty \cdot \omega_1 \left(f, \mu_n \right), \tag{2}$$

where

$$\mu_n := \left\| \left(L_n \left((t - x)^2 \right) \right) (x) \right\|_\infty^{\frac{1}{2}}. \tag{3}$$

Shisha–Mond inequality generated and inspired a lot of research done by many authors worldwide on the rate of convergence of a sequence of positive linear operators to the unit operator, always producing similar inequalities, however, in many different directions.

The author (see [1]) in his 1993 research monograph produces in many directions best upper bounds for $|(L_n f) (x_0) - f (x_0)|$, $x_0 \in Q \subseteq \mathbb{R}^n$, $n \geq 1$, compact and convex, which lead for the first time to sharp/attained inequalities of Shisha–Mond type. The method of proving is probabilistic from the theory of moments. His pointwise approach is closely related to the study of the weak convergence with rates of a sequence of finite positive measures to the unit measure at a specific point.

The author in [3, pp. 383–412] continued this work in an abstract setting: Let X be a normed vector space, Y be a Banach lattice; $M \subset X$ is a compact and convex subset. Consider the space of continuous functions from M into Y, denoted by $C (M, Y)$; also consider the space of bounded functions $B (M, Y)$. He studied the rate of the uniform convergence of lattice homomorphisms $T : C (M, Y) \to C (M, Y)$ or $T : C (M, Y) \to B (M, Y)$ to the unit operator I. See also [2].

Also the author in [4, pp. 175–188] continued the last abstract work for bounded linear operators that are bounded by companion real positive linear operators. Here the involved functions are from $[a, b] \subset \mathbb{R}$ into $(X, \|\cdot\|)$ a Banach space.

All the above have inspired and motivated the work of this chapter. Our results are of Shisha–Mond type, i.e., of Korovkin type.

Namely here let K be a convex and compact subset of $\mathbb{C}$ and L be a linear operator from $C (K, \mathbb{C})$ into itself, and let $\tilde{L}$ be a positive linear operator from $C (K, \mathbb{R})$ into itself, such that $|L (f)| \leq \lambda \tilde{L} (|f|)$, $\forall f \in C (K, \mathbb{C})$, where $\lambda > 0$.

Clearly then L is a bounded linear operator. Here we create a complete quantitative Korovkin type theory over the last described setting.

2 Preparation and Motivation

We need

Theorem 1 ([5]) *Let $K \subseteq (\mathbb{C}, |\cdot|)$ and f a function from K into $\mathbb{C}$. Consider the first complex modulus of continuity*

$$\omega_1 (f, \delta) := \sup_{\substack{x,y \in K \\ |x-y| < \delta}} |f (x) - f (y)|, \ \delta > 0. \tag{4}$$

We have:

(1)′ If K is open convex or compact convex, then $\omega_1 (f, \delta) < \infty$, $\forall \, \delta > 0$, where $f \in UC (K, \mathbb{C})$ (uniformly continuous functions).

(2)′ If K is open convex or compact convex, then $\omega_1 (f, \delta)$ is continuous on $\mathbb{R}_+$ in δ, for $f \in UC (K, \mathbb{C})$.

(3)′ If K is convex, then

$$\omega_1 (f, t_1 + t_2) \le \omega_1 (f, t_1) + \omega_1 (f, t_2), \ t_1, t_2 > 0, \tag{5}$$

that is, the subadditivity property is true. Also it holds

$$\omega_1 (f, n\delta) \le n\omega_1 (f, \delta) \tag{6}$$

and

$$\omega_1 (f, \lambda\delta) \le \lceil \lambda \rceil \omega_1 (f, \delta) \le (\lambda + 1) \omega_1 (f, \delta), \tag{7}$$

where $n \in \mathbb{N}$, $\lambda > 0$, $\delta > 0$, $\lceil \cdot \rceil$ is the ceiling of the number.

(4)′ Clearly in general $\omega_1 (f, \delta) \ge 0$ and is increasing in $\delta > 0$ and $\omega_1 (f, 0) = 0$.

(5)′ If K is open or compact, then $\omega_1 (f, \delta) \to 0$ as $\delta \downarrow 0$, iff $f \in UC (K, \mathbb{C})$.

(6)′ It holds

$$\omega_1 (f + g, \delta) \le \omega_1 (f, \delta) + \omega_1 (g, \delta), \tag{8}$$

for $\delta > 0$, any $f, g : K \to \mathbb{C}$, $K \subset \mathbb{C}$ is arbitrary.

Next we give examples that motivate our main assumptions in this chapter.

Example 1 Let $K \subset \mathbb{C}$ be a compact and convex set, $l : C (K, \mathbb{C}) \to \mathbb{C}$ a linear functional, and $\widetilde{l} : C (K, \mathbb{R}) \to \mathbb{R}$ a positive linear functional. If $f \in C (K, \mathbb{C})$, then $|f| \in C (K, \mathbb{R})$. We want to see that

$$|l (f)| \le \lambda \widetilde{l} (|f|), \text{ where } \lambda > 0 \tag{9}$$

is possible.

Also, we want to see that

$$l(cg) = c\widetilde{l}(g), \ \forall \, g \in C(K, \mathbb{R}), \ \forall \, c \in \mathbb{C}, \tag{10}$$

is also possible.

So here is a concrete example of $l, \widetilde{l}$.

Take $K = [a_1, b_1] \times [a_2, b_2] \subset \mathbb{C}$ a rectangle. Here $z = x + iy \in \mathbb{C}$, and $f(z) = f_1(x, y) + i f_2(x, y)$. We have that $f \in C(K, \mathbb{C})$ iff $f_1, f_2 \in C(K, \mathbb{R})$. Define the following linear functional

$$l(f) := \int_{a_1}^{b_1} \int_{a_2}^{b_2} f_1(x, y) \, dxdy + i \int_{a_1}^{b_1} \int_{a_2}^{b_2} f_2(x, y) \, dxdy, \ \forall \, f \in C(K, \mathbb{C}). \tag{11}$$

This is a linear functional from $C(K, \mathbb{C}) \to \mathbb{C}$.

Let now $g \in C(K, \mathbb{R})$, then

$$l(g) = \int_{a_1}^{b_1} \int_{a_2}^{b_2} g(x, y) \, dxdy \in \mathbb{R}, \tag{12}$$

so that $\widetilde{l} := l|_{C(K, \mathbb{R})}$ is a positive linear functional from $C(K, \mathbb{R})$ into $\mathbb{R}$.

Let $c \in K$, then $c = a + ib$, hence $cg = (a + ib) g = ag + ibg$, thus

$$l(cg) = \int_{a_1}^{b_1} \int_{a_2}^{b_2} (ag(x, y)) \, dxdy + i \int_{a_1}^{b_1} \int_{a_2}^{b_2} (bg(x, y)) \, dxdy = \tag{13}$$

$$a \int_{a_1}^{b_1} \int_{a_2}^{b_2} g(x, y) \, dxdy + ib \int_{a_1}^{b_1} \int_{a_2}^{b_2} g(x, y) \, dxdy =$$

$$(a + ib) \int_{a_1}^{b_1} \int_{a_2}^{b_2} g(x, y) \, dxdy = c \int_{a_1}^{b_1} \int_{a_2}^{b_2} g(x, y) \, dxdy = c\widetilde{l}(g).$$

Thus

$$l(cg) = c\widetilde{l}(g) \tag{14}$$

is true, where

$$\widetilde{l}(g) := \int_{a_1}^{b_1} \int_{a_2}^{b_2} g(x, y) \, dxdy, \ \forall \, g \in C(K, \mathbb{R}).$$

Next, we notice that

$$|l(f)| \leq \int_{a_1}^{b_1} \int_{a_2}^{b_2} |f_1(x, y)| \, dx \, dy + \int_{a_1}^{b_1} \int_{a_2}^{b_2} |f_2(x, y)| \, dx \, dy =$$

$$\int_{a_1}^{b_1} \int_{a_2}^{b_2} (|f_1(x, y)| + |f_2(x, y)|) \, dx \, dy \leq \qquad (15)$$

$$\sqrt{2} \int_{a_1}^{b_1} \int_{a_2}^{b_2} \sqrt{(f_1(x, y))^2 + (f_2(x, y))^2} \, dx \, dy =$$

$$\sqrt{2} \int_{a_1}^{b_1} \int_{a_2}^{b_2} |f(z)| \, dx \, dy = \sqrt{2} \tilde{l}(|f|).$$

That is,

$$|l(f)| \leq \sqrt{2} \tilde{l}(|f|), \ \forall \ f \in C(K, \mathbb{C}) \qquad (16)$$

is valid.

Relations (14) and (16) motivate our major assumptions of our theory here.

We continue with a more general example.

Example 2 Let K be a compact and convex subset of $\mathbb{C}$, and $f \in C(K, \mathbb{C})$, which is $f(z) = u(x, y) + iv(x, y) = u + iv$, where $z = x + iy$, $z \in K$; $x, y \in \mathbb{R}$.

All linearities here are over the field of $\mathbb{R}$.

Consider $\tilde{L} : C(K, \mathbb{R}) \to C(K, \mathbb{R})$ a positive linear operator. And consider $L : C(K, \mathbb{C}) \to C(K, \mathbb{C})$ the linear operator such that:

$$L(f)(z) := \tilde{L}(u)(x, y) + i\tilde{L}(v)(x, y), \qquad (17)$$

indeed L is a linear operator.

Notice from $|u| \leq |u| \Leftrightarrow -|u| \leq u \leq |u| \Leftrightarrow -\tilde{L}(|u|) \leq \tilde{L}(u) \leq \tilde{L}(|u|) \Leftrightarrow |\tilde{L}(u)| \leq \tilde{L}(|u|)$.

Thus

$$|L(f)(z)| \leq |\tilde{L}(u)(x, y)| + |\tilde{L}(v)(x, y)| \leq$$

$$\tilde{L}(|u|)(x, y) + \tilde{L}(|v|)(x, y) = \tilde{L}(|u| + |v|)(x, y) \leq$$

$$\sqrt{2} \tilde{L}\left(\sqrt{u^2 + v^2}\right)(x, y) = \sqrt{2} \tilde{L}(|f(z)|) = \sqrt{2} \tilde{L}(|f|)(z). \qquad (18)$$

We have proved that

$$(1) \qquad |L(f)|(z) \le \sqrt{2}\widetilde{L}(|f|)(z), \ \forall z \in K. \qquad (19)$$

Next, let $g \in C(K, \mathbb{R})$, and $c \in \mathbb{C}$, i.e. $c = a + bi$; $a, b \in \mathbb{R}$. Then $cg = ag + ibg$. Clearly $L(cg) = \widetilde{L}(ag) + i\widetilde{L}(bg) = a\widetilde{L}(g) + ib\widetilde{L}(g) = c\widetilde{L}(g)$.

That is true

$$(2) \qquad L(cg) = c\widetilde{L}(g), \ \forall c \in \mathbb{C} \text{ and } \forall g \in C(K, \mathbb{R}). \qquad (20)$$

Properties (1) and (2), see (19), (20), justify our theory here. Notice that $f \in C(K, \mathbb{C})$, iff $u, v \in C(K, \mathbb{R})$.

Application 2 *Take* $K := [0, 1]^2$, $z \in K$ ($z = x + iy$), $x, y \in [0, 1]$. *Let* $g \in C([0, 1]^2, \mathbb{R})$, *then the two-dimensional Bernstein polynomials are*

$$B_{n_1, n_2}(g)(x, y) :=$$

$$\sum_{k_1=0}^{n_1} \sum_{k_2=0}^{n_2} g\left(\frac{k_1}{n_1}, \frac{k_2}{n_2}\right) \binom{n_1}{k_1} \binom{n_2}{k_2} x^{k_1} (1-x)^{n_1-k_1} y^{k_2} (1-y)^{n_2-k_2}, \qquad (21)$$

and they converge uniformly to g, for $n_1, n_2 \to \infty$.

Thus, for $f \in C([0, 1]^2, \mathbb{C})$, *we define*

$$B_{n_1, n_2}^{\mathbb{C}}(f)(z) := B_{n_1, n_2}(u)(x, y) + i B_{n_1, n_2}(v)(x, y), \qquad (22)$$

the complex Bernstein operators.

Indeed it is

$$B_{n_1, n_2}^{\mathbb{C}}(f)(z) =$$

$$\sum_{k_1=0}^{n_1} \sum_{k_2=0}^{n_2} u\left(\frac{k_1}{n_1}, \frac{k_2}{n_2}\right) \binom{n_1}{k_1} \binom{n_2}{k_2} x^{k_1} (1-x)^{n_1-k_1} y^{k_2} (1-y)^{n_2-k_2} +$$

$$i \sum_{k_1=0}^{n_1} \sum_{k_2=0}^{n_2} v\left(\frac{k_1}{n_1}, \frac{k_2}{n_2}\right) \binom{n_1}{k_1} \binom{n_2}{k_2} x^{k_1} (1-x)^{n_1-k_1} y^{k_2} (1-y)^{n_2-k_2} =$$

$$\sum_{k_1=0}^{n_1} \sum_{k_2=0}^{n_2} \left[u\left(\frac{k_1}{n_1}, \frac{k_2}{n_2}\right) + iv\left(\frac{k_1}{n_1}, \frac{k_2}{n_2}\right)\right] \binom{n_1}{k_1} \binom{n_2}{k_2} x^{k_1} (1-x)^{n_1-k_1} y^{k_2} (1-y)^{n_2-k_2},$$

$$(23)$$

a complex linear operator.

Notice that

$$\left| B_{n_1,n_2}^{\mathbb{C}} (f) (z) - f (z) \right| =$$

$$\left| \left(B_{n_1,n_2} (u) (x, y) - u (x, y) \right) + i \left(B_{n_1,n_2} (v) (x, y) - v (x, y) \right) \right| =$$

$$\sqrt{ \left(B_{n_1,n_2} (u) (x, y) - u (x, y) \right)^2 + \left(B_{n_1,n_2} (v) (x, y) - v (x, y) \right)^2 } =: (*). \quad (24)$$

We have that $\left| B_{n_1,n_2} (u) (x, y) - u (x, y) \right| < \varepsilon_1$, $\forall\, x, y \in [0, 1]^2$, $\forall\, n_1, n_2 \geq N_1$, *and* $\left| B_{n_1,n_2} (v) (x, y) - v (x, y) \right| < \varepsilon_2$, $\forall\, x, y \in [0, 1]^2$, $\forall\, n_1, n_2 \geq N_2$; $N_1, N_2 \in \mathbb{N}$, *where* $\varepsilon_1, \varepsilon_2 > 0$.

Thus, it holds

$$(*) \leq \sqrt{\varepsilon_1^2 + \varepsilon_2^2} =: \varepsilon, \quad (25)$$

$\forall\, x, y \in [0, 1]^2$, $\forall\, n_1, n_2 \geq \max (N_1, N_2) =: N^*$, $\varepsilon > 0$.

Hence

$$\left| B_{n_1,n_2}^{\mathbb{C}} (f) (z) - f (z) \right| \leq \varepsilon, \, \forall\, z \in [0, 1]^2, \, \forall\, n_1, n_2 \geq N^* \in \mathbb{N}, \textit{ where } \varepsilon > 0.$$

Therefore $B_{n_1,n_2}^{\mathbb{C}} (f) \to f$, *uniformly convergent, as* $n_1, n_2 \to \infty$.

3   Main Results

Let K be a compact and convex subset of $\mathbb{C}$. Consider $L : C (K, \mathbb{C}) \to C (K, \mathbb{C})$ a linear operator and $\widetilde{L} : C (K, \mathbb{R}) \to C (K, \mathbb{R})$ a positive linear operator (i.e. for $f_1 . f_2 \in C (K, \mathbb{R})$ with $f_1 \geq f_2$ we get $\widetilde{L} (f_1) \geq \widetilde{L} (f_2)$) both over the field of $\mathbb{R}$.

We assume that

$$|L (f)| \leq \lambda \widetilde{L} (|f|), \; \forall\, f \in C (K, \mathbb{C}), \text{ where } \lambda > 0, \quad (27)$$

(i.e. $|L (f) (z)| \leq \lambda \widetilde{L} (|f|) (z), \forall\, z \in K$).

We call $\widetilde{L}$ the companion operator of L.

Let $z_0 \in K$. Clearly, then $L (\cdot) (z_0)$ is a linear functional from $C (K, \mathbb{C})$ into $\mathbb{C}$, and $\widetilde{L} (\cdot) (z_0)$ is a positive linear functional from $C (K, \mathbb{R})$ into $\mathbb{R}$. Notice $L (f) (z) \in \mathbb{C}$ and $\widetilde{L} (|f|) (z) \in \mathbb{R}$, $\forall\, f \in C (K, \mathbb{C})$ (thus, $|f| \in C (K, \mathbb{R})$). Here $L (f) \in C (K, \mathbb{C})$, and $\widetilde{L} (|f|) \in C (K, \mathbb{R})$, $\forall\, f \in C (K, \mathbb{C})$.

Notice that $C (K, \mathbb{C}) = UC (K, \mathbb{C})$, also $C (K, \mathbb{R}) = UC (K, \mathbb{R})$ (uniformly continuous functions).

By [3, p. 388], we have that $\widetilde{L} \left(| \cdot - z_0 |^r \right) (z_0)$, $r > 0$, is a continuous function in $z_0 \in K$.

We have the following approximation result with rates of Korovkin type.

Theorem 3 *Here K is a convex and compact subset of $\mathbb{C}$ and L_n is a sequence of linear operators from $C(K, \mathbb{C})$ into itself, $n \in \mathbb{N}$. There is a sequence of companion positive linear operators $\widetilde{L}_n$ from $C(K, \mathbb{R})$ into itself, such that*

$$|L_n(f)| \leq \lambda \widetilde{L}_n(|f|), \ \lambda > 0, \ \forall \ f \in C(K, \mathbb{C}), \ \forall \ n \in \mathbb{N} \tag{28}$$

(i.e. $|L_n(f)(z_0)| \leq \lambda \left(\widetilde{L}_n(|f|)\right)(z_0)$, $\forall \ z_0 \in K$).
    Additionally, we assume that

$$L_n(cg) = c\widetilde{L}_n(g), \quad \forall \ g \in C(K, \mathbb{R}), \ \forall \ c \in \mathbb{C} \tag{29}$$

(i.e. $(L_n(cg))(z_0) = c\left(\widetilde{L}_n(g)\right)(z_0)$, $\forall \ z_0 \in K$).
    Then, for any $f \in C(K, \mathbb{C})$, we have

$$|(L_n(f))(z_0) - f(z_0)| \leq |f(z_0)| \left|\widetilde{L}_n(1(\cdot))(z_0) - 1\right| +$$

$$\lambda \left(\widetilde{L}_n(1(\cdot))(z_0) + 1\right) \omega_1\left(f, \widetilde{L}_n(|\cdot - z_0|)(z_0)\right), \tag{30}$$

$\forall \ z_0 \in K, \forall \ n \in \mathbb{N}$.
    If $\widetilde{L}_n(1(\cdot))(z_0) = 1$, $\forall \ z_0 \in K$, then

$$|(L_n(f))(z_0) - f(z_0)| \leq 2\lambda\omega_1\left(f, \widetilde{L}_n(|\cdot - z_0|)(z_0)\right), \tag{31}$$

$\forall \ z_0 \in K, \forall \ n \in \mathbb{N}$.
    If $\widetilde{L}_n(1(\cdot))(z_0) \to 1$, and $\widetilde{L}_n(|\cdot - z_0|)(z_0) \to 0$, as $n \to \infty$, then $L_n(f)(z_0) \to f(z_0)$, $\forall \ f \in C(K, \mathbb{C})$. Here $\widetilde{L}_n(1(\cdot))(z_0)$ is bounded.

Proof We notice that

$$|(L_n(f))(z_0) - f(z_0)| =$$

$$|(L_n(f))(z_0) - L_n(f(z_0)(\cdot))(z_0) + L_n(f(z_0)(\cdot))(z_0) - f(z_0)| \overset{(29)}{=}$$

$$\left|(L_n(f))(z_0) - L_n(f(z_0)(\cdot))(z_0) + f(z_0)\widetilde{L}_n(1(\cdot))(z_0) - f(z_0)\right| \leq$$

$$|(L_n(f))(z_0) - L_n(f(z_0)(\cdot))(z_0)| + |f(z_0)| \left|\widetilde{L}_n(1(\cdot))(z_0) - 1\right| = \tag{32}$$

$$|L_n(f(\cdot) - f(z_0))(z_0)| + |f(z_0)| \left|\widetilde{L}_n(1(\cdot))(z_0) - 1\right| \overset{(28)}{\leq}$$

$$|f(z_0)| \left|\widetilde{L}_n(1(\cdot))(z_0) - 1\right| + \lambda \left(\widetilde{L}_n(|f(\cdot) - f(z_0)|)\right)(z_0) \leq$$

$$|f(z_0)| \left|\widetilde{L}_n(1(\cdot))(z_0) - 1\right| + \lambda \left(\widetilde{L}_n\left(\omega_1\left(f, \frac{\delta|\cdot - z_0|}{\delta}\right)\right)\right)(z_0) \leq$$

$$\left| f\left(z_0\right)\right| \left|\widetilde{L}_n\left(1\left(\cdot\right)\right)\left(z_0\right) - 1\right| + \lambda \left(\widetilde{L}_n\left(\omega_1\left(f, \delta\right)\left(1\left(\cdot\right) + \frac{1}{\delta}\left|\cdot - z_0\right|\right)\right)\right)\left(z_0\right) =$$

$$\left| f\left(z_0\right)\right| \left|\widetilde{L}_n\left(1\left(\cdot\right)\right)\left(z_0\right) - 1\right| + \lambda\omega_1\left(f, \delta\right)\left[\widetilde{L}_n\left(1\left(\cdot\right)\right)\left(z_0\right) + \frac{1}{\delta}\widetilde{L}_n\left(\left|\cdot - z_0\right|\right)\left(z_0\right)\right] =$$
(33)

$$\left| f\left(z_0\right)\right| \left|\widetilde{L}_n\left(1\left(\cdot\right)\right)\left(z_0\right) - 1\right| + \lambda\omega_1\left(f, \widetilde{L}_n\left(\left|\cdot - z_0\right|\right)\left(z_0\right)\right)\left[\widetilde{L}_n\left(1\left(\cdot\right)\right)\left(z_0\right) + 1\right],$$
(34)

by choosing

$$\delta := \widetilde{L}_n\left(\left|\cdot - z_0\right|\right)\left(z_0\right),$$
(35)

if $\widetilde{L}_n\left(\left|\cdot - z_0\right|\right)\left(z_0\right) > 0$.

Next we consider the case of

$$\widetilde{L}_n\left(\left|\cdot - z_0\right|\right)\left(z_0\right) = 0.$$
(36)

By Riesz representation theorem ([8, p. 304]) there exists a positive finite measure μ_{z_0} such that

$$\widetilde{L}_n\left(g\right)\left(z_0\right) = \int_K g\left(t\right) d\mu_{z_0}\left(t\right), \quad \forall\, g \in C\left(K, \mathbb{R}\right).$$
(37)

That is,

$$\int_K \left|t - z_0\right| d\mu_{z_0}\left(t\right) = 0,$$

which implies $\left|t - z_0\right| = 0$, a.e., hence $t - z_0 = 0$, a.e., and $t = z_0$, a.e. on K. Consequently $\mu_{z_0}\left(\{t \in K : t \neq z_0\}\right) = 0$.

That is, $\mu_{z_0} = \delta_{z_0} M$ (where $0 < M := \mu_{z_0}\left(K\right) = \widetilde{L}_n\left(1\left(\cdot\right)\right)\left(z_0\right)$). Hence, in that case $\widetilde{L}_n\left(g\right)\left(z_0\right) = g\left(z_0\right) M$. Consequently, it holds $\omega_1\left(f, \widetilde{L}_n\left(\left|\cdot - z_0\right|\right)\left(z_0\right)\right) = 0$, and the right-hand side of (30) equals $\left|f\left(z_0\right)\right|\left|M - 1\right|$.

Also, it is $\widetilde{L}_n\left(\left|f\left(\cdot\right) - f\left(z_0\right)\left(\cdot\right)\right|\right)\left(z_0\right) = \left|f\left(z_0\right) - f\left(z_0\right)\right| M = 0$.

And by (28) we obtain

$$\left|\left(L_n\left(f\left(\cdot\right) - f\left(z_0\right)\left(\cdot\right)\right)\right)\left(z_0\right)\right| = 0,$$

that is,

$$\left|L_n\left(f\right)\left(z_0\right) - L_n\left(f\left(z_0\right)\left(\cdot\right)\right)\left(z_0\right)\right| = 0.$$

The last says that

$$L_n\left(f\right)\left(z_0\right) = L_n\left(f\left(z_0\right)\left(\cdot\right)\right)\left(z_0\right) \overset{(29)}{=} f\left(z_0\right)\widetilde{L}_n\left(1\left(\cdot\right)\right)\left(z_0\right) = Mf\left(z_0\right).$$

Consequently the left-hand side of (30) becomes

$$\left|L_n\left(f\right)\left(z_0\right) - f\left(z_0\right)\right| = \left|Mf\left(z_0\right) - f\left(z_0\right)\right| = \left|f\left(z_0\right)\right|\left|M - 1\right|.$$

So that (30) becomes an equality, and both sides equal $\left|f\left(z_0\right)\right|\left|M - 1\right|$ in the extreme case of $\widetilde{L}_n\left(\left|\cdot - z_0\right|\right)\left(z_0\right) = 0$. Thus, inequality (30) is proved completely in both cases.

A similar result follows:

Theorem 4 *Here all as in Theorem 3. Then, for any $f \in C\left(K, \mathbb{C}\right)$, we have*

$$\left|\left(L_n\left(f\right)\right)\left(z_0\right) - f\left(z_0\right)\right| \le \left|f\left(z_0\right)\right|\left|\widetilde{L}_n\left(1\left(\cdot\right)\right)\left(z_0\right) - 1\right| + \tag{38}$$

$$\lambda\left(\widetilde{L}_n\left(1\left(\cdot\right)\right)\left(z_0\right) + 1\right)\omega_1\left(f, \left(\widetilde{L}_n\left(\left|\cdot - z_0\right|^2\right)\left(z_0\right)\right)^{\frac{1}{2}}\right),$$

$\forall\, z_0 \in K, \forall\, n \in \mathbb{N}$.
   If $\widetilde{L}_n\left(1\left(\cdot\right)\right)\left(z_0\right) = 1, \forall\, z_0 \in K$, then

$$\left|\left(L_n\left(f\right)\right)\left(z_0\right) - f\left(z_0\right)\right| \le 2\lambda\omega_1\left(f, \widetilde{L}_n\left(\left|\cdot - z_0\right|^2\right)\left(z_0\right)\right)^{\frac{1}{2}}, \tag{39}$$

$\forall\, z_0 \in K, \forall\, n \in \mathbb{N}$.

Remark 1 (To Theorem 4) If $\widetilde{L}_n\left(1\left(\cdot\right)\right)\left(z_0\right) \to 1$, and $\widetilde{L}_n\left(\left|\cdot - z_0\right|^2\right)\left(z_0\right) \to 0$, as $n \to \infty$, we get that $L_n\left(f\right)\left(z_0\right) \to f\left(z_0\right), \forall\, f \in C\left(K, \mathbb{C}\right)$. Here $\widetilde{L}_n\left(1\left(\cdot\right)\right)\left(z_0\right)$ is bounded.
   For $t, z_0 \in K$ with $t = t_1 + it_2$ and $z_0 = z_{01} + iz_{02}$ we have

$$\widetilde{L}_n\left(\left|t - z_0\right|^2\right)\left(z_0\right) = \widetilde{L}_n\left(\left(t_1 - z_{01}\right)^2 + \left(t_2 - z_{02}\right)^2\right)\left(z_0\right) = \tag{40}$$

$$\widetilde{L}_n\left(\left(t_1 - z_{01}\right)^2\right)\left(z_0\right) + \widetilde{L}_n\left(\left(t_2 - z_{02}\right)^2\right)\left(z_0\right).$$

so if $\widetilde{L}_n\left(\left(t_1 - z_{01}\right)^2\right)\left(z_0\right)$ and $\widetilde{L}_n\left(\left(t_2 - z_{02}\right)^2\right)\left(z_0\right)$ converge to zero, as $n \to \infty$, we get that $\widetilde{L}_n\left(\left|t - z_0\right|^2\right)\left(z_0\right) \to 0$.
   We also notice that

$$\widetilde{L}_n\left(\left|t - z_0\right|^2\right)\left(z_0\right) = \left(\widetilde{L}_n\left(t_1^2\right)\left(z_0\right) - z_{01}^2\right) + \left(\widetilde{L}_n\left(t_2^2\right)\left(z_2\right) - z_{02}^2\right) + \tag{41}$$

$$\left|z\right|^2\left(\widetilde{L}_n\left(1\left(\cdot\right)\right)\left(z_0\right) - 1\right) - 2z_{01}\left(\widetilde{L}_n\left(t_1\right)\left(z_0\right) - z_{01}\right) - 2z_{02}\left(\widetilde{L}_n\left(t_2\right)\left(z_0\right) - z_{02}\right).$$

Thus, if $\widetilde{L}_n\left(1\left(\cdot\right)\right)\left(z_0\right) \rightarrow 1$, $\widetilde{L}_n\left(t_1\right)\left(z_0\right) \rightarrow z_{01}$, $\widetilde{L}_n\left(t_2\right)\left(z_0\right) \rightarrow z_{02}$, $\widetilde{L}_n\left(t_1^2\right)\left(z_0\right) \rightarrow z_{01}^2$ and $\widetilde{L}_n\left(t_2^2\right)\left(z_0\right) \rightarrow z_{02}^2$, as $n \rightarrow \infty$, then we get that $L_n\left(f\right)\left(z_0\right) \rightarrow f\left(z_0\right)$, $\forall\, f \in C\left(K, \mathbb{C}\right)$.

Proof of Theorem 4 Let $t, z_0 \in K$ and $\delta > 0$. If $|t - z_0| > \delta$, then

$$|f\left(t\right) - f\left(z_0\right)| \le \omega_1\left(f, |t - z_0|\right) = \omega_1\left(f, |t - z_0|\,\delta^{-1}\delta\right) \le \qquad (42)$$

$$\left(1 + \frac{|t - z_0|}{\delta}\right)\omega_1\left(f, \delta\right) \le \left(1 + \frac{|t - z_0|^2}{\delta^2}\right)\omega_1\left(f, \delta\right).$$

The estimate

$$|f\left(t\right) - f\left(z_0\right)| \le \left(1 + \frac{|t - z_0|^2}{\delta^2}\right)\omega_1\left(f, \delta\right) \qquad (43)$$

also holds trivially when $|t - z_0| \le \delta$.

So (43) is true always, $\forall\, t \in K$, for any $z_0 \in K$.

As in the proof of Theorem we have

$$|\left(L_n\left(f\right)\right)\left(z_0\right) - f\left(z_0\right)| \le \ldots \le |f\left(z_0\right)|\left|\widetilde{L}_n\left(1\left(\cdot\right)\right)\left(z_0\right) - 1\right|$$

$$+\lambda\left(\widetilde{L}_n\left(|f\left(\cdot\right) - f\left(z_0\right)|\right)\right)\left(z_0\right) \overset{(43)}{\le} |f\left(z_0\right)|\left|\widetilde{L}_n\left(1\left(\cdot\right)\right)\left(z_0\right) - 1\right| +$$

$$\lambda\left(\widetilde{L}_n\left(\left(1\left(\cdot\right) + \frac{|\cdot - z_0|^2}{\delta^2}\right)\omega_1\left(f, \delta\right)\right)\right)\left(z_0\right) =$$

$$|f\left(z_0\right)|\left|\widetilde{L}_n\left(1\left(\cdot\right)\right)\left(z_0\right) - 1\right| + \lambda\omega_1\left(f, \delta\right)\left[\widetilde{L}_n\left(1\left(\cdot\right)\right)\left(z_0\right) + \frac{1}{\delta^2}\widetilde{L}_n\left(|\cdot - z_0|^2\right)\left(z_0\right)\right] =$$

$$|f\left(z_0\right)|\left|\widetilde{L}_n\left(1\left(\cdot\right)\right)\left(z_0\right) - 1\right| + \lambda\omega_1\left(f, \left(\widetilde{L}_n\left(|\cdot - z_0|^2\right)\left(z_0\right)\right)^{\frac{1}{2}}\right)\left[\widetilde{L}_n\left(1\left(\cdot\right)\right)\left(z_0\right) + 1\right],$$

$$\qquad (44)$$

by choosing

$$\delta := \left(\widetilde{L}_n\left(|\cdot - z_0|^2\right)\left(z_0\right)\right)^{\frac{1}{2}}, \qquad (45)$$

if $\widetilde{L}_n\left(|\cdot - z_0|^2\right)\left(z_0\right) > 0$.

Next we consider the case of

$$\widetilde{L}_n\left(|\cdot - z_0|^2\right)\left(z_0\right) = 0.$$

By Riesz representation theorem there exists a positive finite measure μ_{z_0} such that

$$\tilde{L}_n (g) (z_0) = \int_K g(t) \, d\mu_{z_0}(t), \quad \forall \, g \in C(K, \mathbb{R}). \tag{46}$$

That is,

$$\int_K |t - z_0|^2 \, d\mu_{z_0}(t) = 0,$$

which implies $|t - z_0|^2 = 0$, a.e., hence $|t - z_0| = 0$, a.e., thus $t - z_0 = 0$, a.e., and $t = z_0$, a.e. on K. Consequently $\mu_{z_0} (\{t \in K : t \neq z_0\}) = 0$.

That is, $\mu_{z_0} = \delta_{z_0} M$ (where $0 < M := \mu_{z_0}(K) = \tilde{L}_n (1 (\cdot)) (z_0)$). Hence, in that case $\tilde{L}_n (g) (z_0) = g(z_0) M$.

Consequently, it holds $\omega_1 \left(f, \left(\tilde{L}_n \left(|\cdot - z_0|^2 \right) (z_0) \right)^{\frac{1}{2}} \right) = 0$, and the right-hand side of (38) equals $|f(z_0)| \, |M - 1|$. Also, it is $\tilde{L}_n (|f(\cdot) - f(z_0) (\cdot)|) (z_0) = |f(z_0) - f(z_0)| M = 0$.

And by (28) we obtain

$$|(L_n (f(\cdot) - f(z_0) (\cdot))) (z_0)| = 0,$$

that is,

$$|L_n (f) (z_0) - L_n (f(z_0) (\cdot)) (z_0)| = 0.$$

The last says that

$$L_n (f) (z_0) = L_n (f(z_0) (\cdot)) (z_0) \stackrel{(29)}{=} f(z_0) \tilde{L}_n (1 (\cdot)) (z_0) = Mf(z_0).$$

Consequently the left-hand side of (38) becomes

$$|L_n (f) (z_0) - f(z_0)| = |Mf(z_0) - f(z_0)| = |f(z_0)| \, |M - 1|.$$

So that (38) becomes an equality, and both sides equal $|f(z_0)| \, |M - 1|$ in the extreme case of $\tilde{L}_n \left(|\cdot - z_0|^2 \right) (z_0) = 0$. Thus, inequality (38) is proved completely in both cases.

We give

Corollary 1 *All as in Theorem 3, $z_0 \in K$. Then*

$$\|L_n (f) - f\|_\infty \leq \|f\|_\infty \left\| \tilde{L}_n (1 (\cdot)) - 1 \right\|_\infty + \tag{47}$$

$$\lambda \left\| \tilde{L}_n (1 (\cdot)) + 1 \right\|_\infty \omega_1 \left(f, \left\| \tilde{L}_n (|\cdot - z_0|) (z_0) \right\|_{\infty, z_0} \right),$$

$\forall\, n \in \mathbb{N}.$

If $\widetilde{L}_n\,(1\,(\cdot)) = 1$, *then*

$$\|L_n\,(f) - f\|_\infty \leq 2\lambda\omega_1\left(f, \left\|\widetilde{L}_n\,(|\cdot - z_0|)\,(z_0)\right\|_{\infty, z_0}\right), \tag{48}$$

$\forall\, n \in \mathbb{N}.$

As $\widetilde{L}_n\,(1) \overset{u}{\rightarrow} 1$, *and* $\widetilde{L}_n\,(|\cdot - z_0|)\,(z_0) \overset{u}{\rightarrow} 0$ *(u is uniformly), as $n \to \infty$, then* $L_n\,(f) \overset{u}{\rightarrow} f,\, \forall\, f \in C\,(K, \mathbb{C})$. *Notice* $\widetilde{L}_n\,(1)$ *is bounded, and all suprema in (47) are finite.*

Corollary 2 *All as in Theorem 4, $z_0 \in K$. Then*

$$\|L_n\,(f) - f\|_\infty \leq \|f\|_\infty \left\|\widetilde{L}_n\,(1\,(\cdot)) - 1\right\|_\infty + \tag{49}$$

$$\lambda\left\|\widetilde{L}_n\,(1\,(\cdot)) + 1\right\|_\infty \omega_1\left(f, \left\|\widetilde{L}_n\left(|\cdot - z_0|^2\right)(z_0)\right\|_{\infty, z_0}^{\frac{1}{2}}\right),$$

$\forall\, n \in \mathbb{N}.$

If $\widetilde{L}_n\,(1\,(\cdot)) = 1$, *then*

$$\|L_n\,(f) - f\|_\infty \leq 2\lambda\omega_1\left(f, \left\|\widetilde{L}_n\left(|\cdot - z_0|^2\right)(z_0)\right\|_{\infty, z_0}^{\frac{1}{2}}\right), \tag{50}$$

$\forall\, n \in \mathbb{N}.$

As $\widetilde{L}_n\,(1) \overset{u}{\rightarrow} 1$, *and* $\widetilde{L}_n\left(|\cdot - z_0|^2\right)(z_0) \overset{u}{\rightarrow} 0$, *then* $L_n\,(f) \overset{u}{\rightarrow} f$, *as $n \to \infty$, $\forall$* $f \in C\,(K, \mathbb{C})$.

We need

Theorem 5 ([5]) *Let $K \subseteq \mathbb{C}$ convex, $x_0 \in K^0$ (interior of K) and $f : K \to \mathbb{C}$ such that $|f\,(t) - f\,(x_0)|$ is convex in $t \in K$. Furthermore let $\delta > 0$ so that the closed disk $D\,(x_0, \delta) \subset K$. Then*

$$|f\,(t) - f\,(x_0)| \leq \frac{\omega_1\,(f, \delta)}{\delta}\,|t - x_0|,\ \forall\, t \in K. \tag{51}$$

We present a convex Korovkin type result:

Theorem 6 *Here all as in Theorem 3. Let a fixed $z_0 \in K^0$ and assume that $|f\,(t) - f\,(z_0)|$ is convex in $t \in K$. Assume the closed disk $D\left(z_0, \widetilde{L}_n\,(|\cdot - z_0|)\,(z_0)\right) \subset K$. Then*

$$|(L_n\,(f))\,(z_0) - f\,(z_0)| \leq |f\,(z_0)|\left|\widetilde{L}_n\,(1\,(\cdot))\,(z_0) - 1\right|$$

$$+ \lambda\omega_1\left(f, \widetilde{L}_n\,(|\cdot - z_0|)\,(z_0)\right),\ \forall\, n \in \mathbb{N}. \tag{52}$$

As $\widetilde{L}_n (1 (\cdot)) (z_0) \to 1$, and $\widetilde{L}_n (|\cdot - z_0|) (z_0) \to 0$, then $(L_n (f)) (z_0) \to f (z_0)$, as $n \to \infty$.

Proof As in the proof of Theorem 3 we have

$$|(L_n (f)) (z_0) - f (z_0)| \leq$$

$$|f (z_0)| \left| \widetilde{L}_n (1 (\cdot)) (z_0) - 1 \right| + \lambda \left(\widetilde{L}_n (|f (\cdot) - f (z_0)|) \right) (z_0) \overset{(51)}{\leq}$$

$(\delta > 0 : D (z_0, \delta) \subset K)$

$$|f (z_0)| \left| \widetilde{L}_n (1 (\cdot)) (z_0) - 1 \right| + \frac{\lambda \omega_1 (f, \delta)}{\delta} \widetilde{L}_n (|\cdot - z_0|) (z_0) =$$

$$|f (z_0)| \left| \widetilde{L}_n (1 (\cdot)) (z_0) - 1 \right| + \lambda \omega_1 \left(f, \widetilde{L}_n (|\cdot - z_0|) (z_0) \right), \tag{53}$$

by choosing

$$\delta := \widetilde{L}_n (|\cdot - z_0|) (z_0),$$

if $\widetilde{L}_n (|\cdot - z_0|) (z_0) > 0$.

The case $\widetilde{L}_n (|\cdot - z_0|) (z_0) = 0$ is treated similarly as in the proof of Theorem 3. The theorem is proved.

We make

Remark 2 Let $f : D \subseteq \mathbb{C} \to \mathbb{C}$ be an analytic function on the convex domain D and $y, x \in D$, then we have the following Taylor's expansion with integral remainder

$$f (y) = \sum_{k=0}^{N-1} \frac{f^{(k)} (x)}{k!} (y - x)^k +$$

$$\frac{1}{(N - 1)!} (y - x)^N \int_0^1 f^{(N)} [(1 - s) x + sy] (1 - s)^{N-1} ds, \tag{54}$$

for $N \in \mathbb{N}$, see [10, p. 8].

Clearly then

$$f (y) = \sum_{k=0}^{N} \frac{f^{(k)} (x)}{k!} (y - x)^k +$$

$$\frac{1}{(N - 1)!} (y - x)^N \int_0^1 \left[f^{(N)} [(1 - s) x + sy] - f^{(N)} (x) \right] (1 - s)^{N-1} ds, \tag{55}$$

for $N \in \mathbb{N}$.

Call the remainder of (55) as

$$R_N(x, y) := \frac{(y-x)^N}{(N-1)!} \int_0^1 \left[f^{(N)} [(1-s)x + sy] - f^{(N)}(x) \right] (1-s)^{N-1} ds.$$
(56)

We have that

$$|R_N(x, y)| \le \frac{|y-x|^N}{(N-1)!} \int_0^1 \left| f^{(N)} [(1-s)x + sy] - f^{(N)}(x) \right| (1-s)^{N-1} ds =: (*),$$
(57)

$N \in \mathbb{N}$.

Next assume $f^{(N)} \in UC(D, \mathbb{C})$.

We observe that

$$(*) \le \frac{|y-x|^N}{(N-1)!} \int_0^1 \omega_1 \left(f^{(N)}, \frac{\delta s |y-x|}{\delta} \right) (1-s)^{N-1} ds \le$$

$$\frac{|y-x|^N}{(N-1)!} \omega_1 \left(f^{(N)}, \delta \right) \int_0^1 \left[1 + \frac{s|y-x|}{\delta} \right] (1-s)^{N-1} ds =$$

$$\frac{|y-x|^N}{(N-1)!} \omega_1 \left(f^{(N)}, \delta \right) \left[\int_0^1 (1-s)^{N-1} ds + \frac{|y-x|}{\delta} \int_0^1 (1-s)^{N-1} (s-0)^{2-1} ds \right] =$$

$$\frac{|y-x|^N}{(N-1)!} \omega_1 \left(f^{(N)}, \delta \right) \left[\frac{1}{N} + \frac{|y-x|}{\delta} \frac{1}{N(N+1)} \right] =$$

$$\frac{|y-x|^N}{N!} \omega_1 \left(f^{(N)}, \delta \right) \left[1 + \frac{|y-x|}{\delta(N+1)} \right].$$
(59)

We have proved

$$|R_N(x, y)| \le \frac{|y-x|^N}{N!} \omega_1 \left(f^{(N)}, \delta \right) \left[1 + \frac{|y-x|}{\delta(N+1)} \right],$$
(60)

$N \in \mathbb{N}, \delta > 0$.

The last means that

$$\left| f(y) - \sum_{k=0}^N \frac{f^{(k)}(x)}{k!} (y-x)^k \right| \le$$

$$\omega_1 \left(f^{(N)}, \delta \right) \frac{|y-x|^N}{N!} \left[1 + \frac{|y-x|}{\delta(N+1)} \right],$$
(61)

$N \in \mathbb{N}, \delta > 0, \forall x, y \in D$, where $f^{(N)} \in UC(D, \mathbb{C})$.

We make

Remark 3 Let $f : K \subseteq \mathbb{C} \to \mathbb{C}$ be an analytic function on the convex and compact set K, and $z_0 \in K$, where $\delta > 0$.

Then, as in (61), we get

$$\left| f(\cdot) - \sum_{k=0}^{N} \frac{f^{(k)}(z_0)}{k!} (\cdot - z_0)^k \right| \leq$$

$$\omega_1 \left(f^{(N)}, \delta \right) \frac{|\cdot - z_0|^N}{N!} \left[1 + \frac{|\cdot - z_0|}{\delta(N+1)} \right], \tag{62}$$

$\forall N \in \mathbb{N}$. Here ω_1 is on K.

Above we mean that $f : D \subseteq \mathbb{C} \to \mathbb{C}$ is analytic on the convex domain D, where $K \subseteq D$. For convenience we set and use $f = f|_K$.

We have proved

Theorem 7 *Let $f : K \subseteq \mathbb{C} \to \mathbb{C}$ be an analytic function on the convex and compact set K; $z_0 \in K$, $\delta > 0$, and $f^{(k)}(z_0) = 0$, $k = 1, 2, \ldots, N$. Then*

$$|f(\cdot) - f(z_0)| \leq \frac{\omega_1 \left(f^{(N)}, \delta \right)}{N!} \left[|\cdot - z_0|^N + \frac{|\cdot - z_0|^{N+1}}{\delta(N+1)} \right], \tag{63}$$

over K, $N \in \mathbb{N}$.

We present higher order of approximation:

Theorem 8 *Here K is a convex and compact subset of $\mathbb{C}$ and L_n is a sequence of linear operators from $C(K, \mathbb{C})$ into itself, $n \in \mathbb{N}$. There is a sequence of companion positive linear operators $\widetilde{L}_n$ from $C(K, \mathbb{R})$ into itself, such that*

$$|L_n(f)| \leq \lambda \widetilde{L}_n(|f|), \ \lambda > 0, \forall f \in C(K, \mathbb{C}), \ \forall n \in \mathbb{N}. \tag{64}$$

Additionally, we assume that

$$L_n(cg) = c\widetilde{L}_n(g), \ \forall g \in C(K, \mathbb{R}), \ \forall c \in \mathbb{C}. \tag{65}$$

Here we consider $f : K \to \mathbb{C}$ that are analytic, so that $f^{(k)}(z_0) = 0$, $k = 1, 2, \ldots, N$, where $z_0 \in K$.

Then

$$|(L_n(f))(z_0) - f(z_0)| \leq |f(z_0)| \left|\tilde{L}_n(1(\cdot))(z_0) - 1\right| +$$

$$\frac{\lambda\omega_1\left(f^{(N)}, \left((\tilde{L}_n(|\cdot - z_0|^{N+1}))(z_0)\right)^{\frac{1}{(N+1)}}\right)}{N!}\left(\left(\tilde{L}_n(|\cdot - z_0|^{N+1})\right)(z_0)\right)^{\left(\frac{N}{N+1}\right)}$$

$$\left[\left(\tilde{L}_n(1(\cdot))(z_0)\right)^{\frac{1}{(N+1)}} + \frac{1}{(N+1)}\right], \tag{66}$$

$\forall\, n \in \mathbb{N}$.

If $\tilde{L}_n(1(\cdot))(z_0) = 1$, *then*

$$|(L_n(f))(z_0) - f(z_0)| \leq \frac{\lambda(N+2)\,\omega_1\left(f^{(N)}, \left((\tilde{L}_n(|\cdot - z_0|^{N+1}))(z_0)\right)^{\frac{1}{(N+1)}}\right)}{(N+1)!}$$

$$\left(\left(\tilde{L}_n(|\cdot - z_0|^{N+1})\right)(z_0)\right)^{\left(\frac{N}{N+1}\right)}, \tag{67}$$

$\forall\, n \in \mathbb{N}$.

If $\tilde{L}_n(1(\cdot))(z_0) \to 1$ *and* $\tilde{L}_n(|\cdot - z_0|^{N+1})(z_0) \to 0$, *then* $(L_n(f))(z_0) \to f(z_0)$, *as* $n \to \infty$. *Here* $\tilde{L}_n(1(\cdot))(z_0)$ *is bounded.*

Proof We notice that

$$|(L_n(f))(z_0) - f(z_0)| =$$

$$|(L_n(f))(z_0) - L_n(f(z_0)(\cdot))(z_0) + L_n(f(z_0)(\cdot))(z_0) - f(z_0)| \overset{(65)}{=}$$

$$\left|(L_n(f))(z_0) - L_n(f(z_0)(\cdot))(z_0) + f(z_0)\tilde{L}_n(1(\cdot))(z_0) - f(z_0)\right| \leq$$

$$|(L_n(f))(z_0) - L_n(f(z_0)(\cdot))(z_0)| + |f(z_0)|\left|\tilde{L}_n(1(\cdot))(z_0) - 1\right| = \tag{68}$$

$$|L_n(f(\cdot) - f(z_0))(z_0)| + |f(z_0)|\left|\tilde{L}_n(1(\cdot))(z_0) - 1\right| \overset{(64)}{\leq}$$

$$|f(z_0)|\left|\tilde{L}_n(1(\cdot))(z_0) - 1\right| + \lambda\left(\tilde{L}_n(|f(\cdot) - f(z_0)|)\right)(z_0) \overset{(63)}{\leq}$$

$$|f(z_0)|\left|\tilde{L}_n(1(\cdot))(z_0) - 1\right| +$$

$$\lambda\frac{\omega_1\left(f^{(N)}, \delta\right)}{N!}\left[\tilde{L}_n\left(|\cdot - z_0|^N\right)(z_0) + \frac{1}{\delta(N+1)}\tilde{L}_n\left(|\cdot - z_0|^{N+1}\right)(z_0)\right] =: (*).$$

By Hölder's inequality and Riesz representation theorem we obtain

$$\tilde{L}_n \left(|\cdot - z_0|^N\right)(z_0) \le \left(\left(\tilde{L}_n \left(|\cdot - z_0|^{N+1}\right)\right)(z_0)\right)^{\left(\frac{N}{N+1}\right)} \left(\tilde{L}_n (1 (\cdot)) (z_0)\right)^{\frac{1}{N+1}}.$$
(70)

Therefore

$$(*) \le |f(z_0)| \left|\tilde{L}_n (1(\cdot))(z_0) - 1\right| +$$

$$\lambda \frac{\omega_1 \left(f^{(N)}, \delta\right)}{N!} \left[\left(\tilde{L}_n \left(|\cdot - z_0|^{N+1}\right)(z_0)\right)^{\left(\frac{N}{N+1}\right)} \left(\tilde{L}_n (1(\cdot))(z_0)\right)^{\frac{1}{N+1}}\right.$$

$$\left. + \frac{1}{\delta(N+1)} \tilde{L}_n \left(|\cdot - z_0|^{N+1}\right)(z_0)\right] =: (\xi).$$
(71)

We choose

$$\delta := \left(\left(\tilde{L}_n \left(|\cdot - z_0|^{N+1}\right)\right)(z_0)\right)^{\frac{1}{(N+1)}},$$
(72)

in case of $\tilde{L}_n \left(|\cdot - z_0|^{N+1}\right)(z_0) > 0$.

Then it holds

$$(\xi) = |f(z_0)| \left|\tilde{L}_n (1(\cdot))(z_0) - 1\right| +$$

$$\frac{\lambda \omega_1 \left(f^{(N)}, \left(\left(\tilde{L}_n \left(|\cdot - z_0|^{N+1}\right)\right)(z_0)\right)^{\frac{1}{(N+1)}}\right)}{N!}$$

$$\left[\delta^N \left(\tilde{L}_n (1(\cdot))(z_0)\right)^{\frac{1}{N+1}} + \frac{\delta^N}{(N+1)}\right] =$$

$$|f(z_0)| \left|\tilde{L}_n (1(\cdot))(z_0) - 1\right| +$$

$$\frac{\lambda \omega_1 \left(f^{(N)}, \left(\left(\tilde{L}_n \left(|\cdot - z_0|^{N+1}\right)\right)(z_0)\right)^{\frac{1}{(N+1)}}\right)}{N!} \left(\left(\tilde{L}_n \left(|\cdot - z_0|^{N+1}\right)\right)(z_0)\right)^{\left(\frac{N}{N+1}\right)}$$

$$\left[\left(\tilde{L}_n (1(\cdot))(z_0)\right)^{\frac{1}{(N+1)}} + \frac{1}{(N+1)}\right].$$
(73)

Next we treat the case of

$$\tilde{L}_n \left(|\cdot - z_0|^{N+1}\right)(z_0) = 0.$$

By Riesz representation theorem there exists a positive finite measure μ_{z_0} such that

$$\widetilde{L}_n(g)(z_0) = \int_K g(t)\, d\mu_{z_0}(t), \quad \forall\, g \in C(K, \mathbb{R}). \tag{74}$$

That is,

$$\int_K |t - z_0|^{N+1}\, d\mu_{z_0}(t) = 0,$$

which implies $|t - z_0|^{N+1} = 0$, a.e., hence $|t - z_0| = 0$, a.e., thus $t - z_0 = 0$, a.e., and $t = z_0$, a.e. on K. Consequently $\mu_{z_0}(\{t \in K : t \neq z_0\}) = 0$.

That is, $\mu_{z_0} = \delta_{z_0} M$ (where $0 < M := \mu_{z_0}(K) = \widetilde{L}_n(1(\cdot))(z_0)$). Hence, in that case $\widetilde{L}_n(g)(z_0) = g(z_0) M$.

Consequently, it holds $\omega_1\left(f^{(N)}, \left(\widetilde{L}_n\left(|\cdot - z_0|^{N+1}\right)(z_0)\right)^{\frac{1}{(N+1)}}\right) = 0$, and the right-hand side of (66) equals $|f(z_0)|\,|M - 1|$. Also, it is $\widetilde{L}_n(|f(\cdot) - f(z_0)(\cdot)|)(z_0) = |f(z_0) - f(z_0)|\, M = 0$.

And by (64) we obtain

$$|(L_n(f(\cdot) - f(z_0)(\cdot)))(z_0)| = 0,$$

that is,

$$|L_n(f)(z_0) - L_n(f(z_0)(\cdot))(z_0)| = 0.$$

The last says that

$$L_n(f)(z_0) = L_n(f(z_0)(\cdot))(z_0) \overset{(65)}{=} f(z_0)\,\widetilde{L}_n(1(\cdot))(z_0) = M f(z_0).$$

Consequently the left-hand side of (66) becomes

$$|L_n(f)(z_0) - f(z_0)| = |M f(z_0) - f(z_0)| = |f(z_0)|\,|M - 1|.$$

So that (66) becomes an equality, and both sides equal $|f(z_0)|\,|M - 1|$ in the extreme case of $\widetilde{L}_n\left(|\cdot - z_0|^{N+1}\right)(z_0) = 0$. Thus, inequality (66) is proved completely in both cases.

We give

Corollary 3 *All as in Theorem 8. Here* $N = 1$, *i.e.* $f'(z_0) = 0$. *Then*

$$|(L_n(f))(z_0) - f(z_0)| \leq |f(z_0)|\,\left|\widetilde{L}_n(1(\cdot))(z_0) - 1\right| +$$

$$\lambda\omega_1\left(f', \left(\left(\widetilde{L}_n\left(|\cdot - z_0|^2\right)\right)(z_0)\right)^{\frac{1}{2}}\right)\left(\left(\widetilde{L}_n\left(|\cdot - z_0|^2\right)\right)(z_0)\right)^{\frac{1}{2}}$$

$$\left[\left(\widetilde{L}_n \left(1 \left(\cdot \right) \right) \left(z_0 \right) \right)^{\frac{1}{2}} + \frac{1}{2} \right], \quad \forall\, n \in \mathbb{N}. \tag{75}$$

If $\widetilde{L}_n \left(1 \left(\cdot \right) \right) \left(z_0 \right) = 1$, *then*

$$\left| \left(L_n \left(f \right) \right) \left(z_0 \right) - f \left(z_0 \right) \right| \leq \frac{3\lambda\omega_1 \left(f', \left(\left(\widetilde{L}_n \left(\left| \cdot - z_0 \right|^2 \right) \right) \left(z_0 \right) \right)^{\frac{1}{2}} \right)}{2}$$

$$\left(\left(\widetilde{L}_n \left(\left| \cdot - z_0 \right|^2 \right) \right) \left(z_0 \right) \right)^{\frac{1}{2}}, \quad \forall\, n \in \mathbb{N}. \tag{76}$$

If $\widetilde{L}_n \left(1 \left(\cdot \right) \right) \left(z_0 \right) \rightarrow 1$ *and* $\widetilde{L}_n \left(\left| \cdot - z_0 \right|^2 \right) \left(z_0 \right) \rightarrow 0$, *as* $n \rightarrow \infty$, *we get that* $\left(L_n \left(f \right) \right) \left(z_0 \right) \rightarrow f \left(z_0 \right)$.

We make

Remark 4 Let $f : D \subseteq \mathbb{C} \rightarrow \mathbb{C}$ be an analytic function on the convex domain D and K be a compact and convex subset of D and $t, z_0 \in K$, with $z_0 \in K^0$ (interior of K), then we have the following modified Taylor's expansion with integral remainder:

$$f \left(t \right) = \sum_{k=0}^{N} \frac{f^{(k)} \left(z_0 \right)}{k!} \left(t - z_0 \right)^k +$$

$$\frac{\left(t - z_0 \right)^N}{\left(N - 1 \right)!} \int_0^1 \left[f^{(N)} \left[\left(1 - s \right) z_0 + st \right] - f^{(N)} \left(z_0 \right) \right] \left(1 - s \right)^{N-1} ds, \tag{77}$$

for $N \in \mathbb{N}$.

Assuming $f^{(k)} \left(z_0 \right) = 0, k = 1, \ldots, N$, we get

$$f \left(t \right) - f \left(z_0 \right) = \frac{\left(t - z_0 \right)^N}{\left(N - 1 \right)!} \int_0^1 \left[f^{(N)} \left[\left(1 - s \right) z_0 + st \right] - f^{(N)} \left(z_0 \right) \right] \left(1 - s \right)^{N-1} ds, \tag{78}$$

$N \in \mathbb{N}$.

We have that

$$\left| f \left(t \right) - f \left(z_0 \right) \right| \leq$$

$$\frac{\left| t - z_0 \right|^N}{\left(N - 1 \right)!} \int_0^1 \left| f^{(N)} \left[\left(1 - s \right) z_0 + st \right] - f^{(N)} \left(z_0 \right) \right| \left(1 - s \right)^{N-1} ds =: \left(* \right). \tag{79}$$

We assume that $\left| f^{(N)} \left(t \right) - f^{(N)} \left(z_0 \right) \right|$ is convex in $t \in K$. Let $\delta > 0$ such that the closed disk $D \left(z_0, \delta \right) \subset K$. Then, by Theorem 5, we obtain that

$$\left|f^{(N)}(t) - f^{(N)}(z_0)\right| \le \frac{\omega_1\left(f^{(N)}, \delta\right)}{\delta} |t - z_0|, \quad \forall\, t \in K. \tag{80}$$

Notice that by convexity of K, $(1-s) z_0 + st \in K, 0 \le s \le 1$. Therefore

$$(*) \overset{(80)}{\le} \frac{|t - z_0|^N}{(N-1)!} \frac{\omega_1\left(f^{(N)}, \delta\right)}{\delta} \int_0^1 s\, |t - z_0| (1 - s)^{N-1}\, ds =$$

$$\frac{|t - z_0|^{N+1}}{(N-1)!} \frac{\omega_1\left(f^{(N)}, \delta\right)}{\delta} \int_0^1 (1 - s)^{N-1} (s - 0)^{2-1}\, ds =$$

$$\frac{|t - z_0|^{N+1}}{(N-1)!} \frac{\omega_1\left(f^{(N)}, \delta\right)}{\delta} \frac{1}{N(N+1)} = \frac{|t - z_0|^{N+1}}{(N+1)!} \frac{\omega_1\left(f^{(N)}, \delta\right)}{\delta}. \tag{81}$$

We have proved that

$$|f(t) - f(z_0)| \le \frac{\omega_1\left(f^{(N)}, \delta\right)}{\delta\,(N+1)!} |t - z_0|^{N+1}, \tag{82}$$

$\forall\, t \in K, N \in \mathbb{N}.$

We have proved

Theorem 9 *Let K be a compact and convex subset of the convex domain $D \subseteq \mathbb{C}$, $z_0 \in K^0$. Here $f : K \to \mathbb{C}$ is analytic such that $f^{(k)}(z_0) = 0$, $k = 1, 2, \ldots, N \in \mathbb{N}$. We assume that $\left|f^{(N)}(\cdot) - f^{(N)}(z_0)\right|$ is convex over K. Let $\delta > 0$ such that the closed disk $D(z_0, \delta) \subset K$. Then*

$$|f(\cdot) - f(z_0)| \le \frac{\omega_1\left(f^{(N)}, \delta\right)}{\delta\,(N+1)!} |\cdot - z_0|^{N+1}, \tag{83}$$

over K.

The convex analog of Theorem 8 follows:

Theorem 10 *All as in Theorem 8. Additionally we assume that: $z_0 \in K^0$, $\left|f^{(N)}(\cdot) - f^{(N)}(z_0)\right|$ is convex over K, and that the closed disk $D\left(z_0, \widetilde{L}_n\left(|\cdot - z_0|^{N+1}\right)(z_0)\right) \subset K$. Then*

$$|(L_n(f))(z_0) - f(z_0)| \le |f(z_0)|\left|\widetilde{L}_n(1(\cdot))(z_0) - 1\right| +$$

$$\frac{\lambda\omega_1\left(f^{(N)}, \widetilde{L}_n\left(|\cdot - z_0|^{N+1}\right)(z_0)\right)}{(N+1)!}, \tag{84}$$

$\forall\, n \in \mathbb{N}.$

If $\widetilde{L}_n (1 (\cdot)) (z_0) \to 1$ and $\widetilde{L}_n (|\cdot - z_0|^{N+1}) (z_0) \to 0$, as $n \to \infty$, then $L_n (f) (z_0) \to f (z_0)$.

Proof As in the proof of Theorem 8 we have

$$|(L_n (f)) (z_0) - f (z_0)| \leq \dots \leq$$

$$|f (z_0)| |\widetilde{L}_n (1 (\cdot)) (z_0) - 1| + \lambda \left(\widetilde{L}_n (|f (\cdot) - f (z_0)|) \right) (z_0) \overset{(83)}{\leq} \tag{83}$$

$$|f (z_0)| |\widetilde{L}_n (1 (\cdot)) (z_0) - 1| +$$

$$\lambda \frac{\omega_1 \left(f^{(N)}, \delta \right)}{\delta (N + 1)!} \widetilde{L}_n \left(|\cdot - z_0|^{N+1} \right) (z_0) = \tag{85}$$

$$|f (z_0)| |\widetilde{L}_n (1 (\cdot)) (z_0) - 1| + \frac{\lambda}{(N + 1)!} \omega_1 \left(f^{(N)}, \widetilde{L}_n \left(|\cdot - z_0|^{N+1} \right) (z_0) \right),$$

by choosing

$$\delta := \widetilde{L}_n \left(|\cdot - z_0|^{N+1} \right) (z_0),$$

if $\widetilde{L}_n \left(|\cdot - z_0|^{N+1} \right) (z_0) > 0$.

If $\widetilde{L}_n \left(|\cdot - z_0|^{N+1} \right) (z_0) = 0$, then this case is treated similarly to the proof of Theorem 8.

We give

Corollary 4 (To Theorem 10, Case of $N = 1$) *All as in Theorem 10. Assume* $|f' (\cdot) - f' (z_0)|$ *is convex over K, and the closed disk $D \left(z_0, \widetilde{L}_n (|\cdot - z_0|^2) (z_0) \right) \subset K$. Then*

$$|(L_n (f)) (z_0) - f (z_0)| \leq |f (z_0)| |\widetilde{L}_n (1 (\cdot)) (z_0) - 1| +$$

$$\frac{\lambda \omega_1 \left(f', \widetilde{L}_n (|\cdot - z_0|^2) (z_0) \right)}{2}, \tag{86}$$

$\forall n \in \mathbb{N}$.

If $\widetilde{L}_n (1 (\cdot)) (z_0) \to 1$ and $\widetilde{L}_n (|\cdot - z_0|^2) (z_0) \to 0$, as $n \to \infty$, then $L_n (f) (z_0) \to f (z_0)$.

4 Illustration

Here we go according to Example 2 and Application 2. We will study the quantitative uniform convergence of complex Bernstein operators $B^{\mathbb{C}}_{n_1,n_2}(f)$ to $f \in C\left([0, 1]^2, \mathbb{C}\right)$. Indeed we have

$$\left|B^{\mathbb{C}}_{n_1,n_2}(f)\right|(z) \leq \sqrt{2} B_{n_1,n_2}(|f|)(z), \quad \forall z \in [0, 1]^2 \text{ and } \forall f \in C\left([0, 1]^2, \mathbb{C}\right), \tag{87}$$

and

$$B^{\mathbb{C}}_{n_1,n_2}(cg) = c B_{n_1,n_2}(g), \quad \forall c \in \mathbb{C} \text{ and } \forall g \in C\left([0, 1]^2, \mathbb{R}\right). \tag{88}$$

Clearly $B^{\mathbb{C}}_{n_1,n_2}$ maps $C\left([0, 1]^2, \mathbb{C}\right)$ into itself and B_{n_1,n_2} maps $C\left([0, 1]^2 R\right)$ into itself. Notice that $B_{n_1,n_2}(1(\cdot))(x, y) = 1$.

Hence by Theorem 4 (39) we get:

$$\left|B^{\mathbb{C}}_{n_1,n_2}(f)(z_0) - f(z_0)\right| \leq 2\sqrt{2}\omega_1\left(f, \sqrt{B_{n_1,n_2}\left(|\cdot - z_0|^2\right)(z_0)}\right), \tag{89}$$

$\forall z_0 \in [0, 1]^2, \forall n_1, n_2 \in \mathbb{N}$.

Here $z_0 = z_{01} + i z_{02}$, $z_{01}, z_{02} \in [0, 1]$, and $t = t_1 + i t_2$, where $t_1, t_2 \in [0, 1]$. We notice that

$$B_{n_1,n_2}\left(|t - z_0|^2\right)(z_0) = B_{n_1,n_2}\left((t_1 - z_{01})^2 + (t_2 - z_{02})^2\right)(z_{01}, z_{02}) =$$

$$B_{n_1,n_2}\left((t_1 - z_{01})^2\right)(z_{01}, z_{02}) + B_{n_1,n_2}\left((t_2 - z_{02})^2\right)(z_{01}, z_{02}) = \tag{90}$$

$$\left(B_{n_1}\left((t_1 - z_{01})^2\right)\right)(z_{01}) + \left(B_{n_2}\left((t_2 - z_{02})^2\right)\right)(z_{02}) =$$

$$\frac{z_{01}(1 - z_{01})}{n_1} + \frac{z_{02}(1 - z_{02})}{n_2}, \tag{91}$$

where B_{n_1}, B_{n_2} are the basic univariate Bernstein operators over $[0, 1]$.

That is,

$$B_{n_1,n_2}\left(|t - z_0|^2\right)(z_0) = \frac{z_{01}(1 - z_{01})}{n_1} + \frac{z_{02}(1 - z_{02})}{n_2}, \tag{92}$$

$\forall z_0 \in [0, 1]^2$.

Therefore we find

$$B_{n_1,n_2}\left(|t-z_0|^2\right)(z_0) \le \frac{1}{4}\left(\frac{1}{n_1}+\frac{1}{n_2}\right), \tag{93}$$

$\forall z_0 \in [0,1]^2$.

That is,

$$\sqrt{B_{n_1,n_2}\left(|\cdot-z_0|^2\right)(z_0)} \le \frac{1}{2}\sqrt{\frac{1}{n_1}+\frac{1}{n_2}}, \tag{94}$$

$\forall z_0 \in [0,1]^2$.

By (89), finally, we obtain

$$\left\|B_{n_1,n_2}^{\mathbb{C}}(f)-f\right\|_\infty \le 2\sqrt{2}\omega_1\left(f,\frac{1}{2}\sqrt{\frac{1}{n_1}+\frac{1}{n_2}}\right), \tag{95}$$

$\forall n_1, n_2 \in \mathbb{N}$.

Consequently, as $n_1, n_2 \to \infty$, we get that $B_{n_1,n_2}^{\mathbb{C}}(f) \overset{u}{\to} f$, uniformly, $\forall f \in C\left([0,1]^2, \mathbb{C}\right)$.

Many other examples as above could be given but we choose to omit this task.

References

1. G.A. Anastassiou, *Moments in Probability and Approximation Theory*. Pitman Research Notes in Mathematics, vol. 287 (Longman Scientific & Technical, Harlow, 1993)
2. G.A. Anastassiou, Lattice homomorphism - Korovkin type inequalities for vector valued functions. Hokkaido Math. J. **26**, 337–364 (1997)
3. G.A. Anastassiou, *Quantitative Approximations* (Chapman & Hall/CRC, Boca Raton, 2001)
4. G.A. Anastassiou, *Intelligent Computations: Abstract Fractional Calculus, Inequalities, Approximations* (Springer, Heidelberg, 2018)
5. G.A. Anastassiou, Complex Korovkin theory. J. Comput. Anal. Appl. **28**(6), 981–996 (2020)
6. P.P. Korovkin, *Linear Operators and Approximation Theory* (Hindustan, Delhi, 1960)
7. R.G. Mamedov, On the order of the approximation of functions by linear positive operators. Dokl. Akad. Nauk USSR **128**, 674–676 (1959)
8. H.L. Royden, *Real Analysis*, 2nd edn. (Macmillan, New York, 1968)
9. O. Shisha, B. Mond, The degree of convergence of sequences of linear positive operators. Natl. Acad. Sci. **60**, 1196–1200 (1968)
10. Z.X. Wang, D.R. Guo, *Special Functions* (World Scientific, Teaneck, 1989)

Hyperstability of a Linear Functional Equation on Restricted Domains

Jaeyoung Chung, John Michael Rassias, Bogeun Lee, and Chang-Kwon Choi

Abstract Let X, Y be real Banach spaces, $f : X \to Y$ and $\mathcal{H}$ be a subset of X such that $\mathcal{H}^c$ is of the first category. Using the Baire category theorem we prove the Ulam–Hyers stability of the linear functional equation

$$f(ax + by + \alpha) = Af(x) + Bf(y) + C$$

for all $x, y \in \mathcal{H}$, such that $\|x\| + \|y\| \geq d$ with $d > 0$, where a, b, A, B are nonzero real numbers and $\alpha \in X$ is fixed. As a consequence we solve the hyperstability problem associated to

$$\| f(ax + by + \alpha) - Af(x) - Bf(y) - C \| \leq \delta\psi(x, y)$$

for all $x, y \in \mathcal{K}$, where $\mathcal{K}$ is a subset of $\mathbb{R}$ with Lebesgue measure zero and $\psi(x, y) = |x|^p + |y|^q$, $p, q < 0$; or $\psi(x, y) = |x|^p|y|^q$, $p + q < 0$; or $\psi(x, y) = |x|^p|y|^q$, $pq < 0$.

2010 Mathematics Subject Classification 39B82

J. Chung
Department of Mathematics, Kunsan National University, Gunsan, Republic of Korea
e-mail: jychung@kunsan.ac.kr

J. M. Rassias
National and Kapodistrian University of Athens, Pedagogical Department E. E.,
Section of Mathematics and Informatics, Athens, Greece
e-mail: jrassias@primedu.uoa.gr; jrass@otenet.gr

B. Lee
Department of Mathematics and Institute of Pure and Applied Mathematics,
Chonbuk National University, Jeonju, Republic of Korea
e-mail: akiles@jbnu.ac.kr

C.-K. Choi (✉)
Department of Mathematics and Liberal Education Institute, Kunsan National University,
Gunsan, Republic of Korea
e-mail: ck38@kunsan.ac.kr

© Springer Nature Switzerland AG 2019
G. A. Anastassiou and J. M. Rassias (eds.), *Frontiers in Functional Equations and Analytic Inequalities*, https://doi.org/10.1007/978-3-030-28950-8_2

1   Introduction

Throughout this paper we denote by X, Y a real normed space and a Banach space, respectively. A mapping $f : X \to Y$ is called *an additive mapping* if f satisfies $f(x+y) = f(x) + f(y)$ for all $x, y \in X$. The Hyers–Ulam stability problems for functional equations have been originated by Ulam in 1940 (see [22]). One of the first assertions to be obtained is the following result, essentially due to Hyers [8] that gives an answer to the fundamental question of Ulam.

Theorem 1.1 *Suppose that* $f : X \to Y$ *satisfies the inequality*

$$\|f(x+y) - f(x) - f(y)\| \le \epsilon$$

for all $x, y \in X$. *Then there exists a unique additive mapping* $g : X \to Y$ *such that*

$$\|f(x) - g(x)\| \le \epsilon$$

for all $x \in X$.

In 1950 Aoki [1] generalized the above result and in 1978 Rassias [17] generalized the result by allowing the Cauchy difference to be unbounded (see also [3]). This stability concept is also applied to the case of other functional equations. In particular, the Hyers' result was generalized to the general linear functional equation

$$f(ax + by + \alpha) = Af(x) + Bf(y) + C \tag{1.1}$$

for all $x, y \in X$, where a, b, A, B are nonzero real numbers and $\alpha \in X$ is fixed (see [5, 12, 14–16] for the pertinent results).

Among the numerous results, Skof [21] solved the Hyers–Ulam problem for additive mappings on a restricted domain. Jung [9, 11] investigated the Hyers–Ulam stability for additive and quadratic mappings on restricted domains and Rassias [18, 19] investigated the Hyers–Ulam stability of mixed type mappings on restricted domains. For more results on functional equations or inequalities satisfied on restricted domains or satisfied under restricted conditions, we refer the reader to [2, 4, 6, 7, 10, 20, 21].

In this paper we first consider the stability of the Eq. (1.1) in some abstract domains satisfying a certain condition which we denote by (C) (see the beginning of the Sect. 2 for the condition). Secondly, using the Baire category theorem, we prove the stability of the Eq. (1.1) on restricted domains of form $\mathcal{H}^2 \cap \{(x, y) \in X^2 : \|x\| + \|y\| \ge d\}$ with $d > 0$, where $\mathcal{H}$ is a subset of X such that $\mathcal{H}^c$ is of the first category. Thirdly, as a consequence we show that for some particular forms of $\psi(x, y)$, the functional equation is ψ-hyperstable, i.e., each $f : X \to Y$ satisfying the inequality

$$\|f(ax + by + \alpha) - Af(x) - Bf(y) - C\| \le \delta\psi(x, y) \tag{1.2}$$

has the form $f(x) = g(x) + f(0)$, where g is an additive function (see [4, 14, 15] for hyperstabilities of functional equations). Finally, using the fact that the set $\mathbb{R}$ of real number can be partitioned as $\mathbb{R} = \mathcal{K} \cup (\mathbb{R} \setminus \mathcal{K})$, where $\mathcal{K}$ is of Lebesgue measure zero and $\mathbb{R} \setminus \mathcal{K}$ is of the first category, we prove the hyperstability of the equation on a set $\Gamma \subset \{(x, y) \in \mathbb{R}^2 : |x| + |y| \ge d\}$ of Lebesgue measure zero when f is defined on $\mathbb{R}$.

2 Stability on Abstract Restricted Domains

In this section consider the Ulam–Hyers stability of the functional equation on restricted domains $\Omega \subset X \times X$ satisfying some of the conditions: Let a, b be fixed real numbers with $ab \ne 0$ and $\alpha \in X$ be fixed. Throughout this section we assume that $\Omega \subset X \times X$ satisfies the following condition (C): For any $x, y \in X$ and $M > 0$, there exists $t \in X$ with $\|t\| \ge M$ such that

(C) $\quad \left\{(x - bt, y + at), (-bt, y + at), \left(x - bt, -\dfrac{\alpha}{b} + at\right), \left(-bt, -\dfrac{\alpha}{b} + at\right)\right\} \subset \Omega.$

We obtain the following stability theorem on restricted domain Ω.

Theorem 2.1 *Suppose that $f : X \to Y$ satisfies the inequality*

$$\|f(ax + by + \alpha) - Af(x) - Bf(y) - C\| \le \delta \tag{2.1}$$

for all $x, y \in \Omega$. Then there exists a unique additive mapping $g : X \to Y$ such that

$$g(ax) = Ag(x) \text{ and } g(bx) = Bg(x) \tag{2.2}$$

for all $x \in X$, and inequality

$$\|f(x) - g(x) - f(0)\| \le 4\delta \tag{2.3}$$

for all $x \in X$. Furthermore, if there exists an unbounded function f satisfying (2.1), then each of the couples $\{(a, A), (b, B)\}$ consists either of algebraic numbers which are algebraic conjugates each other or of transcendental numbers.

Proof Let the following difference:

$$D(x, y) = f(ax + by + \alpha) - Af(x) - Bf(y) - C \tag{2.4}$$

for all $x, y \in X$. Then we clearly have

$$D\left(\frac{x}{a} - bt, \frac{y-\alpha}{b} + at\right) = f(x+y) - Af\left(\frac{x}{a} - bt\right) - Bf\left(\frac{y-\alpha}{b} + at\right) - C,$$

$$D\left(\frac{x}{a} - bt, -\frac{\alpha}{b} + at\right) = f(x) - Af\left(\frac{x}{a} - bt\right) - Bf\left(-\frac{\alpha}{b} + at\right) - C,$$

$$D\left(-bt, \frac{y-\alpha}{b} + at\right) = f(y) - Af(-bt) - Bf\left(\frac{y-\alpha}{b} + at\right) - C,$$

$$D\left(-bt, -\frac{\alpha}{b} + at\right) = f(0) - Af(-bt) - Bf\left(-\frac{\alpha}{b} + at\right) - C$$

and we can write

$$f(x+y) - f(x) - f(y) + f(0)$$

$$= D\left(\frac{x}{a} - bt, \frac{y-\alpha}{b} + at\right) - D\left(\frac{x}{a} - bt, -\frac{\alpha}{b} + at\right) - D\left(-bt, \frac{y-\alpha}{b} + at\right)$$

$$+ D\left(-bt, -\frac{\alpha}{b} + at\right) \tag{2.5}$$

for all $x, y, t \in X$.

Since Ω satisfies the said condition (C), for given $\frac{x}{a}, \frac{y-\alpha}{b} \in X$ we can choose $t \in X$ such that

$$\left\|D\left(\frac{x}{a} - bt, \frac{y-\alpha}{b} + at\right)\right\| \le \delta, \quad \left\|D\left(\frac{x}{a} - bt, -\frac{\alpha}{b} + at\right)\right\| \le \delta, \tag{2.6}$$

$$\left\|D\left(-bt, \frac{y-\alpha}{b} + at\right)\right\| \le \delta, \quad \left\|D\left(-bt, -\frac{\alpha}{b} + at\right)\right\| \le \delta.$$

Thus from (2.5) and (2.6) we have

$$\|f(x+y) - f(x) - f(y) + f(0)\| \le 4\delta \tag{2.7}$$

for all $x, y \in X$. Thus, replacing f by $f - f(0)$ in Theorem 1.1 we get (2.3). Now, we prove (2.2). Since Ω satisfies the condition (C), for given $x, y \in X$ there exists $t \in X$ such that

$$\|f(ax + by + \alpha) - Af(x - bt) - Bf(y + at) - C\| \le \delta, \tag{2.8}$$

$$\left\|-f(0) + Af(-bt) + Bf\left(-\frac{\alpha}{b} + at\right) + C\right\| \le \delta. \tag{2.9}$$

From (2.7) we have

$$\|Af(x - bt) - Af(x) - Af(-bt) + Af(0)\| \le 4|A|\delta, \tag{2.10}$$

$$\|Bf(y + at) - Bf(y) - Bf(at) + Bf(0)\| \le 4|B|\delta, \tag{2.11}$$

$$\left\| -Bf\left(-\frac{\alpha}{b}+at\right) + Bf\left(-\frac{\alpha}{b}\right) + Bf\left(at\right) - Bf(0)\right\| \le 4|B|\delta \qquad (2.12)$$

for all $x, y, t \in X$. From (2.8)–(2.10), using the triangle inequality we get

$$\|f(ax + by + \alpha) - Af(x) - Bf(y)\|$$

$$\le (2 + 4|A| + 8|B|)\delta + |Af(0)| + \left|Bf\left(-\frac{\alpha}{b}\right)\right| := M_1 \qquad (2.13)$$

for all $x, y \in X$. From (2.3) we have

$$|f(x) - g(x)| \le 4\delta + |f(0)| := M_2 \qquad (2.14)$$

for all $x \in X$. Using (2.13), (2.14) and the triangle inequality we have

$$\|g(ax + by + \alpha) - Ag(x) - Bg(y)\| \le M_1 + (1 + |A| + |B|)M_2 := M_3 \qquad (2.15)$$

for all $x, y \in X$. Putting $y = -\frac{\alpha}{b}$ in (2.15) and using the triangle inequality we have

$$\|g(ax) - Ag(x)\| \le M_3 + \left|Bg\left(-\frac{\alpha}{b}\right)\right| \qquad (2.16)$$

for all $x \in X$. Putting $x = -\frac{c}{a}$ in (2.15) and using the triangle inequality we have

$$\|g(by) - Bg(y)\| \le M_3 + \left|Ag\left(-\frac{c}{a}\right)\right| \qquad (2.17)$$

for all $y \in X$. Since g is additive, from (2.16) and (2.17) we get $g(ax) = Ag(x)$ and $g(bx) = Bg(x)$ for all $x \in X$. Now, we prove the final statement. If f is unbounded, then from (2.3) we get $g \ne 0$. Now, it suffices to show for one couple, say (a, A). Assume that a is algebraic. Let $p(x)$ be the irreducible polynomial such that $p(a) = 0$. Then from the relation $g(ax) = Ag(x)$ for all $x \in X$ we have $p(A)g(x) = g(p(a)x) = g(0) = 0$ for all $x \in X$, which implies $p(A) = 0$ since $g \ne 0$. Conversely, if $p(A) = 0$, then we have $g(p(a)x) = 0$ and hence $p(a) = 0$. Now, the proof is complete. $\qquad \square$

Assume that Ω, in addition to the condition (C), satisfies the following two conditions: For any $x, y \in X$ and $M > 0$, there exists $t \in X$ with $\|t\| \ge M$ such that

$$(C)' \quad \left\{(x + y, t), \left(x, \frac{a}{b}y + t\right), (y, t), \left(0, \frac{a}{b}y + t\right)\right\} \subset \Omega,$$

$$(C)'' \quad \left\{(t, x + y), \left(\frac{a}{b}x + t, y\right), (t, x), \left(\frac{a}{b}x + t, 0\right)\right\} \subset \Omega.$$

Then the constant 4δ in (2.3) can be replaced by $\min\left\{4\delta, \frac{4\delta}{|A|}, \frac{4\delta}{|B|}\right\}$.

Theorem 2.2 *Let Ω satisfy the three conditions: (C), (C)$'$, and (C)$''$. Suppose that $f : X \to Y$ satisfies the inequality (2.1) for all $x, y \in \Omega$. Then there exists a unique additive mapping $g : X \to Y$ such that both relations (2.2) hold for all $x \in X$, and*

$$\|f(x) - g(x) - f(0)\| \le \min\left\{4\delta, \frac{4\delta}{|A|}, \frac{4\delta}{|B|}\right\} \qquad (2.18)$$

for all $x \in X$.

Proof Similarly, from the following four equalities

$$D(x + y, t) = f(ax + ay + bt + \alpha) - Af(x + y) - Bf(t) - C,$$

$$D\left(x, \frac{a}{b}y + t\right) = f(ax + ay + bt + \alpha) - Af(x) - Bf\left(\frac{a}{b}y + t\right) - C,$$

$$D(y, t) = f(ay + bt + \alpha) - Af(y) - Bf(t) - C,$$

$$D\left(0, \frac{a}{b}y + t\right) = f(ay + bt + \alpha) - Af(0) - Bf\left(\frac{a}{b}y + t\right) - C,$$

we have

$$Af(x + y) - Af(x) - Af(y) + Af(0) \qquad (2.19)$$

$$= -D(x + y, t) + D\left(x, \frac{a}{b}y + t\right) + D(y, t) - D\left(0, \frac{a}{b}y + t\right)$$

for all $x, y, t \in X$. Since Ω satisfies (C)$'$, for given $x, y \in X$ we can choose $t \in X$ such that

$$\|D(x + y, t)\| \le \delta, \ \left\|D\left(x, \frac{a}{b}y + t\right)\right\| \le \delta, \ \|D(y, t)\| \le \delta, \ \left\|D\left(0, \frac{a}{b}y + t\right)\right\| \le \delta. \qquad (2.20)$$

Thus from (2.19) and (2.20) we have

$$\|f(x + y) - f(x) - f(y) + f(0)\| \le \frac{4\delta}{|A|} \qquad (2.21)$$

for all $x, y \in X$. Thus, there exists an additive mapping $g_1 : X \to Y$ such that

$$\|f(x) - g_1(x) - f(0)\| \le \frac{4\delta}{|A|} \qquad (2.22)$$

for all $x \in X$. Now, from the following equalities

$$D(t, x + y) = f(at + bx + by + \alpha) - Af(t) - Bf(x + y) - C,$$

$$D\left(\frac{b}{a}x + t, y\right) = f(at + bx + by + \alpha) - Af\left(\frac{b}{a}x + t\right) - Bf(y) - C,$$

$$D(t, x) = f(at + bx + \alpha) - Af(t) - Bf(x) - C,$$

$$D\left(\frac{b}{a}x + t, 0\right) = f(at + bx + \alpha) - Af\left(\frac{b}{a}x + t\right) - Bf(0) - C,$$

we have

$$Bf(x + y) - Bf(x) - Bf(y) + Bf(0) \tag{2.23}$$

$$= -D(t, x + y) + D\left(\frac{a}{b}x + t, y\right) + D(t, x) - D\left(\frac{a}{b}x + t, 0\right)$$

for all $x, y, t \in X$. Since Ω satisfies (C)″, for given $x, y \in X$ we can choose $t \in X$ such that

$$\|D(t, x + y)\| \le \delta, \ \left\|D\left(\frac{a}{b}x + t, y\right)\right\| \le \delta, \ \|D(t, x)\| \le \delta, \ \left\|D\left(\frac{a}{b}x + t, 0\right)\right\| \le \delta. \tag{2.24}$$

Thus from (2.23) and (2.24) we have

$$\|f(x + y) - f(x) - f(y) + f(0)\| \le \frac{4\delta}{|B|} \tag{2.25}$$

for all $x, y \in X$. Thus, there exists an additive $g_2 : X \to Y$ such that

$$|f(x) - g_2(x) - f(0)| \le \frac{4\delta}{|B|} \tag{2.26}$$

for all $x \in X$. From three inequalities (2.3), (2.22), and (2.26), and using the triangle inequality we have

$$|g(x) - g_1(x)| \le 4\delta + \frac{4\delta}{|A|}, \tag{2.27}$$

$$|g(x) - g_2(x)| \le 4\delta + \frac{4\delta}{|B|} \tag{2.28}$$

for all $x \in X$. Since g, g_1, and g_2 are additive mappings, from (2.27) we have $g = g_1$, and from (2.27) we have $g = g_2$. Thus from (2.3), (2.22), and (2.26) we get (2.18). Now, the proof is complete. □

Remark 2.3 It is easy to see that for each $d > 0$, the set $\Omega = \{(x, y) : \|x\| + \|y\| \geq d\}$ satisfies the conditions (C). Thus, if $f : X \to Y$ satisfies the inequality (2.1) for all $x, y \in X$ such that $\|x\| + \|y\| \geq d$, then the result in Theorem 2.2 holds true.

Now, let $\Omega_n = \{(x, y) : \|x\| \geq n, \|y\| \geq n\}$ for positive integers $n \in \mathbb{N}$. Then Ω_n satisfies the condition (C) for all $n \in \mathbb{N}$. Thus, as a consequence of Theorem 2.1, we have the following Corollary 2.4.

Corollary 2.4 *Let $p, q < 0$. Suppose that $f : X \to Y$ satisfies the inequality*

$$\|f(ax + by + \alpha) - Af(x) - Bf(y) - C\| \leq \delta(\|x\|^p + \|y\|^q) \tag{2.29}$$

for all $x, y \in X$. Then there exists a unique additive mapping $g : X \to Y$ satisfying

$$g(ax) = Ag(x), \quad g(bx) = Bg(x) \tag{2.30}$$

for all $x \in X$, and

$$f(x) = g(x) + f(0) \tag{2.31}$$

for all $x \in X$. Furthermore, if f is not constant, then each of the two couples $\{(a, A), (b, B)\}$ consists either of algebraic numbers which are algebraic conjugates each other or of transcendental numbers.

Proof For each positive integer $m \in \mathbb{N}$, we can choose $n \in \mathbb{N}$ such that $\delta(\|x\|^p + \|y\|^p) \leq \frac{1}{m}$ if $(x, y) \in \Omega_n$. Since Ω_n satisfies (C), by Theorem 2.1, there exists an additive mapping $g : X \to Y$ such that

$$\|f(x) - g(x) - f(0)\| \leq \frac{4}{m} \tag{2.32}$$

for all $x \in X$, where g is independent of m. Now, letting $m \to \infty$ in (2.32) we get $f(x) = g(x) + f(0)$. Thus, using the triangle inequality with (2.29) we have

$$\|g(ax+by+\alpha)-Ag(x)-Bg(y)\| \leq \delta(\|x\|^p+\|y\|^q)+\|f(0)-Af(0)-Bf(0)-C\| \tag{2.33}$$

for all $x, y \in X$. Putting $y = -\frac{\alpha}{b}$ in (2.33) and using the triangle inequality we have

$$\|g(ax) - Ag(x)\| \leq \delta\|x\|^p + M \tag{2.34}$$

for all $x, y \in X$, where $M = \delta \left\| \frac{\alpha}{b} \right\|^q + \left\| Bg\left(\frac{\alpha}{b}\right) \right\| + \|f(0) - Af(0) - Bf(0) - C\|$.
For each $x \in X$ with $x \neq 0$, choose $k \in \mathbb{N}$, so that $\|kx\| \geq 1$. Then we have

$$\|g(ax) - Ag(x)\| = \frac{1}{k}\|g(akx) - Ag(kx)\| \leq \frac{1}{k}(\delta\|kx\|^p + M) \leq \frac{1}{k}(\delta + M). \tag{2.35}$$

Letting $k \to \infty$ in (2.35) we obtain $g(ax) = Ag(x)$ for all $x \neq 0$, and $g(0) = Ag(0) = 0$. Putting $x = -\frac{\alpha}{a}$ in (2.34) and using the same approach we get $g(bx) = Bg(x)$ for all $x \in X$. The last statement can be proved by the same methods as in the proof of Theorem 2.1. □

Remark 2.5 Let $\phi(x, y) = \|x + y\|^p + \|x - y\|^q + \|y\|^r$, or $\|x + y\|^p + \|x - y\|^q + \|x\|^r$, where $p, q, r < 0$. Assume that $f : X \to Y$ satisfies the inequality.

$$\|f(ax + by + \alpha) - Af(x) - Bf(y) - C\| \leq \delta\phi(x, y) \tag{2.36}$$

for all $x, y \in X$ instead of (2.29). Then using (2.22) and (2.26) we can show that the result in Theorem 2.4 holds true.

3 Stability on Concrete Restricted Domains

Throughout this section we assume that X is complete. By constructing subsets $\Omega \subset X \times X$ satisfying the three conditions (C), (C)$'$, (C)$''$ we prove the Hyers–Ulam stability of the functional equation (1.1) satisfied on restricted domains of form $\mathcal{H}^2 \cap \{(x, y) \in X^2 : \|x\| + \|y\| \geq d\}$ with $d > 0$, where $\mathcal{H}$ is a subset of X such that $\mathcal{H}^c$ is of the first category. As a consequence we obtain a stability theorem of the equation on a set of Lebesgue measure zero when $X = \mathbb{R}$.

Recall that a subset K of a topological space E is said to be of the first category if K is a countable union of nowhere dense subsets of E, and otherwise it is said to be of the second category. As named *Baire category theorem* it is well known that every nonempty open subset of a compact Hausdorff space or a complete metric space is of the second category.

The proof of the following lemma can be found in [7]. For the reader we give the proof.

Lemma 3.1 *Let $\mathcal{H}$ be a subset of X such that $\mathcal{H}^c := X \setminus \mathcal{H}$ is of the first category. Then, for any countable subsets $U \subset X$, $\Gamma \subset \mathbb{R} \setminus \{0\}$ and $M > 0$, there exists $t \in X$ with $\|t\| \geq M$ such that*

$$U + \Gamma t = \{u + \gamma t : u \in U, \gamma \in \Gamma\} \subset \mathcal{H}. \tag{3.1}$$

Proof Let $\mathcal{H}^c_{u,\gamma} = \gamma^{-1}(\mathcal{H}^c - u)$, $u \in U, \gamma \in \Gamma$. Then, since $\mathcal{H}^c$ is of the first category, $\mathcal{H}^c_{u,\gamma}$ are also of the first category for all $u \in U, \gamma \in \Gamma$. Since each $\mathcal{H}^c_{u,\gamma}$ consists of a countable union of nowhere dense subsets of X, by the Baire category theorem, the countable union of all $\{\mathcal{H}^c_{u,\gamma} : u \in U, \gamma \in \Gamma\}$ cannot cover $X_0 := \{t \in X : \|t\| \geq M\}$, i.e.,

$$X_0 \not\subset \bigcup_{(u,\gamma) \in U \times \Gamma} \mathcal{H}^c_{u,\gamma}.$$

Choose a $t \in X_0$ such that $t \notin \mathcal{H}^c_{u,\gamma}$ for all $u \in U$, $\gamma \in \Gamma$. Then we have $u + \gamma t \in \mathcal{H}$ for all $u \in U$, $\gamma \in \Gamma$. Now, the proof is complete.   □

From now on we identify $\mathbb{R}^2$ with $\mathbb{C}$.

Lemma 3.2 *Let* $P = \{(p_j + \gamma_j t, q_j + \lambda_j t) : j = 1, 2, \ldots, r\}$, *where* $p_j, q_j, t \in X$, $\gamma_j, \lambda_j \in \mathbb{R}$ *with* $\gamma_j^2 + \lambda_j^2 \neq 0$ *for all* $j = 1, 2, \ldots, r$. *Then there exists a* $\theta \in [0, 2\pi)$ *such that* $e^{-i\theta} P := \{(p'_j + \gamma'_j t, q'_j + \lambda'_j t) : j = 1, 2, \ldots, r\}$ *satisfies* $\gamma'_j \lambda'_j \neq 0$ *for all* $j = 1, 2, \ldots, r$.

From now on we let $P = P(x, y, t) := S_1 \cup S_2 \cup S_3$, where

$$S_1 = \left\{ (x - bt, y + at), (-bt, y + at), \left(x - bt, -\frac{\alpha}{b} + at \right) \left(-bt, -\frac{\alpha}{b} + at \right) \right\},$$

$$S_2 = \left\{ (x + y, t), \left(x, \frac{a}{b}y + t \right), (y, t), \left(0, \frac{a}{b}y + t \right) \right\},$$

$$S_3 = \left\{ (t, x + y), \left(\frac{a}{b}x + t, y \right), (t, x), \left(\frac{a}{b}x + t, 0 \right) \right\}.$$

Lemma 3.3 *Let* $\mathcal{H}$ *be a subset of* X *such that* $\mathcal{H}^c$ *is of the first category. Then there exists a* $\theta \in [0, 2\pi)$ *such that* $\Omega_{\theta,d} := (e^{i\theta} \mathcal{H}^2) \cap \{(x, y) \in X^2 : \|x\| + \|y\| \geq d\}$ *satisfies the conditions* (C), (C)′, (C)″ *for all* $d > 0$.

Proof Let θ be the angle of rotation in Lemma 3.2 for $P = S_1 \cup S_2 \cup S_3$. It suffices to show that for given $x, y \in X$ there exists $t \in X$ such that

$$e^{-i\theta} P(x, y, t) \subset \mathcal{H}^2 \text{ and } P(x, y, t) \subset \{(u, v) : \|u\| + \|v\| \geq d\}. \qquad (3.2)$$

Let $e^{-i\theta} P(x, y, t) := \{(p'_j + \gamma'_j t, q'_j + \lambda'_j t) : j = 1, 2, \ldots, r\}$. Then by Lemma 3.2, we have $\gamma'_j \lambda'_j \neq 0$ for all $j = 1, 2, \ldots, r$. Let $U = \{p'_j, q'_j : j = 1, 2, \ldots, r\}$, $\Gamma = \{\gamma'_j, \lambda'_j : j = 1, 2, \ldots, r\}$. Then we have

$$\{u, v : (u, v) \in e^{-i\theta} P(x, y, t)\} \subset U + \Gamma t. \qquad (3.3)$$

Now, by Lemma 3.1, there exists $t \in X$ with $\|t\| \geq \max_{1 \leq j \leq r} (|\gamma_j| + |\lambda_j|)^{-1} (|p_j| + |q_j| + d)$ such that

$$U + \Gamma t \subset \mathcal{H}. \qquad (3.4)$$

From (3.3) and (3.4) we have

$$e^{-i\theta} P \subset \mathcal{H}^2.$$

By the choice of t, we have $P(x, y, t) \subset \{(u, v) : \|u\| + \|v\| \geq d\}$. This completes the proof.   □

Remark 3.4 In view of the proof of Lemma 3.3 we can see that $\Omega_{0,d} := \mathcal{H}^2 \cap$ $\{(x, y) \in X^2 : \|x\| + \|y\| \geq d\}$ satisfies (C) for all $d > 0$.

Thus, as a consequence of Theorem 2.1 we obtain the following Theorem 3.5.

Theorem 3.5 *Let $d > 0$. Suppose that $f : X \to Y$ satisfies the inequality*

$$\|f(ax + by + \alpha) - Af(x) - Bf(y) - C\| \leq \delta \tag{3.5}$$

for all $x, y \in \mathcal{H}$ with $\|x\| + \|y\| \geq d$. Then there exists a unique additive mapping $g : X \to Y$ satisfying

$$g(ax) = Ag(x), \quad g(bx) = Bg(x) \tag{3.6}$$

for all $x \in X$, and

$$\|f(x) - g(x) - f(0)\| \leq 4\delta \tag{3.7}$$

for all $x \in X$. Furthermore, if there exists an unbounded function f satisfying (3.5), then each of the two couples $\{(a, A), (b, B)\}$ consists either of algebraic numbers which are algebraic conjugates each other or of transcendental numbers.

As a consequence of Theorem 2.2 and Lemma 3.3 we obtain the following.

Theorem 3.6 *Suppose that $f : X \to Y$ satisfies the inequality (3.5) for all $x, y \in \Omega_{\theta,d}$. Then there exists a unique additive mapping $g : X \to Y$ satisfying*

$$g(ax) = Ag(x) \text{ and } g(bx) = Bg(x) \tag{3.8}$$

for all $x \in X$, and

$$\|f(x) - g(x) - f(0)\| \leq \min\left\{4\delta, \frac{4\delta}{|A|}, \frac{4\delta}{|B|}\right\} \tag{3.9}$$

for all $x \in X$. Furthermore, if there exists an unbounded function f satisfying (3.5) for all $x, y \in \Omega_{\theta,d}$, then each of the two couples $\{(a, A), (b, B)\}$ consists either of algebraic numbers which are algebraic conjugates each other or of transcendental numbers.

Remark 3.7 The set $\mathbb{R}$ of real numbers can be partitioned as follows:

$$\mathbb{R} = \mathcal{K} \cup (\mathbb{R} \setminus \mathcal{K}),$$

where $\mathcal{K}$ is of Lebesgue measure zero and $\mathbb{R} \setminus \mathcal{K}$ is of the first category [13, Theorem 1.6]. Thus, in view of Lemma 3.3, $\Gamma_d := (e^{i\theta}\mathcal{K}^2) \cap \{(x, y) \in X^2 : |x| + |y| \geq d\}$ is of Lebesgue measure zero satisfying (C), (C)$'$, (C)$''$. Now, as a consequence of Theorem 3.5 we obtain the following.

Corollary 3.8 *For any $d > 0$, there exists a subset $\Gamma_d \subset \{(x, y) \in X^2 : |x| + |y| \geq d\}$ of Lebesgue measure zero such that if $f : \mathbb{R} \to Y$ satisfies the functional inequality*

$$\| f(ax + by + \alpha) - Af(x) - Bf(y) - C\| \leq \delta$$

for all $(x, y) \in \Gamma_d$, then there exists a unique additive mapping $g : \mathbb{R} \to Y$ satisfying

$$g(ax) = Ag(x), \quad g(bx) = Bg(x) \tag{3.10}$$

for all $x \in X$, and

$$\| f(x) - g(x) - f(0)\| \leq \min \left\{ 4\delta, \frac{4\delta}{|A|}, \frac{4\delta}{|B|} \right\}$$

for all $x \in X$. Furthermore, if there exists an unbounded function f satisfying (2.15), then each of the couples $\{(a, A), (b, B)\}$ consists either of algebraic numbers which are algebraic conjugates each other or of transcendental numbers.

Letting $\delta = 0$ in Theorem 3.6 we have the following Corollary 3.9.

Corollary 3.9 *For any $d > 0$, there exists a subset $\Gamma_d \subset \{(x, y) \in X^2 : |x| + |y| \geq d\}$ of Lebesgue measure zero such that if $f : \mathbb{R} \to Y$ satisfies the functional equation*

$$f(ax + by + \alpha) - Af(x) - Bf(y) - C = 0 \tag{3.11}$$

for all $(x, y) \in \Gamma_d$, then the functional equation (3.11) holds for all $x, y \in X$.

4 Hyperstability on Restricted Domains and on a Set of Lebesgue Measure Zero

In this section we consider the hyperstability of the functional equation (1.1) in the restricted domain $\mathcal{H}^2$, where $\mathcal{H}^c$ is of the first category. Let us consider the functional inequalities

$$\| f(ax + by + \alpha) - Af(x) - Bf(y) - C\| \leq \delta\psi(\|x\|, \|y\|) \tag{4.1}$$

for all $(x, y) \in e^{i\theta}\mathcal{H}^2$, where $\delta \geq 0$ and ψ is one of the following:

$$\text{(Case 1)} \qquad \psi(s, t) = s^p + t^q, \quad p, q < 0;$$

$$\text{(Case 2)} \qquad \psi(s, t) = s^p t^q, \quad p + q < 0;$$

$$\text{(Case 3)} \qquad \psi(s, t) = s^p t^q, \quad pq < 0.$$

We first prove the hyperstability theorem for Case 1 and Case 2. In particular, the following result for Case 2 is a generalization of Theorem 2.1 of [15].

Theorem 4.1 *Suppose that* $f : X \to Y$ *satisfies the inequality* (4.1) *for all* $x, y \in \mathcal{H}$ *when* $\psi(s,t) = s^p + t^q$, $p, q < 0$ *or* $s^p t^q$, $p + q < 0$. *Then there exists a unique additive mapping* $g : X \to Y$ *satisfying*

$$g(ax) = Ag(x) \text{ and } g(bx) = Bg(x) \tag{4.2}$$

for all $x \in X$, *and*

$$f(x) = g(x) + f(0) \tag{4.3}$$

for all $x \in X$. *Furthermore, if* f *not constant, then each of the couples* $\{(a, A), (b, B)\}$ *consists either of algebraic numbers which are algebraic conjugates each other or of transcendental numbers.*

Proof First, we consider Case 1 when $\psi(s,t) = s^p + t^q$, $p, q < 0$. Let $\Omega_d = \mathcal{H}^2 \cap \{(x, y) : \|x\| \geq d, \|y\| \geq d\}$ for $d > 0$. Then by Lemma 3.1, for any $U = \{0, x, y, -\alpha/b\}$, $\Gamma = \{a, -b\}$ and $M > 0$ there exists $t \in X$ with $\|t\| \geq M$ such that $U + \Gamma t \subset \mathcal{H}$. Thus, for any $x, y \in X$ and $M > 0$ there exists $t \in X$ with $\|t\| \geq M$ such that

$$P(x, y, t) := \left\{ (x - bt, y + at), (-bt, y + at), \right.$$
$$\left. \left(x - bt, -\frac{\alpha}{b} + at \right) \left(-bt, -\frac{\alpha}{b} + at \right) \right\} \subset \Omega_d. \tag{4.4}$$

Thus, Ω_d satisfies condition (C). For each positive integer $m \in \mathbb{N}$, we can choose $d_m > 0$ such that $\|x\|^p + \|y\|^q \leq \frac{1}{m}$ for all $(x, y) \in \Omega_{d_m}$. Thus, from (4.1) we get

$$\|f(ax + by + \alpha) - Af(x) - Bf(y) - C\| \leq \frac{\delta}{m} \tag{4.5}$$

for all $(x, y) \in \Omega_{d_m}$. Since Ω_{d_m} satisfies (C), by Theorem 2.1, there exists an additive mapping $g : X \to Y$ satisfying (4.2) and

$$\|f(x) - g(x) - f(0)\| \leq \frac{4\delta}{m} \tag{4.6}$$

for all $x \in X$, where g is independent of m. Now, letting $m \to \infty$ in (4.6) we get $f(x) = g(x) + f(0)$.

Since Ω_{d_m} satisfies (C), for given $x, y \in X$ and $M > 0$, there exists $t \in X$ with $\|t\| \geq M$ such that

$$\|f(ax + by + \alpha) - Af(x - bt) - Bf(y + at) - C\| \le \frac{4\delta}{m}, \tag{4.7}$$

$$\left\| - f(0) + Af(-bt) + Bf\left(-\frac{\alpha}{b} + at\right) + C\right\| \le \frac{4\delta}{m}. \tag{4.8}$$

Since $f - f(0)$ is additive we have

$$Af(x - bt) = Af(x) + Af(-bt) - Af(0), \tag{4.9}$$

$$Bf(y + at) = Bf(y) + Bf(at) - Bf(0), \tag{4.10}$$

$$Bf\left(-\frac{\alpha}{b} + at\right) = Bf\left(-\frac{\alpha}{b}\right) + Bf(at) - Bf(0) \tag{4.11}$$

for all $x, y, t \in X$. From (4.7)~(4.9), we get

$$\left\| f(ax + by + \alpha) - Af(x) - Bf(y) - f(0) + Af(0) + Bf\left(-\frac{\alpha}{b}\right)\right\| \le \frac{8\delta}{m} \tag{4.12}$$

for all $x, y \in X$. Letting $m \to \infty$ in (4.12) we get

$$f(ax + by + \alpha) - Af(x) - Bf(y) - f(0) + Af(0) + Bf\left(-\frac{\alpha}{b}\right) = 0 \tag{4.13}$$

for all $x, y \in X$. Since $f(x) = g(x) + g(0)$, putting $y = -\frac{\alpha}{b}$ in (4.13) we have $g(ax) = Ag(x)$ for all $x \in X$. Putting $x = -\frac{\alpha}{a}$, $y = 0$ in (4.13) we have

$$-f(0) + Bf(0) + Af\left(-\frac{\alpha}{a}\right) = -f(0) + Af(0) + Bf\left(-\frac{\alpha}{b}\right). \tag{4.14}$$

Putting $y = -\frac{\alpha}{a}$ in (4.13) and using (4.14) we have $g(bx) = Bg(x)$ for all $x \in X$.

Secondly, we consider Case 2 when $\psi(s, t) = s^p t^q$, $p + q < 0$. Let $\Omega_d = \mathcal{H}^2 \cap \{(x, y) : d \le \frac{1}{2}\left|\frac{a}{b}\right| \|x\| \le \|y\| \le 2\left|\frac{a}{b}\right| \|x\|\}$ for $d > 0$. For given $x, y \in X$, if we choose $t \in X$ such that $\|t\|$ is sufficiently large, then $d \le \frac{1}{2}\left|\frac{a}{b}\right| \|u\| \le \|v\| \le 2\left|\frac{a}{b}\right| \|u\|$ for all $(u, v) \in P(x, y, t)$ in (4.5). Thus, as in the first case we can see that Ω_d satisfies the condition (C). Since there exists $c > 0$ such that $\|x\|^p \|y\|^q \le c\|x\|^{p+q}$ for all $(x, y) \in \Omega_d$, for each positive integer $m \in \mathbb{N}$, we can choose $d_m > 0$ such that $\|x\|^p \|y\|^q \le c\|x\|^{p+q} \le \frac{1}{m}$ for all $(x, y) \in \Omega_{d_m}$. Thus, from (4.2) we get (4.6) for all $(x, y) \in \Omega_{d_m}$. Now, the remaining part of the proof is similar to the first case. Finally, the last statement can be proved by the same methods as in the proof of Theorem 2.1. Now, the proof is complete.   □

Now, we prove the hyperstability theorem for Case 3 when $\psi(s, t) = s^p t^q$, $pq < 0$.

Theorem 4.2 *Let $p, q \in \mathbb{R}$ with $pq < 0$. Then there exists a $\theta \in [0, 2\pi)$ such that if $f : X \to Y$ satisfies the inequality*

$$\|f(ax + by + \alpha) - Af(x) - Bf(y) - C\| \leq \delta \|x\|^p \|y\|^q \tag{4.15}$$

for all $(x, y) \in e^{i\theta}\mathcal{H}^2$, *then* $g = f - f(0)$ *is an additive mapping.*

Proof We may assume that $p > 0$, $q < 0$. For each $m \in \mathbb{N}$, let $\Omega_{\theta,m} = e^{i\theta}\mathcal{H}^2 \cap \{(x, y) \in X : \|x\| \geq m\}$, where θ is the angle of rotation in Lemma 3.3. Then $\Omega_{\theta,m}$ satisfies the condition (C)'. Thus, from (1.19) in Theorem 2.2, for given $x, y \in X$ and $m \in \mathbb{N}$, we can choose $t_m \in X$ with $\|t_m\| \geq m$ such that

$$\|Af(x + y) - Af(x) - Af(y) + Af(0)\| \tag{4.16}$$

$$= \| - D(x + y, t_m)\| + \left\|D\left(x, \frac{a}{b}y + t_m\right)\right\| + \|D(y, t_m)\| + \left\|-D\left(0, \frac{a}{b}y + t_m\right)\right\|$$

$$\leq \|x + y\|^p \|t_m\|^q + \|x\|^p \left\|\frac{a}{b}y + t_m\right\|^q + \|y\|^p \|t_m\|^q,$$

where $D(x, y) = f(ax + by + \alpha) - Af(x) - Bf(y) - C$. Letting $m \to \infty$ in (4.19) we get

$$f(x + y) - f(x) - f(y) + f(0) = 0 \tag{4.17}$$

for all $x \in X$. Letting $g = f - f(0)$ we get (4.17). Now, the proof is complete. $\square$

Remark 4.3 We have no idea if $g = f - f(0)$ in Theorem 4.2 satisfies some of the relations (4.2).

Let $\mathcal{K}$ be a subset of $\mathbb{R}$ with Lebsegue measure zero such that $\mathcal{K}^c$ is of the first category(see Remark 3.7). Then as a consequence of Theorem 4.1 we obtain the following.

Corollary 4.4 *Suppose that* $f : \mathbb{R} \to Y$ *satisfies the inequality*

$$\|f(ax + by + \alpha) - Af(x) - Bf(y) - C\| \leq \delta(|x|^p + |y|^q),$$

$$p, q < 0 \ (or \ \delta|x|^p|y|^q, \ p + q < 0)$$

for all $x, y \in \mathcal{K}$. *Then* $g = f - f(0)$ *is an additive mapping satisfying*

$$g(ax) = Ag(x) \ \text{and} \ g(bx) = Bg(x)$$

for all $x \in \mathbb{R}$. *Furthermore, if* f *is not constant, then each of the two couples* $\{(a, A), (b, B)\}$ *consists either of algebraic numbers which are algebraic conjugates each other or of transcendental numbers.*

As a consequence of Theorem 4.2 we obtain the following.

Corollary 4.5 *There exists an angle* $\theta \in [0, 2\pi)$ *such that if* $f : \mathbb{R} \to Y$ *satisfies the inequality*

$$\|f(ax + by + \alpha) - Af(x) - Bf(y) - C\| \le \delta |x|^p |y|^q, \quad pq < 0 \qquad (4.18)$$

for all $(x, y) \in e^{i\theta} \mathcal{K}^2$, *then* $g = f - f(0)$ *is an additive mapping.*

Acknowledgements This research was completed with the help of Professor Jaeyoung Chung. After finishing this work, Professor Jaeyoung Chung tragically passed away. Pray for the bliss of dead.

References

1. T. Aoki, On the stability of the linear transformation in Banach spaces. J. Math. Soc. Jpn. **2**, 64–66 (1950)
2. B. Batko, Stability of an alternative functional equation. J. Math. Anal. Appl. **339**, 303–311 (2008)
3. D.G. Bourgin, Classes of transformations and bordering transformations. Bull. Am. Math. Soc. **57**, 223–237, (1951)
4. J. Brzdęk, K. Ciepliński, Hyperstability and superstability. Abstr. Appl. Anal. **2013**, Art. ID 401756, 13 pages (2013)
5. J. Brzdęk, A. Pietrzyk, A note on stability of the general linear equation. Aequationes Math. **75**(3), 267–270 (2008)
6. J. Brzdęk, J. Sikorska, A conditional exponential functional equation and its stability. Nonlinear Anal. TMA **72**, 2929–2934 (2010)
7. J. Chung, J.M. Rassias, Quadratic functional equations in a set of Lebesgue measure zero. J. Math. Anal. Appl. **419**, 1065–1075 (2014)
8. D.H. Hyers, On the stability of the linear functional equations. Proc. Natl. Acad. Sci. U. S. A. **27**, 222–224 (1941)
9. S.-M. Jung, On the Hyers-Ulam stability of the functional equations that have the quadratic property. J. Math. Anal. Appl. **222**, 126–137 (1998)
10. S.-M. Jung, *Hyers-Ulam-Rassias Stability of Functional Equations in Nonlinear Analysis* (Springer, New-York, 2011)
11. S.-M. Jung, M.S. Moslehian, P.K. Sahoo, Stability of a generalized Jensen equation on restricted domains. J. Math. Inequal. **4**, 191–206 (2010)
12. D. Molaei, A. Najati, Hyperstability of the general linear equation on restricted domains. Acta Math. Hungar. **149**, 238–253 (2016)
13. J.C. Oxtoby, *Measure and Category* (Springer, New-York, 1980)
14. M. Piszczek, Remark on hyperstability of the general linear equation. Aequationes Math. **88**(1–2), 163–168 (2014)
15. M. Piszczek, Hyperstability of the general linear functional equation. Bull. Korean Math. Soc. **52**(6), 1827–1838 (2015)
16. D. Popa, On the stability of the general linear equation. Results Math. **53**(3–4), 383–389 (2009)
17. T. M. Rassias, On the stability of the linear mapping in Banach spaces. Proc. Am. Math. Soc. **72**(2), 297–300 (1978)
18. J.M. Rassias, On the Ulam stability of mixed type mappings on restricted domains. J. Math. Anal. Appl. **281**, 747–762 (2002)
19. J.M. Rassias, Asymptotic behavior of mixed type functional equations. Aust. J. Math. Anal. Appl. **1**(1), Article 10, 1–21 (2004)
20. J. Sikorska, On two conditional Pexider functional equations and their stabilities. Nonlinear Anal. TMA **70**, 2673–2684 (2009)
21. F. Skof, Sull' approssimazione delle applicazioni localmente δ-additive. Atti della Accademia delle Scienze di Torino **117**, 377–389 (1983)
22. S.M. Ulam, *Problems in Modern Mathematics*, vol. VI, Science edn. (Wiley, New York, 1940)

Part II
Functional Equations

Hyers–Ulam's Stability Results to a Three-Point Boundary Value Problem of Nonlinear Fractional Order Differential Equations

Kamal Shah, Zamin Gul, Yongjin Li, and Rahmat Ali Khan

Abstract This research is devoted to investigate the existence and multiplicity results of boundary value problem (BVP) for nonlinear fractional order differential equation (FDEs). To obtain the required results, we use some fixed point theorems due to Leggett–Williams and Banach. Further in this paper, we introduce different types of Ulam's stability concepts for the aforesaid problem of nonlinear FDEs. The concerned types of Ulam's stability are devoted to Ulam–Hyers (UH), generalized Ulam–Hyers (GUH) stability and Ulam–Hyers–Rassias (UHR), generalized Ulam–Hyers–Rassias (GUHR) stability. Finally the whole analysis is verified by some adequate examples.

2010 Mathematics Subject Classification 26A33, 34A08, 35B40

1 Introduction

Fractional differential equations (FDEs) received considerable attention due to its large numbers of applications in various disciplines of sciences and technology like physics, mechanics, chemistry, engineering, signal and image processing phenomenons, etc. There has been a significant development in the theory of initial and BVPs for nonlinear FDEs. Researchers have been paid much attention to investigate different aspects of the aforesaid problems which are included qualitative theory, numerical analysis, optimization, and stability analysis. It is a well-known fact that in the last few years, many researchers focused their attention to develop the existence theory of solutions to nonlinear FDEs, see, for example, [1–5, 35] and the references therein. This is because of numerous applications of BVPs of

K. Shah · Z. Gul · R. A. Khan
Department of Mathematics, University of Malakand, Chakadara Dir (L), Khyber Pakhtunkhwa, Pakistan

Y. Li (✉)
Department of Mathematics, Sun Yat-sen University, Guangzhou, China
e-mail: stslyj@mail.sysu.edu.cn

© Springer Nature Switzerland AG 2019
G. A. Anastassiou and J. M. Rassias (eds.), *Frontiers in Functional Equations and Analytic Inequalities*, https://doi.org/10.1007/978-3-030-28950-8_3

differential and integral equations in physics, engineering, and technology. As far as
we know that the mentioned aspect has been very well studied for BVPs of ordinary
differential equations and plenty of research articles, books, and monograph can be
found on this topic in literature. Similarly, the mentioned aspect has been very well
explored for nonlinear FDEs with boundary conditions.

For instance, Bai and Lü [2] investigated the existence theory of solutions to BVP
recalled as

$$D_{0+}^p u(t) + f(t, u(t)) = 0, \ t \in I, \ 1 < p \leq 2,$$

$$u(0) = u(1) = 0,$$

where D_{0+}^p is the Riemann–Liouville derivative of arbitrary order and f : I $\times$
$[0, \infty) \to [0, \infty)$ is continuous function.

In same line Kaufmann and Mboumi [12] investigated appropriate results for the
existence of solutions to the BVP given by

$$D_{0+}^p u(t) + \alpha(t) f(u(t)) = 0, \ t \in I, \ 1 < p \leq 2,$$

$$u(0) = 0, \ u'(1) = 0,$$

where $\alpha(t)$ is a positive and continuous function on I.

Following the aforesaid investigation, Li et al. [17], considered the following
nonlinear FDE as

$$D_{0+}^p u(t) + f(t, u(t)) = 0, \quad t \in I,$$

$$u(0) = 0, \quad D_{0+}^q u(1) = a D^q u(\xi), \ 0 < q < 1,$$

where $1 < p \leq 2$, D_{0+}^p is the Riemann–Liouville derivative and f : I $\times [0, \infty) \to$
$[0, \infty)$ is continuous function. In the last few years some other remarkable work
which is devoted to existence theory of nonlinear FDEs has been done, few of them
we refer as [14, 22, 29, 36].

One of the important area of differential equations is devoted to investigate
BVPs, because such problems arise in various disciplines like physics, engineering,
fluid mechnics, etc. In this regard from applications point of view, here we refer
some famous BVPs of differential equations which are the wave equation, like
the computation of the normal modes, the Sturm–Liouville problems, and Dirichlet
problem, etc. For usability purposes, a BVP should be well posed which implies that
a unique solution exists corresponding to the input which depends continuously on
the input. In thermal sciences BVPs have significant applications, for instance to find
the temperature at all points of an iron bar with one end kept at lowest energy level
and the other end at the freezing point of water. Due to these importance applications
researchers studied BVPs of both classical and arbitrary order differential equations
from different aspects. Besides from the existence theory another aspect which is
recently considered by many authors is known as stability of solutions. In fact
it is very difficult job to solve every nonlinear problem of differential or integral

equations for their exact solution, therefore in such a situation one need some iterative methods to find approximate or numerical solutions. Now for the numerical solutions stability is a demanding task. Researchers investigated different kinds of stability for differential, integral, and functional equations like exponential, Mittag–Leffler, and Lyapunov stability, for details, see [18, 23, 25]. Recently some authors explored another form of stability known as UH and GUH stability for the solutions of FDEs, see [19, 23, 26, 27, 30]. This form of stability was pointed out by Ulam in 1940 which was later explained by Hyers in a more brilliant way in 1941. The mentioned stability has been improved further in 1970 by Rassias to some other forms known as UHR and GUHR stability, see [21]. The aforesaid stability has been very well studied for initial value problems and simple two-point BVPs of linear and nonlinear FDEs, see [7, 14, 32]. In the last few years the said stability analysis has been extensively considered by many authors, we refer some work as [6, 8–11, 19, 24, 28, 31, 33, 34]. The concerned stability is very rarely investigated for the multi-point BVPs of FDEs. Here we remark that nonlocal BVPs of FDEs are of key impotence for engineers, physics, etc. The stable solutions of the aforesaid problems help us in understanding the phenomenon which has the differential equations.

Inspired from the aforesaid work, we consider the following three-point BVP of nonlinear FDEs as

$$D_{0+}^{\mathbf{p}} u(t) + f(t, u(t)) = 0, \qquad t \in \mathbf{I}, \tag{1}$$

$$D_{0+}^{\mathbf{q}} u(0) = 0, \quad D_{0+}^{\mathbf{q}} u(1) = \lambda u(\xi), \ 0 \le \mathbf{q} < 1, \ \mathbf{p} - \mathbf{q} \ge 1,$$

where $D_{0+}^{\mathbf{p}}$ is the standard Riemann–Liouville fractional derivative of order $1 < \mathbf{p} \le 2$, and $\lambda \Gamma(\mathbf{p} - \mathbf{q})\xi^{\mathbf{p}-2} \le \Gamma(\mathbf{p})(1 - \mathbf{q})$, $f : \mathbf{I} \times [0, \infty) \to [0, \infty)$ is continuous function. By using Leggett–Williams and Banach fixed point theorems we develop the required results. Further different kinds of Ulam's stability results are also investigated. Here we remark that investigating the UH and GUH stability for the solutions of nonlinear FDEs involving Riemann–Liouville fractional derivative is very rarely studied in literature. Most of the investigations about the stability are related to Caputo type derivative. Further we also studied UHR and GUHR stability results for the considered BVP of FDEs. The whole analysis is justified by providing some problems.

2 Axillary Results

This part of the paper is devoted to some necessary notations, definitions, and results related to fractional calculus and functional analysis for which reader can see [13, 15, 20]. These definitions and lemmas will be used throughout this manuscript.

Definition 1 The fractional order integral of arbitrary order $\mathbf{p} > 0$ for the function $f \in L([a, b], \mathbf{R})$ is recalled as

$$\mathscr{I}_a^{\mathbf{p}} f(t) = \frac{1}{\Gamma(\mathbf{p})} \int_a^t (t-s)^{\mathbf{p}-1} f(s) ds,$$

where the integration on the right-hand side is pointwise defined on $(0, \infty)$.

Definition 2 If $\mathbf{p} > 0$, then the Riemann–Liouville fractional derivatives for a continuous function $f \in ((0, \infty), \mathbf{R})$, is given by

$$D_{0+}^{\mathbf{p}} f(t) = \frac{1}{\Gamma(n-\mathbf{p})} \left(\frac{d}{dt}\right)^n \int_0^t (t-s)^{n-\mathbf{p}-1} f(s) ds,$$

where $n = [\mathbf{p}] + 1$, provided that the integral on the right is pointwise on $(0, \infty)$.

Definition 3 A concave functional $\theta \geq 0$ on a cone $\mathscr{P}$ of a real Banach space $\mathbf{X}$ for all $0 \leq \lambda \leq 1$ is defined by

$$\theta(\lambda u + (1-\lambda)v) \geq \lambda\theta(u) + (1-\lambda)\theta(v) \quad \text{for all} \quad u, v \in \mathscr{P}.$$

Lemma 1 ([12]) *The solution of FDE*

$$D_{0+}^{\mathbf{p}} u(t) = 0$$

is given by

$$u(t) = b_1 t^{\mathbf{p}-1} + b_2 t^{\mathbf{p}-2} + \ldots + b_n t^{\mathbf{p}-n},$$

for some $b_i \in \mathbf{R}$, $i = 1, 2, \ldots, n$, where $n = [\mathbf{p}] + 1$.

Lemma 2 *Using the above Lemma 1, we get the result given by*

$$\mathscr{I}_{0+}^{\mathbf{p}} D_{0+}^{\mathbf{p}} u(t) = u(t) + b_1 t^{\mathbf{p}-1} + b_2 t^{\mathbf{p}-2} + \ldots + b_n t^{\mathbf{p}-n},$$

for some $b_i \in \mathbf{R}$, $i = 1, 2, \ldots, n$, where $N = [\mathbf{p}] + 1$.

Lemma 3 *Let $f \in L(\mathbf{I})$ and $\mathbf{p}, \mathbf{q}$ are two constants such that $\mathbf{p} > 1 > \mathbf{q} \geq 0$, then*

$$D_{0+}^{\mathbf{q}} \mathscr{I}_{0+}^{\mathbf{p}} f(t) = \mathscr{I}_{0+}^{\mathbf{p}-\mathbf{q}} f(t)$$

$$= \frac{1}{\Gamma(\mathbf{p}-\mathbf{q})} \int_0^t (t-s)^{\mathbf{p}-\mathbf{q}-1} f(s) ds.$$

To get our main results of existence, we present the following lemmas from fixed point theory.

Theorem 1 ([12]) *Let $\mathbf{X}$ be a Banach space, $\mathscr{P} \subset \mathbf{X}$ is a cone, and $\mathbf{A}_1$, $\mathbf{A}_2$ are open sets with $0 \in \mathbf{A}_1$, $\overline{\mathbf{A}_1} \subset \mathbf{A}_2$, and let $\mathscr{T} : \mathscr{P} \cap (\overline{\mathbf{A}_2} \setminus \mathbf{A}_1) \to \mathscr{P}$ be a completely continuous operator such that either*

(i) $\|\mathscr{T}u\|_{\mathbf{X}} \leq \|u\|_{\mathbf{X}}, \; u \in \mathscr{P} \cap \partial \mathbf{A}_1, \; and \; \|\mathscr{T}u\|_{\mathbf{X}} \geq \|u\|_{\mathbf{X}}, \; u \in \mathscr{P} \cap \partial \mathbf{A}_2, or$

(ii) $\|\mathscr{T}u\|_{\mathbf{X}} \geq \|u\|_{\mathbf{X}}, \; u \in \mathscr{P} \cap \partial \mathbf{A}_1, \; and \; \|\mathscr{T}u\|_{\mathbf{X}} \leq \|u\|_{\mathbf{X}}, \; u \in \mathscr{P} \cap \partial \mathbf{A}_2,$

then $\mathscr{T}$ has a fixed point in $\mathscr{P} \cap (\overline{\mathbf{A}_2} \setminus \mathbf{A}_1)$.

Theorem 2 ([12]) *Let* $\mathbf{X}$ *be a Banach space with* $\mathscr{P} \subset \mathbf{X}$ *is closed and convex set. Assume that* $\mathbf{U}$ *is a relatively open subset of* $\mathscr{P}$ *with* $0 \in \mathbf{U}$ *and* $\mathscr{T} : \overline{\mathbf{U}} \to \mathscr{P}$ *is completely continuous. Then either*

1. $\mathscr{T}$ *has a fixed in* $\overline{\mathbf{U}}$, *or*
2. *there exist* $u \in \partial \mathbf{U}$ *and* $\gamma \in (0, 1)$ *with* $u = \gamma \mathscr{T}u$.

Theorem 3 (Leggett–Williams's fixed point theorem, [16]) *Let* $\mathscr{P}$ *be a cone in a real Banach space* $\mathbf{X}$, $\mathscr{P}_c = \{u; \|u\|_{\mathbf{X}} < c\}$, θ *is a nonnegative continuous concave functional on* $\mathscr{P}$ *such that* $\theta(u) \leq \|u\|_{\mathbf{X}}$, *for all* $u \in \bar{\mathscr{P}}_c$, *and* $\mathscr{P}(\theta, b, d) = \{u \in \mathscr{P}; b \leq \theta(u), \|u\|_{\mathbf{X}} \leq d\}$. *Suppose that* $\mathscr{T} : \bar{\mathscr{P}}_c \to \bar{\mathscr{P}}_c$ *is completely continuous and there exist positive constants* $0 < a < b < d \leq c$ *such that*

(C_1) $\{u \in \mathscr{P}(\theta, b, d) : \theta(u) > b\} \neq \emptyset$ *and* $\theta(\mathscr{T}u) > b$, *for* $u \in \mathscr{P}(\theta, b, d)$,
(C_2) $\|\mathscr{T}u\|_{\mathbf{X}} < a$ *for* $u \in \bar{\mathscr{P}}_a$,
(C_2) $\theta(\mathscr{T}u) > b$ *for* $u \in \mathscr{P}(\theta, b, c)$ *with* $\|\mathscr{T}u\|_{\mathbf{X}} > d$.

Then $\mathscr{T}$ *has at least three fixed points* u_1, u_2, *and* u_3 *with*

$$\|u\|_{\mathbf{X}} < a, \quad b < \theta(u_2), \quad a < \|u_3\|_{\mathbf{X}}, \quad with \; \theta(u_3) < b.$$

3 Existence Theory

Lemma 4 *Let* $y \in L(\mathbf{I})$ *and* $1 < \mathbf{p} \leq 2$, *then FDE*

$$D^{\mathbf{p}}_{+0}u(t) + y(t) = 0, \; t \in \mathbf{I}, \tag{2}$$

$$D^{\mathbf{q}}_{+0}u(0) = 0, \; D^{\mathbf{q}}_{+0}u(1) = \lambda u(\xi), \quad 0 \leq \mathbf{q} < 1, \quad \mathbf{p} - \mathbf{q} \geq 1,$$

where $\lambda, \xi \in (0, 1)$ *and* $\lambda \Gamma(\mathbf{p} - \mathbf{q})\xi^{\mathbf{p}-2} \leq \Gamma(\mathbf{p})(1 - \mathbf{q})$ *has a solution given as*

$$u(t) = \int_0^1 \mathbf{K}(t, s)y(s)ds,$$

where

$$\mathbf{K}(t,s) = \frac{1}{\Gamma(\mathbf{p})} \begin{cases} d\Gamma(\mathbf{p})\,t^{\mathbf{p}-1}(1-s)^{\mathbf{p}-\mathbf{q}-1} \\ \qquad -dt^{\mathbf{p}-1}\lambda\Gamma(\mathbf{p}-\mathbf{q})(\xi-s)^{\mathbf{p}-1} - (t-s)^{\mathbf{p}-1}, \\ \qquad if\ 0 \le s \le \min\{\xi,t\} < 1 \\[2mm] d\Gamma(\mathbf{p})t^{\mathbf{p}-1}(1-s)^{\mathbf{p}-\mathbf{q}-1} - (t-s)^{\mathbf{p}-1}, \ \ if\ 0 \le \xi \le s \le t \le 1 \\[2mm] d\Gamma(\mathbf{p})t^{\mathbf{p}-1}(1-s)^{\mathbf{p}-\mathbf{q}-1} - dt^{\mathbf{p}-1}\lambda\Gamma(\mathbf{p}-\mathbf{q})(\xi-s)^{\mathbf{p}-1}, \\ \qquad if\ 0 \le t \le s \le \xi < 1 \\[2mm] d\Gamma(\mathbf{p})t^{\mathbf{p}-1}(1-s)^{\mathbf{p}-\mathbf{q}-1}, \ \ if\ \max\{\xi,t\} \le s \le 1, \end{cases}$$

and $d = (\Gamma(\mathbf{p}) - \Gamma(\mathbf{p}-\mathbf{q})\lambda\xi^{\mathbf{p}-1})^{-1}$.
Or

$$\mathbf{K}(t,s) = \begin{cases} g_1(t,s), & if\ 0 \le s \le \min\{\xi,t\} < 1 \\[2mm] g_2(t,s), & if\ 0 \le \xi \le s \le t \le 1 \\[2mm] g_3(t,s), & if\ 0 \le t \le s \le \xi < 1 \\[2mm] g_4(t,s), & if\ \max\{\xi,t\} \le s \le 1. \end{cases}$$

Proof Applying Lemma 2 to the linear BVP (2), we get

$$u(t) = b_1 t^{\mathbf{p}-1} + b_2 t^{\mathbf{p}-2} - \mathscr{I}^{\mathbf{p}} y(t),$$

for some real constants b_1 and b_2. Also by using definition 2, we have

$$D_{0+}^{\mathbf{q}} u(t) = b_1 \frac{\Gamma(\mathbf{p})}{\Gamma(\mathbf{p}-\mathbf{q})} t^{\mathbf{p}-\mathbf{q}-1} + b_2 \frac{\Gamma(\mathbf{p}-1)}{\Gamma(\mathbf{p}-\mathbf{q}-1)} t^{\mathbf{p}-\mathbf{q}-2} - \mathscr{I}^{\mathbf{p}-\mathbf{q}} y(t),$$

Multiplying both sides by $t^{2+\mathbf{q}-\mathbf{p}}$ and taking $t \to 0$, we get $b_2 = 0$, therefore

$$u(t) = b_1 t^{\mathbf{p}-1} - \mathscr{I}^{\mathbf{p}} y(t),$$

and

$$D_{+0}^{\mathbf{q}} u(t) = b_1 \frac{\Gamma(\mathbf{p})}{\Gamma(\mathbf{p}-\mathbf{q})} t^{\mathbf{p}-\mathbf{q}-1} - \mathscr{I}^{\mathbf{p}-\mathbf{q}} y(t).$$

Using boundary condition $D_{+0}^{\mathbf{q}} u(1) = \lambda u(\xi)$, we get

$$b_1 = \frac{\Gamma(\mathbf{p}-\mathbf{q})}{\Gamma(\mathbf{p}) - \Gamma(\mathbf{p}-\mathbf{q})\lambda\xi^{\mathbf{p}-1}}(\mathscr{I}^{\mathbf{p}-\mathbf{q}}y(1) - \lambda\mathscr{I}^{\mathbf{p}}y(\xi)).$$

Hence, BVP (2) has unique solution given as

$$u(t) = \frac{d}{\Gamma(\mathbf{p})}\Gamma(\mathbf{p})t^{\mathbf{p}-1}\int_0^1 (1-s)^{\mathbf{p}-\mathbf{q}-1}y(s)ds$$

$$+ \frac{1}{\Gamma(\mathbf{p})}\left[-dt^{\mathbf{p}-1}\lambda\Gamma(\mathbf{p}-\mathbf{q})\int_0^\xi (\xi-s)^{\mathbf{p}-1}y(s)ds - \int_0^t (t-s)^{\mathbf{p}-1}y(s)ds \right]$$

$$= \int_0^1 \mathbf{K}(t,s)y(s)ds.$$

Lemma 5 *The Green's function* $\mathbf{K}(t,s)$ *obeys the following hypothesis:*

1. $\mathbf{K}(t,s)$ *is continuous on* $\mathbf{I} \times \mathbf{I}$,
2. $\mathbf{K}(t,s) > 0$ *for any* $s,\ t \in (0,1)$.

Proof Clearly $\mathbf{K}(t,s)$ is continuous function on $\mathbf{I} \times \mathbf{I}$. To prove the second part we prove that each $g_i(t,s) > 0$ for $i = 1,2,3,4$. Now taking

$$g_1(t,s) = \frac{t^{\mathbf{p}-1}}{\Gamma(\mathbf{p})}h(t,s),$$

where

$$h(t,s) = d\Gamma(\mathbf{p})(1-s)^{\mathbf{p}-\mathbf{q}-1} - d\lambda\Gamma(\mathbf{p}-\mathbf{q})(\xi-s)^{\mathbf{p}-1} - (1-\frac{s}{t})^{\mathbf{p}-1}$$

implies that for all $t \in [s,1]$,

$$\frac{\partial h(t,s)}{\partial t} = -(\mathbf{p}-1)(1-\frac{s}{t})^{\mathbf{p}-2}(\frac{s}{t^2}) \leq 0,$$

Showing that $h(t,s)$ is decreasing function at $t \in [s,1]$, but always non-negative. Because for maximum value of $t = 1$,

$$h(1,s) = d\Gamma(\mathbf{p})(1-s)^{\mathbf{p}-\mathbf{q}-1} - d\lambda\Gamma(\mathbf{p}-\mathbf{q})(\xi-s)^{\mathbf{p}-1} - (1-s)^{\mathbf{p}-1}$$

so

$$h(1,0) = d\Gamma(\mathbf{p}) - d\lambda\Gamma(\mathbf{p}-\mathbf{q})\xi^{\mathbf{p}-1} - 1$$

$$= d(\Gamma(\mathbf{p}) - \lambda\Gamma(\mathbf{p}-\mathbf{q})\xi^{\mathbf{p}-1}) - 1$$

$$= 0.$$

Also

$$h(1, \xi) = d\Gamma(\mathbf{p})(1 - \xi)^{\mathbf{p}-\mathbf{q}-1} - (1 - \xi)^{\mathbf{p}-1}$$
$$\geq d(1 - \xi)^{\mathbf{p}-1}(\Gamma(\mathbf{p})(1 + \mathbf{q}\xi) - (\Gamma(\mathbf{p}) - \lambda\Gamma(\mathbf{p} - \mathbf{q})\xi^{\mathbf{p}-1})$$
$$= d(1 - \xi)^{\mathbf{p}-1}\xi(\mathbf{q}\Gamma(\mathbf{p}) + \lambda\Gamma(\mathbf{p} - \mathbf{q})\xi^{\mathbf{p}-2})$$
$$> 0.$$

Further

$$\frac{\partial h(1, s)}{\partial s} = \quad -d\Gamma(\mathbf{p})(\mathbf{p} - \mathbf{q} - 1)(1 - s)^{\mathbf{p}-\mathbf{q}-2}$$
$$+\lambda d(\mathbf{p} - 1)(\xi - s)^{\mathbf{p}-2} + (\mathbf{p} - 1)(1 - s)^{\mathbf{p}-2}.$$

Let at $s = s^*$, $\frac{\partial h(1,s)}{\partial s} = 0$, which implies that

$$\lambda d(\xi - s^*)^{\mathbf{p}-2} = \frac{d\Gamma(\mathbf{p})(\mathbf{p} - \mathbf{q} - 1)(1 - s^*)^{\mathbf{p}-\mathbf{q}-2} - (\mathbf{p} - 1)(1 - s^*)^{\mathbf{p}-2}}{\mathbf{p} - 1}.$$

Therefore

$$h(1, s^*) = d\Gamma(\mathbf{p})(1 - s^*)^{\mathbf{p}-\mathbf{q}-1}$$
$$-\frac{d\Gamma(\mathbf{p})(\mathbf{p} - \mathbf{q} - 1)(1 - s^*)^{\mathbf{p}-\mathbf{q}-2} - (\mathbf{p} - 1)(1 - s^*)^{\mathbf{p}-2}}{\mathbf{p} - 1}(\xi - s^*)$$
$$-(1 - s^*)^{\mathbf{p}-1}$$
$$> d\Gamma(\mathbf{p})(1 - s^*)^{\mathbf{p}-\mathbf{q}-1} - d\Gamma(\mathbf{p})(1 - s^*)^{\mathbf{p}-\mathbf{q}-2}(\xi - s^*)$$
$$+(1 - s^*)^{\mathbf{p}-2}(\xi - s^*)$$
$$-(1 - s^*)^{\mathbf{p}-1}$$
$$> d(1 - \xi)(1 - s^*)^{\mathbf{p}-2}(\Gamma(\mathbf{p})\mathbf{q}s^* + \lambda\Gamma(\mathbf{p} - \mathbf{q})\xi^{\mathbf{p}-1})$$
$$> 0.$$

Hence we have

$$\min_{0 \leq s \leq \xi} h(1, s) = \min\{h(1, 0), h(1, s^*), h(1, \xi)\} = 0.$$

Also

$$\frac{\partial h(1, 0)}{\partial s} = -d\Gamma(\mathbf{p})(\mathbf{p} - \mathbf{q} - 1) + \lambda d\Gamma(\mathbf{p} - \mathbf{q})(\mathbf{p} - 1)\xi^{\mathbf{p}-2} + \mathbf{p} - 1.$$

But

$$\lambda d\Gamma(\mathbf{p} - \mathbf{q})(\mathbf{p} - 1)\xi^{\mathbf{p}-2} \geq d\Gamma(\mathbf{p})(\mathbf{p} - \mathbf{q} - 1),$$

if $\mathbf{p} \leq 2$ and $\lambda\Gamma(\mathbf{p} - \mathbf{q})\xi^{\mathbf{p}-2} \leq \mathbf{p}\Gamma(\mathbf{p})(1 - \mathbf{q})$, which show that $\frac{\partial h(1,0)}{\partial s} > 0$. Hence we concluded that $h(1, s) > 0$ for all $s \in (0, \xi]$, or $g_1(t, s) > 0$ for all $0 \leq s \leq \min\{t, \xi\} < 1$. By using similar argument we can prove that each $g_i(t, s)$ for all $i = 2, 3, 4$, are positive in their corresponding interval. Therefore $\mathbf{K}(t, s) > 0$ for all $0 < s, t < 1$.

Lemma 6 *The function* $\mathbf{K}(t, s)$ *satisfies the following hypothesis*

1. $\mathbf{K}(t, s) \leq \mathbf{K}(s, s)$, *for all* $s, t \in \mathbf{I}$,
2. *There exists a positive function* $\gamma(s) \in C(0, 1)$ *such that*

$$\min_{\xi \leq t \leq 1} \mathbf{K}(t, s) \geq \gamma(s) \max_{0 \leq t \leq 1} \mathbf{K}(t, s) = \gamma(s)\mathbf{K}(s, s), \quad for \quad 0 < s < 1.$$

Proof For part first we will prove that $g_1(t, s)$ and $g_2(t, s)$ are both decreasing with respect to $t \in [s, 1]$, while $g_3(t, s)$ and $g_4(t, s)$ are both increasing with respect to $t \in [0, s]$.
Consider

$$g_1(t, s) = \frac{h(t, s)}{\Gamma(\mathbf{p})} \quad \text{if} \quad 0 \leq s \leq \{t, \xi\} < 1,$$

and

$$\frac{\partial h(t, s)}{\partial t} = d\Gamma(\mathbf{p})(\mathbf{p} - 1)t^{\mathbf{p}-2}(1 - s)^{\mathbf{p}-\mathbf{q}-1} - d(\mathbf{p} - 1)t^{\mathbf{p}-2}\lambda\Gamma(\mathbf{p} - \mathbf{q})(\xi - s)^{\mathbf{p}-1}$$

$$-(\mathbf{p} - 1)(t - s)^{\mathbf{p}-2}$$

$$\leq (\mathbf{p} - 1)t^{\mathbf{p}-2}\left[d\Gamma(\mathbf{p})(1 - s)^{\mathbf{p}-\mathbf{q}-1} - (1 - s)^{\mathbf{p}-2} \right]$$

$$\leq (\mathbf{p} - 1)d(t(1 - s))^{\mathbf{p}-2}\xi\left[-\Gamma(\mathbf{p})(1 - \mathbf{q}) + \Gamma(\mathbf{p} - \mathbf{q})\lambda\xi^{\mathbf{p}-2} \right]$$

$$\leq 0.$$

From which we see that $g_1(t, s)$ is decreasing function for t. Similarly one can prove that $g_2(t, s)$ is decreasing function for $t \in [s, 1]$.
Further

$$\frac{\partial g_3(t, s)}{\partial t} = \frac{d(\mathbf{p} - 1)t^{\mathbf{p}-2}}{\Gamma(\mathbf{p})}(\Gamma(\mathbf{p})(1 - s)^{\mathbf{p}-\mathbf{q}-1} - \lambda\Gamma(\mathbf{p} - \mathbf{q})(\xi - s)^{\mathbf{p}-1})$$

$$= \frac{d(\mathbf{p}-1)t^{\mathbf{p}-2}}{\Gamma(\mathbf{p})}r(s),$$

where

$$r(s) = \Gamma(\mathbf{p})(1-s)^{\mathbf{p}-\mathbf{q}-1} - \lambda\Gamma(\mathbf{p}-\mathbf{q})(\xi-s)^{\mathbf{p}-1} \text{ for } 0 \le s \le \xi.$$

Now

$$r(0) = (\Gamma(\mathbf{p}) - \lambda\Gamma(\mathbf{p}-\mathbf{q})\xi^{\mathbf{p}-1})$$
$$= d^{-1} > 0,$$

and

$$r(\xi) = \Gamma(\mathbf{p})(1-\xi)^{\mathbf{p}-\mathbf{q}-1} \ge 0.$$

Also

$$r'(s) = -\Gamma(\mathbf{p})(\mathbf{p}-\mathbf{q}-1)(1-s)^{\mathbf{p}-\mathbf{q}-2} + \lambda\Gamma(\mathbf{p}-\mathbf{q})(\mathbf{p}-1)(\xi-s)^{\mathbf{p}-2},$$

Setting $r'(s^*) = 0$ which implies that

$$\lambda\Gamma(\mathbf{p}-\mathbf{q})(\mathbf{p}-1)(\xi-s^*)^{\mathbf{p}-2} = (\mathbf{p}-\mathbf{q}-1)\Gamma(\mathbf{p})(1-s^*)^{\mathbf{p}-\mathbf{q}-2},$$

or

$$\lambda\Gamma(\mathbf{p}-\mathbf{q})(\xi-s^*)^{\mathbf{p}-2} = \frac{\mathbf{p}-\mathbf{q}-1}{\mathbf{p}-1}\Gamma(\mathbf{p})(1-s^*)^{\mathbf{p}-\mathbf{q}-2}.$$

Therefore

$$r(s^*) = \Gamma(\mathbf{p})(1-s^*)^{\mathbf{p}-\mathbf{q}-1} - \Gamma(\mathbf{p})(\frac{\mathbf{p}-\mathbf{q}-1}{\mathbf{p}-1})(1-s^*)^{\mathbf{p}-\mathbf{q}-2}(\xi-s^*)$$

$$> \Gamma(\mathbf{p})(1-s^*)^{\mathbf{p}-\mathbf{q}-2}\left[1-s^*-\xi+s^*\right]$$

$$> 0.$$

Hence $\min r(s) > 0$ which implies that $\frac{\partial g_3(t,s)}{\partial t} > 0$, showing that $g_3(t,s)$ is increasing function for $t \in [s, 1]$. Finally $g_4(t, s)$ is also increasing function for $t \in [s, 1]$, because

$$\frac{\partial g_4(t, s)}{\partial t} = d(\mathbf{p}-1)t^{\mathbf{p}-2}(1-s)^{\mathbf{p}-\mathbf{q}-1} > 0.$$

Hence we concluded that $\mathbf{K}(t, s)$ is increasing function for $s \leq t$, and decreasing for $s \geq t$.

So $\mathbf{K}(t, s) \leq \mathbf{K}(s, s)$ for $s, t \in \mathbf{I}$. Also, one has

$$
\min_{\xi \leq t \leq 1} \mathbf{K}(t, s) =
\begin{cases}
\min_{\xi \leq t \leq 1}\{g_1(t, s), g_3(t, s)\}, & \text{if } 0 \leq s \leq \xi, \\[2mm]
\min_{\xi \leq t \leq 1}\{g_2(t, s), g_4(t, s)\}, & \text{if } \xi \leq s \leq 1,
\end{cases}
$$

or

$$
\min_{\xi \leq t \leq 1} \mathbf{K}(t, s) =
\begin{cases}
g_1(t, s), & \text{if } 0 \leq s \leq \xi, \\[2mm]
\psi(s), & \text{if } \xi \leq s \leq 1,
\end{cases}
$$

where $\psi(s) = \min\{g_2(1, s), g_4(\xi, s)\}$. From which we have

$$
\frac{\min_{\xi \leq t \leq 1} \mathbf{K}(t, s)}{\mathbf{K}(s, s)} =
\begin{cases}
\frac{g_1(t,s)}{\mathbf{K}(s,s)}, & \text{if } 0 \leq s \leq \xi, \\[2mm]
\frac{\psi(s)}{\mathbf{K}(s,s)}, & \text{if } \xi \leq s \leq 1,
\end{cases}
$$

where

$$
\mathbf{K}(s, s) =
\begin{cases}
\dfrac{d\Gamma(\mathbf{p})s^{\mathbf{p}-1}(1-s)^{\mathbf{p}-\mathbf{q}-1} - d\lambda\Gamma(\mathbf{p}-\mathbf{q})s^{\mathbf{p}-1}(\xi-s)^{\mathbf{p}-1}}{\Gamma(\mathbf{p})}, & \text{if } 0 \leq s \leq \xi, \\[3mm]
ds^{\mathbf{p}-1}(1 - s)^{\mathbf{p}-\mathbf{q}-1}, & \text{if } \xi \leq s \leq 1.
\end{cases}
$$

Therefore taking

$$
\gamma(s) =
\begin{cases}
\frac{g_1(t,s)}{\mathbf{K}(s,s)}, & \text{if } 0 \leq s \leq \xi, \\[2mm]
\frac{\psi(s)}{\mathbf{K}(s,s)}, & \text{if } \xi \leq s \leq 1,
\end{cases}
$$

we have

$$
\min_{\xi \leq t \leq 1} \mathbf{K}(t, s) \leq \gamma(s)\mathbf{K}(s, s) = \gamma(s) \max_{0 \leq t \leq 1} \mathbf{K}(t, s), \tag{3}
$$

where $0 < s < 1$ is the required result.

Theorem 4 *If there exists a real valued function $h \in L(\mathbf{I})$ such that $|f(t, u) - f(t, v)| \leq h(t)|u - v|$, for $t \in \mathbf{I}$ and for all $u, v \in [0, \infty)$ and moreover if*

$$0 < \int_0^1 \mathbf{K}(s,s)h(s)ds < 1,$$

then there exists a unique positive solution of FDE (2) on **I**.

Proof Let $\mathbf{X} = C(\mathbf{I}, \mathbf{R})$ be the Banach space of all continuous functions with $\|u\|_{\mathbf{X}} = \sup_{0 \le t \le 1} |u(t)|$ and $\mathscr{P}$ be the cone such that

$$\mathscr{P} = \{u \in \mathbf{X} : u(t) \ge 0, t \in \mathbf{I}\}.$$

We prove that the mapping

$$\mathscr{T} : \mathscr{P} \to \mathscr{P}$$

defined as

$$\mathscr{T}u(t) = \int_0^1 \mathbf{K}(t,s)\mathrm{f}(s,u(s))ds.$$

is contraction mapping. We consider $u, v \in \mathscr{P}$ such that

$$|\mathscr{T}u(t) - \mathscr{T}v(t)| \le \int_0^1 \mathbf{K}(t,s)|\mathrm{f}(s,u(s)) - \mathrm{f}(s,v(s))|)ds$$

$$\le \|u - v\|_{\mathbf{X}} \int_0^1 \mathbf{K}(s,s)h(s)ds.$$

This implies that

$$\|\mathscr{T}u - \mathscr{T}v\|_{\mathbf{X}} \le \varphi \|u - v\|_{\mathbf{X}},$$

where $\varphi = \int_0^1 \mathbf{K}(s,s)h(s)ds \in (0,1)$. Hence considered FDE (2) has unique solution.

Theorem 5 *Let there exist two nonnegative real valued function* $\alpha, \beta \in L(\mathbf{I})$ *with* $\mathrm{f}(t,u) \le \alpha(t) + \beta(t)u$, *for every* $(t,u) \in \mathbf{I} \times [0, \infty)$. *Then the mapping* $\mathscr{T} : \mathscr{P} \to \mathscr{P}$ *defined by* $\mathscr{T}u(t) = \int_0^1 \mathbf{K}(t,s)\mathrm{f}(s,u(s))ds$ *is completely continuous.*

Proof

Step 1 First we will show that operator $\mathscr{T} : \mathscr{P} \to \mathscr{P}$ is continuous. Let $u_n(t)$ be converged to $u \in \mathscr{P}$ then by continuity of $\mathrm{f}(t,u(s))$, $\lim_{n \to \infty} \mathrm{f}(t,u_n(t)) = \mathrm{f}(t,u(t))$ for $t \in \mathbf{I}$.

So we have as $n \to \infty$

$$\sup_{s \in \mathbf{I}} \|\mathrm{f}(s,u_n(s)) - \mathrm{f}(t,u(s))\|_{\mathbf{X}} \to 0.$$

Now consider

$$|(\mathscr{T}u_n)(t) - (\mathscr{T}u)(t)| = \left| \int_0^1 \mathbf{K}(t, s)[f(s, u_n(s)) - f(s, u(s))]ds \right|$$

$$\leq \sup_{s \in \mathbf{I}} \left| f(s, u_n(s)) - f(t, u(s)) \right| \int_0^1 \mathbf{K}(s, s)ds.$$

From which it is clear that

$$\|\mathscr{T}u_n - \mathscr{T}u\|_\mathbf{X} \to 0, \ n \to \infty.$$

Hence $\mathscr{T}$ is continuous.

Step 2 To show $\mathscr{T}(u)$ is bounded for every bounded set u of $\mathscr{P}$. Let $\|u\|_\mathbf{X} \leq r$, so we have

$$|\mathscr{T}u(t)| = \left| \int_0^1 \mathbf{K}(t, s)f(s, u(s))ds \right|$$

$$\leq \int_0^1 \mathbf{K}(t, s)\alpha(s)ds + \int_0^1 \mathbf{K}(t, s)\beta(s)\|u\|_\mathbf{X}ds$$

$$\leq \int_0^1 \mathbf{K}(s, s)\alpha(s)ds + r \int_0^1 \mathbf{K}(s, s)\beta(s)ds$$

$$= l,$$

which yields

$$\|\mathscr{T}u\|_\mathbf{X} \leq l.$$

Step 3 To show $\mathscr{T}$ maps every bounded set of $\mathscr{P}$ into equi-continuous set of $\mathscr{P}$. Let u belong to a bounded subset of $\mathscr{P}$ such that $\|u\|_\mathbf{X} \leq r$, then we have

$$|\mathscr{T}u(t_2) - \mathscr{T}u(t_1)| = \left| \int_0^1 |\mathbf{K}(t_2, s) - \mathbf{K}(t_1, s)|f(s, u(s))ds \right|$$

$$\leq \int_0^1 |\mathbf{K}(t_2, s) - \mathbf{K}(t_1, s)|(\alpha(s) + \beta(s)\|u(s)\|_\mathbf{X})ds$$

$$\leq l \int_0^1 |\mathbf{K}(t_2, s) - \mathbf{K}(t_1, s)|(\alpha(s) + \beta(s)r)ds.$$

Since $\mathbf{K}(t, s)$ is continuous on $\mathbf{I} \times \mathbf{I}$, so also uniformly continuous on $\mathbf{I} \times \mathbf{I}$. Hence for any $\epsilon > 0$, there exist $\delta > 0$ such that

$$|\mathbf{K}(t_2, s) - \mathbf{K}(t_1, s)| < \frac{\epsilon}{l \times \int_0^1 (\alpha(s) + r\beta(s))ds},$$

whenever $|t_2 - t_1| < \delta$,
which implies that

$$||\mathscr{T}u(t_2) - \mathscr{T}u(t_1)||_{\mathbf{X}} < \epsilon,$$

whenever $|t_2 - t_1| < \delta$.
Hence $\{\mathscr{T}u\}$ is equi-continuous. Therefore by Arzelá–Ascoli theorem the operator $\mathscr{T} : \mathscr{P} \to \mathscr{P}$ is completely continuous.

Theorem 6 *Assume that there exist two nonnegative real valued function* $\alpha, \beta \in L(\mathbf{I})$ *such that* $f(t, x) \leq \alpha(t) + \beta(t)x$, *for every* $(t, u) \in \mathbf{I} \times [0, \infty)$. *If*

$$\int_0^1 \mathbf{K}(s, s)\beta(s)ds < 1,$$

then the FDE (2) has at least one positive solution.

Proof Let

$$\mathbf{U} = \{u \in \mathscr{P} : ||u||_{\mathbf{X}} < r\},$$

where

$$r = \frac{\int_0^1 \mathbf{K}(s, s)\alpha(s)ds}{1 - \int_0^1 \mathbf{K}(s, s)\beta(s)ds} > 0,$$

and for a completely continuous operator

$$\mathscr{T} : \overline{\mathbf{U}} \to \mathscr{P},$$

defined by

$$\mathscr{T}u(t) = \int_0^1 \mathbf{K}(t, s)f(s, u(s))ds,$$

assume that there exist $u \in \mathscr{P}$ and $\gamma \in (0, 1)$ such that $u = \gamma \mathscr{T}u$, we claim that $||u||_{\mathbf{X}} \neq r$. So we get

$$u(t) = \gamma \int_0^1 \mathbf{K}(t, s)\mathbf{f}(s, u(s))ds$$

$$\leq \gamma \int_0^1 \mathbf{K}(t, s)(\alpha(s) + \beta(s)u(s))ds$$

$$\leq \gamma \left[\int_0^1 \mathbf{K}(s, s)\alpha(s)ds + \int_0^1 \mathbf{K}(s, s)\beta(s)\|u\|_\mathbf{X} ds \right],$$

or

$$\|u\|_\mathbf{X} \leq \gamma \left[\int_0^1 \mathbf{K}(s, s)\alpha(s)ds + r \int_0^1 \mathbf{K}(s, s)\beta(s)ds \right]$$

$$< \int_0^1 \mathbf{K}(s, s)\alpha(s)ds + r \int_0^1 \mathbf{K}(s, s)\beta(s)ds = r$$

or in other words u is not in $\partial \mathbf{U}$. Hence $\mathscr{T}$ has a fixed point $u \in \overline{\mathbf{U}}$. Therefore by Theorem 2, the FDE(2) has at least one positive solution.

Theorem 7 *Assume that there exist*

1. *Two positive real valued functions* $\alpha, \beta \in L(\mathbf{I})$, *such that* $\mathbf{f}(t, u) \leq \alpha(t) + \beta(t)u$, *for every* $t \in \mathbf{I}$ *and for* $u \in [0, \infty)$,
2. $\mathbf{f}(t, u) \leq \mathscr{M} r_2$, *for* $(t, s) \in \mathbf{I} \times [0, r_2]$,
3. $\mathbf{f}(t, u) \geq \mathscr{N} r_1$, *for* $(t, s) \in \mathbf{I} \times [0, r_1]$,

where r_1 and r_2 are positive constants with $0 < r_1 < r_2$ and

$$\mathscr{M} = \left(\int_0^1 \mathbf{K}(s, s)\beta(s)ds \right)^{-1}, \quad \mathscr{N} = \left(\int_\xi^1 \mathbf{K}(s, s)n(s)ds \right)^{-1}.$$

Then the FDE (2) has at least a positive solution.

Proof Given that there exist two positive real valued functions $\alpha, \beta \in L(\mathbf{I})$ with $\mathbf{f}(t, u) \leq \alpha(t) + \beta(t)u$, for every $(t, u) \in \mathbf{I} \times [0, \infty)$. Therefore the mapping

$$\mathscr{T} : \mathscr{P} \to \mathscr{P}$$

defined by

$$\mathscr{T}u(t) = \int_0^1 \mathbf{K}(t, s)\mathbf{f}(s, u(s))ds,$$

is completely continuous. Also

(a) Let $\mathbf{A}_2 = \{u \in \mathbf{X} : \|u\|_\mathbf{X} < r_2\}$. For $u \in \mathscr{P} \cap \partial \mathbf{A}_2$, $\|u\|_\mathbf{X} = r_2$ which yields that $0 \leq u(t) \leq r_2$ for each $t \in \mathbf{I}$. Then by condition (2) we have

$$|\mathscr{T}u(t)| = \left| \int_0^1 \mathbf{K}(t, s)\mathrm{f}(s, u(s))ds \right|$$

$$\leq \int_0^1 \mathbf{K}(s, s)\mathscr{M}r_2 ds$$

$$\leq r_2,$$

which means that

$$\|\mathscr{T}u\|_{\mathbf{X}} \leq \|u\|_{\mathbf{X}}.$$

(b) Let $\mathbf{A}_1 = \{u \in \mathbf{X} : \|u\|_{\mathbf{X}} < r_1\}$. For $u \in \mathscr{P} \cap \partial\mathbf{A}_1$, $\|u\|_{\mathbf{X}} = r_1$ then for every $t \in [\xi, 1]$ we have

$$|\mathscr{T}u(t)| = \left| \int_0^1 \mathbf{K}(t, s)\mathrm{f}(s, u(s))ds \right|$$

$$\geq \int_0^1 \gamma(s)\mathbf{K}(s, s)\mathscr{N}r_1 ds$$

$$\geq \mathscr{N}r_1 \int_\xi^1 \gamma(s)\mathbf{K}(s, s)ds = r_1,$$

which produces

$$\|\mathscr{T}u\|_{\mathbf{X}} \geq \|u\|_{\mathbf{X}}.$$

Hence by Lemma 1, $\mathscr{T}$ has a fixed point $u \in \mathscr{P} \cap (\overline{\mathbf{A}_2} \setminus \mathbf{A}_1)$, which is the required solution of FDE (2).

Theorem 8 *Assume that there exist two positive real valued functions $\alpha, \beta \in L(\mathbf{I})$ such that $\mathrm{f}(t, u) \leq \alpha(t) + \beta(t)u$, for every $t \in \mathbf{I}$ and*

1. $\mathrm{f}(t, u) < \mathscr{M}a_1$, for $(t, u) \in \mathbf{I} \times [0, a_1]$,
2. $\mathrm{f}(t, u) \leq \mathscr{M}c$, for $(t, u) \in \mathbf{I} \times [0, c]$,
3. $\mathrm{f}(t, u) \geq \mathscr{N}b$, for $(t, u) \in [\xi, 1] \times [b, c]$,

where a_1, b, and c are constants such that $0 < a_1 < b < c$, then the FDE (2) has at least three positive solutions $u_1, u_2,$ and u_3 with

$$\|u_1\|_{\mathbf{X}} < a_1, \quad b < \theta(u_2) < \|u_2\|_{\mathbf{X}} < c, \quad a_1 < \|u_3\|_{\mathbf{X}}, \quad \theta(u_3) < b.$$

Proof To derive all the assumptions of Lemma 3, we follow as if $u \in \overline{\mathscr{P}_c}$, then $\|u\|_{\mathbf{X}} \leq c$. So $0 \leq u(t) \leq c$ for $t \in \mathbf{I}$. We have

$$|\mathscr{T}u(t)| = \left| \int_0^1 \mathbf{K}(t,s)\mathbf{f}(s,u(s))ds \right|$$

$$\leq \int_0^1 \mathbf{K}(s,s)\mathscr{M}c\,ds$$

$$= \mathscr{M}c \int_0^1 \mathbf{K}(s,s)ds = c,$$

which implies that

$$\|\mathscr{T}u\|_{\mathbf{X}} \leq c, \quad u \in \overline{\mathscr{P}_c}.$$

Hence $\mathscr{T} : \overline{\mathscr{P}_c} \to \overline{\mathscr{P}_c}$ is operator which is completely continuous.

Following similar fashion one claims that if $u \in \overline{\mathscr{P}_{a_1}}$, then $\|\mathscr{T}u\|_{\mathbf{X}} \leq a_1$. Further we choose $u(t) = \frac{b+c}{2}$, $t \in \mathbf{I}$, it is clear that $u(t) = \frac{b+c}{2} \in \mathscr{P}(\theta, b, c)$, $\theta(u) = \frac{b+c}{2} > b$. Therefore $\{u \in \mathscr{P}(\theta, b, c) : \theta(u) > b\} \neq \emptyset$. Also, if $u \in \mathscr{P}(\theta, b, c)$ then $b \leq u(t) \leq c$, $t \in [\xi, 1]$. And by last condition, we have

$$\mathbf{f}(t, u(t)) \geq \mathscr{N}b.$$

Hence, we get

$$\theta(\mathscr{T}u) = \min_{\xi \leq t \leq 1} |\mathscr{T}u(t)|$$

$$= \min_{\xi \leq t \leq 1} \left| \int_0^1 \mathbf{K}(t,s)\mathbf{f}(s,u(s))ds \right|$$

$$\geq \min_{\xi \leq t \leq 1} \int_0^1 \mathbf{K}(t,s)\mathscr{N}b\,ds$$

$$> \int_\xi^1 \gamma(s)\mathbf{K}(s,s)\mathscr{N}b\,ds$$

$$> \mathscr{N}b \int_\xi^1 \gamma(s)\mathbf{K}(s,s)ds = b,$$

which means that $\theta(\mathscr{T}u) > b$, with $u \in (\theta, b, c)$.

Thus, thanks to Theorem 3, there exist at least three solutions u_1, u_2, and u_3 for the FDE (2) such that

$$\|u_1\|_{\mathbf{X}} < a_1, \ b < \theta(u_2) < \|u_2\|_{\mathbf{X}} < c, \ a_1 < \|u_3\|_{\mathbf{X}}, \ \theta(u_3) < b.$$

Hence we get the required result.

4 Different Kinds of Ulam's Stability for the Solutions of BVP (1)

This portion of the paper is devoted to the UH, GUH, UHR, and GUHR stability of the solutions to consider BVP (1) of nonlinear FDEs. To come across the required result, we give the following axillary results needed.

Definition 4 The solution $u \in C(\mathbf{I}, \mathbf{R})$ of the considered problem (1) is UH stable if we can find a real number $\hat{C}_f > 0$ with the property that for every $\varpi > 0$ and for each solution $u \in C(\mathbf{I}, \mathbf{R})$ of the inequality

$$|^c\mathrm{D}^\mathbf{p}u(t) - \mathrm{f}(t, u(t))| \leq \varpi, \ t \in \mathbf{I}, \tag{4}$$

there exists a unique solution $v \in C(\mathbf{I}, \mathbf{R})$ of the proposed BVP (1) with a constant $\hat{C}_f > 0$ with

$$\|u - v\|_\mathbf{X} \leq \hat{C}_f\varpi > 0, \ t \in \mathbf{I}.$$

Definition 5 The solution $u \in C(\mathbf{I}, \mathbf{R})$ of the proposed BVP (1) is called to be GUH stable, if we can find

$$\Theta_f : (0, \infty) \to \mathbf{R}^+, \ with \ \Theta_f(0) = 0,$$

such that for each solution $v \in C(\mathbf{I}, \mathbf{R})$ of the inequality (4), we can find a unique solution $v \in C(\mathbf{I}, \mathbf{R})$ of the considered BVP (1) with

$$\|u - v\|_\mathbf{X} \leq \hat{C}_f\Theta_f(\varpi), \ t \in \mathbf{I}.$$

Remark 1 A function $u \in (\mathbf{I}, \mathbf{R})$ is said to be the solution of inequality given in (4) if and only if, we can find a function $\aleph \in (\mathbf{I}, \mathbf{R})$ depends on u only, then

(i) $|\aleph(t)| \leq \varpi, \ for \ all \ t \in \mathbf{I}$;
(ii) $^c\mathrm{D}^\mathbf{p}u(t) = \mathrm{f}(t, u(t)) + \aleph(t), \ for \ all \ t \in \mathbf{I}.$

Next we recall the definitions of UHR and GUHR stability [21] for our considered problem (1) as below:

Definition 6 FDE (1) is said to be UHR stable with respect to $\chi \in C(\mathbf{I}, \mathbf{R})$ if there exists a real constant $\hat{C}_f > 0$ such that for each $\varpi > 0$ and for every solution $u \in C\mathbf{I}$ of the inequality

$$|^c\mathrm{D}^\mathbf{p}u(t) - \mathrm{f}(t, u(t))| \leq \varpi\chi(t), \ t \in \mathbf{I}, \tag{5}$$

there exists a solution $v \in C(\mathbf{I})$ of the Eq. (1), such that

$$|u(t) - v(t)| \leq \hat{C}_{\phi,\varphi}\epsilon\chi(t), \ t \in \mathbf{I}.$$

Definition 7 Equation (1) is said to be GUHR stable with respect to $\chi \in C(\mathbf{I})$, if there exists a real number $\hat{C}_f > 0$ such that for each solution $u \in C\mathbf{I}$ of the inequality

$$|{}^cD^pu(t) - f(t, u(t))| \leq \chi(t), \ t \in \mathbf{I}, \tag{6}$$

there exists a solution $w \in C(\mathbf{I})$ of the Eq. (1) such that $|w(t) - x(t)| \leq \hat{C}_{\chi,\ell,\Upsilon}\chi(t)$, $t \in \mathbf{I}$.

Lemma 7 *Under assumption $h \in L(\mathbf{I})$ such that $|f(t, u) - f(t, v)| \leq h(t)|u - v|$, for $t \in (\mathbf{I})$, the solution $u \in (\mathbf{I}, \mathbf{R})$ of the considered problem is given by*

$$D_{0+}^p u(t) + f(t, u(t)) = \aleph(t), \quad 0 < t < 1, \tag{7}$$

$$D_{0+}^q u(0) = 0, \ D_{0+}^q u(1) = \lambda u(\xi), \ 0 \leq q < 1, \ p - q \geq 1,$$

satisfying the relation given by

$$\left| u(t) - \int_0^1 K(t, s)f(s, u(s))ds \right| \leq K\varpi, \ for \ all \ t \in \mathbf{I}. \tag{8}$$

Proof Thanks to Lemma 4 , we get the solution of the considered problem (12) as

$$u(t) = \int_0^1 K(t, s)f(s, u(s))ds + \int_0^1 K(t, s)\aleph(s)ds, \ t \in \mathbf{I}, \tag{9}$$

where $K(t, s)$ is the same Green's function defined in Lemma 4. From (14), we may write as

$$\left| u(t) - \int_0^1 K(t, s)f(s, u(s))ds \right| = \left| \int_0^1 K(t, s)\aleph(s)ds \right|$$

$$\leq \max_{t \in \mathbf{I}} \int_0^1 |K(t, s)||\aleph(s)|ds$$

$$\leq K^*\varpi, \ for \ all \ t \in \mathbf{I}, \ K^* = \max_{t \in \mathbf{I}} \int_0^1 |K(t, s)|ds.$$

Theorem 9 *Under the assumption $h \in L(\mathbf{I})$ such that $|f(t, u) - f(t, v)| \leq h(t)|u - v|$, for all $t \in \mathbf{I}$ and Lemma 7 together with the condition $\Upsilon = \max_{t \in \mathbf{I}} \int_0^1 K(t, s)h(s)ds \neq 1$, the solution of FDE (1) is UH stable and consequently GUH stable.*

Proof Let $u \in C(\mathbf{I}, \mathbf{R})$ be any solution of FDE (1) and $v \in C(\mathbf{I}, \mathbf{R})$ be the unique solution of the considered problem, then consider

$$|v(t) - u(t)| = \left| v(t) - \int_0^1 \mathbf{K}(t,s)\mathbf{f}(s,v(s))ds + \int_0^1 \mathbf{K}(t,s)\mathbf{f}(s,v(s))ds \right.$$

$$\left. - \int_0^1 \mathbf{K}(t,s)\mathbf{f}(s,u(s))ds \right|$$

$$\leq \left| v(t) - \int_0^1 \mathbf{K}(t,s)\mathbf{f}(s,v(s))ds \right| + \left| \int_0^1 \mathbf{K}(t,s)\mathbf{f}(s,v(s))ds \right.$$

$$\left. - \int_0^1 \mathbf{K}(t,s)\mathbf{f}(s,u(s))ds \right|$$

$$\leq \mathbf{K}^*\varpi + \int_0^1 |\mathbf{K}^*(t,s)| |h(s)| |v - u| ds,$$

which on simplification yields that

$$\|v - u\|_{\mathbf{X}} \leq \mathbf{K}^*\varpi + \max_{t \in \mathbf{I}} \int_0^1 \mathbf{K}(t,s) h(s) ds \|v - u\|_{\mathbf{X}}$$

which further gives

$$\|y - x\|_{\mathbf{X}} \leq \frac{\mathbf{K}^*\varpi}{1 - \Upsilon}, \; for \; all \; t \in \mathbf{I}, \; where \; \Upsilon = \max_{t \in \mathbf{I}} \int_0^1 \mathbf{K}(t,s) h(s) ds. \quad (10)$$

Hence the solution of the considered problem (1) is UH stable . Also if we let $\Theta(\varpi) = \varpi$ and $\hat{C}_f = \frac{\mathbf{K}^*\varpi}{1-\Upsilon}$, then (15) can be written as

$$\|y - x\|_{\mathbf{X}} \leq \hat{C}_f \Theta(\varpi), \; for \; all \; t \in \mathbf{I}$$

It is clear that $\Theta(0) = 0$. Hence the solution of the proposed problem (1) is GUH stable.

Let the following inequality hold for a nondecreasing function $\chi : (0, \infty) \to \mathbf{R}$

$$\mathscr{I}^{\mathbf{p}} \chi(t) \leq \lambda_\chi \chi(t). \quad (11)$$

Lemma 8 *Under the inequality (11) and assumption $h \in L(\mathbf{I})$ with $|\mathbf{f}(t,u) - \mathbf{f}(t,v)| \leq h(t)|u - v|$, for all $t \in \mathbf{I}$, the solution $u \in (\mathbf{I}, \mathbf{R})$ of the considered problem is given by*

$$D_{0+}^{\mathbf{p}} u(t) + \mathbf{f}(t, u(t)) = \aleph(t), \quad 0 < t < 1,$$

$$D_{0+}^{\mathbf{q}} u(0) = 0, \; D_{0+}^{\mathbf{q}} u(1) = \lambda u(\xi), \; 0 \leq \mathbf{q} < 1, \; \mathbf{p} - \mathbf{q} \geq 1, \quad (12)$$

satisfying the following relation

$$\left| u(t) - \int_0^1 \mathbf{K}(t,s)f(s,u(s))ds \right| \leq \mathbf{K}\lambda_\chi \varpi \chi(t), \ for \ all \ t \in \mathbf{I}. \tag{13}$$

Proof Thanks to Lemma 4 , we get the solution of the considered problem (12) as

$$u(t) = \int_0^1 \mathbf{K}(t,s)f(s,u(s))ds + \int_0^1 \mathbf{K}(t,s)\aleph(s)ds, \ t \in \mathbf{I}, \tag{14}$$

where $\mathbf{K}(t,s)$ is the same Green's function defined in Lemma 4. From (14), we may write as

$$\begin{aligned}
\left| u(t) - \int_0^1 \mathbf{K}(t,s)f(s,u(s))ds \right| &= \left| \frac{1}{\Gamma(\mathbf{p})} \left[d\Gamma(\mathbf{p})t^{\mathbf{p}-1} \int_0^1 (1-s)^{\mathbf{p}-\mathbf{q}-1} y(s)ds \right. \right. \\
&\quad \left. \left. -dt^{\mathbf{p}-1}\lambda\Gamma(\mathbf{p}-\mathbf{q}) \int_0^\xi (\xi-s)^{\mathbf{p}-1} y(s)ds \right. \right. \\
&\quad \left. \left. - \int_0^t (t-s)^{\mathbf{p}-1} y(s)ds \right] \right| \\
&\leq d\lambda\Gamma(p-q)\lambda_\chi \chi(t) + \lambda_\chi \chi(t) \\
&= \lambda_{\chi,\ell}\chi(t), \ where \ \ell = \lambda_\chi(1+d\lambda\Gamma(p-q)).
\end{aligned}$$

Theorem 10 *Under the assumption $h \in L(\mathbf{I})$ such that $|f(t,u) - f(t,v)| \leq h(t)|u-v|$, for all $t \in \mathbf{I}$ and Lemma 8 together with the condition $\Upsilon = \max_{t \in \mathbf{I}} \int_0^1 \mathbf{K}(t,s)h(s)ds < 1$, the solution of FDE (1) is UHR and consequently GUHR stable.*

Proof Let $u \in C(\mathbf{I},\mathbf{R})$ be any solution of FDE (1) and $v \in C(\mathbf{I},\mathbf{R})$ be the unique solution of the considered problem, then consider

$$\begin{aligned}
|v(t) - u(t)| &= \left| v(t) - \int_0^1 \mathbf{K}(t,s)f(s,v(s))ds + \int_0^1 \mathbf{K}(t,s)f(s,v(s))ds \right. \\
&\quad \left. - \int_0^1 \mathbf{K}(t,s)f(s,u(s))ds \right| \\
&\leq \left| v(t) - \int_0^1 \mathbf{K}(t,s)f(s,v(s))ds \right| + \left| \int_0^1 \mathbf{K}(t,s)f(s,v(s))ds \right. \\
&\quad \left. - \int_0^1 \mathbf{K}(t,s)f(s,u(s))ds \right| \\
&\leq \lambda_{\chi,\ell}\chi(t) + \int_0^1 |\mathbf{K}^*(t,s)|h(s)|v-u|ds,
\end{aligned}$$

which on simplification yields that

$$\|v - u\|_X \leq \lambda_{\chi,\ell}\chi(t) + \max_{t \in I}\int_0^1 K(t,s)h(s)ds\|v - u\|_X$$

which further gives

$$\|v - u\|_X \leq \hat{C}_{\chi,\ell,\Upsilon}\chi(t), \ for \ all \ t \in I, \ where \ \hat{C}_{\chi,\ell,\Upsilon} = \frac{\lambda_{\chi,\ell}}{1 - \Upsilon}. \quad (15)$$

Hence the solution of the considered problem (1) is UHR stable. On a similar computation, we can easily derive that the solution of the proposed problem (1) is GUHR stable.

5 Example

Example 1 Consider the fractional order boundary value differential equation

$$D_{0+}^{\mathbf{p}}u(t) + \frac{e^{2t}u(t)}{4(2 + e^{2t})(2 + u(t))} + cos^2 t + 1 = 0, \ for \ 0 < t < 1,$$

$$D_{0+}^{\mathbf{q}}(0) = 0, \ D_{0+}^{\mathbf{q}}u(1) = \lambda u(\xi), \quad (16)$$

where $\mathbf{p} = \frac{3}{2}$, $p = \frac{1}{20}$, $\xi = \frac{9}{10}$, $\lambda = \frac{1}{200}$.

Let $f(t, u) = \frac{e^{2t}u(t)}{4(2 + e^{2t})(2 + u(t))} + cos^2 t + 1$, $(t, u) \in I \times [0, \infty)$.

Taking $h(t) = \frac{e^{2t}}{2(2 + e^{2t})}$, one has

$$\left|f(t, u_2(t)) - f(t, u_1(t))\right| = \frac{e^{2t}}{4(2 + e^{2t})}\left|\frac{u_2(t)}{2 + u_2(t)} - \frac{u_1(t)}{2 + u_1(t)}\right|$$

$$\leq \frac{e^{2t}}{2(2 + e^{2t})}\|u_2 - u_1\|_X$$

$$\leq h(t)\|u_2 - u_1\|_X \ for \ all \ (t, u) \in I \times [0, \infty).$$

Also, we get

$$0 < \int_0^1 K(s, s)h(s)ds \leq \int_0^1 ds^{\mathbf{p}-1}(1 - s)^{\mathbf{p}-\mathbf{q}-1}ds$$

$$\leq d\beta(\mathbf{p}, \mathbf{p} - \mathbf{q})$$

$$\leq \frac{\Gamma(\mathbf{p})\Gamma(\mathbf{p} - \mathbf{q})}{\Gamma(2\mathbf{p} - p)(\Gamma(\mathbf{p}) - \Gamma(\mathbf{p} - \mathbf{q})\lambda\xi^{\mathbf{p}-1})}$$

$$\leq 0.464248 < 1.$$

Thus, thanks to Theorem 4, the FDE (16) has a unique solution. Further, it is easy to compute that $\Upsilon = \max_{t \in I} \int_0^1 \frac{e^{2s}}{2(2+e^{2s})} K(t, s) ds \neq 1$. Thus by Theorem 9, the solution of BVP (16) is UH stable and consequently GUH stable. Further taking $\chi(t) = t$, one can obtain the conditions of UHR and GUHR stability of the solutions for the BVP (16).

Example 2

$$D_{0+}^{p} u(t) + \frac{t^{\frac{3}{2}} u(t)}{2 + 5u(t)} + te^{2t} + 1 = 0, \text{ for } 0 < t < 1,$$

$$D_{0+}^{q}(0) = 0, \quad D_{0+}^{q} u(1) = \lambda u(\xi), \tag{17}$$

where $\mathbf{p} = \frac{3}{2}$, $\mathbf{q} = \frac{1}{2}$, $\xi = \frac{4}{10}$, $\lambda = \frac{1}{200}$.

Let $f(t, u) = \frac{t^{\frac{3}{2}} u(t)}{2+5u(t)} + te^{2t} + 1$, $(t, u) \in I \times [0, \infty)$, Taking $\beta(t) = te^{2t} + 1$ and $\alpha(t) = t^{\frac{3}{2}}$ we have

$$f(t, u(t)) = \frac{u(t)}{2 + 5u(t)} \alpha(t) + \beta(t)$$

$$\leq \alpha(t) + \beta(t) \text{ for all } (t, u) \in I \times [0, \infty).$$

Further we have

$$0 < \int_0^1 K(s, s)\beta(s) ds \leq \int_0^1 ds^{p-1}(1-s)^{p-q-1} s^{\frac{3}{2}} ds$$

$$\leq d\beta(3, 1)$$

$$\leq \frac{\Gamma(3)\Gamma(1)}{\Gamma(4)(\Gamma(1.5) - \Gamma(1)(0.05)(0.02))}$$

$$\leq 0.376527 < 1.$$

Thanks to Theorem 6, the FDE (17) has at least one positive solution. Also, it is easy to compute that $\Upsilon = \max_{t \in I} \int_0^1 s^{\frac{3}{2}} K(t, s) ds \neq 1$. Thus by Theorem 9, the solution of FDE (17) is UH stable and consequently GUH-stable. Further on an easy computation we can easily obtain the results of UHR and GUHR stability by taking $\chi(t) = t$.

Example 3

$$D_{0+}^{p} u(t) + \frac{u^2(t)}{5} + \frac{\cos^2 t}{20} + 1 = 0, \text{ for } 0 < t < 1,$$

$$D_{0+}^{q}(0) = 0, \quad D_{0+}^{q} u(1) = \lambda u(\xi), \tag{18}$$

where $\mathbf{p} = \frac{3}{2}$, $\mathbf{q} = 0$, $\xi = \frac{1}{2}$, $\lambda = \frac{1}{2}$
Let

$$f(t, u) = \frac{u^2(t)}{5} + \frac{cos^2 t}{20} + 1, \text{ for all } (t, u) \in \mathbf{I} \times [0, \infty).$$

Now $d = (\Gamma(\mathbf{p}) - \Gamma(\mathbf{p} - \mathbf{q})\lambda\xi^{\mathbf{p}-1})^{-1} = 1.5045$, $\int_0^1 \mathbf{K}(s, s)ds = 0.59077$, and

$$\int_\xi^1 \gamma(s)\mathbf{K}(s, s)ds = \int_{\frac{1}{2}}^1 g_2(1, s)ds$$

$$= 0.37598 \ldots \int_{\frac{1}{2}}^1 (1 - s)^{\frac{1}{2}} ds$$

$$= 0.08862 \ldots .$$

Hence

$$\mathscr{M} = \left(\int_0^1 \mathbf{K}(s, s)ds \right)^{-1} = 1.6927, \quad \mathscr{N} = \left(\int_\xi^1 \gamma(s)\mathbf{K}(s, s)ds \right)^{-1} = 2.6597.$$

Choosing $r_1 = \frac{1}{3}$ and $r_2 = 1$, we have

$$f(t, u) = \frac{u^2(t)}{5} + \frac{cos^2(t)}{20} + 1 \leq 1.25 \leq \mathscr{M}r_2, \text{ for all } (t, u) \in \mathbf{I} \times [0, r_2],$$

$$f(t, u) = \frac{u^2(t)}{5} + \frac{cos^2(t)}{20} + 1 \geq 1 \geq \mathscr{N}r_1, \text{ for all } (t, u) \in \mathbf{I} \times [0, r_1].$$

Hence all conditions of Theorem 7 are satisfied, thus the FDE (18) has at least one positive solution u such that $\frac{1}{3} \leq \|u\|_\mathbf{X} \leq 1$. It is easy to verify the results of Theorem 9 by taking $h(t) = 1$ to derive $\Upsilon \neq 1$, by taking $h(t) = 1$.

Example 4 Consider the problem

$$D_{0+}^{\mathbf{p}} u(t) + f(t, u) = 0, \text{ for } 0 < t < 1,$$

$$D_{0+}^{\mathbf{q}}(0) = 0, \; D_{0+}^{\mathbf{q}} u(1) = \lambda u(\xi), \tag{19}$$

where $\mathbf{p} = \frac{3}{2}$, $\mathbf{q} = 0$, $\xi = \frac{1}{2}$, $\lambda = \frac{1}{2}$.
Let

$$f(t, u) = \begin{cases} \frac{t+3}{25} + u^2, & \text{if } (t, u) \in \mathbf{I} \times \mathbf{I}, \\ \\ \frac{3t}{25} + u + 2, & \text{if } (t, u) \in \mathbf{I} \times (0, \infty). \end{cases}$$

By computing we have

$$\mathcal{M} = \left(\int_{\xi}^{1} \gamma(s) \mathbf{K}(s,s) ds \right)^{-1} = 1.6927 \text{ and}$$

$$\mathcal{N} = \left(\int_{\xi}^{1} \gamma(s) \mathbf{K}(s,s) ds \right)^{-1} = 2.6597.$$

Taking $a_1 = \frac{1}{5}$, $b = \frac{3}{4}$, $c = 5$, we have

$$f(t,u) = \frac{t+3}{25} + u^2 \le 0.2 \le 0.3385 = \mathcal{M} a_1, \ (t,u) \in \mathbf{I} \times [0, \frac{1}{5}],$$

$$f(t,u) = \frac{3t}{25} + u + 2 \le 7.12 \le 8.4635 = \mathcal{M} c, \ (t,u) \in \mathbf{I} \times [0,5],$$

$$f(t,u) = \frac{3t}{25} + u + 2 \ge 2.75 \ge 1.9947 = \mathcal{N} b, \ (t,u) \in [\frac{1}{2},1] \times [\frac{3}{4},5].$$

Since assumptions of Theorem 8 hold, so the FDE (19) has at least three positive solutions $u_1, u_2,$ and u_3 with

$$\|u_1\|_{\mathbf{X}} < \frac{1}{5}, \quad \frac{3}{4} < \theta(u_2), \quad \frac{1}{5} < \|u_3\|_{\mathbf{X}}, \quad \text{where } \theta(u_3) < \frac{3}{4}.$$

for all the three solutions. One can easily show by taking $h(t) = \frac{t+3}{25}$ and $h(t) = \frac{3t}{25}$ in respective cases that $\Upsilon \ne 1$. Hence by Theorem 9, the solutions u_1, u_2, u_3 are UH and GUH stable. The results of UHR and GUHR stability are obvious for the BVP (19), if we consider $\chi(t) = t$.

6 Conclusion

In the current work, we have developed some results for qualitative theory of solutions by using classical fixed point theorems due to Leggett–Williams and Banach. Further stability analysis which is very important for numerical and optimization purposes has been investigated. The respective analysis has been very rarely investigated for the nonlinear FDEs involving Riemann–Liouville fractional derivative. The whole analysis has been demonstrated by providing several examples.

Acknowledgements The authors declare that there does not exist any conflict of interest. Further, this work has been supported by the National Natural Science Foundation of China (11571378).

References

1. A. Babakhani, V.D. Gejji, Existence of positive solutions of nonlinear fractional differential equations. J. Math. Anal. Appl. **278**, 434–442 (2003)
2. Z.B. Bai, H.S. Lu, Positive solutions for boundary value problem of nonlinear fractional differential equation. J. Math. Anal. Appl. **311**, 495–505 (2005)
3. M. Benchohra, S. Hamani, S.K. Ntouyas, Boundary value problems for differential equations with fractional order. Surv. Appl. Math. **3**, 1–12 (2008)
4. B. Benchohra, J.R. Graef, S. Hamani, Existence results for boundary value problem with nonlinear fractional differential equation. Appl. Anal. **87**, 851–863 (2008)
5. R. Hilfer, *Applications of Fractional Calculus in Physics* (World Scientific, Singapore, 2000)
6. D.S. Cimpean, D. Popa, Hyers-Ulam stability of Euler's equation. Appl. Math. Lett. **24**, 1539–1543 (2011)
7. F. Haq, K. Shah, G. Rahman, M. Shahzad, Hyers-Ulam stability to a class of fractional differential equations with boundary conditions. Int. J. Appl. Comput. Math. **2017**, 1–13 (2017)
8. D.H. Hyers, G. Isac, T.M. Rassias, *Stability of Functional Equations in Several Variables* (Birkhäuser, Basel, 1998)
9. S.M. Jung, *Hyers-Ulam-Rassias Stability of Functional Equations in Nonlinear Analysis* (Springer, New York, 2011)
10. S.M. Jung, *Hyers-Ulam-Rassias Stability of Functional Equations in Mathematical Analysis* (Hadronic Press, Palm Harbor, 2001)
11. S.M. Jung, Hyers-Ulam stability of linear differential equations of first order. Appl. Math. Lett. **17**, 1135–1140 (2004)
12. E.R. Kaufmann, E. Mboumi, Positive solutions of boundary value problem for a nonlinear fractional differential equation. Electron. J. Qual. Theory Differ. Equ. **2008**, 1–11 (2008)
13. A.A. Kilbas, H.M. Srivastava, J.J. Trujillo, *Theory and Application of Fractional Differential Equation*. North-Holland Mathematics studies, vol. 204 (Elsevier Science B.V., Amsterdam, 2006)
14. P. Kumama, A. Ali, K. Shah, R.A. Khan, Existence results and Hyers-Ulam stability to a class of nonlinear arbitrary order differential equations. J. Nonlinear Sci. Appl. **10**, 2986–2997 (2017)
15. V. Lakshmikantham, S. Leela, J. Vasuandhara Devi, *Theory of Fractional Dynamic* (Cambridge Scientific Publishers, Cambridge, 2009)
16. R.W. Leggett, L.R. William, Multiple positive fixed points of nonlinear operators on ordered Banach space. Indiana Univ. Math. J. **28**, 673–688 (1979)
17. C.F. Li, X.N. Luo, Y. Zhou, Existence of positive solution of the boundary value problem for nonlinear fractional differential equations. Comput. Math. Appl. **59**, 1363–1375 (2010)
18. G. Lijun, D. Wang, G. Wang, Further results on exponential stability for impulsive switched nonlinear time-delay systems with delayed impulse effects. Appl. Math. Comput. **268**, 186–200 (2015)
19. M. Obloza, Hyers-Ulam stability of the linear differential equation. Rocznik Nauk.-Dydakt. Prace Mat. **13**, 259–270 (1993)
20. I. Podlubny, *Fractional Differential Equation* (Academic Press, New York, 1999)
21. T. M. Rassias, on the stability of the linear mapping in Banach spaces. Proc. Am. Math. Soc. **72**, 297–300 (1978)
22. M. Shoaib, M. Sarwara, K. Shaha, P. Kumam, Fixed point results and its applications to the systems of non-linear integral and differential equations of arbitrary order. J. Nonlinear Sci. Appl. **9**, 4949–4962 (2016)
23. I. Stamova, Mittag-Leffler stability of impulsive differential equations of fractional order. Q. Appl. Math. **73**(3), 525–535 (2015)
24. S.-E. Takahasi, T. Miura, S. Miyajima, On the Hyers-Ulam stability of the Banach space-valued differential equation $y' = \lambda y$. Bull. Korean Math. Soc. **39**, 309–315 (2002)

25. J.C. Trigeassou, et al., A Lyapunov approach to the stability of fractional differential equations. Signal Process. **91**(3), 437–445 (2011)
26. S.M. Ulam, *Problems in Modern Mathematics* (Wiley, New York, 1940)
27. S.M. Ulam, *A Collection of Mathematical Problems* (Interscience, New York, 1960)
28. J. Vanterler da C. Sousa, E. Capelas de Oliveira, Ulam–Hyers stability of a nonlinear fractional Volterra integro-differential equation. Appl. Math. Lett. **81**, 50–56 (2018)
29. A.Vinodkumar, K. Malar, M. Gowrisankar, P. Mohankumar, Existence, uniqueness and stability of random impulsive fractional differential equations. Acta Math. Sci. **36**(2), 428–442 (2016)
30. J. Wang, X. Li, Ulam-Hyers stability of fractional Langevin equations. Appl. Math. Comput. **258**, 72–83 (2015)
31. J. Wang, Y. Zhou, Mittag-Leffler-Ulam stabilities of fractional evolution equations. Appl. Math. Lett. **25**, 723–728 (2012)
32. J. Wang, L. Lv, Y. Zhou, Ulam stability and data dependence for fractional differential equations with Caputo derivative. Electron. J. Qual. Theory Differ. Equ. **63**, 1–10 (2011)
33. J. Wang, M. Feckan, Y. Zhou, Ulam's type stability of impulsive ordinary differential equations. J. Math. Anal. Appl. **395**, 258–264 (2012)
34. A. Zada, S. Ali, Y. Li, Ulam's type stability for a class of implicit fractional differential equations with non-instantaneous integral impulses and boundary condition. Adv. Difference Equ. **2017**, 317 (2017)
35. S.Q. Zhang, Existence of positive solution for some class of a nonlinear fractional differential. J. Math. Anal. Appl. **278**, 136–148 (2003)
36. S. Zhang, Positive solutions for boundary value problem of nonlinear fractional differential equations. Electron. J. Differ. Equ. **2006**, 1–12 (2006)

Topological Degree Theory and Ulam's Stability Analysis of a Boundary Value Problem of Fractional Differential Equations

Amjad Ali, Kamal Shah, and Yongjin Li

Abstract In this article, we study the existence and uniqueness of positive solution to a class of nonlinear fractional order differential equations with boundary conditions. By using fixed point theorems on contraction mapping together with topological degree theory, we investigate some sufficient conditions in order to obtain the existence and uniqueness of positive solution for the considered problem. Further we also investigate different kinds of Ulam stability for the considered problem. Moreover, we also provide an example to justify the whole results.

2010 Mathematics Subject Classification 26A33, 26A42, 34A08

1 Introduction

The study of fractional differential equations is an important area for research in recent time, because of its wide range of applications in describing the real-word problems. These applications can be found in various scientific and engineering disciplines such as physics, chemistry, optimization theory, biology, viscoelasticity, control theory, signal processing, etc. For details, we refer [1–8]. Due to large number of applications of fractional differential equations, researchers are giving much attention to study fractional order differential equations, we refer the readers to [9–13] and the references therein for the recent development in the theory of fractional differential equations. It is worthwhile to mention that Caputo's fractional

A. Ali
Department of Mathematics, Government Postgraduate Jehanzeb College
Saidu Sharif Mingora Swat, Saidu Sharif, Khyber Pakhtunkhwa, Pakistan

K. Shah
Department of Mathematics, University of Malakand, Chakadara Dir (L),
Khyber Pakhtunkhwa, Pakistan

Y. Li (✉)
Department of Mathematics, Sun Yat-sen University, Guangzhou, China
e-mail: stslyj@mail.sysu.edu.cn

© Springer Nature Switzerland AG 2019
G. A. Anastassiou and J. M. Rassias (eds.), *Frontiers in Functional Equations and Analytic Inequalities*, https://doi.org/10.1007/978-3-030-28950-8_4

order derivative plays important roles in applied problems as it provides known physical interpretation for initial and boundary value problems of differential equations of arbitrary order. On the other hand, the Riemann–Liouville derivative of fractional order does not provide physical interpretations in most of the cases for initial and boundary problems. Existence theory of differential equations of classical as well as arbitrary order with multi-point boundary conditions has attracted the attention of many researchers and is a rapidly growing area of research, because such problems occurred in applications, we refer the readers to [14–19]. The area devoted to study boundary value problems of classical order differential equations has been studied and plenty of work is available on it by means of degree theory, however, for differential equations of fractional order, the area is quite recent and very few papers are available on it. As in [20], the authors studied the following problems by using topological degree theory

$$^{c}D^{q}u(t) = f(t, u(t)), \ 0 < q < 1, \ t \in [0, T],$$

$$u(0) + g(u) = u_0,$$

and

$$^{c}D^{q}u(t) = f(t, u(t)), \ 0 < q < 1, \ t \in [0, T],$$

$$au(0) + bu(T) = c, \ a + b \neq 0,$$

$$^{c}D^{q}u(t) = f(t, u(t)), \ 0 < q < 1, \ t \in [0, T],$$

$$u(0) = u_0, \ \Delta u(t_k) = I_k(u(t_k)), k = 1, 2, \cdots n.$$

where $g \in C([0, T], \mathbf{R}), f \in C([0, T] \times \mathbf{R}, \mathbf{R}), I_k : \mathbf{R} \to \mathbf{R}$ is continuous function called impulse. Similarly in [21], the authors studied the following multi-point boundary value problems by topological degree theory given by

$$^{c}D^{q}u(t) = f(t, u(t)), \ 1 < q \leq 2, \ t \in [0, 1],$$

$$u(0) = g(u), \ \sum_{k=0}^{m-2} \lambda_k u(\eta_k) + h(u) = u(1).$$

where $g, h \in C([0, 1], \mathbf{R}), f \in C([0, 1] \times \mathbf{R}, \mathbf{R})$. In very recent times, Shah et al. [22] developed sufficient conditions for the existence and uniqueness of positive solution to a coupled system with four-point boundary conditions via topological degree.

Motivated by the above work, in this article, we study the following class of nonlinear fractional order differential equations with given boundary conditions as

$$^{c}D^{q}u(t) = f(t, u(t)), \ 1 < q \leq 2, \ t \in J = [0, 1],$$

$$\lambda_1 u(0) + \mu_1 u(1) = g_1(u),$$

$$\lambda_2 u'(0) + \mu_2 u'(1) = g_1(u).$$

where $g_k : C(J, \mathbf{R}) \to \mathbf{R}$ *for* $k = 1, 2$, are continuous functions and $f : J \times \mathbf{R} \to \mathbf{R}$ is nonlinear continuous function and $\lambda_k, \mu_k (k = 1, 2)$ are real constants with $\lambda_k + \mu_k \neq 0, k = 1, 2$.

Here we remark that existence theory together with stability analysis is very important from numerical as well as optimization point of view. Beside from existence theory of solutions to the nonlinear fractional differential equations, the aspect devoted to stability analysis has been attracted the attention, see [23–26]. Different kinds of stability including exponential, Mittag–Leffler, and Lyapunov stability have been studied for the said differential equations, for details see [27–29]. Another kind of stability which greatly attracted the researchers' attention has been recently considered for nonlinear and linear fractional differential equations, we refer [30–33]. This important form of stability was first pointed out by Ulam in 1940 and was brilliantly explained by Hyers in 1940. After that valuable contributions have been done in this regard. In 1997, Rassias extended the aforementioned stability to some other forms known as Ulam–Hyers–Rassias and generalized Ulam–Hyers–Rassias stability. The concerned stability results have been investigated recently for fractional differential equations, ordinary and functional equations, see [34]. The aforementioned stability has been investigated for functional, integral, and differential equations very well, see [35–37]. In the last few years significant contribution has been done in the aforementioned aspects. Problems devoted to integral, functional, and differential equations have been evaluated for the aforesaid stability, see [38–50, 55, 56].

Therefore in this work, the considered class of differential equations of fractional order is investigated for positive solutions by means of contraction mapping principle coupled with topological degree theory. Sufficient conditions are developed under which the considered class of boundary value problem has at least one and unique solution. Then using nonlinear analysis we develop sufficient conditions for different kinds of Ulam stability.

We present the rest of the paper in four sections, in Sect. 2, we present some of the basic results and theorems, which are helpful in this paper. Also, we give some assumptions which are needed for this study. Section 3 is devoted to the main results. In Sect. 4, we provide a detailed analysis for stability theory. In Sect. 5, we give an example for verification of the established results. In the last section, we give a brief conclusion.

2 Preliminaries Results

In this section, we recall some definitions and basic results which are helpful throughout in this article, for details see [51–54].

The notation $C(J, \mathbf{R})$ is used for Banach space for all continuous function defined for J into $\mathbf{R}$ with norm

$$\|u\|_c = \sup\{|u(t)| : 0 \le t \le 1\}.$$

We denote $X = C[0, 1]$, we recall the following results from degree theory.

Definition 1 Let $C \subset X$ and $T : C \to X$ be a continuous bounded map, then T is α-condensing if $\alpha(T(V)) < \alpha(V)$ for all V bounded subset of C with $\alpha(V) > 0$.

The following theorem given by Isaia is important for our main results

Theorem 1 *Let $T : V \to X$ be α-condensing and*

$$V = \{u \in X : there exists \quad \lambda \in [0, 1] \quad such that \quad u = \lambda T u\}.$$

If V is a bounded subset of X and there exists $r > 0$ such that $V \subset \mathcal{U}_r(0)$, then the degree

$$D(I - \lambda T, \mathcal{U}_r(0), 0) = 1, \quad for all \quad \lambda \in [0, 1].$$

Consequently, T has at least one fixed point and the set of the fixed points of T lies in $\mathcal{U}_r(0)$.

The following propositions are needed.

Proposition 1 *If $T_1, T_2 : V \to X$ are α-Lipschitz maps with constants κ_1 and κ_2 respectively, then $T_1 + T_2 : V \to X$ are α-Lipschitz with constants $\kappa_1 + \kappa_2$.*

Proposition 2 *If $T_1 : V \to X$ is compact, then T is α-Lipschitz with constant $\kappa = 0$.*

Proposition 3 *If $T_1 : V \to X$ is Lipschitz with constant κ, then T_1 is α-Lipschitz with the same constant κ.*

Definition 2 The fractional(arbitrary) order integral of a function $u \in L^1([0, b], \mathbf{R})$ of order $q \in \mathbf{R}^+$ is defined by

$$I^q u(t) = \frac{1}{\Gamma(q)} \int_0^t (t - s)^{q-1} u(s) ds, \quad n - 1 < q \leq n.$$

Definition 3 The Caputo's fractional order derivative of a function u on the interval $[0, b]$ is defined by

$${}^c D^q u(t) = \frac{1}{\Gamma(n - q)} \int_0^t (t - s)^{n-q-1} u^{(n)}(s) \, ds, \ n = [q] + 1,$$

where $[q]$ represents integer part of q.

For the existence of solutions to the considered problem, we need the following results:

Theorem 2 *The fractional order differential eqnarray of order $q > 0$ of the form*

$${}^c D^q u(t) = 0, \ n - 1 < q \leq n,$$

has a solution of the form

$$u(t) = c_0 + c_1 t + c_2 t^2 + \ldots + c_{n-1} t^{n-1},$$

where $c_i \in \mathbf{R}$, for $i = 0, 1, \ldots, n-1$.

Theorem 3 *The following result holds for a fractional order differential equation q*

$$I^q[^c D^q u](t) = u(t) + c_0 + c_1 t + c_2 t^2 + \ldots + c_{n-1} t^{n-1},$$

for arbitrary $c_i \in \mathbf{R}$, for $i = 0, 1, 2, \ldots, n-1$.

The consequence of Theorems 2 and 3 leads us to the following useful result.

Theorem 4 *Let $u \in CJ$ and $y \in C(J \times \mathbf{R}, \mathbf{R})$, then the solution of linear fractional differential equation*

$$^c D^q u(t) = f(t, u(t)), \ 1 < q \le 2, \ t \in [0, 1],$$

$$\lambda_1 u(0) + \mu_1 u(1) = g_1(u), \tag{1}$$

$$\lambda_2 u'(0) + \mu_2 u'(1) = g_1(u).$$

where $g_k (k = 1, 2) : C(J, \mathbf{R}) \to \mathbf{R}$ are nonlocal continuous functions and the real constant λ_k, μ_k satisfy the relations $\lambda_k + \mu_k \ne 0$, for $k = 1, 2$, is given by

$$u(t) = g(u) + \int_0^1 \mathscr{G}(t, s) f(s, u(s)) ds,$$

where

$$g(u) = \frac{1}{\lambda_1 + \mu_1} g_1(u) + \frac{1}{\lambda_2 + \mu_2} [t - \mu_1] g_2(u)$$

and $\mathscr{G}(t, s)$ is the Green's function provided by

$$\mathscr{G}(t, s) = \begin{cases} \frac{(t-s)^{q-1}}{\Gamma(q)} + \frac{\mu_1 (1-s)^{q-1}}{(\lambda_1+\mu_1)\Gamma(q)} + \frac{\mu_2}{\lambda_2+\mu_2} \left(\frac{\mu_1}{\lambda_1+\mu_1} - t \right) \frac{(1-s)^{q-2}}{\Gamma(q-1)}, & 0 \le s \le t \le 1, \\ \frac{\mu_1 (1-s)^{q-1}}{(\lambda_1+\mu_1)\Gamma(q)} + \frac{\mu_2}{\lambda_2+\mu_2} \left(\frac{\mu_1}{\lambda_1+\mu_1} - t \right) \frac{(1-s)^{q-2}}{\Gamma(q-1)}, & 0 \le t \le s \le 1, \end{cases}$$

$$\tag{2}$$

Proof Consider the following linear problem of FDES subject to the given boundary condition for $y \in C(J, \mathbf{R})$

$$^c D^q u(t) = y(t), \ 1 < q \le 2, \ t \in [0, 1],$$

$$\lambda_1 u(0) + \mu_1 u(1) = g_1(u), \tag{3}$$

$$\lambda_2 u'(0) + \mu_2 u'(1) = g_1(u).$$

In view of Lemma 2, (3) can be written as

$$u(t) = c_0 + c_1 t - I^q y(t), \quad c_0, \ c_1 \in \mathbf{R}, \tag{4}$$

using $\lambda_1 u(0) + \mu_1 u(1) - g_1(u) = 0$ in (4), we get

$$\lambda_1 c_0 + \mu_1 I^q y(t) + \mu_1 c_0 + \mu_1 c_1 = g_1(u)$$

which yields

$$c_0 = -\frac{\mu_1}{\lambda_1 + \mu_1} c_1 - \frac{\mu_1}{\lambda_1 + \mu_1} I^q y(t) + \frac{1}{\lambda_1 + \mu_1}.$$

Now using $\lambda_2 u'(0) + \mu_2 u'(1) - g_2(u) = 0$ in (4), we get

$$\lambda_2 c_1 + \mu_2 I^{q-1} y(1) + \mu_2 c_1 = g_2(u)$$

implies that

$$c_1 = \frac{1}{\lambda_2 + \mu_2} \left[g_2(u) - \mu_2 I^{q-1} y(1) \right].$$

By simple calculation, we get

$$c_0 = \frac{1}{\lambda_1 + \mu_1} \left[g_1(u) - \frac{\mu_1}{\lambda_2 + \mu_2} g_2(u) \right] + \frac{\mu_1}{\lambda_1 + \mu_1} \left[\frac{\mu_2}{\lambda_2 + \mu_2} I^{q-1} y(1) - I^q y(1) \right].$$

Hence (4) becomes

$$u(t) = \frac{1}{\lambda_1 + \mu_1} g_1(u) + \frac{1}{\lambda_2 + \mu_2} (t - \mu_1) g_2(u) + \frac{\mu_1}{\lambda_1 + \mu_1} I^q y(1)$$
$$+ \frac{\mu_2}{\lambda_2 + \mu_2} \left(\frac{\mu_1}{\lambda_1 + \mu_1} - t \right) I^{q-1} y(1) + I^q y(t).$$

hencewehave $u(t) = g(u) + \displaystyle\int_0^1 \mathscr{G}(t, s) f(s, u(s)) ds,$ \hfill (5)

where

$$g(u) = \frac{1}{\lambda_1 + \mu_1} g_1(u) + \frac{1}{\lambda_2 + \mu_2} \left[t - \mu_1 \right] g_2(u) \tag{6}$$

Thus in view of (5), our considered problem (1) is written as in the form of Fredholm integral eqnarray given by

$$u(t) = g(u) + \int_0^1 \mathscr{G}(t, s) f(s, u(s)) ds, \ t \in [0, 1], \tag{7}$$

where $\mathscr{G}(t, s)$ is Green's function defined as in (2) and $g(u)$ is defined in (6).

In other words, we need the following assumptions to be hold, which are needed for our main results:

(B_1) For $u, v \in C[0, 1]$, there exist $k_g \in [0, 1)$, such that

$$|g(u) - g(v)| \leq k_g \|u - v\|_c;$$

(B_2) For arbitrary $u \in C(J, \mathbf{R})$, there exist $C_g, M_g > 0$ and $r_1 \in [0, 1)$,

$$|f(u)| \leq C_g \|u\|_c^{r_1} + M_g;$$

(B_3) For arbitrary $u \in C(J, \mathbf{R})$, there exist $C_f, M_f > 0$ and $r_2 \in [0, 1)$,

$$|f(t, u)| \leq C_f \|u\|_c^{r_2} + M_f;$$

(B_4) There exists a constant $L_f > 0$, such that

$$|f(t, u) - f(t, \bar{u})| \leq L_f \|u - \bar{u}\|_c, \ for any \ u, \bar{u} \in \mathbf{R}.$$

Let an operator $T : C(J, \mathbf{R}) \longrightarrow C(J, \mathbf{R}) be defined$. Then 7 in the form of operator equation as

$$Tu(t) = Fu(t) + Gu(t), \tag{8}$$

where

$$Fu(t) = \frac{1}{\lambda_1 + \mu_1} g_1(u) + \frac{1}{\lambda_2 + \mu_2} [t - \mu_1] g_2(u), \ Gu(t) = \int_0^1 \mathscr{G}(t, s) f(s, u(s)) ds.$$

The solution of operator equation (8) is the corresponding solution of the considered problem (1).

3   Main Results

Theorem 5 *The operator $F : C(J, \mathbf{R}) \to C(J, \mathbf{R})$ is Lipschitz with constant $k_g \in [0, 1)$. Consequently F is α-Lipschitz with constant k_g. Moreover F obeys the growth condition given by*

$$\|Fu\|_c \leq C_g \|u\|_c^{r_1} + M_g, \quad for every \ u \in C(J, \mathbf{R}). \tag{9}$$

Proof By (B_1)

$$\|Fu - Fv\|_c = \sup\left|\frac{1}{\lambda_1 + \mu_1}(g_1(u) - g_1(v)) + \left(\frac{t}{\lambda_2 + \mu_2} - \frac{\mu_1}{\lambda_2 + \mu_2}\right)(g_2(u) - g_1(v))\right|$$

$$\leq \frac{1}{|\lambda_1 + \mu_1|}|g_1(u) - g_1(v)| + \left|\frac{1}{\lambda_2 + \mu_2} - \frac{\mu_1}{\lambda_2 + \mu_2}\right||g_2(u) - g_2(v)|,$$

using $t \leq 1$

$$\|Fu - Fv\|_c \leq \frac{k_{g_1}}{|\lambda_1 + \mu_1|}\|u - v\|_c + \frac{k_{g_2}}{|\lambda_2 + \mu_2|}\|u - v\|_c \ using \ k_{g_1}, \ k_{g_2} \in [0, 1)$$

$$\leq \left[\frac{k_{g_1}}{|\lambda_1 + \mu_1|} + \frac{k_{g_2}}{|\lambda_2 + \mu_2|}\right]\|u - v\|_c,$$

using

$$\left[\frac{k_{g_1}}{|\lambda_1 + \mu_1|} + \frac{k_{g_2}}{|\lambda_2 + \mu_2|}\right] = k_g.$$

Thus

$$\|Fu - Fv\| \leq k_g \|u - v\|_c.$$

Hence in view of Proposition 1, F is α-Lipschitz with constant k_g. For growth condition, consider

$$\|Fu\|_c = \sup\left|\frac{1}{\lambda_1 + \mu_1}g_1(u) + \frac{1}{\lambda_2 + \mu_2}(t - \mu_1)g_2(u)\right|$$

$$\leq \sup\left|\frac{1}{\lambda_1 + \mu_1}g_1(u)\right| + \sup\left|\frac{t - \mu_1}{\lambda_2 + \mu_2}\right||g_2(u)|$$

$$\leq \frac{C_{g_1}}{|\lambda_1 + \mu_1|}\|u\|_c^{r_1} + M_{g_1} + \frac{1}{|\lambda_2 + \mu_2|}C_{g_2}\|u\|_c^{r_1} + M_{g_2}$$

$$= \left[\frac{C_{g_1}}{|\lambda_1 + \mu_1|} + \frac{C_{g_2}}{|\lambda_2 + \mu_2|}\right]\|u\|_c^{r_1} + M_{g_1} + M_{g_2}$$

which implies that

$$\|Fu\|_c \leq C_g \|u\|_c^{r_1} + M_g, \ C_g = \frac{C_{g_1}}{|\lambda_1 + \mu_1|} + \frac{C_{g_2}}{|\lambda_2 + \mu_2|}, \ M_g = M_{g_1} + M_{g_2}.$$

which is the growth condition (9).

Theorem 6 *The operator* $G : C(J, \mathbf{R}) \longrightarrow C(J, \mathbf{R})$ *is continuous, moreover G satisfies the following growth condition:*

$$\|Gu_n\|_c \leq M \frac{C_f \|u\|_c^{r2} + M_f}{\Gamma(q)}, \ for every \ u \in C(J, \mathbf{R}),$$

where $M = 1 + \left| \frac{\mu_1}{\lambda_1 + \mu_1} \right| + \left| \frac{\mu_2}{\lambda_2 + \mu_2} \left(\frac{\mu_1}{\lambda_1 + \mu_1} \right) \right|.$

Proof Consider that $\{u_n\}$ be the sequence of bounded set $\mathscr{B}_k = \{\|u\|_c \leq k : u \in C(J, \mathbf{R})\}$.
Where $\mathscr{B}_k \subseteq C(J, \mathbf{R})$ and $u_n \to u$ as $n \to \infty$ in $\mathscr{B}_k$. We have to show that $\|Gu_n - Gu\|_c \to 0$ as $n \to \infty$.
Consider

$$|Gu_n(t) - Gu(t)| \leq \int_0^1 |\mathscr{G}(t, s)| \, |f(s, u_n(s)) - f(s, u(s))| \, ds$$

$$\leq \int_0^t \frac{(t - s)^{q-1}}{\Gamma(q)} |f(s, u_n(s)) - f(s, u(s))| \, ds$$

$$+ \left| \frac{\mu_1}{\lambda_1 + \mu_1} \right| \int_0^1 \frac{(1 - s)^{q-1}}{\Gamma(q)} |f(s, u_n(s)) - f(s, u(s))| \, ds$$

$$+ \left| \frac{\mu_2}{\lambda_2 + \mu_2} \right| \left[\frac{\mu_1}{\lambda_1 + \mu_1} \right] \int_0^1 \frac{(1 - s)^{q-2}}{\Gamma(q - 1)}$$
$$\times |f(s, u_n(s)) - f(s, u(s))| \, ds.$$

In view of continuity of f, we have

$$f(t, u_n(s)) \to f(t, u(s)) \ as \ n \to \infty, \ for each \ t \in J.$$

Applying (B_3), and using Lebesgue dominated convergent theorem, we have

$$\int_0^t \frac{(t - s)^{q-1}}{\Gamma(q)} \left[C_f \|u\|_c^{r2} + M_f \right] ds \to 0 \ as \ n \to \infty.$$

Consequently, $\left| \frac{\mu_1}{\lambda_1 + \mu_1} \right| \int_0^1 \frac{(1-s)^{q-1}}{\Gamma(q)} \left[C_f \|u\|_c^{r2} + M_f \right] ds \to 0 \ as \ n \to \infty$
and

$$\left| \frac{\mu_2}{\lambda_2 + \mu_2} \right| \left[\frac{\mu_1}{\lambda_1 + \mu_1} \right] \int_0^1 \frac{(1 - s)^{q-2}}{\Gamma(q - 1)} \left[C_f \|u\|_c^{r2} + M_f \right] ds \to 0 \ as \ n \to \infty.$$

From which it is followed that $\|Gu_n - Gu\| \to 0 \ as \ n \to \infty$. Thus G is continuous.

For growth condition, consider

$$|Gu_n(t)| = \left| \int_0^1 \mathscr{G}(t,s) f(s, u_n(s)) ds \right|$$

$$\leq \int_0^1 \mathscr{G}(t,s) |f(s, u_n(s))| \, ds$$

$$\leq \int_0^1 \frac{(t-s)^{q-1}}{\Gamma(q)} \left[C_f \|u\|_c^{r_2} + M_f \right] ds$$

$$+ \left| \frac{\mu_1}{\lambda_1 + \mu_1} \right| \int_0^1 \frac{(1-s)^{q-1}}{\Gamma(q)} \left[C_f \|u\|_c^{r_2} + M_f \right] ds$$

$$+ \frac{1}{\Gamma(q-1)} \left| \frac{\mu_2}{\lambda_2 + \mu_2} \right| \left(\left| \frac{\mu_1}{\lambda_1 + \mu_1} - 1 \right| \right)$$

$$\times \int_0^1 (1-s)^{q-2} \left[C_f \|u\|_c^{r_2} + M_f \right] ds.$$

From which, we have

$$\|Gu_n\| \leq \frac{C_f \|u\|_c^{r_2} + M_f}{\Gamma(q)} \left[1 + \left| \frac{\mu_1}{\lambda_1 + \mu_1} \right| \left(\left| \frac{\mu_2}{\lambda_2 + \mu_2} \right| + 1 \right) \right]. \tag{10}$$

Hence

$$\|Gu_n\| \leq \frac{M}{\Gamma(q)} \left(C_f \|u\|_c^{r_2} + M_f \right). \tag{11}$$

Theorem 7 *The operator* $G : C(J, \mathbf{R}) \to C(J, \mathbf{R})$ *is completely continuous and* α-*Lipschitz with constant zero.*

Proof For the compactness of G, we consider $\mathscr{D} \subseteq \mathscr{B}_k \subseteq C(J, \mathbf{R})$ is bounded set. We have to show that $G(\mathscr{D})$ is relatively compact in $C(J, \mathbf{R})$ with the help of Arzelà Ascali theorem.

Let $\{u_n\}$ be sequence in $\mathscr{D} \subseteq \mathscr{B}_k$ for every $u_n \in \mathscr{D}$. Then from Growth condition (11), it is obvious that $G(\mathscr{D})$ is bounded in $C(J, \mathbf{R})$.

Let $0 \leq t_1 \leq t_2 \leq 1$, then for equi-continuity, we discuss two cases from Green's function (2) as:

Case I $0 \leq s \leq t \leq 1$.

$$|Gu_n(t_1) - Gu_n(t_2)| \leq \frac{1}{\Gamma(q)} \int_0^{t_1} \left[(t_1 - s)^{q-1} - (t_2 - s)^{q-1} \right] |f(s, u_n(s))| \, ds$$

$$+ \frac{1}{\Gamma(q)} \int_{t_1}^{t_2} (t_2 - s)^{q-1} |f(s - u_n(s))| \, ds$$

$$+(t_2 - t_1) \left| \frac{\mu_2}{\lambda_2 + \mu_2} \right| \frac{1}{\Gamma(q-1)} \int_0^1 (1-s)^{q-2} |f(s, u_n(s))| \, ds$$

$$\leq \frac{1}{\Gamma(q)} \int_0^{t_1} \left[(t_1 - s)^{q-1} - (t_2 - s)^{q-1} \right] \left(C_f \|u_n\|_c^{r_2} + M_f \right) ds$$

$$+ \frac{1}{\Gamma(q)} \int_{t_1}^{t_2} (t_2 - s)^{q-1} \left(C_f \|u_n\|_c^{r_2} + M_f \right) ds$$

$$+ (t_2 - t_1) \frac{|\mu_2|}{|\lambda_2 + \mu_2|} \frac{1}{\Gamma(q-1)} \int_0^1 (1-s)^{q-2} \left(C_f \|u_n\|_c^{r_2} + M_f \right) ds.$$

Therefore

$$\|Gu_n(t_1) - Gu_n(t_2)\| = \left[\frac{(t_1^q - t_2^q)}{\Gamma(q+1)} + \frac{(t_2 - t_1)^q}{\Gamma(q+1)} + \frac{(t_2 - t_1)|\mu_2|}{|\lambda_2 + \mu_2| \Gamma(q)} \right]$$
$$\times \left(C_f \|u\|_c^{r_2} + M_f \right). \tag{12}$$

Clearly $t_1 \to t_2$, then the right-hand side of (12) tends to zero. So

$$\|Gu_n(t_1) - Gu_n(t_2)\| \to 0 \ \text{ as } \ t_1 \to t_2.$$

Thus in this case G is equi-continuous.

Case II   if $0 \leq t \leq s \leq 0$, then

$$|Gu_n(t_1) - Gu_n(t_2)| \leq \frac{\mu_2 (t_1 - t_2)}{|\lambda_2 + \mu_2| \Gamma(q-1)} \int_0^1 (1-s)^{q-2} |f(s, u_n(s))| \, ds$$

$$\leq \frac{\mu_2 (t_1 - t_2)}{|\lambda_2 + \mu_2| \Gamma(q)} \left(C_f \|u\|_c^{r_2} + M_f \right) \to 0, \ \text{as } t_1 \to t_2.$$

So G in this case is also equi-continuous. Hence G is equi-continuous and $G(\mathscr{D}) \subseteq C(J, \mathbf{R})$, which satisfies the hypothesis of Arzela Ascali theorem. So $G(\mathscr{D})$ is relatively compact in $C(J, \mathbf{R})$. G is completely continuous. It is easy to show that G is α-Lipschitz with constant zero by using Proposition 2.

Theorem 8 *Assume that (B_1)–(B_3) hold, then boundary value problem (1) has at least one positive solution $u \in C(J, \mathbf{R})$ and the set of the solutions is bounded in $C(J, \mathbf{R})$.*

Proof As $F, G, T : C(J, \mathbf{R}) \to C(J, \mathbf{R})$ have been defined previously are continuous in view of continuity of f, g. Moreover F and G are α-Lipschitz. Thus T is strict α-contraction. Consider

$$W_0 = \{u \in C(J, \mathbf{R}) : there exist \ \lambda \in [0, 1], \lambda u = \lambda Tu \}.$$

To prove that W_0 is bounded subset of $C(J, \mathbf{R})$. Let $u \in W_0$ such that $u = \lambda Tu$, one can see

$$\|u\|_c = \|\lambda Tu\|_c = \lambda \left(\|Fu + Gu\|_c\right)$$
$$\leq \lambda \left(\|Fu\|_c + \|Gu\|_c\right)$$

In view of Growth conditions of F, G, we get

$$\|u\|_c \leq \left(C_g \|u\|_c^{r_1} + M_g + M \frac{C_f \|u\|_c^{r_2} + M_f}{\Gamma(q)}\right), \quad r_1, r_2 \in [0, 1). \quad (13)$$

Thus W_0 is bounded. If not, let $\mathscr{R} = \|u\|_c$, taking $\mathscr{R} = \|u\|_c$ such that $\mathscr{R} \to \infty$. Then from (13), we have

$$1 \leq \lim_{\mathscr{R} \to \infty} \lambda \left[\frac{C_g \|u\|_c^{r_1} + M_g}{\mathscr{R}} + M \frac{C_f \|u\|_c^{r_2} + M_f}{\mathscr{R}\Gamma(q)}\right] = 0.$$

which is contraction.

This implies that W_0 is bounded and T has at least one fixed point by means of Theorem 1, which is the corresponding positive solution of boundary value problem therefore (1).

Theorem 9 *Under the assumption (B_1) to (B_4), boundary value problem (1) has a unique solution if $G^* < 1$, where*

$$G^* = \frac{k_{g_1}}{|\lambda_1 + \mu_1|} + \frac{k_{g_2}}{|\lambda_2 + \mu_2|} + L_f \int_0^1 \mathscr{G}(t, s)ds.$$

Proof From (B_1)–(B_3), we have

$$\|Tu - Tv\| \leq \frac{1}{|\lambda_1 + \mu_1|} \|g_1(u) - g_1(v)\|_c + \frac{1}{|\lambda_2 + \mu_2|} \|g_2(u) - g_2(v)\|_c$$

$$+ \int_0^1 \mathscr{G}(t, s) \|f(s, u) - f(s, v)\| ds$$

$$\leq \frac{k_{g_1}}{|\lambda_1 + \mu_1|} \|u - v\|_c + \frac{k_{g_2}}{|\lambda_2 + \mu_2|} \|u - v\|_c$$

$$+ L_f \int_0^1 \mathscr{G}(t, s) \|u - v\|_c ds$$

$$\leq \left[\frac{k_{g_1} + k_{g_2}}{|\lambda_1 + \mu_1| + |\lambda_2 + \mu_2|} + L_f \int_0^1 \mathscr{G}(1, s)ds\right] \|u - v\|_c$$

$$\|Tu - Tv\|_c \leq G^* \|u - v\|_c$$

where

$$G^* = \frac{k_{g_1}}{|\lambda_1 + \mu_1|} + \frac{k_{g_2}}{|\lambda_2 + \mu_2|} + L_f \int_0^1 \mathcal{G}(1, s)ds < 1.$$

Hence T has a unique fixed point, which is the corresponding positive solution to the considered problem (1).

4 Ulam's Stability Analysis of Boundary Value Problem (1)

In this section, we prove necessary and sufficient conditions for various types of Ulam's stability like Ulam–Hyers, generalized Ulam–Hyers stability, Ulam–Hyers–Rassias, and generalized Ulam–Hyers–Rassias stability of the solutions to the considered problem (1) of nonlinear fractional differential equations. In this regard we review the following definitions and results for further analysis.

Definition 4 The solution $u \in C([0, 1])$ of the fractional differential equation given by

$$^cD^q u(t) = f(t, u(t)), \ t \in J, \tag{14}$$

is Ulam–Hyers stable if we can find a real number $\hat{C}_{L_f, k_g, \mathcal{G}^*} > 0$ with the property that for every $\epsilon > 0$ and for every solution $u \in C[0, 1]$ of the inequality

$$\left| {}^cD^q u(t) - f(t, u(t)) \right| \leq \epsilon, \ t \in [0, 1], \tag{15}$$

there exists unique solution $v \in C[0, 1]$ of the given fractional differential equation (1) with a constant $\hat{C}_{L_f, k_g, \mathcal{G}^*} > 0$ with

$$\|u - v\|_c \leq \hat{C}_{L_f, k_g, \mathcal{G}^*}\epsilon.$$

Definition 5 The solution $u \in C[0, 1]$ of the fractional differential equation (1) is called to be generalized Ulam–Hyers stable , if we can find

$$\theta_{f,q} : (0, \infty) \to \mathbf{R}^+, \ \theta_{f,q}(0) = 0,$$

such that for each solution $u \in C[0, 1]$ of the inequality (15), we can find a unique solution $v \in C[0, 1]$ of the fractional differential equation (1) with

$$\|u - v\|_c \leq \hat{C}_{L_f, k_g, \mathcal{G}^*}\theta_{f,q}.$$

Next we recall the definitions of Ulam–Hyers–Rassias and generalized Ulam–Hyers–Rassias stability [34] for our considered problem (1) as below:

Definition 6 Fractional differential equation (1) is said to be Ulam–Hyers–Rassias stable with respect to $\varphi \in C([0, 1], \mathbf{R})$ if there exists a nonzero positive real constant $\hat{C}_{L_f, k_g, \mathscr{G}^*}$ such that for each $\epsilon > 0$ and for every solution $u \in C[0, 1]$ of the inequality

$$\left| {}^c D^q u(t) - f(t, u(t)) \right| \leq \varphi(t) \epsilon, \ t \in [0, 1], \tag{16}$$

there exists a solution $v \in C[0, 1]$ of the Eq. (1), such that

$$|u(t) - v(t)| \leq \hat{C}_{L_f, k_g, \mathscr{G}^*} \epsilon \varphi(t), \ t \in [0, 1].$$

Definition 7 Equation (1) is said to be generalized Ulam–Hyers–Rassias stable with respect to $\varphi \in C[0, 1]$, if there exists a real number $\hat{C}_{L_f, k_g, \mathscr{G}^*} > 0$ such that for each solution $u \in C[0, 1]$ of the inequality

$$\left| {}^c D^q u(t) - f(t, u(t)) \right| \leq \varphi(t), \ t \in [0, 1], \tag{17}$$

there exists a solution $v \in C[0, 1]$ of the Eq. (1) such that $|u(t) - v(t)| \leq \hat{C}_{L_f, k_g, \mathscr{G}^*} \theta(\epsilon) \varphi(t), \ t \in [0, 1]$.

Remark 1 A function $u \in C[0, 1]$ is said to be the solution of inequality given in (15) if and only if there exists a function $\varpi \in C[0, 1]$ that depends on u only such that

(i) $|\varpi(t)| \leq \epsilon, \ for \, all \ t \in [0, 1]$;
(ii) ${}^c D^q u(t) = f(t, u(t)) + \varpi(t), \quad for \, all \ t \in [0, 1]$.

Lemma 1 *Under the assumption given in Remark 1, the solution $u \in C[0, 1]$ of the boundary value problem given by*

$$\begin{cases} {}^c D^q u(t) = f(t, \ u(t)) + \varpi(t), \ 1 < q \leq 2, \ t \in [0, 1], \\ \lambda_1 u(0) + \mu_1 u(1) = g_1(u), \\ \lambda_2 u'(0) + \mu_2 u'(1) = g_1(u) \end{cases} \tag{18}$$

satisfies the following relation:

$$\left| u(t) - \left(g(u) + \int_0^1 \mathscr{G}(t, s) f(s, u(s)) ds \right) \right| \leq \epsilon \mathscr{G}^*, \ where \ \max_{t \in [0,1]} \int_0^1 |\mathscr{G}(t, s)| ds = \mathscr{G}^*. \tag{19}$$

Proof In view of Theorem 4, the solution of the problem (18) is given by

$$u(t) = g(u) + \int_0^1 \mathscr{G}(t, s) f(s, u(s)) ds + \int_0^1 \mathscr{G}(t, s) \varpi(s) ds,$$

where $\mathscr{G}$ is the Green's function as defined in Theorem 4. Using Remark 1 we can easily get the result given in (19).

Theorem 10 *Under the assumption* (B_1), (B_4) *and Lemma 1, the solution of the considered problem (1) is Ulam's stable and consequently generalized Ulam–Hyers stable if the condition* $[k_g + L_f \mathscr{G}^*] < 1$ *holds.*

Proof Let $u \in C[0, 1]$ be any solution of boundary value problem (1) and $v \in C[0, 1]$ be the unique solution of the considered problem (1), then take

$$|u(t) - v(t)| = \left| u(t) - \left(g(v) + \int_0^1 \mathscr{G}(t, s) f(s, v(s)) ds \right) \right|$$

$$\leq \left| u(t) - \left(g(u) + \int_0^1 \mathscr{G}(t, s) f(s, u(s)) ds \right) \right|$$

$$+ \left| g(u) - g(v) + \int_0^1 \mathscr{G}(t, s) [f(s, u(s)) - f(s, v(s))] \right|$$

$$\leq \epsilon \mathscr{G}^* + k_g \|u - v\|_c + L_f \mathscr{G}^* \|u - v\|_c.$$

From which we have

$$\|u - v\|_c \leq \epsilon \mathscr{G}^* + [k_g + L_f \mathscr{G}^*] \|u - v\|_c,$$

where k_g is defined in Theorem 5 which yields

$$\|u - v\|_c \leq \hat{C}_{L_f, k_g, \mathscr{G}^*} \epsilon, \quad where \quad \frac{\mathscr{G}^*}{1 - [k_g + L_f \mathscr{G}^*]} = \hat{C}_{L_f, k_g, \mathscr{G}^*}. \tag{20}$$

Hence the solution of the considered problem (1) is Ulam–Hyers stable. Further if we set $\theta(\epsilon) = \epsilon$ such that $\theta(0) = 0$, then we get

$$\|u - v\|_c \leq \hat{C}_{L_f, k_g, \mathscr{G}^*} \theta(\epsilon) \tag{21}$$

which implies that the solution of the proposed problem is generalized Ulam–Hyers stable.

(B_5) Let for $\delta_\varphi > 0$ there exists a nondecreasing function $\varphi \in ([0, 1], \mathbf{R}^+)$ such that

$$\int_0^t \frac{(t - s)^{q-1}}{\Gamma(q)} \varphi(s) ds \leq \delta_\varphi \varphi(t), \quad for \; t \in [0, 1].$$

Theorem 11 *Under the assumptions* (B_1), (B_4), (B_5), *the solution of the considered problem (1) is Ulam–Hyers–Rassias stable if* $k_g + L_f \mathscr{G}^* < 1$.

Proof Let $u \in C[0, 1]$ be any solution of the inequality (16) and $v \in C[0, 1])$ be the unique solution of the problem (1), then the solution of

$$^c D^q u(t) = f(t, u(t)) + \varpi(t), \ 1 < q \leq 2, \ t \in [0, 1],$$

$$\lambda_1 u(0) + \mu_1 u(1) = g_1(u),$$

$$\lambda_2 u'(0) + \mu_2 u'(1) = g_1(u)$$

is given by

$$u(t) = \frac{1}{\lambda_1 + \mu_1} g_1(u) + \frac{1}{\lambda_2 + \mu_2} (t - \mu_1) g_2(u) + \frac{\mu_1}{\lambda_1 + \mu_1} \int_0^1 \frac{(1-s)^{q-1}}{\Gamma(q)} f(s, u(s))$$

$$+ \frac{\mu_2}{\lambda_2 + \mu_2} \left(\frac{\mu_1}{\lambda_1 + \mu_1} - t \right) \int_0^1 \frac{(1-s)^{q-2}}{\Gamma(q-1)} f(s, u(s)) + \int_0^t \frac{(t-s)^{q-1}}{\Gamma(q)} f(s, u(s))$$

$$+ \frac{\mu_1}{\lambda_1 + \mu_1} \int_0^1 \frac{(1-s)^{q-1}}{\Gamma(q)} \varpi(s) + \frac{\mu_2}{\lambda_2 + \mu_2} \left(\frac{\mu_1}{\lambda_1 + \mu_1} - t \right)$$

$$\times \int_0^1 \frac{(1-s)^{q-2}}{\Gamma(q-1)} \varpi(s) + \int_0^t \frac{(t-s)^{q-1}}{\Gamma(q)} \varpi(s)$$

$$u(t) = g(u) + \int_0^1 \mathscr{G}(t, s) f(s, u(s)) ds \qquad (22)$$

$$+ \frac{\mu_1}{\lambda_1 + \mu_1} \int_0^1 \frac{(1-s)^{q-1}}{\Gamma(q)} \varpi(s) + \frac{\mu_2}{\lambda_2 + \mu_2} \left(\frac{\mu_1}{\lambda_1 + \mu_1} - t \right)$$

$$\times \int_0^1 \frac{(1-s)^{q-2}}{\Gamma(q-1)} \varpi(s) + \int_0^t \frac{(t-s)^{q-1}}{\Gamma(q)} \varpi(s).$$

Then from (22) we have

$$\left| u(t) - g(u) - \int_0^1 \mathscr{G}(t, s) f(s, u(s)) ds \right|$$

$$= \left| \frac{\mu_1}{\lambda_1 + \mu_1} \int_0^1 \frac{(1-s)^{q-1}}{\Gamma(q)} \varpi(s) + \frac{\mu_2}{\lambda_2 + \mu_2} \left(\frac{\mu_1}{\lambda_1 + \mu_1} - t \right) \right.$$

$$\left. \times \int_0^1 \frac{(1-s)^{q-2}}{\Gamma(q-1)} \varpi(s) + \int_0^t \frac{(t-s)^{q-1}}{\Gamma(q)} \varpi(s) \right|$$

$$\leq \epsilon \delta_\varphi \varphi(t).$$

Then using the same fashion as in Theorem 10, we have

$$|u(t) - v(t)| \leq \left| u(t) - \left(g(u) + \int_0^1 \mathscr{G}(t, s) f(s, u(s)) ds \right) \right|$$

$$+ |g(u) - g(v)| + \left| \int_0^1 \mathscr{G}(t, s) f(s, u(s)) ds \right.$$

$$\left. - \int_0^1 \mathscr{G}(t, s) f(s, v(s)) ds \right|$$

$$\leq \epsilon \delta_\varphi \varphi(t) + k_g \|u - v\|_c + L_f \mathscr{G}^* \|u - v\|_c$$

which further gives $\|u - v\|_c \leq \epsilon \delta_\varphi \varphi(t) + [k_g + L_f \mathscr{G}^*] \|u - v\|_c.$ (23)

Hence we have

$$\|u - v\|_c \leq \hat{C}_{L_f, k_g, \mathscr{G}^*} \epsilon \delta_\varphi \varphi(t), \ t \in [0, 1], \ where \ \hat{C}_{L_f, k_g, \mathscr{G}^*} = \frac{1}{1 - [k_g + L_f \mathscr{G}^*]}.$$
(24)

Hence from (24) we concluded that the solution of the considered problem (1) is Ulam–Hyers–Rassias stable. Further it is easy to prove that the solution of the considered problem (1) is generalized Ulam–Hyers Rassias stable.

5 Example

Example 1 Consider the boundary value problem

$$^c D^{\frac{3}{2}} u(t) = \frac{|u(t)|}{(1 + e^t)(1 + 9u(t))}, \ t \in [0, 1],$$

$$u(0) + u(1) = g_1(u) = \sum_{k=1}^5 \delta_k u(t_k), \ t_k \in (0, 1), \ \sum_{k=1}^5 \delta_k \leq \frac{1}{20}, \quad (25)$$

$$\frac{1}{2} \acute{u}(0) + \frac{1}{2} \acute{u}(1) = g_2(u) = \sum_{k=1}^3 \acute{\delta}_k u(t_k), \ t_k \in (0, 1), \ \sum_{k=1}^3 \acute{\delta}_k \leq \frac{1}{10}.$$

Then $\lambda_1 = \mu_1 = 1, \lambda_2 = \mu_2 = \frac{1}{2}, \ g_1(u) = \sum_{k=1}^5 \delta_k u(t_k), g_2(u) = \sum_{k=1}^3 \acute{\delta}_k u(t_k),$
and $f(t, u) = \frac{|u(t)|}{(1 + e^t)(1 + 9u(t))}.$
We have

$$|f(t, u) - f(t, v)| \leq \left| \frac{|u|(1 + 9|v|) - |v|(1 + 9|u|)}{(1 + e^t)(1 + 9|u|)(1 + 9|v|)} \right| \leq \frac{1}{200} |u - v|.$$

Clearly

$$M_f = 0, C_f = \frac{1}{2}, \; k_{g_1} = \frac{1}{20}, \; k_{g_2} = \frac{1}{10}, \; from \; which, \; we \; have$$

$$k_g = \frac{3}{20}, \; r_1 = r_2 = \frac{1}{2}, L_f = \frac{1}{200}, q = \frac{3}{2}.$$

By simple computation, one can show that

$$G^* = \frac{\frac{1}{20}}{2} + \frac{\frac{1}{10}}{1} + \frac{1}{200} \int_0^1 \mathscr{G}(1,s)ds = \frac{1}{40} + \frac{1}{10} + \frac{1}{200} \int_0^1 \mathscr{G}(1,s)ds < 1.$$

Thus in view of Theorem 9, (25) has unique solution. Further, it is easy to show that the set of solution is bounded by using Theorem 8. Further the condition $k_g + L_f \mathscr{G}^* < 1$ obviously holds so by Theorem 10, the solution of the given problem is Ulam–Hyers stable and consequently generalized Ulam–Hyers stable. Let $\varphi(t) = t$, then the conditions of Ulam–Hyers Rassias and generalized Ulam–Hyers Rassias stability can be easily received by using Theorem 11.

6 Conclusion

Considering the Caputo fractional derivative we have successfully established existence theory of at least one solution to a boundary value problem of fractional differential equations by using topological degree theory. Further by using nonlinear functional analysis we have developed appropriate conditions for different kinds of Ulam stability theory including Ulam–Hyers, generalized Ulam–Hyers, Ulam–Hyers Rassias, and generalized Ulam–Hyers–Rassias stability. The whole results have been demonstrated by a proper example.

Acknowledgements The authors declare that there does not exist any conflict of interest. Further, this work has been supported by the National Natural Science Foundation of China (11571378).

References

1. R. Hilfer, *Applications of Fractional Calculus in Physics* (World Scientific, Singapore, 2000)
2. A.A. Kilbas, O.I. Marichev, S.G. Samko, *Fractional Integrals and Derivatives (Theory and Applications)* (Gordon and Breach, Switzerland, 1993)
3. A.A. Kilbas, H.M. Srivastava, J.J. Trujillo, *Theory and Applications of Fractional Differential Equations*. North-Holland Mathematics Studies, vol. 204 (Elsevier, Amsterdam, 2006)
4. V. Lakshmikantham, S. Leela, J. Vasundhara, *Theory of Fractional Dynamic Systems* (Cambridge Academic, Cambridge, 2009)

5. K.S. Miller, B. Ross, *An Introduction to the Fractional Calculus and Fractional Differential Equations* (Wiley, New York, 1993)
6. I. Podlubny, *Fractional Differential Equations*. Mathematics in Science and Engineering (Academic, New York, 1999)
7. V.E. Tarasov, *Fractional Dynamics: Application of Fractional Calculus to Dynamics of Particles, Fields and Media* (Springer, HEP, Heidelberg, Germany, 2010)
8. I. Podlubny, *Fractional differential equations. An introduction to fractional derivative, fractional differential equations, to methods of their solution and some of their applications.* Mathematics in Science and Engineering, vol. 198 (Academic, San Diego, 1999)
9. M. Benchohra, S. Hamani, S.K. Ntouyas, Boundary value problems for differential equations with fractional order and nonlocal conditions. Nonlinear Anal. **71**, 2391–2396 (2009)
10. B. Ahmad, S. Sivasundaram, On four-point nonlocal boundary value problems of nonlinear integro-differential equations of fractional order. Appl. Math. Comput. **217**, 480–487 (2010)
11. Z. Bai, On positive solutions of a nonlocal fractional boundary value problem. Nonlinear Anal. **72**, 916–924 (2010)
12. M. El-Shahed, J.J. Nieto, Nontrivial solutions for a nonlinear multi-point boundary value problem of fractional order. Comput. Math. Appl. **59**, 3438–3443 (2010)
13. A. Belarbi, M. Benchohra, A. Ouahab, Existence results for functional differential equations of fractional order. Appl. Anal. **85**, 1459–1470 (2006)
14. R.P. Agarwal, M. Benchohra, S. Hamani, Boundary value problems for differential inclusions with fractional order. Adv. Stud. Contemp. Math. **12**(2), 181–196 (2008)
15. M. Benchohra, J.R. Graef, S. Hamani, Existence results for boundary value problems with nonlinear fractional differential equations. Appl. Anal. **87**, 851–863 (2008)
16. M. Benchohra, S. Hamani, S.K. Ntouyas, Boundary value problems for differential equations with fractional order. Surv. Math. Appl. **3**, 1–12 (2008)
17. K. Shah, R.A. Khan, Existence and uniqueness of positive solutions to a coupled system of nonlinear fractional order differential equations with anti periodic boundary conditions. Differ. Equ. Appl. **7**(2), 245–262 (2015)
18. K. Shah, H. Khalil, R.A. Khan, Investigation of positive solution to a coupled system of impulsive boundary value problems for nonlinear fractional order differential equations. J. Chaos Solitons Fractals **77**, 240–246 (2015)
19. C.F. Li, X.N. Luo, Y. Zhou, Existence of positive solutions of the boundary value problem for nonlinear fractional differential equations. Comput. Math. Appl. **59**, 1363–1375 (2010)
20. J. Wang , Y. Zhou, W. Wei, Study in fractional differential equations by means of topological degree methods. Numer. Funct. Anal. Optim. **33**(2), 216–238 (2012)
21. R.A. Khan, K. Shah, Existence and uniqueness of positive solutions to fractional order multi-point boundary value problems. Commun. Appl. Anal. **19**, 515–526 (2015)
22. K. Shah, A. Ali, R.A. Khan, Degree theory and existence of positive solutions to coupled systems of multi-point boundary value problems. Bound. Value Probl. **2016**, 43 (2016)
23. I.A. Rus, Ulam stabilities of ordinary differential equations in a Banach space. Carpathian J. Math. **26**, 103–107 (2010)
24. M. Obloza, Hyers stability of the linear differential equation. Rocznik Nauk.-Dydakt. Prace Mat. **13**, 259–270 (1993)
25. J. Wang, X. Li, Ulam-Hyers stability of fractional Langevin equations. Appl. Math. Comput. **258**, 72–83 (2015)
26. S. Tang, A. Zada, S. Faisal, M.M.A. El-Sheikh, T. Li, Stability of higher order nonlinear impulsive differential equations. J. Nonlinear Sci. Appl. **9**, 4713–4721 (2016)
27. G. Lijun, D. Wang, G. Wang, Further results on exponential stability for impulsive switched nonlinear time-delay systems with delayed impulse effects. Appl. Math. Comput. **268**, 186–200 (2015)
28. I. Stamova, Mittag-Leffler stability of impulsive differential equations of fractional order. Q. Appl. Math. **73**(3), 525–535 (2015)
29. J.C. Trigeassou, et al., A Lyapunov approach to the stability of fractional differential equations. Signal Process. **91**(3), 437–445 (2011)

30. J. Wang, L. Lv, W. Zhou, Ulam stability and data dependence for fractional differential equations with Caputo derivative. Electron. J. Qual. Theory Differ. Equ. **63**, 1–10 (2011)
31. K. Shah, D. Vivek, K. Kanagarajan, Dynamics and stability of Ψ-fractional pantograph equations with boundary conditions. Bol. Soc. Paran. Mat. **22**(2), 1–13 (2018)
32. A. Zada, S. Faisal, Y. Li, On the Hyers–Ulam stability of first order impulsive delay differential equations. J. Funct. Spaces **2016**, 6 pages (2016)
33. A. Zada, O. Shah, R. Shah, Hyers–Ulam stability of non-autonomous systems in terms of boundedness of Cauchy problems. Appl. Math. Comput. **271**, 512–518 (2015)
34. T.M. Rassias, On the stability of the linear mapping in Banach spaces. Proc. Am. Math. Soc. **72**, 297–300 (1978)
35. S.M. Ulam, *Problems in Modern Mathematics* (Wiley, New York, 1940)
36. S.M. Ulam, *A Collection of Mathematical Problems* (Interscience, New York, 1960)
37. P. Kumama, A. Ali, K. Shah, R.A. Khan, Existence results and Hyers-Ulam stability to a class of nonlinear arbitrary order differential equations. J. Nonlinear Sci. Appl. **10**, 2986–2997 (2017)
38. J. Wang, L. Lv, Y. Zhou, Ulam stability and data dependece for fractional differential equations with Caputo derivative. Electron. J. Qual. Theory Differ. Equ. **63**, 1–10 (2011)
39. F. Haq, K. Shah, G. Rahman, M. Shahzad, Hyers-Ulam stability to a class of fractional differential equations with boundary conditions. Int. J. Appl. Comput. Math. **2017**, 1–13 (2017)
40. J. Vanterler da C. Sousa, E. Capelas de Oliveira, Ulam–Hyers stability of a nonlinear fractional Volterra integro-differential equation. Appl. Math. Lett. **81**, 50–56 (2018)
41. A. Zada, S. Ali, Y. Li, Ulam's type stability for a class of implicit fractional differential equations with non-instantaneous integral impulses and boundary condition. Adv. Difference Equ. **2017**, 317 (2017)
42. J. Wang, M. Feckan, Y. Zhou, Ulam's type stability of impulsive ordinary differential equations. J. Math. Anal. Appl. **395**, 258–264 (2012)
43. S.M. Jung, *Hyers-Ulam-Rassias Stability of Functional Equations in Nonlinear Analysis* (Springer, New York, 2011)
44. J. Wang, M. Feckan, Y. Zhou, Ulam's type stability of impulsive ordinary differential equations. J. Math. Anal. Appl. **395**, 258–264 (2012)
45. D.H. Hyers, G. Isac, T.M. Rassias, *Stability of Functional Equations in Several Variables* (Birkhäuser, Basel, 1998)
46. S.M. Jung, *Hyers-Ulam-Rassias Stability of Functional Equations in Mathematical Analysis* (Hadronic Press, Palm Harbor, 2001)
47. S.M. Jung, Hyers-Ulam stability of linear differential equations of first order. Appl. Math. Lett. **17**, 1135–1140 (2004)
48. S.E. Takahasi, T. Miura, S. Miyajima, On the Hyers-Ulam stability of the Banach space-valued differential equation $y' = \lambda y$. Bull. Korean Math. Soc. **39**, 309–315 (2002)
49. D.S. Cimpean, D. Popa, Hyers-Ulam stability of Euler's equation. Appl. Math. Lett. **24**, 1539–1543 (2011)
50. J. Wang, Y. Zhou, Mittag-Leffler-Ulam stabilities of fractional evolution equations. Appl. Math. Lett. **25**, 723–728 (2012)
51. J. Wang, L. Lv, Y. Zhou, Boundary value problems for fractional differential equations involving Caputo derivative in Banach spaces. J. Appl. Math. Comput. **38**(1–2), 209–224 (2012)
52. K. Deimling, *Nonlinear Functional Analysis* (Springer, New York, 1985)
53. K. Diethelm, *The Analysis of Fractional Differential Equations*. Lecture Notes in Mathematics (Springer, New York, 2010)
54. F. Isaia, On a nonlinear integral equation without compactness. Acta. Math. Univ. Comenianae. **75**, 233–240 (2006)
55. S.M. Ulam, *A Collection of the Mathematical Problems* (Interscience, New York, 1960)
56. T.M. Rassias, On the stability of functional equations and a problem of Ulam. Acta. Appl. Math. **62**, 23–130 (2000)

On a Variant of μ-Wilson's Functional Equation with an Endomorphism

K. H. Sabour, A. Charifi, and S. Kabbaj

Abstract The main goal of this chapter is to find the solutions (f, g) of the generalized variant of μ-d'Alembert's functional equation

$$f(xy) + \mu(y)f(\varphi(y)x) = 2f(x)f(y),$$

and μ-Wilson's functional equation

$$f(xy) + \mu(y)f(\varphi(y)x) = 2f(x)g(y),$$

in the setting of semigroups, monoids, and groups, where φ is an endomorphism not necessarily involutive and μ is a multiplicative function. We prove that their solutions can be expressed in terms of multiplicative and additive functions. Many consequences of these results are presented.

1 Notation and Terminology

To formulate our results we recall the following notations and assumptions that will be used throughout the chapter.

Let S be a semigroup, i.e., a set equipped with an associative operation. A monoid M is a semigroup with an identity element that we denote e. Let G be a group.

The map $\sigma : S \rightarrow S$ denotes an involutive automorphism. That σ is involutive means that $\sigma(\sigma(x)) = x$ for all $x \in S$. The map $\varphi : M \rightarrow M$ denotes an endomorphism where M is a possibly non-abelian group or monoid.

A multiplicative function on M is a map $\chi : M \rightarrow \mathbb{C}$ such that $\chi(xy) = \chi(x)\chi(y)$ for all $x, y \in M$. A character on a group G is a homomorphism from G into the multiplicative group of non-zero complex numbers. While a non-zero multiplicative function on a group can never take the value 0, it is possible for a

K. H. Sabour (✉) · A. Charifi · S. Kabbaj
Department of Mathematics, Faculty of Sciences, University of Ibn Tofail, Kenitra, Morocco

© Springer Nature Switzerland AG 2019

G. A. Anastassiou and J. M. Rassias (eds.), *Frontiers in Functional Equations and Analytic Inequalities*, https://doi.org/10.1007/978-3-030-28950-8_5

multiplicative function on a monoid to take the value 0 on a proper, non-empty subset of S. If $\chi : S \to \mathbb{C}$ is multiplicative and $\chi \neq 0$, then

$$I_\chi = \{x \in S \mid \chi(x) = 0\}$$

is either empty or a proper subset of S. The fact that χ is multiplicative establishes that I_χ is a two-sided ideal in S if not empty (for us an ideal is never the empty set). It follows also that $S \backslash I_\chi$ is a subsemigroup of S. These ideals play an essential role in our discussion of Eq. (5) on monoids.

A function $A : S \to \mathbb{C}$ is called additive, if it satisfies $A(xy) = A(x) + A(y)$ for all $x, y \in S$.

If S is a topological space, then we let $C(S)$ denote the algebra of continuous functions from S into $\mathbb{C}$.

2 Introduction

The functional equation

$$g(x + y) + g(x - y) = 2g(x)g(y), \quad x, y \in \mathbb{R}, \tag{1}$$

is known as the d'Alembert's functional equation. It has a long history going back to d'Alembert [2]. As the name suggests this functional equation was introduced by d'Alembert in connection with the composition of forces and plays a central role in determining the sum of two vectors in Euclidean and non-Euclidean geometries. The continuous solutions of (1) were determined by Cauchy in 1821 (see [1]).

D'Alembert's functional equation was generalized in another direction by Wilson [10], viz. to the functional equation

$$f(x + y) + f(x - y) = 2f(x)g(y), \quad x, y \in \mathbb{R}, \tag{2}$$

that contains the two unknown functions f and g.

D'Alembert's functional equation (1) possesses periodic and non-periodic solutions. To exclude the non-periodic solutions Kannappan, in 1968, modified (1) to the functional equation

$$f(x - y + z_0) + f(x + y + z_0) = 2f(x)f(y), \quad x, y \in \mathbb{R}, \tag{3}$$

where z_0 is a non-zero real constant. Kannappan proved that any solution $f : \mathbb{R} \to \mathbb{C}$ of (3) has the form $f(x) = g(x - z_0)$, where $g : \mathbb{R} \to \mathbb{C}$ is a periodic solution of (1) with period $2z_0$. This enabled him to find all Lebesgue measurable solutions (see [7]).

Equations (1)–(3) have been extended to abelian groups: You just replace the domain of definition $\mathbb{R}$ by an abelian group $(G, +)$. They have been solved in that setting. For more details concerning (1)–(3), and their further generalizations we refer to the monographs [8, 9], and the references therein.

The purpose of the present chapter is first to solve the μ-d'Alembert's functional equation

$$g(xy) + \mu(y)g(\varphi(y)x) = 2g(x)g(y), \quad x, y \in S, \tag{4}$$

where $\mu : S \to \mathbb{C}\backslash\{0\}$ is a multiplicative function such that $\mu(x\varphi(x)) = 1$ for all $x \in S$. This equation provides a common generalization of (1), the symmetrized multiplicative Cauchy equation

$$g(xy) + g(yx) = 2g(x)g(y), \quad x, y \in S,$$

and the variant of d'Alembert's functional equation

$$g(xy) + g(\varphi(y)x) = 2g(x)g(y), \quad x, y \in S,$$

solved by Fadli et al. [6], and secondly to use that to find the solutions of the following generalization of μ-Wilson's functional equation

$$f(xy) + \mu(y)f(\varphi(y)x) = 2f(x)g(y), \quad x, y \in M, \tag{5}$$

where M is a possibly non-abelian group or monoid (that is, a semigroup with identity), $\mu : M \to \mathbb{C}\backslash\{0\}$ is a multiplicative function such that $\mu(x\varphi(x)) = 1$ for all $x \in M$, and $\varphi : M \to M$ is an endomorphism, for unknown functions $f, g : M \to \mathbb{C}$. This equation, in the case where φ is an involutive automorphism, has been introduced and solved by Elqorachi et al. in [5].

By various elementary methods we find all solutions of (5) on monoids that are generated by their squares and on groups, in terms of multiplicative and additive functions. This contrasts the solutions of the functional equation $f(xy) + f(y^{-1}x) = 2f(x)g(y)$, where the non-abelian phenomena like 2-dimensional irreducible representations may occur (see [4]). Our results constitute a natural extension of earlier results of, e.g., [6].

As other important results in this chapter, we solve the following functional equations

$$f(xyz_0) + \mu(y)f(\varphi(y)xz_0) = 2f(x)g(y), \quad x, y \in G,$$
$$f(xyz_0) + \mu(y)f(\varphi(y)xz_0) = 2g(x)f(y), \quad x, y \in G,$$
$$f(xyz_0) + \mu(y)f(\varphi(y)xz_0) = 2f(x)f(y), \quad x, y \in G,$$
$$f(xy) + \mu(y)f(\varphi(y)x) = 2g(x)h(y), \quad x, y \in G,$$

where G is a group, $z_0 \in G$ is a fixed element, and $\varphi : G \to G$ is an endomorphism.

Finally, we note that the sine addition law on semigroups given in [3] is a key ingredient of the proof of our main results (Theorems (1)–(3)).

3   Main Results

The following lemma will be used in the proof of our main results (Theorems (1)–(3)).

Lemma 1 *Let $\mu : S \to \mathbb{C}\backslash\{0\}$ be a multiplicative function such that $\mu(x\varphi(x)) = 1$ for all $x \in S$. Let the pair $\phi, \psi : S \to \mathbb{C}$ be a solution of the functional equation*

$$\phi(xy) + \mu(y)\phi(\varphi(y)x) = 2\phi(x)\psi(y), \quad x, y \in S. \tag{6}$$

Then we have

$$\phi(xyz) - \phi(x)\psi(yz) = [\phi(xy) - \phi(x)\psi(y)]\psi(z)$$
$$+ [\phi(xz) - \phi(x)\psi(z)]\psi(y). \tag{7}$$

Proof Let $x, y, z \in S$. If we replace x by xy and y by z in (6), we get

$$\phi(xyz) + \mu(z)\phi(\varphi(z)xy) = 2\phi(xy)\psi(z). \tag{8}$$

On the other hand, if we replace x by $\varphi(z)x$ in (6), we obtain

$$\phi(\varphi(z)xy) + \mu(y)\phi(\varphi(yz)x) = 2\phi(\varphi(z)x)\psi(y)$$
$$= 2\mu(\varphi(z))[\mu(z)\phi(\varphi(z)x)]\psi(y)$$
$$= 2[2\mu(\varphi(z))\phi(x)\psi(z)$$
$$- \mu(\varphi(z))\phi(xz)]\psi(y).$$

Also

$$\mu(y)\phi(\varphi(yz)x) = \mu(y)\mu(\varphi(yz))[\mu(yz)\phi(\varphi(yz)x)]$$
$$= \mu(\varphi(z))[2\phi(x)\psi(yz) - \phi(xyz)].$$

So by using $\mu(z\varphi(z)) = 1$, we have

$$\mu(z)\phi(\varphi(z)xy) + [2\phi(x)\psi(yz) - \phi(xyz)] = 2[2\phi(x)\psi(z) - \phi(xz)]\psi(y).$$

Subtracting this functional equation from (8) we get

$$\phi(xyz) - \phi(x)\psi(yz) = [\phi(xy) - \phi(x)\psi(y)]\psi(z)$$
$$+ [\phi(xz) - \phi(x)\psi(z)]\psi(y).$$

The following theorem leads to the solution of the functional equation (4) on an arbitrary semigroup.

Theorem 1 *Let* $\mu : S \to \mathbb{C}\backslash\{0\}$ *be a multiplicative function such that* $\mu(x\varphi(x)) = 1$ *for all* $x \in S$. *The function* $f : S \to \mathbb{C}$ *satisfies* (4) *if and only if it has one of the following forms:*

a. $f = 0$.

b. *There exists a non-zero multiplicative function* χ *of* S *with* $\chi \circ \varphi = 0$, *such that* $f = \frac{1}{2}\chi$.

c. *There exists a multiplicative function* χ *of* S *with* $\chi \circ \varphi^2 = \chi$, *such that*

$$f = \frac{\chi + \mu\chi \circ \varphi}{2}.$$

Moreover, if S *is a topological semigroup and* $f \in C(S)$, *then* $\chi, \mu\chi \circ \varphi \in C(S)$.

Proof If $f = 0$, then (*a*) is the case. From now on we assume that $f \neq 0$.

Using Lemma 1 with $\Phi = \psi = f$, we obtain

$$f(xyz) - f(x)f(yz) = [f(xy) - f(x)f(y)]f(z)$$
$$+ [f(xz) - f(x)f(z)]f(y). \tag{9}$$

With the notation $f_a(b) := f(ab) - f(a)f(b)$ we can reformulate (9) to

$$f_x(yz) = f_x(y)f(z) + f_x(z)f(y).$$

This shows that the pair (f_x, f) satisfies the sine addition low for any $x \in S$.

Case 1 Suppose that $f_x = 0$ for all $x \in S$. By the definition of f_x, we see that f is multiplicative. Substituting f into (4), we conclude that $\mu f \circ \varphi = f$. We may thus write it as $f = (f + \mu f \circ \varphi)/2$, which is the form claimed in the theorem.

Case 2 We now suppose that $f_x \neq 0$ for some $x \in S$.

From Lemma 3.4 in [3], we see that there exist two multiplicative functions $\chi_1, \chi_2 : S \to \mathbb{C}$ such that

$$f = \frac{\chi_1 + \chi_2}{2}.$$

We may assume that they are different, because we otherwise are back to the multiplicative case already treated. Substituting f into (4) we find after a reduction that

$$\chi_1(x)[\chi_2(y) - \mu(y)\chi_1 \circ \varphi(y)] + \chi_2(x)[\chi_1(y) - \mu(y)\chi_2 \circ \varphi(y)] = 0$$

for all $x, y \in S$. Since $\chi_1 \neq \chi_2$ we get from the theory of multiplicative functions (see, for instance, [9, Theorem 3.18(d)]) that both terms are 0. Therefore

$$
\begin{cases}
\chi_1(x)[\chi_2(y) - \mu(y)\chi_1 \circ \varphi(y)] = 0 \\
\chi_2(x)[\chi_1(y) - \mu(y)\chi_2 \circ \varphi(y)] = 0
\end{cases}
\tag{10}
$$

for all $x, y \in S$. Since $\chi_1 \neq \chi_2$, then at least $\chi_1 \neq 0$ or $\chi_2 \neq 0$.

If $\chi_2 = 0$, then $\chi_1 \neq 0$ and hence $\chi_1 \circ \varphi = 0$. Thus $f = \frac{1}{2}\chi_1$.
So we are in Case (b). The same is true for $\chi_1 = 0$ and $\chi_2 \neq 0$.
If $\chi_1 \neq 0$ and $\chi_2 \neq 0$, then (10) becomes

$$
\chi_1 = \mu\chi_2 \circ \varphi = \chi_1 \circ \varphi^2,
$$

which yields the desired formula with $\chi = \chi_1$.

The rest of the proof is trivial.

The continuity statement follows from [9, Theorem 3.18(d)].

The following theorem solves the functional equation (5) on an arbitrary group.

Theorem 2 *Let $\mu : G \to \mathbb{C}$ be a character such that $\mu(x\varphi(x)) = 1$ for all $x \in G$. The pair $f, g : G \to \mathbb{C}$ satisfies (5) if and only if it has one of the following forms:*

a. *$f = 0$ and g is arbitrary.*
b. *There exists a character χ of G such that*

$$
f = \alpha\chi \quad and \quad g = \frac{\chi + \mu\chi \circ \varphi}{2},
$$

for some $\alpha \in \mathbb{C}\backslash\{0\}$.
c. *There exists a character χ of G with $\chi = \chi \circ \varphi^2$, such that*

$$
g = \frac{\chi + \mu\chi \circ \varphi}{2}.
$$

Furthermore, we have

i. *If $\chi \neq \mu\chi \circ \varphi$, then*

$$
f = \alpha\chi + \beta\mu\chi \circ \varphi,
$$

for some $\alpha, \beta \in \mathbb{C}\backslash\{0\}$.
ii. *If $\chi = \mu\chi \circ \varphi$, then there exists a non-zero additive function $A : G \to \mathbb{C}$ with $A \circ \varphi = -A$ such that*

$$
f = (\alpha + A)\chi,
$$

for some $\alpha \in \mathbb{C}$.

Moreover, if G is a topological group, $f \neq 0$ and $f, g \in C(G)$, then χ, $\mu\chi \circ \varphi$, $A \in C(G)$.

The monoid version (Theorem 3) differs from Theorem 2 only when $\chi = \mu\chi \circ \varphi$ (case $(c)(ii)$), where the formulations are more complicated. The conclusions of the two versions agree if χ vanishes nowhere, which is the case on groups.

Theorem 3 *Let M be a monoid which is generated by its squares and let $\mu : M \to \mathbb{C}\setminus\{0\}$ be a multiplicative function such that $\mu(x\varphi(x)) = 1$ for all $x \in M$. The pair $f, g : M \to \mathbb{C}$ satisfies (5) if and only if it has one of the following forms:*

a. $f = 0$ and g is arbitrary.
b. There exists a non-zero multiplicative function $\chi : M \to \mathbb{C}$ such that

$$f = \alpha\chi \quad and \quad g = \frac{\chi + \mu\chi \circ \varphi}{2},$$

for some $\alpha \in \mathbb{C}\setminus\{0\}$.
c. There exists a non-zero multiplicative function $\chi : M \to \mathbb{C}$ with $\chi = \chi \circ \varphi^2$, such that

$$g = \frac{\chi + \mu\chi \circ \varphi}{2}.$$

Furthermore, we have

i. If $\chi \neq \mu\chi \circ \varphi$, then

$$f = \alpha\chi + \beta\mu\chi \circ \varphi,$$

for some $\alpha, \beta \in \mathbb{C}\setminus\{0\}$.
ii. If $\chi = \mu\chi \circ \varphi$, then there exists a non-zero additive function $A : M\setminus I_\chi \to \mathbb{C}$ with $A \circ \varphi = -A$ such that

$$f(x) = \begin{cases} (\alpha + A(x))\chi(x) & for \quad x \in M\setminus I_\chi \\ 0 & for \quad x \in I_\chi \end{cases}$$

for some $\alpha \in \mathbb{C}$.

Moreover, if M is a topological monoid generated by its squares, $f \neq 0$, and $f, g \in C(M)$, then χ, $\mu\chi \circ \varphi \in C(M)$, while $A \in C(M\setminus I_\chi)$.

Proof The case $f = 0$ is trivial, so we will assume from now on that $f \neq 0$. Using Lemma 1 with $\Phi = f$ and $\psi = g$, we find that

$$f(xyz) - f(x)g(yz) = [f(xy) - f(x)g(y)]g(z) + [f(xz) - f(x)g(z)]g(y),$$

i.e.

$$h_x(yz) = h_x(y)g(z) + h_x(z)g(y),$$

where $h_a(b) := f(ab) - f(a)g(b)$. So, the pair (h_x, g) is a solution of the sine addition law for any $x \in S$. In particular for $x = e$, we have

$$f = h_e + f(e)g.$$

Case 1 Suppose that $h_e = 0$. Then $f = f(e)g$ and hence $f(e) \neq 0$. Indeed, $f(e) = 0$ would imply $f = 0$, contradicting our assumption. So, g is a solution of the functional equation (4). According to Theorem 1, we have only the following two cases:

Subcase 1.1 There exists a non-zero multiplicative function $\chi : M \to \mathbb{C}$ with $\chi \circ \varphi = 0$ such that $g = \frac{1}{2}\chi$. So $g = (\chi + \mu\chi \circ \varphi)/2$. It is Case (b) of our theorem.

Subcase 1.2 There exists a multiplicative function $\chi : M \to \mathbb{C}$ with $\chi = \chi \circ \varphi^2$ such that $g = (\chi + \mu\chi \circ \varphi)/2$. Since $g \neq 0$ (because $f \neq 0$), we have $\chi \neq 0$. So we are in Case (c)(i).

Case 2 Let us suppose that $h_e \neq 0$. From Lemma 3.4 in [3], we see that there exist two multiplicative functions $\chi_1, \chi_2 : M \to \mathbb{C}$, such that

$$g = \frac{\chi_1 + \chi_2}{2}.$$

Subcase 2.1 Let us assume that $\chi_1 \neq \chi_2$. By Lemma 3.4 in [3], we have $h_e = c(\chi_1 - \chi_2)$ for some constant $c \in \mathbb{C}\backslash\{0\}$. So

$$f = c(\chi_1 - \chi_2) + \frac{1}{2}f(e)(\chi_1 + \chi_2)$$

$$= \alpha\chi_1 + \beta\chi_2,$$

where $\alpha = c + \frac{1}{2}f(e)$ and $\beta = -c + \frac{1}{2}f(e)$. Substituting f into (5), we find after a reduction that

$$\alpha\chi_1(x)[\chi_2(y) - \mu\chi_1 \circ \varphi(y)] + \beta\chi_2(x)[\chi_1(y) - \mu\chi_2 \circ \varphi(y)] = 0,$$

for all $x, y \in M$. Since $\chi_1 \neq \chi_2$ we get that

$$\begin{cases} \alpha\chi_1(x)[\chi_2(y) - \mu\chi_1 \circ \varphi(y)] = 0 \\ \beta\chi_2(x)[\chi_1(y) - \mu\chi_2 \circ \varphi(y)] = 0 \end{cases} \qquad (11)$$

If $\chi_2 = 0$, then $\chi_1 \neq 0$ and hence $\mu\alpha\chi_1 \circ \varphi = 0$. If $\alpha = 0$, then $f = 0$ but $f \neq 0$ by assumption, so that $\alpha \neq 0$ and hence $\mu\chi_1 \circ \varphi = 0$. Thus $f = \alpha\chi_1$ and $g = \frac{1}{2}\chi_1$. So we are in Case (b). The same is true for $\chi_1 = 0$ and $\chi_2 \neq 0$.

If $\chi_1 \neq 0$ and $\chi_2 \neq 0$, then (11) becomes

$$\begin{cases} \alpha[\chi_2(y) - \mu\chi_1 \circ \varphi(y)] = 0 \\ \beta[\chi_1(y) - \mu\chi_2 \circ \varphi(y)] = 0 \end{cases} \tag{12}$$

Suppose that $\beta = 0$. Then $\alpha = 2c \neq 0$. From (12), we see that $\chi_2 = \mu\chi_1 \circ \varphi$ and arrive at the solution in Case (b) with $\chi = \chi_1$. The same is true for $\alpha = 0$ with the multiplicative function χ_2 replacing χ.

We now suppose that $\alpha \neq 0$ and $\beta \neq 0$. From (12), we see that $\chi_2 = \mu\chi_1 \circ \varphi$ and $\mu\chi_2 \circ \varphi = \chi_1$. So $\chi_1 \neq \mu\chi_1 \circ \varphi$ (because $\chi_1 \neq \chi_2$), $\chi_1 = \mu\chi_2 \circ \varphi = \chi_1 \circ \varphi^2$, and we have

$$f = \alpha\chi_1 + \beta\chi_2 = \alpha\chi_1 + \beta\mu\chi_1 \circ \varphi,$$
$$g = \frac{\chi_1 + \chi_2}{2} = \frac{\chi_1 + \mu\chi_1 \circ \varphi}{2}.$$

This is Case $(c)(i)$ with $\chi = \chi_1$.

Subcase 2.2 Assume that $\chi_1 = \chi_2 = \chi$. If M is a group, we get from Lemma 3.4 in [3] that $h_e = \chi A$ for some additive function $A : M \to \mathbb{C}$. So

$$g = \chi \quad \text{and} \quad f = \chi A + f(e)\chi = (\alpha + A)\chi,$$

where $\alpha = f(e)$. Substituting f into (5), we get

$$[A(y) - A(x) - \alpha]\chi(y) + [A \circ \varphi(y) + A(x) + \alpha]\mu(y)\chi \circ \varphi(y) = 0, \tag{13}$$

for all $x, y \in M$. Using (13) and the fact that $A \neq 0$ (because $h_e \neq 0$), we infer that $\chi = \mu\chi \circ \varphi$. This implies that $A \circ \varphi = -A$. So, we are in case $(c)(ii)$ of Theorem 2.

If M is a monoid which is generated by its squares, we get from Lemma 3.4 in [3] that there exists an additive function $A : M \backslash I_\chi \to \mathbb{C}$ for which

$$h_e(x) = \begin{cases} \chi(x)A(x) & \text{for} \quad x \in M \backslash I_\chi \\ 0 & \text{for} \quad x \in I_\chi \end{cases}$$

Since $f = h_e + f(e)g$ and $\chi(x) = 0$ for $x \in I_\chi$ we have

$$f(x) = \begin{cases} (\alpha + A(x))\chi(x) & \text{for} \quad x \in M \backslash I_\chi \\ 0 & \text{for} \quad x \in I_\chi \end{cases}$$

where $\alpha = f(e)$. Substitution into (5) gives that $\chi = \mu\chi \circ \varphi$ and $A \circ \varphi = -A$. So, we are in case $(c)(ii)$ of Theorem 3.

Conversely, simple computations prove that the formulas above for f and g define solutions of (5).

The continuity statements follow from Lemma 3.4 in [3] and [9, Theorem 3.18(d)].

Remark 1 When the condition $\chi = \chi \circ \varphi^2$ in Case (c) of Theorem 2 is always satisfied, the Cases (b) and (c) of Theorem 2 can be reduced to the following:

d. There exists a character χ of G such that

$$g = \frac{\chi + \mu\chi \circ \varphi}{2}.$$

Furthermore, we have

i. If $\chi \neq \mu\chi \circ \varphi$, then

$$f = \alpha\chi + \beta\mu\chi \circ \varphi,$$

for some $(\alpha, \beta) \in \mathbb{C}^2 \backslash \{(0,0)\}$.
ii. If $\chi = \mu\chi \circ \varphi$, then there exists an additive function $A : G \to \mathbb{C}$ with $A \circ \varphi = -A$ such that

$$f = (\alpha + A)\chi,$$

for some $\alpha \in \mathbb{C}$.

The same idea is valid for Theorem 3.

Example 1 As a non-abelian example on a group, we consider the 3-dimensional Heisenberg group $G = H_3(\mathbb{R})$ described in [9, Example A.17(a)]), and we take as the endomorphism

$$\varphi \begin{pmatrix} 1 & x & z \\ 0 & 1 & y \\ 0 & 0 & 1 \end{pmatrix} = \begin{pmatrix} 1 & rx & r^2z \\ 0 & 1 & ry \\ 0 & 0 & 1 \end{pmatrix}$$

where r is a real constant. In the case where $r > 0$ the map φ is known in the literature as the dilation.

The continuous characters on $H_3(\mathbb{R})$ are parametrized by $(\alpha, \beta) \in \mathbb{C}^2$ as follows (see, e.g., [9, Example 3.14]).

$$\chi_{\alpha,\beta} \begin{pmatrix} 1 & x & z \\ 0 & 1 & y \\ 0 & 0 & 1 \end{pmatrix} = e^{\alpha x + \beta y} \text{ for } x, y, z \in \mathbb{R}.$$

The continuous additive functions on $H_3(\mathbb{R})$ are parametrized by $(\lambda, \nu) \in \mathbb{C}^2$ as follows (see, e.g., [9, Example 2.11]).

$$A_{\lambda,\nu} \begin{pmatrix} 1 & x & z \\ 0 & 1 & y \\ 0 & 0 & 1 \end{pmatrix} = \lambda x + \nu y \text{ for } x, y, z \in \mathbb{R}.$$

Let μ be a continuous character in $H_3(\mathbb{R})$. Then there exist constants $\alpha_0, \beta_0 \in \mathbb{C}$ such that

$$\mu \begin{pmatrix} 1 & x & z \\ 0 & 1 & y \\ 0 & 0 & 1 \end{pmatrix} = e^{\alpha_0 x + \beta_0 y} \text{ for } x, y, z \in \mathbb{R}.$$

In this example, we have dealt with the following three cases:

Case 1 Suppose that $r = 1$. Then $\varphi = id$. So that $\mu(X\varphi(X)) = 1$ for all $X \in H_3(\mathbb{R})$ if and only if $\mu = 1$ and $A_{\lambda,\nu} \circ \varphi = -A_{\lambda,\nu}$ if and only if $A_{\lambda,\nu} = 0$. In conclusion, by help of Theorem 2 combined with Remark 1, the continuous solutions $f, g : H_3(\mathbb{R}) \to \mathbb{C}$ of (5) are the following:

1. $f = 0$ and g is arbitrary in $C(H_3(\mathbb{R}))$.
2. There exist constants $c \in \mathbb{C} \setminus \{0\}$ and $\alpha, \beta \in \mathbb{C}$ such that

$$f \begin{pmatrix} 1 & x & z \\ 0 & 1 & y \\ 0 & 0 & 1 \end{pmatrix} = c e^{\alpha x + \beta y} \text{ for } x, y, z \in \mathbb{R},$$

$$g \begin{pmatrix} 1 & x & z \\ 0 & 1 & y \\ 0 & 0 & 1 \end{pmatrix} = e^{\alpha x + \beta y} \text{ for } x, y, z \in \mathbb{R}.$$

Case 2 Suppose that $r = -1$. Simple computations prove that $\mu(X\varphi(X)) = 1$ for all $X \in H_3(\mathbb{R})$, $\chi_{\alpha,\beta} \circ \varphi^2 = \chi_{\alpha,\beta}$ and $A_{\lambda,\nu} \circ \varphi = -A_{\lambda,\nu}$.

We compute that $\mu \chi_{\alpha,\beta} \circ \varphi = \chi_{\alpha,\beta}$ if and only if $\alpha = \frac{\alpha_0}{2}$ and $\beta = \frac{\beta_0}{2}$, and in that case

$$\chi_{\frac{\alpha_0}{2}, \frac{\beta_0}{2}} \begin{pmatrix} 1 & x & z \\ 0 & 1 & y \\ 0 & 0 & 1 \end{pmatrix} = e^{\frac{\alpha_0}{2} x + \frac{\beta_0}{2} y} \text{ for } x, y, z \in \mathbb{R}.$$

In conclusion, by help of Theorem 2 combined with Remark 1, the continuous solutions $f, g : H_3(\mathbb{R}) \to \mathbb{C}$ of (5) are the following:

1. $f = 0$ and g is arbitrary in $C(H_3(\mathbb{R}))$.
2. There exist constants $(c, c_1) \in \mathbb{C} \setminus \{(0, 0)\}$ and $\alpha, \beta \in \mathbb{C}$ such that

$$f \begin{pmatrix} 1 & x & z \\ 0 & 1 & y \\ 0 & 0 & 1 \end{pmatrix} = ce^{\alpha x + \beta y} + c_1 e^{(\alpha - \alpha_0)x + (\beta - \beta_0)y} \quad \text{for } x, y, z \in \mathbb{R},$$

$$g \begin{pmatrix} 1 & x & z \\ 0 & 1 & y \\ 0 & 0 & 1 \end{pmatrix} = \frac{e^{\alpha x + \beta y} + e^{(\alpha - \alpha_0)x + (\beta - \beta_0)y}}{2} \quad \text{for } x, y, z \in \mathbb{R}.$$

3. There exist constants $c, \lambda, \nu \in \mathbb{C}$ such that

$$f \begin{pmatrix} 1 & x & z \\ 0 & 1 & y \\ 0 & 0 & 1 \end{pmatrix} = (c + \lambda x + \nu y)(e^{\frac{\alpha_0 x + \beta_0 y}{2}}) \quad \text{for } x, y, z \in \mathbb{R},$$

$$g \begin{pmatrix} 1 & x & z \\ 0 & 1 & y \\ 0 & 0 & 1 \end{pmatrix} = e^{\frac{\alpha_0 x + \beta_0 y}{2}} \quad \text{for } x, y, z \in \mathbb{R}.$$

Case 3 Now we suppose that $r \neq \mp 1$. Then $\mu(X\varphi(X)) = 1$ for all $X \in H_3(\mathbb{R})$ if and only if $\mu = 1$. We compute that $\chi_{\alpha, \beta} \circ \varphi^2 = \chi_{\alpha, \beta}$ if and only if $\alpha = \beta = 0$, and $A_{\lambda, \nu} \circ \varphi = -A_{\lambda, \nu}$ if and only if $A_{\lambda, \nu} = 0$. So point (c) of Theorem 2 does not occur. In conclusion, by help of Theorem 2, the continuous solutions $f, g : H_3(\mathbb{R}) \to \mathbb{C}$ of (5) are the following:

1. $f = 0$ and g is arbitrary in $C(H_3(\mathbb{R}))$.
2. There exist constants $c \in \mathbb{C} \setminus \{0\}$ and $(\alpha, \beta) \in \mathbb{C}$ such that

$$f \begin{pmatrix} 1 & x & z \\ 0 & 1 & y \\ 0 & 0 & 1 \end{pmatrix} = ce^{\alpha x + \beta y} \quad \text{for } x, y, z \in \mathbb{R},$$

$$g \begin{pmatrix} 1 & x & z \\ 0 & 1 & y \\ 0 & 0 & 1 \end{pmatrix} = \frac{e^{\alpha x + \beta y} + e^{r(\alpha x + \beta y)}}{2} \quad \text{for } x, y, z \in \mathbb{R}.$$

4 Applications

As immediate consequences of Theorems 2 and 3, combined with Remark 1, we obtain the following two corollaries:

Corollary 1 *Let $\mu : G \to \mathbb{C}$ be a character such that $\mu(x\sigma(x)) = 1$ for all $x \in G$. The pair $f, g : G \to \mathbb{C}$ satisfies the equation*

$$f(xy) + \mu(y)f(\sigma(y)x) = 2f(x)g(y), \quad x, y \in G \qquad (14)$$

if and only if it has one of the following forms:

a. *$f = 0$ and g is arbitrary.*
b. *There exists a character χ of G such that*

$$g = \frac{\chi + \mu\chi \circ \sigma}{2}.$$

Furthermore, we have

  i. *If $\chi \neq \mu\chi \circ \sigma$, then*

$$f = \alpha\chi + \beta\mu\chi \circ \sigma,$$

  for some $(\alpha, \beta) \in \mathbb{C}^2 \backslash \{(0, 0)\}$.
  ii. *If $\chi = \mu\chi \circ \sigma$, then there exists an additive function $A : G \to \mathbb{C}$ with $A \circ \sigma = -A$ such that*

$$f = (\alpha + A)\chi,$$

  for some $\alpha \in \mathbb{C}$.

Moreover, if G is a topological group, $f \neq 0$, and $f, g \in C(G)$, then $\chi, \mu\chi\circ\sigma, A \in C(G)$.

Corollary 2 *Let M be a monoid which is generated by its squares and let $\mu : M \to \mathbb{C}\backslash\{0\}$ be a multiplicative function such that $\mu(x\sigma(x)) = 1$ for all $x \in M$. The pair $f, g : M \to \mathbb{C}$ satisfies (14) if and only if it has one of the following forms:*

a. *$f = 0$ and g is arbitrary.*
b. *There exists a non-zero multiplicative function $\chi : M \to \mathbb{C}$ such that*

$$g = \frac{\chi + \mu\chi \circ \sigma}{2}.$$

Furthermore, we have

  i. *If $\chi \neq \mu\chi \circ \sigma$, then*

$$f = \alpha\chi + \beta\mu\chi \circ \sigma,$$

  for some $(\alpha, \beta) \in \mathbb{C}^2 \backslash \{(0, 0)\}$.

ii. If $\chi = \mu\chi \circ \sigma$, then there exists an additive function $A : M\backslash I_\chi \to \mathbb{C}$ with $A \circ \sigma = -A$ such that

$$f(x) = \begin{cases} (\alpha + A(x))\chi(x) & for \quad x \in M\backslash I_\chi \\ 0 & for \quad x \in I_\chi \end{cases}$$

for some $\alpha \in \mathbb{C}$.

Moreover, if M is a topological monoid generated by its squares, $f \neq 0$, and $f, g \in C(M)$, then $\chi, \mu\chi \circ \sigma \in C(M)$, while $A \in C(M\backslash I_\chi)$.

In the rest of the chapter let G be a group, $\varphi : G \to G$ be an endomorphism, let $\mu : G \to \mathbb{C}$ be a character such that $\mu(x\varphi(x)) = 1$ for all $x \in G$, and let $z_0 \in G$ be a fixed element.

In view of Theorem 2, we obtain the following result which is an extension of the result of [9, Exercise 11.6].

Corollary 3 *Let the pair $f, g : G \to \mathbb{C}$ be a solution of the functional equation*

$$f(xyz_0) + \mu(y)f(\varphi(y)xz_0) = 2f(x)g(y), \quad x, y \in G. \tag{15}$$

Then we have the following possibilities:

a. *$f = 0$ and g is arbitrary.*
b. *There exists a character χ of G such that*

$$f = \alpha\chi \quad and \quad g = \frac{\chi(z_0)}{2}(\chi + \mu\chi \circ \varphi),$$

for some $\alpha \in \mathbb{C}\backslash\{0\}$.
c. *There exists a character χ of G with $\chi = \chi \circ \varphi^2$ and $\mu(z_0)\chi \circ \varphi(z_0) = \chi(z_0)$ such that*

$$g = \frac{\chi(z_0)}{2}(\chi + \mu\chi \circ \varphi).$$

Furthermore, we have

i. *If $\chi \neq \mu\chi \circ \varphi$, then*

$$f = \alpha\chi + \beta\mu\chi \circ \varphi,$$

for some $\alpha, \beta \in \mathbb{C}\backslash\{0\}$.

ii. If $\chi = \mu\chi \circ \varphi$, then there exists a non-zero additive function $A : G \to \mathbb{C}$ with $A \circ \varphi = -A$ and $A(z_0) = 0$ such that

$$f = (\alpha + A)\chi,$$

for some $\alpha \in \mathbb{C}$.

Conversely, the functions given with these properties satisfy the functional equation (15).

Moreover, if G is a topological group, $f \neq 0$, and $f, g \in C(G)$, then $\chi, \mu\chi \circ \varphi, A \in C(G)$.

Proof If $f = 0$, then (a) is the case. From now on we assume that $f \neq 0$.

By putting $y = e$ in (15) we get that

$$f(xz_0) = g(e)f(x), \quad x \in G. \tag{16}$$

Since $f \neq 0$, we get immediately $g(e) \neq 0$. So, using (16), we can reformulate the form of Eq. (15) as

$$f(xy) + \mu(y)f(\varphi(y)x) = 2f(x)\frac{g(y)}{g(e)}, \quad x, y \in G.$$

So the pair $(f, \frac{g}{g(e)})$ is a solution of (5). Since $f \neq 0$, we know from Theorem 2 that there are only the following three cases:

Case 1 There exist a character χ of G and a constant $\alpha \in \mathbb{C}\backslash\{0\}$ such that

$$f = \alpha\chi \quad \text{and} \quad g = \frac{g(e)}{2}(\chi + \mu\chi \circ \varphi).$$

Simple computations based on (16) show that $g(e) = \chi(z_0)$. So we are in Case (b) of our statement.

Case 2 There exist a character χ of G with $\chi \neq \mu\chi \circ \varphi$ and $\chi = \chi \circ \varphi^2$, and constants $\alpha, \beta \in \mathbb{C}\backslash\{0\}$ such that

$$f = \alpha\chi + \beta\mu\chi \circ \varphi \quad \text{and} \quad g = \frac{g(e)}{2}(\chi + \mu\chi \circ \varphi).$$

Simple computations based on (16) show that

$$\alpha[g(e) - \chi(z_0)]\chi + \beta[g(e) - \mu(z_0)\chi \circ \varphi(z_0)]\mu\chi \circ \varphi = 0.$$

By the linear independence of different characters we infer that

$$\begin{cases} \alpha[g(e) - \chi(z_0)] = 0 \\ \beta[g(e) - \mu(z_0)\chi \circ \varphi(z_0)] = 0 \end{cases}$$

Since $\alpha, \beta \neq 0$ we get that $g(e) = \chi(z_0) = \mu(z_0)\chi \circ \varphi(z_0)$. So we arrive at the solution in Case $(c)(i)$.

Case 3 There exist a character χ of G with $\chi = \mu\chi \circ \varphi$, an additive function $A : G \to \mathbb{C}$ with $A \circ \varphi = -A$, and a constant $\alpha \in \mathbb{C}$ such that

$$f = (\alpha + A)\chi \quad \text{and} \quad g = g(e)\chi.$$

Using (16), we get that

$$[g(e) - \chi(z_0)]A = A(z_0)\chi(z_0) - \alpha[g(e) - \chi(z_0)].$$

Since A is additive, the last equality can hold only if $g(e) = \chi(z_0)$ and $A(z_0) = 0$. So we are in Case (b) or $(c)(ii)$. This finishes the necessity assertion.

Conversely, simple computations prove that the formulas above for (f, g) define solutions of (15).

The continuity statements follow from Theorem [9, Theorem 3.18(d)].

As another consequence of Theorem 2, we have the following result.

Corollary 4 *The solutions $f, g : G \to \mathbb{C}$ of the functional equation*

$$f(xyz_0) + \mu(y)f(\varphi(y)xz_0) = 2g(x)f(y), \quad x, y \in G, \tag{17}$$

are the following:

a. $f = 0$ and g is arbitrary.
b. There exist a character χ of G and a constant $c \in \mathbb{C}\backslash\{0\}$ such that

$$f = c\frac{\chi + \mu\chi \circ \varphi}{2} \quad \text{and} \quad g = \frac{\chi(z_0)}{2}\chi + \frac{\mu(z_0)\chi \circ \varphi(z_0)}{2}\mu\chi \circ \varphi.$$

Proof We leave out the simple verifications that the formulas of (a) and (b) define solutions of (17). It is thus smug to prove that any solution (f, g) of (17) falls into one of these two categories. We note that the proof is similar to the proof of Corollary 3.

The first case is obvious, so we suppose that $f \neq 0$. By putting $y = e$ in (17) we get that

$$f(xz_0) = f(e)g(x), \quad x \in G. \tag{18}$$

Since $f \neq 0$, we have $f(e) \neq 0$. So, using (18), we can reformulate the form of Eq. (17) as

$$g(xy) + \mu(y)g(\varphi(y)x) = 2g(x)\frac{f(y)}{f(e)}, \quad x, y \in G.$$

So the pair $(g, \frac{f}{f(e)})$ is a solution of (5). Since $f \neq 0$, we know from Theorem 2 that there are only the following two cases:

Case 1 There exists a character χ of G such that

$$g = \alpha\chi \quad \text{and} \quad f = \frac{f(e)}{2}(\chi + \mu\chi \circ \varphi),$$

for some $\alpha \in \mathbb{C}\backslash\{0\}$.

Simple computations based on (18) show that

$$[\chi(z_0) - 2\alpha]\chi + \mu(z_0)\chi \circ \varphi(z_0)\mu\chi \circ \varphi = 0.$$

Since $\mu(z_0)\chi \circ \varphi(z_0) \neq 0$, the last equality can hold only if $\chi = \mu\chi \circ \varphi$ and $\alpha = \chi(z_0)$. So we are in Case (*b*) above with $c = f(e)$.

Case 2 There exist a character χ of G with $\chi \circ \varphi^2 = \chi$ and $\chi \neq \mu\chi \circ \varphi$, and constants $\alpha, \beta \in \mathbb{C}\backslash\{0\}$ such that

$$f = \frac{f(e)}{2}(\chi + \mu\chi \circ \varphi) \quad \text{and} \quad g = \alpha\chi + \beta\mu\chi \circ \varphi.$$

Simple computations based on (18) show that

$$[\chi(z_0) - 2\alpha]\chi + [\mu(z_0)\chi \circ \varphi(z_0) - 2\beta]\mu\chi \circ \varphi = 0.$$

By the linear independence of different characters we infer that $\alpha = \frac{1}{2}\chi(z_0)$ and $\beta = \frac{1}{2}\mu(z_0)\chi \circ \varphi(z_0)$. So we are in Case (*b*) above with $c = f(e)$.

Case 3 There exist a character χ of G with $\chi = \mu\chi \circ \varphi$, a non-zero additive function $A : G \to \mathbb{C}$ with $A \circ \varphi = -A$, and a constant $\alpha \in \mathbb{C}$ such that

$$f = f(e)\chi \quad \text{and} \quad g = (\alpha + A)\chi.$$

Using (18), we get that

$$\alpha + A = \chi(z_0).$$

Since A is additive, the last equality can hold only if $A = 0$ and $\alpha = \chi(z_0)$. This case does not apply, because $A \neq 0$ by assumption. This finishes the proof.

As a consequence of Corollary 4, we have the following result which is a natural extension of Kannappan's functional equation (3).

Corollary 5 *The solutions* $f : G \to \mathbb{C}$ *of the functional equation*

$$f(xyz_0) + \mu(y)f(\varphi(y)xz_0) = 2f(x)f(y), \quad x, y \in G, \tag{19}$$

are either $f \equiv 0$ *or*

$$f = \chi(z_0)\frac{\chi + \mu\chi \circ \varphi}{2},$$

where χ *is a character of* G *satisfying* $\mu(z_0)\chi \circ \varphi(z_0) = \chi(z_0)$.

Remark 2 By using Theorem 3, we can get the solutions of the functional equations (15), (17), and (19) on monoids that are generated by their squares.

In the following lemma, we give a characterization of the solutions of

$$f(xy) + \mu(y)f(\varphi(y)x) = 2g(x)h(y), \quad x, y \in G. \tag{20}$$

Lemma 2 *The triple* $f, g, h : G \to \mathbb{C}$ *be a solution of the functional equation (20). Then we have the following possibilities:*

1. $f = 0$, $g = 0$, h *is arbitrary.*
2. $f = 0$, $h = 0$, g *is arbitrary.*
3. $f = h(e)g$, *where* $h(e) \neq 0$, *and* g *is a solution of (5) with companion function* $h/h(e)$, *i.e.,*

$$g(xy) + g(\varphi(y)x) = 2g(x)\frac{h(y)}{h(e)}, \quad x, y \in G.$$

Proof The first two cases are obvious, so we suppose that $f \neq 0$. Taking $y = e$ in (20) we get $f = h(e)g$. If $h(e) = 0$, we get $f = 0$ contradicting our assumption. Hence $h(e) \neq 0$. Replacing f by $h(e)g$ in (20), we obtain the identity in (3).

In view of Theorem 2 and Lemma 2 we find the complete solution of (20) on an arbitrary group.

Theorem 4 *The triple* $f, g, h : G \to \mathbb{C}$ *satisfies (20) if and only if it has one of the following forms:*

1. $f = 0$, $g = 0$, h *is arbitrary.*
2. $f = 0$, $h = 0$, g *is arbitrary.*
3. *There exists a character* χ *of* G *such that*
   $h = \frac{\gamma}{2}(\chi + \mu\chi \circ \varphi)$, $g = \alpha\chi$ *and* $f = \gamma\alpha\chi$,
   for some $\alpha, \gamma \in \mathbb{C}\backslash\{0\}$.

4. *There exists a character χ of G with $\chi = \chi \circ \varphi^2$ such that*

$$h = \frac{\gamma}{2}(\chi + \mu\chi \circ \varphi),$$

for some $\gamma \in \mathbb{C}\backslash\{0\}$. Furthermore, we have

a. *If $\chi \neq \mu\chi \circ \varphi$, then*

$$g = \alpha\chi + \beta\mu\chi \circ \varphi,$$
$$f = \gamma(\alpha\chi + \beta\mu\chi \circ \varphi),$$

for some $\alpha, \beta, \gamma \in \mathbb{C}\backslash\{0\}$.
b. *If $\chi = \mu\chi \circ \varphi$, then there exists a non-zero additive function $A : G \to \mathbb{C}$ with $A \circ \varphi = -A$ such that*

$$g = (\alpha + A)\chi,$$
$$f = \gamma(\alpha + A)\chi,$$

for some $\alpha, \gamma \in \mathbb{C}$.

Moreover, if G is a topological group, and $f, g, h \in C(G)$, then $\chi, \chi \circ \varphi, A, A \circ \varphi \in C(G)$.

References

1. A.L. Cauchy, in *Cours d'Analyse de l'ecole Polytechnique*. Analyse Algebrique, V., Paris, vol. 1 (1821)
2. J. d'Alembert, Addition au Mémoire sur la courbe que forme une corde tendue mise en vibration. Hist. Acad. Berlin **1750**, 355–360 (1750)
3. B.R. Ebanks, H. Stetkær, d'Alembert's other functional equation on monoids with an involution. Aequationes Math. **89**, 187–206 (2015)
4. B.R. Ebanks, H. Stetkær, On Wilson's functional equations. Aequationes Math. **89**(2), 339–354 (2015)
5. E. Elqorachi, A. Redouani, *Solutions and Stability of Variant of Wilson's Functional Equation* (2015). arXiv:1505.06512v1 [math.CA]
6. B. Fadli, S. Kabbaj, K.H. Sabour, D. Zeglami, Functional equation on semigroups with an endomorphism. Acta Math. Hung. **150**(2) (2016). https://doi.org/10.1007/s10474-016-0635-9
7. P.L. Kannappan, A functional equation for the cosine. Can. Math. Bull. **2**, 495–498 (1968)
8. P.L. Kannappan, *Functional Equations and Inequalities with Applications* (Springer, New York, 2009)
9. H. Stetkær, *Functional Equations on Groups* (World Scientific Publishing Co, Singapore, 2013)
10. W.H. Wilson, On certain related functional equations. Bull. Am. Math. Soc. **26**, 300–312 (1919–1920). Fortschr.: **47**, 320 (1919–20)

On the Additivity of Maps Preserving Triple Jordan Product $A^*B + \lambda B^*A$ on $*$-Algebras

Vahid Darvish, Mojtaba Nouri, and Mehran Razeghi

Abstract Suppose that $\mathcal{A}$ and $\mathcal{B}$ are $*$-algebras and $\Phi : \mathcal{A} \longrightarrow \mathcal{B}$ is a unital bijective map such that

$$\Phi(P \bullet A \bullet P) = \Phi(P) \bullet \Phi(A) \bullet \Phi(P)$$

for all $A \in \mathcal{A}$ and $P \in \{I_{\mathcal{A}}, P_1, I_{\mathcal{A}} - P_1\}$ where P_1 is a projection in $\mathcal{A}$. The operation $\bullet_\lambda$ between two arbitrary elements S and T is defined as $S \bullet_\lambda T = S^*T + \lambda T^*S$ for $\lambda \in \{-1, 1\}$. Then, Φ is additive.

1 Introduction

Let $\mathcal{R}$ be a $*$-ring and $A \bullet B = AB + BA^*$ and $[A, B]_* = AB - BA^*$ for $A, B \in \mathcal{R}$. These are two new products. These products play an important role in some research topics and attract many authors' attention recently (for example, see [1, 3]).

The authors in [5] reduced the assumption of the two above-mentioned results. They proved that if $\mathcal{A}$ and $\mathcal{B}$ are two C^*-algebras and the map Φ from $\mathcal{A}$ onto $\mathcal{B}$ is bijective, unital, and satisfies

$$\Phi(A \bullet_\lambda P) = \Phi(A) \bullet_\lambda \Phi(P),$$

for all $A \in \mathcal{A}$ and $P \in \{P_1, I_{\mathcal{A}} - P_1\}$ where P_1 is a nontrivial projection in $\mathcal{A}$ and $\lambda \in \{-1, 1\}$, then, Φ is $*$-additive.

V. Darvish (✉)
School of Mathematics and Statistics, Nanjing University of Information Science and Technology, Nanjing, China

M. Nouri · M. Razeghi
Department of Mathematics, Faculty of Mathematical Sciences, University of Mazandaran, Babolsar, Iran

© Springer Nature Switzerland AG 2019
G. A. Anastassiou and J. M. Rassias (eds.), *Frontiers in Functional Equations and Analytic Inequalities*, https://doi.org/10.1007/978-3-030-28950-8_6

In [4], they considered $\mathcal{A}$ is a von Neumann algebra with no central abelian projections and $\mathcal{B}$ is a *-algebra. Suppose that a bijective map $\Phi : \mathcal{A} \to \mathcal{B}$ satisfies

$$\Phi([[A, B]_*, C]_*) = [[\Phi(A), \Phi(B)]_*, \Phi(C)]_*$$

for all $A, B, C \in \mathcal{A}$ where $[A, B]_* = AB - BA^*$ is the skew Lie product. They showed that the following hold:

1. $\Phi(I)$ is self-adjoint central element in $\mathcal{B}$ with $\Phi(I)^2 = I$.
2. Let $\Psi(A) = \Phi(I)\Phi(A)$ for all $A \in \mathcal{A}$, then there exists a central projection $E \in \mathcal{A}$ such that the restriction of Ψ to $\mathcal{A}E$ is a linear *-isomorphism and the restriction of Ψ to $\mathcal{A}(I - E)$ is a conjugate linear *-isomorphism.

Also, the authors in [2] by considering the same assumptions on Φ as above which holds in the following condition

$$\Phi(A \bullet B \bullet C) = \Phi(A) \bullet \Phi(B) \bullet \Phi(C)$$

where $A \bullet B = AB + BA^*$ obtained the same results.

Motivated by the above results we show that if $\Phi : \mathcal{A} \to \mathcal{B}$ which is bijective and satisfies

$$\Phi(P \bullet_\lambda A \bullet_\lambda P) = \Phi(P) \bullet_\lambda \Phi(A) \bullet_\lambda \Phi(P),$$

for all $A \in \mathcal{A}$ and $P \in \{I_\mathcal{A}, P_1, I_\mathcal{A} - P_1\}$ where P_1 is a projection in $\mathcal{A}$. The operation $\bullet_\lambda$ between two arbitrary elements S and T is defined as $S \bullet_\lambda T = S^*T + \lambda T^*S$ for $\lambda \in \{-1, 1\}$. Then, Φ is additive.

It is well known that ring $\mathcal{A}$ is prime, in the sense that $A\mathcal{A}B = 0$ for $A, B \in \mathcal{A}$ which implies either $A = 0$ or $B = 0$. Also, for the real and imaginary part of T we will use $\Re(T)$ and $\Im(T)$, respectively.

2 Main Results

We need the following lemma for proving our theorems.

Lemma 1 *Suppose $\mathcal{A}$ and $\mathcal{B}$ are two *-algebras and $\Phi : \mathcal{A} \longrightarrow \mathcal{B}$ is a map that satisfies*

$$\Phi(P \bullet_\lambda A \bullet_\lambda P) = \Phi(P) \bullet_\lambda \Phi(A) \bullet_\lambda \Phi(P)$$

for $\lambda \in \{-1, 1\}$ and $P \in \{I, I - P_1, P_1\}$. If for $T, A, B \in \mathcal{A}$, we have

$$\Phi(T) = \Phi(A) + (B).$$

Then,

$$\Phi(P \bullet_\lambda T \bullet_\lambda P) = \Phi(P \bullet_\lambda A \bullet_\lambda P) + \Phi(P \bullet_\lambda B \bullet_\lambda P).$$

Lemma 2 *It is easy to show that $\Phi(0) = 0$.*

Theorem 1 *Let $\mathcal{A}$ and $\mathcal{B}$ be two prime $*$-algebras with a nontrivial projection and for each $A, B \in \mathcal{A}$, we have*

$$A \bullet_\lambda B = A^* B + \lambda B^* A$$

for $\lambda \in \{-1, 1\}$. If a unital bijective map $\Phi : \mathcal{A} \to \mathcal{B}$ satisfies the following condition

$$\Phi(P \bullet_\lambda A \bullet_\lambda P) = \Phi(P) \bullet_\lambda \Phi(A) \bullet_\lambda \Phi(P),$$

then Φ is additive.

Proof Let P_1 be a nontrivial projection in $\mathcal{A}$ and $P_2 = I_{\mathcal{A}} - P_1$. Denote $\mathcal{A}_{ij} = P_i \mathcal{A} P_j$, $i, j = 1, 2$, then $\mathcal{A} = \sum_{i,j=1}^{2} \mathcal{A}_{ij}$. For every $A \in \mathcal{A}$, we may write $A = A_{11} + A_{12} + A_{21} + A_{22}$. In all that follow, when we write A_{ij}, it indicates that $A_{ij} \in \mathcal{A}_{ij}$. For showing additivity of Φ on $\mathcal{A}$, we use the above partition of $\mathcal{A}$ and give some steps that prove Φ is additive on each $\mathcal{A}_{ij}$, $i, j = 1, 2$.

We prove the above theorem by several steps.

Step 1 For $A_{11} \in \mathcal{A}_{11}$ and $A_{21} \in \mathcal{A}_{21}$ we have

$$\Phi(A_{11} + A_{21}) = \Phi(A_{11}) + \Phi(A_{21}).$$

Since Φ is surjective, then we have $T \in \mathcal{A}$ such that

$$\Phi(T) = \Phi(A_{11}) + \Phi(A_{22}). \tag{1}$$

By applying Lemma 1 for P_2 in relation (1), we have

$$\Phi(P_2 \bullet_\lambda T \bullet_\lambda P_2) = \Phi(P_2 \bullet_\lambda A_{11} \bullet_\lambda P_2) + \Phi(P_2 \bullet_\lambda A_{21} \bullet_\lambda P_2)$$
$$= \Phi(A_{21}^* + \lambda A_{21}).$$

Since Φ is injective, we have

$$T_{22}^* + T_{21}^* + 2\lambda T_{22} + \lambda T_{21} + \lambda^2 T_{22}^* = A_{21}^* + \lambda A_{21}.$$

For $\lambda \in \{-1, 1\}$, we obtain $\Re T_{22} = \Im T_{22} = 0$, $\Re T_{21} = \Re A_{21}$ and $\Im T_{21} = \Im A_{21}$. Hence, $T_{22} = 0$, $A_{21} = T_{21}$.

By applying Lemma 1 for P_1 in relation (1), we have

$$\Phi(P_1 \bullet_\lambda T \bullet_\lambda P_1) = \Phi(P_1 \bullet_\lambda A_{11} \bullet_\lambda P_1) + \Phi(P_1 \bullet_\lambda A_{21} \bullet_\lambda P_1)$$
$$= \Phi(A_{11}^* + 2\lambda A_{11} + \lambda^2 A_{11}^*).$$

Since Φ is injective, we have

$$T_{11}^* + T_{12}^* + 2\lambda T_{11} + \lambda T_{12} + \lambda^2 T_{11}^* = A_{11}^* + 2\lambda A_{11} + \lambda^2 A_{11}^*.$$

For $\lambda \in \{-1, 1\}$, we can write $T_{11} = A_{11}$ and $T_{12} = 0$.

Step 2 For $A_{12} \in \mathcal{A}_{12}$ and $A_{21} \in \mathcal{A}_{21}$, we have

$$\Phi(A_{12} + A_{21}) = \Phi(A_{12}) + \Phi(A_{21}).$$

Since Φ is surjective, we can find $T \in \mathcal{A}$ such that

$$\Phi(T) = \Phi(A_{12}) + \Phi(A_{21}). \tag{2}$$

By applying Lemma 1 for P_1 and P_2 in Eq. (1), we have

$$\Phi(P_1 \bullet_\lambda T \bullet_\lambda P_1) = \Phi(P_1 \bullet_\lambda A_{12} \bullet_\lambda P_1) + \Phi(P_1 \bullet_\lambda A_{21} \bullet_\lambda P_1) = \Phi(A_{12}^* + \lambda A_{12}),$$

and

$$\Phi(P_2 \bullet_\lambda T \bullet_\lambda P_2) = \Phi(P_2 \bullet_\lambda A_{12} \bullet_\lambda P_2) + \Phi(P_2 \bullet_\lambda A_{21} \bullet_\lambda P_2) = \Phi(A_{21}^* + \lambda A_{21}).$$

Hence, Φ is injective and we have

$$T_{11}^* + T_{12}^* + 2\lambda T_{11} + \lambda T_{12} + \lambda^2 T_{11}^* = A_{12}^* + \lambda A_{12}^*,$$

and

$$T_{22}^* + T_{21}^* + 2\lambda T_{22} + \lambda T_{21} + \lambda^2 T_{22}^* = A_{21}^* + \lambda A_{21}^*.$$

For $\lambda \in \{-1, 1\}$, we obtain $\Re(T_{12}) = \Re(A_{12})$, $\Im(T_{12}) = \Im(A_{12})$, $\Re(T_{21}) = \Re(A_{21})$, $\Im(T_{21}) = \Im(A_{21})$. So, $A_{21} = T_{21}$, $A_{12} = T_{12}$ and $T_{11} = T_{22} = 0$.

Step 3 For each $A \in \mathcal{A}$, we have

$$\Phi(4\Re(A)) = 4\Re(\Phi(A))$$

and

$$\Phi(-4\Im(A)) = -4\Im(\Phi(A)).$$

$$\begin{aligned}
\Phi(2A + 2A^*) &= \Phi(I \bullet_\lambda A \bullet_\lambda I) \\
&= \Phi(I) \bullet_\lambda \Phi(A) \bullet_\lambda \Phi(I) \\
&= (\Phi(I)^* \Phi(A) + \Phi(A)^* \Phi(I)) \bullet_\lambda \Phi(I) \\
&= 2\Phi(A) + 2\Phi(A)^*,
\end{aligned}$$

then $\Phi(4\Re(A)) = 4\Re(\Phi(A))$ and similarly

$$\Phi(2A^* - 2A) = \Phi(I \bullet_\lambda A \bullet_\lambda I) = \Phi(I) \bullet_\lambda \Phi(A) \bullet_\lambda \Phi(I)$$
$$= (\Phi(I)^*\Phi(A) - \Phi(A)^*\Phi(I)) \bullet_\lambda \Phi(I) = 2\Phi(A)^* - 2\Phi(A)$$

then $\Phi(-4\Im(A)) = -4\Im(\Phi(A))$.

Step 4 For $A_{11} \in \mathcal{A}_{11}$ and $A_{12} \in \mathcal{A}_{12}$, we have

$$\Phi(A_{11} + A_{12}) = \Phi(A_{11}) + \Phi(A_{12}).$$

There exists $T \in \mathcal{A}$ such that

$$\Phi(2T) = \Phi(2A_{11}) + \Phi(2A_{12}). \tag{3}$$

By applying Lemma 1 for P_2 in Eq. (3), we have

$$\Phi(P_2 \bullet_\lambda 2T \bullet_\lambda P_2) = \Phi(P_2 \bullet_\lambda 2A_{11} \bullet_\lambda P_2) + \Phi(P_2 \bullet_\lambda 2A_{12} \bullet_\lambda P_2) = 0.$$

Since Φ is injective, we have

$$T_{22}^* + T_{21}^* + 2\lambda T_{22} + \lambda T_{21} + \lambda^2 T_{22}^* = 0.$$

For $\lambda \in \{-1, 1\}$, we have $\Re T_{22} = \Im T_{22} = 0$ and $T_{21} = 0$. Hence, $T_{22} = T_{21} = 0$.

By applying Lemma 1 for P_1 and $\lambda = 1$ in Eq. (3) and Steps 1 and 3, we have

$$\Phi(P_1 \bullet_1 2T \bullet_1 P_1) = \Phi(P_1 \bullet_1 2A_{11} \bullet_1 P_1) + \Phi(P_1 \bullet_1 2A_{12} \bullet_1 P_1)$$

$$= \Phi(4A_{11}^* + 4A_{11}) + \Phi(2A_{12} + 2A_{12}^*)$$

$$= \Phi(4\Re(2A_{11})) + \Phi(4\Re(A_{12}^*)) = 4\Re(\Phi(2A_{11})) + 4\Re(\Phi(A_{12}^*))$$

$$= 4\Re(\Phi(2A_{11} + A_{12}^*)) = \Phi(8\Re(A_{11}) + 4\Re(A_{12})).$$

From injectivity of Φ, we have

$$8\Re T_{11} + 4\Re T_{12} = 8\Re A_{11} + 4\Re A_{12}. \tag{4}$$

Similarly, by applying Lemma 1 for P_1 and $\lambda = -1$ in relation (3) Steps 1 and 3, we have

$$\Phi(P_1 \bullet_{-1} 2T \bullet_{-1} P_1) = \Phi(P_1 \bullet_{-1} 2A_{11} \bullet_{-1} P_1) + \Phi(P_1 \bullet_{-1} 2A_{12} \bullet_{-1} P_1)$$

$$= \Phi(4A_{11}^* - 4A_{11}) + \Phi(2A_{12}^* - 2A_{12}) = \Phi(-4i\Im(2A_{11}))$$

$$+\Phi(-4i\Im(A_{12}^*)) = -4i\Im(\Phi(2A_{11})) + 4i\Im(\Phi(A_{12}^*))$$

$$= -4i\Im(\Phi(2A_{11} + A_{12}^*)) = \Phi(-8i\Im(A_{11}) - 4i\Im(A_{12}^*))$$

$$= \Phi(-8i\Im(A_{11}) - 4i\Im(A_{12})).$$

From injectivity of Φ we have

$$8\Im T_{11} + 4i\Im T_{12} = 8i\Im A_{11} + 4i\Im A_{12}. \tag{5}$$

By summing up relations (4) and (5), we have $T_{11} = A_{11}$ and $T_{12} = A_{12}$.
Similarly, we can show that

$$\Phi(A_{21} + A_{22}) = \Phi(A_{21}) + \Phi(A_{22})$$

for $A_{21} \in \mathcal{A}_{21}$ and $A_{22} \in \mathcal{A}_{22}$.

Step 5 For $A_{ij}, B_{ij} \in \mathcal{A}_{ij}$ for $i \neq j$, we have

$$\Phi(A_{ij} + B_{ij}) = \Phi(A_{ij}) + \Phi(B_{ij}).$$

Equivalently, we prove

$$\Phi(2A_{ij} + 2B_{ij}) = \Phi(2A_{ij}) + \Phi(2B_{ij}).$$

There exists $T \in \mathcal{A}$ such that

$$\Phi(2T) = \Phi(2A_{ij}) + \Phi(2B_{ij}). \tag{6}$$

By applying Lemma 1 for P_j in relation (6), we can write

$$\Phi(P_j \bullet_\lambda 2T \bullet_\lambda P_j) = \Phi(P_j \bullet_\lambda 2A_{ij} \bullet_\lambda P_j) + \Phi(P_j \bullet_\lambda 2B_{ij} \bullet_\lambda P_j) = 0.$$

Since Φ is injective, we have

$$T_{jj}^* + T_{ji}^* + 2\lambda T_{jj} + \lambda T_{ji} + \lambda^2 T_{jj}^* = 0.$$

For $\lambda \in \{-1, 1\}$, we have $T_{jj} = 0$ and $T_{ji} = 0$. By applying Lemma 1 for P_i in relation (6), for $\lambda = 1$ we have

$$\Phi(P_i \bullet_1 2T \bullet_1 P_i) = \Phi(P_i \bullet_1 2A_{ij} \bullet_1 P_i) + \Phi(P_i \bullet_1 2B_{ij} \bullet_1 P_i)$$

$$= \Phi(2A_{ij}^* + 2A_{ij}) + \Phi(2B_{ij}^* + 2B_{ij})$$

$$= \Phi(4\Re(A_{ij})) + \Phi(4\Re(B_{ij}^*))$$

$$= 4\Re\Phi(A_{ij}) + 4\Re\Phi(B_{ij}^*)$$

$$= 4\Re\Phi(A_{ij} + B_{ij}^*) = \Phi(4\Re A_{ij} + 4\Re B_{ij})$$
$$= \Phi(4\Re(A_{ij} + 4\Re B_{ij})).$$

From injectivity of Φ, we have

$$8\Re(T_{ii}) + 4\Re(T_{ij}) = 4\Re(A_{ij}) + 4\Re(B_{ij}). \tag{7}$$

Similarly, by applying Lemma 1 for P_i in relation (6), for $\lambda = -1$, we have

$$8i\Im(T_{ii}) + 4i(\Im T_{ij}) = 4i\Im(A_{ij}) + 4i\Im(B_{ij}). \tag{8}$$

By summing up relations (7) and (8), we have $T_{ii} = 0$ and $T_{ij} = A_{ij} + B_{ij}$.

Step 6 For $A_{11} \in \mathcal{A}_{11}$, $A_{12} \in \mathcal{A}_{12}$, $A_{21} \in \mathcal{A}_{21}$ and $A_{22} \in \mathcal{A}_{22}$, we have

$$\Phi(A_{11} + A_{12} + A_{21} + A_{22}) = \Phi(A_{11}) + \Phi(A_{12}) + \Phi(A_{21}) + \Phi(A_{22}).$$

Equivalently, we prove

$$\Phi(2A_{11} + 2A_{12} + 2A_{21} + 2A_{22}) = \Phi(2A_{11}) + \Phi(2A_{12}) + \Phi(2A_{21}) + \Phi(2A_{22}).$$

Since Φ is surjective, we have

$$\Phi(2T) = \Phi(2A_{11}) + \Phi(2A_{12}) + \Phi(2A_{21}) + \Phi(2A_{22}). \tag{9}$$

By applying Lemma 1 for P_1 in relation (9), we have

$$\Phi(P_1 \bullet_1 2T \bullet_1 P_1) = \Phi(P_1 \bullet_1 2A_{11} \bullet_1 P_1) + \Phi(P_1 \bullet_1 2A_{12} \bullet_1 P_1)$$
$$+ \Phi(P_1 \bullet_1 2A_{22} \bullet_1 P_1) + \Phi(P_1 \bullet_1 2A_{21} \bullet_1 P_1)$$
$$= \Phi(4A_{11}^* + 4A_{11}) + \Phi(2A_{12}^* + 2A_{12})$$
$$= \Phi(4\Re(2A_{11})) + \Phi(4\Re(A_{12}))$$
$$= 4\Re(\Phi(2A_{11})) + 4\Re(\Phi(A_{12})) = 4\Re(\Phi(2A_{11} + A_{12}))$$
$$= \Phi(8\Re(A_{11}) + 4\Re(A_{12})).$$

From injectivity of Φ, we obtain

$$8\Re(T_{11}) + 4\Re(T_{12}) = 8\Re(A_{11}) + 4\Re(A_{12}). \tag{10}$$

By applying Lemma 1 for P_1 in relation (10) for $\lambda = -1$ and applying Steps 3 and 4, we have

$$\Phi(P_1 \bullet_{-1} 2T \bullet_{-1} P_1) = \Phi(P_1 \bullet_{-1} 2A_{11} \bullet_{-1} P_1) + \Phi(P_1 \bullet_{-1} 2A_{12} \bullet_{-1} P_1)$$

$$+\Phi(P_1 \bullet_{-1} 2A_{22} \bullet_{-1} P_1) + \Phi(P_1 \bullet_{-1} 2A_{21} \bullet_{-1} P_1)$$

$$= \Phi(4A_{11}^* - 4A_{11}) + \Phi(2A_{12}^* - 2A_{12})$$

$$= \Phi(-4i(\Im 2A_{11})) + \Phi(-4i\Im(A_{12}))$$

$$= -4i\Im(\Phi(2A_{11})) - 4i\Im(\Phi(A_{12}))$$

$$= -4i\Im(\Phi(2A_{11} + A_{12}))$$

$$= \Phi(-8i\Im(A_{11}) - 4i\Im(A_{12})).$$

Since Φ is injective, we have

$$8i\Im(T_{11}) + 4i\Im(T_{12}) = 8i\Im(A_{11}) + 4i\Im(A_{12}). \tag{11}$$

By summing up relations (10) and (11), we have $T_{11} = A_{11}$ and $T_{12} = A_{12}$.

Similarly, by applying Lemma 1 for P_2 in relation (9), we have $T_{22} = A_{22}$ and $T_{21} = A_{21}$.

Step 7 Φ is the preserver of the orthogonal projection P_1 and P_2 in the both sides.

We have

$$\Phi(I \bullet_{-1} P_i \bullet_{-1} I) = \Phi(I) \bullet_{-1} \Phi(P_i) \bullet_{-1} \Phi(I) = 0.$$

Hence, we obtain

$$2\Phi(P_i)^* = 2\Phi(P_i). \tag{12}$$

For each $A \in \mathcal{A}$, we have

$$\Phi(P_i \bullet_\lambda A \bullet_\lambda P_i) = \Phi(P_i) \bullet_\lambda \Phi(A) \bullet_\lambda \Phi(P_i).$$

For $\lambda \in \{-1, 1\}$, we have

$$\Phi(A^*P_i + P_iAP_i + P_iA + P_iA^*P_i) = \Phi(A)^*\Phi(P_i)^2 + \Phi(P_i)^*\Phi(A)\Phi(P_i)$$
$$+ (\Phi(P_i)^*)^2\Phi(A) + \Phi(P_i)^*\Phi(A)^*\Phi(P_i)$$

and

$$\Phi(A^*P_i - P_iAP_i - P_iA + P_iA^*P_i) = \Phi(A)^*\Phi(P_i)^2 - \Phi(P_i)^*\Phi(A)\Phi(P_i)$$
$$+ (\Phi(P_i)^*)^2\Phi(A) - \Phi(P_i)^*\Phi(A)^*\Phi(P_i).$$

We subtract the above two relations from each other, to obtain

$$
\begin{aligned}
2\Phi(P_i)^*\Phi(A)\Phi(P_i) + 2(\Phi(P_i)^*)^2\Phi(A) &= \Phi(A^*P_i + P_i A P_i + P_i A + P_i A^* P_i) \\
&\quad -\Phi(A^*P_i - P_i A P_i - P_i A + P_i A^* P_i) \\
&= \Phi(2A_{ii}^* + 2A_{ii} + A_{ij} + A_{ij}^*) \\
&\quad -\Phi(2A_{ii}^* - 2A_{ii} + A_{ij}^* - A_{ij}) \\
&= \Phi(4\Re A_{ii}) - \Phi(-4i\Im A_{ii}) + 2\Phi(A_{ij}) \\
&= 4\Re\Phi(A_{ii}) + 4i\Im\Phi(A_{ii}) + 2\Phi(A_{ij}) \\
&= 4\Phi(A_{ii}) + 2\Phi(A_{ij}) \\
&= 2\Phi(A_{ii}) + 2\Phi(A_{ii} + A_{ij}) \\
&= 2\Phi(A_{ii}) + 2\Phi(A P_i).
\end{aligned}
$$

Now, from (12), we have

$$
\Phi(P_i)^*\Phi(A)\Phi(P_i) + (\Phi(P_i)^*)^2\Phi(A) = \Phi(P_i A P_i) + \Phi(A P_i).
$$

Put $A = I$. From (10) we have

$$
\Phi(P_i)^2 = \Phi(P_i).
$$

Also, one can easily show that

$$
\Phi(P_i)\Phi(P_j) = \Phi(P_i)(I - \Phi(P_i)) = 0.
$$

Step 8 We show that

$$
\Phi(A_{ij}) = B_{ij},
$$

for $i \neq j$.
For each $A_{ij} \in \mathcal{A}_{ij}$, we have

$$
\Phi(P_i \bullet_\lambda A_{ij} \bullet_\lambda P_i) = \Phi(P_i) \bullet_\lambda \Phi(A_{ij}) \bullet_\lambda \Phi(P_i).
$$

So, we have

$$
\begin{aligned}
\Phi(A_{ij}^* + A_{ij}) &= \Phi(A_{ij})^*\Phi(P_i)^2 + \Phi(P_i)^*\Phi(A_{ij})\Phi(P_i) \\
&\quad +(\Phi(P_i)^*)^2\Phi(A_{ij}) + \Phi(P_i)^*\Phi(A_{ij})^*\Phi(P_i)
\end{aligned}
$$

and

$$
\begin{aligned}
\Phi(A_{ij}^* - A_{ij}) &= \Phi(A_{ij})^*\Phi(P_i)^2 - \Phi(P_i)^*\Phi(A_{ij})\Phi(P_i) \\
&\quad -(\Phi(P_i)^*)^2\Phi(A_{ij}) + \Phi(P_i)^*\Phi(A_{ij})^*\Phi(P_i).
\end{aligned}
$$

By subtracting the above two relations, we have

$$\Phi(A_{ij}) = \Phi(P_i)\Phi(A_{ij})\Phi(P_i) + \Phi(P_i)\Phi(A_{ij}). \tag{13}$$

From relation (13), we have

$$\Phi(P_i)\Phi(A_{ij})\Phi(P_i) = 0$$
$$\Phi(P_j)\Phi(A_{ij})\Phi(P_i) = 0$$
$$\Phi(P_j)\Phi(A_{ij})\Phi(P_j) = 0.$$

Hence, $\Phi(\mathcal{A}_{ij}) \subseteq \mathcal{B}_{ij}$. Since Φ^{-1} has the properties of Φ, we have the result.

Step 9 We show that

$$\Phi(\mathcal{A}_{ii}) = \mathcal{B}_{ii}$$

for $1 \leq i \leq 2$.

For each $A_{ii} \in \mathcal{A}_{ii}$, we have

$$\Phi(P_i \bullet_\lambda A_{ii} \bullet_\lambda P_i) = \Phi(P_i) \bullet_\lambda \Phi(A_{ii}) \bullet_\lambda \Phi(P_i).$$

Since $\lambda \in \{-1, 1\}$, we have

$$\Phi(2A_{ii}^* + 2A_{ii}) = \Phi(P_i)^*\Phi(A_{ii})\Phi(P_i) + \Phi(A_{ii})^*\Phi(P_i)^2$$
$$+(\Phi(P_i)^*)^2\Phi(A_{ii}) + \Phi(P_i)^*\Phi(A_{ii})^*\Phi(P_i)$$

and

$$\Phi(2A_{ii}^* - 2A_{ii}) = \Phi(P_i)^*\Phi(A_{ii})^*\Phi(P_i) + \Phi(A_{ii})^*\Phi(P_i)^2$$
$$-\Phi(P_i)^2\Phi(A_{ii}) - \Phi(P_i)^*\Phi(A_{ii})^*\Phi(P_i).$$

We subtract the above two relations from each other and obtain

$$2\Phi(A_{ii}) = \Phi(P_i)\Phi(A_{ii})\Phi(P_i) + \Phi(P_i)\Phi(A_{ii}). \tag{14}$$

From (14), we have

$$\Phi(P_i)\Phi(A_{ii})\Phi(P_j) = 0$$
$$\Phi(P_j)\Phi(A_{ii})\Phi(P_j) = 0$$
$$\Phi(P_j)\Phi(A_{ii})\Phi(P_i) = 0.$$

Hence, we have

$$\Phi(A_{ii}) = \Phi(P_i)\Phi(A_{ii})\Phi(P_i) + \Phi(P_j)\Phi(A_{ii})\Phi(P_j)$$
$$+\Phi(P_j)\Phi(A_{ii})\Phi(P_i) + \Phi(P_i)\Phi(A_{ii})\Phi(P_j)$$
$$= \Phi(P_i)\Phi(A_{ii})\Phi(P_i) \in \mathcal{B}_{ii}.$$

So, $\Phi(\mathcal{A}_{ii}) \subseteq \mathcal{B}_{ii}$.
Since Φ^{-1} has the properties of Φ, $\Phi(\mathcal{A}_{ii}) = \mathcal{B}_{ii}$

Step 10 For $A_{ii}, B_{ii} \in \mathcal{A}_{ii}$, we have

$$\Phi(A_{ii} + B_{ii}) = \Phi(A_{ii}) + \Phi(B_{ii}).$$

First, we will prove that $\Phi(P_i A + P_i B) = \Phi(P_i A) + \Phi(P_i B)$ for every $A, B \in \mathcal{A}$.
By Step 5 and for every $\Phi(T)_{ji} \in \mathcal{B}_{ji}$ such that $i \neq j$, we obtain

$$\Phi(P_i)\left(\Phi(P_i A + P_i B) - \Phi(P_i A) - \Phi(P_i B)\right)\Phi(T)_{ji}$$
$$= \Phi(P_i)\left(\Phi(P_i A + P_i B) - \Phi(P_i A) - \Phi(P_i B)\right)Q_j\Phi(T)Q_i$$
$$= \Phi(P_i)\Phi(P_i A + P_i B)Q_j - \Phi(P_i)\Phi(P_i A)Q_j$$
$$\quad -\Phi(P_i)\Phi(P_i B)Q_j)\Phi(T)Q_i$$
$$= (\Phi(P_i A P_j + P_i B P_j) - \Phi(P_i A P_j)$$
$$\quad -\Phi(P_i B P_j))\Phi(T)Q_i \quad \text{by Step 8}$$
$$= \left(\Phi(A_{ij} + B_{ij}) - \Phi(A_{ij}) - \Phi(B_{ij})\right)\Phi(T)Q_i$$
$$= \left(\Phi(A_{ij}) + \Phi(B_{ij}) - \Phi(A_{ij}) - \Phi(B_{ij})\right)\Phi(T)Q_i \quad \text{by Step 5}$$
$$= 0.$$

Since $\mathcal{B}$ is prime and $\Phi(T)_{ij} \in \mathcal{B}_{ij}$,

$$\Phi(P_i)\left(\Phi(P_i A + P_i B) - \Phi(P_i A) - \Phi(P_i B)\right)\Phi(P_j) = 0.$$

Also, we know that $\Phi(P_i)$ and $\Phi(P_j)$ are strictly positive, then the above equation leads us to

$$\Phi(P_i A + P_i B) = \Phi(P_i A) + \Phi(P_i B).$$

Multiplying the both sides of the above equation by $\Phi(P_i)$ and applying Step 9, we have

$$\Phi(A_{ii} + B_{ii}) = \Phi(A_{ii}) + \Phi(B_{ii}).$$

The additivity of Φ comes from the above steps.

Acknowledgements The first author is supported by the Talented Young Scientist Program of Ministry of Science and Technology of China (Iran-19-001).

References

1. J. Cui, C.L. Li, Maps preserving product $XY - YX^*$ on factor von Neumann algebras. Linear Algebra Appl. **431**, 833–842 (2009)
2. C. Li, F. Lu, Nonlinear maps preserving the Jordan triple 1-∗-product on von Neumann algebras. Complex Anal. Oper. Theory **11**, 109–117 (2017)
3. C. Li, F. Lu, X. Fang, Nonlinear mappings preserving product $XY+YX^*$ on factor von Neumann algebras. Linear Algebra Appl. **438**, 2339–2345 (2013)
4. C. Li, F. Lu, T. Wang, Nonlinear maps preserving the Jordan triple ∗-product on von Neumann algebras. Ann. Funct. Anal. **7**, 496–507 (2016)
5. A. Taghavi, V. Darvish, H. Rohi, Additivity of maps preserving products $AP \pm PA^*$ on C^*-algebras. Math. Slovaca **67**, 213–220 (2017)

General Solution and Hyers–Ulam Stability of DuoTrigintic Functional Equation in Multi-Banach Spaces

Murali Ramdoss and Antony Raj Aruldass

Abstract In this paper, we introduce the general form of a new duotrigintic functional equation. Then, we find the general solution and study the generalized Hyers–Ulam stability of such functional equation in multi-Banach spaces by employing fixed point technique. Also, we give an example for non-stability cases for this new functional equation.

1 Introduction

Stability problem of a functional equation was first posed by Ulam [36] and that was partially answered by Hyers [14] and then generalized by Aoki [1] and Rassias [26] for additive mappings and linear mappings, respectively. In 1994, a generalization of Rassias theorem was obtained by Gâvruta [13], who replaced $\epsilon \left(\|x\|^p + \|y\|^p \right)$ by a general control function $\phi(x, y)$. This idea is known as generalized Hyers–Ulam–Rassias stability. After that, the general stability problems of various functional equations such as quadratic [8], cubic [3, 5, 17, 28], quartic [3, 4, 27], quintic [39], sextic [39], septic and octic [38], nonic [6, 29, 30], decic [2], undecic [32], quattuordecic [33], hexadecic [22], octadecic [23], vigintic [25], viginticduo [19], quattuorvigintic [15, 24, 31], octavigintic [16] and trigintic [7] functional equations have been investigated by a number of authors with more general domains and co-domains.

2 Preliminaries

In this section, we recall some basic concepts concerning Multi-Banach Spaces. The Multi-Banach Spaces were first investigated by Dales and Polyakov [10]. Theory of Multi-Banach Spaces is similar to operator sequence space and has some

M. Ramdoss (✉) · A. R. Aruldass
PG and Research Department of Mathematics, Sacred Heart College (Autonomous), Tirupattur, Tamil Nadu, India

© Springer Nature Switzerland AG 2019
G. A. Anastassiou and J. M. Rassias (eds.), *Frontiers in Functional Equations and Analytic Inequalities*, https://doi.org/10.1007/978-3-030-28950-8_7

connections with operator spaces and Banach Spaces. In 2007 Dales and Moslehian [9] first proved the stability of mappings and also gave some examples on multi-normed spaces. The asymptotic aspects of the quadratic functional equations in multi-normed spaces were investigated by Moslehian et al. [21] in 2009. In the last two decades, the stability of functional equations on multi-normed spaces was proved by many mathematicians (see [12, 18, 34, 35, 37, 40]). Let $(\wp, \|\cdot\|)$ be a complex normed space, and let $k \in \mathbb{N}$. We denote by $\wp^k$ the linear space $\wp \oplus \wp \oplus \wp \oplus \ldots \oplus \wp$ consisting of k- tuples $(x_1, \ldots, x_k)$ where $x_1, \ldots, x_k \in \wp$. The linear operations on $\wp^k$ are defined coordinate wise. The zero element of either $\wp$ or $\wp^k$ is denoted by 0. We denote by $\mathbb{N}_k$ the set $\{1, 2, \ldots, k\}$ and by Ψ_k the group of permutations on k symbols.

Definition 1 ([9]) A multi-norm on $\{\wp^k : k \in \mathbb{N}\}$ is a sequence $(\|.\|) = (\|.\|_k : k \in \mathbb{N})$ such that $\|.\|_k$ is a norm on $\wp^k$ for each $k \in \mathbb{N}$, $\|x\|_1 = \|x\|$ for each $x \in \wp$, and the following axioms are satisfied for each $k \in \mathbb{N}$ with $k \geq 2$:

1. $\left\|\left(x_{\sigma(1)}, \ldots, x_{\sigma(k)}\right)\right\|_k = \|(x_1 \ldots x_k)\|_k$, for $\sigma \in \Psi_k, x_1, \ldots, x_k \in \wp$;
2. $\|(\alpha_1 x_1, \ldots, \alpha_k x_k)\|_k \leq \left(\max_{i \in \mathbb{N}_k} |\alpha_i|\right) \|(x_1 \ldots x_k)\|_k$
   for $\alpha_1 \ldots \alpha_k \in \mathbb{C}, x_1, \ldots, x_k \in \wp$;
3. $\|(x_1, \ldots, x_{k-1}, 0)\|_k = \|(x_1, \ldots, x_{k-1})\|_{k-1}$, for $x_1, \ldots, x_{k-1} \in \wp$;
4. $\|(x_1, \ldots, x_{k-1}, x_{k-1})\|_k = \|(x_1, \ldots, x_{k-1})\|_{k-1}$ for $x_1, \ldots, x_{k-1} \in \wp$.

In this case, we say that $((\wp^k, \|.\|_k) : k \in \mathbb{N})$ is a multi-normed space.

Suppose that $((\wp^k, \|.\|_k) : k \in \mathbb{N})$ is a multi-normed space, and take $k \in \mathbb{N}$. We need the following two properties of multi-norms. They can be found in [9].

(a) $\|(x, \ldots x)\|_k = \|x\|, \forall x \in \wp,$

(b) $\max_{i \in \mathbb{N}_k} \|x_i\| \leq \|(x_1, \ldots, x_k)\|_k \leq \sum_{i=1}^{k} \|x_i\| \leq k \max_{i \in \mathbb{N}_k} \|x_i\|, \forall x_1, \ldots, x_k \in \wp.$

It follows from (b) that if $(\wp, \|\cdot\|)$ is a Banach Space, then $(\wp^k, \|.\|_k)$ is a Banach Space for each $k \in \mathbb{N}$;

In this case, $((\wp^k, \|.\|_k) : k \in \mathbb{N})$ is a multi-Banach space.

3 The Fixed Point Method

The fixed point method is one of the most dynamic areas of research during the last 60 years with lot of applications in various fields of pure and applied mathematics. Let X be a nonempty set. A function $d : X \times X \to [0, \infty)$ is called a generalized metric on X if d satisfies

1. $d(x, y) = 0$ if and only if $x = y$;
2. $d(x, y) = d(y, x)$ for all $x, y \in X$;
3. $d(x, z) \leq d(x, y) + d(y, z)$ for all $x, y, z \in X$.

The following fixed point theorem proved by Diaz and Margolis [11] plays an important role in proving our theorem:

Theorem 1 ([11]) *Let (X, d) be a complete generalized metric space and let $\mathcal{J}$: $X \to X$ be a strictly contractive mapping with Lipschitz constant $\mathcal{L} < 1$. Then for each given element $x \in X$, either*

$$d(\mathcal{J}^n x, \mathcal{J}^{n+1} x) = \infty$$

for all nonnegative integers n or there exists a positive integer n_0 such that

(i) $d(\mathcal{J}^n x, \mathcal{J}^{n+1} x) < \infty$ for all $n \geq n_0$;
(ii) The sequence $\{\mathcal{J}^n x\}$ is convergent to a fixed point y^* of $\mathcal{J}$;
(iii) y^* is the unique fixed point of T in the set $Y = \{y \in X : d(\mathcal{J}^{n_0} x, y) < \infty\}$;
(iv) $d(y, y^*) \leq \frac{1}{1-\mathcal{L}} d(y, \mathcal{J}y)$ for all $y \in Y$.

Let X and Y be real vector spaces. For convenience, we use the following abbreviation for a mapping $f : X \to Y$

$Df(x, y) = f(x + 16y) - 32f(x + 15y) + 496f(x + 14y) - 4960f(x + 13y)$
$+35960f(x + 12y) - 201376f(x + 11y) + 906192f(x + 10y) - 3365856f(x + 9y)$
$+10518300f(x + 8y) - 28048800f(x + 7y) + 64512240f(x + 6y)$
$-129024480f(x + 5y) + 225792840f(x + 4y) - 347373600f(x + 3y)$
$+471435600f(x + 2y) - 565722720f(x + y) + 601080390f(x)$
$-565722720f(x - y) + 471435600f(x - 2y) - 347373600f(x - 3y)$
$+225792840f(x - 4y) - 129024480f(x - 5y) + 64512240f(x - 6y)$
$-28048800f(x - 7y) + 10518300f(x - 8y) - 3365856f(x - 9y)$
$+906192f(x - 10y) - 201376f(x - 11y) + 35960f(x - 12y)$

$$-4960f(x - 13y) + 496f(x - 14y) - 32f(x - 15y) + f(x - 16y) - 32!f(y) \quad (1)$$

for all $x, y \in X$, where $32! = 2.631308369 \times 10^{35}$.

In this paper, we introduce the Duotrigintic functional equation:

$$Df(x, y) = 0.$$

for all $x, y \in X$. Moreover, we prove the stability of the Duotrigintic functional equation (1) in Multi-Banach Spaces by using fixed point method. It is easy to show that the function $f(x) = x^{32}$ satisfies the functional equation (1), which is called as duotrigintic functional equation and every solution of the duotrigintic functional equation is said to be a duotrigintic mapping.

4 General Solution of Duotrigintic Functional Equation in (1)

Theorem 2 *Let X and Y be vector spaces. If $f : X \to Y$ is the function (1) for all $x, y \in X$, then f is a Duotrigintic mapping.*

Proof Substituting $x = 0$ and $y = 0$ in (1), we obtain that $f(0) = 0$. Substituting (x, y) with (x, x) and $(x, -x)$ in (1), respectively, and subtracting two resulting equations, we can arrive at $f(-x) = f(x)$, that is to say, f is an even function.

Letting (x, y) by $(16x, x)$ and $(0, 2x)$ respectively in (1), and subtracting the two resulting equations, we arrive at

$32f(31x) - 528f(30x) + 4960f(29x) - 35464f(28x) + 201376f(27x)$
$-911152f(26x) + 3365856f(25x) - 10482340f(24x)$
$+28048800f(23x) - 64713616f(22x) + 129024480f(21x)$
$-224886648f(20x) + 347373600f(19x) - 474801456f(18x)$
$+565722720f(17x) - 590562090f(16x) + 565722720f(15x)$
$-499484400f(14x) + 347373600f(13x) - 161280600f(12x) + 129024480f(11x)$
$-193536720f(10x) + 28048800f(9x) + 215274540f(8x) + 3365856f(7x)$
$-348279792f(6x) + 201376f(5x) + 471399640f(4x) + 4960f(3x)$

$$-\frac{32!}{2}f(2x) + 32!f(x) = 0 \tag{2}$$

for all $x \in X$. Substituting $x = 15x$ and $y = x$ in (1), further multiplying the resulting equation by 32, and subtracting the obtained result from (2), we get

$496f(30x) - 10912f(29x) + 123256f(28x) - 949344f(27x) + 5532880f(26x)$
$-25632288f(25x) + 97225052f(24x) - 308536800f(23x) + 852847984f(22x)$
$-1935367200f(21x) + 3903896712f(20x) - 6877997280f(19x)$
$+1.064115374 \times 10^{10}f(18x) - 1.452021648 \times 10^{10}f(17x)$
$+1.751256495 \times 10^{10}f(16x) - 1.866884976 \times 10^{10}f(15x)$
$+1.760364264 \times 10^{10}f(14x) - 1.47385656 \times 10^{10}f(13x)$
$+1.09546746 \times 10^{10}f(12x) - 7096346400f(11x) + 3935246640f(10x)$
$-2036342880f(9x) + 1112836146f(8x) - 333219744f(7x)$
$-240572400f(6x) - 28796768f(5x) + 477843672f(4x) - 1145760f(3x)$

$$-\frac{32!}{2}f(2x) + 32!(33)f(x) = 0 \tag{3}$$

for all $x \in X$. Replacing (x, y) with $(14x, x)$ in (1), further multiplying the resulting equation by 496, and subtracting the obtained result from (3), we have

$4960f(29x) - 122760f(28x) + 1510816(27x) - 12303280f(26x)$
$+74250208f(25x) - 352246180f(24x) + 1360927776f(23x)$
$-43842288f(22x) + 1.19768376 \times 10^{10}f(21x)$
$-2.809417433 \times 10^{10}f(20x) + 5.71181448 \times 10^{10}f(19x)$
$-1.013520949 \times 10^{11}f(18x) + 1.577770891 \times 10^{11}f(17x)$

$-2.163194927 \times 10^{11} f(16x) + 2.619296193 \times 10^{11} f(15x)$
$-2.805322308 \times 10^{11} f(14x) + 2.658599035 \times 10^{11} f(13x)$
$-2.22877383 \times 10^{11} f(12x) + 1.652009592 \times 10^{11} f(11x)$
$-1.08058002 \times 10^{11} f(10x) + 6.19597992 \times 10^{10} f(9x)$
$-3.08852349 \times 10^{10} f(8x) + 1.357898506 \times 10^{10} f(7x)$
$-5457649200 f(6x) + 1640667808 f(5x) + 28372440 f(4x) + 98736736 f(3x)$

$$-\frac{32!}{2} f(2x) + 32!(529) f(x) = 0 \tag{4}$$

for all $x \in X$. Replacing (x, y) with $(13x, x)$ in (1), further multiplying the resulting equation by 4960, and subtracting the obtained result from (4), we get

$35960 f(28x) - 949344 f(27x) + 12298320 f(26x) - 104111392 f(25x)$
$+646578780 f(24x) - 3133784544 f(23x)$
$+1.231041694 \times 10^{10} f(22x) - 4.01939304 \times 10^{10} f(21x)$
$+1.110278737 \times 10^{11} f(20x) - 2.628625656 \times 10^{11} f(19x)$
$+5.386093259 \times 10^{11} f(18x) - 9.621553973 \times 10^{11} f(17x)$
$+1.506653563 \times 10^{12} f(16x) - 2.076390957 \times 10^{12} f(15x)$
$+2.52545246 \times 10^{12} f(14x) - 2.715498831 \times 10^{12} f(13x)$
$+2.583107308 \times 10^{12} f(12x) - 2.173119617 \times 10^{12} f(11x)$
$+1.614915054 \times 10^{12} f(10x) - 1.057972687 \times 10^{12} f(9x)$
$+6.090761859 \times 10^{11} f(8x) - 3.064017253 \times 10^{11} f(7x)$
$+1.336643988 \times 10^{11} f(6x) - 5.053010019 \times 10^{10} f(5x)$
$+1.67230182 \times 10^{10} f(4x) - 4395980544 f(3x)$

$$-\frac{32!}{2} f(2x) + 32!(5489) f(x) = 0 \tag{5}$$

for all $x \in X$. Plugging (x, y) into $(12x, x)$ in (1), further multiplying the resulting equation by 35960, and subtracting the obtained result from (5), we have

$201376 f(27x) - 5537840 f(26x) + 74250208 f(25x)$
$-646542820 f(24x) + 4107696416 f(23x)$
$-2.027624738 \times 10^{10} f(22x) + 8.084225136 \times 10^{10} f(21x)$
$-2.672101943 \times 10^{11} f(20x) + 7.457722824 \times 10^{11} f(19x)$
$-1.781250825 \times 10^{12} f(18x) + 3.677564904 \times 10^{12} f(17x)$
$-6.612856963 \times 10^{12} f(16x) + 1.04151637 \times 10^{13} f(15x)$
$-1.442737172 \times 10^{13} f(14x) + 1.762789018 \times 10^{13} f(13x)$
$-1.903174351 \times 10^{13} f(12x) + 1.817026939 \times 10^{13} f(11x)$
$-1.533790913 \times 10^{13} f(10x) + 1.43358197 \times 10^{13} f(9x)$
$-7.51043434 \times 10^{12} f(8x) + 4.333318575 \times 10^{12} f(7x)$
$-2.186195751 \times 10^{12} f(6x) + 9.581047478 \times 10^{11} f(5x)$
$-3.615150858 \times 10^{11} f(4x) + 1.16641352 \times 10^{11} f(3x)$

$$-\frac{32!}{2} f(2x) + 32!(41449) f(x) = 0 \tag{6}$$

for all $x \in X$. Plugging (x, y) into $(11x, x)$ in (1), further multiplying the resulting equation by 201376, and subtracting the obtained result from (6), we arrive at

$906192 f(26x) - 25632288 f(25x) + 352282140 f(24x) - 3133784544 f(23x)$
$+2.0276046 \times 10^{10} f(22x) - 1.016430688 \times 10^{11} f(21x)$
$+4.105924235 \times 10^{11} f(20x) - 1.372360898 \times 10^{12} f(19x)$
$+3.867104325 \times 10^{12} f(18x) - 9.313651939 \times 10^{12} f(17x)$
$+1.936957672 \times 10^{13} f(16x) - 3.505409525 \times 10^{13} f(15x)$
$+5.552533436 \times 10^{13} f(14x) - 7.730792521 \times 10^{13} f(13x)$
$+9.489123499 \times 10^{13} f(12x) - 1.028728952 \times 10^{14} f(11x)$
$+9.858506937 \times 10^{13} f(10x) - 8.350223342 \times 10^{13} f(9x)$
$+6.244227173 \times 10^{13} f(8x) - 4.113594037 \times 10^{13} f(7x)$
$+2.379623793 \times 10^{13} f(6x) - 1.203311299 \times 10^{13} f(5x)$
$+5.286846507 \times 10^{12} f(4x) - 2.001591711 \times 10^{12} f(3x)$

$$- \frac{32!}{2} f(2x) + 32!(242825) f(x) = 0 \tag{7}$$

for all $x \in X$. Plugging (x, y) into $(10x, x)$ in (1), further multiplying the resulting equation by 906192, and subtracting the obtained result from (7), we have

$3365856 f(25x) - 97189092 f(24x) + 1360927776 f(23x)$
$-1.231061832 \times 10^{10} f(22x) + 8.084225136 \times 10^{10} f(21x)$
$-4.105915173 \times 10^{11} f(20x) + 1.677750882 \times 10^{12} f(19x)$
$-5.664494989 \times 10^{12} f(18x) + 1.610394623 \times 10^{13} f(17x)$
$-3.909089907 \times 10^{13} f(16x) + 8.186685635 \times 10^{13} f(15x)$
$-1.490863309 \times 10^{14} f(14x) + 2.374792521 \times 10^{14} f(13x)$
$-3.323199342 \times 10^{14} f(12x) + 4.097805079 \times 10^{14} f(11x)$
$-4.461091714 \times 10^{14} f(10x) + 4.291511697 \times 10^{14} f(9x)$
$-3.647688975 \times 10^{14} f(8x) + 2.73651237 \times 10^{14} f(7x)$
$-1.808154283 \times 10^{14} f(6x) + 1.048878683 \times 10^{14} f(5x)$
$-5.317407875 \times 10^{13} f(4x) + 2.342050117 \times 10^{13} f(3x)$

$$- \frac{32!}{2} f(2x) + 32!(1149017) f(x) = 0 \tag{8}$$

for all $x \in X$. Replacing (x, y) with $(9x, x)$ in (1), further multiplying the resulting equation by 3365856, and subtracting the obtained result from (8), we get

$10518300 f(24x) - 308536800 f(23x) + 4384027440 f(22x) - 4.01939304 \times 10^{10} f(21x) + 2.672111005 \times 10^{11} f(20x) - 1.372360898 \times 10^{12} f(19x)$
$+5.664491624 \times 10^{12} f(18x) - 1.929913693 \times 10^{13} f(17x)$
$+5.53173227 \times 10^{13} f(16x) - 1.352720538 \times 10^{14} f(15x)$
$+2.851914892 \times 10^{14} f(14x) - 5.225069332 \times 10^{14} f(13x)$
$+8.368895818 \times 10^{14} f(12x) - 1.177003885 \times 10^{15} f(11x)$
$+1.45803204 \times 10^{15} f(10x) - 1.593998867 \times 10^{15} f(9x)$
$+1.539372314 \times 10^{15} f(8x) - 1.313133109 \times 10^{15} f(7x)$
$+9.883941957 \times 10^{14} f(6x) - 6.550999864 \times 10^{14} f(5x)$

$$+3.811204361 \times 10^{14} f(4x) - 1.938394451 \times 10^{14} f(3x)$$

$$-\frac{32!}{2} f(2x) + 32!(4514873) f(x) = 0 \tag{9}$$

for all $x \in X$. Replacing (x, y) with $(8x, x)$ in (1), further multiplying the resulting equation by 10518300, and subtracting the obtained result from (9), we arrive at

$28048800 f(23x) - 833049360 f(22x) 1.19768376 \times 10^{10} f(21x)$
$-1.110269675 \times 10^{11} f(20x) + 7.457722824 \times 10^{11} f(19x)$
$-3.867107693 \times 10^{12} f(18x) + 1.610394623 \times 10^{13} f(17x)$
$-5.53173122 \times 10^{13} f(16x) + 1.597536392 \times 10^{14} f(15x)$
$-3.933676047 \times 10^{14} f(14x) + 8.346112548 \times 10^{14} f(13x)$
$-1.538067247 \times 10^{15} f(12x) + 2.476775902 \times 10^{15} f(11x)$
$-3.500669031 \times 10^{15} f(10x) + 4.356442419 \times 10^{15} f(9x)$
$-4.782971563 \times 10^{15} f(8x) + 4.637308513 \times 10^{15} f(7x)$
$-3.970312093 \times 10^{15} f(6x) + 2.998731922 \times 10^{15} f(5x)$
$-1.994214631 \times 10^{15} f(4x) + 1.165396876 \times 10^{15} f(3x)$

$$-\frac{32!}{2} f(2x) + 32!(15033173) f(x) = 0 \tag{10}$$

for all $x \in X$. Replacing (x, y) with $(7x, x)$ in (1), further multiplying the resulting equation 28048800, and subtracting the obtained result from (10), we arrive at

$64512240 f(22x) - 1935367200 f(21x) + 2.809508052 \times 10^{10} f(20x)$
$-2.6286625656 \times 10^{11} f(19x) + 1.781247459 \times 10^{12} f(18x) - 9.313651939 \times 10^{12} f(17x)$
$+3.909090958 \times 10^{13} f(16x) - 1.352720538 \times 10^{14} f(15x)$
$+3.933675767 \times 10^{14} f(14x) - 9.748796622 \times 10^{14} f(13x)$
$+2.080914588 \times 10^{15} f(12x) - 3.856442309 \times 10^{15} f(11x)$
$+6.242743601 \times 10^{15} f(10x) - 8.866760471 \times 10^{15} f(9x)$
$+1.108487277 \times 10^{16} f(8x) - 1.222228904 \times 10^{16} f(7x)$
$+1.189767046 \times 10^{16} f(6x) - 1.022547957 \times 10^{16} f(5x)$
$+7.754846356 \times 10^{15} f(4x) - 5.193238933 \times 10^{15} f(3x)$

$$-\frac{32!}{2} f(2x) + 32!(43081973) f(x) = 0 \tag{11}$$

for all $x \in X$. Setting (x, y) by $(6x, x)$ in (1), further multiplying the resulting equation by 64512240, and subtracting the obtained result from (11), we arrive at

$129024480 f(21x) - 3902990520 f(20x) + 5.71181448 \times 10^{10} f(19x)$
$-5.386126918 \times 10^{11} f(18x) + 3.677564904 \times 10^{12} f(17x)$
$-1.936956622 \times 10^{13} f(16x) + 8.186685633 \times 10^{13} f(15x)$
$-2.851915174 \times 10^{14} f(14x) + 8.346112548 \times 10^{14} f(13x)$
$-2.080914522 \times 10^{15} f(12x) + 4.467215911 \times 10^{15} f(11x)$
$-8.323658349 \times 10^{15} f(10x) + 1.354309065 \times 10^{16} f(9x)$
$-1.93285258 \times 10^{16} f(8x) + 2.427407082 \times 10^{16} f(7x)$

$-2.688169178 \times 10^{16} f(6x) + 2.628355153 \times 10^{16} f(5x)$
$-2.271698069 \times 10^{16} f(4x) + 1.743374903 \times 10^{16} f(3x)$

$$-\frac{32!}{2} f(2x) + 32!(107594213) f(x) = 0 \qquad (12)$$

for all $x \in X$. Replacing (x, y) with $(5x, x)$ in (1), further multiplying the resulting equation by 129024480, and subtracting the obtained result from (12), we have

$225792840 f(20x) - 6877997280 f(19x)1.013487293 \times 10^{11} f(18x)$
$-9.621553973 \times 10^{11} f(17x) + 6.612867481 \times 10^{12} f(16x)$
$-3.505409525 \times 10^{13} f(15x) + 1.490863029 \times 10^{14} f(14x)$
$-5.225069332 \times 10^{14} f(13x) + 1.538067313 \times 10^{15} f(12x)$
$-3.856442438 \times 10^{15} f(11x) + 8.323662221 \times 10^{15} f(10x)$
$-1.558977711 \times 10^{16} f(9x) + 2.549181227 \times 10^{16} f(8x)$
$-3.655730204 \times 10^{16} f(7x) + 4.613637043 \times 10^{16} f(6x)$
$-5.138745418 \times 10^{16} f(5x) + 5.07093769 \times 10^{16} f(4x)$
$-4.47501023 \times 10^{16} f(3x)$

$$-\frac{32!}{2} f(2x) f(2x) + 32!(236618693) f(x) = 0 \qquad (13)$$

for all $x \in X$. Plugging (x, y) into $(4x, x)$ in (1), further multiplying the resulting equation by 225792840, and subtracting the obtained result from (13), we have

$347373600 f(19x) - 1.064451964 \times 10^{10} f(18x)$
$1.577770891 \times 10^{11} f(17x) - 1.506643045 \times 10^{12} f(16x) + 1.041516365 \times 10^{13} f(15x)$
$-5.552536241 \times 10^{13} f(14x) + 2.374792521 \times 10^{14} f(13x)$
$-8.36889742 \times 10^{14} f(12x) + 2.476782998 \times 10^{15} f(11x)$
$-6.242851659 \times 10^{15} f(10x) + 1.354414659 \times 10^{16} f(9x)$
$-2.549871384 \times 10^{16} f(8x) + 4.19226389 \times 10^{16} f(7x)$
$-6.051502427 \times 10^{16} f(6x) + 7.710867162 \times 10^{16} f(5x)$
$-8.73852283 \times 10^{16} f(4x) + 8.93192555 \times 10^{16} f(3x)$

$$-\frac{32!}{2} f(2x) + 32!(462411533) f(x) = 0 \qquad (14)$$

for all $x \in X$. Replacing (x, y) with $(3x, x)$ in (1), further multiplying the resulting equation by 347373600, and subtracting the obtained result from (14), we have

$471435600 f(18x) - 1.45202165 \times 10^{10} f(17x) + 2.16330011 \times 10^{11} f(16x)$
$-2.076390956 \times 10^{12} f(15x) + 1.442734367 \times 10^{13} f(14x)$
$-7.73082726 \times 10^{13} f(13x) + 3.3233089 \times 10^{14} f(12x)$
$-1.77169036 \times 10^{15} f(11x) + 3.502283946 \times 10^{15} f(10x)$
$-8.87819402 \times 10^{15} f(9x) + 1.939093697 \times 10^{16} f(8x)$
$-3.682661996 \times 10^{16} f(7x) + 6.132260323 \times 10^{16} f(6x)$
$-9.030938968 \times 10^{16} f(5x) + 1.88753222 \times 10^{17} f(4x)$
$-1.418900525 \times 10^{17} f(3x)$

$$-\frac{32!}{2}f(2x) + 32!(809785133)f(x) = 0 \qquad (15)$$

for all $x \in X$. Replacing (x, y) with $(2x, x)$ in (1), further multiplying the resulting equation by 471435600, and subtracting the obtained result from (15), we get

$565722700 f(17x) - 1.75020466 \times 10^{10} f(16x) + 2.619296198 \times 10^{11} f(15x)$
$-2.525951944 \times 10^{12} f(14x) + 1.764262872 \times 10^{13} f(13x)$
$-9.51141113 \times 10^{13} f(12x) + 4.11953627 \times 10^{14} f(11x)$
$-1.47336995 \times 10^{15} f(10x) + 4.43994465 \times 10^{15} f(9x)$
$-1.144964077 \times 10^{16} f(8x) + 2.558689753 \times 10^{16} f(7x)$
$-5.008288087 \times 10^{16} f(6x) + 8.667539472 \times 10^{16} f(5x)$
$-1.337895693 \times 10^{17} f(4x) + 1.856385106 \times 10^{17} f(3x)$

$$-\frac{32!}{2}f(2x) + 32!(1281220733)f(x) = 0 \qquad (16)$$

for all $x \in X$. Replacing (x, y) with (x, x) in (1), further multiplying the resulting equation by 565722720, and subtracting the obtained result from (16), we arrive

$601080390 f(16x) - 1.923457248 \times 10^{10} f(15x)$
$+2.981358734 \times 10^{11} f(14x) - 2.981358734 \times 10^{12} f(13x)$
$+2.161485082 \times 10^{13} f(12x) - 1.210431646 \times 10^{14} f(11x)$
$+5.446942408 \times 10^{14} f(10x) - 2.023150037 \times 10^{15} f(9x)$
$+6.32234387 \times 10^{15} f(8x) - 1.685958364 \times 10^{16} f(7x)$
$+3.877704238 \times 10^{16} f(6x) - 7.755678478 \times 10^{16} f(5x)$
$+1.357196483 \times 10^{17} f(4x) - 2.08799458 \times 10^{17} f(3x)$

$$-\frac{32!}{2}f(2x) + 32!(1846943453)f(x) = 0 \qquad (17)$$

for all $x \in X$. Taking $x = 0$ and $y = x$ in (1), further multiplying the resulting equation by 601080390, and subtracting the obtained result from (17), we arrive

$$-\frac{32!}{2}f(2x) + 32!(2147483648)f(x) = 0 \qquad (18)$$

for all $x \in X$. It follows from (18), we get

$$f(2x) = 2^{32} f(x) \qquad (19)$$

for all $x \in X$. □

5   Hyers–Ulam Stability of Functional Equation (1) in Multi-Banach Spaces

Theorem 3 *Let X be a linear space and let $\big((Y^k, \|.\|_k) : k \in \mathbb{N}\big)$ be a Multi-Banach Space. Suppose that δ is a non-negative real number and $f : X \to Y$ be a function fulfills*

$$\sup_{k \in \mathbb{N}} \|(f(x_1, y_1), \ldots, f(x_k, y_k))\|_k \leq \delta \tag{20}$$

$\forall x_1, \ldots, x_k, y_1, \ldots, y_k \in X$. *Then there exists a unique Duotrigintic mapping* $\mathcal{T} : X \to Y$ *such that*

$$\sup_{k \in \mathbb{N}} \|(f(x_1) - \mathcal{T}(x_1), \ldots, f(x_k) - \mathcal{T}(x_k))\|_k \leq \frac{4294967297}{32!(4294967295)}\delta. \tag{21}$$

$\forall x_i \in X$, *where* $i = 1, 2, \ldots, k$.

Proof Letting (x_i, y_i) by $(16x_i, x_i)$ and $(0, 2x_i)$ respectively in (20), and subtracting the two resulting equations, we arrive at

$\sup_{k \in \mathbb{N}} \|(32f(31x_1) - 528f(30x_1) + 4960f(29x_1) - 35464f(28x_1)$
$+201376f(27x_1) - 911152f(26x_1) + 3365856f(25x_1) - 10482340f(24x_1)$
$+28048800f(23x_1) - 64713616f(22x_1) + 129024480f(21x_1)$
$-224886648f(20x_1) + 347373600f(19x_1) - 474801456f(18x_1)$
$+565722720f(17x_1) - 590562090f(16x_1) + 565722720f(15x_1)$
$-499484400f(14x_1)$
$+347373600f(13x_1) - 161280600f(12x_1) + 129024480f(11x_1)$
$-193536720f(10x_1) + 28048800f(9x_1) + 215274540f(8x_1) + 3365856f(7x_1)$
$-348279792f(6x_1) + 201376f(5x_1) + 471399640f(4x_1) + 4960f(3x_1)$
$-\dfrac{32!}{2}f(2x_1) + 32!f(x_1), \ldots, 32f(31x_k) - 528f(30x_k) + 4960f(29x_k)$
$-35464f(28x_k)$
$+201376f(27x_k) - 911152f(26x_k) + 3365856f(25x_k)$
$-10482340f(24x_k) + 28048800f(23x_k) - 64713616f(22x_k)$
$+129024480f(21x_k)$
$-224886648f(20x_k) + 347373600f(19x_k) - 474801456f(18x_k)$
$+565722720f(17x_k) - 590562090f(16x_k) + 565722720f(15x_k)$
$-499484400f(14x_k) + 347373600f(13x_k) - 161280600f(12x_k)$
$+129024480f(11x_k)$
$-193536720f(10x_k) + 28048800f(9x_k) + 215274540f(8x_k) + 3365856f(7x_k)$
$-348279792f(6x_k) + 201376f(5x_k) + 471399640f(4x_k)$

$$+4960f(3x_k) - \frac{32!}{2}f(2x_k) + 32!f(x_k)\bigg)\bigg\|_k \leq \frac{3}{2}\delta \tag{22}$$

for all $x_i \in X$, where $i = 1, 2, .., k$. Substituting $x_i = 15x_i$ and $y_i = x_i$ in (20), further multiplying the resulting equation by 32, and subtracting the obtained result from (22), we arrive at

$$\sup_{k \in \mathbb{N}} \| (496 f(30x_1) - 10912 f(29x_1) + 123256 f(28x_1) - 949344 f(27x_1)$$
$$+5532880 f(26x_1) - 25632288 f(25x_1) + 97225052 f(24x_1)$$
$$-308536800 f(23x_1) \quad +852847984 f(22x_1) - 1935367200 f(21x_1)$$
$$+3903896712 f(20x_1)$$
$$-6877997280 f(19x_1) + 1.064115374 \times 10^{10} f(18x_1)$$
$$-1.452021648 \times 10^{10} f(17x_1) + 1.751256495 \times 10^{10} f(16x_1)$$
$$-1.866884976 \times 10^{10} f(15x_1) + 1.760364264 \times 10^{1} 0 f(14x_1)$$
$$-1.47385656 \times 10^{10} f(13x_1) + 1.09546746 \times 10^{10} f(12x_1)$$
$$-7096346400 f(11x_1) + 3935246640 f(10x_1) - 2036342880 f(9x_1)$$
$$+1112836146 f(8x_1) - 333219744 f(7x_1) - 240572400 f(6x_1)$$
$$-28796768 f(5x_1) + 477843672 f(4x_1) - 1145760 f(3x_1) - \frac{32!}{2} f(2x_1)$$
$$+32!(33) f(x_1), ..,$$
$$496 f(30x_k) - 10912 f(29x_k) + 123256 f(28x_k) - 949344 f(27x_k)$$
$$+5532880 f(26x_k) - 25632288 f(25x_k) + 97225052 f(24x_k)$$
$$-308536800 f(23x_k)$$
$$+852847984 f(22x_k) - 1935367200 f(21x_k) + 3903896712 f(20x_k)$$
$$-6877997280 f(19x_k) + 1.064115374 \times 10^{10} f(18x_k)$$
$$-1.452021648 \times 10^{10} f(17x_k) + 1.751256495 \times 10^{10} f(16x_k)$$
$$-1.866884976 \times 10^{10} f(15x_k) + 1.760364264 \times 10^{1} 0 f(14x_k)$$
$$-1.47385656 \times 10^{10} f(13x_k) + 1.09546746 \times 10^{10} f(12x_k)$$
$$-7096346400 f(11x_k) + 3935246640 f(10x_k) - 2036342880 f(9x_k)$$
$$+1112836146 f(8x_k) - 333219744 f(7x_k) - 240572400 f(6x_k)$$
$$-28796768 f(5x_k) + 477843672 f(4x_k) - 1145760 f(3x_k)$$

$$-\frac{32!}{2} f(2x_k) + 32!(33) f(x_k) \Big) \Big\|_k \leq \frac{67}{2} \delta \tag{23}$$

for all $x_i \in X$, where $i = 1, 2, \ldots, k$. Applying the same procedure of Theorem 2 and using (19), we get

$$\sup_{k \in \mathbb{N}} \left\| \left(f(x_1) - \frac{1}{2^{32}} f(2x_1), \ldots, f(x_k) - \frac{1}{2^{32}} f(2x_k) \right) \right\|_k \leq \frac{4294967297}{(32!)(4294967295)} \delta \tag{24}$$

for all $x_i \in X$, where $i = 1, 2, \ldots, k$.

Let $\Lambda = \{ g : X \to Y | g(0) = 0 \}$ and introduce the generalized metric d defined on Λ by

$$d(u, v) = \inf \left\{ \lambda \in [0, \infty] | \sup_{k \in \mathbb{N}} \| (u(x_1) - v(x_1), \ldots, u(x_k) - v(x_k)) \|_k \leq \lambda \right\}$$

forall $x_1, \ldots, x_k \in X$. Then it is easy to show that (Λ, d) is a generalized complete metric space. See [20].

We define an operator $\mathcal{J} : \Lambda \to \Lambda$ by

$$\mathcal{J}u(x) = \frac{1}{2^{32}}u(2x) \ \forall x \in X.$$

We assert that $\mathcal{J}$ is a strictly contractive operator. Given $u, v \in \Lambda$, let $\lambda \in (0, \infty)$ be an arbitrary constant with $d(u, v) \leq \lambda$. By the definition of d, it follows that

$$\sup_{k \in \mathbb{N}} \|(u(x_1) - v(x_1), \ldots, u(x_k) - v(x_k))\|_k \leq \lambda,$$

for all $x_1, \ldots, x_k \in X$. Therefore,

$$\sup_{k \in \mathbb{N}} \|(\mathcal{J}u(x_1) - \mathcal{J}v(x_1), \ldots, \mathcal{J}u(x_k) - \mathcal{J}v(x_k))\|_k$$

$$\leq \sup_{k \in \mathbb{N}} \left\|\left(\frac{1}{2^{32}}u(2x_1) - \frac{1}{2^{32}}v(2x_1), \ldots, \frac{1}{2^{32}}u(2x_k) - \frac{1}{2^{32}}v(2x_k)\right)\right\|_k$$

$$\leq \frac{1}{2^{32}}\lambda$$

for all $x_1, \ldots, x_k \in X$. Hence, it holds that $d(\mathcal{J}u, \mathcal{J}v) \leq \frac{1}{2^{32}}\lambda$ i.e., $d(\mathcal{J}u, \mathcal{J}v) \leq \frac{1}{2^{32}}d(u, v) \ \forall u, v \in \Lambda$. This means that $\mathcal{J}$ is strictly contractive operator on Λ with the Lipschitz constant $\mathcal{L} = \frac{1}{2^{32}}$.

By (24), we have $d(\mathcal{J}h, h) \leq \frac{4294967297}{(32!)(4294967296)}\delta$. According to Theorem 1, we deduce the existence of a fixed point of $\mathcal{J}$ that is the existence of mapping $\mathcal{T} : X \to Y$ such that

$$\mathcal{T}(2x) = 2^{32}\mathcal{T}(x) \quad \forall x \in X.$$

Moreover, we have $d\left(\mathcal{J}^n h, \mathcal{T}\right) \to 0$, which implies

$$\mathcal{T}(x) = \lim_{n \to \infty} \mathcal{J}^n h(x) = \lim_{n \to \infty} \frac{h(2^n x)}{2^{32n}}$$

for all $x \in X$.

Also, $d(h, \mathcal{T}) \leq \frac{1}{1 - \mathcal{L}}d(\mathcal{J}h, h)$ implies the inequality

$$d(h, \mathcal{T}) \le \frac{1}{1 - \dfrac{1}{2^{32}}} d(\mathcal{J}h, h)$$

$$\le \frac{4294967297}{(32!)(4294967295)} \delta.$$

Setting $x_1 = \ldots = x_k = 2^n x$, $y_1 =, \ldots, = y_k = 2^n y$ in (20) and divide both sides by 2^{32^n}. Then, using property (a) of multi-norms, we obtain

$$\|D\mathcal{T}(x, y)\| = \lim_{n \to \infty} \frac{1}{2^{32^n}} \left\| Dh \left(2^n x, 2^n y \right) \right\| = 0$$

for all $x, y \in X$. Hence $\mathcal{T}$ is Duotrigintic mapping.

The uniqueness of $\mathcal{T}$ follows from the fact that $\mathcal{T}$ is the unique fixed point of $\mathcal{J}$ with the property that there exists $\ell \in (0, \infty)$ such that

$$\sup_{k \in \mathbb{N}} \|(f(x_1) - \mathcal{T}(x_1), \ldots, f(x_k) - \mathcal{T}(x_k))\|_k \le \ell$$

for all $x_1, \ldots, x_k \in X$.

This completes the proof of the theorem. □

Corollary 1 *Let X be a linear space, and let $(Y^k, \|.\|_k)$ be a Multi-Banach space. Let $\theta > 0, 0 < p < 32$ and $f : X \to Y$ be a mapping satisfying $f(0) = 0$*

$$\sup_{k \in \mathbb{N}} \|\mathcal{D}f(x_1, y_1), \ldots, \mathcal{D}f(x_k, y_k)\|_k \le \theta \left(\|x_1\|^p + \|y_1\|^p, \ldots, \|x_k\|^p + \|y_k\|^p \right)$$

$$(25)$$

for all $x_1, \ldots, x_k, y_1 \ldots, y_k \in X$. Then there exists a unique mapping $\mathcal{T} : X \to Y$ such that

$$\sup_{k \in \mathbb{N}} \|(f(x_1) - \mathcal{T}(x_1), \ldots, f(x_k) - \mathcal{T}(x_k))\| \le \frac{1}{2^{32} - 2^p} \Psi(\|x_1\|^p, \ldots, \|x_k\|^p)$$

$$(26)$$

where

$$\Psi = \frac{2}{32!} \theta \left[\frac{1}{2} 2^p + (16^p + 1) + 32(15^p + 1) + 496(14^p + 1) + 4960(13^p + 1) \right.$$

$$+ 35960(12^p + 1) + 201376(11^p + 1) + 906192(10^p + 1) + 3365856(9^p + 1)$$

$$+ 10518300(8^p + 1) + 28048800(7^p + 1)$$

$$+ 64512240(6^p + 1) + 129024480(5^p + 1) + 225792840(4^p + 1)$$

$$+ 347373600(3^p + 1) + 471435600(2^p + 1) + 866262915 \Big]$$

Proof The proof is similar to that of Theorem 3, replacing δ by $\theta \left(\|x_1\|^p + \|y_1\|^p, \ldots, \|x_k\|^p + \|y_k\|^p \right)$.   □

Corollary 2 *Let X be a linear space, and let $(Y^k, \|.\|_k)$ be a Multi-Banach Space. Let $\theta > 0, 0 < r + s = p < 32$ and $f : X \to Y$ be a mapping satisfying $f(0) = 0$*

$$\sup_{k \in \mathbb{N}} \|\mathcal{D}f(x_1, y_1), \ldots, \mathcal{D}f(x_k, y_k)\|_k \le \theta \left(\|x_1\|^r \cdot \|y_1\|^s, \ldots, \|x_k\|^r \cdot \|y_k\|^s \right)$$

(27)

for all $x_1, \ldots, x_k, y_1 \ldots, y_k \in X$. Then there exists a unique mapping $\mathcal{T} : X \to Y$ such that

$$\sup_{k \in \mathbb{N}} \|(f(x_1) - \mathcal{T}(x_1), \ldots, f(x_k) - \mathcal{T}(x_k))\| \le \frac{1}{2^{32} - 2^p} \Psi_{32}(\|x_1\|^{r+s}, \ldots, \|x_k\|^{r+s})$$

(28)

where

$$
\begin{aligned}
\Psi_{32} = \frac{2}{32!}\theta \big[& 16^r + 32(15^r) + 496(14^r) + 4960(13^r) + 35960(12^r) \\
& + 201376(11^r) + 906192(10^r) + 3365856(9^r) + 10518300(8^r) + 28048800(7^r) \\
& + 64512240(6^r) + 129024480(5^r) + 225792840(4^r) + 347373600(3^r) \\
& + 471435600(2^r) + 565722720 \big]
\end{aligned}
$$

Proof The proof is similar to that of Theorem 3, replacing δ by $\theta \left(\|x_1\|^r \cdot \|y_1\|^s, \ldots, \|x_k\|^r \cdot \|y_k\|^s \right)$.   □

Example Let $k \in \mathbb{N}$. We define $\phi : \mathbb{R} \to \mathbb{R}$, by

$$\phi(x) = \begin{cases} 1 & x \in [1, \infty) \\ x^{32} & x \in (-\infty, \infty) \\ -1 & x \in (-\infty, -1]. \end{cases}$$

We consider the function $f : \mathbb{R} \to \mathbb{R}$ defined by

$$f(x) = \sum_{n=0}^{\infty} \frac{\phi(4^n x)}{4^{32n}}, \quad (x \in \mathbb{R}).$$

Then f satisfies the following functional inequality:

$$\|Df(x_1, y_1)), \ldots, Df(x_k, y_k)\|_k \le \frac{2^{32} + 32!}{4^{32} - 1} 4^{96}$$

$$\left(|x_1|^{32} +, \ldots, |x_k|^{32} + |y_1|^{32} +, \ldots, |y_k|^{32} \right)$$

for all $x_1, \ldots, x_k, y_1, \ldots, y_k \in \mathbb{R}$.

Proof We have

$$|f(x)| \leq \frac{4^{32}}{4^{32} - 1}$$

for all $x \in \mathbb{R}$. Therefore, we see that f is bounded. Let $x, y \in \mathbb{R}$. If $|x|^{32} + |y|^{32} = 0$ or $|x|^{32} + |y|^{32} \geq \frac{1}{4^{32}}$, then

$$|Df(x, y)| \leq \frac{(2^{32} + 32!) \, 4^{32}}{4^{32} - 1} \leq \frac{(2^{32} + 32!) \, 4^{32}}{4^{32} - 1} 4^{32} \left(|x|^{32} + |y|^{32}\right).$$

Now, suppose that $0 < |x|^{32} + |y|^{32} \leq \frac{1}{4^{32}}$. Then there exists a non-negative integer k such that

$$\frac{1}{4^{32(k+2)}} \leq |x|^{32} + |y|^{32} < \frac{1}{4^{32(k+1)}}.$$

Hence, $4^k x < \frac{1}{4}$ and $4^k y < \frac{1}{4}$, and
$4^n(x + 16y), 4^n(x + 15y), 4^n(x + 14y), 4^n(x + 13y), 4^n(x + 12y),$
$4^n(x + 11y), 4^n(x + 10y), 4^n(x + 9y), 4^n(x + 8y), 4^n(x + 7y)$
$4^n(x + 6y), 4^n(x + 5y), 4^n(x + 4y), 4^n(x + 3y), 4^n(x + 2y)$
$4^n(x + y)4^n(x), 4^n(x - y), 4^n(x - 2y), 4^n(x - 3y),$
$4^n(x - 4y), 4^n(x - 5y), 4^n(x - 6y), 4^n(x - 7y),$
$4^n(x - 8y), 4^n(x - 9y), 4^n(x - 10y), 4^n(x - 11y), 4^n(x - 12y)$
$4^n(x - 13y), 4^n(x - 14y), 4^n(x - 15y), 4^n(x - 16y) \in (-1, 1)$
for all $n = 0, 1, \ldots, k - 1$. Thus we get

$$\frac{|Df(x, y)|}{|x|^{32} + |y|^{32}} \leq \sum_{n=k}^{\infty} \frac{2^{32} + 32!}{4^{32n}(|x|^{32} + |y|^{32})}$$

$$\leq \sum_{n=0}^{\infty} \frac{2^{32} + 32!}{4^{32n} 4^{32(k+2)}(|x|^{32} + |y|^{32})} 4^{64}$$

$$\leq \sum_{n=0}^{\infty} \frac{2^{32} + 32!}{4^{32n}} 4^{64} = \frac{2^{32} + 32!}{4^{32} - 1} 4^{96},$$

or

$$|Df(x, y)| \leq \leq \frac{2^{32} + 32!}{4^{32} - 1} 4^{96}(|x|^{32} + |y|^{32}).$$

$\square$

References

1. T. Aoki, On the stability of the linear transformation in Banach spaces. J. Math. Soc. Jpn. **2**, 64–66 (1950)
2. M. Arunkumar, A. Bodaghi, J.M. Rassias, E. Sathiya, The general solution and approximations of a decic type functional equation in various normed spaces. J. Chungcheong Math. Soc. **29**(2) 287–328 (2016)
3. A. Bodaghi, Intuitionistic fuzzy stability of the generalized forms of cubic and quartic functional equations. J. Intel. Fuzzy Syst. **30**(4), 2309–2317 (2016)
4. A. Bodaghi, Stability of a mixed type additive and quartic function equation. Filomat **28**(8), 1629–1640 (2014)
5. A. Bodaghi, S.M. Moosavi, H. Rahimi, The generalized cubic functional equation and the stability of cubic Jordan *-derivations. Ann. Univ. Ferrara **59**(2), 235–250 (2013)
6. A. Bodaghi, C. Park, J.M. Rassias, Fundamental stabilities of the nonic functional equation in intuitionistic fuzzy normed spaces. Commun. Korean Math. Soc. **31**(4), 729–743 (2016)
7. P. Choonkil, R. Murali, A. Antony Raj, Stability of trigintic functional equation in multi-banach spaces: a fixed point approach. Korean J. Math. **26**, 615–628 (2018)
8. S. Czerwik, On the stability of the quadratic mapping in normed spaces. Abh. Math. Sem. Univ. Hamb. **62**(1), 59–64 (1992)
9. H.G. Dales, M. Moslehian, Stability of mappings on multi-normed spaces. Glasg. Math. J. **49**, 321–332 (2007)
10. H.G. Dales, M.E. Polyakov, Multi-normed spaces and multi-Banach algebras (2011). arXiv:1112.5148
11. J.B. Diaz, B. Margolis, A fixed point theorem of the alternative for contractions on generalized complete metric space. Bull. Am. Math. Soc. **74**, 305–309 (1968)
12. M. Fridoun, Approximate Euler-Lagrange-Jensen type additive mapping in multi-banach spaces: a fixed point approach. Commun. Korean Math. Soc. **28**, 319–333 (2013)
13. P. Gâvruta, A generalization of the Hyers-Ulam-Rassias stability of approximately additive mappings. J. Math. Anal. Appl. **184**, 431–436 (1994)
14. D.H. Hyers, On the stability of the linear functional equation. Proc. Natl. Acad. Sci. USA **27**, 222–224 (1941)
15. R. John Michael, R. Murali, J.R. Matina, A. Antony Raj, General solution, stability and non-stability of quattuorvigintic functional equation in multi-banach spaces. Int. J. Math. Appl. **5**(2-A), 181–194 (2017)
16. R. John Michael, R. Murali, A. Antony Raj, Stability of octavigintic functional equation in multi-banach spaces: a fixed point approach (submitted)
17. K.W. Jun, H.M. Kim, The generalized Hyers-Ulam-Rassias stability of a cubic functional equation. J. Math. Anal. Appl. **274**(2), 867–878 (2002)
18. W. Liguang, L. Bo, B. Ran, Stability of a mixed type functional equation on multi-Banach spaces: a fixed point approach. Fixed Point Theory Appl. **2010**(1), 9 (2010)
19. R. Murali, P. Sandra, A. Antony Raj, General solution and a fixed point approach to the Ulam-Hyers stability of Viginti Duo functional equation in multi-banach spaces. IOSR J. Math. **13**(4), 48–59 (2017)
20. D. Mihet, V. Radu, On the stability of the additive Cauchy functional equation in random normed spaces. J. Math. Anal. Appl. **343**, 567–572 (2008)
21. M.S. Moslehian, K. Nikodem, D. Popa, Asymptotic aspects of the quadratic functional equations in multi-normed spaces. J. Math. Anal. Appl. **355**, 717–724 (2009)
22. R. Murali, P. Sandra, A. Antony Raj, Ulam- Hyers stability of hexadecic functional equations in multi-Banach spaces. Analysis **34**(4) (2017)
23. M. Nazarianpoor, J.M. Rassias, G. Sadeghi, Stability and nonstability of octadecic functional equation in multi-normed spaces. Arab. J. Math. **7**(3) (2017)
24. M. Nazarianpoor, J.M. Rassias, G. Sadeghi, Solution and Stability of Quattuorvigintic Functional Equation in Intuitionistic Fuzzy Normed Spaces. Iran. J. Fuzzy Syst. **15**(4), 13–30 (2018)

25. M. Ramdoss, B. Abasalt, A. Antony Raj, General solution and Ulam-Hyers stability of Viginti functional equations in multi-banach spaces. J. Chungcheong Math. Soc. **31**(2) (2018)
26. T.M. Rassias, On the stability of the linear mapping in Banach spaces. Proc. Am. Math. Soc. **72**, 297–300 (1978)
27. J.M. Rassias, Solution of the Ulam stability problem for quartic mappings. Glas. Mat. Ser. III **34**(2), 243–252 (1999)
28. J.M. Rassias, Solution of the Ulam stability problem for cubic mappings. Glas. Mat. Ser. III **36**(1), 63–72 (2001)
29. J.M. Rassias, M. Eslamian, Fixed points and stability of nonic functional equation in quasi-β-normed spaces. Contin. Anal. Appl. Math. **3**, 293–309 (2015)
30. J.M. Rassias, M. Arunkumar, E. Sathya, T. Namachivayam, Various generalized Ulam-Hyers Stabilities of a nonic functional equations. Tbilisi Math. J. **9**(1), 159–196 (2016)
31. J.M. Rassias, R. Murali, M.J. Rassias, V. Vithya, A. Antony Raj, General solution and stability of Quattuorvigintic functional equation in matrix normed spaces. Tbilisi Math. J. **11**(2), 97–109 (2018)
32. K. Ravi, J.M. Rassias, B.V. Senthil Kumar, Ulam-Hyers stability of undecic functional equation in quasi-β normed spaces fixed point method. Tbilisi Math. Sci. **9**(2), 83–103 (2016)
33. K. Ravi, J.M. Rassias, S. Pinelas, S. Suresh, General solution and stability of quattuordecic functional equation in quasi β-normed spaces. Adv. Pure Math. **6**(12) 921–941 (2016)
34. A. Sattar, M. Fridoun, Approximate a quadratic mapping in multi-banach spaces, a fixed point approach. Int. J. Nonlinear Anal. Appl. **7**, 63–75 (2016)
35. X. Tian Zhou, R. John Michael, X. Wan Xin, Generalized Ulam-Hyers stability of a general mixed AQCQ functional equation in Multi-Banach Spaces: a fixed point approach. Eur. J. Pure Appl. Math. **3**, 1032–1047 (2010)
36. S.M. Ulam, *A Collection of the Mathematical Problems* (Interscience, New York, 1960)
37. W. Xiuzhong, C. Lidan, L. Guofen, Orthogonal stability of mixed Additive-Quadratic Jenson type functional equation in multi-banach spaces. Adv. Pure Math. **5**, 325–332 (2015)
38. T.Z. Xu, J.M. Rassias, Approximate septic and octic mappings in quasi-β-normed spaces. J. Comput. Anal. Appl. **15**(6), 1110–1119 (2013)
39. T.Z. Xu, J.M. Rassias, M.J. Rassias, W.X. Xu, A fixed point approach to the stability of quintic and sextic functional equations in quasi-β-normed spaces. J. Inequal. Appl. **2010**, 23 (2010)
40. W. Zhihua, L. Xiaopei, T.M. Rassias, Stability of an additive-cubic-quartic functional equation in multi-Banach spaces. Abstr. Appl. Anal. **2011**, 11 (2011)

Stabilities of MIQD and MIQA Functional Equations via Fixed Point Technique

B. V. Senthil Kumar, S. Sabarinathan, and M. J. Rassias

Abstract In this chapter, we investigate the stabilities of multiplicative inverse quadratic difference and multiplicative inverse quadratic adjoint functional equations in the setting of non-Archimedean fields via fixed point method.

1 Introduction and Preliminaries

The question posed by Ulam [15] in 1940 is the basis for the theory of stability of functional equations. The question raised by Ulam was answered by Hyers [5] which made a cornerstone in the study of stability of functional equations. The result obtained by Hyers is termed as Hyers–Ulam stability or ϵ-stability of functional equation. Then, Hyers' result was generalized by Aoki [1]. Also, Hyers' result was modified by Rassias [10] considering the upper bound as sum of powers of norms (Hyers–Ulam–T. Rassias stability). Rassias [11] established Hyers' result by taking the upper bound as product of powers of norms (Ulam–Gavruta–J. Rassias stability). In 1994, the stability result was further generalized into simple form by Gavruta [4] by replacing the upper bound by a general control function (generalized Hyers–Ulam stability).

In recent times, Ravi and Suresh [12] have investigated the generalized Hyers–Ulam stability of multiplicative inverse quadratic functional equation in two variables of the form

B. V. Senthil Kumar (✉)
Section of Mathematics, Department of Information Technology, Nizwa College of Technology, Nizwa, Oman
e-mail: senthilkumar@nct.edu.om

S. Sabarinathan
Department of Mathematics, SRM Institute of Science & Technology, Kattankulathur, Tamil Nadu, India

M. J. Rassias
Department of Statistical Science, University College London, London, UK
e-mail: matina@stats.ucl.ac.uk

© Springer Nature Switzerland AG 2019
G. A. Anastassiou and J. M. Rassias (eds.), *Frontiers in Functional Equations and Analytic Inequalities*, https://doi.org/10.1007/978-3-030-28950-8_8

$$R(u + v) = \frac{R(u)R(v)}{R(u) + R(v) + 2\sqrt{R(u)R(v)}} \tag{1}$$

in the setting of real numbers. It is easy to verify that the multiplicative inverse quadratic function $R(u) = \frac{1}{u^2}$ is a solution of Eq. (1). For further stability results using fixed point method concerning different types of functional equations and rational functional equations, one may refer [2, 6, 7, 9, 13, 14].

Here, we evoke a few fundamental notions of non-Archimedean field and fixed point alternative theorem in non-Archimedean spaces. Throughout this chapter, let us assume that N and R are the sets of natural numbers and real numbers, respectively.

Definition 1 Let F be a field with a mapping (valuation) $| \cdot |$ from F into $[0, \infty)$. Then F is said to be a non-Archimedean field if the upcoming requirements persist:

(i) $|k| = 0$ if and only if $k = 0$;
(ii) $|k_1 k_2| = |k_1||k_2|$;
(iii) $|k_1 + k_2| \leq max\{|k_1|, |k_2|\}$

for all $k, k_1, k_2 \in F$.

It is evident that $|1| = |-1| = 1$ and $|k| \leq 1$ for all $k \in N$. Furthermore, we presume that $| \cdot |$ is non-trivial, that is, there exists an $\alpha_0 \in F$ such that $|\alpha_0| \neq 0, 1$.

Suppose E is a vector space over a scalar field F with a non-Archimedean non-trivial valuation $| \cdot |$. A function $|| \cdot || : E \longrightarrow R$ is a *non-Archimedean norm* (*valuation*) if it satisfies the ensuing requirements:

(i) $||u|| = 0$ if and only if $u = 0$;
(ii) $||\rho u|| = |\rho|||u||$ $(\rho \in F, u \in E)$;
(iii) the strong triangle inequality (ultrametric); namely,

$$||u + v|| \leq max\{||u||, ||v||\} \quad (u, v \in E).$$

Then $(E, || \cdot ||)$ is known as a non-Archimedean space. By virtue of the inequality

$$||u_\ell - u_k|| \leq max\left\{||u_{j+1} - u_j|| : k \leq j \leq \ell - 1\right\} \quad (\ell > k),$$

a sequence $\{u_k\}$ is Cauchy if and only if $\{u_{k+1} - u_k\}$ converges to 0 in a non-Archimedean space. If every Cauchy sequence is convergent in the space, then it is called as complete non-Archimedean space.

Definition 2 Assume H is a nonempty set. Suppose $d : H \times H \to [0, \infty]$ satisfies the following properties:

(i) $d(\alpha, \beta) = 0$ if and only if $\alpha = \beta$;
(ii) $d(\alpha, \beta) = d(\beta, \alpha)$ (symmetry);
(iii) $d(\alpha, \gamma) \leq max\{d(\alpha, \beta), d(\beta, \gamma)\}$ (strong triangle inequality)

for all $\alpha, \beta, \gamma \in H$. Then (H, d) is called a generalized non-Archimedean metric space. Suppose every Cauchy sequence in A is convergent, then (A, d) is called complete.

Example 1 Let F be a non-Archimedean field. Assume A and B are two non-Archimedean spaces over F. If B is complete and $\phi : A \longrightarrow [0, \infty)$, for every $s, t : A \longrightarrow B$, define

$$d(s, t) = \inf\{\delta > 0 : |s(u) - t(u)| \le \delta\phi(u), \text{ for all } u \in A\}.$$

Using Theorem 2.5 [3], Mirmostafaee [8] proposed new version of the alternative fixed point principle in the setting of non-Archimedean space as follows:

Theorem 1 ([8] (Alternative Fixed Point Principle in Non-Archimedean Scheme)) *Suppose (H, d) is a non-Archimedean generalized metric space. Let a mapping $\Lambda : H \longrightarrow H$ be a strictly contractive, (that is, $d(\Lambda(u), \Lambda(v)) \le \rho d(v, u)$, for all $u, v \in H$ and a Lipschitz constant $\rho < 1$), then either*

(i) $d\left(\Lambda^p(u), \Lambda^{p+1}u\right) = \infty$ *for all $p \ge 0$, or;*

(ii) there exists some $p_0 \ge 0$ such that $d\left(\Lambda^p(u), \Lambda^{p+1}(u)\right) < \infty$ for all $p \ge p_0$;

the sequence $\{\Lambda^p(u)\}$ is convergent to a fixed point $u^\star$ of Λ; $u^\star$ is the distinct invariant point of Λ in the set $Y = \{y \in X : d(\Lambda^{p_0}(u), v) < \infty\}$ and $d(v, u^\star) \le d(v, \Lambda(v))$ for all v in this set.

In this chapter, we consider the following functional equations

$$R_q\left(\frac{u+v}{2}\right) - R_q(u+v) = \frac{3R_q(u)R_q(v)}{R_q(u) + R_q(v) + 2\sqrt{R_q(u)R_q(v)}} \tag{2}$$

and

$$R_q\left(\frac{u+v}{2}\right) + R_q(u+v) = \frac{5R_q(u)R_q(v)}{R_q(u) +_q (v) + 2\sqrt{R_q(u)R_q(v)}}. \tag{3}$$

Clearly, the multiplicative inverse quadratic function $R_q(u) = \frac{1}{u^2}$ satisfies Eqs. (2) and (3). Hence, Eqs. (2) and (3) are called as Multiplicative Inverse Quadratic Difference (MIQD) functional equation and Multiplicative Inverse Quadratic Adjoint (MIQA) functional equation, respectively. We prove the stabilities of the above Eqs. (2) and (3) in non-Archimedean fields by fixed point approach.

Let us presume that E and F are a non-Archimedean field and a complete non-Archimedean field, respectively, in this chapter. In the sequel, we represent $E^* = E\backslash\{0\}$, where E is a non-Archimedean field. For the sake of easy computation, we describe the difference operators $\Delta_1 R_q, \Delta_2 R_q : E^* \times E^* \longrightarrow F$ by

$$\Delta_1 R_q(u, v) = R_q\left(\frac{u+v}{2}\right) - R_q(u+v) - \frac{3R_q(u)R_q(v)}{R_q(u) + R_q(v) + 2\sqrt{R_q(u)R_q(v)}}$$

and

$$\Delta_2 R_q(u, v) = R_q\left(\frac{u+v}{2}\right) + R_q(u+v) - \frac{5R_q(u)R_q(v)}{R_q(u) + R_q(v) + 2\sqrt{R_q(u)R_q(v)}}$$

for all $u, v \in E^*$.

2 Solution of Eqs. (2) and (3)

In this section, we attain the solution of functional equations (2) and (3). In the following, we denote $R\backslash\{0\}$ by R^*.

Theorem 2 *A mapping $R_q : R^* \longrightarrow R$ satisfies Eq. (1) if and only if $R_q : R^* \longrightarrow R$ satisfies Eq. (2) if and only if $R_q : R^* \longrightarrow R$ satisfies Eq. (3). Therefore, every solution of Eqs. (2) and (3) is also a multiplicative inverse quadratic mapping.*

Proof Let $R_q : R^* \longrightarrow R$ satisfy Eq. (1). Switching v into u in (1), we obtain

$$R_q(2u) = \frac{1}{4}R_q(u) \tag{4}$$

for all $u \in R^*$. Now, letting u to $\frac{u}{2}$ in (4), one finds

$$R_q\left(\frac{u}{2}\right) = 4R_q(u) \tag{5}$$

for all $u \in R^*$. Again, substituting (u, v) by $(\frac{u}{2}, \frac{v}{2})$ in (1) and applying (5), we obtain

$$R_q\left(\frac{u+v}{2}\right) = \frac{4R_q(u)R_q(v)}{R_q(u) + R_q(v) + 2\sqrt{R_q(u)R_q(v)}} \tag{6}$$

for all $u, v \in R^*$. Subtracting (1) from (6), we arrive at (3).

Now, suppose $R_q : R^* \longrightarrow R$ satisfies Eq. (3). Plugging v by u in (3), we obtain

$$R_q(2u) = \frac{1}{4}R_Q(u) \tag{7}$$

for all $u \in R^*$. Now, replacing u by $\frac{u}{2}$ in (7), we get

$$R_q\left(\frac{u}{2}\right) = 4R_q(u) \tag{8}$$

for all $u \in R^*$. Using (8) in (3), we obtain

$$R_q(u+v) = \frac{R_q(u)R_q(v)}{R_q(u) + R_q(v) + \sqrt{R_q(u)R_q(v)}} \tag{9}$$

for all $u, v \in R^*$. Now, summing (9) with (3), we lead to (3).

Lastly, suppose $R_q : R^* \longrightarrow R$ satisfies Eq. (3). Letting $v = u$ in (3), we obtain

$$R_q(2u) = \frac{1}{4}R_q(u) \tag{10}$$

for all $u \in R^*$. Replacing u by $\frac{u}{2}$ in (10), we obtain

$$R_q\left(\frac{u}{2}\right) = 4R_q(u) \tag{11}$$

for all $u \in R^*$. In lieu of (11) and (3), we arrive at (1), which completes the proof.

3 Stabilities of Eqs. (2) and (3)

In this section, we investigate stabilities of Eqs. (2) and (3) via fixed point method in non-Archimedean fields.

Theorem 3 *Assume a mapping $R_q : E^* \longrightarrow F$ satisfies the inequality*

$$\left|\Delta_1 R_q(u, v)\right| \le \varphi(u, v) \tag{12}$$

for all $u, v \in E^*$, where $\varphi : E^* \times E^* \longrightarrow F$ is a given function. If $0 < L < 1$,

$$|2|^{-2}\varphi\left(2^{-1}u, 2^{-1}v\right) \le L\varphi(u, v) \tag{13}$$

for all $u, v \in E^*$, then there exists a unique multiplicative inverse quadratic mapping $r_d : E^* \longrightarrow F$ satisfying Eq. (2) and

$$|R_q(u) - r_d(u)| \le L|2|^2\varphi(u, u) \tag{14}$$

for all $u \in E^*$.

Proof Plugging (u, v) by $\left(\frac{u}{2}, \frac{u}{2}\right)$ in (12), we obtain

$$\left|R_q(u) - 2^{-2}R_q\left(2^{-1}u\right)\right| \le \varphi\left(2^{-1}u, 2^{-1}u\right) \tag{15}$$

for all $u \in E^*$. Let $\mathcal{A} = \{p | p : E^* \longrightarrow F\}$, and define

$$d(p, q) = \inf\{\gamma > 0 : |p(u) - q(u)| \le \gamma\varphi(u, u), \text{ for all } u \in E^*\}.$$

In lieu of Example 1, we find that d turns into a complete generalized non-Archimedean complete metric on $\mathcal{A}$. Let $\Gamma : \mathcal{A} \longrightarrow \mathcal{A}$ be a mapping defined by

$$\Gamma(p)(u) = 2^{-2} p \left(2^{-1} u \right)$$

for all $u \in E^*$ and $p \in \mathcal{A}$. Then Γ is strictly contractive on $\mathcal{A}$, in fact if $|p(u) - q(u)| \le \gamma \varphi(u, u), \ (u \in E^*)$, then by (13), we obtain

$$|\Gamma(p)(u) - \Gamma(q)(u)| = |2|^{-2} \left| p \left(2^{-1} u \right) - q \left(2^{-1} u \right) \right|$$

$$\le \gamma |2|^{-2} \varphi \left(2^{-1} u, 2^{-1} u \right)$$

$$\le \gamma L \varphi(u, u) \quad (u \in E^*).$$

From the above, we conclude that

$$(\Gamma(p), \Gamma(q)) \le L d(p, q) \ (p, q \in \mathcal{A}).$$

Consequently, the mapping d is strictly contractive with Lipschitz constant L. Using (15), we have

$$|\rho(u)(s) - u(s)| = \left| 3^{-11} u \left(3^{-1} s \right) - u(s) \right|$$

$$\le \zeta \left(\frac{s}{3}, \frac{s}{3} \right) \le |3|^{11} L \zeta(s, s) \quad (s \in G^*).$$

This indicates that $d(\Gamma(R_q), R_q) \le L|2|^2$. Due to Theorem 1 (ii), Γ has a distinct invariant point $r_d : E^* \longrightarrow F$ in the set $G = \{g \in F : d(u, g) < \infty\}$ and for each $u \in E^*$, $r_d(u) = \lim_{s \to \infty} \Gamma^s R_q(u) = \lim_{s \to \infty} 2^{-2s} R_q \left(2^{-s} u \right) \ (u \in E^*)$. Therefore, for all $u, v \in E^*$,

$$|\Delta_1 r_d(u, v)| = \lim_{s \to \infty} |2|^{-2s} \left| \Delta_1 R_q \left(2^{-s} u, 2^{-s} v \right) \right|$$

$$\le \lim_{s \to \infty} |2|^{-2s} \varphi \left(2^{-s} u, 2^{-s} v \right)$$

$$\le \lim_{s \to \infty} L^s \varphi(u, v) = 0$$

which shows that r_d is multiplicative inverse quadratic. Theorem 1 (ii) implies $d(R_q, r_d(u)) \le d(\Gamma(R_q), R_q)$, that is, $|R_q(u) - r_d(u)| \le |2|^2 L \varphi(u, u)$ $(u \in E^*)$. Let $r'_d : E^* \longrightarrow F$ be a multiplicative inverse quadratic mapping which satisfies (14), then r'_d is a fixed point of Γ in $\mathcal{A}$. However, by Theorem 1, Γ has only one invariant in G. This completes the distinctiveness allegation of the theorem.

The following theorem is dual of Theorem 3. We skip the proof as it is analogous to Theorem 3.

Theorem 4 *Suppose the mapping $R_q : E^* \longrightarrow F$ satisfies the inequality (13). If $0 < L < 1$,*

$$|2|^2 \varphi(2u, 2v) \leq L\varphi(u, v),$$

for all $u, v \in E^*$, then there exists a unique multiplicative inverse quadratic mapping $r_d : E^* \longrightarrow F$ satisfying Eq. (2) and

$$|R_q(u) - r_d(u)| \leq L\varphi\left(\frac{u}{2}, \frac{u}{2}\right),$$

for all $u \in E^*$.

The following corollaries follow directly from Theorems 3 and 4. In the following corollaries, we assume that $|2| < 1$ for a non-Archimedean field E.

Corollary 1 *Let ϵ(independent of u, v)≥ 0 be a constant exists for a mapping $R_q : E^* \longrightarrow F$ such that the functional inequality satisfies*

$$\left|\Delta_1 R_q(u, v)\right| \leq \epsilon,$$

for all $u, v \in E^*$. Then there exists a unique multiplicative inverse quadratic mapping $r_d : E^* \longrightarrow F$ satisfying Eq. (2) and

$$|R_q(u) - r_d(u)| \leq \epsilon,$$

for all $u \in E^*$.

Proof Assuming $\varphi(u, v) = \epsilon$ and selecting $L = |2|^{-2}$ in Theorem 3, we get the desired result.

Corollary 2 *Let $\lambda \neq -2$ and $c_1 \geq 0$ be real numbers exists for a mapping $R_q : E^* \longrightarrow F$ such that the following inequality holds*

$$\left|\Delta_1 R_q(u, v)\right| \leq c_1 \left(|u|^\lambda + |v|^\lambda\right),$$

for all $u, v \in E^*$. Then there exists a unique multiplicative inverse quadratic mapping $r_d : E^* \longrightarrow F$ satisfying Eq. (2) and

$$|R_q(u) - r_d(u)| \leq \begin{cases} \frac{|2|c_1}{|2|^\lambda}|u|^\lambda, & \lambda > -2 \\ |2|^3 c_1 |u|^\lambda, & \lambda < -2 \end{cases}$$

for all $u \in E^*$.

Proof Consider $\varphi(u, v) = c_1 \left(|u|^\lambda + |v|^\lambda \right)$ in Theorems 3 and 4 and then assume $L = |2|^{-\lambda-2}, \lambda > -2$ and $L = |2|^{\lambda+2}, \lambda < -2$, respectively, the proof follows directly.

Corollary 3 *Let $c_2 \geq 0$ and $\lambda \neq -2$ be real numbers, and $R_q : E^* \longrightarrow F$ be a mapping satisfying the functional inequality*

$$\left| \Delta_1 R_q(u, v) \right| \leq c_2 |u|^{\lambda/2} |v|^{\lambda/2},$$

for all $u, v \in E^*$. Then there exists a unique multiplicative inverse quadratic mapping $r_d : E^* \longrightarrow F$ satisfying Eq. (2) and

$$|R_q(u) - r_d(u)| \leq \begin{cases} \frac{c_2}{|2|^\lambda} |u|^\lambda, & \lambda > -2 \\ |2|^2 c_2 |u|^\lambda, & \lambda < -2 \end{cases}$$

for all $u \in E^*$.

Proof It is easy to prove this corollary, by taking $\varphi(u, v) = c_2 |u|^{\lambda/2} |v|^{\lambda/2}$ and then choosing $L = |2|^{-\lambda-2}, \lambda > -2$ and $L = |2|^{\lambda+2}, \lambda < -2$, respectively in Theorems 3 and 4.

In the sequel, using fixed point technique, we investigate the stabilities of Eq. (3) in the framework of non-Archimedean fields. Since the proof of the subsequent results is akin to the results of Eq. (2), for the sake of completeness, we state only theorems and skip their proofs.

Theorem 5 *Let $R_q : E^* \longrightarrow F$ be a mapping satisfying the inequality*

$$\left| \Delta_2 R_q(u, v) \right| \leq \xi(u, v) \tag{16}$$

for all $u, v \in E^*$, where $\xi : E^* \times E^* \longrightarrow [0, \infty)$ is an arbitrary function. If $0 < L < 1$,

$$|2|^{-2} \xi \left(2^{-1} u, 2^{-1} v \right) \leq L \xi(u, v),$$

for every $u, v \in E^*$, then there exists a unique multiplicative inverse quadratic mapping $r_a : E^* \longrightarrow F$ gratifying Eq. (3) and

$$|R_q(u) - r_a(u)| \leq L |2|^2 \xi(u, u),$$

for each $u \in E^*$.

Theorem 6 *Let $R_q : E^* \longrightarrow F$ be a mapping satisfying the inequality (16). If $0 < L < 1$,*

$$|2|^2 \xi(2u, 2v) \leq L \xi(u, v),$$

for every $u, v \in E^*$, *then there exists a unique multiplicative inverse quadratic mapping* $r_a : E^* \longrightarrow F$ *satisfying Eq. (3) and*

$$|R_q(u) - r_a(u)| \leq L\xi \left(\frac{u}{2}, \frac{u}{2}\right)$$

for each $u \in E^*$.

Corollary 4 *Let* θ *(independent of* u, v)≥ 0 *be a constant. Suppose a mapping* $R_q : E^* \longrightarrow F$ *satisfies the inequality*

$$\left|\Delta_2 R_q(u, v)\right| \leq \theta$$

for every $u, v \in E^*$. *Then there exists a unique multiplicative inverse quadratic mapping* $r_a : E^* \longrightarrow F$ *satisfying Eq. (3) and*

$$|R_q(u) - r_a(u)| \leq \theta,$$

for each $u \in E^*$.

Corollary 5 *Let* $\alpha \neq -2$ *and* $\delta_1 \geq 0$ *be real numbers. If* $R_q : E^* \longrightarrow F$ *is a mapping satisfying the inequality*

$$\left|\Delta_2 R_q(u, v)\right| \leq \delta_1 \left(|u|^\alpha + |v|^\alpha\right),$$

for every $u, v \in E^*$, *then there exists a unique multiplicative inverse quadratic mapping* $r_a : E^* \longrightarrow F$ *satisfying Eq. (3) and*

$$|R_q(u) - r_a(u)| \leq \begin{cases} \frac{|2|\delta_1}{|2|^\alpha}|u|^\alpha, & \alpha > -2 \\ |2|^3\delta_1|u|^\alpha, & \alpha < -2 \end{cases}$$

for each $u \in E^*$.

Corollary 6 *Let* $R_q : E^* \longrightarrow F$ *be a mapping and* $\delta_2 \geq 0$ *and* $\alpha \neq -2$ *be real numbers. If the mapping* R_q *satisfies the functional inequality*

$$\left|\Delta_2 R_q(u, v)\right| \leq \delta_2|u|^{\alpha/2}|v|^{\alpha/2}$$

for every $u, v \in E^*$, *then there exists a unique multiplicative inverse quadratic mapping* $r_a : E^* \longrightarrow F$ *satisfying Eq. (3) and*

$$|R_q(u) - r_a(u)| \leq \begin{cases} \frac{\delta_2}{|2|^\alpha}|u|^\alpha, & \alpha > -2 \\ |2|^2\delta_2|u|^\alpha, & \alpha < -2 \end{cases}$$

for each $u \in E^*$.

References

1. T. Aoki, On the stability of the linear transformation in Banach spaces. J. Math. Soc. Jpn. **2**, 64–66 (1950)
2. L. Cădariu, V. Radu, Fixed points and the stability of Jensen's functional equation. J. Inequ. Pure Appl. Math. **4**(1), Art. 4 (2003)
3. L. Cădariu, V. Radu, On the stability of the Cauchy functional equation: a fixed point approach. Grazer Math. Ber. **346**, 43–52 (2006)
4. P. Găvruta, A generalization of the Hyers-Ulam-Rassias stability of approximately additive mappings. J. Math. Anal. Appl. **184**, 431–436 (1994)
5. D.H. Hyers, On the stability of the linear functional equation. Proc. Natl. Acad. Sci. USA **27**, 222–224 (1941)
6. S.M. Jung, A fixed point approach to the stability of the equation $f(x + y) = \frac{f(x)f(y)}{f(x)+f(y)}$. Aust. J. Math. Anal. Appl. **6**(1), Art. 8, 1–6 (1998)
7. B. Margolis, J. Diaz, A fixed point theorem of the alternative for contractions on a generalized complete metric space. Bull. Am. Math. Soc. **74**, 305–309 (1968)
8. A.K. Mirmostafaee, Non-Archimedean stability of quadratic equations. Fixed Point Theory **11**(1), 67–75 (2010)
9. V. Radu, The fixed point alternative and the stability of functional equations. Fixed Point Theory **4**(1), 91–96 (2003)
10. T.M. Rassias, On the stability of the linear mapping in Banach spaces. Proc. Am. Math. Soc. **72**, 297–300 (1978)
11. J.M. Rassias, On approximation of approximately linear mappings by linear mappings. J. Funct. Anal. **46**, 126–130 (1982)
12. K. Ravi, S. Suresh, Solution and generalized Hyers–Ulam stability of a reciprocal quadratic functional equation. Int. J. Pure Appl. Math. **117**(2), Art. No. AP2017-31-4927 (2017)
13. K. Ravi, J.M. Rassias, B.V. Senthil Kumar, A fixed point approach to the generalized Hyers–Ulam stability of reciprocal difference and adjoint functional equations. Thai J. Math. **8**(3), 469–481 (2010)
14. B.V. Senthil Kumar, A. Bodaghi, Approximation of Jensen type reciprocal functional equation using fixed point technique. Boletim da Sociedade Paranaense de Mat. **38**(3) (2018). https://doi.org/10.5269/bspm.v38i3.36992
15. S.M. Ulam, *Problems in Modern Mathematics*. Chapter VI (Wiley-Interscience, New York, 1964)

Hyers–Ulam Stability of First Order Differential Equation via Integral Inequality

S. Tamilvanan, E. Thandapani, and J. M. Rassias

Abstract In this chapter, first we derive a nonlinear integral inequality of Gollwitzer type, and as an application we investigate the Hyers–Ulam stability of nonlinear differential equation

$$y'(t) = f(t, y(t)), \ t \geq a,$$

where f is a given function. The obtained results are new to the literature.

2010 Mathematics Subject Classification 34K10

1 Introduction

The Hyers–Ulam stability of functional equations received great attention in the last few years, see, for example, [9–11] and the references cited therein. C. Alsina and R. Ger [1] were the first authors who investigated the Hyers–Ulam stability of a differential equation. In fact, they proved that if a differentiable function $y : I \to \mathbb{R}$ satisfies $|y'(t) - y(t)| \leq \epsilon$ for all $t \in I$, then there exists a differentiable function

S. Tamilvanan
Department of Mathematics, SRM Institute of Science and Technology, Kattankulathur, Tamil Nadu, India
e-mail: tamilvanan.s@ktr.srmuniv.ac.in

E. Thandapani (✉)
Ramanujan Institute for Advanced Study in Mathematics, University of Madras, Chennai, India

J. M. Rassias
Pedagogical Department E.E., Section of Mathematics and Informatics,
National and Kapodistrian University of Athens, Athens, Greece
e-mail: jrassias@primedu.uoa.gr

© Springer Nature Switzerland AG 2019
G. A. Anastassiou and J. M. Rassias (eds.), *Frontiers in Functional Equations and Analytic Inequalities*, https://doi.org/10.1007/978-3-030-28950-8_9

$g : I \to \mathbb{R}$ satisfying $g'(t) = g(t)$ for any $t \in I$ such that $|y(t) - g(t)| \leq 3\epsilon$ for all $t \in I$.

The above result has been generalized and extended in different directions by many researchers, see, for example, [4–7, 12] and the references therein. In all these papers, the authors discussed the Hyers–Ulam stability of first or second order linear differential equations and it seems that no such result is available in the literature dealing with Hyers–Ulam stability for the nonlinear differential equations. This observation motivated us to study the Hyers–Ulam stability of the following first order nonlinear differential equation

$$y'(t) = f(t, y(t)), \ t \in I = [a, b] \tag{1}$$

where $y \in C'(I)$, $f \in C(I, \mathbb{R})$ and $-\infty < a, b < \infty$.

Next, we give the definition of Hyers–Ulam stability for differential equations.

Definition 1 We say that Eq. (1) has the Hyers–Ulam stability if the following property holds: for every $\epsilon > 0$, $y \in C'(I)$, if

$$|y'(t) - f(t, y(t))| \leq \epsilon$$

then there exists some $u \in C'(I)$ satisfying

$$u'(t) = f(t, u(t))$$

such that $|y(t) - u(t)| \leq K(\epsilon)$ for every $t \in I$, where $K(\epsilon)$ is an expression of ϵ only.

If the above statement is also true when we replace ϵ and $K(\epsilon)$ by $\phi(t)$ and $\Phi(t)$, where $\phi, \Phi : I \to [0, \infty)$ are functions not depending on y and u explicitly, then we say that the corresponding differential equation has the Hyers–Ulam–Rassias stability (or the generalized Hyers–Ulam stability). For more detailed definitions of the Hyers–Ulam stability and Hyers–Ulam–Rassias stability, one can refer to [9–11]. In this chapter, first we present nonlinear integral inequality of Gollwitzer type, and as an application we discuss the Hyers–Ulam stability of nonlinear differential equation (1).

2 Gollwitzer Type Integral Inequality

First we derive a nonlinear Gollwitzer type integral inequality that provides us a powerful tool for proving the Hyers–Ulam stability of a nonlinear first order differential equations. We begin with the following result which can be found in [3, Theorem 41, p.39].

Lemma 1 If $x > 0$ and $0 < \alpha \leq 1$, then

$$x^{\alpha} \leq \alpha x + (1 - \alpha), \tag{2}$$

and the equality holds if $\alpha = 1$.

Theorem 1 *Let u, f, g and h be nonnegative real-valued continuous functions defined on I, and*

$$u(t) \leq f(t) + g(t) \int_a^t h(s) u^\alpha(s)\, ds, \quad t \in I \tag{3}$$

where $0 < \alpha \leq 1$. Then

$$u(t) \leq f(t) + g(t) \int_a^t h(s)(\alpha f(s) + (1 - \alpha)) \cdot \exp\left(\int_s^t \alpha h(\sigma) g(\sigma)\, d\sigma\right) ds. \tag{4}$$

Proof Define a function $R(t)$ by

$$R(t) = \int_a^t h(s) u^\alpha(s)\, ds, \quad t \in I$$

then $R(a) = 0$, and $u(t) \leq f(t) + g(t) R(t)$. Then using Lemma 1, we obtain

$$R'(t) = h(t) u^\alpha(t) \leq h(t)(f(t) + g(t) R(t))^\alpha$$
$$\leq (\alpha h(t) f(t) + (1 - \alpha)h(t)) + \alpha h(t) g(t) R(t). \tag{5}$$

Multiplying (5) by $\exp\left(-\int_a^t \alpha h(\sigma) g(\sigma)\, d\sigma\right)$ we have

$$\frac{d}{dt}\left[R(t) \exp\left(-\int_a^t \alpha h(\sigma) g(\sigma)\, d\sigma\right)\right]$$
$$\leq (\alpha h(t) f(t) + (1 - \alpha)h(t)) \exp\left(-\int_a^t \alpha h(\sigma) g(\sigma)\, d\sigma\right). \tag{6}$$

By setting $t = s$ in the last inequality and integrating it with respect to s from a to t, we obtain

$$R(t) \exp\left(-\int_a^t \alpha h(\sigma) g(\sigma)\, d\sigma\right)$$
$$\leq \int_a^t (\alpha h(s) f(s) + (1 - \alpha)h(s)) \exp\left(-\int_a^s \alpha h(\sigma) g(\sigma)\, d\sigma\right) ds. \tag{7}$$

Using the bound on $R(t)$ from (7) in $u(t) \leq f(t) + g(t) R(t)$, we have the desired inequality (4). This completes the proof.

Corollary 1 *Let $u(t)$ and $p(t)$ be real-valued nonnegative continuous functions defined on I such that*

$$u(t) \leq u_0 + \int_a^t p(s)u^\alpha(s)\, ds, \quad t \in I \tag{8}$$

where $u_0 \geq 0$ and $0 < \alpha \leq 1$. Then

$$u(t) \leq A(t) \exp\left(\alpha \int_a^t p(s)\, ds\right) \tag{9}$$

where $A(t) = u_0 + (1-\alpha) \int_a^t p(s)\, ds$.

Proof The proof follows from Theorem 1 by taking $f(t) = u_0 \geq 0$, $g(t) \equiv 1$ and $h(t) = p(t)$ and the details are omitted.

Remark 1 If we take $\alpha = 1$, then Theorem 1 reduced to the well-known Gollwitzer inequality [2]. However for $0 < \alpha < 1$, the result obtained in Theorem 1 is new to the literature.

Remark 2 If we take $\alpha = 1$, then Corollary 1 reduced to the well-known Gronwall inequality [8].

3 Hyers–Ulam Stability

As an application of the integral inequality established in Sect. 2 we investigate the Hyers–Ulam stability of Eq. (1).

Theorem 2 *Assume that $p(t) \in C(I)$ such that*

$$|f(t, u) - f(t, v)| \leq p(t)|u - v|^\alpha \tag{10}$$

where $0 < \alpha \leq 1$, and

$$\int_a^\infty p(t)\, dt < \infty. \tag{11}$$

If for $\phi(t) \in C(I)$ such that $\int_a^\infty \phi(t)\, dt < \infty$, and

$$|y' - f(t, y)| \leq \phi(t) \tag{12}$$

then there exists a $u \in C'(I)$ and $K > 0$ satisfying

$$u' = f(t, u) \tag{13}$$

such that $|y(t) - u(t)| \leq K$; that is, Eq. (1) has the Hyers–Ulam stability.

Proof From the inequality (12), we have

$$y(t) \le y(a) + \int_a^t f(s, y(s))\, ds + \int_a^t \phi(s)\, ds, \tag{14}$$

and from Eq. (13) we obtain

$$u(t) = u(a) + \int_a^t f(s, u(s))\, ds. \tag{15}$$

Combining (14) and (15) yields

$$|y(t) - u(t)| \le |y(a) - u(a)| + \int_a^t |f(s, y(s)) - f(s, u(s))|\, ds + \int_a^t \phi(s)\, ds.$$

Using (10) in the last inequality, one obtains

$$|y(t) - u(t)| \le M_1 + \int_a^t p(s)|y(s) - u(s)|^\alpha\, ds + M_2 \tag{16}$$

where $M_1 = |y(a) - u(a)|$ and $\int_a^\infty \phi(t)\, dt \le M_2$, by hypothesis.
Applying Corollary 1 in (16), we obtain

$$|y(t) - u(t)| \le \left[M_1 + M_2 + (1 - \alpha) \int_a^t p(s)\, ds \right] \exp\left(\alpha \int_a^t p(s)\, ds \right). \tag{17}$$

It follows from (11) that there is a constant $M_3 > 0$ such that $\int_a^\infty p(t)\, dt \le M_3$, and using this in (17), we have

$$|y(t) - u(t)| \le K$$

where $K = [M_1 + M_2 + (1 - \alpha)M_3] \exp(\alpha M_3)$. This completes the proof.

4 Conclusion

In this chapter, first we have obtained a new nonlinear integral inequality and as an application we present a Hyers–Ulam stability of a nonlinear differential equation. In this approach, we don't need the explicit form of the solution of the studied equation, whereas in [1, 4–7, 12] the authors used an explicit form of the solutions to prove their results.

References

1. C. Alsina, A. Ger, On some inequalities and stability results related to the exponential function. J. Inequal. Appl. **2**, 373–380 (1998)
2. H.E. Gollwitzer, A note on functional inequality. Proc. Am. Math. Soc. **23**, 642–647 (1969)
3. G.H. Hardy, J.E. Littlewood, G. Polya, *Inequalities* (Cambridge University Press, Cambridge, 1934)
4. Y. Hi, Y. Shen, Hyers-Ulam stability of linear differential equations of second order. Appl. Math. Lett. **23**, 306–309 (2010)
5. S.M. Jung, Hyers-Ulam stability of linear differential equations of first order. Appl. Math. Lett. **17**,1135–1140 (2004)
6. S.M. Jung, Hyers-Ulam stability of linear differential equations of first order-III. J. Math. Appl. **311**, 139–146 (2005)
7. S.M. Jung, Hyers-Ulam stability of linear differential equations of first order-II. Appl. Math. Lett. **19**, 854–858 (2006)
8. B.G. Pachpatte, *Inequalities for Differential and Integral Equations* (Academic Press, New York, 1998)
9. J.M. Rassias, Solution of a problem of Ulam. J. Approx. Theory **57**, 268–273 (1989)
10. J.M. Rassias, E. Thandapani, K. Ravi, B.V. Senthilkumar, *Functional Equations and Inequalities: Solutions and Stability Results*. Series on Concrete and applicable mathematics (World Scientific Publishing Co. Pte. Ltd., Singapore, 2017)
11. K. Ravi, M. Arunkumar, J.M. Rassias, Ulam stability for the orthogonally general Euler-Lagrange type functional equation. Int. J. Math. Stat. **3**, 36–46 (2008)
12. G. Wang, M. Zhou, L. Sun, Hyers-Ulam stability of linear differential equations of first order. Appl. Math. Lett. **21**, 1024–1028 (2008)

Stability of an n-Dimensional Functional Equation in Banach Space and Fuzzy Normed Space

Sandra Pinelas, V. Govindan, and K. Tamilvanan

Abstract In this paper, the authors investigate the general solution of a new additive functional equation

$$f\left(\sum_{i=1}^{n} x_i\right) + \sum_{j=1; i\neq j}^{n} f\left(-x_j - x_i + \sum_{1\leq i<j<k\leq n} x_k\right) = \left(\frac{n^2 - 5n + 6}{2}\right) \sum_{i=1}^{n} f(x_i)$$

where n is a positive integer with $\mathbb{N} - \{1, 2, 3, 4\}$ and discuss its generalized Hyers–Ulam stability in Banach spaces and stability in fuzzy normed spaces using two different methods.

1 Introduction

In 1940, Ulam [26] raised the following question. Under what conditions does there exist an additive mapping near an approximately addition mapping? The case of approximately additive functions was solved by Hyers [11] under the assumption that for $\epsilon > 0$ and $f : E_1 \rightarrow E_2$ be such that $\|f(x + y) - f(x) - f(y)\| \leq \epsilon$ for all $x, y \in E_1$ then there exists a unique additive mapping $T : E_1 \rightarrow E_2$ such that $\|f(x) - T(x)\| \leq \epsilon$ for all $x \in E_1$.

In 1978, a generalized version of the theorem of Hyers for approximately linear mapping was given by Rassias [20]. He proved that for a mapping $f : E_1 \rightarrow E_2$ be

S. Pinelas (✉)
Departmento de Ciências Exatas e Engenharia, Academia Militar, Lisboa, Portugal

V. Govindan
Department of Mathematics, Sri Vidya Mandir Arts and Science College, Uthangarai, Tamil Nadu, India

K. Tamilvanan
Department of Mathematics, Government Arts and Science College (for Men), Krishnagiri, Tamil Nadu, India

© Springer Nature Switzerland AG 2019
G. A. Anastassiou and J. M. Rassias (eds.), *Frontiers in Functional Equations and Analytic Inequalities*, https://doi.org/10.1007/978-3-030-28950-8_10

such that $f(tx)$ is continuous in $t \in \mathbb{R}$ and for each fixed $x \in E_1$ assume that there exist constant $\epsilon > 0$ and $p \in [0, 1)$ with

$$\| f(x + y) - f(x) - f(y) \| \le \epsilon(\|x\|^p + \|y\|^p) \tag{1}$$

for all $x, y \in E_1$ then there exists a unique R-Linear mapping $T : E_1 \to E_2$ such that

$$\| f(x) - T(x) \| \le \frac{2\epsilon}{2 - 2^p} \|x\|^p \tag{2}$$

for all $x \in E_1$.

A number of mathematicians were attracted by the result of Rassias. The stability concept that was introduced and investigated by Rassias is called the Hyers–Ulam–Rassias stability.

During the last decades, the stability problems of several functional equations have been extensively investigated by a number of authors [1, 5, 8, 12, 23, 24].

In 1982–1989, Rassias [21, 22] replaced the sum appeared in the right-hand side of the Eq. (1) by the product of powers of norms. In modelling applied problems only partial information may be known (or) there may be a degree of uncertainty in the parameters used in the model or some measurements may be imprecise. Due to such features, we are tempted to consider the study of functional equations in the fuzzy setting.

For the last 40 years, fuzzy theory has become a very active area of research and a lot of development has been made in the theory of fuzzy sets to find the fuzzy analogues of the classical set theory. This branch finds a wide range of applications in the field of science and engineering.

Katsaras [13] introduced an idea of fuzzy norm on a linear space in 1984, in the same year Wu and Fang [27] introduced a notion of fuzzy normed space to give a generalization of the Kolmogoroff normalized theorem for fuzzy topological linear spaces. In 1991, Biswas [4] defined and studied fuzzy inner product spaces in linear space. In 1991, Felbin [7] introduced an alternative definition of a fuzzy norm on a linear topological structures of a fuzzy normed linear spaces. In 1994, Cheng and Mordeson [6] introduced a definition of fuzzy norm on a linear space in such a manner that the corresponding induced fuzzy metric is of Kramosil and Michalek [14]. In 2003, Bag and Samanta [2] modified the definition of Cheng and Mordeson [6] by removing a regular condition. Recently various results have been investigated by numerous authors, one can refer to [3, 9, 10, 15–19, 25].

Before we proceed to the main theorems, we will introduce some definitions and an example to illustrate the idea of fuzzy norm.

Definition 1 Let X be a real linear space. A function $N : X \times \mathbb{R} \longrightarrow [0, 1]$ is said to be fuzzy norm on X if for all $x, y \in X$ and $a, b \in \mathbb{R}$.

(N_1) $N(x, a) = 0$ for $a \le 0$;

(N_2) $x = 0$ iff $N(x, a) = 1$ for all $a > 0$;

(N_3) $N(ax, b) = N\left(x, \frac{b}{|a|}\right)$ if $a \neq 0;$

(N_4) $N(x + y, a + b) \geq min\{N(x, a), N(y, b)\};$

(N_5) $N(x, .)$ is a non-decreasing function on $\mathbb{R}$ and $lim_{a \longrightarrow \infty} N(x, a) = 1.$

(N_6) For $x \neq 0$, $N(x, .)$ is continuous on $\mathbb{R}$.

The pair (X, N) is called a fuzzy normed linear space. One may regard $N(x, a)$ as the truth value of the statement the norm of x is less than or equal to the real number a.

Definition 2 Let (X, N) be a fuzzy normed linear space. Let x_n be a sequence in X. Then x_n is said to be convergent if there exists $x \in X$ such that $lim_{n \to \infty} N(x_n - x, t) = 1$ for all $t > 0$. In that case, x is called the limit of the sequence x_n and we denote it by $N - lim_{n \to \infty} x_n = x.$

Definition 3 A sequence x_n in X is called Cauchy if for each $\epsilon > 0$ and each $t > 0$ there exists n_0 such that for all $n \geq n_0$ and all $p > 0$, we have $N\left(x_{n+p} - x_n, t\right) > 1 - \epsilon.$

Definition 4 Every convergent sequence in a fuzzy normed space is Cauchy. If each Cauchy sequence is convergent, then the fuzzy norm is said to be complete and the fuzzy normed space is called a fuzzy Banach space.

Definition 5 A mapping $f : X \to Y$ between fuzzy normed spaces X and Y is continuous at a point x_0 if for each sequence $\{x_n\}$ converging to x_0 in X, the sequence $f\{x_n\}$ converges to $f\{x_0\}$. If f is continuous at each point of $x_0 \in X$, then f is said to be continuous on X.

Example Let $(X, \|.\|)$ be a normed linear space. Then

$$N(x, a) = \begin{cases} \frac{a}{a + \|x\|}, & a > 0, \ x \in X \\ 0, & a \leq 0, \ x \in X \end{cases}$$

is a fuzzy norm on X.

In the following we will suppose that $N(x, .)$ is left continuous for every x. A fuzzy normed linear space is a pair(X, N), where X is a real linear space and N is a fuzzy norm on X. Let (X, N) be a fuzzy normed linear space. A sequence $\{x_n\}$ in X is said to be convergent if there exist $x \in X$ such that $lim_{n \to \infty} N(x_n - x, t) = 1$ $(t > 0)$. In that case, x is called the limit of the sequence $\{x_n\}$ and we write $N - lim_{n \to \infty} x_n = x$. A sequence $\{x_n\}$ in fuzzy normed space (X, N) is called cauchy if for each $\epsilon > 0$ and $\delta > 0$, there exist $n_0 \in N$ such that

$$N(x_m - x_n, \delta) > 1 - \epsilon, \quad (m, n \geq n_0).$$

If each cauchy sequence is convergent, then the fuzzy norm is said to be complete and the fuzzy normed space is called a fuzzy Banach space.

In this paper, the authors investigate the general solution and generalized Hyers–Ulam stability of a new type of n-dimensional functional equation of the form

$$f\left(\sum_{i=1}^{n} x_i\right) + \sum_{j=1; i\neq j}^{n} f\left(-x_j - x_i + \sum_{1\leq i<j<k\leq n} x_k\right) = \left(\frac{n^2 - 5n + 6}{2}\right)\sum_{i=1}^{n} f(x_i)$$

(3)

where n is a positive integer with $\mathbb{N} - \{1, 2, 3, 4\}$, in the setting of Banach space and fuzzy normed space using direct and fixed point methods.

Theorem 1 (Banach's Contraction Principle) *Let (X, d) be a complete metric space and consider a mapping $T : X \longrightarrow X$ which is strictly contractive mapping, that is*

(A1) $d(Tx, Ty) \leq Ld(x, y)$ for some (Lipschitz constant) $L < 1$, then

> *(i) The mapping T has one and only fixed point $x^* = T(x^*)$;*
> *(ii) The fixed point for each given element $x^*$is globally attractive that is*

(A2) $\lim_{n\longrightarrow\infty} T^n x = x^*$, for any starting point $x \in X$;

> *(iii) One has the following estimation inequalities:*

(A3) $(T^n x, x^*) \leq \frac{1}{1-L} d(T^n x, T^{n+1} x)$, for all $n \geq 0, x \in X$.
(A4) $(x, x^*) \leq \frac{1}{1-L} d(x, x^*), \forall x \in X$.

Theorem 2 (The Alternative of Fixed Point) *Suppose that for a complete generalized metric space (X, d) and a strictly contractive mapping $T : X \longrightarrow Y$ with Lipschitz constant L. Then, for each given element $x \in X$ either*

(B1) $(T^n x, T^{n+1} x) = +\infty$, for all $n \geq 0$, or
(B2)   there exists natural number n_0 such that:

> *(i) $d(T^n x, T^{n+1} x) < \infty$ for all $n \geq n_0$;*
> *(ii) The sequence $(T^n x)$ is convergent to a fixed point y^* of T;*
> *(iii) y^* is the unique fixed point of T in the set $Y = \{y \in X; d(T^{n_0} x, y) < \infty\}$;*
> *iv) $d(y^*, y) \leq \frac{1}{1-L} d(y, Ty)$ for all $y \in L$.*

2   General Solution of the Functional Equation (3)

In this section, we obtain the general solution of the functional equation (3). Throughout this section , let X and Y be real vector spaces.

Theorem 3 *A function $f : X \longrightarrow Y$ satisfies the functional equation (3) then $f : X \longrightarrow Y$ satisfies the functional equation (1).*

Proof Let $f : X \longrightarrow Y$ satisfy the functional equation (3). Replacing $(x_1, x_2, x_3, \ldots, x_n)$ by $(0, 0, \ldots, 0)$, $(x, 0, \ldots, 0)$ and $(x, x, 0 \ldots, 0)$ in (3) we obtain

$$f(0) = 0, \ f(-x) = -f(x) \ and \ f(2x) = 2f(x) \tag{4}$$

for all $x \in X$. It is easy to verify from (3) that

$$f\left(\frac{x}{2^i}\right) = \frac{1}{2^i} f(x), i = 1, 2, 3, \ldots, n \tag{5}$$

for all $x \in X$. Setting $(x_1, x_2, x_3, \ldots, x_n)$ by $(x, y, 0, \ldots, 0)$ in (3) and using oddness of f, we obtain the result of (1).

Define a mapping $f : X \to Y$ by

$$D_f(x_1, x_2, \ldots, x_n) = f(\sum_{i=1}^{n} x_i) + \sum_{j=1; i \neq j}^{n} f\left(-x_j - x_i + \sum_{1 \leq i < j < k \leq n} x_k\right)$$
$$- \left(\frac{n^2 - 5n + 6}{2}\right) \sum_{i=1}^{n} f(x_i)$$

for all $x_1, x_2, \ldots, x_n \in X$.

2.1 Stability Result for (3) in Banach Space Using Direct Method

In this section, we consider X to be a real vector space and Y to be a Banach space, we present the Hyers–Ulam stability of the functional equation (3).

Theorem 4 *Let $\psi : X^n \longrightarrow [0, \infty)$ be a function such that* $\sum_{k=0}^{\infty} \frac{\psi(2^{kj} x, 2^{kj} x, 0, \ldots, 0)}{2^{kj}}$ *converges in $\mathbb{R}$ and*

$$\lim_{k \to \infty} \frac{\psi(2^{kj} x_1, 2^{kj} x_2, \ldots, 2^{kj} x_n)}{2^{kj}} = 0 \tag{6}$$

for all $x_1, x_2, \ldots, x_n \in X$. If a function $f : X \longrightarrow Y$ satisfies

$$\|D_f(x_1, x_2, \ldots, x_n)\| \leq \psi(x_1, x_2, x_3, \ldots, x_n) \tag{7}$$

for all $x_1, x_2, \ldots, x_n \in X$, then there exists a unique additive function $A : X \longrightarrow Y$ which satisfies the functional equation (3) and

$$\|f(x) - A(x)\| \leq \frac{1}{(n^2 - 5n + 6)} \sum_{k=0}^{\infty} \frac{\psi(2^{kj} x, 2^{kj} x, 0, \ldots, 0)}{2^{kj}} \tag{8}$$

for all $x \in X$. The function A is given by

$$A(x) = \lim_{k \to \infty} \frac{f(2^{kj}x)}{2^{kj}} \tag{9}$$

for all $x \in X$.

Proof Setting $(x_1, x_2, x_3, \ldots, x_n)$ by $(x, x, 0, \ldots, 0)$ in (7), we obtain

$$\left\| \left(\frac{n^2 - 5n + 6}{2} \right) f(2x) - (n^2 - 5n + 6) f(x) \right\| \leq \psi(x, x, 0, \ldots, 0) \tag{10}$$

for all $x \in X$. It follows from (10) that

$$\left\| \frac{f(2x)}{2} - f(x) \right\| \leq \frac{\psi(x, x, 0, \ldots, 0)}{(n^2 - 5n + 6)} \tag{11}$$

for all $x \in X$. Setting x by $2x$ in (11), we obtain

$$\left\| \frac{f(2^2 x)}{2} - f(2x) \right\| \leq \frac{\psi(2x, 2x, 0, \ldots, 0)}{(n^2 - 5n + 6)} \tag{12}$$

for all $x \in X$. It follows from (12) we get

$$\left\| \frac{f(2^2 x)}{2^2} - \frac{f(2x)}{2} \right\| \leq \frac{\psi(2x, 2x, 0, \ldots, 0)}{2(n^2 - 5n + 6)} \tag{13}$$

for all $x \in X$. It follows from (11) and (13) that

$$\left\| \frac{f(2^2 x)}{2^2} - f(x) \right\| \leq \frac{1}{(n^2 - 5n + 6)} \left[\psi(x, x, 0, \ldots, 0) + \frac{\psi(2x, 2x, 0, \ldots, 0)}{2} \right] \tag{14}$$

for all $x \in X$. Generalizing, we get

$$\left\| \frac{f(2^n x)}{2^n} - f(x) \right\| \leq \frac{1}{(n^2 - 5n + 6)} \sum_{k=0}^{n-1} \frac{\psi(2^k x, 2^k x, 0, \ldots, 0)}{2^k}$$

$$\leq \frac{1}{(n^2 - 5n + 6)} \sum_{k=0}^{\infty} \frac{\psi(2^k x, 2^k x, 0, \ldots, 0)}{2^k} \tag{15}$$

for all $x \in X$. Now we have to prove that the sequence $\left\{ \frac{f(2^k x)}{2^k} \right\}$ is a cauchy sequence for all $x \in X$. For every positive integer n, m and for all $x \in X$, consider

$$\left\|\frac{f(2^{n+m}x)}{2^{n+m}} - \frac{f(2^n x)}{2^n}\right\| = \frac{1}{2^n}\left\|f(2^n x) - \frac{f(2^{n+m}x)}{2^m}\right\|$$

$$\leq \frac{1}{(n^2 - 5n + 6)}\sum_{i=0}^{m-1}\frac{\psi(2^{i+n}x, 2^{i+n}x, 0, \ldots, 0)}{2^{i+n}}$$

$$\leq \frac{1}{(n^2 - 5n + 6)}\sum_{i=0}^{\infty}\frac{\psi(2^{i+n}x, 2^{i+n}x, 0, \ldots, 0)}{2^{i+n}} \tag{16}$$

for all $x \in X$. By condition (6), the right-hand side approaches 0 as $n \to \infty$. Thus, the sequence is a cauchy sequence due to the completeness of the Banach space Y

$$A(x) = \lim_{k \to \infty}\frac{f(2^k x)}{2^k} \quad \forall x \in X,$$

is well-defined. We can see that (9) holds. To show that A satisfies (3), we set $(x, y) = (2^n x_1, 2^n x_2, \ldots, 2^n x_n)$ in (7) and divide the resulting equation by 2^n, we obtain

$$\frac{1}{2^k}\|D_f(2^k x_1, 2^k x_2, \ldots, 2^k x_n)\| \leq \frac{1}{2^k}\psi(2^k x_1, 2^k x_2, \ldots, 2^k x_n).$$

Taking the limit as $n \to \infty$, using (6) and (9), A satisfies (3). To prove the uniqueness of A, suppose that there exist another cubic function $B : X \to Y$ such that B satisfies (3) and (8), we have

$$\|A(x) - B(x)\| \leq \frac{1}{2^l}\|A(2^l x) - f(2^l x)\| + \|f(2^l x) - B(2^l x)\|$$

$$\leq \frac{2}{(n^2 - 5n + 6)}\sum_{k=0}^{\infty}\frac{\psi(2^{k+1}x, 2^{k+1}x, 0, \ldots, 0)}{2^{k+1}} \quad \forall x \in X.$$

By condition (6), the right-hand side approaches 0 as $n \to \infty$, and it follows that $A(x) = B(x)$ for all $x \in X$. Hence, A is unique. This completes the proof of the theorem.

The following corollary is an immediate consequence of Theorem 4, concerning the stability of (3).

Corollary 1 *Let λ and s be a non-negative real numbers. Let $f : X \longrightarrow Y$ be a function satisfying the inequality*

$$\|D_f(x_1, x_2, x_3, \ldots, x_n)\| \leq \begin{cases} \lambda \\ \lambda(\sum_{i=1}^{n}\|x_i\|^s) \\ \lambda(\prod_{i=1}^{n}\|x_i\|^s + \sum_{i=1}^{n}\|x_i\|^{ns}) \end{cases}$$

for all $x_1, x_2, \ldots, x_n \in X$. Then there exists a unique additive function $A : X \longrightarrow Y$
such that

$$\|f(x) - A(x)\| \leq \begin{cases} \dfrac{|2|\lambda}{(n^2-5n+6)} \\ \dfrac{4\lambda\|x\|^s}{(n^2-5n+6)|2-2^s|} & ; \quad s \neq 1 \\ \dfrac{4\lambda\|x\|^{ns}}{(n^2-5n+6)|2-2^{ns}|} & ; \quad s \neq \dfrac{1}{n} \end{cases}$$

2.1.1 Stability Result for (3) in Banach Space Using Fixed Point Method

In this segment, the authors presented the generalized Ulam–Hyers stability of the functional equation (3) in Banach space and using fixed point method.

Theorem 5 *Let $f : X \longrightarrow Y$ be a mapping for which there exists a function*
$\psi : X^n \longrightarrow [0, \infty)$ with the condition

$$lim_{k \longrightarrow \infty} \frac{\psi(\eta_i^k x_1, \eta_i^k x_2, \ldots, \eta_i^k x_n)}{\eta_i^k} = 0 \tag{17}$$

where

$$\eta_i = \begin{cases} 2, \text{ if } & i = 0 \\ \frac{1}{2}, \text{ if } & i = 1 \end{cases}$$

such that the functional inequality

$$\|D_f(x_1, x_2, \ldots, x_n)\| \leq \psi(x_1, x_2, \ldots, x_n) \tag{18}$$

for all $x_1, x_2, \ldots, x_n \in W$. If there exists $L = L(i)$ such that the function

$$x \longrightarrow \beta(x) = \frac{\psi(x/2, x/2, 0, \ldots, 0)}{(n^2 - 5n + 6)}$$

has the property,

$$\frac{1}{\eta_i} \beta(\eta_i x) = L\beta(x) \tag{19}$$

for all $x \in X$. Then there exists a unique additive function $A : X \longrightarrow Y$ satisfying
the functional equation (3) and

$$\|f(x) - A(x)\| \leq \frac{L^{1-i}}{1-L} \beta(x) \tag{20}$$

holds for all $x \in X$.

Proof Consider the set $d = \{u/u : X \longrightarrow Y, u(0) = 0)\}$ and introduce the generalized metric on M. $d(u, v) = inf \{k \in (0, \infty) : ||u(x) - v(x)|| \leq k\beta(x), x \in X\}$. It is easy to see that (M, d) is complete. Define $T : M \longrightarrow M$ by

$$Tu(x) = \frac{1}{\eta_i} u(\eta_i x)$$

for all $x \in M$. Now $u, v \in M$,

$$d(u, v) \leq k \Rightarrow ||u(x) - v(x)|| \leq k\beta(x) \quad \forall x \in X;$$

$$\Rightarrow ||\frac{1}{\eta_i} u(\eta_i x) - \frac{1}{\eta_i} v(\eta_i x)|| \leq \frac{1}{\eta_i} k\beta(\eta_i x) \quad \forall x \in X;$$

$$\Rightarrow ||Tu(x) - Tv(x)|| \leq k\beta(x) \quad \forall x \in X;$$

$$\Rightarrow d(Tu, Tv) \leq Lk$$

This implies $d(Tu, Tv) \leq Ld(u, v)$ for all $u, v \in M$. (i.e.,) T is strictly contractive mapping on with Lipschitz constant L. Replacing $(x_1, x_2, x_3, \ldots, x_n)$ by $(x, x, 0, \ldots, 0)$ in (18), we obtain

$$||\frac{(n^2 - 5n + 6)}{2} f(2x) - (n^2 - 5n + 6) f(x)|| \leq \psi(x, x, 0, \ldots, 0) \qquad (21)$$

for all $x \in X$. It follows from (21) that

$$||f(x) - \frac{f(2x)}{2}|| \leq \frac{\psi(x, x, 0, \ldots, 0)}{(n^2 - 5n + 6)} \qquad (22)$$

for all $x \in X$. Using (19) for the case $i = 0$, it reduces to

$$||f(x) - \frac{f(2x)}{2}|| \leq \beta(x)$$

for all $x \in X$.

$$i.e., d(f, Tf) \leq L \Rightarrow d(f, Tf) \leq 1 = L = L^0 < \infty.$$

Again replacing $x = \frac{x}{2}$ in (21) and (22), we get

$$||\left(\frac{n^2 - 5n + 6}{2}\right) f(x) - (n^2 - 5n + 6) f\left(\frac{x}{2}\right)|| \leq \psi(\frac{x}{2}, \frac{x}{2}, 0, \ldots, 0)$$

and

$$\| f(x) - 2f\left(\frac{x}{2}\right) \| \leq \frac{2}{(n^2 - 5n + 6)} \psi\left(\frac{x}{2}, \frac{x}{2}, 0, \ldots, 0\right) \tag{23}$$

for all $x \in X$. Using (19) for the case $i = 0$, it reduces to

$$\| f(x) - \frac{f(2x)}{2} \| \leq L\beta(x) \tag{24}$$

for all $x \in X$. (i.e.,) $d(f, Tf) \leq 2 \Rightarrow d(f, Tf) \leq 2 = L^0 < \infty$. In the above case, we arrive

$$d(f, Tf) \leq L^{1-i}.$$

Therefore $(B_2(i))$holds. By $(B_2(ii))$, it follows that there exists a fixed point A of T in X, such that

$$A(x) = \lim_{k \longrightarrow \infty} \frac{f(\eta_i^k x)}{\eta_i^k} \tag{25}$$

for all $x \in X$. In order to prove $A : X \longrightarrow Y$ is additive. Replacing $(x_1, x_2, \ldots, x_n)$ by $(\eta_i^k x_1, \eta_i^k x_2, \ldots, \eta_i^k x_n)$ in (18) and dividing by η_i^k, it follows from (17) and (25), we see that A satisfies (3) for all $x_1, x_2, \ldots, x_n \in X$. Hence A satisfies the functional equation (3). By $(B_2(iii))$, A is the unique fixed point of T in the set, $Y = \{f \in M; d(Tf, A) < \infty\}$. Using the fixed point alternative result, A is the unique function such that

$$\| f(x) - A(x) \| \leq k\beta(x)$$

for all $x \in W$ and $k > 0$.Finally by $(B_2(iv))$, we obtain

$$d(f, A) \leq \frac{1}{1 - L} d(f, Tf)$$

$$(i.e.,) \, d(f, A) \leq \frac{L^{1-i}}{1 - L}.$$

Hence, we conclude that

$$\| f(x) - A(x) \| \leq \frac{L^{1-i}}{1 - L} \beta(x)$$

for all $x \in X$. This completes the proof of the theorem.

The following corollary is an immediate consequence of Theorem 5 concerning the stability of (3).

Corollary 2 *Let $f : X \longrightarrow Y$ be a mapping and there exist real numbers λ and s such that*

$$
\|D_f(x_1, x_2, \ldots, x_n)\| \leq \begin{cases} \lambda \\ \lambda(\sum_{i=1}^{n} \|x_i\|^s) \\ \lambda(\prod_{i=1}^{n} \|x_i\|^s + \sum_{i=1}^{n} \|x_i\|^{ns}) \end{cases}
$$

for all $x_1, x_2, \ldots, x_n \in X$. Then there exists a unique additive function $A : X \longrightarrow Y$ such that

$$
\|f(x) - A(x)\| \leq \begin{cases} \frac{|2|\lambda}{(n^2-5n+6)} \\ \frac{4\lambda\|x\|^s}{(n^2-5n+6)|2-2^s|} & ; \quad s \neq 1 \\ \frac{4\lambda\|x\|^{ns}}{(n^2-5n+6)|2-2^{ns}|} & ; \quad s \neq \frac{1}{n} \end{cases}
$$

for all $x \in X$.

Proof Setting

$$
\psi(x_1, x_2, x_3, \ldots, x_n) \leq \begin{cases} \lambda \\ \lambda(\sum_{i=1}^{n} \|x_i\|^s) \\ \lambda(\prod_{i=1}^{n} \|x_i\|^s + \sum_{i=1}^{n} \|x_i\|^{ns}) \end{cases}
$$

for all $x_1, x_2, \ldots, x_n \in X$. Now

$$
\frac{\psi(\eta_i^k x_1, \eta_i^k x_2, \ldots, \eta_i^k x_n)}{\eta_i^k} \leq \begin{cases} \frac{\lambda}{\eta_i^k} \\ \frac{\lambda}{\eta_i^k}\{\sum_{i=1}^{n} \|\eta_i^k x_i\|^s\} \\ \frac{\lambda}{\eta_i^k}\{\prod_{i=1}^{n} \|\eta_i^k x_i\|^{ns} + \sum_{i=1}^{n} \|\eta_i^k x_i\|^{ns}\} \end{cases}
$$

$$
= \begin{cases} \longrightarrow & 0 \quad as \quad k \longrightarrow \infty \\ \longrightarrow & 0 \quad as \quad k \longrightarrow \infty \\ \longrightarrow & 0 \quad as \quad k \longrightarrow \infty \end{cases}
$$

i.e., (21) holds. But we have $\beta(x) = \frac{2}{(n^2-5n+6)}\psi(\frac{x}{2}, \frac{x}{2}, 0, \ldots, 0)$.

Hence

$$\beta(x) = \frac{1}{(n^2 - 5n + 6)} \psi\left(\frac{x}{2}, \frac{x}{2}, 0, \ldots, 0\right) = \begin{cases} \frac{2\lambda}{(n^2-5n+6)} \\ \frac{4\lambda||x||^s}{(n^2-5n+6)2^s} \\ \frac{4\lambda||x||^{ns}}{(n^2-5n+6)2^{ns}} \end{cases}$$

$$\frac{1}{\eta_i}\beta(\eta_i x) = \begin{cases} \frac{1}{\eta_i}\frac{2\lambda}{(n^2-5n+6)} \\ \frac{1}{\eta_i}\frac{4\lambda||x||^s}{(n^2-5n+6)2^s} \\ \frac{1}{\eta_i}\frac{4\lambda||x||^n s}{(n^2-5n+6)2^{ns}} \end{cases}$$

$$= \begin{cases} \eta_i^{-1}\beta(x) \\ \eta_i^{s-1}\beta(x) \\ \eta_i^{ns-1}\beta(x) \end{cases}$$

for all $x \in X$. Hence the inequality (3) holds for

$L = 2^{-1}$ if $i = 0$ and $L = \frac{1}{2^{-1}}$ if $i = 1$
$L = 2^{s-1}$ for $s < 1$ if $i = 0$ and $L = \frac{1}{2^{s-1}}$ for $s > 1$ if $i = 1$.
$L = 2^{ns-1}$ for $s < \frac{1}{n}$ if $i = 0$ and $L = \frac{1}{2^{ns-1}}$ for $s > \frac{1}{n}$ if $i = 1$.

Now, from (21) we prove the following cases:

Case1: $L = 2^{-1}$ if $i = 0$
$$||f(x) - A(x)| \leq \frac{L^{1-i}}{1-L}\beta(x) = \frac{(2^{-1})}{1-2^{-1}}\frac{2\lambda}{(n^2-5n+6)} = \frac{2\lambda}{(n^2-5n+6)}$$
Case2: $L = \frac{1}{3^{-1}}$ if $i = 1$
$$||f(x) - A(x)|| \leq \frac{L^{1-i}}{1-L}\beta(x) = \frac{1}{1-2}\frac{2\lambda}{(n^2-5n+6)} = \frac{-2\lambda}{(n^2-5n+6)}$$
Case3: $L = 2$ for $s < 1$ if $i = 0$
$$||f(x) - A(x)|| \leq \frac{L^{1-i}}{1-L}\beta(x) = \frac{2^{s-1}}{1-2^{s-1}}\frac{4\lambda||x||^s}{(n^2-5n+6)2^s} = \frac{4\lambda||x||^s}{(n^2-5n+6)(2-2^s)}$$
Case4: $L = \frac{1}{2^{s-1}}$ for $s > 1$ if $i = 1$
$$||f(x) - A(x)|| \leq \frac{L^{1-i}}{1-L}\beta(x) = \frac{1}{1-\frac{1}{2^{s-1}}}\frac{4\lambda||x||^s}{(n^2-5n+6)2^s} = \frac{4\lambda||x||^s}{(n^2-5n+6)(2^s-2)}$$
Case5: $L = 2^{ns-1}$ for $s < 1$ if $i = 0$
$$||f(x) - A(x)|| \leq \frac{L^{1-i}}{1-L}\beta(x) = \frac{2^{ns-1}}{1-2^{ns-1}}\frac{4\lambda||x||^{ns}}{(n^2-5n+6)2^{ns}} = \frac{4\lambda||x||^{ns}}{(n^2-5n+6)(2-2^{ns})}$$
Case6: $L = \frac{1}{2^{ns-1}}$ for $s > \frac{1}{n}$ if $i = 1$
$$||f(x) - A(x)|| \leq \frac{L^{1-i}}{1-L}\beta(x) = \frac{1}{1-\frac{1}{2^{ns-1}}}\frac{4\lambda||x||^{ns}}{(n^2-5n+6)2^{ns}} = \frac{4\lambda||x||^{ns}}{(n^2-5n+6)(2^{ns}-2)}.$$

3 Stability Result for (3) in Fuzzy Normed Space Using Direct Method

Throughout this section, assume that X, $\left(Z, N'\right)$, (Y, N) are linear space, Banach space, and fuzzy normed space, respectively, we now investigate the fuzzy stability of the functional equation (3).

Theorem 6 *Let $\beta \in \{1, -1\}$ be fixed and let $\psi : X^n \longrightarrow Z$ be a mapping such that for some $d > 0$ with $0 < \left(\frac{d}{2}\right)^\beta < 1$.*

$$N'\left(\psi(2^\beta x, 2^\beta x, 0, \ldots, 0), r\right) \geq N'(d^\beta \psi(x, x, 0, \ldots, 0), r) \qquad (26)$$

for all $x \in X$ and all $r > 0, d > 0$, and

$$lim_{k \longrightarrow \infty} N'\left(\psi(2^{\beta k} x_1, 2^{\beta k} x_2, \ldots, 2^{\beta k} x_n), 2^{\beta k} r\right) = 1 \qquad (27)$$

for all $x_1, x_2, \ldots, x_n \in X$ and all $r > 0$. Suppose an odd mapping $f : X \longrightarrow Y$ with $f(0) = 0$ satisfies the inequality

$$N(D_f(x_1, x_2, \ldots, x_n), r) \geq N'(\psi(x_1, x_2, \ldots, x_n), r) \qquad (28)$$

for all $r > 0$ and all $x_1, x_2, \ldots, x_n \in X$. Then the limit

$$A(x) = N - lim_{k \longrightarrow \infty} \frac{f(2^{\beta k} x)}{2^{\beta k}} \qquad (29)$$

exists for all $x \in X$ and the mapping $A : X \longrightarrow Y$ is the unique additive mapping such that

$$N(f(x) - A(x), r) \geq N'\left(\psi(x, x, 0, \ldots, 0), \frac{(n^2 - 5n + 6)}{2} r|2 - d|\right) \qquad (30)$$

for all $x \in X$ and for all $r > 0$.

Proof Let $\beta = 1$. Replacing $(x_1, x_2, x_3, \ldots, x_n)$ by $(x, x, 0, \ldots, 0)$ in (28), we get

$$N\left((n^2 - 5n + 6)f(x) - \frac{(n^2 - 5n + 6)}{2} f(2x), r\right) \geq N'(\psi(x, x, 0, \ldots, 0), r) \qquad (31)$$

for all $x \in X$ and all $r > 0$. Replacing x by $2^k x$ in (31), we obtain

$$N\left(\frac{f(2^{k+1} x)}{2} - f(2^k x), \frac{r}{(n^2 - 5n + 6)}\right) \geq N'(\psi(2^k x, 2^k x, 0, \ldots, 0), r) \qquad (32)$$

for all $x \in X$ and for all $r > 0$. Using (26), we get

$$N\left(\frac{f(2^{k+1}x)}{2} - f(2^k x), \frac{r}{(n^2 - 5n + 6)}\right) \geq N'(\psi(x, x, 0, \ldots, 0), \frac{r}{d^k}) \qquad (33)$$

for all $x \in X$ and for all $r > 0$. It is easy to verify from (33) that

$$N\left(\frac{f(2^{k+1}x)}{2^{k+1}} - \frac{f(2^k x)}{2^k}, \frac{r}{(n^2 - 5n + 6)2^k}\right) \geq N'(\psi(x, x, 0, \ldots, 0), \frac{r}{d^k}) \qquad (34)$$

holds for all $x \in X$ and for all $r > 0$. Replacing r by $d^k r$ in (34),we get

$$N\left(\frac{f(2^{k+1}x)}{2^{k+1}} - \frac{f(2^k x)}{2^k}, \frac{d^k r}{(n^2 - 5n + 6)2^k}\right) \geq N'(\psi(x, x, 0, \ldots, 0), r) \qquad (35)$$

for all $x \in X$ and for all $r > 0$. It follows from

$$\frac{f(2^k x)}{2^k} - f(x) = \sum_{i=0}^{k-1}\left[\frac{f(2^{i+1}x)}{2^{i+1}} - \frac{f(2^i x)}{2^i}\right] \qquad (36)$$

and (35) that

$$N\left(\frac{f(2^k x)}{2^k} - f(x), \sum_{i=0}^{k-1}\frac{d^i r}{(n^2 - 5n + 6)2^i}\right)$$

$$\geq min\left\{N\left(\frac{f(2^{i+1}x)}{2^{i+1}} - \frac{f(2^i x)}{2^i}, \frac{d^i r}{(n^2 - 5n + 6)2^i}\right) : i = 0, 1, 2, \ldots, k-1\right\}$$

$$\geq N'(\psi(x, x, 0, \ldots, 0), r) \qquad (37)$$

for all $x \in X$ and for all $r > 0$. Replacing x by $2^m x$ in (37), we get

$$N\left(\frac{f(2^{k+m}x)}{2^{k+m}} - \frac{f(2^m x)}{2^m}, \sum_{i=m}^{m+k-1}\frac{d^i r}{(n^2 - 5n + 6)2^i}\right) \geq N'(\psi(x, x, 0, \ldots, 0), \frac{r}{d^m}) \qquad (38)$$

for all $x \in X$ and for all $r > 0$ and all $m, k \geq 0$. Replacing r by $d^m r$ in (38), we get

$$N\left(\frac{f(2^{k+m}x)}{2^{k+m}} - \frac{f(2^m x)}{2^m}, \sum_{i=0}^{k-1}\frac{d^i r}{(n^2 - 5n + 6)2^i}\right) \geq N'(\psi(x, x, 0, \ldots, 0), r) \qquad (39)$$

for all $x \in X$ and for all $r > 0$ and all $m, k \geq 0$. Using (N_3) in (38), we obtain

$$N\left(\frac{f(2^{k+m}x)}{2^{k+m}} - \frac{f(2^m x)}{2^m}, r\right) \geq N'\left(\psi(x, x, 0, \ldots, 0), \frac{r}{\sum_{i=m}^{m+k-1}\frac{d^i}{(n^2-5n+6)2^i}}\right) \qquad (40)$$

for all $x \in X, r > 0$ and all $m, k \geq 0$. Since $0 < d < 2$ and $\sum_{i=0}^{k} \left(\frac{d}{2}\right)^i < \infty$, the Cauchy criterion for convergence and (N_5) implies that $\left\{\frac{f(2^k x)}{2^k}\right\}$ is a Cauchy sequence in (Y, N). Since (Y, N) is a fuzzy Banach space, this sequence converges to some point $A(x) \in Y$. So one can define the mapping $A : X \longrightarrow Y$ by

$$A(x) := N - \lim_{k \longrightarrow \infty} \frac{f(2^k x)}{2^k}$$

for all $x \in X$. Letting $m = 0$ in (40), we get

$$N\left(\frac{f(2^k x)}{2^k} - f(x), r\right) \geq N'\left(\psi(x, x, 0, \ldots, 0), \frac{r}{\sum_{i=0}^{k-1} \frac{d^i}{(n^2 - 5n + 6)2^i}}\right) \quad (41)$$

for all $x \in X$. Taking the limits as $k \longrightarrow \infty$ and using (N_6), we arrive

$$N(f(x) - A(x), r) \geq N'(\psi(x, x, 0, \ldots, 0), (n^2 - 5n + 6)r.(2 - d))$$

for all $x \in X$ and for all $r > 0$. Now, we claim that A is additive. Replacing $(x_1, x_2, x_3, \ldots, x_n)$ by $(2^k x_1, 2^k x_2, \ldots, 2^k x_n)$ in (28), respectively, we get

$$N\left(\frac{1}{2^k} D_f(2^k x_1, 2^k x_2, \ldots, 2^k x_n), r\right) \geq N'(\psi(2^k x_1, 2^k x_2, \ldots, 2^k x_n), 2^k r)$$

$$(42)$$

for all $r > 0$ and for all $x_1, x_2, \ldots, x_n \in X$. Since

$$\lim_{k \longrightarrow \infty} N'\left(\psi(2^{\beta k} x_1, 2^{\beta k} x_2, \ldots, 2^{\beta k} x_n), 2^{\beta k} r\right) = 1.$$

A satisfies the additive functional equation (3). Hence $A : X \to Y$ is additive. To prove the uniqueness of A, let A' be another additive mapping satisfying (30). Fix $x \in X$, clearly $A(2^n x) = 2^n A(x)$ and $A'(2^n x) = 2^n A'(x)$ for all $x \in X$ and all $n \in N$. It follows from (30) that $N(A(x) - A'(x), r) = N\left(\frac{A(2^k x)}{2^k} - \frac{A'(2^k x)}{2^k}, r\right)$

$$\geq min\left\{N\left(\frac{A(2^k x)}{2^k} - \frac{f(2^k x)}{2^k}, \frac{r}{2}\right), N\left(\frac{f(2^k x)}{2^k} - \frac{A'(2^k x)}{2^k}, \frac{r}{2}\right)\right\}$$

$$\geq N'\left(\psi(2^k x, 2^k x, 0, \ldots, 0), \frac{r(n^2 - 5n + 6)2^k(2 - d)}{2}\right)$$

$$\geq N'\left(\psi(2^k x, 2^k x, 0, \ldots, 0), \frac{r(n^2 - 5n + 6)2^k(2 - d)}{2d^k}\right)$$

for all $x \in X$ and $r > 0$. Since $lim_{k \to \infty} \frac{r(n^2 - 5n + 6)2^k(2-d)}{2d^k} = \infty$, we obtain

$$lim_{k \to \infty} N'\left(\psi(x, x, 0, \ldots, 0), \frac{r(n^2 - 5n + 6)2^k(2-d)}{2d^k}\right) = 1.$$

Thus $N(A(x) - A'(x), r) = 1$ for all $x \in X$ and $r > 0$ and so $A(x) = A'(x)$. For $\beta = -1$, we can prove the result by a similar method.

The following corollary is an immediate consequence of Theorem 6, concerning the stability for the functional equation (3).

Corollary 3 *Suppose that the function* $f : X \longrightarrow Y$ *satisfies the inequality*

$$N(Df(x_1.x_2, \ldots, x_n), r) \geq \begin{cases} N'(\theta, r) \\ N'(\theta \sum_{i=1}^{n} ||x_i||^s, r) \\ N'(\theta(\sum_{i=1}^{n} ||x_i||^{ns} + \Pi_{i=1}^{n} ||x_i||^s), r) \end{cases}$$

for all $x_1, x_2, \ldots, x_n \in X$ *and all* $r > 0$, *where* θ, s *are constants then there exists a unique additive mapping* $A : X \to Y$ *such that*

$$N(f(x) - A(x), r) \geq \begin{cases} N'(\theta, \frac{r(n^2 - 5n + 6)}{|2|}) \\ N'\left(2\theta ||x||^s, \frac{r(n^2 - 5n + 6)|2 - 2^s|}{2}\right) & ; s \neq 1 \\ N'\left(2\theta ||x||^{ns}, \frac{r(n^2 - 5n + 6)|2 - 2^{ns}|}{2}\right) & ; s \neq \frac{1}{n} \end{cases}$$

3.1 Stability Result for (3) in Fuzzy Normed Space Using Fixed Point Method

Throughout this section, the authors investigated the generalized Ulam–Hyers stability of the functional equation (3) in fuzzy normed space using fixed point method.

To prove the stability result, we define the following μ_i is a constant such that

$$\eta_i = \begin{cases} 2 & if \quad i = 0 \\ \frac{1}{2} & if \quad i = 1 \end{cases}$$

and Ω is the set such that $\Omega = \{t/t : W \longrightarrow B, t(0) = 0\}$.

Theorem 7 *Let* $f : X \longrightarrow Y$ *be a mapping for which there exists a function* $\psi : X^n \longrightarrow Z$ *with condition*

$$lim_{k \to \infty} N'\left(\psi(\eta^k x_1, \eta^k x_2, \ldots, \eta^k x_n), \eta^k r\right) = 1 \qquad (43)$$

for all $x_1, x_2, \ldots, x_n \in X$ and all $r > 0$ and satisfying the inequality

$$N(D_f(x_1, x_2, \ldots, x_n), r) \geq N'(\psi(x_1, x_2, \ldots, x_n), r) \tag{44}$$

for all $x \in X$ and $r > 0$. If there exist $L = L[i]$ such that the function $x \longrightarrow$ $\beta(x) = \frac{1}{(n^2 - 5n + 6)} \psi \left(\frac{x}{2}, \frac{x}{2}, 0, \ldots, 0 \right)$ has the property

$$N' \left(L \frac{1}{\eta_i} \beta(\eta_i x), r \right) = N'(\beta(x), r) \tag{45}$$

for all $x \in X$ and $r > 0$, then there exists a unique additive function $A : X \longrightarrow Y$ satisfying the functional equation (3) and

$$N(f(x) - A(x), r) \geq N' \left(\frac{L^{1-i}}{1 - L} \beta(x), r \right)$$

for all $x \in X$ and $r > 0$.

Proof Let d be a general metric on Ω such that

$$d(t, u) = \inf \left\{ k \in (0, \infty) \mid N(t(x) - u(x), r) \geq N'(\beta(x), kr), x \in X, r > 0 \right\}$$

It is easy to see that (Ω, d) is complete. Define $T : \Omega \longrightarrow \Omega$ by $Tt(x) = \frac{1}{\eta_i} t(\eta_i x)$ for all $x \in X$, for $t, u \in \Omega$, we have

$$d(t, u) = k \Rightarrow N(t(x) - u(x), r) \geq N'(\beta(x), kr)$$

$$\Rightarrow N \left(\frac{t(\eta_i x)}{\eta_i} - \frac{u(\eta_i x)}{\eta_i}, r \right) \geq N'(\beta(\eta_i x), k\eta_i r) \tag{46}$$

$$\Rightarrow N(Tt(x) - Tu(x), r) \geq N'(\beta(\eta_i x), k\eta_i r)$$

$$\Rightarrow N(Tt(x) - Tu(x), r) \geq N'(\beta(x), kLr)$$

$$\Rightarrow d(Tt(x) - Tu(x)) \geq kL$$

$$\Rightarrow d(Tt - Tu, r) \geq Ld(t, u)$$

for all $t, u \in \Omega$. Therefore T is strictly contractive mapping on Ω with Lipschitz constant L, replacing $(x_1, x_2, x_3, \ldots, x_n)$ by $(x, x, 0, \ldots, 0)$ in (44), we get

$$N \left(\frac{(n^2 - 5n + 6)}{2} f(2x) - (n^2 - 5n + 6) f(x), r \right) \geq N'(\psi(x, x, 0, \ldots, 0), r) \tag{47}$$

for all $x \in X$ and $r > 0$. Using (N_3) in (47), we arrive

$$N\left(\frac{f(2x)}{2} - f(x), r\right) \geq N'\left(\frac{\psi(x, x, 0, \ldots, 0)}{(n^2 - 5n + 6)}, r\right) \tag{48}$$

for all $x \in X$ and $r > 0$ with the help of (45) when $i = 0$, it follows from (48) that

$$\Rightarrow N\left(\frac{f(2x)}{2} - f(x), r\right) \geq N'(L\beta(x), r)$$

$$\Rightarrow d(Tf, f) \geq L = L^1 = L^{1-i}. \tag{49}$$

Replacing x by $\frac{x}{2}$ in (47), we obtain

$$N\left(f(x) - 2f\left(\frac{x}{2}\right), r\right) \geq N'\left(\frac{2}{(n^2 - 5n + 6)}\psi\left(\frac{x}{2}, \frac{x}{2}, 0, \ldots, 0\right), r\right)$$

for all $x \in X$ and $r > 0$, when $i = 1$, it follows from (49), we get

$$\Rightarrow N\left(f(x) - 2f\left(\frac{x}{2}\right), r\right) \geq N'(\beta(x), r)$$

$$\Rightarrow T(f, Tf) \leq 1 = L^0 = L^{1-i}. \tag{50}$$

Then from (49) and (50), we can conclude

$$\Rightarrow T(f, Tf) \leq L^{1-i} < \infty.$$

Now from the fixed point alternative in both cases, it follows that there exists a fixed point A of T in Ω such that

$$A(x) = N - \lim_{k \longrightarrow \infty} \frac{f(\eta_i^k x)}{\eta_i^k}$$

for all $x \in W$ and $r > 0$. Replacing $(x_1, x_2, \ldots, x_n)$ by $(\eta_i^k x_1, \eta_i^k x_2, \ldots, \eta_i^k x_n)$ in (44), we arrive

$$N\left(\frac{1}{\eta_i^k}Df(\eta_i^k x_1, \eta_i^k x_2, \ldots, \eta_i^k x_n), r\right) \geq N'(\psi(\eta_i^k x_1, \eta_i^k x_2, \ldots, \eta_i^k x_n), \eta_i^k r)$$

for all $r > 0$ and all $x_1, x_2, \ldots, x_n \in X$. By proceeding the same procedure of the Theorem 5.1 , we can prove the function $A : X \longrightarrow Y$ is additive and it satisfies the functional equation (3). By a fixed point alternative, since A is a unique fixed point of T in the set

$$\Delta = \{f \in \Omega / d(f, A) < \infty\}.$$

Therefore A is a unique function such that

$$N(f(x) - A(x), r) \geq N'(\beta(x), kr)$$

for all $x \in W$ and $r > 0$. Again using the fixed point alternative, we obtain

$$d(f, A) \leq \frac{1}{1 - L} d(f, Tf)$$

$$\Rightarrow d(f, A) \leq \frac{L^{1-i}}{1 - L}$$

$$\Rightarrow N(f(x) - A(x), r) \geq N'\left(\beta(x)\frac{L^{1-i}}{1 - L}, r\right)$$

for all $x \in X$ and $r > 0$. This completes the proof of the theorem.

The following corollary is an immediate consequence of Theorem 7 concerning the stability of (3).

Corollary 4 *Suppose a function* $f : X \longrightarrow Y$ *satisfies the inequality*

$$N(D_f(x_1, x_2, \ldots, x_n), r) \geq \begin{cases} N'(\theta, r) \\ N'(\theta \sum_{i=1}^{n} ||x_i||^s, r) \\ N'(\theta(\sum_{i=1}^{n} ||x_i||^{ns} + \Pi_{i=1}^{n} ||x_i||^s), r) \end{cases}$$

for all $x_1, x_2, \ldots, x_n \in X$ *and* $r > 0$, *where* θ, s *are constants with* $\theta > 0$. *Then there exists a unique additive mapping* $A : X \longrightarrow Y$ *such that*

$$N(f(x) - A(x), r) \geq \begin{cases} N'\left(\theta, \frac{r(n^2 - 5n + 6)}{|2|}\right) & \\ N'\left(2\theta||x||^s, \frac{r(n^2 - 5n + 6)|2 - 2^s|}{2}\right) & ; \ s \neq 1 \\ N'\left(2\theta||x||^{ns}, \frac{r(n^2 - 5n + 6)|2 - 2^{ns}|}{2}\right) & ; \ s \neq \frac{1}{n} \end{cases}$$

for all $x \in X$ *and* $r > 0$.

Proof Setting

$$\psi(x_1, x_2, x_3, \ldots, x_n) \leq \begin{cases} \theta \\ \theta(\sum_{i=1}^{n} ||x_i||^s) \\ \theta(\Pi_{i=1}^{n} ||x_i||^s + \sum_{i=1}^{n} ||x_i||^{ns}) \end{cases}$$

for all $x_1, x_2, \ldots, x_n \in X$. Then

$$N'\left(\psi\left(\eta_i^k x_1, \eta_i^k x_2, \ldots, \eta_i^k x_n\right), \eta_i^k r\right) = \begin{cases} N'(\theta, \eta_i^k r) \\ N'\left(\theta \sum_{i=1}^n ||x_i||^s, \eta_i^{(1-s)k} r\right) \\ N'\left(\theta(\sum_{i=1}^n ||x_i||^{ns} + \Pi_{i=1}^n ||x_i||^s), \eta_i^{(1-ns)k} r\right) \end{cases}$$

$$= \begin{cases} \longrightarrow & 1 \quad as \quad k \longrightarrow \infty, \\ \longrightarrow & 1 \quad as \quad k \longrightarrow \infty, \\ \longrightarrow & 1 \quad as \quad k \longrightarrow \infty. \end{cases}$$

Thus, (6) holds. But we have

$$\beta(x) = \frac{2}{(n^2 - 5n + 6)} \psi\left(\frac{x}{2}, \frac{x}{2}, 0, \ldots, 0\right)$$

has the property

$$N'\left(L\frac{1}{\eta_i}\beta(\eta_i x), r\right) \geq N'(\beta(x), r)$$

for all $x \in X$ and $r > 0$. Hence

$$N'(\beta(x), r) = N'\left(\psi\left(\frac{x}{2}, \frac{x}{2}, 0, \ldots, 0\right), (n^2 - 5n + 6)r\right)$$

$$= \begin{cases} N'(\theta, r(n^2 - 5n + 6)) \\ N'\left(\frac{2}{2^s}\theta||x||^s, r(n^2 - 5n + 6)\right) \\ N'\left(\frac{2}{2^{ns}}\theta||x||^{ns}, r(n^2 - 5n + 6)\right). \end{cases}$$

Now,

$$N'\left(\frac{1}{\eta_i}\beta(\eta_i x), r\right) = \begin{cases} N'\left(\frac{\theta}{\eta_i}, r(n^2 - 5n + 6)\right) \\ N'\left(\frac{\theta}{\eta_i}\left(\frac{2}{2^s}\right)||\eta_i x||^s, r(n^2 - 5n + 6)\right) \\ N'\left(\frac{\theta}{\eta_i}\left(\frac{2}{2^{ns}}\right)||\eta_i x||^{ns}, r(n^2 - 5n + 6)\right) \end{cases}$$

$$= \begin{cases} N'(\eta_i^{-1}\beta(x), r) \\ N'(\eta_i^{s-1}\beta(x), r) \\ N'(\eta_i^{ns-1}\beta(x), r) \end{cases}$$

Now from the following cases for the conditions (i) and (ii)

Case(i): $L = 2^{-1}$ *for* $s = 0$ *if* $i = 0$

$$N(f(x) - A(x), r) \geq N'\left(\frac{L^{1-i}}{1-L}\beta(x), r\right) \geq N'\left(\frac{2^{-1}}{1-2^{-1}}\frac{2\theta}{(n^2-5n+6)}, r\right) \geq$$

$$N'\left(\theta, \frac{r(n^2-5n+6)}{2}\right)$$

Case(ii): $L = \left(\frac{1}{2}\right)^{-1}$ *for* $s = 0$ *if* $i = 1$

$$N(f(x) - A(x), r) \geq N'\left(\frac{L^{1-i}}{1-L}\beta(x), r\right) \geq N'\left(\frac{1}{1-\left(\frac{1}{2}\right)^{-1}}\frac{2\theta}{(n^2-5n+6)}, r\right) \geq$$

$$N'\left(\theta, \frac{-r(n^2-5n+6)}{2}\right)$$

Case(iii): $L = (2)^{s-1}$ *for* $s < 1$ *if* $i = 0$

$$N(f(x) - A(x), r) \geq N'\left(\frac{L^{1-i}}{1-L}\beta(x), r\right)$$

$$\geq N'\left(\frac{2^{s-1}}{1-2^{s-1}}\frac{2\theta||x||^s}{(n^2-5n+6)2^s}, r\right)$$

$$\geq N'\left(2\theta||x||^s, \frac{r(n^2-5n+6)(2-2^s)}{2}\right)$$

Case(iv): $L = (2)^{1-s}$ *for* $s > 1$ *if* $i = 1$

$$N(f(x) - A(x), r) \geq N'\left(\frac{L^{1-i}}{1-L}\beta(x), r\right)$$

$$\geq N'\left(\frac{2^{1-s}}{1-2^{1-s}}\frac{2\theta||x||^s}{(n^2-5n+6)2^s}, r\right)$$

$$\geq N'\left(2\theta||x||^s, \frac{r(n^2-5n+6)(2^s-2)}{2}\right)$$

Case(v): $L = (2)^{ns-1}$ *for* $s < \frac{1}{n}$ *if* $i = 0$

$$N(f(x) - A(x), r) \geq N'\left(\frac{L^{1-i}}{1-L}\beta(x), r\right)$$

$$\geq N'\left(\frac{2^{ns-1}}{1-2^{ns-1}}\frac{2\theta||x||^{ns}}{(n^2-5n+6)2^{ns}}, r\right)$$

$$\geq N'\left(2\theta||x||^{ns}, \frac{r(n^2-5n+6)(2-2^{ns})}{2}\right)$$

Case(vi): $L = (2)^{1-ns} \quad for \quad s < \frac{1}{n} \quad if \quad i = 1$

$$N(f(x) - A(x), r) \geq N'\left(\frac{L^{1-i}}{1-L}\beta(x), r\right)$$

$$\geq N'\left(\frac{2^{1-ns}}{1-2^{1-ns}}\frac{2\theta||x||^{ns}}{(n^2 - 5n + 6)2^{ns}}, r\right)$$

$$\geq N'\left(2\theta||x||^{ns}, \frac{r(n^2 - 5n + 6)(2^{ns} - 2)}{2}\right)$$

Hence the proof is completed.

References

1. T. Aoki, On the stability of the linear transformation in Banach space. Int. J. Math. Soc. **2**(1–2), 64–66 (1950)
2. T. Bag, S.K. Samanta, Finite dimensional fuzzy normed linear spaces. J. Fuzzy Math. **11**(3), 687–705 (2003)
3. T. Bag, S.K. Samanta, Fuzzy bounded linear operators. Fuzzy Sets Syst. **151**, 513–547 (2005)
4. R. Biswas, Fuzzy inner product space and fuzzy norm functions. Inf. Sci. **53**, 185–190 (1991)
5. I.S. Chang, H.M. Kim, On the Hyers-Ulam stability of quadratic functional equations. J. Inequal. Pure Appl. Math. **3**, 33 (2002)
6. S.C. Cheng, J.N. Mordeson, Fuzzy linear operator and fuzzy normed linear spaces. Bull. Calcuta Math. Soc. **86**, 429–436 (1994)
7. C. Felbin, Finite dimensional fuzzy normed space. Fuzzy Sets Syst. **48**, 239–248 (1992)
8. P. Gavruta, A generalization of the Hyers-Ulam-Rassias stability of approximately additive mappings. J. Math. Anal. Appl. **184**, 431–436 (1992)
9. V. Govindan, K. Tamilvanan, Stability of functional equation in Banach space: using two different methods. Int. J. Math. Appl. **6**(1-C), 527–536 (2018)
10. V. Govindan, S. Murthy, M. Saravanan, Solution and stability of New type of (aaq,bbq,caq,daq) mixed type functional equation in various normed spaces: using two different methods. Int. J. Math. Appl. 5(1-B), 187–211 (2017)
11. D.H. Hyers, On the stability of the linear functional equation. Proc. Natl. Acad. Sci. **27**, 222–224 (1941)
12. P. L. Kannappan, Quadratic functional equation and inner product spaces. Results Math. **27**, 368–372 (1995)
13. A.K. Katsaras, Fuzzy topological vector spaces II. Fuzzy Sets Syst. **12**, 143–154 (1984)
14. I. Kramosil, J. Michalek, Fuzzy metric and statistical metric spaces. Kybernetica **11**, 326–334 (1975)
15. A.K. Mirmostafee, M.S. Moslehian, Fuzzy versions of Hyers-Ulam-Rassias theorem. Fuzzy Sets Syst. **159**, 720–729 (2008)
16. A.K. Mirmostafee, M.S. Moslehian, Fuzzy almost quadratic functions. Results Math. https://doi.org/10.1007/s00025-007-0278-9
17. S. Murthy, V. Govindan, General solution and generalized HU (Hyers-Ulam) stability of new dimension cubic functional equation in Fuzzy ternary Banach algebras: using two different methods. Int. J. Pure Appl. Math. **113**(6), 394–403 (2017)
18. S. Pinelas, V. Govindan, K. Tamilvanan, Stability of a quartic functional equation. J. Fixed Point Theory Appl. **20**(148), 10pp. (2018). https://doi.org/10.007/s11784-018-0629-z

19. S. Pinelas, V. Govindan, K. Tamilvanan, Stability of non- additive functional equation. IOSR J. Math. **14**(2 - I), 60–78 (2018)
20. T.M. Rassias, On the stability of the linear mapping in Banach spaces. Proc. Am. Math. Soc. **72**, 297–300 (1978)
21. J.M. Rassias, On approximation of approximately linear mappings by linear mapping. J. Funct. Anal. **46**(1), 126–130 (1982)
22. J.M. Rassias, On approximation of approximately linear mappings by linear mappings. Bull. Sci. Math. **108**(4), 445–446 (1984)
23. K. Ravi, P. Narasimman, R. Kishore Kumar, Generalized Hyers-Ulam-Rassias stability and J. M. Rassias stability of a quadratic functional equation. Int. J. Math. Sci. Eng. Appl. **3**(2), 79–94 (2009)
24. K. Ravi, R. Kodandan, P. Narasimman, Ulam stability of a quadratic functional equation. Int. J. Pure Appl. Math. **51**(1), 87–101 (2009)
25. B. Shieh, Infinite fuzzy relation equations with continuous t-norms. Inf. Sci. **178**, 1961–1967 (2008)
26. S.M. Ulam, *A Collection of the Mathematical Problems* (Interscience, New York, 1960).
27. C. Wu, J. Fang, Fuzzy generalization of Klomogoroffs theorem. J. Harbin Inst. Technol. **1**, 1–7 (1984)

Measure Zero Stability Problem for Drygas Functional Equation with Complex Involution

Ahmed Nuino, Muaadh Almahalebi, and Ahmed Charifi

Abstract In this chapter, we discuss the Hyers–Ulam stability theorem for the σ-Drygas functional equation

$$f(x + y) + f\big(x + \sigma(y)\big) = 2f(x) + f(y) + f\big(\sigma(y)\big)$$

for all $(x, y) \in \Omega \subset \mathbb{C}^2$ for Lebesgue measure $m(\Omega) = 0$, where $f : \mathbb{C} \to Y$ and $\sigma : X \to X$ is an involution.

1 Introduction

The study of stability problems of functional equations was motivated by a question of S. M. Ulam asked in 1940 [35]. The first result giving answer to this question is due to Hyers [21]. Subsequently, his result was extended and generalized in several ways by many authors worldwide.

Characterizing quasi-inner product spaces, Drygas considers in [15] the functional equation

$$f(x + y) + f(x - y) = 2f(x) + f(y) + f(-y), \quad x, y \in \mathbb{R} \tag{1}$$

which is a generalization of an important quadratic functional equation

$$f(x + y) + f(x - y) = 2f(x) + 2f(y), \quad x, y \in \mathbb{R}. \tag{2}$$

The functional equation (1) is now known in the literature as *Drygas equation*. The general solution of Drygas equation was given by Ebanks, Kannappan, and Sahoo in [16]. It has the form

A. Nuino (✉) · M. Almahalebi · A. Charifi
Department of Mathematics, Faculty of Sciences, University of Ibn Tofail, Kenitra, Morocco

© Springer Nature Switzerland AG 2019
G. A. Anastassiou and J. M. Rassias (eds.), *Frontiers in Functional Equations and Analytic Inequalities*, https://doi.org/10.1007/978-3-030-28950-8_11

$$f(x) = A(x) + Q(x),$$

where $A : \mathbb{R} \longrightarrow \mathbb{R}$ is an additive function and $Q : \mathbb{R} \longrightarrow \mathbb{R}$ is a quadratic function. In 2002, Jung and Sahoo [24] considered the stability problem of the following functional equation:

$$f(x + y) + f(x - y) = 2f(x) + f(y) + g(2y), \tag{3}$$

and as a consequence they obtained the stability theorem of functional equation of Drygas (1) where f and g are functions from a real vector space X to a Banach space Y.

Here we state a slightly modified version of the results in [24].

Theorem 1 *Let $\varepsilon \geq 0$ be fixed and let X be a real vector space and Y a Banach space. If a function $f : X \longrightarrow Y$ satisfies the inequality*

$$\| f(x + y) + f(x - y) - 2f(x) - f(y) - f(-y)\| \leq \varepsilon, \tag{4}$$

for all $x, y \in X$, then there exists a unique additive mapping $A : X \longrightarrow Y$ and a unique quadratic mapping $Q : X \longrightarrow Y$ such that $S = A + Q$ is a solution of (1) such that

$$\| f(x) - S(x)\| \leq \frac{25}{3}\varepsilon, \tag{5}$$

for all $x \in X$.

This result was improved first by Yang in [36] and later by Sikorska in [31]. In this chapter we use the Sikorska's result as a basic tool in the main result. So, we need to present the following theorem.

Theorem 2 ([31]) *Let $(X, +)$ be a group and Y be a Banach space. Given an $\varepsilon > 0$, assume that $f : X \to Y$ satisfies the condition*

$$\| f(x + y) + f(x - y) - 2f(x) - f(y) - f(-y)\| \leq \varepsilon, \qquad x, y \in X.$$

Then there exists a uniquely determined function $g : X \to Y$ such that

$$g(x) = \frac{2}{9}g(3x) - \frac{1}{9}g(-3x), \qquad x \in X,$$

and

$$\| f(x) - g(x)\| \leq \varepsilon \qquad x \in X.$$

Moreover, if X is Abelian, then g satisfies

$$g(x + y) + g(x - y) = 2g(x) + g(y) + g(-y), \qquad x, y \in X.$$

The stability and solution of the Drygas equation under some additional conditions was also studied by Forti and Sikorska in [19] in the case when X and Y are amenable groups.

It is a very natural subject to consider functional equations or inequalities satisfied on restricted domains or satisfied under restricted conditions [2–10, 13, 18–23, 25, 26, 28–30]. Among the results, S. M. Jung and J. M. Rassias proved the Hyers–Ulam stability of the quadratic functional equations in a restricted domain [22, 27].

It is very natural to ask if the restricted domain $D := \{(x, y) \in X^2 : \|x\| + \|y\| \geq d\}$ can be replaced by a much smaller subset $\Omega \subset D$, i.e., a subset of measure 0 in a measure space X. In 2013, J. Chung considered the stability of the Cauchy functional equation

$$f(x + y) = f(x) + f(y) \tag{6}$$

in a set $\Omega \subset \{(x, y) \in \mathbb{R}^2 : |x| + |y| \geq d\}$ of measure $m(\Omega) = 0$ when $f : \mathbb{R} \longrightarrow \mathbb{R}$. In 2014, J. Chung and J. M. Rassias proved the stability of the quadratic functional equation in a set of measure zero.

Let E be a real vector space, $G = (G, +)$ be an arbitrary semigroup. We say that a function $f : G \longrightarrow E$ satisfies the σ-Drygas equation if

$$f(x + y) + f(x + \sigma(y)) = 2f(x) + f(y) + f(\sigma(y)) \tag{7}$$

for all $x, y \in G$, where σ be an involution of G (which means that $\sigma(x + y) = \sigma(x) + \sigma(y)$ and $\sigma(\sigma(x)) = x$ for all $x, y \in G$).

If $\sigma(x) = -x$ and G is an abelian group in Eq. (7), then Eq. (7) reduces to the classic Drygas functional equation (1).

The solutions of Drygas equation in abelian group are obtained by Stetkær in [32] and [33]. Various authors studied the Drygas equation, for example Szabo [34], Ebanks et al. [16], Jung and Sahoo [24], Yang [36], Faïziev and Sahoo [17]. There are several functional equations reduced to those of the Drygas functional equation (1), i.e. the mixed type additive, quadratic, Jensen and Pexidered equations, we refer, for example, to [1, 3, 13, 14].

In this chapter, our aim is to prove the Hyers–Ulam stability on $\Omega \subset X^2$ of Lebesgue measure 0 for the σ-Drygas functional equation (7), where $f : X \rightarrow Y$ and $\sigma : X \rightarrow X$ is a complex involution. We also obtain an asymptotic behavior of this equation.

2   Stability of Eq. (7) in Set of Measure Zero

Throughout this section, assume that X is a complex normed space and Y is a complex Banach space. If an additive mapping $\sigma : X \to Y$ satisfies $\sigma(\sigma(x)) = x$ for all $x \in X$, then σ is called complex involution on X.

For given $x, y, t \in X$, we define

$$P_{x,y,t} := \left\{ (x + y, t), (x + \sigma(x), t), (x, y + t), (x, y + \sigma(t)), (y, t), (\sigma(y), \sigma(t)) \right\}.$$

Throughout this section, we assume that $\Omega \subset X^2$ satisfies the following condition:

For given $x, y \in X$, there exists $t \in X$ such that

$$(C) \qquad\qquad P_{x,y,t} \subset \Omega$$

where $\sigma : X \to X$ is a complex involution. In the following theorem, we prove the Hyers–Ulam stability theorem for the Drygas functional equation (7) in Ω.

Theorem 3 *Let $\varepsilon \geq 0$ be fixed. Suppose that $f : X \longrightarrow Y$ satisfies the functional inequality*

$$\| f(x + y) + f(x + \sigma(y)) - 2f(x) - f(y) - f(\sigma(y)) \| \leq \varepsilon \qquad (8)$$

for all $(x, y) \in \Omega$, where $\sigma : X \to X$ is a complex involution. Then there exists a unique mapping $S : X \longrightarrow Y$ such that S is a solution of (7) and

$$\| f(x) - S(x) \| \leq 25\varepsilon \qquad (9)$$

for all $x \in X$.

Proof Assume that $f : X \to Y$ be a function satisfying (7) for all $(x, y) \in \Omega$ and let $\sigma : X \to X$. Define $D : X \times X \to Y$ by

$$D(x, y) := f(x + y) + f(x + \sigma(y)) - 2f(x) - f(y) - f(\sigma(y)), \quad (x, y) \in \Omega.$$

Since Ω satisfies (C), for given $x, y \in X$, there exists $t \in X$ such that

$$\| D(x+y, t) \| \leq \varepsilon, \qquad \| D(x+\sigma(y), t) \| \leq \varepsilon, \qquad \| D(x, y+t) \| \leq \varepsilon,$$

$$\| D(x, y+\sigma(t)) \| \leq \varepsilon, \qquad \| D(y, t) \| \leq \varepsilon, \qquad \| D(\sigma(y), \sigma(t)) \| \leq \varepsilon.$$

Thus, using the triangle inequality we have

$$\left\| f(x+y) + f\big(x+\sigma(y)\big) - 2f(x) - f(y) - f\big(\sigma(y)\big) \right\|$$

$$= \left\| -\frac{1}{2}D(x+y,t) - \frac{1}{2}D\big(x\sigma(y),t\big) \right.$$

$$\left. + \frac{1}{2}D(x,y+t) + \frac{1}{2}D\big(x,y\sigma(t)\big) + \frac{1}{2}D(y,t) + \frac{1}{2}D\big(\sigma(y),\sigma(t)\big) \right\| \leq 3\varepsilon$$

for all $x, y \in X$. Next, according to Theorem 1, there exists a unique additive mapping $A : X \longrightarrow Y$ and a unique quadratic mapping $Q : X \longrightarrow Y$ such that $S = A + Q$ is a solution of (7) such that

$$\| f(x) - S(x) \| \leq 25\,\varepsilon$$

for all $x \in X$. This completes the proof.

The following corollary is a particular case of Theorem 3, where $\varepsilon = 0$.

Corollary 1 *Suppose that $f : X \longrightarrow Y$ satisfies the functional equation*

$$f(x+y) + f(x+\sigma(y)) = 2f(x) + f(y) + f(\sigma(y)) \qquad (10)$$

for all $(x, y) \in \Omega$, where $\sigma : X \to X$ is a complex involution. Then, (10) holds for all $x, y \in X$.

3 Construction of a Set Ω of Lebesgue Measure Zero

In this section, we construct a set Ω of measure zero satisfying the condition (C) when $X = \mathbb{C}$. From now on, we identify $\mathbb{R}^2$ with $\mathbb{C}$ and we suppose that $\sigma : \mathbb{C} \to \mathbb{C}$ is an involution. The following lemma is a crucial key of our construction [[27], Theorem 1.6].

Lemma 1 ([27]) *The set $\mathbb{R}$ of real numbers can be partitioned as $\mathbb{R} = F \cup K$ where F is of first Baire category, i.e., F is a countable union of nowhere dense subsets of $\mathbb{R}$, and K is of Lebesgue measure 0.*

The following lemma was proved by Chung and Rassias in [11] and [12].

Lemma 2 ([11, 12]) *Let K be a subset of $\mathbb{R}$ of measure 0 such that $K^c := \mathbb{R} \setminus K$ is of first Baire category. Then, for any countable subsets $U \subset \mathbb{R}$, $V \subset \mathbb{R} \setminus \{0\}$ and $M > 0$, there exists $\lambda \geq M$ such that*

$$U + \lambda V = \{u + \lambda v : u \in U, v \in V\} \subset K. \qquad (11)$$

We study two cases for the involution $\sigma : \mathbb{C} \to \mathbb{C}$. First, we suppose that $\sigma(z) = -\bar{z}$, where $\bar{z}$ is the conjugate of z for all $z \in \mathbb{C}$. It is easy to prove the following lemma.

Lemma 3 *The general solution $f : \mathbb{C} \to Y$ of the functional equation*

$$f(x + y) + f(x - \bar{y}) = 2f(x) + f(y) + f(-\bar{y}), \quad x, y \in \mathbb{C},$$

is

$$f(x) = A(x) + B(x, x), \quad x \in \mathbb{C},$$

where $B : \mathbb{C} \times \mathbb{C} \to Y$ is an arbitrary symmetric bi-additive function with $B(a, ib) = B(ia, ib) = 0$ and $A : \mathbb{C} \to Y$ is an arbitrary additive function with $A(ib) = 0$ for all $a, b \in \mathbb{R}$.

In this case, the condition (C) can be reduced to the following condition:

For given $x, y \in \mathbb{C}$, there exists $t \in \mathbb{C}$ such that

(C_1) $\quad \left\{ (x+y, t), (x-\bar{x}, t), (x, y+t), (x, y-\bar{t}), (y, t), (-\bar{y}, -\bar{t}) \right\} \subset \Omega.$ $\quad$ (12)

In the following theorem, we give the construction of a set $\Omega \subset \mathbb{C}^2$ of Lebesgue measure zero satisfying (12).

Theorem 4 *Let K be the set defined in Lemma 2, R be the rotation*

$$R = \begin{pmatrix} \frac{\sqrt{3}}{2} & 0 & -\frac{1}{2} & 0 \\ 0 & \frac{\sqrt{3}}{2} & 0 & -\frac{1}{2} \\ \frac{1}{2} & 0 & \frac{\sqrt{3}}{2} & 0 \\ 0 & \frac{1}{2} & 0 & \frac{\sqrt{3}}{2} \end{pmatrix} \quad (13)$$

and $\Omega = R^{-1}(K \times K \times K \times K)$. Then Ω satisfies (12) and has four-dimensional Lebesgue measure 0.

Proof Let $x, y, t \in \mathbb{C}$ such that $x = a + ib$, $y = c + id$ and $t = u + iv$ where $a, b, c, d, u, v \in \mathbb{R}$ and let

$$Q_{x,y,t} = \big\{ (a + c, b + d, u, v), (0, 2b, u, v), (a, b, c + u, d + v),$$

$$\times (a, b, c - u, d + v), (c, d, u, v), (-c, d, -u, v) \big\}.$$

Then Ω satisfies (12) if and only if, for every $x = a + ib$, $y = c + id \in \mathbb{C}$, there exists $t = u + iv \in \mathbb{C}$ such that

$$R(Q_{x,y,t}) \subset K \times K \times K \times K. \quad (14)$$

The inclusion (14) is equivalent to

$$S_{x,y,t} := \left\{ \frac{\sqrt{3}}{2}p_1 - \frac{1}{2}p_3, \; \frac{\sqrt{3}}{2}p_2 - \frac{1}{2}p_4, \; \frac{1}{2}p_1 + \frac{\sqrt{3}}{2}p_3, \; \frac{1}{2}p_2 \right.$$
$$\left. + \frac{\sqrt{3}}{2}p_4 : (p_1, p_2, p_3, p_4) \in Q_{x,y,t} \right\} \subset K.$$

If we choose $\alpha \in \mathbb{R}$ such that $v = \alpha u$, then we can easily check that the set $S_{x,y,u+i\alpha u}$ is included in a set of form $U + uV$, where

$$U = \left\{ 0, \frac{1}{2}a, b, \sqrt{3}b, \pm\frac{1}{2}c, \pm\frac{\sqrt{3}}{2}c, \frac{1}{2}d, \frac{\sqrt{3}}{2}d, \frac{1}{2}(a+c), \frac{\sqrt{3}}{2}(a+c), \right.$$
$$\left. \times \frac{1}{2}(b+d), \frac{\sqrt{3}}{2}(b+d), \left(\frac{\sqrt{3}}{2}a - \frac{1}{2}c\right), \left(\frac{\sqrt{3}}{2}b - \frac{1}{2}d\right), \left(\frac{1}{2}b + \frac{\sqrt{3}}{2}d\right) \right\}$$

and

$$V = \left\{ \pm\frac{1}{2}u, \pm\frac{\sqrt{3}}{2}u, -\frac{1}{2}\alpha u, \frac{\sqrt{3}}{2}\alpha u \right\}.$$

According to (11) in Lemma 2, for each $x = a + ib$, $y = c + id \in \mathbb{C}$ and $M > 0$, there exists $u \geq M$ such that

$$S_{x,y,u+i\alpha u} \subset U + uV \subset K.$$

Thus, Ω satisfies (12). This completes the proof.

In the second case, we assume that $\sigma(z) = -i\overline{z}$ for all $z \in \mathbb{C}$. We obtain the following results.

Lemma 4 *The general solution $f : \mathbb{C} \to Y$ of the functional equation*

$$f(x+y) + f(x - i\overline{y}) = 2f(x) + f(y) + f(-i\overline{y}), \quad x, y \in \mathbb{C},$$

is

$$f(x) = A(x) + B(x, x), \quad x \in \mathbb{C},$$

where $B : \mathbb{C} \times \mathbb{C} \to Y$ is an arbitrary symmetric bi-additive function with $B(a, b) = B(a, ib) = B(ia, ib) = 0$ and $A : \mathbb{C} \to Y$ is an arbitrary additive function with $A(ib) = A(b)$ for all $a, b \in \mathbb{R}$.

Here, the condition (C) reduces to the following condition:
For given $x, y \in \mathbb{C}$, there exists $t \in \mathbb{C}$ such that

$$(C_1) \quad \left\{ (x+y, t), (x - i\bar{x}, t), (x, y+t), (x, y - i\bar{t}), (y, t), (-i\bar{y}, -i\bar{t}) \right\} \subset \Omega.$$
$$(15)$$

By virtue of Theorem 3, it suffices to construct a set $\Omega \in \mathbb{C}^2$ of measure zero satisfying (15).

Theorem 5 *Let K be the set defined in Lemma 2, R be the rotation*

$$R = \begin{pmatrix} \frac{\sqrt{3}}{2} & 0 & -\frac{1}{2} & 0 \\ 0 & \frac{\sqrt{3}}{2} & 0 & -\frac{1}{2} \\ \frac{1}{2} & 0 & \frac{\sqrt{3}}{2} & 0 \\ 0 & \frac{1}{2} & 0 & \frac{\sqrt{3}}{2} \end{pmatrix} \qquad (16)$$

and $\Omega = R^{-1}(K \times K \times K \times K)$. Then Ω satisfies (15) and has four-dimensional Lebesgue measure 0.

Proof Similarly to the proof of the Theorem 4, let $x, y, t \in \mathbb{C}$ such that $x = a + ib$, $y = c + id$ and $t = u + iv$ where $a, b, c, d, u, v \in \mathbb{R}$ and let

$$Q'_{x,y,t} = \{ (a+c, b+d, u, v), (a-b, b-a, u, v), (a, b, c+u, d+v),$$

$$(a, b, c-v, d-u), (c, d, u, v), (-d, -c, -v, -u) \}.$$

Then, we get

$$S'_{x,y,t} := \left\{ \frac{\sqrt{3}}{2} p_1 - \frac{1}{2} p_3, \ \frac{\sqrt{3}}{2} p_2 - \frac{1}{2} p_4, \ \frac{1}{2} p_1 + \frac{\sqrt{3}}{2} p_3, \right.$$

$$\left. \frac{1}{2} p_2 + \frac{\sqrt{3}}{2} p_4 \ : \ (p_1, p_2, p_3, p_4) \in Q_{x,y,t} \right\} \subset U + uV \subset K, \qquad (17)$$

for some $u \in \mathbb{R}$, where

$$U = \left\{ \frac{1}{2} c, \ \frac{\sqrt{3}}{2} c, \ \pm\frac{\sqrt{3}}{2} d, \ \pm\frac{1}{2} d, \ \frac{\sqrt{3}}{2}(a+b), \ \frac{\sqrt{3}}{2}(a-b), \ \frac{\sqrt{3}}{2}(b+d), \ \frac{\sqrt{3}}{2}(b-a), \ \frac{1}{2}(a+c), \right.$$

$$\left. \frac{1}{2}(a-b), \frac{1}{2}(b+d), \frac{1}{2}(b-a), \left(\frac{\sqrt{3}}{2}a - \frac{1}{2}c \right), \left(\frac{\sqrt{3}}{2}b - \frac{1}{2}d \right), \left(\frac{1}{2}a + \frac{\sqrt{3}}{2}c \right), \left(\frac{1}{2}b + \frac{\sqrt{3}}{2}d \right) \right\}$$

and

$$V = \left\{ \pm \frac{1}{2}u, \pm \frac{1}{2}\alpha u, \pm \frac{\sqrt{3}}{2}u, \pm \frac{\sqrt{3}}{2}\alpha u \right\}.$$

Thus, Ω satisfies (15). This completes the proof.

In the following corollaries, we consider σ as the previous cases.

Corollary 2 *Let $\varepsilon \geq 0$ be a constant and let $f : \mathbb{C} \to Y$ satisfy*

$$\| f(x + y) + f(x + \sigma(y)) - 2f(x) - f(y) - f(\sigma(y)) \| \leq \varepsilon$$

for all $(x, y) \in \Omega$ where $\Omega \in \mathbb{C}^2$ of Lebesgue measure zero. Then there exist a unique arbitrary additive mapping $A : \mathbb{C} \to Y$ and a unique arbitrary symmetric bi-additive mapping $B : \mathbb{C} \times \mathbb{C} \to Y$ such that

$$\| f(x) - A(x) - B(x, x) \| \leq 25\varepsilon$$

for all $x \in \mathbb{C}$.

Corollary 3 *Suppose that $f : \mathbb{C} \to Y$ satisfies*

$$\| f(x + y) + f(x + \sigma(y)) - 2f(x) - f(y) - f(\sigma(y)) \| \to 0 \qquad (18)$$

as $(x, y) \in \Omega, |x| + |y| \to \infty$. Then f is a drygas mapping.

Proof The condition (18) implies that for each $n \in \mathbb{N}^*$, there exists $d_n > 0$ such that

$$\| f(x + y) + f(x + \sigma(y)) - 2f(x) - f(y) - f(\sigma(y)) \| \leq \frac{1}{n} \qquad (19)$$

for all $(x, y) \in \Omega_{d_n} := \{ (x, y) \in \Omega : |x| + |y| \geq d_n \}$. In view of the proof of theorems 4 and 5, the inclusions (14) and (17) imply that for every $x, y \in \mathbb{C}$ and $M > 0$ there exists $u \geq M$ such that

$$S_{x,y,t} = S_{x,y,u+i\alpha u} \subset \Omega \quad \text{and} \quad S'_{x,y,t} = S'_{x,y,u+i\alpha u} \subset \Omega \qquad (20)$$

For given $x, y \in \mathbb{C}$, if we take $M = d_n + |x| + |y|$ and if $u \geq M$, then we get

$$S_{x,y,u+i\alpha u} \subset \{ (p, q) : |p| + |q| \geq d_n \} \quad \text{and} \quad S'_{x,y,u+i\alpha u} \subset \{ (p, q) : |p| + |q| \geq d_n \}. \qquad (21)$$

It follows from (20) and (21) that for each x, $y \in \mathbb{C}$ there exists $u \in \mathbb{R}$ such that

$$S_{x,y,u+i\alpha u} \subset \Omega_{d_n} \quad \text{and} \quad S'_{x,y,u+i\alpha u} \subset \Omega_{d_n}. \tag{22}$$

So, Ω_{d_n} satisfies the conditions (12) and (15). Thus, by Theorem 3, there exists a unique mapping $S_n : \mathbb{C} \to Y$ such that S_n is a solution of (7) and

$$\| f(x) - S_n(x) \| \leq \frac{25}{n} \tag{23}$$

for all $x \in \mathbb{C}$. Now, replacing $n \in \mathbb{N}^*$ by $m \in \mathbb{N}^*$ in (23) and using the triangle inequality, we get

$$\| S_m(x) - S_n(x) \| \leq \| S_m(x) - f(x) + f(x) - S_n(x) \| \leq \frac{25}{m} + \frac{25}{n} \leq 50 \tag{24}$$

for all $m, n \in \mathbb{N}^*$ and all $x \in \mathbb{C}$. Hence, $S_m - S_n$ is bounded, so we conclude that $S_m = S_n$ for all $m, n \in \mathbb{N}^*$. Finally, letting $n \to \infty$ in (23), we get the result.

References

1. M. Ait Sibaha, B. Bouikhalene, E. Elqorachi, Hyers-Ulam-Rassias stability of the K-quadratic functional equation. J. Inequal. Pure Appl. Math. **8**(3) (2007)
2. C. Alsina, J.L. Garcia-Roig, On a conditional Cauchy equation on rhombuses, in *Functional Analysis, Approximation Theory and Numerical Analysis*, ed. by J.M. Rassias (World Scientific, Singapore, 1994)
3. L.M. Arriola, W.A. Beyer, Stability of the Cauchy functional equation over p-adic fields. Real Anal. Exch. **31**(1), 125–132 (2005)
4. A. Bahyrycz, J. Brzdęk, On solutions of the d'Alembert equation on a restricted domain. Aequationes Math. **85**, 169–183 (2013)
5. B. Batko, Stability of an alternative functional equation. J. Math. Anal. Appl. **339**, 303–311 (2008)
6. B. Batko, On approximation of approximate solutions of Dhombres' equation. J. Math. Anal. Appl. **340**, 424–432 (2008)
7. J. Brzdęk, On the quotient stability of a family of functional equations. Nonlinear Anal. **71**, 4396–4404 (2009)
8. J. Brzdęk, On a method of proving the Hyers-Ulam stability of functional equations on restricted domains. Aust. J. Math. Anal. Appl. **6**, 1–10 (2009)
9. J. Brzdęk, J. Sikorska, A conditional exponential functional equation and its stability. Nonlinear Anal. **72**, 2929–2934 (2010)
10. J. Chung, Stability of functional equations on restricted domains in a group and their asymptotic behaviors. Comput. Math. Appl. **60**, 2653–2665 (2010)
11. J. Chung, Stability of a conditional Cauchy equation on a set of measure zero. Aequationes Math. (2013). http://dx.doi.org/10.1007/s00010-013-0235-5
12. J. Chung, J.M. Rassias, Quadratic functional equations in a set of Lebesgue measure zero. J. Math. Anal. Appl. **419**(2), 1065–1075 (2014)
13. S. Czerwik, *Stability of Functional Equations of Ulam-Hyers-Rassias Type* (Hadronic Press, Inc., Palm Harbor, 2003)

14. D.Ž. Djoković, A representation theorem for $(X_1 - 1)(X_2 - 1)\ldots(X_n - 1)$ and its applications. Ann. Polon. Math. **22**(2), 189–198 (1969)
15. H. Drygas, Quasi-inner products and their applications, in *Advances in Multivariate Statistical Analysis*, ed. by A.K. Gupta (Reidel Publishing Company, Boston, 1987), pp. 13–30
16. B.R. Ebanks, P.L. Kannappan, P.K. Sahoo, A common generalization of functional equations characterizing normed and quasi-inner-product spaces. Can. Math. Bull. **35**, 321–327 (1992)
17. V.A. Faĭziev, P.K. Sahoo, On Drygas functional equation on groups. Int. J. Appl. Math. Stat. **7**, 59–69 (2007)
18. M. Fochi, An alternative functional equation on restricted domain. Aequationes Math. **70**, 201–212 (2005)
19. G.L. Forti, J. Sikorska, Variations on the Drygas equations and its stability. Nonlinear Anal. **74**, 343–350 (2011)
20. R. Ger, J. Sikorska, On the Cauchy equation on spheres. Ann. Math. Sil. **11**, 89–99 (1997)
21. D.H. Hyers, On the stability of the linear functional equation. Proc. Natl. Acad. Sci. USA **27**, 222–224 (1941)
22. S.-M. Jung, On the Hyers-Ulam stability of the functional equations that have the quadratic property. J. Math. Anal. Appl. **222**, 126–137 (1998)
23. S.-M. Jung, *Hyers-Ulam-Rassias Stability of Functional Equations in Nonlinear Analysis* (Springer, New York, 2011)
24. S.-M. Jung, P.K. Sahoo, Stability of functional equation of Drygas. Aequationes Math. **64**, 263–273 (2002)
25. M. Kuczma, Functional equations on restricted domains. Aequationes Math. **18**, 1–34 (1978)
26. Y.-H. Lee, Hyers-Ulam-Rassias stability of a quadratic-additive type functional equation on a restricted domain. Int. J. Math. Anal. **7**(55), 2745–2752 (2013)
27. J.C. Oxtoby, *Measure and Category* (Springer, New York, 1980)
28. J.M. Rassias, On the Ulam stability of mixed type mappings on restricted domains. J. Math. Anal. Appl. **281**, 747–762 (2002)
29. J.M. Rassias, M.J. Rassias, On the Ulam stability of Jensen and Jensen type mappings on restricted domains. J. Math. Anal. Appl. **281**, 516–524 (2003)
30. J. Sikorska, On two conditional Pexider functional equations and their stabilities. Nonlinear Anal. **70**, 2673–2684 (2009)
31. J. Sikorska, On a direct method for proving the Hyers-Ulam stability of functional equations. J. Math. Anal. Appl. **372**, 99–109 (2010)
32. H. Stetkær, Functional equations on abelian groups with involution. II. Aequationes Math. **55**, 227–240 (1998)
33. H. Stetkær, Functional equations involving means of functions on the complex plane. Aequationes Math. **55**, 47–62 (1998)
34. Gy. Szabo, Some functional equations related to quadratic functions. Glasnik Math. **38**, 107–118 (1983)
35. S.M. Ulam, A Collection of Mathematical Problems (Interscience Publication, New York, 1961). Problems in Modern Mathematics, Wiley, New York 1964
36. D. Yang, Remarks on the stability of Drygas equation and the Pexider-quadratic equation. Aequationes Math. **68**, 108–116 (2004)

Fourier Transforms and Ulam Stabilities of Linear Differential Equations

Murali Ramdoss and Ponmana Selvan Arumugam

Abstract The purpose of this paper is to study the Hyers–Ulam stability and Generalized Hyers–Ulam stability of the general Linear Differential Equations of first order and second order with constant coefficients using Fourier Transform method. Moreover, the Hyers–Ulam stability constants of these differential equations are obtained. Some examples are given to illustrate the main results.

AMS Subject Classification 35B35, 34K20, 26D10, 44A10, 39B82

1 Introduction

We say that a functional equation is stable, if for every approximate solution, there exists an exact solution near to it. A simulating and famous talk presented by Ulam [44] in 1940 motivated the study of stability problems for various functional equations. He gave a wide range of talk before a Mathematical Colloquium at the University of Wisconsin in which he presented a list of unsolved problems. Among those was the following question concerning the stability of homomorphisms.

Theorem 1 (Ulam [44]) *Let G_1 be a group and let G_2 be a group endowed with a metric ρ. Given $\epsilon > 0$, does there exist a $\delta > 0$ such that if $f : G_1 \rightarrow G_2$ satisfies $\rho(f(xy), f(x) \, f(y)) < \delta$, for all $x, y \in G$, then we can find a homomorphism $h : G_1 \rightarrow G_2$ exists with $\rho(f(x), h(x)) < \epsilon$ for all $x \in G_1$?*

If the answer is affirmative, we say that the functional equation for homomorphisms is stable. In 1941, Hyers [11] was the first mathematician to present the result concerning the stability of functional equations. He brilliantly answered the question of Ulam, the problem for the case of approximately additive mappings,

M. Ramdoss (✉) · P. S. Arumugam
PG and Research Department of Mathematics, Sacred Heart College (Autonomous), Tirupattur, Tamil Nadu, India

© Springer Nature Switzerland AG 2019
G. A. Anastassiou and J. M. Rassias (eds.), *Frontiers in Functional Equations and Analytic Inequalities*, https://doi.org/10.1007/978-3-030-28950-8_12

when G_1 and G_2 are assumed to be Banach spaces. The result of Hyers is stated in the following celebrated theorem.

Theorem 2 (Hyers [11]) *Assume that G_1 and G_2 are Banach spaces. If a function $f : G_1 \to G_2$ satisfies the inequality $\|f(x + y) - f(x) - f(y)\| \leq \epsilon$ for some $\epsilon > 0$ and for all $x, y \in G_1$, then the limit*

$$A(x) = \lim_{n \to \infty} 2^{-n} f(2^n x)$$

exists for each $x \in G_1$ and $A : G_1 \to G_2$ is the unique additive function such that

$$\|f(x) - A(x)\| \leq \epsilon \tag{1}$$

for all $x \in G_1$. Moreover, if $f(tx)$ is continuous in t for each fixed $x \in G_1$, then A is linear.

Taking the above fact into account, the additive functional equation

$$f(x + y) = f(x) + f(y)$$

is said to have *Hyers–Ulam stability* on (G_1, G_2). In the above theorem, an additive function A satisfying the inequality (1) is constructed directly from the given function f and it is the most powerful tool to study the stability of several functional equations. In course of time, the theorem formulated by Hyers was generalized by Rassias [33], Aoki [4], and Bourgin [6] for additive mappings.

In 1982, Rassias [34] gave a further generalization of the result of D.H. Hyers and proved a theorem using weaker conditions controlled by a product of different powers of norms. His theorem is presented as follows:

Theorem 3 (Rassias [34]) *Let $f : X \to Y$ be a mapping from a Normed Vector space X into a Banach space Y subject to the inequality*

$$\|f(x + y) - f(x) - f(y)\| \leq \epsilon \|x\|^p \|y\|^p \tag{2}$$

for all $x, y \in X$, where ϵ and p are constants with $\epsilon > 0$ and $0 \leq p < \frac{1}{2}$. Then the limit

$$A(x) = \lim_{n \to \infty} \frac{1}{2^n} f(2^n x)$$

exists for all $x \in X$ and $A : X \to Y$ is the unique additive mapping which satisfies

$$\|f(x) - A(x)\| \leq \frac{\epsilon}{2 - 2^{2p}} \|x\|^{2p} \tag{3}$$

for all $x \in X$. If $p < 0$, then the inequality (2) holds for $x, y \neq 0$ and (3) for $x \neq 0$. If $p > 0$, then the inequality (2) holds for all $x, y \in X$ and the limit

$$A(x) = \lim_{n \to \infty} 2^n f\left(\frac{x}{2^n}\right)$$

exists for all $x \in X$. If in addition $f : X \to Y$ is a mapping such that the transformation $t \to f(tx)$ is continuous in $t \in \mathbb{R}$ for each fixed $x \in X$, then A is $\mathbb{R}$-linear mapping.

This type of stability involving a product of powers of norms is called Hyers–Ulam–Gavruta stability by Bouikhalence and Elquorachi [5], Nakmahachalasint [26, 27], Park and Nataji [31], Pietrzyk [32], and Sibaha et al. [42]. Since then, almost many mathematicians are studied the related Ulam stability problems on different types of functional equations or abstract spaces. (See, for example, [7–10, 17, 18, 28, 35, 36, 39, 40, 46]).

A generalization of Ulam's problem was recently proposed by replacing functional equations with differential equations: The differential equation

$$\phi\left(f, x, x', x'', \ldots x^{(n)}\right) = 0$$

has the Hyers–Ulam stability if for a given $\epsilon > 0$ and a function x such that

$$\left|\phi\left(f, x, x', x'', \ldots x^{(n)}\right)\right| \le \epsilon,$$

there exists a solution x_a of the differential equation such that $|x(t) - x_a(t)| \le K(\epsilon)$ and

$$\lim_{\epsilon \to 0} K(\epsilon) = 0.$$

If the preceding statement is also true when we replace ϵ and $K(\epsilon)$ by $\phi(t)$ and $\varphi(t)$, where ϕ, φ are appropriate functions not depending on x and x_a explicitly, then we say that the corresponding differential equation has the generalized Hyers–Ulam stability or Hyers–Ulam–Rassias stability.

Obloza seems to be the first author who has investigated the Hyers–Ulam stability of linear differential equations [29, 30]. Thereafter, in 1998, Alsina and Ger [3] were the first authors who investigated the Hyers–Ulam stability of differential equations. They proved in [3] the following theorem.

Theorem 4 *Assume that a differentiable function $f : I \to \mathbb{R}$ is a solution of the differential inequality $\|x'(t) - x(t)\| \le \epsilon$, where I is an open subinterval of $\mathbb{R}$. Then there exists a solution $g : I \to \mathbb{R}$ of the differential equation $x'(t) = x(t)$ such that for any $t \in I$, we have $\|f(t) - g(t)\| \le 3\epsilon$.*

This result of Alsina and Ger [3] has been generalized by Takahasi [43]. They proved in [43] that the Hyers–Ulam stability holds true for the Banach Space valued differential equation $y'(t) = \lambda y(t)$. Indeed, the Hyers–Ulam stability has been proved for the first-order linear differential equations in more general settings [13–15, 20–23].

Using the approach as in [44], Miura et al. [22], Miura [23], Takahasi et al. [43], and Miura et al. [20] proved that the Hyers–Ulam stability holds true for the differential equation $x' = \lambda x$, while Jung [13] proved a similar result for the differential equation $\phi(t)x'(t) = x$.

In 2006, Jung [16] investigated the Hyers–Ulam stability of a system of first-order linear differential equations with constant coefficients by using matrix method. In 2007, Wang et al. [45] studied the Hyers–Ulam stability of a class of first-order linear differential equations. Rus [41] discussed four types of Ulam stability: Ulam–Hyers stability, Generalized Ulam–Hyers stability, Ulam–Hyers-Rassias stability, and Generalized Ulam–Hyers–Rassias stability of the Ordinary Differential Equation $u'(t) = A(u(t)) + f(t, u(t)), t \in [a, b]$. In 2014, Alqifiary and Jung [1] proved the Generalized Hyers–Ulam stability of linear differential equation of the form

$$x^{(n)}(t) + \sum_{k=0}^{n-1} \alpha_k \, x^{(k)}(t) \ = \ f(t)$$

by using the Laplace Transform method, where α_k are scalars and x and f are n times continuously differentiable function and of the exponential order, respectively. Recently, the Hyers–Ulam stability of differential equations has been investigated in a series of paper [2, 12, 19, 24, 25, 37, 38, 47] and the investigation is ongoing. Motivated and connected by the above discussions, our main intention is by applying Fourier Transform method to investigate the Hyers–Ulam stability and Hyers–Ulam–Rassias stability of the first-order homogeneous linear differential equation of the form

$$x'(t) + l \, x(t) = 0 \tag{4}$$

and the non-homogeneous linear differential equation

$$x'(t) + l \, x(t) = r(t) \tag{5}$$

where l is a scalar, $x(t)$ and $r(t)$ are the continuously differentiable functions. Also, by using Fourier Transforms, we establish the Hyers–Ulam stability and Hyers–Ulam–Rassias stability of the second order homogeneous linear differential equation

$$x''(t) + l \, x'(t) + m \, x(t) = 0 \tag{6}$$

and the non-homogeneous second order differential equation

$$x''(t) + l \, x'(t) + m \, x(t) = r(t) \tag{7}$$

where l and m are scalars, $x(t)$ is a twice continuously differentiable function, and $r(t)$ is a continuously differentiable function.

2 Preliminaries

In this section, we introduce some standard notations, definitions, and theorems, it will be very useful to prove our main results.

Throughout this paper, $\mathbb{F}$ denotes the real field $\mathbb{R}$ or the complex field $\mathbb{C}$. A function $f : (0, \infty) \to \mathbb{F}$ of exponential order if there exists constants $A, B \in \mathbb{R}$ such that $|f(t)| \le Ae^{tB}$ for all $t > 0$.

For each function $f : (0, \infty) \to \mathbb{F}$ of exponential order. Let g denote the Fourier Transform of f so that

$$g(u) = \int_{-\infty}^{\infty} f(t) e^{-itu} \, dt.$$

Then, at points of continuity of f, we have

$$f(x) = \frac{1}{2\pi} \int_{-\infty}^{\infty} g(u) e^{-ixu} \, du,$$

this is called the inverse Fourier transforms. The Fourier transform of f is denoted by $\mathcal{F}(\xi)$. We also introduce a notion, the convolution of two functions.

Definition 1 (Convolution) Given two functions f and g, both are Lebesgue integrable on $(-\infty, +\infty)$. Let S denote the set of x for which the Lebesgue integral

$$h(x) = \int_{-\infty}^{\infty} f(t) g(x - t) \, dt$$

exists. This integral defines a function h on S called the convolution of f and g. We also write $h = f * g$ to denote this function.

Theorem 5 *The Fourier transform of the convolution of $f(x)$ and $g(x)$ is the product of the Fourier transform of $f(x)$ and $g(x)$. That is,*

$$\mathcal{F}\{f(x) * g(x)\} = \mathcal{F}\{f(x)\} \, \mathcal{F}\{g(x)\} = F(s) \, G(s)$$

or

$$\mathcal{F}\left\{ \int_{-\infty}^{\infty} f(t) g(x - t) \, dt \right\} = F(s) \, G(s),$$

where $F(s)$ and $G(s)$ are Fourier transform of $f(x)$ and $g(x)$, respectively.

Now, we give the definition of Hyers–Ulam stability and Generalized Hyers–Ulam stability of the differential equations (4), (5), (6), and (7).

Definition 2 The linear differential equation (4) is said to have the Hyers–Ulam stability, if there exists a constant $K > 0$ having the following properties: For every $\epsilon > 0$, there exists $x(t)$ is a continuously differentiable function satisfying the inequality $|x'(t) + l\,x(t)| \leq \epsilon$. Then there exists some $y : (0, \infty) \to \mathbb{F}$ satisfying the differential equation (4) such that $|x(t) - y(t)| \leq K\epsilon$, for any $t > 0$. We call such K as the Hyers–Ulam stability constant for (4).

Definition 3 We say that the non-homogeneous linear differential equation (5) has the Hyers–Ulam stability, if there exists $x(t)$ is a continuously differentiable function satisfying the following condition: For every $\epsilon > 0$ there exists a positive constant K such that $|x'(t) + l\,x(t) - r(t)| \leq \epsilon$. Then there exists a solution $y : (0, \infty) \to \mathbb{F}$ satisfies the differential equation (5) such that $|x(t) - y(t)| \leq K\epsilon$, for any $t > 0$. We call such K as the Hyers–Ulam stability constant for the differential equation (5).

Definition 4 We say that the homogeneous linear differential equation (6) has the Hyers–Ulam stability property, if there exists a real constant $K > 0$ satisfying the following properties: For every $\epsilon > 0$, there exists $x(t)$ is a twice continuously differentiable function satisfying $|x''(t) + l\,x'(t) + m\,x(t)| \leq \epsilon$. Then there exists $y : (0, \infty) \to \mathbb{F}$ satisfying the differential equation (6) such that $|x(t) - y(t)| \leq K\epsilon$, for any $t > 0$. We call such K as the Hyers–Ulam stability constant for (6).

Definition 5 The non-homogeneous linear differential equation (7) is said to have the Hyers–Ulam stability, if there exists a positive real constant $K > 0$ satisfying the following properties: For every $\epsilon > 0$, there exists $x(t)$ is a twice continuously differentiable function satisfying $|x''(t) + l\,x'(t) + m\,x(t) - r(t)| \leq \epsilon$. Then there exists some $y : (0, \infty) \to \mathbb{F}$ satisfying (7) such that $|x(t) - y(t)| \leq K\epsilon$, for any $t > 0$. We call such K as the Hyers–Ulam stability constant for (7).

Definition 6 We say that the homogeneous linear differential equation (4) has the Generalized Hyers–Ulam stability, if there exists a constant $K > 0$ having the following properties: For every $\epsilon > 0$ and $x(t)$ is a continuously differentiable function, if there exists $\phi : (0, \infty) \to (0, \infty)$ satisfies the inequality $|x'(t) + l\,x(t)| \leq \phi(t)\epsilon$. Then there exists some $y : (0, \infty) \to \mathbb{F}$ satisfying the differential equation (4) such that $|x(t) - y(t)| \leq K\,\phi(t)\epsilon$, for any $t > 0$. We call such K as Generalized Hyers–Ulam stability constant for (4).

Definition 7 The linear differential equation (5) is said to have the Generalized Hyers–Ulam stability, if there exists a positive constant K satisfying the following conditions: For every $\epsilon > 0$, there exists $x(t)$ is a continuously differentiable function and $\phi : (0, \infty) \to (0, \infty)$ satisfying the inequality

$$|x'(t) + l\,x(t) - r(t)| \leq \phi(t)\epsilon.$$

Then there exists a solution $y : (0, \infty) \to \mathbb{F}$ satisfying the differential equation (5) such that $|x(t) - y(t)| \leq K\,\phi(t)\epsilon$, for any $t > 0$. We call such K as the Generalized Hyers–Ulam stability constant for the differential equation (5).

Definition 8 We say that the homogeneous linear differential equation (6) has the Generalized Hyers–Ulam stability, if there exists a constant $K > 0$ having the following properties: For every $\epsilon > 0$ and $x(t)$ is a twice continuously differentiable function, if there exists $\phi : (0, \infty) \to (0, \infty)$ satisfies the inequality

$$|x''(t) + l\, x'(t) + m\, x(t)| \leq \phi(t)\epsilon.$$

Then there is a solution $y : (0, \infty) \to \mathbb{F}$ of the differential equation (6) such that $|x(t) - y(t)| \leq K\, \phi(t)\epsilon$, for any $t > 0$. We call such K as Generalized Hyers–Ulam stability constant for (6).

Definition 9 The non-homogeneous linear differential equation (7) is said to have the Generalized Hyers–Ulam stability, if there exists a positive constant K satisfying the following: For every $\epsilon > 0$, there exists $x(t)$ is a twice continuously differentiable function and $\phi : (0, \infty) \to (0, \infty)$ satisfying the inequality

$$|x''(t) + l\, x'(t) + m\, x(t) - r(t)| \leq \phi(t)\epsilon.$$

Then there exists a solution $y : (0, \infty) \to \mathbb{F}$ satisfying the differential equation (7) such that $|x(t) - y(t)| \leq K\, \phi(t)\epsilon$, for any $t > 0$. We call such K as the Generalized Hyers–Ulam stability constant for the differential equation (7).

3 Hyers–Ulam Stability

In the following theorems, we prove the Hyers–Ulam stability of the homogeneous and non-homogeneous linear differential equations (4), (5), (6), and (7). Firstly, we prove the Hyers–Ulam stability of first-order homogeneous differential equation (4).

Theorem 6 *Let l be a constant in $\mathbb{F}$. For every $\epsilon > 0$, there exists a positive constant K such that $x : (0, \infty) \to \mathbb{F}$ is a continuously differentiable function satisfying the inequality*

$$|x'(t) + l\, x(t)| \leq \epsilon \tag{8}$$

for all $t > 0$. Then there exists a solution $y : (0, \infty) \to \mathbb{F}$ of the differential equation (4) such that $|x(t) - y(t)| \leq K\epsilon$, for any $t > 0$.

Proof Assume that $x(t)$ is a continuously differentiable function that satisfies the inequality (8). Let us define a function $p : (0, \infty) \to \mathbb{F}$ such that $p(t) =: x'(t) + l\, x(t)$ for each $t > 0$. In view of (8), we have $|p(t)| \leq \epsilon$. Now, taking Fourier transform to $p(t)$, we have

$$\mathcal{F}\{p(t)\} = \mathcal{F}\{x'(t) + l\, x(t)\}$$

$$P(\xi) = \mathcal{F}\{x'(t)\} + l\, \mathcal{F}\{x(t)\} = -i\xi X(\xi) + l\, X(\xi) = (l - i\xi)X(\xi)$$

$$X(\xi) = \frac{P(\xi)}{(l - i\xi)}.$$

Thus

$$\mathcal{F}\{x(t)\} = X(\xi) = \frac{P(\xi)\,(l + i\xi)}{l^2 - \xi^2}. \tag{9}$$

Taking $Q(\xi) = \dfrac{1}{(l - i\xi)}$, then we have

$$\mathcal{F}\{q(t)\} = \frac{1}{(l - i\xi)} \quad \Rightarrow \quad q(t) = \mathcal{F}^{-1}\left\{\frac{1}{(l - i\xi)}\right\}.$$

Now, we set $y(t) = e^{-lt}$ and taking Fourier transform on both sides, we get

$$\mathcal{F}\{y(t)\} = Y(\xi) = \int_{-\infty}^{\infty} e^{-lt}\, e^{ist}\, dt = \int_{-\infty}^{0} e^{-lt}\, e^{ist}\, dt + \int_{0}^{\infty} e^{-lt}\, e^{ist}\, dt = 0$$

$$\tag{10}$$

Now,

$$\mathcal{F}\{y'(t) + l\,y(t)\} = \mathcal{F}\{y'(t)\} + l\,\mathcal{F}\{y(t)\} = -i\xi Y(\xi) + l\,Y(\xi) = (l - i\xi)Y(\xi)$$

Then by using (10), we have $\mathcal{F}\{y'(t) + l\,y(t)\} = 0$, since $\mathcal{F}$ is one-to-one operator, thus $y'(t) + l\,y(t) = 0$. Hence $y(t)$ is a solution of the differential equation (4). Then by using (9) and (10) we can obtain

$$\mathcal{F}\{x(t)\} - F\{y(t)\} = X(\xi) - Y(\xi) = \frac{P(\xi)\,(l + i\xi)}{l^2 - \xi^2}$$

$$= P(\xi)\,Q(\xi) = \mathcal{F}\{p(t)\}\,\mathcal{F}\{q(t)\}$$

$$\Rightarrow \qquad \mathcal{F}\{x(t) - y(t)\} = \mathcal{F}\{p(t) * q(t)\}$$

The operator $\mathcal{F}$ is one-to-one and linear, which gives $x(t) - y(t) = p(t) * q(t)$. Taking modulus on both sides, we have

$$|x(t) - y(t)| = |p(t) * q(t)|$$

$$= \left|\int_{-\infty}^{\infty} p(t)\,q(t - x)\,dx\right| \le |p(t)| \left|\int_{-\infty}^{\infty} q(t - x)\,dx\right| \le K\epsilon.$$

where $K = \left| \int\limits_{-\infty}^{\infty} q(t - x)\, dx \right|$ and the integral exists for each value of t. Then by the virtue of Definition 2 the homogeneous linear differential equation (4) has the Hyers–Ulam stability. $\qquad\square$

Now, we prove the Hyers–Ulam stability of the non-homogeneous linear differential equation (5) using Fourier transform method.

Theorem 7 *Let l be a constant in $\mathbb{F}$. For every $\epsilon > 0$, there exists a positive constant K such that $x : (0, \infty) \to \mathbb{F}$ is a continuously differentiable function that satisfies the inequality*

$$|x'(t) + l\, x(t) - r(t)| \leq \epsilon \tag{11}$$

for all $t > 0$. Then there exists a solution $y : (0, \infty) \to \mathbb{F}$ of the non-homogeneous differential equation (5) such that $|x(t) - y(t)| \leq K\epsilon$, for any $t > 0$.

Proof Assume that $x(t)$ is a continuously differentiable function that satisfies the inequality (11). Let us define a function $p : (0, \infty) \to \mathbb{F}$ such that $p(t) =: x'(t) + l\, x(t) - r(t)$ for each $t > 0$. In view of (11), we have $|p(t)| \leq \epsilon$. Now, taking Fourier transform to $p(t)$, we have

$$\mathcal{F}\{p(t)\} = \mathcal{F}\{x'(t) + l\, x(t) - r(t)\}$$

$$P(\xi) = \mathcal{F}\{x'(t)\} + l\, \mathcal{F}\{x(t)\} - \mathcal{F}\{r(t)\}$$

$$= -i\xi X(\xi) + l\, X(\xi) - R(\xi) = (l - i\xi)X(\xi) - R(\xi)$$

$$X(\xi) = \frac{P(\xi) + R(\xi)}{(l - i\xi)}.$$

Thus

$$\mathcal{F}\{x(t)\} = X(\xi) = \frac{\{P(\xi) + R(\xi)\}\,(l + i\xi)}{l^2 - \xi^2}. \tag{12}$$

Let us choose $Q(\xi)$ as $\dfrac{1}{(l - i\xi)}$, then we have

$$\mathcal{F}\{q(t)\} = \frac{1}{(l - i\xi)} \quad \Rightarrow \quad q(t) = \mathcal{F}^{-1}\left\{\frac{1}{(l - i\xi)}\right\}.$$

Now, we set $y(t) = e^{-lt} + (r(t) * q(t))$ and taking Fourier transform on both sides, we get

$$\mathcal{F}\{y(t)\} = Y(\xi) = \int\limits_{-\infty}^{\infty} e^{-lt}\, e^{ist}\, dt + \frac{R(\xi)}{(l - i\xi)} = \frac{R(\xi)}{(l - i\xi)} \tag{13}$$

Now,

$$\mathcal{F}\{y'(t) + l\, y(t)\} = -i\xi Y(\xi) + l\, Y(\xi) = R(\xi)$$

Then by using (13), we have $\mathcal{F}\{y'(t) + l\, y(t)\} = F\{r(t)\}$, since $\mathcal{F}$ is one-to-one operator, thus $y'(t) + l\, y(t) = r(t)$. Hence $y(t)$ is a solution of the differential equation (5). Then by using (12) and (13) we can obtain

$$\mathcal{F}\{x(t)\} - F\{y(t)\} = X(\xi) - Y(\xi) = \frac{\{P(\xi) + R(\xi)\}\,(l + i\xi)}{l^2 - \xi^2} - \frac{R(\xi)}{(l - i\xi)}$$

$$= P(\xi)\, Q(\xi) = \mathcal{F}\{p(t)\}\, \mathcal{F}\{q(t)\}$$

$$\Rightarrow \qquad \mathcal{F}\{x(t) - y(t)\} = \mathcal{F}\{p(t) * q(t)\}$$

The operator $\mathcal{F}$ is one-to-one and linear, which gives $x(t) - y(t) = p(t) * q(t)$. Taking modulus on both sides, we have

$$|x(t) - y(t)| = |p(t) * q(t)| = \left|\int_{-\infty}^{\infty} p(t)\, q(t - x)\, dx\right| \le |p(t)| \left|\int_{-\infty}^{\infty} q(t - x)\, dx\right| \le K\epsilon.$$

where $K = \left|\int_{-\infty}^{\infty} q(t - x)\, dx\right|$ and the integral exists for each value of t. Hence, by the virtue of Definition 3, the non-homogeneous differential equation (5) has the Hyers–Ulam stability. $\qquad \square$

Theorem 8 *Let l, m be a constant in $\mathbb{F}$ such that there exists μ, $\nu \in \mathbb{F}$ with $\mu\nu = m$, $\mu + \nu = -l$, and $\mu \ne \nu$. For every $\epsilon > 0$, there exists a positive constant K such that $x : (0, \infty) \to \mathbb{F}$ is a twice continuously differentiable function satisfying the inequality*

$$|x''(t) + l\, x'(t) + m\, x(t)| \le \epsilon \tag{14}$$

for all $t > 0$. Then there exists a solution $y : (0, \infty) \to \mathbb{F}$ of the homogeneous differential equation (6) such that $|x(t) - y(t)| \le K\epsilon$, for any $t > 0$.

Proof Assume that $x(t)$ be a continuously differentiable function satisfying the inequality (14). Let us define a function $p : (0, \infty) \to \mathbb{F}$ such that $p(t) =: x''(t) + l\, x'(t) + m\, x(t)$ for each $t > 0$. In view of (14), we have $|p(t)| \le \epsilon$. Now, taking Fourier transform to $p(t)$, we have

$$\mathcal{F}\{p(t)\} = \mathcal{F}\{x''(t) + l\, x'(t) + m\, x(t)\}$$

$$P(\xi) = \mathcal{F}\{x''(t)\} + l\, \mathcal{F}\{x'(t)\} + m\, \mathcal{F}\{x(t)\} = (\xi^2 - i\xi l + m)\, X(\xi)$$

$$X(\xi) = \frac{P(\xi)}{\xi^2 - i\xi l + m}.$$

Since l, m be a constant in $\mathbb{F}$ such that there exists $\mu, \nu \in \mathbb{F}$ with $\mu + \nu = -l$, $\mu\nu = m$, and $\mu \neq \nu$, we have $(\xi^2 - i\xi l + m) = (i\xi - \mu)(i\xi - \nu)$. Thus

$$\mathcal{F}\{x(t)\} = X(\xi) = \frac{P(\xi)}{(i\xi - \mu)(i\xi - \nu)}. \tag{15}$$

Let $Q(\xi) = \dfrac{1}{(i\xi - \mu)(i\xi - \nu)}$, then we have

$$\mathcal{F}\{q(t)\} = \frac{1}{(i\xi - \mu)(i\xi - \nu)} \quad \Rightarrow \quad q(t) = \mathcal{F}^{-1}\left\{\frac{1}{(i\xi - \mu)(i\xi - \nu)}\right\}.$$

Now, setting $y(t)$ as $\dfrac{\mu e^{-\mu t} - \nu e^{-\nu t}}{\mu - \nu}$ and taking Fourier transform, we obtain

$$\mathcal{F}\{y(t)\} = Y(\xi) = \int_{-\infty}^{\infty} \frac{\mu e^{-\mu t} - \nu e^{-\nu t}}{\mu - \nu} e^{ist} \, dt = 0. \tag{16}$$

Now,

$$\mathcal{F}\{y''(t) + l\, y'(t) + m\, y(t)\} = (\xi^2 - i\xi l + m)\, Y(\xi).$$

Then by using (16), we have $\mathcal{F}\{y''(t) + l\, y'(t) + m\, y(t)\} = 0$, since $\mathcal{F}$ is one-to-one operator, thus $y''(t) + l\, y'(t) + m\, y(t) = 0$. Hence $y(t)$ is a solution of the differential equation (6). Then by using (15) and (16) we can obtain

$$\mathcal{F}\{x(t)\} - F\{y(t)\} = X(\xi) - Y(\xi) = \frac{P(\xi)}{\xi^2 - i\xi l + m} = P(\xi)\, Q(\xi)$$

$$= \mathcal{F}\{p(t)\}\, \mathcal{F}\{q(t)\}$$

$$\Rightarrow \qquad \mathcal{F}\{x(t) - y(t)\} = \mathcal{F}\{p(t) * q(t)\}$$

The operator $\mathcal{F}$ is one-to-one and linear, which gives $x(t) - y(t) = p(t) * q(t)$. Taking modulus on both sides, we have

$$|x(t) - y(t)| = |p(t) * q(t)| = \left| \int_{-\infty}^{\infty} p(t)\, q(t-x)\, dx \right| \leq |p(t)| \left| \int_{-\infty}^{\infty} q(t-x)\, dx \right| \leq K\epsilon.$$

where $K = \left| \int_{-\infty}^{\infty} q(t-x)\, dx \right|$ and the integral exists for each value of t. Then by the virtue of Definition 4 the homogeneous linear differential equation (6) has the Hyers–Ulam stability. $\qquad\square$

Theorem 9 *Let l, m be a constant in $\mathbb{F}$ such that there exists μ, $\nu \in \mathbb{F}$ with $\mu\nu = m$, $\mu + \nu = -l$, and $\mu \neq \nu$. For every $\epsilon > 0$, there exists a positive constant K such that $x : (0, \infty) \to \mathbb{F}$ which is a twice continuously differentiable function satisfying the inequality*

$$|x''(t) + l\,x'(t) + m\,x(t) - r(t)| \leq \epsilon \qquad (17)$$

for all $t > 0$. Then there exists a solution $y : (0, \infty) \to \mathbb{F}$ of the non-homogeneous differential equation (7) such that $|x(t) - y(t)| \leq K\epsilon$, for any $t > 0$.

Proof Assume that $x(t)$ be a continuously differentiable function satisfying the inequality (17). Let us define a function $p : (0, \infty) \to \mathbb{F}$ such that $p(t) =: x''(t) + l\,x'(t) + m\,x(t) - r(t)$ for each $t > 0$. In view of (17), we have $|p(t)| \leq \epsilon$. Now, taking Fourier transform to $p(t)$, we have

$$\mathcal{F}\{p(t)\} = \mathcal{F}\{x''(t) + l\,x'(t) + m\,x(t) - r(t)\}$$

$$P(\xi) = \mathcal{F}\{x''(t)\} + l\,\mathcal{F}\{x'(t)\} + m\,\mathcal{F}\{x(t)\} - \mathcal{F}\{r(t)\}$$

$$= (\xi^2 - i\xi l + m)\,X(\xi) - R(\xi)$$

$$X(\xi) = \frac{P(\xi) + R(\xi)}{\xi^2 - i\xi l + m}.$$

Since l, m be a constant in $\mathbb{F}$ such that there exists μ, $\nu \in \mathbb{F}$ with $\mu + \nu = -l$, $\mu\nu = m$ and $\mu \neq \nu$, we have $(\xi^2 - i\xi l + m) = (i\xi - \mu)(i\xi - \nu)$. Thus

$$\mathcal{F}\{x(t)\} = X(\xi) = \frac{P(\xi) + R(\xi)}{(i\xi - \mu)(i\xi - \nu)}. \qquad (18)$$

Taking $Q(\xi) = \mathcal{F}\{q(t)\} = \dfrac{1}{(i\xi - \mu)(i\xi - \nu)}$ and set

$$y(t) = \frac{\mu e^{-\mu t} - \nu e^{-\nu t}}{\mu - \nu} + (r(t) * q(t))$$

and taking Fourier transform on both sides, we get

$$\mathcal{F}\{y(t)\} = Y(\xi) = \int\limits_{-\infty}^{\infty} \frac{\mu e^{-\mu t} - \nu e^{-\nu t}}{\mu - \nu}\,e^{ist}\,dt + \frac{R(\xi)}{(i\xi - \mu)(i\xi - \nu)}$$

$$= \frac{R(\xi)}{(i\xi - \mu)(i\xi - \nu)}. \qquad (19)$$

Now,

$$\mathcal{F}\{y''(t) + l\,y'(t) + m\,y(t)\} = \mathcal{F}\{y''(t)\} + l\,\mathcal{F}\{y'(t)\} + m\,\mathcal{F}\{y(t)\}$$

$$= (\xi^2 - i\xi l + m)\,Y(\xi) = R(\xi).$$

Then by using (19), we have $\mathcal{F}\{y''(t) + l\ y'(t) + m\ y(t)\} = 0\mathcal{F}\{r(t)\}$, since $\mathcal{F}$ is one-to-one operator, thus $y''(t) + l\ y'(t) + m\ y(t) = r(t)$. Hence $y(t)$ is a solution of the differential equation (7). Then by using (18) and (19) we can obtain

$$\mathcal{F}\{x(t)\} - F\{y(t)\} = X(\xi) - Y(\xi) = \frac{P(\xi) + R(\xi)}{(i\xi - \mu)\ (i\xi - \nu)} - \frac{R(\xi)}{(i\xi - \mu)\ (i\xi - \nu)}$$

$$= P(\xi)\ Q(\xi) = \mathcal{F}\{p(t)\}\ \mathcal{F}\{q(t)\}$$

$$\Rightarrow \qquad \mathcal{F}\{x(t) - y(t)\} = \mathcal{F}\{p(t) * q(t)\}$$

The operator $\mathcal{F}$ is one-to-one and linear, which gives $x(t) - y(t) = p(t) * q(t)$. Taking modulus on both sides, we have

$$|x(t) - y(t)| = |p(t) * q(t)| = \left| \int\limits_{-\infty}^{\infty} p(t)\ q(t-x)\ dx \right| \le |p(t)| \left| \int\limits_{-\infty}^{\infty} q(t-x)\ dx \right| \le K\epsilon.$$

where $K = \left| \int\limits_{-\infty}^{\infty} q(t-x)\ dx \right|$ and the integral exists for each value of t. Then by the virtue of Definition 5 the non-homogeneous linear differential equation (7) has the Hyers–Ulam stability. $\qquad\qquad\square$

4 Generalized Hyers–Ulam Stability

In the following theorems, we prove the Generalized Hyers–Ulam stability of the differential equations (4), (5), (6), and (7). Firstly, we prove the Generalized Hyers–Ulam stability of first-order homogeneous differential equation (4).

Theorem 10 *Let l be a constant in $\mathbb{F}$. For every $\epsilon > 0$, there exists a positive constant K such that $x : (0, \infty) \to \mathbb{F}$ is a continuously differentiable function and $\phi : (0, \infty) \to (0, \infty)$ be an integrable function satisfying*

$$|x'(t) + l\ x(t)| \le \phi(t)\epsilon \tag{20}$$

for all $t > 0$. Then there exists a solution $y : (0, \infty) \to \mathbb{F}$ of the homogeneous differential equation (4) such that $|x(t) - y(t)| \le K\ \phi(t)\epsilon$, for any $t > 0$.

Proof Assume that $x(t)$ is a continuously differentiable function satisfying the inequality (20). Let us define a function $p : (0, \infty) \to \mathbb{F}$ such that $p(t) =: x'(t) + l\ x(t)$ for each $t > 0$. In view of (20), we have $|p(t)| \le \phi(t)\epsilon$. Now, taking Fourier transform to $p(t)$, we have

$$\mathcal{F}\{x(t)\} = X(\xi) = \frac{P(\xi)\ (l + i\xi)}{l^2 - \xi^2}. \tag{21}$$

Choosing $Q(\xi) = \dfrac{1}{(l - i\xi)}$, then we have $q(t) = \mathcal{F}^{-1}\left\{\dfrac{1}{(l - i\xi)}\right\}$. Now, we set $y(t) = e^{-lt}$ and taking Fourier transform on both sides, we get

$$\mathcal{F}\{y(t)\} = Y(\xi) = \int\limits_{-\infty}^{\infty} e^{-lt}\, e^{ist}\, dt = 0. \tag{22}$$

Hence

$$\mathcal{F}\{y'(t) + l\, y(t)\} = -i\xi Y(\xi) + l\, Y(\xi) = (l - i\xi)Y(\xi)$$

Then by using (22), we have $\mathcal{F}\{y'(t) + l\, y(t)\} = 0$, since $\mathcal{F}$ is one-to-one operator, thus $y'(t) + l\, y(t) = 0$. Hence $y(t)$ is a solution of the differential equation (4). Then by using (21) and (22) we can obtain

$$\mathcal{F}\{x(t) - y(t)\} = \mathcal{F}\{p(t) * q(t)\}$$

The operator $\mathcal{F}$ is one-to-one and linear, which gives $x(t) - y(t) = p(t) * q(t)$. Taking modulus on both sides, we have

$$|x(t) - y(t)| = |p(t) * q(t)| = \left|\int\limits_{-\infty}^{\infty} p(t)\, q(t - x)\, dx\right| \leq |p(t)| \left|\int\limits_{-\infty}^{\infty} q(t - x)\, dx\right| \leq K\phi(t)\epsilon.$$

where $K = \left|\int\limits_{-\infty}^{\infty} q(t - x)\, dx\right|$, the integral exists for each value of t and $\phi(t)$ is an integrable function. Then by the virtue of Definition 6 the differential equation (4) has the Generalized Hyers–Ulam stability. $\qquad\square$

Now, we prove the Hyers–Ulam stability of the non-homogeneous linear differential equation (5) using Fourier transform method.

Theorem 11 *Let l be a constant in $\mathbb{F}$. For every $\epsilon > 0$, there exists a positive constant K such that $x : (0, \infty) \to \mathbb{F}$ is a continuously differentiable function and $\phi : (0, \infty) \to (0, \infty)$ be an integrable function satisfying the condition*

$$|x'(t) + l\, x(t) - r(t)| \leq \phi(t)\epsilon \tag{23}$$

for all $t > 0$. Then there exists a solution $y : (0, \infty) \to \mathbb{F}$ of the non-homogeneous differential equation (5) such that $|x(t) - y(t)| \leq K\, \phi(t)\epsilon$, for any $t > 0$.

Proof Assume that $x(t)$ is a continuously differentiable function satisfying the inequality (23). Let us define a function $p : (0, \infty) \to \mathbb{F}$ such that $p(t) =: x'(t) + l\, x(t) - r(t)$ for each $t > 0$. In view of (23), we have $|p(t)| \leq \phi(t)\epsilon$. Now, taking Fourier transform to $p(t)$, we have

$$\mathcal{F}\{x(t)\} = X(\xi) = \frac{\{P(\xi) + R(\xi)\}\ (l + i\xi)}{l^2 - \xi^2}. \tag{24}$$

Now, let us take $Q(\xi)$ as $\dfrac{1}{(l - i\xi)}$, then we have

$$\mathcal{F}\{q(t)\} = \frac{1}{(l - i\xi)} \quad \Rightarrow \quad q(t) = \mathcal{F}^{-1}\left\{\frac{1}{(l - i\xi)}\right\}.$$

We set $y(t) = e^{-lt} + (r(t) * q(t))$ and taking Fourier transform on both sides, we get

$$\mathcal{F}\{y(t)\} = Y(\xi) = \int_{-\infty}^{\infty} e^{-lt}\, e^{ist}\, dt + \frac{R(\xi)}{(l - i\xi)} = \frac{R(\xi)}{(l - i\xi)} \tag{25}$$

Now,

$$\mathcal{F}\{y'(t) + l\, y(t)\} = \mathcal{F}\{y'(t)\} + l\, \mathcal{F}\{y(t)\} = -i\xi Y(\xi) + l\, Y(\xi) = R(\xi)$$

Then by using (25), we have $\mathcal{F}\{y'(t) + l\, y(t)\} = F\{r(t)\}$, since $\mathcal{F}$ is one-to-one operator, thus $y'(t) + l\, y(t) = r(t)$. Hence $y(t)$ is a solution of the differential equation (5). Then by using (24) and (25) we can obtain

$$\mathcal{F}\{x(t) - y(t)\} = \mathcal{F}\{p(t) * q(t)\}$$

The operator $\mathcal{F}$ is one-to-one and linear, which gives $x(t) - y(t) = p(t) * q(t)$. Taking modulus on both sides, we have

$$|x(t) - y(t)| = |p(t) * q(t)| = \left| \int_{-\infty}^{\infty} p(t)\, q(t - x)\, dx \right| \le |p(t)| \left| \int_{-\infty}^{\infty} q(t - x)\, dx \right| \le K\, \phi(t)\epsilon.$$

where $K = \left| \int_{-\infty}^{\infty} q(t - x)\, dx \right|$, the integral exists for each value of t and $\phi(t)$ is an integrable function. Hence by the virtue of Definition 7 the differential equation (5) has the Generalized Hyers–Ulam stability. $\qquad \square$

Now, we are going to establish the Generalized Hyers–Ulam stability of the second order homogeneous differential equation (6).

Theorem 12 *Let l, m be a constant in $\mathbb{F}$ such that there exists $\mu, \nu \in \mathbb{F}$ with $\mu\nu = m$, $\mu + \nu = -l$, and $\mu \ne \nu$. For every $\epsilon > 0$, there exists a positive constant K such that $x : (0, \infty) \to \mathbb{F}$ is a twice continuously differentiable function and $\phi : (0, \infty) \to (0, \infty)$ be an integrable function satisfying the inequality*

$$|x''(t) + l\, x'(t) + m\, x(t)| \le \phi(t)\epsilon \tag{26}$$

for all $t > 0$. Then there exists a solution $y : (0, \infty) \to \mathbb{F}$ of the homogeneous differential equation (6) such that $|x(t) - y(t)| \le K\phi(t)\epsilon$, for any $t > 0$.

Proof Assume that $x(t)$ is a continuously differentiable function satisfying the inequality (26). Let us define a function $p : (0, \infty) \to \mathbb{F}$ such that $p(t) =: x''(t) + l\, x'(t) + m\, x(t)$ for each $t > 0$. In view of (26), we have $|p(t)| \le \phi(t)\epsilon$. Now, taking Fourier transform to $p(t)$, we have

$$P(\xi) = \mathcal{F}\{x''(t)\} + l\, \mathcal{F}\{x'(t)\} + m\, \mathcal{F}\{x(t)\} = (\xi^2 - i\xi l + m)\, X(\xi)$$

$$X(\xi) = \frac{P(\xi)}{\xi^2 - i\xi l + m}.$$

Since l, m be a constant in $\mathbb{F}$ such that there exists $\mu, \nu \in \mathbb{F}$ with $\mu + \nu = -l$, $\mu\nu = m$ and $\mu \ne \nu$, we have $(\xi^2 - i\xi l + m) = (i\xi - \mu)(i\xi - \nu)$. Thus

$$\mathcal{F}\{x(t)\} = X(\xi) = \frac{P(\xi)}{(i\xi - \mu)(i\xi - \nu)}. \tag{27}$$

Choosing $Q(\xi)$ as $\dfrac{1}{(i\xi - \mu)(i\xi - \nu)}$, then we have $\mathcal{F}\{q(t)\} = \dfrac{1}{(i\xi - \mu)(i\xi - \nu)}$ and we define a function $y(t) = \dfrac{\mu e^{-\mu t} - \nu e^{-\nu t}}{\mu - \nu}$ and taking Fourier transform on both sides, we get

$$\mathcal{F}\{y(t)\} = Y(\xi) = \int\limits_{-\infty}^{\infty} \frac{\mu e^{-\mu t} - \nu e^{-\nu t}}{\mu - \nu}\, e^{ist}\, dt = 0. \tag{28}$$

Now, $\mathcal{F}\{y''(t) + l\, y'(t) + m\, y(t)\} = (\xi^2 - i\xi l + m)\, Y(\xi)$. Then by using (28), we have $\mathcal{F}\{y''(t) + l\, y'(t) + m\, y(t)\} = 0$, since $\mathcal{F}$ is one-to-one operator, thus $y''(t) + l\, y'(t) + m\, y(t) = 0$. Hence $y(t)$ is a solution of the differential equation (6). Then by using (27) and (28) we can obtain

$$\mathcal{F}\{x(t)\} - F\{y(t)\} = X(\xi) - Y(\xi) = \frac{P(\xi)}{\xi^2 - i\xi l + m}$$

$$= P(\xi)\, Q(\xi) = \mathcal{F}\{p(t)\}\, \mathcal{F}\{q(t)\}$$

$$\Rightarrow \quad \mathcal{F}\{x(t) - y(t)\} = \mathcal{F}\{p(t) * q(t)\}$$

The operator $\mathcal{F}$ is one-to-one and linear, which gives $x(t) - y(t) = p(t) * q(t)$. Taking modulus on both sides, we have

$$|x(t) - y(t)| = |p(t) * q(t)| = \left| \int_{-\infty}^{\infty} p(t)\, q(t - x)\, dx \right|$$

$$\leq |p(t)| \left| \int_{-\infty}^{\infty} q(t - x)\, dx \right| \leq K\phi(t)\epsilon.$$

where $K = \left| \int_{-\infty}^{\infty} q(t - x)\, dx \right|$ exists for each value of t and $\phi(t)$ is an integrable function. Then by the virtue of Definition 8 the homogeneous linear differential equation (6) has the Generalized Hyers–Ulam stability. □

Finally, we are going to investigate the Generalized Hyers–Ulam stability of the second order non-homogeneous differential equation (7).

Theorem 13 *Let l, m be a constant in $\mathbb{F}$ such that there exists $\mu, v \in \mathbb{F}$ with $\mu v = m$, $\mu + v = -l$, and $\mu \neq v$. For every $\epsilon > 0$, there exists a positive constant K such that $x : (0, \infty) \to \mathbb{F}$ is a twice continuously differentiable function and $\phi : (0, \infty) \to (0, \infty)$ be an integrable function satisfying the inequality*

$$|x''(t) + l\, x'(t) + m\, x(t) - r(t)| \leq \phi(t)\epsilon \tag{29}$$

for all $t > 0$. Then there exists a solution $y : (0, \infty) \to \mathbb{F}$ of the non-homogeneous differential equation (7) such that $|x(t) - y(t)| \leq K\phi(t)\epsilon$, for any $t > 0$.

Proof Assume that $x(t)$ is a continuously differentiable function satisfying the inequality (29). Let us define a function $p : (0, \infty) \to \mathbb{F}$ such that $p(t) =: x''(t) + l\, x'(t) + m\, x(t) - r(t)$ for each $t > 0$. In view of (29), we have $|p(t)| \leq \phi(t)\epsilon$. Now, taking Fourier transform to $p(t)$, we have

$$P(\xi) = \mathcal{F}\{x''(t)\} + l\, \mathcal{F}\{x'(t)\} + m\, \mathcal{F}\{x(t)\} - \mathcal{F}\{r(t)\}$$

$$= (\xi^2 - i\xi l + m)\, X(\xi) - R(\xi)$$

$$X(\xi) = \frac{P(\xi) + R(\xi)}{\xi^2 - i\xi l + m}.$$

Since l, m be a constant in $\mathbb{F}$ such that there exists $\mu, v \in \mathbb{F}$ with $\mu + v = -l$, $\mu v = m$ and $\mu \neq v$, we have $(\xi^2 - i\xi l + m) = (i\xi - \mu)(i\xi - v)$. Thus

$$\mathcal{F}\{x(t)\} = X(\xi) = \frac{P(\xi) + R(\xi)}{(i\xi - \mu)(i\xi - v)}. \tag{30}$$

Assuming $Q(\xi) = \mathcal{F}\{q(t)\} = \dfrac{1}{(i\xi - \mu)(i\xi - v)}$ and let us define a function

$$y(t) = \frac{\mu e^{-\mu t} - \nu e^{-\nu t}}{\mu - \nu} + (r(t) * q(t))$$

and taking Fourier transform on both sides, we get

$$\mathcal{F}\{y(t)\} = Y(\xi) = \int_{-\infty}^{\infty} \frac{\mu e^{-\mu t} - \nu e^{-\nu t}}{\mu - \nu} e^{ist} \, dt + \frac{R(\xi)}{(i\xi - \mu)(i\xi - \nu)}$$

$$= \frac{R(\xi)}{(i\xi - \mu)(i\xi - \nu)}. \tag{31}$$

Now, $\mathcal{F}\{y''(t) + l\, y'(t) + m\, y(t)\} = (\xi^2 - i\xi l + m)\, Y(\xi) = R(\xi)$. Then by using (31), we have $\mathcal{F}\{y''(t) + l\, y'(t) + m\, y(t)\} = 0\mathcal{F}\{r(t)\}$, since $\mathcal{F}$ is one-to-one operator, thus $y''(t) + l\, y'(t) + m\, y(t) = r(t)$. Hence $y(t)$ is a solution of the differential equation (7). Then by using (30) and (31) we can obtain

$$\mathcal{F}\{x(t)\} - F\{y(t)\} = \frac{P(\xi) + R(\xi)}{(i\xi - \mu)(i\xi - \nu)} - \frac{R(\xi)}{(i\xi - \mu)(i\xi - \nu)}$$

$$= P(\xi)\, Q(\xi) = \mathcal{F}\{p(t)\}\, \mathcal{F}\{q(t)\}$$

$$\Rightarrow \quad \mathcal{F}\{x(t) - y(t)\} = \mathcal{F}\{p(t) * q(t)\}$$

The operator $\mathcal{F}$ is one-to-one and linear, which gives $x(t) - y(t) = p(t) * q(t)$. Taking modulus on both sides, we have

$$|x(t) - y(t)| = |p(t) * q(t)| = \left| \int_{-\infty}^{\infty} p(t)\, q(t - x)\, dx \right| \le |p(t)| \left| \int_{-\infty}^{\infty} q(t - x)\, dx \right| \le K\phi(t)\epsilon.$$

where $K = \left| \int_{-\infty}^{\infty} q(t - x)\, dx \right|$ and the integral exists for each value of t. Then by the virtue of Definition 9 the non-homogeneous linear differential equation (7) has the Generalized Hyers–Ulam stability. □

5 Applications

In this section, we investigate some examples to illustrate the main results.

Example 1 Consider the non-homogeneous differential equation

$$x'(t) + x(t) = 2\cos t. \tag{32}$$

Using Theorem 7, we have $|x'(t) + x(t) - 2\cos t| \leq \epsilon$, where x is a continuously differentiable function. Let $p(t) = x'(t) + x(t) - 2\cos t$ for each $t > 0$ and we have $|p(t)| \leq \epsilon$. Now, taking Fourier transform to $p(t)$,

$$P(\xi) = (-i\xi + 1)X(\xi) + \mathcal{F}\{2\cos t\} = (1 - i\xi)X(\xi) - \delta(w - 1) - \delta(w + 1)$$

$$X(\xi) = \frac{P(\xi) + \delta(w - 1) + \delta(w + 1)}{(1 - i\xi)}.$$

where $\mathcal{F}\{\cos t\} = \dfrac{\delta(w - 1) + \delta(w + 1)}{2}$, $\delta(t)$ is a delta function, and w is a frequency of 1 cycle/second. Let $Q(\xi)$ as $\dfrac{1}{(1 - i\xi)}$, then we have $\mathcal{F}\{q(t)\} = \dfrac{1}{(1 - i\xi)}$. Since, we have a solution function $y(t) = e^{-t} + [(2\cos t) * q(t)]$ and taking Fourier transform, we get

$$\mathcal{F}\{y(t)\} = Y(\xi) = \frac{\delta(w - 1) + \delta(w + 1)}{(1 - i\xi)}.$$

Also, $\mathcal{F}\{y'(t) + y(t)\} = (1 - i\xi)\,Y(\xi) = 2\mathcal{F}\{\cos t\}$, since $\mathcal{F}$ is one-to-one operator, thus $y'(t) + y(t) = 2\cos t$. Hence $y(t)$ is a solution of the differential equation (32). Then by Theorem 7, we obtain that $|x(t) - y(t)| \leq K\epsilon$. Hence, the non-homogeneous differential equation (32) has the Hyers–Ulam stability.

Example 2 Let us consider the non-homogeneous differential equation

$$x''(t) - 21x'(t) + 90x(t) = e^{-|t|}, \tag{33}$$

where $x(t)$ is a twice continuously differentiable function satisfying the inequality

$$\left| x''(t) - 21x'(t) + 90x(t) - e^{-|t|} \right| \leq \epsilon.$$

Take $p(t) = x''(t) - 21x'(t) + 90x(t) - e^{-|t|}$, then $|p(t)| \leq \epsilon$. Now, taking Fourier transform, we get

$$P(\xi) = (\xi^2 + 21i\xi + 90) - \mathcal{F}\{e^{-|t|}\} = (\xi^2 + 21i\xi + 90)\,X(\xi) - \frac{2}{1 + \xi^2}$$

$$X(\xi) = \frac{P(\xi) + \dfrac{2}{1 + \xi^2}}{\xi^2 + 21i\xi + 90}.$$

Since l, m be a constant in $\mathbb{F}$ such that there exists $\mu, \nu \in \mathbb{F}$ with $\mu + \nu = -l$, $\mu\nu = m$, and $\mu \neq \nu$, we have $(\xi^2 + 21i\xi + 90) = (i\xi + 6)(i\xi + 15)$. Taking

$$Q(\xi) = \mathcal{F}\{q(t)\} = \frac{1}{(i\xi + 6)\,(i\xi + 15)}$$

and set $y(t) = \dfrac{-6e^{6t} + 15e^{15t}}{9} + \left(e^{-|t|} * q(t)\right)$ and taking Fourier transform on

both sides, we get $Y(\xi) = \dfrac{2}{(1 + \xi^2)\,(i\xi + 6)\,(i\xi + 15)}$. Now,

$$\mathcal{F}\{y''(t) - 21\,y'(t) + 90\,y(t)\} = (\xi^2 + 21i\xi + 90)\,Y(\xi) = \frac{2}{1 + \xi^2}.$$

then by using the Theorem 9, we can have $y(t)$ satisfying the differential equation (33) and $X(\xi) - Y(\xi) = \dfrac{P(\xi)}{(i\xi + 6)\,(i\xi + 15)}$ gives that $x(t) - y(t) = p(t) * q(t)$.
Hence by Theorem 9, we obtain $|x(t) - y(t)| \le K\epsilon$. Thus the non-homogeneous linear differential equation (33) has the Hyers–Ulam stability.

Example 3 Consider the differential equation

$$x''(t) - x(t) = e^{-a|t|}, \tag{34}$$

$a > 0$ and $x(t)$ is a twice continuously differentiable function satisfying the inequality

$$\left| x''(t) - x(t) - e^{-a|t|} \right| \le \epsilon.$$

Take $p(t) = x''(t) - x(t) - e^{-a|t|}$, then $|p(t)| \le \epsilon$. Now, taking Fourier transform, we get

$$\mathcal{F}\{p(t)\} = \mathcal{F}\left\{ x''(t) - x(t) - e^{-a|t|} \right\}$$

$$P(\xi) = (\xi^2 - 1) - \mathcal{F}\{e^{-a|t|}\} = (\xi^2 - 1)\,X(\xi) - \frac{2a}{a^2 + \xi^2}$$

$$X(\xi) = \frac{P(\xi) + \dfrac{2a}{a^2 + \xi^2}}{\xi^2 - 1} = \frac{P(\xi) + \dfrac{2a}{a^2 + \xi^2}}{(i\xi + 1)\,(i\xi - 1)}.$$

Letting $Q(\xi) = \mathcal{F}\{q(t)\} = \dfrac{1}{(i\xi + 1)\,(i\xi - 1)}$ and set $y(t) = e^{-t} + \left(e^{-a|t|} * q(t)\right)$ and taking Fourier transform on both sides, we get $Y(\xi) = \dfrac{2a}{(a^2 + \xi^2)\,(i\xi + 1)\,(i\xi - 1)}$. Now,

$$\mathcal{F}\{y''(t) - y(t)\} = (\xi^2 - 1)\,Y(\xi) = \frac{2a}{a^2 + \xi^2}.$$

then by using Theorem 9, we can see that $y(t)$ satisfies the differential equation (34) and $X(\xi) - Y(\xi) = \dfrac{P(\xi)}{(i\xi + 1)(i\xi - 1)}$ gives that $x(t) - y(t) = p(t) * q(t)$. Hence by Theorem 9, we obtain $|x(t) - y(t)| \le K\epsilon$. Thus the differential equation (34) has the Hyers–Ulam stability.

Remark 1 The above examples are also true when we replace ϵ and $K\epsilon$ with $\phi(t)\epsilon$ and $K\phi(t)\epsilon$, respectively, where $\phi(t)$ does not depend on x and y explicitly, then we say that the corresponding differential equations has the Generalized Hyers–Ulam stability.

Conclusion We have proved the Hyers–Ulam stability and Generalized Hyers–Ulam stability of the linear differential equations of the first and second order with constant coefficients using Fourier Transform method. That is, we established the sufficient criteria for Hyers–Ulam stability and Generalized Hyers–Ulam stability of the linear differential equation of the first and second order with constant coefficients using Fourier Transform method. Additionally, this paper also provides another method to study the Hyers–Ulam stability of differential equations. Also, this paper shows that the Fourier Transform method is more convenient to study the Hyers–Ulam stability and Generalized Hyers–Ulam stability of the linear differential equation with constant coefficients.

References

1. Q.H. Alqifiary, S.M. Jung, Laplace transform and generalized Hyers–Ulam stability of differential equations. Electron. J. Differ. Equ. **2014**(80), 1–11 (2014)
2. Q.H. Alqifiary, J.K. Miljanovic, Note on the stability of system of differential equations $\dot{x}(t) = f(t, x(t))$. Gen. Math. Notes **20**(1), 27–33 (2014)
3. C. Alsina, R. Ger, On Some inequalities and stability results related to the exponential function. J. Inequal. Appl. **2**, 373–380 (1998)
4. T. Aoki, On the stability of the linear transformation in Banach spaces. J. Math. Soc. Jpn. **2**, 64–66 (1950)
5. B. Bouikhalene, E. Elquorachi, Ulam–Gavruta–Rassias stability of the Pexider functional equation. Int. J. Appl. Math. Stat. **7**, 7–39 (2007)
6. D.G. Bourgin, Classes of transformations and bordering transformations. Bull. Am. Math. Soc. **57**, 223–237 (1951)
7. M. Burger, N. Ozawa, A. Thom, On Ulam stability. Israel J. Math. **193**, 109–129 (2013)
8. M. Eshaghi Gordji, A. Javadian, J.M. Rassias, Stability of systems of bi-quadratic and additive-cubic functional equations in Frechet's spaces. Funct. Anal. Approx. Comput. **4**(1), 85–93 (2012)
9. L. Gang, X. Jun, L. Qi, J. Yuanfeng, Hyers–Ulam stability of derivations in fuzzy Banach space. J. Nonlinear Sci. Appl. **9**(12), 5970–5979 (2016)
10. P. Gavruta, L. Gavruta, A New method for the generalized Hyers–Ulam–Rassias stability. Int. J. Non Linear Anal. **2**, 11–18 (2010)
11. D.H. Hyers, On the stability of a linear functional equation. Proc. Natl. Acad. Sci. U.S.A. **27**, 222–224 (1941)
12. J. Huang, S.M. Jung, Y. Li, On Hyers–Ulam stability of nonlinear differential equations. Bull. Korean Math. Soc. **52**, 685–697 (2015)

13. S.M. Jung, Hyers–Ulam stability of linear differential equation of first order. Appl. Math. Lett. **17**, 1135–1140 (2004)
14. S.M. Jung, Hyers–Ulam stability of linear differential equations of first order (III). J. Math. Anal. Appl. **311**, 139–146 (2005)
15. S.M. Jung, Hyers–Ulam stability of linear differential equations of first order (II). Appl. Math. Lett. **19**, 854–858 (2006)
16. S.M. Jung, Hyers–Ulam stability of a system of first order linear differential equations with constant coefficients. J. Math. Anal. Appl. **320**, 549–561 (2006)
17. S.M. Jung, *Hyers–Ulam–Rassias Stability of Functional Equation in Nonlinear Analysis* (Springer, New York, 2011)
18. Y.H. Lee, K.W. Jun, A generalization of Hyers–Ulam–Rassias stability of Jensen's equation. J. Math. Anal. Appl. **238**, 305–315 (1999)
19. T. Li, A. Zada, S. Faisal, Hyers–Ulam stability of nth order linear differential equations. J. Nonlinear Sci. Appl. **9**, 2070–2075 (2016)
20. T. Miura, On the Hyers–Ulam stability of a differentiable map. Sci. Math. Jpn. **55**, 17–24 (2002)
21. T. Miura, S.M. Jung, S.E. Takahasi, Hyers–Ulam–Rassias stability of the Banach space valued linear differential equation $y' = \lambda y$. J. Korean Math. Soc. **41**, 995–1005 (2004)
22. T. Miura, S. Miyajima, S.E. Takahasi, A characterization of Hyers–Ulam stability of first order linear differential operators. J. Math. Anal. Appl. **286**, 136–146 (2003)
23. T. Miura, S.E. Takahasi, H. Choda, On the Hyers–Ulam stability of real continuous function valued differentiable map. Tokyo J. Math. **24**, 467–476 (2001)
24. R. Murali, A. Ponmana Selvan, On the generalized Hyers–Ulam stability of linear ordinary differential equations of higher order. Int. J. Pure Appl. Math. **117**(12), 317–326 (2017)
25. R. Murali, A. Ponmana Selvan, Hyers–Ulam–Rassias stability for the linear ordinary differential equation of third order. Kragujevac J. Math. **42**(4), 579–590 (2018)
26. P. Nakmahachalasint, On the generalized Ulam–Gavruta–Rassias stability of a mixed type linear and Euler-Lagrange-Rassias functional equation. Int. J. Math. Math. Sci. **2007**, 63239 (2007). http://dx.doi.org/10.1155/2007/63239
27. P. Nakmahachalasint, On the generalized Hyers–Ulam–Rassias and Ulam–Gavruta–Rassias stability of an additive functional equation in several variables. Int. J. Math. Math. Sci. **2007**, 13437 (2007). http://dx.doi.org/10.1155/2007/13437
28. M. Nazarianpoor, J.M. Rassias, G.H. Sadeghi, Solution and stability of quattuorvigintic functional equation in intuitionistic fuzzy normed spaces. Iran. J. Fuzzy Syst. **15**(4), 13–30 (2018)
29. M. Obloza, Hyers stability of the linear differential equation. Rockznik Nauk-Dydakt. Prace Math. **13**, 259–270 (1993)
30. M. Obloza, Connection between Hyers and Lyapunov stability of the ordinary differential equations. Rockznik Nauk-Dydakt. Prace Math. **14**, 141–146 (1997)
31. C. Park, A. Najati, Homomorphisms and derivations in C^* Algebras. Abstr. Appl. Anal. **2007**, 80630 (2007) http://dx.doi.org/10.1155/2007/80630
32. A. Pietrzyk, Stability if the Euler-Lagrange-Rassias functional equation. Demonstr. Math. **39**(3), 523–530 (2006)
33. Th.M. Rassias, On the stability of the linear mappings in Banach spaces. Proc. Am. Math. Soc. **72**, 297–300 (1978)
34. J.M. Rassias, On approximation of approximately linear mappings by linear mappings. J. Funct. Anal. **46**, 126–130 (1982)
35. J.M. Rassias, R. Murali, M.J. Rassias, A. Antony Raj, General solution, stability and nonstability of quattuorvigintic functional equation in multi-Banach spaces. Int. J. Math. Appl. **5**, 181–194 (2017)
36. K. Ravi, J.M. Rassias, B.V. Senthil Kumar, Generalized Ulam–Hyers stability of the Harmonic Mean functional equation in two variables. Int. J. Anal. Appl. **1**(1), 1–17 (2013)

37. K. Ravi, R. Murali, A. Ponmana Selvan, Hyers–Ulam stability of nth order linear differential equation with initial and boundary condition. Asian J. Math. Comput. Res. **11**(3), 201–207 (2016)
38. K. Ravi, R. Murali, A. Ponmana Selvan, Ulam stability of a General nth order linear differential equation with constant coefficients. Asian J. Math. Comput. Res. **11**(1), 61–68 (2016)
39. K. Ravi, J.M. Rassias, B.V. Senthil Kumar, Ulam–Hyers stability of undecic functional equation in quasi-β-normed spaces fixed point method. Tbilisi Math. Sci. **9**(2), 83–103 (2016)
40. K. Ravi, J.M. Rassias, S. Pinelas, S. Suresh, General solution and stability of Quattuordecic functional equation in quasi-β-normed spaces. Adv. Pure Math. **6**, 921–941 (2016) http://dx.doi.org/10.4236/apm.2016.612070
41. I.A. Rus, Ulam Stabilities of ordinary differential equations in a Banach space. Carpathian J. Math. **26**(1), 103–107 (2010)
42. A. Sibaha, B. Bouikhalene, E. Elquorachi, Ulam–Gavruta–Rassias stability for a linear functional equation. Int. J. Appl. Math. Stat. **7**, 157–168 (2007)
43. S.E. Takahasi, T. Miura, S. Miyajima, On the Hyers–Ulam stability of the Banach space-valued differential equation $y' = \alpha y$. Bull. Korean Math. Soc. **39**, 309–315 (2002)
44. S.M. Ulam, in *Problem in Modern Mathematics*. Chapter IV, Science Editors (Willey, New York, 1960)
45. G. Wang, M. Zhou, L. Sun, Hyers–Ulam stability of linear differential equations of first order. Appl. Math. Lett. **21**, 1024–1028 (2008)
46. T.Z. Xu, J.M. Rassias, W.X. Xu, A fixed point approach to the stability of a general mixed additive-cubic functional equation in quasi fuzzy normed spaces. Int. J. Phys. Sci. **6**(2), 313–324 (2011)
47. J. Xue, Hyers–Ulam stability of linear differential equations of second order with constant coefficient. Ital. J. Pure Appl. Math. **32**, 419–424 (2014)

A Class of Functional Equations of Type d'Alembert on Monoids

Belaid Bouikhalene and Elhoucien Elqorachi

Abstract Recently, the solutions of the functional equation $f(xy) - f(\sigma(y)x) = g(x)h(y)$ obtained, where σ is an involutive automorphism and f, g, h are complex-valued functions, in the setting of a group G and a monoid S. Our main goal is to determine the general complex-valued solutions of the following version of this equation, viz. $f(xy) - \mu(y)f(\sigma(y)x) = g(x)h(y)$ where $\mu : G \longrightarrow \mathbb{C}$ is a multiplicative function such that $\mu(x\sigma(x)) = 1$ for all $x \in G$. As an application we find the complex-valued solutions (f, g, h) on groups of equation $f(xy) + \mu(y)g(\sigma(y)x) = h(x)h(y)$ on monoids.

1 Introduction

We recall that a semigroup S is a non-empty set equipped with an associative operation. We write the operation multiplicatively. A monoid is a semigroup S with identity element that we denote e. A function $\mu : S \longrightarrow \mathbb{C}$ is said to be multiplicative if $\mu(xy) = \mu(x)\mu(y)$ for all $x, y \in S$.

Let S be a semigroup and $\sigma : S \longrightarrow G$ a homomorphism involutive, that is $\sigma(xy) = \sigma(x)\sigma(y)$ and $\sigma(\sigma(x)) = x$ for all $x, y \in S$.

Recently, Stetkær [21] obtained the complex-valued solutions of the following variant of d'Alembert's functional equation

$$f(xy) + f(\sigma(y)x) = 2f(x)f(y), \ x, y \in S. \tag{1}$$

B. Bouikhalene (✉)
Laboratory LIMATI, Polydisciplinary Faculty, Sultan Moulay Slimane University, Beni Mellal, Morocco
e-mail: b.bouikhalene@usms.ma

E. Elqorachi
Department of Mathematics, Faculty of Sciences, Ibn Zohr University, Agadir, Morocco

© Springer Nature Switzerland AG 2019
G. A. Anastassiou and J. M. Rassias (eds.), *Frontiers in Functional Equations and Analytic Inequalities*, https://doi.org/10.1007/978-3-030-28950-8_13

They are the functions of the form

$$f(x) = \frac{\chi + \chi \circ \sigma}{2},$$

where $\chi : S \longrightarrow \mathbb{C}$ is multiplicative.

At a later stage, that is in 2015, Ebanks and Stetkær [8] obtained the complex-valued solutions on monoids of the following d'Alembert's other functional equation

$$f(xy) - f(\sigma(y)x) = g(x)h(y), \quad x, y \in S. \tag{2}$$

This functional equation contains, among others, an equation of d'Alembert [1–5, 9, 12, 14, 20, 22, 23]

$$f(x + y) - f(x - y) = g(x)h(y), \quad x, y \in \mathbb{R} \tag{3}$$

whose general solutions are known on abelian groups, and a functional equation

$$f(x + y) - f(x + \sigma(y)) = g(x)h(y), \quad x, y \in G \tag{4}$$

studied by Stetkær [[15], Corollary III.5] on abelian group G.

There are various ways of extending functional equations from abelian groups to non-abelian groups. The μ-d'Alembert functional equation

$$f(xy) + \mu(y)f(x\sigma(y)) = 2f(x)f(y), \quad x, y \in G \tag{5}$$

which is an extension of d'Alembert functional equation

$$f(xy) + f(x\sigma(y)) = 2f(x)f(y), \quad x, y \in G, \tag{6}$$

where in this case σ is an involutive of G, is closely related to pre-d'Alembert function. It occurs in the literature. See Parnami et al. [11], Davison [6, Proposition 2.11], Ebanks and Stetkær [7], Stetkær [16, Lemma IV.4], and Yang [24, Proposition 4.2]. The functional equation (5) has been treated systematically by Stetkær [18] and [19]. The non-zero solutions of (5) are the normalized traces of certain representation of G on $\mathbb{C}^2$. Davison proved this via his work [6] on the pre-d'Alembert functional equation on monoids.

The variant Wilson's functional equation

$$f(xy) + \mu(y)f(\sigma(y)x) = 2f(x)g(y), \quad x, y \in G \tag{7}$$

with $\mu \neq 1$ was recently studied on groups by Elqorachi and Redouani [10].

The complex-valued solutions of Eq. (7) with $\sigma(x) = x^{-1}$ and $\mu(x) = 1$ for all $x \in G$ are obtained on groups by Ebanks and Stetkær [7].

The present paper complements and contains the existing results of (2) by finding the solutions f, g, h of the extension

$$f(xy) - \mu(y)f(\sigma(y)x) = g(x)h(y), \quad x, y \in S \tag{8}$$

of it to monoids that need not be abelian.

As in the previous results [8] one of the main ideas is to relate the functional equation (2) to a sine subtraction laws on monoids. In our case we need the solutions of the following version of the sine subtraction law

$$\mu(z)f(y\sigma(z)) = f(y)g(z) - f(z)g(y), \quad x, y \in S. \tag{9}$$

These results are obtained in Theorem 1. We need also the solutions of Eq. (7) on monoids that are not in the literature, the results are derived in Theorem 3.

In Sect. 4 we obtain the main results of the present paper. Furthermore, as an application we find the complex-valued solutions (f, g, h) of the functional equation

$$f(xy) + \mu(y)g(\sigma(y)x) = h(x)h(y), \quad x, y \in G \tag{10}$$

on groups and monoids in terms of multiplicative and additive functions.

1.1   Notation and Preliminary

Throughout this paper G denotes a group and S a semigroup. A mapping $\mu : S \longrightarrow \mathbb{C}$ is a multiplicative function such that $\mu(x\sigma(x)) = 1$ for all $x \in G$. Let $\chi : S \longrightarrow \mathbb{C}$ be a multiplicative function such that $\chi \neq 0$, then $I_\chi = \{x \in S \mid \chi(x) = 0\}$ is either empty or a proper subset of S. Furthermore, I_χ is a two-sided ideal in S if not empty and $S \backslash I_\chi$ is a subsemigroup of S. If S is a topological space, then we let $C(S)$ denote the algebra of continuous functions from S into $\mathbb{C}$.

For later use we need the following results.

Proposition 1 ([9, 13]) *Let G be a group, and suppose $f, g : G \longrightarrow \mathbb{C}$ satisfy the sine addition law*

$$f(xy) = f(x)g(y) + f(y)g(x), \quad x, y \in G \tag{11}$$

with $f \neq 0$. Then there exist multiplicative functions $\chi_1, \chi_2 : G \longrightarrow \mathbb{C}$ and a constant $c \in \mathbb{C} \setminus \{0\}$ such that:

(i) $g = \frac{\chi_1 + \chi_2}{2}$ and $f = c(\chi_1 - \chi_2)$.
(ii) $g = \chi_1$ and $f = \chi_1 A$ where $A : G \longrightarrow \mathbb{C}$ is an additive function such that $A \neq 0$.

Furthermore, if G is a topological group and $f, g \in C(G)$, then $\chi_1, \chi_2 \in C(G)$.

The next proposition corresponds to lemma 3.4 in [8]

Proposition 2 *Let S be a semigroup, and suppose* $f, g : S \longrightarrow \mathbb{C}$ *satisfy the sine addition law*

$$f(xy) = f(x)g(y) + f(y)g(x), \ x, y \in S \tag{12}$$

with $f \neq 0$. *Then there exist multiplicative functions* $\chi_1, \chi_2 : S \longrightarrow \mathbb{C}$ *such that*

$$g = \frac{\chi_1 + \chi_2}{2}.$$

Additionally we have the following

(i) If $\chi_1 \neq \chi_2$, *then* $f = c(\chi_1 - \chi_2)$ *for some constant* $c \in \mathbb{C} \setminus \{0\}$.
(ii) If $\chi_1 = \chi_2$, *then letting* $\chi := \chi_1$ *we have* $g = \chi$. *If S is a semigroup such that* $S = \{xy \in S : x, y \in S\}$ *(for instance, a monoid), then* $\chi \neq 0$.

If S is a group, then there is an additive function $A : S \longrightarrow \mathbb{C}$, $A \neq 0$, *such that* $f = \chi A$.

If S is a semigroup which is generated by its squares, then there exists an additive function $A : S \setminus I_\chi \longrightarrow \mathbb{C}$ *for which*

$$f(x) = \begin{cases} \chi(x)A(x) & for \ x \in S \setminus I_\chi \\ 0 & for \ x \in I_\chi \end{cases}$$

Furthermore, if S is a topological group, or if S is a topological semigroup generated by its squares, and $f, g \in C(S)$, *then* $\chi_1, \chi_2, \chi \in C(S)$. *In the group case* $A \in C(S)$ *and in the second case* $A \in C(S \setminus I_\chi)$.

2 μ-Sine Subtraction Law on a Group and on a Monoid

In this section we deal with a new version of the sine subtraction law

$$\mu(z)k(y\sigma(z)) = k(y)l(z) - k(z)l(y), \ x, y \in S \tag{13}$$

where k, l are complex valued functions and μ is a multiplicative function. We shall say that k satisfies μ-sine subtraction law with companion function l. As in Lemma 3.1 in [6] if S is a topological semigroup and k, l satisfy (13) such that $k \neq 0$ and k is a continuous function then l is also a continuous function. In the case where $\mu = 1$ and G is a topological group, the functional equation (13) was solved in [8].

Here we focus exclusively on (13), and we include nothing about other extension of cosine, sine addition, and subtraction formulas:

$$\mu(z)k(y\sigma(z))=k(y)l(z)+k(z)l(y), \quad \mu(z)k(y\sigma(z))=k(y)k(z)-l(z)l(y), \quad y,z \in S.$$

The next theorem is the analogous of Theorem 3.2 in [8].

Theorem 1 *Let G be a group and let $\sigma : G \longrightarrow \mathbb{C}$ be an involutive automorphism. Let $\mu : G \longrightarrow \mathbb{C}$ be a multiplicative function such that $\mu(x\sigma(x)) = 1$ for all $x \in S$. The solutions $k,l : G \longrightarrow \mathbb{C}$ of the μ-sine subtraction law (13) with $k \neq 0$ are the following pairs of functions, where $\chi : G \longrightarrow \mathbb{C} \setminus \{0\}$ denotes a character and $c_1 \in \mathbb{C}$, $c_2 \in \mathbb{C} \setminus \{0\}$ are constants.*

(i) If $\chi \neq \mu\chi \circ \sigma$, then

$$k = c_2\frac{\chi - \mu\chi \circ \sigma}{2}, \quad l = \frac{\chi + \mu\chi \circ \sigma}{2} + c_1\frac{\chi - \mu\chi \circ \sigma}{2}.$$

(ii) If $\chi = \mu\chi \circ \sigma$, then

$$k = \chi A, \quad l = \chi(1 + c_1 A)$$

where $A : G \longrightarrow \mathbb{C} \setminus \{0\}$ is an additive function such that $A \circ \sigma = -A \neq 0$.

Furthermore, if G is a topological group and $k \in C(G)$, then $l, \chi, \mu\chi \circ \sigma, A \in C(G)$.

Proof By using some ideas from [8] we get by interchanging x and y that $\mu(x)k(y\sigma(x)) = -\mu(y)k(x\sigma(y))$ for all $x, y \in G$. By setting $y = e$ we get that $k(x) = -\mu(x)k(\sigma(x))$ for all $x \in G$. Using this and Eq. (13) we get for all $x, y \in G$ that

$$k(x)[l(y) - \mu(y)l(\sigma(y))] - k(y)[l(x) - \mu(x)l(\sigma(x))]$$
$$= \mu(y)k(x\sigma(y)) - \mu(y)k(x\sigma(y))$$
$$= 0.$$

So that we get for all $x, y \in G$ that

$$k(x)[l(y) - \mu(y)l(\sigma(y))] = k(y)[l(x) - \mu(x)l(\sigma(x))]. \tag{14}$$

Let $l^+(x) = \frac{l(x)+\mu(x)l(\sigma(x))}{2}$ and $l^-(x) = \frac{l(x)-\mu(x)l(\sigma(x))}{2}$ for all $x \in G$. We have $l = l^+ + l^-$, $l^+(\sigma(x)) = \mu(\sigma(x))l^+(x)$ and $l^-(\sigma(x)) = -\mu(\sigma(x))l^-(x)$ for all $x \in G$. From (13) we have for all $x, y \in G$ that $k(x)l^-(y) = k(y)l^-(x)$. Since $k \neq 0$ we assume that there exists an $x_0 \in G$ such that $k(x_0) \neq 0$. So that we have $l^-(y) = \frac{l^-(x_0)}{k(x_0)}f(y) = cf(y)$ for all $y \in G$. Then $l^- = ck$. On the other hand, by subsisting the fact that $l = l^+ + l^-$ in (13) and by using $k(x)l^-(y) = k(y)l^-(x)$ we get that

$$\mu(y)k(x\sigma(y)) = k(y)l^+(z) - k(z)l^+(y), \quad x, y \in G \tag{15}$$

Replacing y by $\sigma(y)$ in (15) we get that

$$k(xy) = k(y)l^+(z) + k(z)l^+(y), \quad x, y \in G. \tag{16}$$

According to Proposition 1 we have

(i) $k = c_1(\chi_1 - \chi_2)$ and $l^+ = \frac{\chi_1 + \chi_2}{2}$ where $c_1 \in \mathbb{C} \setminus \{0\}$. Since $k(x) = -\mu(x)k(\sigma(x))$ and $l^+(x) = \mu(\sigma(x))k(\sigma(x))$ for all $x \in G$ we get that $c_2(\chi_1 - \chi_2) = -c_2\mu(\chi_1 \circ \sigma - \chi_2 \circ \sigma)$ and $\mu(\chi_1 \circ \sigma + \chi_2 \circ \sigma) = \chi_1 + \chi_2$. Then $\chi_2 = \mu\chi_1 \circ \sigma$, $l^+ = \frac{\chi_1 + \mu\chi_1 \circ \sigma}{2}$ and $k = c_2\frac{\chi_1 - \mu\chi_1 \circ \sigma}{2}$. Since $l^- = ck = cc_2\frac{\chi_1 - \mu\chi_1 \circ \sigma}{2} = c_1\frac{\chi_1 - \mu\chi_1 \circ \sigma}{2}$ where $c_1 \in \mathbb{C} \setminus \{0\}$. By using the fact $l = l^- + l^+$ we get (i).

(ii) we have $l^+ = \chi$ and $k = \chi A$. Since $l^+(\sigma(x)) = \mu(\sigma(x))l^+(x)$ for all $x \in G$ we get that $\chi = \mu\chi \circ \sigma$. From $k(x) = -\mu(x)k(\sigma(x))$ for all $x \in G$ we get that $A = -A \circ \sigma$. By using the fact $l = l^- + l^+ = c\chi A + \chi = \chi(cA + 1)$. This ends the proof.

□

In the next proposition we extend Theorem 1 to monoids.

Proposition 3 *Let S be a monoid and let $\sigma : S \longrightarrow S$ be an involutive automorphism. Let $\mu : G \longrightarrow \mathbb{C}$ be a multiplicative function such that $\mu(x\sigma(x)) = 1$ for all $x \in S$. The solutions $k, l : G \longrightarrow \mathbb{C}$ of the μ-sine subtraction law (13) with $k \neq 0$ are the following pairs of functions, where $\chi : G \longrightarrow \mathbb{C} \setminus \{0\}$ denotes a multiplicative function and $\chi(e) = 1$ and $c_1 \in \mathbb{C}$, $c_2 \in \mathbb{C} \setminus \{0\}$ are constants:*

(i) If $\chi \neq \mu\chi \circ \sigma$, then

$$k = c_1\frac{\chi - \mu\chi \circ \sigma}{2}, \quad l = \frac{\chi + \mu\chi \circ \sigma}{2} + c_2\frac{\chi - \mu\chi \circ \sigma}{2}.$$

(ii) If $\chi = \mu\chi \circ \sigma$ and S is generated by its squares, then

$$k(x) = \chi(x)A(x), \quad l(x) = \chi(x)(1 + c_1A(x)) \; \forall x \in S \setminus I_\chi$$

$$l(x) = k(x) = 0, \quad \forall x \in S \setminus I_\chi$$

where $A : G \longrightarrow \mathbb{C} \setminus \{0\}$ is an additive function such that $A \circ \sigma = -A \neq 0$.

Furthermore, if S is a topological monoid, and $k \in C(S)$, then $l, \chi, \mu\chi \circ \sigma \in C(S)$ and $A \in C(S \setminus I_\chi)$.

Proof By the same way in Proposition 3.6 in [8] and by using Theorem 1 and the Proposition 2 we get the proof. □

3 Variant of Wilson's Functional Equation on Monoids

The solutions of the functional equation (7) on groups are obtained in [10]. More precisely, Elqorachi and Redouani proved the following theorem.

Theorem 2 *Let G be a group, let $\sigma: G \longrightarrow G$ a homomorphism such that $\sigma \circ \sigma = I$, where I denotes the identity map, and $\mu: G \longrightarrow \mathbb{C}$ be a multiplicative function such that $\mu(x\sigma(x)) = 1$ for all $x \in G$. The solutions f, g of the functional equation (7) are the following pairs of functions, where $\chi: G \longrightarrow \mathbb{C}$ denotes a function multiplicative and $c, \alpha \in \mathbb{C}^*$*

  (i) $f = 0$ and g arbitrary.
  (ii) $g = \frac{\chi + \mu\chi\circ\sigma}{2}$ and $f = \alpha g$.
  (iii) $g = \frac{\chi + \mu\chi\circ\sigma}{2}$ and $f = (c+\alpha/2)\chi - (c-\alpha/2)\chi\circ\sigma$ with $(\mu-1)\chi = (\mu-1)\chi\circ\sigma$.
  (iv) $g = \chi$ and $f = \chi(a + \alpha)$, where $\chi = \mu\chi \circ \sigma$ and a is an additive map which satisfies $a \circ \sigma + a = 0$.

We shall now extend this result to monoids.

Theorem 3 *Let S be a monoid, let $\sigma: S \longrightarrow S$ a homomorphism such that $\sigma \circ \sigma = I$, where I denotes the identity map, and $\mu: S \longrightarrow \mathbb{C}$ be a multiplicative function such that $\mu(x\sigma(x)) = 1$ for all $x \in S$. The solutions f, g of the functional equation (7) are the following pairs of functions, where $\chi: S \longrightarrow \mathbb{C}$ denotes a function multiplicative and $c, \alpha \in \mathbb{C}^*$*

  (i) $f = 0$ and g arbitrary.
  (ii) $g = \frac{\chi + \mu\chi\circ\sigma}{2}$ and $f = \alpha g$.
  (iii) $g = \frac{\chi + \mu\chi\circ\sigma}{2}$ and $f = (c+\alpha/2)\chi - (c-\alpha/2)\chi\circ\sigma$ with $(\mu-1)\chi = (\mu-1)\chi\circ\sigma$.
  (iv) $g = \chi$. Furthermore, if S is a monoid which is generated by its squares, then $\chi = \mu\chi \circ \sigma$, there exists an additive function $a: S\backslash I_\chi \longrightarrow \mathbb{C}$ for which $a \circ \sigma + a = 0$,

$$f(x) = \begin{cases} \chi(x)(a(x) + \alpha) & for\ x \in S \setminus I_\chi \\ 0 & for\ x \in I_\chi \end{cases}$$

Indeed, if S is a topological group, or S is a topological monoid generated by its squares, $f, g, \mu \in C(S)$, and $\sigma: G \longrightarrow G$ is continuous, then $\chi \in C(S)$. In the group case $a \in C(S)$ and in the second case $a \in C(S\backslash I_\chi)$.

Proof Verifying that the stated pairs of functions constitute solutions consists of simple computations. To see the converse, i.e., that any solution f, g of (7) is contained in one of the cases below, we will use [8, Lemma 3.4] and [10,

Theorem 3.1]]. All, except the last paragraphs of part (iv) and the continuity statements, are in Theorem 3.1 in [10]. Now, we assume that S is a monoid generated by its squares. We use the notation used in the proof of Theorem 3.1 in [10], in particular for the last paragraphs of part (iv) we have

$$f_e(xy) = f_e(x)\chi(y) + f_e(y)\chi(x) \tag{17}$$

for all $x, y \in S$ and with $f = f_e + f(e)\chi$. So, from [[8], Lemma 3.4] we get $f(x) = 0 + f(e)\chi(x) = 0 + f(e)0 = 0$ if $x \in I_\chi$ and $f(x) = \chi(x)(a(x) + f(e))$ if $x \in S\backslash I_\chi$ and where a is an additive function of $S\backslash I_\chi$.

Now, we will verify that $\chi = \mu\chi \circ \sigma$ and $a \circ \sigma = -a$.

Since f, g are solution of Eq. (7) we have

$$f(xy) + \mu(y)f(\sigma(y)x) = 2f(x)\chi(y) \tag{18}$$

for all $x, y \in G$. By using the new expression of f and the fact that I_χ is an ideal, we get after an elementary computation that $f(xy) = f(yx)$ for all $x, y \in G$. So, Eq. (18) can be written as follows

$$f(xy) + \mu(y)f(x\sigma(y)) = 2f(x)\chi(y), \quad x, y \in G. \tag{19}$$

By replacing y by $\sigma(y)$ in (19) and multiplying the result obtained by $\mu(y)$ we get

$$f(xy) + \mu(y)f(x\sigma(y)) = 2f(x)\mu(y)\chi(\sigma(y)), \quad x, y \in G. \tag{20}$$

Finally, by comparing (19), (20) and using $f \neq 0$ we get $\chi(y) = \mu(y)\chi(\sigma(y))$ for all $y \in G$.

By substituting the expression of f into (19) and using $\chi(y) = \mu(y)\chi(\sigma(y))$ and $\mu(y\sigma(y)) = 1$ for all $y \in G$, we find after reduction that $\chi(x)\chi(y)[a(y) + a(\sigma(y))] = 0$ for all $x, y \in S\backslash I_\chi$. Since $\chi \neq 0$ we get $a(y) + a(\sigma(y)) = 0$ for all $y \in S\backslash I_\chi$.

For the topological statement we use [[17], Theorem 3.18(d)]. This completes the proof. $\square$

4   Solutions of (8) on Groups and Monoids

In this section we solve the functional equation

$$f(xy) - \mu(y)f(\sigma(y)x) = g(x)h(y), \quad x, y \in S \tag{21}$$

where S is a monoid.

In the next proposition we show that h satisfies the μ-sine subtraction law

Proposition 4 *Let S be a monoid, let σ be an involutive automorphism on S, let μ be a multiplicative function on S and suppose $f, g, h : S \longrightarrow \mathbb{C}$ satisfy the functional equation (13). Suppose also that $g \neq 0$ and $h \neq 0$. Then*

- *(i) $h(\sigma(y)) = -\mu(\sigma(x))h(x))$ for all $x \in S$.*
- *(ii) $h(xy) = h(yx)$ for all $x, y \in S$.*
- *(iii) h satisfies the μ-sine subtraction law (13).*
- *(iv) If $g(e) = 0$, then $g = bh$ for some $b \in \mathbb{C} \setminus \{0\}$.*
- *(vi) If $g(e) \neq 0$, then h satisfies the μ-sine subtraction law with companion function $\frac{g}{g(e)}$.*

Moreover, if S is a topological monoid, and $h \in C(S)$, then the companion function in case (i) is also continuous.

Proof We follow the path of the proof of Proposition 3.1 in [6].

By substituting (x, yz), $(\sigma(y), \sigma(z)x)$ and $(z, \sigma(xy))$ and $(z, \sigma(xy))$ in (21) we obtain

$$f(xyz) - \mu(xyz)f(\sigma(yz)x) = g(x)h(yz), \tag{22}$$

$$f(\sigma(yz)x) - \mu(\sigma(z)x)f(z\sigma(xy)) = g(\sigma(y))h(\sigma(z)x), \tag{23}$$

$$f(z\sigma(xy)) - \mu(\sigma(xy))f(xyz) = g(z)h(\sigma(xy)). \tag{24}$$

By multiplying (22) by $\mu(\sigma(xy))$ we obtain that

$$\mu(\sigma(xy))f(xyz) - \mu(\sigma(x)z)f(\sigma(yz)x) = \mu(\sigma(xy))g(x)h(yz). \tag{25}$$

By adding (25) and (24) we obtain

$$f(z\sigma(xy)) - \mu(\sigma(x)z)f(\sigma(yz)x) = g(z)h(\sigma(xy)) + \mu(\sigma(xy))g(x)h(yz). \tag{26}$$

By multiplying (26) by $\mu(\sigma(z)x)$ we obtain

$$\mu(\sigma(z)x)f(z\sigma(xy)) - f(\sigma(yz)x) = \mu(\sigma(z)x)g(z)h(\sigma(xy)) + \mu(\sigma(zy))g(x)h(yz). \tag{27}$$

By adding (27) and (23) we obtain

$$0 = g(\sigma(y))h(\sigma(z)x) + \mu(\sigma(z)x)g(z)h(\sigma(xy)) + \mu(\sigma(zy))g(x)h(yz). \tag{28}$$

Setting x_0 such that $g(x_0) \neq 0$ and the fact that $\mu(x\sigma(x)) = \mu(x)\mu(\sigma(x)) = 1$ for all $x \in G$, we get that

$$h(yz) = \mu(y)g(\sigma(y))l(z) + g(z)l_1(y), \quad y, z \in S \tag{29}$$

where l, l_1 are complex valued functions on S. Using (29) in (28) we obtain for all $x, y, z \in S$

$$g(x)g(\sigma(y))\{l_1(\sigma(z)) + \mu(\sigma(z))l(z)\} + g(x)g(z)\{\mu(\sigma(z))l(\sigma(y))$$
$$+ \mu(\sigma(zy))l_1(y)\} + g(z)g(\sigma(y))\{\mu(\sigma(z))l(x) + \mu(\sigma(z)x)l_1(\sigma(x))\} = 0. \tag{30}$$

Putting $x = x_0, y = \sigma(x_0)$, Eq. (30) becomes

$$l_1(\sigma(z)) + \mu(\sigma(z))l(z) = c\mu(\sigma(z))g(z), \quad z \in S \tag{31}$$

By putting (31) in (30) we obtain $3\mu(\sigma(z))cg(x)g(\sigma(y))g(z) = 0$ for all $x, y, z \in S$. Since $g \neq 0$ and $\mu(\sigma(z)) \neq 0$ it follows that $c = 0$ and then $l_1(\sigma(z)) = -\mu(\sigma(z))l(z)$ for all $z \in S$. So that Eq. (29) becomes

$$h(yz) = \mu(y)g(\sigma(y))l(z) - \mu(y)g(z)l(\sigma(y)), \quad x, y, z \in S \tag{32}$$

Replacing (y, z) by $(\sigma(z), \sigma(y))$ in (32) we obtain

$$h(\sigma(zy)) = \mu(\sigma(zy))(\mu(y)g(\sigma(y))l(z) - \mu(y)g(z)l(\sigma(y)) - \mu(\sigma(zy))h(z). \tag{33}$$

From which we obtain by putting $y = e$ that

$$h(\sigma(z)) = -\mu(\sigma(z))h(z), \quad z \in S. \tag{34}$$

From (33) and (34) we get that h a central function, i.e. $h(yz) = h(zy)$ for all $y, z \in S$.

Next we consider two cases:

First case: Suppose $g(e) = 0$. Let $z = e$ and $x = x_0$ in (28) give that

$$h(y) = -c\mu(y)g(\sigma(y)), \quad y \in S \tag{35}$$

for some $c \in \mathbb{C} \setminus \{0\}$. By replacing y by $\sigma(y)$ in (35) we obtain that $h(\sigma(y)) = -c\mu(\sigma(y))g(y)$ for all $y \in G$. By using (34) we get that $g = \frac{1}{c}h = bh$ where $b = \frac{1}{c}$ and that

$$g(\sigma(z)) = -\mu(\sigma(z))g(z), \quad z \in S. \tag{36}$$

Using (36) in (32) and setting $m = -bl$ we get

$$h(yz) = \mu(y)g(\sigma(y))l(z) - \mu(y)g(z)l(\sigma(y))$$
$$= -\mu(y)\mu(\sigma(y))g(y)l(z) - \mu(y)g(z)l(\sigma(y))$$
$$= -g(y)l(z) - \mu(y)g(z)l(\sigma(y))$$

$$= -bh(y)l(z) - \mu(y)bh(z)l(\sigma(y))$$
$$= h(y)m(z) + \mu(y)h(z)m(\sigma(y)).$$

By replacing z by $\sigma(z)$ we obtain

$$h(y\sigma(z)) = h(y)(m \circ \sigma)(z) + \mu(y)h(\sigma(z))m(\sigma(y)), \quad y, z \in S. \tag{37}$$

By multiplying (37) by $\mu(z)$ and by setting $n(z) = (m \circ \sigma)(z)\mu(z)$ we obtain the μ-sine subtraction law with the companion function n

$$\mu(z)h(y\sigma(z)) = h(y)n(z) - h(z)n(y). \tag{38}$$

This ends the first case.

Second case: Suppose $g(e) \neq 0$. Then we obtain from (28) with $x = e$ that

$$h(yz) = [\mu(y)g(\sigma(y))h(z) + g(z)h(y)]/g(e). \tag{39}$$

Interchanging y and z in (39) we get

$$h(zy) = [\mu(z)g(\sigma(z))h(y) + g(y)h(z)]/g(e). \tag{40}$$

By replacing z by $\sigma(z)$ in (40) and multiplying it by $\mu(z)$ we get

$$\mu(z)h(\sigma(z)y) = h(y)\frac{g}{g(e)}(z) - \frac{g}{g(e)}(y)h(z), \quad y, z \in S. \tag{41}$$

Since h is central it follows that h satisfies the μ-sine subtraction law with the companion function $\frac{g}{g(e)}$

$$\mu(z)h(y\sigma(z)) = h(y)\frac{g}{g(e)}(z) - \frac{g}{g(e)}(y)h(z), \quad y, z \in S. \tag{42}$$

□

In the next two theorems, by using the result obtained for μ-sine subtraction law, we give solutions of Eq. (21). We will follow the method used in [8].

Let $\mathcal{N}_\mu(\sigma, S)$ be the null space given by

$$\mathcal{N}_\mu(\sigma, S) = \{\theta : S \longrightarrow \mathbb{C} : \theta(xy) - \mu(y)\theta(\sigma(y)x) = 0, \ x, y \in S\}.$$

In the next theorem we consider the group case

Theorem 4 *Let G be a group, let σ be an involutive automorphism on G, let $\mu : G \longrightarrow \mathbb{C}$ be a multiplicative function and suppose $f, g, h : G \longrightarrow \mathbb{C}$ satisfy functional equation (21). Suppose also that $g \neq 0$ and $h \neq 0$. Then there exist a* *character χ of G, constants $c, c_1, c_2 \in \mathbb{C}$, and a function $\theta \in N_\mu(\sigma, S)$ such that one of the following holds:*

(i) If $\chi \neq \mu\chi \circ \sigma$, then

$$h = c_1 \frac{\chi - \mu\chi \circ \sigma}{2}, \quad g = \frac{\chi + \mu\chi \circ \sigma}{2} + c_2 \frac{\chi - \mu\chi \circ \sigma}{2},$$

$$f = \theta + \frac{c_1}{2}[c\frac{\chi - \mu\chi \circ \sigma}{2} + c_2 \frac{\chi + \mu\chi \circ \sigma}{2}]$$

(ii) If $\chi = \mu\chi \circ \sigma$, then

$$h = \chi A, \quad g = \chi(c + c_2 A), \quad f = \theta + \chi A(\frac{c}{2} + \frac{c_2}{4}A).$$

where $A : G \longrightarrow \mathbb{C} \setminus \{0\}$ is an additive function such that $A \circ \sigma = -A \neq 0$. Conversely, the formulas of (i) and (ii) define solutions of (3.1).

Moreover, if G is a topological group, and $f, g, h \in C(G)$, then $\chi, \mu\chi\circ\sigma, A, \theta \in C(G)$, while $A \in C(G)$.

The proof of the Theorem 4 will be integrated into that of Theorem 5 in which we consider the monoid case

Theorem 5 *Let S be a monoid which is generated by its squares, let σ be an involutive automorphism on G, let $\mu : G \longrightarrow \mathbb{C}$ be a multiplicative function and suppose $f, g, h : G \longrightarrow \mathbb{C}$ satisfy functional equation (21). Suppose also that $g \neq 0$ and $h \neq 0$. Then there exist a multiplicative function $\chi : S \longrightarrow \mathbb{C}\chi \neq 0$, constants $c, c1, c_2 \in \mathbb{C}$, and a function $\theta \in N_\mu(\sigma, S)$ such that one of the following holds:*

(i) If $\chi \neq \mu\chi \circ \sigma$, then

$$h = c_1 \frac{\chi - \mu\chi \circ \sigma}{2}, \quad g = \frac{\chi + \mu\chi \circ \sigma}{2} + c_2 \frac{\chi - \mu\chi \circ \sigma}{2},$$

$$f = \theta + \frac{c_1}{2}[c\frac{\chi - \mu\chi \circ \sigma}{2} + c_2 \frac{\chi + \mu\chi \circ \sigma}{2}]$$

(ii) If $\chi = \mu\chi \circ \sigma$, then $h(x) = g(x) = 0$ and $f(x) = \theta(x)$ for $x \in I_\chi$, and

$$h(x) = \chi(x)A(x), \quad g(x) = \chi(x)(c + c_2 A(x)),$$

$$f(x) = \theta(x) + \chi(x)A(x)(\frac{c}{2} + \frac{c_2}{4}A(x))$$

for $x \in S \setminus I_\chi$ where $A : S \setminus I_\chi \longrightarrow \mathbb{C} \setminus \{0\}$ is an additive function such that $A \circ \sigma = -A \neq 0$.

Furthermore, if S is a topological monoid and $k \in C(G)$, then l, χ, $\mu\chi \circ \sigma$, $A \in C(G)$. Conversely, the formulas of (i) and (ii) define solutions of (21).

Moreover, if S is a topological monoid, and $f, g, h \in C(S)$, then $\chi, \mu\chi \circ \sigma, A, \theta \in C(S)$, while $A \in C(S \setminus I_\chi)$.

Proof According to Proposition 4 we have the two following cases:

First case: Suppose that $g(e) = 0$, then h satisfies the μ-sine subtraction law and $g = bh$ where $b \in \mathbb{C} \setminus \{0\}$. According to Theorem 1 and Proposition 3 we get (for S is a group or a monoid) that if $\chi \neq \mu\chi \circ \sigma$, then $h = c_1 \frac{\chi - \mu\chi \circ \sigma}{2}$ where $c_1 \in \mathbb{C}$. Since $g = bh$, then $g = c_2 \frac{\chi - \mu\chi \circ \sigma}{2}$ where $c_2 \in \mathbb{C}$. Subsisting g and h in (21) we get for all $x, y \in S$

$$f(xy) - \mu(y)f(\sigma(y)x) = g(x)h(y)$$
$$= \frac{c_1 c_2}{4}[\chi(x) - \mu(x)\chi(\sigma(x))][\chi(y) - \mu(y)\chi(\sigma(y))]$$
$$= \frac{c_1 c_2}{4}[\chi(xy) - \mu(y)\chi(\sigma(y)x) - \mu(x)\chi(\sigma(\sigma(y)x))$$
$$+ \mu(xy)\chi(\sigma(xy))].$$

Let $\theta = f - \frac{c_1 c_2}{4}(\chi + \mu\chi \circ \sigma)$ we have $\theta \in N_\mu(\sigma, S)$ and $f = \theta + \frac{c_1}{2}[c\frac{\chi - \mu\chi \circ \sigma}{2} + c_2 \frac{\chi + \mu\chi \circ \sigma}{2}]$ with $c = 0$.

Now, if $\chi = \mu\chi \circ \sigma$.

When S is a group then we get from Theorem 1 that $h = \chi A$ where A is an additive function such that $A \circ \sigma = -A \neq 0$. Since $g = bh$, then $g = b\chi A = c_2 \chi A$. Subsisting g and h in 21 we get

$$f(xy) - \mu(y)f(\sigma(y)x) = g(x)h(y)c_2\chi(x)A(x)\chi(y)A(y)$$
$$= \frac{c_2}{4}[\chi(y)A(xy)^2 - \chi(\sigma(y))A(\sigma(y)x)^2].$$

Let $\theta = f - c_2 \frac{\chi A^2}{4}$. Then $\theta \in N_\mu(\sigma, S)$.

When S is a monoid, by Proposition 3 and by the same way as in [6] we have $\theta = \theta_1 \cup \theta_2$ where $\theta_1(x) = f(x) - c_2 \frac{\chi(x)A^2(x)}{4}$ on $S \setminus I_\chi$ and $\theta_2(x) = f(x)$ on I.

Second case: Suppose $g(e) \neq 0$. We have h and $\frac{g}{g(e)}$ play the role of k and l respectively in Theorem 1 or in Proposition 3. If $\chi \neq \mu\chi \circ \sigma$, then $h = c_1 \frac{\chi - \mu\chi \circ \sigma}{2}$ where $c_1 \in \mathbb{C}$ and $g = c\frac{\chi + \mu\chi \circ \sigma}{2} + c_2 \frac{\chi - \mu\chi \circ \sigma}{2}$. By the same way as in [8] we get that $\theta = f - c_1 \frac{[(c+c_2)\chi - (c-c_2)\chi \circ \sigma]}{4} \in N_\mu(\sigma, S)$.

Finally, if $g(e) \neq 0$ and $\chi = \mu\chi \circ \sigma$ we get the remainder by the same way as in [8]. □

5   Applications: Solutions of Eq. (10) on Groups and Monoids

In this section, we use the results obtained in the previous paragraph to solve the functional equation (10) on groups and monoids. We proceed as follows to reduce the equation to the functional equations (7), (8) so that we can apply Theorems 2–5.

Theorem 6 *Let G be a group, and σ a homomorphism involutive of G. Let μ : $G \longrightarrow \mathbb{C}$ be a multiplicative function such that $\mu(x\sigma(x)) = 1$ for all $x \in G$. Suppose that the functions $f, g, h : G \longrightarrow \mathbb{C}$ satisfy the functional equation (10). Suppose also that $f + g \neq 0$. Then there exist a character χ of G, constants $\alpha \in \mathbb{C}^*$, $c_1, c_2 \in \mathbb{C}$, and a function $\theta \in N_\mu(\sigma, S)$ such that one of the following holds:*

(a) *If $\chi \neq \mu\chi \circ \sigma$, then $f = \frac{1}{2}[(1 + \frac{c_1c_2}{2})\frac{\chi+\mu\chi\circ\sigma}{2} + 2c_2\frac{\chi-\mu\chi\circ\sigma}{2} + \theta]$; $g = \frac{1}{2}[(1 - \frac{c_1c_2}{2})\frac{\chi+\mu\chi\circ\sigma}{2} - \theta]$ and $h = \frac{1}{\alpha}[\frac{\chi+\mu\chi\circ\sigma}{2} + c_2\frac{\chi-\mu\chi\circ\sigma}{2}]$.*

(b) *If $\chi = \mu\chi \circ \sigma$ then $f = \frac{c_2\chi(2+A)+\theta+\chi Ac_2(1+\frac{4}{4})}{2}$; $g = \frac{c_2\chi(2+A)-\theta-\chi Ac_2(1+\frac{4}{4})}{2}$ and $h = \frac{1}{\alpha}c_2\chi(2 + A)$.*

Proof Let $f, g, h: G \longrightarrow \mathbb{C}$ satisfy the functional equation (10). The case $g = -f$ was treated in Theorems 2 and 3. From now on, we assume that $f + g \neq 0$. Let $h_o := \frac{h-\mu h\circ\sigma}{2}$ respectively $h_e := \frac{h+\mu h\circ\sigma}{2}$ denote the odd, respectively even, part of h with respect to μ and σ.

Setting $x = e$ in (10) gives us

$$f(y) + \mu(y)g(\sigma(y)) = h(e)h(y) \tag{43}$$

for all $y \in G$. Taking $y = e$ in (10) and using $\mu(e) = 1$ we find

$$f(x) + g(x) = h(e)h(x) \tag{44}$$

for all $x \in G$. So, by comparing (43) with (44) we get

$$g(x) = \mu(x)g(\sigma(x)), \ x \in G. \tag{45}$$

We note that $(f + g)(xy) + \mu(y)(f + g)(\sigma(y)x) = f(xy) + \mu(y)g(\sigma(y)x) + g(xy) + \mu(y)f(\sigma(y)x) = h(x)h(y) + g(xy) + \mu(y)f(\sigma(y)x)$. By using (45) we have $g(xy) = \mu(xy)g(\sigma(x)\sigma(y))$, then we get $g(xy) + \mu(y)f(\sigma(y)x) = \mu(y)f(\sigma(y)x) + \mu(xy)g(\sigma(x)\sigma(y)) = \mu(y)[f(\sigma(y)x) + \mu(x)g(\sigma(x)\sigma(y))] = \mu(y)h(x)h(\sigma(y))$, which implies that

$$(f + g)(xy) + \mu(y)(f + g)(\sigma(y)x) = 2h(x)h_e(y) \tag{46}$$

for all $x, y \in G$. From (44) and the assumption that $f + g \neq 0$ we get $h(e) \neq 0$. So, Eq. (46) can be written as follows:

$$(f+g)(xy) + \mu(y)(f+g)(\sigma(y)x) = 2(f+g)(x)\frac{h_e(y)}{h(e)} \qquad (47)$$

for all $x, y \in G$.

On the other hand, by using similar computation used above, we obtain

$$(f-g)(xy) - \mu(y)(f-g)(\sigma(y)x) = 2h(x)h_o(y) = (f+g)(x)\frac{2h_o(y)}{h(e)} \qquad (48)$$

for all $x, y \in G$. We can now apply Theorems 2–5.

If $h_o = 0$, then $f - g \in N(\sigma, G)$, so there exists $\theta \in N_\mu(\sigma, G)$ such that $f - g = \theta$. Since f, g satisfy (47) then from Theorem 1 we get the only possibility $f + g = \alpha^2 \frac{\chi + \mu\chi\circ\sigma}{2}$ and $h = \alpha\frac{\chi+\mu\chi\circ\sigma}{2}$ for some character $\chi: G \longrightarrow \mathbb{C}$ and a constant $\alpha \in \mathbb{C}$ and we deduce that $f = \frac{1}{2}[\theta + \alpha^2(\frac{\chi+\mu\chi\circ\sigma}{2})]$; $g = \frac{1}{2}[-\theta + \alpha^2(\frac{\chi+\mu\chi\circ\sigma}{2})]$. We deal with case (i).

So for the rest of the proof we will assume that $h_o \neq 0$. The function $f + g$, h_o are solution of Eq. (48) with $f + g \neq 0$ and $h_o \neq 0$, so we know from Theorem 4 that there are only the following two possibilities:

(i) $f - g = \theta + \frac{c_1}{2}[c\frac{\chi-\mu\chi\circ\sigma}{2} + c_2\frac{\chi+\mu\chi\circ\sigma}{2}]$; $f + g = \frac{\chi+\mu\chi\circ\sigma}{2} + c_2\frac{\chi-\mu\chi\circ\sigma}{2}$ for some character χ on G such that $\chi \neq \mu\chi \circ \sigma$, $\theta \in N_\mu(\sigma, G)$ and constants $c, c_1, c_2 \in \mathbb{C}$. So, we have $g = \frac{1}{2}[(1 - \frac{c_1c_2}{2})\frac{\chi+\mu\chi\circ\sigma}{2} + (c_2 - \frac{cc_1}{2})\frac{\chi-\mu\chi\circ\sigma}{2} - \theta]$; $f = \frac{1}{2}[(1 + \frac{c_1c_2}{2})\frac{\chi+\mu\chi\circ\sigma}{2} + (c_2 + \frac{cc_1}{2})\frac{\chi-\mu\chi\circ\sigma}{2} + \theta]$. Since $g = \mu g \circ \sigma$, then we have $c_2 = \frac{cc_1}{2}$ and f, g are as follows: $f = \frac{1}{2}[(1+\frac{c_1c_2}{2})\frac{\chi+\mu\chi\circ\sigma}{2} + 2c_2\frac{\chi-\mu\chi\circ\sigma}{2} + \theta]$; $g = \frac{1}{2}[(1 - \frac{c_1c_2}{2})\frac{\chi+\mu\chi\circ\sigma}{2} - \theta]$. We deal with case (a).

(ii) $f - g = \theta + \chi A(\frac{c}{2} + \frac{c_2}{4}A)$; $f + g = \chi(c + c_2A)$, where χ is a character on G such that $\chi = \mu\chi \circ \sigma$, A is an additive map on G such that $A \circ \sigma = -A$, $\theta \in N_\mu(\sigma, G)$ and $c, c_2 \in \mathbb{C}$. So, we get $g = \frac{\chi(c+c_2A) - \theta - \chi A(\frac{c}{2}+\frac{c}{4})A}{2}$. Since $g = \mu g\circ\sigma$ so, we have $c = 2c_2$. Consequently we have $g = \frac{c_2\chi(2+A) - \theta - \chi Ac_2(1+\frac{A}{4})}{2}$; $f = \frac{c_2\chi(2+A)+\theta+\chi Ac_2(1+\frac{A}{4})}{2}$, where $A: G \longrightarrow \mathbb{C}$ is an additive function such that $A \circ \sigma(x) = -A(x)$ for all $x \in G$. We deal with case (b). This ends this proof. In the next theorem we solve (10) on monoids.

□

Theorem 7 *Let S be a monoid which is generated by its squares, let σ an involutive automorphism on S. Let $\mu : S \longrightarrow \mathbb{C}$ be a multiplicative function such that $\mu(x\sigma(x)) = 1$ for all $x \in S$. Suppose that $f, g, h : S \longrightarrow \mathbb{C}$ satisfy the functional equation (10). Suppose also that $f + g \neq 0$. Then there exist a character χ of G, constants $\alpha \in \mathbb{C}^*$, $c_1, c_2 \in \mathbb{C}$, and a function $\theta \in N_\mu(\sigma, S)$ such that one of the following holds:*

(a) *If $\chi \neq \mu\chi \circ \sigma$, then $f = \frac{1}{2}[(1 + \frac{c_1c_2}{2})\frac{\chi+\mu\chi\circ\sigma}{2} + 2c_2\frac{\chi-\mu\chi\circ\sigma}{2} + \theta]$; $g = \frac{1}{2}[(1 - \frac{c_1c_2}{2})\frac{\chi+\mu\chi\circ\sigma}{2} - \theta]$ and $h = \frac{1}{\alpha}[\frac{\chi+\mu\chi\circ\sigma}{2} + c_2\frac{\chi-\mu\chi\circ\sigma}{2}]$.*

(b) If $\chi = \mu\chi \circ \sigma$, then $f(x) = \frac{\theta(x)}{2}$, $g(x) = \frac{-\theta(x)}{2}$ and $h(x) = 0$ for all $x \in I_\chi$; $f(x) = \frac{c_2\chi(2+A(x))+\theta(x)+\chi(x)Ac_2(1+\frac{A(x)}{4})}{2}$; $g(x) = \frac{c_2\chi(2+A(x))-\theta(x)-\chi(x)Ac_2(1+\frac{A(x)}{4})}{2}$ and $h(x) = \frac{1}{\alpha}c_2\chi(2+A(x))$ for all $x \in S\backslash I_\chi$ and where $A: S\backslash I_\chi \longrightarrow \mathbb{C}$ is an additive function such that $A \circ \sigma(x) = -A(x)$ for all $x \in S\backslash I_\chi$.

References

1. J. Aczél, J. Dhombres, Functional equations in several variables, in *With Applications to Mathematics, Information Theory and to the Natural and Social Sciences*. Encyclopedia of Mathematics and Its Applications, vol 31 (Cambridge University Press, Cambridge, 1989)
2. B. Bouikhalene, E. Elqorachi, An extension of Van Vleck's functional equation for the sine. Acta Math. Hunger. **150**(1), 258–267 (2016)
3. J. d'Alembert, Recherches sur la courbe que forme une corde tendue mise en vibration, I. Hist. Acad. Berlin **1747**, 214–219 (1747)
4. J. d'Alembert, Recherches sur la courbe que forme une corde tendue mise en vibration, II. Hist. Acad. Berlin **1747**, 220–249 (1747)
5. J. d'Alembert, Addition au Mémoire sur la courbe que forme une corde tendue mise en vibration. Hist. Acad. Berlin **1750**, 355–360 (1750)
6. T.M.K. Davison, D'Alembert's functional equation on topological monoids. Publ. Math. Debr. **75**(1/2), 41–66 (2009)
7. B. Ebanks, H. Stetkær, On Wilson's functional equations. Aequationes Math. **89**(2), 339–354 (2015)
8. B.R. Ebanks, H. Stetkær, d'Alembert's other functional equation on monoids with involution. Aequationes Math. **89**, 187–206 (2015)
9. B.R. Ebanks, P.K. Sahoo, W. Sander, *Characterizations of Information Measures* (World Scientific Publishing Company, Singapore, 1998)
10. E. Elqorachi, A. Redouani, Solutions and stability of a variant of Wilson's functional equation. Proyecciones J. Math. **37**(2), 317–344 (2018)
11. J.C. Parnami, H. Singh, H.L. Vasudeva, On an exponential-cosine functional equation. Period. Math. Hung. **19**(4), 287–297 (1988)
12. A.M. Perkins, P.K. Sahoo, On two functional equations with involution on groups related to sine and cosine functions. Aequationes Math. (2014). https://doi.org/10.1007/s00010-014-0309-z
13. T.A. Poulsen, H. Stetkær, On the trigonometric subtraction and addition formulas. Aequationes Math. **59**, 84–92 (2000)
14. P.K. Sahoo, A functional equation with restricted argument related to cosine function. Tamsui Oxford Journal of Information and Mathematical Sciences (TOJIMS) **31**(3), 9–20 (2017)
15. H. Stetkær, Functional equations on abelians groups with involution. Aequationes Math. **54**(1–2), 144–172 (1997)
16. H. Stetkær, On a variant of Wilson's functional equation on groups. Aequationes Math. **68**(3), 160–176 (2004)
17. H. Stetkær, *Functional Equations on Groups* (World Scientific Publishing Company, Singapore, 2013)
18. H. Stetkær, d'Alembert's functional equation on groups, in *Recent Developments in Functional Equations and Inequalities*. Banach Center Publication, vol 99 (Polish Academic Science Institute of Mathematics, Warsaw, 2013), pp. 173–191

19. H. Stetkær, *Functional Equations on Groups* (World Scientific Publishing Company, Singapore, 2013)
20. H. Stetkær, Van Vleck's functional equation for the sine. Aequationes Math. (2014). https://doi.org/10.1007/s00010-015-0349-z
21. H. Stetkær, A variant of d'Alembert's functional equation. Aequationes Math.**89**(3) (2014). https://doi.org/10.1007/s00010-014-0253-y
22. E.B. Van Vleck, A functional equation for the sine. Ann. Math. Second Ser. **11**(4), 161–165 (1910)
23. E.B. Van Vleck, A functional equation for the sine. Additional note. Ann. Math. Second Ser. **13**(1/4), 154 (1911–1912)
24. D. Yang, The symmetrized Sine addition formula. Aequationes Math. **82**(3), 299–318 (2011)

Hyers–Ulam Stability of a Discrete Diamond-Alpha Derivative Equation

Douglas R. Anderson and Masakazu Onitsuka

Abstract We establish the Hyers–Ulam stability (HUS) of a certain first-order linear constant coefficient discrete diamond-alpha derivative equation. In particular, for each parameter value we determine whether the equation has HUS, and if so whether there exists a minimum HUS constant.

1 Introduction

In 1940, Ulam [24, p. 63] posed the following questions:

> When is it true that the solution of an equation differing slightly from a given one, must of necessity be close to the solution of the given equation? Similarly, if we replace a given functional equation by a functional inequality, when can one assert that the solutions of the inequality lie near to the solutions of the strict equation?

The problem for the case of approximately additive mappings was solved by Hyers [7], who proved that the Cauchy equation is stable in Banach spaces, and the result of Hyers was generalized by Rassias [18]. Since then there has been a significant amount of interest in Hyers–Ulam stability (HUS), especially in relation to ordinary differential equations; for example, see [1, 2, 4, 8–14, 21, 25, 26]. See also many pertinent results for functional equations and quadratic functional equations [5, 6, 17, 19, 22]. For some results on time scales, see [23].

For $\lambda \in \mathbb{R}$, the equation

$$x'(t) - \lambda x(t) = 0, \quad t \in \mathbb{R} \tag{1}$$

D. R. Anderson (✉)
Concordia College, Department of Mathematics, Moorhead, MN, USA
e-mail: andersod@cord.edu

M. Onitsuka
Department of Applied Mathematics, Okayama University of Science, Okayama, Japan
e-mail: onitsuka@xmath.ous.ac.jp

© Springer Nature Switzerland AG 2019
G. A. Anastassiou and J. M. Rassias (eds.), *Frontiers in Functional Equations and Analytic Inequalities*, https://doi.org/10.1007/978-3-030-28950-8_14

237

has HUS if and only if there exists a constant $K > 0$ with the following property:

For arbitrary $\varepsilon > 0$, if a function $\phi : \mathbb{R} \to \mathbb{R}$ satisfies $|\phi'(t) - \lambda\phi(t)| \leq \varepsilon$ for all $t \in \mathbb{R}$, then there exists a solution $x : \mathbb{R} \to \mathbb{R}$ of (1) such that $|\phi(t) - x(t)| \leq K\varepsilon$ for all $t \in \mathbb{R}$.

Such a constant K is called an HUS constant for (1) on $\mathbb{R}$. Recently, Onitsuka and Shoji [16] explored the minimum HUS constant for (1). Also, Onitsuka [15] investigated the influence of the constant step size $h > 0$ on HUS for the first-order homogeneous linear difference equation

$$\Delta_h x(t) - \lambda x(t) = 0 \tag{2}$$

on the uniformly discrete time scale $h\mathbb{Z}$, where Δ_h is the forward difference

$$\Delta_h x(t) = \frac{x(t+h) - x(t)}{h}, \quad t \in h\mathbb{Z} := \{hk : k \in \mathbb{Z}\}.$$

We propose to extend these results to the discrete diamond-alpha case, which contains the delta forward difference equation and the nabla backward difference equation [3] as special cases. For more on the diamond-α derivative, see [20] and the references therein.

2 Hyers–Ulam Stability for a Discrete Diamond-Alpha Derivative Equation

Let $\mathbb{Z}$ represent the set of integers. For any nonempty open interval $I \subseteq \mathbb{R}$, let $\mathbb{T} := \mathbb{Z} \cap I$. Define

$$\mathbb{T}_\kappa := \begin{cases} \mathbb{T}\backslash\{\min \mathbb{T}\} & : \min \mathbb{T} \text{ exists}, \\ \mathbb{T} & : \text{otherwise}, \end{cases} \quad \text{and} \quad \mathbb{T}^\kappa := \begin{cases} \mathbb{T}\backslash\{\max \mathbb{T}\} & : \max \mathbb{T} \text{ exists}, \\ \mathbb{T} & : \text{otherwise}, \end{cases}$$

and set $\mathbb{T}_\kappa^\kappa = \mathbb{T}_\kappa \cap \mathbb{T}^\kappa$. In this paper we consider on $\mathbb{T}$ the Hyers–Ulam stability of the first-order linear homogeneous discrete diamond-alpha derivative equation with constant coefficient given by

$$\Diamond_\alpha x(t) - \lambda x(t) = 0, \quad \Diamond_\alpha x(t) := \alpha \Delta x(t) + (1-\alpha)\nabla x(t), \quad \alpha \in [0,1], \tag{3}$$

where $\lambda \in \mathbb{R}$; here, we use the forward difference operator $\Delta x(t) := x(t+1) - x(t)$ and the backward difference operator $\nabla x(t) := x(t) - x(t-1)$ for all $t \in \mathbb{T}_\kappa^\kappa$. Note that if a function x exists on $\mathbb{T}$, then Δx exists on $\mathbb{T}^\kappa$ and ∇x exists on $\mathbb{T}_\kappa$. Thus, for the remainder of the paper, we assume that $\mathbb{T}$ and $\mathbb{T}_\kappa^\kappa$ are nonempty sets in $\mathbb{R}$.

Definition 1 We say that (3) has Hyers–Ulam stability (HUS) on $\mathbb{T}$ if and only if there exists a constant $K > 0$ with the following property. For arbitrary $\varepsilon > 0$, if a function $\phi : \mathbb{T} \to \mathbb{R}$ satisfies

$$|\Diamond_\alpha \phi(t) - \lambda \phi(t)| \le \varepsilon \quad \text{for all} \quad t \in \mathbb{T}_\kappa^\kappa,$$

then there exists a solution $x : \mathbb{T} \to \mathbb{R}$ of (3) such that $|\phi(t) - x(t)| \le K\varepsilon$ for all $t \in \mathbb{T}$. Such a constant K is called an HUS constant for (3) on $\mathbb{T}$.

Given this definition, we would like to know, given $\alpha \in [0, 1]$, for which values of the parameter $\lambda \in \mathbb{R}$ does (3) have HUS, and if it does have HUS, is there a minimum HUS constant?

We start with values for which (3) does not have HUS.

Theorem 1 *For any $\alpha \in [0, 1]$, if $\lambda = 0$ or $\lambda = 2(1 - 2\alpha)$, then (3) does not have Hyers–Ulam stability on $\mathbb{Z}$.*

Proof Let $\lambda = 0$ and $\alpha \in (0, 1]$. Given $\varepsilon > 0$, note that the function $\phi(t) := \varepsilon t$ satisfies $|\Diamond_\alpha \phi(t)| = \varepsilon$ for all $t \in \mathbb{Z}$. As $x(t) = c_1 + c_2 \left(1 - \frac{1}{\alpha}\right)^t$ is the general solution to $\Diamond_\alpha x(t) = 0$, we see that $|\phi(t) - x(t)| \to \infty$ as $t \to \pm\infty$ for any choice of the constants $c_1, c_2 \in \mathbb{R}$, so that (3) does not have HUS on $\mathbb{Z}$ when $\lambda = 0$ and $\alpha \in (0, 1]$. Let $\lambda = 0$ and $\alpha = 0$. Given $\varepsilon > 0$, again see that $\phi(t) := \varepsilon t$ satisfies $|\Diamond_0 \phi(t)| = |\nabla \phi(t)| = \varepsilon$ for all $t \in \mathbb{Z}$. As $x(t) \equiv c$ is the general solution to $\nabla x(t) = 0$, we see that $|\phi(t) - x(t)| \to \infty$ as $t \to \pm\infty$ for any choice of $c \in \mathbb{R}$, so that (3) does not have HUS on $\mathbb{Z}$ when $\lambda = 0$ and $\alpha = 0$.

Next, since $\lambda = 2(1 - 2\alpha)$ for any $\alpha \in [0, 1]$ is equivalent to writing $\alpha = \dfrac{2 - \lambda}{4}$ for $\lambda \in [-2, 2]$, let $\alpha = \dfrac{2 - \lambda}{4}$ for $\lambda \in [-2, 2)$. Given $\varepsilon > 0$, the function $\phi(t) := \varepsilon t (-1)^t$ satisfies

$$\left|\Diamond_{\frac{2-\lambda}{4}} \phi(t) - \lambda \phi(t)\right| = \left|(-1)^{t+1}\varepsilon\right| = \varepsilon$$

for all $t \in \mathbb{Z}$. As

$$x(t) = c_1(-1)^t + c_2 \left(\frac{2+\lambda}{2-\lambda}\right)^t$$

is the general solution to $\Diamond_{\frac{2-\lambda}{4}} x(t) - \lambda x(t) = 0$, we see that $|\phi(t) - x(t)| \to \infty$ as $t \to \pm\infty$ for any choice of the constants $c_1, c_2 \in \mathbb{R}$, so that (3) does not have HUS on $\mathbb{Z}$ when $\alpha = \dfrac{2 - \lambda}{4}$ for $\lambda \in [-2, 2)$. If $\alpha = 0$ and $\lambda = 2$, then using $\phi(t) := \varepsilon t (-1)^t$ and $x(t) = c(-1)^{t/h}$ as above, we see there is no HUS on $\mathbb{Z}$ when $\lambda = 2$ and $\alpha = 0$ either. $\qquad\square$

We now consider cases where (3) does have HUS, given $\alpha \in [0, 1]$ and $\lambda \in \mathbb{R}$ not mentioned in 1 above. We begin with the special cases of $\alpha = 1$ and $\alpha = 0$, respectively, in the next few theorems. The $\alpha = 1$ case is known from [15], while the $\alpha = 0$ is new to the literature.

Theorem 2 ([15, Theorem 1.5]) *Let $\alpha = 1$, so that $\lozenge_\alpha = \Delta$ in (3). Assume $\lambda > -1$ with $\lambda \neq 0$. Let $\varepsilon > 0$ be a given arbitrary constant, and let $\phi : \mathbb{T} \to \mathbb{R}$ satisfy*

$$|\Delta\phi(t) - \lambda\phi(t)| \leq \varepsilon, \qquad t \in \mathbb{T}^\kappa.$$

Then one of the following holds:

(i) *If $\lambda > 0$ and $\tau^* := \max \mathbb{T}$ exists, then any solution x of (3) with $|\phi(\tau^*) - x(\tau^*)| < \frac{\varepsilon}{\lambda}$ satisfies $|\phi(t) - x(t)| < \frac{\varepsilon}{\lambda}$ for all $t \in \mathbb{T}$.*

(ii) *If $\lambda > 0$ and $\max \mathbb{T}$ does not exist, then $\lim_{t \to \infty} \phi(t)(1 + \lambda)^{-t}$ exists, and the function*

$$x(t) := \left(\lim_{t \to \infty} \phi(t)(1 + \lambda)^{-t} \right)(1 + \lambda)^t$$

is the unique solution of (3) such that $|\phi(t) - x(t)| \leq \frac{\varepsilon}{\lambda}$ for all $t \in \mathbb{T}$.

(iii) *If $-1 < \lambda < 0$ and $\tau_* := \min \mathbb{T}$ exists, then any solution x of (3) with $|\phi(\tau_*) - x(\tau_*)| < \frac{\varepsilon}{|\lambda|}$ satisfies $|\phi(t) - x(t)| < \frac{\varepsilon}{|\lambda|}$ for all $t \in \mathbb{T}$.*

(iv) *If $-1 < \lambda < 0$ and $\min \mathbb{T}$ does not exist, then $\lim_{t \to -\infty} \phi(t)(1 + \lambda)^{-t}$ exists, and the function*

$$x(t) := \left(\lim_{t \to -\infty} \phi(t)(1 + \lambda)^{-t} \right)(1 + \lambda)^t$$

is the unique solution of (3) such that $|\phi(t) - x(t)| \leq \frac{\varepsilon}{|\lambda|}$ for all $t \in \mathbb{T}$.

Theorem 3 ([15, Theorem 1.7]) *Let $\alpha = 1$, so that $\lozenge_\alpha = \Delta$ in (3). Assume $\lambda < -1$ with $\lambda \neq -2$. Let $\varepsilon > 0$ be a given arbitrary constant, and let $\phi : \mathbb{T} \to \mathbb{R}$ satisfy*

$$|\Delta\phi(t) - \lambda\phi(t)| \leq \varepsilon, \qquad t \in \mathbb{T}^\kappa.$$

Then one of the following holds:

(i) *If $\lambda < -2$ and $\tau^* := \max \mathbb{T}$ exists, then any solution x of (3) with*

$$|\phi(\tau^*) - x(\tau^*)| < \frac{\varepsilon}{|\lambda + 2|} \qquad satisfies \qquad |\phi(t) - x(t)| < \frac{\varepsilon}{|\lambda + 2|}$$

for all $t \in \mathbb{T}$.

(ii) *If $\lambda < -2$ and $\max \mathbb{T}$ does not exist, then $\lim_{t \to \infty} \phi(t)(1 + \lambda)^{-t}$ exists, and the function*

$$x(t) := \left(\lim_{t \to \infty} \phi(t)(1 + \lambda)^{-t} \right) (1 + \lambda)^t$$

is the unique solution of (3) such that

$$|\phi(t) - x(t)| \le \frac{\varepsilon}{|\lambda + 2|}$$

for all $t \in \mathbb{T}$.

(iii) *If $-2 < \lambda < -1$ and $\tau_* := \min \mathbb{T}$ exists, then any solution x of (3) with*

$$|\phi(\tau_*) - x(\tau_*)| < \frac{\varepsilon}{\lambda + 2} \quad \text{satisfies} \quad |\phi(t) - x(t)| < \frac{\varepsilon}{\lambda + 2}$$

for all $t \in \mathbb{T}$.

(iv) *If $-2 < \lambda < -1$ and $\min \mathbb{T}$ does not exist, then $\lim_{t \to -\infty} \phi(t)(1 + \lambda)^{-t}$ exists, and the function*

$$x(t) := \left(\lim_{t \to -\infty} \phi(t)(1 + \lambda)^{-t} \right) (1 - \lambda)^t$$

is the unique solution of (3) such that

$$|\phi(t) - x(t)| \le \frac{\varepsilon}{\lambda + 2}$$

for all $t \in \mathbb{T}$.

Theorem 4 *Let $\alpha = 0$, so that $\Diamond_\alpha = \nabla$ in (3). Assume $\lambda < 1$ with $\lambda \ne 0$. Let $\varepsilon > 0$ be a given arbitrary constant, and let $\phi : \mathbb{T} \to \mathbb{R}$ satisfy*

$$|\nabla \phi(t) - \lambda \phi(t)| \le \varepsilon, \qquad t \in \mathbb{T}_\kappa.$$

Then one of the following holds:

(i) *If $0 < \lambda < 1$ and $\tau^* := \max \mathbb{T}$ exists, then any solution x of (3) with $|\phi(\tau^*) - x(\tau^*)| < \frac{\varepsilon}{\lambda}$ satisfies $|\phi(t) - x(t)| < \frac{\varepsilon}{\lambda}$ for all $t \in \mathbb{T}$.*

(ii) *If $0 < \lambda < 1$ and $\max \mathbb{T}$ does not exist, then $\lim_{t \to \infty} \phi(t)(1 - \lambda)^t$ exists, and the function*

$$x(t) := \left(\lim_{t \to \infty} \phi(t)(1 - \lambda)^t \right) (1 - \lambda)^{-t}$$

is the unique solution of (3) such that $|\phi(t) - x(t)| \le \frac{\varepsilon}{\lambda}$ for all $t \in \mathbb{T}$.

(iii) *If $\lambda < 0$ and $\tau_* := \min \mathbb{T}$ exists, then any solution x of (3) with $|\phi(\tau_*) - x(\tau_*)| < \frac{\varepsilon}{|\lambda|}$ satisfies $|\phi(t) - x(t)| < \frac{\varepsilon}{|\lambda|}$ for all $t \in \mathbb{T}$.*

(iv) If $\lambda < 0$ and $\min \mathbb{T}$ does not exist, then $\lim\limits_{t \to -\infty} \phi(t)(1 - \lambda)^t$ exists, and the function

$$x(t) := \left(\lim_{t \to -\infty} \phi(t)(1 - \lambda)^t \right) (1 - \lambda)^{-t}$$

is the unique solution of (3) *such that* $|\phi(t) - x(t)| \leq \frac{\varepsilon}{|\lambda|}$ *for all* $t \in \mathbb{T}$.

Proof From $\nabla\phi(t) = \Delta\phi(t - 1) = \phi(t) - \phi(t - 1)$, we have

$$\nabla\phi(t) - \lambda\phi(t) = \Delta\phi(t - 1) - \lambda\left(\Delta\phi(t - 1) + \phi(t - 1)\right)$$

$$= (1 - \lambda)\Delta\phi(t - 1) - \lambda\phi(t - 1)$$

$$= (1 - \lambda)\left(\Delta\phi(t - 1) - \frac{\lambda}{1 - \lambda}\phi(t - 1)\right) \tag{4}$$

for all $t \in \mathbb{T}_\kappa$. Therefore, using the assumption $|\nabla\phi(t) - \lambda\phi(t)| \leq \varepsilon$ for all $t \in \mathbb{T}_\kappa$, we get

$$\left| \Delta\phi(t - 1) - \frac{\lambda}{1 - \lambda}\phi(t - 1) \right| \leq \frac{\varepsilon}{|1 - \lambda|}, \qquad t \in \mathbb{T}_\kappa.$$

That is,

$$\left| \Delta\phi(t) - \frac{\lambda}{1 - \lambda}\phi(t) \right| \leq \frac{\varepsilon}{|1 - \lambda|}, \qquad t \in \mathbb{T}^\kappa. \tag{5}$$

Now, we consider case (i). Suppose that $0 < \lambda < 1$ and $\tau^* := \max \mathbb{T}$ exists. Then $\frac{\lambda}{1-\lambda} > 0$ and $1 - \lambda > 0$ hold. Using Theorem 2 (i) with (5), we have the following: any solution x of

$$\Delta x(t) - \frac{\lambda}{1 - \lambda}x(t) = 0 \tag{6}$$

with

$$|\phi(\tau^*) - x(\tau^*)| < \frac{\frac{\varepsilon}{|1-\lambda|}}{\frac{\lambda}{1-\lambda}} = \frac{\varepsilon}{\lambda}$$

satisfies

$$|\phi(t) - x(t)| < \frac{\frac{\varepsilon}{|1-\lambda|}}{\frac{\lambda}{1-\lambda}} = \frac{\varepsilon}{\lambda}$$

for all $t \in \mathbb{T}$. From (4) and (6) is equivalent to (3) with $\alpha = 0$. Thus, case (i) is now true.

Next, we consider case (ii). Using Theorem 2 (ii) with (5), we see that

$$\lim_{t \to \infty} \phi(t) \left(1 + \frac{\lambda}{1 - \lambda} \right)^{-t} = \lim_{t \to \infty} \phi(t)(1 - \lambda)^t$$

exists, and the function

$$x(t) := \left(\lim_{t \to \infty} \phi(t)(1 - \lambda)^t \right) \left(1 + \frac{\lambda}{1 - \lambda} \right)^t = \left(\lim_{t \to \infty} \phi(t)(1 - \lambda)^t \right) (1 - \lambda)^{-t}$$

is the unique solution of (6) such that $|\phi(t) - x(t)| \le \frac{\varepsilon}{\lambda}$ for all $t \in \mathbb{T}$. From (4), (6) is equivalent to (3) with $\alpha = 0$. Therefore, we conclude that $x(t)$ is the unique solution of (3) with $\alpha = 0$ such that $|\phi(t) - x(t)| \le \frac{\varepsilon}{\lambda}$ for all $t \in \mathbb{T}$. Thus, case (ii) is true.

Since $\lambda < 0$ implies $-1 < \frac{\lambda}{1-\lambda} < 0$, cases (iii) and (iv) can be proved by using Theorem 2 (iii) and (iv). The proof of (iii) and (iv) are the same as the above, thus, the proof is now complete. $\qquad\square$

Theorem 5 *Let $\alpha = 0$, so that $\Diamond_\alpha = \nabla$ in (3). Assume $\lambda > 1$ with $\lambda \neq 2$. Let $\varepsilon > 0$ be a given arbitrary constant, and let $\phi : \mathbb{T} \to \mathbb{R}$ satisfy*

$$|\nabla \phi(t) - \lambda \phi(t)| \le \varepsilon, \qquad t \in \mathbb{T}_\kappa.$$

Then one of the following holds:

(i) If $1 < \lambda < 2$ and $\tau^* := \max \mathbb{T}$ exists, then any solution x of (3) with

$$|\phi(\tau^*) - x(\tau^*)| < \frac{\varepsilon}{2 - \lambda} \quad \text{satisfies} \quad |\phi(t) - x(t)| < \frac{\varepsilon}{2 - \lambda}$$

for all $t \in \mathbb{T}$.

(ii) If $1 < \lambda < 2$ and $\max \mathbb{T}$ does not exist, then $\lim_{t \to \infty} \phi(t)(1 - \lambda)^t$ exists, and the function

$$x(t) := \left(\lim_{t \to \infty} \phi(t)(1 - \lambda)^t \right) (1 - \lambda)^{-t}$$

is the unique solution of (3) such that

$$|\phi(t) - x(t)| \le \frac{\varepsilon}{2 - \lambda}$$

for all $t \in \mathbb{T}$.

(iii) *If* $\lambda > 2$ *and* $\tau_* := \min \mathbb{T}$ *exists, then any solution* x *of* (3) *with*

$$|\phi(\tau_*) - x(\tau_*)| < \frac{\varepsilon}{\lambda - 2} \quad \text{satisfies} \quad |\phi(t) - x(t)| < \frac{\varepsilon}{\lambda - 2}$$

for all $t \in \mathbb{T}$.

(iv) *If* $\lambda > 2$ *and* $\min \mathbb{T}$ *does not exist, then* $\lim_{t \to -\infty} \phi(t)(1 - \lambda)^t$ *exists, and the function*

$$x(t) := \left(\lim_{t \to -\infty} \phi(t)(1 - \lambda)^t \right) (1 - \lambda)^{-t}$$

is the unique solution of (3) *such that*

$$|\phi(t) - x(t)| \leq \frac{\varepsilon}{\lambda - 2}$$

for all $t \in \mathbb{T}$.

Proof Using the same way as in the proof of Theorem 4, it easily turns out that the assertions in Theorem 5 are true. □

By using Theorems 4 and 5, we can establish the following result.

Theorem 6 *Let* $\alpha = 0$, *and assume* $\lambda \neq 0, 1, 2$. *Let* $\varepsilon > 0$ *be a given arbitrary constant, and let* $\phi : \mathbb{T} \to \mathbb{R}$ *satisfy*

$$|\nabla \phi(t) - \lambda \phi(t) - f(t)| \leq \varepsilon, \qquad t \in \mathbb{T}_\kappa,$$

where f *is a real-valued function on* $\mathbb{T}$. *Then one of the following holds:*

(i) *If* $0 < \lambda < 1$, *then there exists a solution* $x : \mathbb{T} \to \mathbb{R}$ *of*

$$\nabla x(t) - \lambda x(t) - f(t) = 0 \tag{7}$$

such that $|\phi(t) - x(t)| \leq \frac{\varepsilon}{\lambda}$ *for all* $t \in \mathbb{T}$.

(ii) *If* $\lambda < 0$, *then there exists a solution* $x : \mathbb{T} \to \mathbb{R}$ *of* (7) *such that* $|\phi(t) - x(t)| \leq \frac{\varepsilon}{|\lambda|}$ *for all* $t \in \mathbb{T}$.

(iii) *If* $1 < \lambda < 2$, *then there exists a solution* $x : \mathbb{T} \to \mathbb{R}$ *of* (7) *such that* $|\phi(t) - x(t)| \leq \frac{\varepsilon}{2 - \lambda}$ *for all* $t \in \mathbb{T}$.

(iv) *If* $\lambda > 2$, *then there exists a solution* $x : \mathbb{T} \to \mathbb{R}$ *of* (7) *such that* $|\phi(t) - x(t)| \leq \frac{\varepsilon}{\lambda - 2}$ *for all* $t \in \mathbb{T}$.

Proof We assume that

$$|\nabla \phi(t) - \lambda \phi(t) - f(t)| \leq \varepsilon$$

for all $t \in \mathbb{T}_\kappa$. Let $u(t) = (1 - \lambda)^{-t} \nabla^{-1} f(t)(1 - \lambda)^{t-1}$ on $\mathbb{T}$, where ∇^{-1} is an anti-backward difference operator. Then $u(t)$ is a solution of (7). Actually, we can check that

$$f(t)(1 - \lambda)^{t-1} = \nabla u(t)(1 - \lambda)^t = u(t)(1 - \lambda)^t - u(t - 1)(1 - \lambda)^{t-1}$$
$$= [(1 - \lambda)u(t) - u(t - 1)](1 - \lambda)^{t-1}$$
$$= [\nabla u(t) - \lambda u(t)](1 - \lambda)^{t-1}$$

holds for all $t \in \mathbb{T}_\kappa$. From this, we see that

$$\nabla(\phi(t) - u(t)) - \lambda(\phi(t) - u(t)) = \nabla \phi(t) - \lambda \phi(t) - (\nabla u(t) - \lambda u(t))$$
$$= \nabla \phi(t) - \lambda \phi(t) - f(t)$$

for all $t \in \mathbb{T}_\kappa$, and thus,

$$|\nabla(\phi(t) - u(t)) - \lambda(\phi(t) - u(t))| = |\nabla \phi(t) - \lambda \phi(t) - f(t)| \leq \varepsilon$$

for all $t \in \mathbb{T}_\kappa$, by the assumption.

First, we consider case (i). Using Theorem 4 (i) and (ii), we can find a solution $v : \mathbb{T} \to \mathbb{R}$ of (3) with $\alpha = 0$ such that $|(\phi(t) - u(t)) - v(t)| \leq \frac{\varepsilon}{\lambda}$ for all $t \in \mathbb{T}$. Let $x(t) = u(t) + v(t)$ for all $t \in \mathbb{T}$. Then we see that

$$\nabla x(t) = \nabla u(t) + \nabla v(t) = \lambda u(t) + f(t) + \lambda v(t) = \lambda x(t) + f(t)$$

holds on $\mathbb{T}_\kappa$. This means that x is a solution of (7) on $\mathbb{T}$. Thus, (i) is true.

Next, we consider case (ii). From Theorem 4 (iii) and (iv), there exists a solution $v : \mathbb{T} \to \mathbb{R}$ of (3) with $\alpha = 0$ such that $|(\phi(t) - u(t)) - v(t)| \leq \frac{\varepsilon}{|\lambda|}$ for all $t \in \mathbb{T}$. Let $x(t) = u(t) + v(t)$ for all $t \in \mathbb{T}$. Using the same argument as above, x is a solution of (7) on $\mathbb{T}$. Thus, (ii) is true.

Cases (iii) and (iv) can be proved by using Theorem 5. The proof of (iii) and (iv) are the same as the above, thus, the proof is now complete. $\qquad \square$

3 Hyers–Ulam Stability for the General Case

In the remainder of the paper, we will explore HUS and HUS constants for (3) in the case that $\alpha \in (0, 1)$ and $\lambda \in \mathbb{R} \backslash \{0\}$. Note that upon expansion, (3) is equivalent to the second-order difference equation

$$\alpha x(t + 1) + (1 - 2\alpha - \lambda)x(t) + (\alpha - 1)x(t - 1) = 0.$$

If we denote the characteristic values of this equation as

$$\Lambda_- := \frac{\lambda + 2\alpha - 1 - \sqrt{1 - 2\lambda + 4\alpha\lambda + \lambda^2}}{2\alpha} \quad \text{and}$$

$$\Lambda_+ := \frac{\lambda + 2\alpha - 1 + \sqrt{1 - 2\lambda + 4\alpha\lambda + \lambda^2}}{2\alpha},$$

then the general solution to (3) is

$$x(t) = a_1 \Lambda_-^t + a_2 \Lambda_+^t, \qquad t \in \mathbb{T},$$

for arbitrary constants $a_1, a_2 \in \mathbb{R}$. Note that $\Lambda_- < 0$ and $\Lambda_+ > 0$ for all $\alpha \in (0, 1)$ and for all $\lambda \in \mathbb{R}$, and that the discriminant satisfies $1 - 2\lambda + 4\alpha\lambda + \lambda^2 > 0$ for all $\alpha \in (0, 1)$ and for all $\lambda \in \mathbb{R}$.

Before presenting the main theorem, we give a lemma.

Lemma 1 *Let $\alpha \in (0, 1)$, $\lambda \in \mathbb{R}$ and $g(\mu) = \alpha\mu^2 - (\lambda + 2\alpha - 1)\mu + \alpha - 1$ for $\mu \in \mathbb{R}$. Then Λ_- and Λ_+ are two different real roots of $g(\mu) = 0$, and satisfy $\Lambda_- < 0 < \Lambda_+$. Furthermore, the following hold:*

(i) If $0 < \lambda < 2(1 - 2\alpha)$, then $\Lambda_+ - 1 > 0$ and $1 < \frac{\Lambda_- - 1}{\Lambda_-} < 2$;

(ii) If $\lambda > 0$ and $\lambda > 2(1 - 2\alpha)$, then $\Lambda_+ - 1 > 0$ and $\frac{\Lambda_- - 1}{\Lambda_-} > 2$;

(iii) If $\lambda < 0$ and $\lambda < 2(1 - 2\alpha)$, then $-1 < \Lambda_+ - 1 < 0$ and $1 < \frac{\Lambda_- - 1}{\Lambda_-} < 2$;

(iv) If $2(1 - 2\alpha) < \lambda < 0$, then $-1 < \Lambda_+ - 1 < 0$ and $\frac{\Lambda_- - 1}{\Lambda_-} > 2$.

Proof From $\alpha \in (0, 1)$, the discriminant $(\lambda + 2\alpha - 1)^2 - 4\alpha(\alpha - 1)$ of $g(\mu) = 0$ is positive, so that Λ_- and Λ_+ are two different real roots of $g(\mu) = 0$. Since $g(0) = \alpha - 1 < 0$, we see that $\Lambda_- < 0 < \Lambda_+$ holds.

First, we consider (i). Since $g(1) = -\lambda < 0$ we have $\Lambda_+ > 1$. On the other hand, from $g(-1) = \lambda + 2(2\alpha - 1) < 0$, we have $\Lambda_- < -1$, so that $2\Lambda_- < \Lambda_- - 1 < \Lambda_-$. Hence we obtain $1 < \frac{\Lambda_- - 1}{\Lambda_-} < 2$.

Next, we consider case (ii). As in the same argument of (i), we have $\Lambda_+ > 1$. From $g(-1) = \lambda + 2(2\alpha - 1) > 0$, we have $-1 < \Lambda_- < 0$, so that $\Lambda_- - 1 < 2\Lambda_-$. Hence we get $\frac{\Lambda_- - 1}{\Lambda_-} > 2$.

Next, we consider case (iii). Since $g(1) = -\lambda > 0$ we have $0 < \Lambda_+ < 1$. Using the same argument of (i), we obtain $1 < \frac{\Lambda_- - 1}{\Lambda_-} < 2$. By using the same arguments, the statement of (iv) is clearly true. □

Theorem 7 *Let $\alpha \in (0, 1)$. Assume $\lambda \neq 0, 2(1 - 2\alpha)$. Let $\varepsilon > 0$ be a given arbitrary constant, and let $\phi : \mathbb{T} \to \mathbb{R}$ satisfy*

$$|\Diamond_\alpha \phi(t) - \lambda\phi(t)| \leq \varepsilon \quad \text{for all} \quad t \in \mathbb{T}_\kappa^\kappa.$$

Then one of the following holds:

(i) *If* $0 < \lambda < 2(1-2\alpha)$, *then* (3) *has Hyers–Ulam stability with an HUS constant*
$$\frac{1}{1-\sqrt{1-2\lambda+4\alpha\lambda+\lambda^2}} \text{ on } \mathbb{T};$$

(ii) *If* $\lambda > 0$ *and* $\lambda > 2(1-2\alpha)$, *then* (3) *has Hyers–Ulam stability with an HUS constant* $\dfrac{1}{\sqrt{1-2\lambda+4\alpha\lambda+\lambda^2}-1} \text{ on } \mathbb{T};$

(iii) *If* $\lambda < 0$ *and* $\lambda < 2(1-2\alpha)$, *then* (3) *has Hyers–Ulam stability with an HUS constant* $\dfrac{1}{\sqrt{1-2\lambda+4\alpha\lambda+\lambda^2}-1} \text{ on } \mathbb{T};$

(iv) *If* $2(1-2\alpha) < \lambda < 0$, *then* (3) *has Hyers–Ulam stability with an HUS constant* $\dfrac{1}{1-\sqrt{1-2\lambda+4\alpha\lambda+\lambda^2}} \text{ on } \mathbb{T}.$

Proof Due to Theorem 1, assume $\lambda \neq 0$ and $\lambda \neq 2(1 - 2\alpha)$ for any $\alpha \in (0, 1)$. We consider the function $g(\mu) = \alpha\mu^2 - (\lambda+2\alpha-1)\mu+\alpha-1$ for $\mu \in \mathbb{R}$. Therefore, by Lemma 1, Λ_- and Λ_+ are two different real roots of $g(\mu) = 0$ with $\Lambda_- < 0 < \Lambda_+$. From this, we see that

$$\Lambda_- + \Lambda_+ = \frac{\lambda + 2\alpha - 1}{\alpha} \quad \text{and} \quad \Lambda_-\Lambda_+ = \frac{\alpha - 1}{\alpha} \tag{8}$$

holds. Now, for arbitrary $\varepsilon > 0$, we assume that a function $\phi : \mathbb{T} \to \mathbb{R}$ satisfies

$$|\lozenge_\alpha\phi(t) - \lambda\phi(t)| \leq \varepsilon$$

for all $t \in \mathbb{T}_\kappa^\kappa$. Let $\psi(t) = \alpha\Lambda_-\nabla\phi(t) - \alpha(\Lambda_- - 1)\phi(t)$ for all $t \in \mathbb{T}_\kappa$. Using (8), we have

$$\Delta\psi(t) - (\Lambda_+ - 1)\psi(t) = \alpha\Lambda_-\Delta(\nabla\phi(t)) - \alpha(\Lambda_- - 1)\Delta\phi(t)$$
$$-\alpha(\Lambda_+ - 1)\Lambda_-\nabla\phi(t) + \alpha(\Lambda_- - 1)(\Lambda_+ - 1)\phi(t)$$
$$= \alpha\Lambda_-\Delta(\phi(t) - \phi(t-1)) - \alpha(\Lambda_- - 1)\Delta\phi(t)$$
$$-\alpha(\Lambda_-\Lambda_+ - \Lambda_-)\nabla\phi(t) + \alpha(\Lambda_-\Lambda_+ - \Lambda_- - \Lambda_+ + 1)\phi(t)$$
$$= \alpha\Lambda_-\Delta\phi(t) - \alpha\Lambda_-\nabla\phi(t) - \alpha(\Lambda_- - 1)\Delta\phi(t)$$
$$-\alpha(\Lambda_-\Lambda_+ - \Lambda_-)\nabla\phi(t) - \lambda\phi(t)$$
$$= \alpha\Delta\phi(t) - \alpha\Lambda_-\Lambda_+\nabla\phi(t) - \lambda\phi(t)$$
$$= \lozenge_\alpha\phi(t) - \lambda\phi(t) \tag{9}$$

for all $t \in \mathbb{T}_\kappa^\kappa$. By the assumption, we see that

$$|\Delta\psi(t) - (\Lambda_+ - 1)\psi(t)| = |\lozenge_\alpha\phi(t) - \lambda\phi(t)| \leq \varepsilon \tag{10}$$

for all $t \in \mathbb{T}_\kappa^\kappa$. The proof can be divided into four cases: (i) $0 < \lambda < 2(1 - 2\alpha)$; (ii) $\lambda > 0$ and $\lambda > 2(1 - 2\alpha)$; (iii) $\lambda < 0$ and $\lambda < 2(1 - 2\alpha)$; (iv) $2(1 - 2\alpha) < \lambda < 0$.

First, we consider case (i) $0 < \lambda < 2(1 - 2\alpha)$. From Lemma 1 (i), we see that

$$\Lambda_+ - 1 > 0 \quad \text{and} \quad 1 < \frac{\Lambda_- - 1}{\Lambda_-} < 2. \tag{11}$$

Using Theorem 2 with (10) and (11), we conclude that there exists a solution $y : \mathbb{T}_\kappa \to \mathbb{R}$ of

$$\Delta y(t) - (\Lambda_+ - 1)y(t) = 0 \tag{12}$$

such that $|\psi(t) - y(t)| \leq \frac{\varepsilon}{\Lambda_+ - 1}$ for all $t \in \mathbb{T}_\kappa$. This inequality implies that

$$\left| \nabla\phi(t) - \frac{\Lambda_- - 1}{\Lambda_-}\phi(t) - \frac{y(t)}{\alpha\Lambda_-} \right| \leq \frac{\varepsilon}{\alpha|\Lambda_-|(\Lambda_+ - 1)} \tag{13}$$

for all $t \in \mathbb{T}_\kappa$. Using Theorem 6 (iii) with (11) and (13), we can find a solution $x : \mathbb{T} \to \mathbb{R}$ of

$$\nabla x(t) - \frac{\Lambda_- - 1}{\Lambda_-}x(t) - \frac{y(t)}{\alpha\Lambda_-} = 0 \tag{14}$$

such that

$$|\phi(t) - x(t)| \leq \frac{\frac{\varepsilon}{\alpha|\Lambda_-|(\Lambda_+ - 1)}}{2 - \frac{\Lambda_- - 1}{\Lambda_-}} = \frac{\varepsilon}{\alpha|\Lambda_- + 1|(\Lambda_+ - 1)}$$

for all $t \in \mathbb{T}$. By using (8), we have

$$\alpha(\Lambda_- + 1)(\Lambda_+ - 1) = -1 + \sqrt{1 - 2\lambda + 4\alpha\lambda + \lambda^2}, \tag{15}$$

and therefore,

$$|\phi(t) - x(t)| \leq \frac{\varepsilon}{1 - \sqrt{1 - 2\lambda + 4\alpha\lambda + \lambda^2}}$$

for all $t \in \mathbb{T}$.

Now, we will show that $x(t)$ is a solution of (3) on $\mathbb{T}$. Recalling (9), we have

$$\Diamond_\alpha x(t) - \lambda x(t) = \alpha\Lambda_-\Delta(\nabla x(t)) - \alpha(\Lambda_- - 1)\Delta x(t)$$
$$-\alpha(\Lambda_+ - 1)\Lambda_-\nabla x(t) + \alpha(\Lambda_- - 1)(\Lambda_+ - 1)x(t)$$

for all $t \in \mathbb{T}_\kappa^\kappa$. Using (12) and (14), we get

$$\Diamond_\alpha x(t) - \lambda x(t) = \Delta\left[\alpha(\Lambda_- - 1)x(t) + y(t)\right] - \alpha(\Lambda_- - 1)\Delta x(t)$$
$$-(\Lambda_+ - 1)\left[\alpha(\Lambda_- - 1)x(t) + y(t)\right] + \alpha(\Lambda_- - 1)(\Lambda_+ - 1)x(t)$$
$$= \Delta y(t) - (\Lambda_+ - 1)y(t) = 0$$

for all $t \in \mathbb{T}_\kappa^\kappa$. Thus, $x(t)$ is a solution of (3) on $\mathbb{T}$. Consequently, (3) has Hyers–Ulam stability with an HUS constant $\dfrac{1}{1-\sqrt{1-2\lambda+4\alpha\lambda+\lambda^2}}$ on $\mathbb{T}$.

Next, we consider case (ii) $\lambda > 0$ and $\lambda > 2(1 - 2\alpha)$. From Lemma 1 (ii), we see that

$$\Lambda_+ - 1 > 0 \quad \text{and} \quad \frac{\Lambda_- - 1}{\Lambda_-} > 2. \tag{16}$$

Repeating the same argument as in the proof of case (i), we see that there exists a solution $y : \mathbb{T}_\kappa \to \mathbb{R}$ of (12) such that $|\psi(t) - y(t)| \le \frac{\varepsilon}{\Lambda_+ - 1}$ for all $t \in \mathbb{T}_\kappa$. That is, (13) holds for all $t \in \mathbb{T}_\kappa$. Using Theorem 6 (iv) with (13) and (16), we can find a solution $x : \mathbb{T} \to \mathbb{R}$ of (14) such that

$$|\phi(t) - x(t)| \le \frac{\frac{\varepsilon}{\alpha|\Lambda_-|(\Lambda_+ - 1)}}{\frac{\Lambda_- - 1}{\Lambda_-} - 2} = \frac{\varepsilon}{\alpha(\Lambda_- + 1)(\Lambda_+ - 1)}$$

for all $t \in \mathbb{T}$. By (15), we obtain

$$|\phi(t) - x(t)| \le \frac{\varepsilon}{\sqrt{1 - 2\lambda + 4\alpha\lambda + \lambda^2} - 1}$$

for all $t \in \mathbb{T}$. Repeating the same argument as in the proof of case (i), $x(t)$ is a solution of (3) on $\mathbb{T}$. Consequently, (3) has Hyers–Ulam stability with an HUS constant $\dfrac{1}{\sqrt{1-2\lambda+4\alpha\lambda+\lambda^2}-1}$ on $\mathbb{T}$.

Next, we consider case (iii) $\lambda < 0$ and $\lambda < 2(1 - 2\alpha)$. From Lemma 1 (iii), we see that

$$-1 < \Lambda_+ - 1 < 0 \quad \text{and} \quad 1 < \frac{\Lambda_- - 1}{\Lambda_-} < 2. \tag{17}$$

Using Theorem 2 with (10) and (17), we conclude that there exists a solution $y : \mathbb{T}_\kappa \to \mathbb{R}$ of (12) such that $|\psi(t) - y(t)| \le \frac{\varepsilon}{|\Lambda_+ - 1|}$ for all $t \in \mathbb{T}_\kappa$. That is,

$$\left| \nabla\phi(t) - \frac{\Lambda_- - 1}{\Lambda_-}\phi(t) - \frac{y(t)}{\alpha\Lambda_-} \right| \le \frac{\varepsilon}{\alpha|\Lambda_-||\Lambda_+ - 1|}$$

holds for all $t \in \mathbb{T}_\kappa$. Using Theorem 6 (iii) with (17), we can find a solution $x : \mathbb{T} \to \mathbb{R}$ of (14) such that

$$|\phi(t) - x(t)| \le \frac{\frac{\varepsilon}{\alpha|\Lambda_-||\Lambda_+ - 1|}}{2 - \frac{\Lambda_- - 1}{\Lambda_-}} = \frac{\varepsilon}{\alpha|\Lambda_- + 1||\Lambda_+ - 1|}$$

for all $t \in \mathbb{T}$. By (15), we obtain

$$|\phi(t) - x(t)| \leq \frac{\varepsilon}{\sqrt{1 - 2\lambda + 4\alpha\lambda + \lambda^2} - 1}$$

for all $t \in \mathbb{T}$. We can easily see that $x(t)$ is a solution of (3) on $\mathbb{T}$. Consequently, (3) has Hyers–Ulam stability with an HUS constant $\frac{1}{\sqrt{1-2\lambda+4\alpha\lambda+\lambda^2}-1}$ on $\mathbb{T}$.

Finally, we consider case (iv) $2(1 - 2\alpha) < \lambda < 0$. From Lemma 1 (iv), we have

$$-1 < \Lambda_+ - 1 < 0 \quad \text{and} \quad \frac{\Lambda_- - 1}{\Lambda_-} > 2.$$

Using this inequalities and the same arguments above, we see that (3) has Hyers–Ulam stability with an HUS constant $\frac{1}{1-\sqrt{1-2\lambda+4\alpha\lambda+\lambda^2}}$ on $\mathbb{T}$. This completes the proof.   □

From Theorems 2, 3, 4, 5, and 7, we obtain the following result.

Theorem 8 *Let $\alpha \in [0, 1]$. If $\lambda \neq 0$, $2(1 - 2\alpha)$, then (3) has Hyers–Ulam stability with an HUS constant $\frac{1}{\left|1-\sqrt{1-2\lambda+4\alpha\lambda+\lambda^2}\right|}$ on $\mathbb{T}$.*

Proof Consider the case $\alpha = 1$. From Theorems 2 and 3, (3) has HUS on $\mathbb{T}$. Moreover, an HUS constant for (3) is $\frac{1}{|1-|\lambda+1||}$. When $\alpha = 1$, we have

$$\frac{1}{\left|1 - \sqrt{1 - 2\lambda + 4\alpha\lambda + \lambda^2}\right|} = \frac{1}{|1 - |\lambda + 1||}.$$

Thus, the assertion is true when $\alpha = 1$.

Next, we consider the case $\alpha = 0$. From Theorems 4 and 5, (3) has HUS on $\mathbb{T}$. Moreover, an HUS constant for (3) is $\frac{1}{|1-|\lambda-1||}$. When $\alpha = 0$, we get

$$\frac{1}{\left|1 - \sqrt{1 - 2\lambda + 4\alpha\lambda + \lambda^2}\right|} = \frac{1}{|1 - |\lambda - 1||}.$$

Thus, the assertion is true when $\alpha = 0$.

By Theorem 7, we can conclude that the case $\alpha \in (0, 1)$ is true immediately.   □

Remark 1 By Theorem 7 (i), if $0 < \lambda < 2(1 - 2\alpha)$, then (3) has Hyers–Ulam stability with an HUS constant of $\frac{1}{1-\sqrt{1-2\lambda+4\alpha\lambda+\lambda^2}}$ on $\mathbb{T}$; consequently, if a minimum HUS constant exists in this case, it is bounded above by this number. Is there a lower bound? Given $\varepsilon > 0$, consider the non-homogeneous diamond-alpha difference equations

$$\Diamond_\alpha y_1(t) - \lambda y_1(t) = -\varepsilon, \quad t \in \mathbb{Z}, \quad 0 < \lambda < 2(1 - 2\alpha) \tag{18}$$

and

$$\lozenge_\alpha y_2(t) - \lambda y_2(t) = (-1)^t \varepsilon, \quad t \in \mathbb{Z}, \quad 0 < \lambda < 2(1 - 2\alpha). \tag{19}$$

Then, for arbitrary constants $c_1, c_2 \in \mathbb{R}$, the function

$$\phi_1(t) = \frac{\varepsilon}{\lambda} + c_1 \Lambda_-^t + c_2 \Lambda_+^t, \quad t \in \mathbb{Z}$$

is the general solution of (18), and the function

$$\phi_2(t) = \frac{(-1)^t \varepsilon}{2 - \lambda - 4\alpha} + c_1 \Lambda_-^t + c_2 \Lambda_+^t, \quad t \in \mathbb{Z}$$

is the general solution of (19). Since $x(t) = c_1 \Lambda_-^t + c_2 \Lambda_+^t$ is a solution of (3),

$$|\phi_1(t) - x(t)| = \frac{\varepsilon}{\lambda}$$

and

$$|\phi_2(t) - x(t)| = \frac{\varepsilon}{2 - \lambda - 4\alpha}, \quad \alpha \in \left(0, \frac{2 - \lambda}{4}\right), \quad \lambda \in (0, 2)$$

for $t \in \mathbb{Z}$. Thus, the minimum HUS constant $K^\dagger$ for (3), if it exists, satisfies

$$K^\dagger \in \begin{cases} \left[\frac{1}{\lambda}, \frac{1}{1 - \sqrt{1 - 2\lambda + 4\alpha\lambda + \lambda^2}} \right] & : \lambda \in (0, 1), \alpha \in \left(0, \frac{1-\lambda}{2}\right], \\[2ex] \left[\frac{1}{2 - 4\alpha - \lambda}, \frac{1}{1 - \sqrt{1 - 2\lambda + 4\alpha\lambda + \lambda^2}} \right] & : \lambda \in (0, 1), \alpha \in \left[\frac{1-\lambda}{2}, \frac{2-\lambda}{4}\right]; \\[2ex] & \quad \lambda \in [1, 2), \alpha \in \left(0, \frac{2-\lambda}{4}\right) \end{cases}$$

for these values of λ and α, leaving a gap. It remains an open question whether one can do better than this in this case. ◇

Remark 2 Consider Theorem 7 (ii), and let $\alpha = \frac{1}{2}$ and $\lambda = \frac{33}{56}$. Then $\lambda > 0$ and $\lambda > 2(1 - 2\alpha)$, so by Theorem 7 (ii), (3) has Hyers–Ulam stability with an HUS constant

$$\frac{1}{\sqrt{1 - 2\lambda + 4\alpha\lambda + \lambda^2} - 1} = \frac{56}{9}$$

on $\mathbb{T}$. Consider $\phi : \mathbb{T} \to \mathbb{R}$ given by

$$\phi(t) := \frac{56\varepsilon}{9(-2)^t} + c_1 \left(\frac{-4}{7}\right)^t + c_2 \left(\frac{7}{4}\right)^t, \quad \mathbb{T} = \mathbb{N}_0 := \{0, 1, 2, 3, \cdots\},$$

which satisfies

$$\left| \lozenge_{1/2}\phi(t) - \frac{33}{56}\phi(t) \right| = \left| \frac{\varepsilon}{(-2)^t} \right| \le \varepsilon \quad \text{for all} \quad t \in \mathbb{T}_\kappa^\kappa = \mathbb{N}$$

for any given arbitrary constant $\varepsilon > 0$ and for arbitrary constants $c_1, c_2 \in \mathbb{R}$. Clearly $x(t) = c_1 \left(\frac{-4}{7} \right)^t + c_2 \left(\frac{7}{4} \right)^t$ is a solution of (3) for these values of α and λ, yielding

$$|\phi(t) - x(t)| = \frac{56\varepsilon}{9(2^t)} \le \frac{56\varepsilon}{9}, \quad t \in \mathbb{T}.$$

Thus

$$\frac{1}{\sqrt{1 - 2\lambda + 4\alpha\lambda + \lambda^2} - 1} = \frac{56}{9}$$

is the minimum HUS constant in this case. ◇

Remark 3 Consider Theorem 7 (iii), and let $\alpha = \frac{1}{2}$ and $\lambda = \frac{-15}{8}$. Then $\lambda < 0$ and $\lambda < 2(1 - 2\alpha)$, so by Theorem 7 (iii), (3) has Hyers–Ulam stability with an HUS constant

$$\frac{1}{\sqrt{1 - 2\lambda + 4\alpha\lambda + \lambda^2} - 1} = \frac{8}{9}$$

on $\mathbb{T}$. For $\phi : \mathbb{T} \to \mathbb{R}$ given by

$$\phi(t) := \frac{8\varepsilon}{9}(2^{-t}) + c_1(4^{-t}) + c_2(-4)^t, \quad \mathbb{T} = \mathbb{N}_0 := \{0, 1, 2, 3, \cdots\},$$

ϕ satisfies

$$\lozenge_{1/2}\phi(t) + \frac{15}{8}\phi(t) = 2^{-t}\varepsilon \le \varepsilon \quad \text{for all} \quad t \in \mathbb{T}_\kappa^\kappa = \mathbb{N}$$

for any given arbitrary constant $\varepsilon > 0$ and for arbitrary constants $c_1, c_2 \in \mathbb{R}$. Clearly $x(t) = c_1(4^{-t}) + c_2(-4)^t$ is a solution of (3) for these values of α and λ, yielding

$$|\phi(t) - x(t)| = \frac{8\varepsilon}{9}(2^{-t}) \le \frac{8\varepsilon}{9}, \quad t \in \mathbb{T}.$$

Thus

$$\frac{1}{\sqrt{1 - 2\lambda + 4\alpha\lambda + \lambda^2} - 1} = \frac{8}{9}$$

is the minimum HUS constant in this case. ◇

Remark 4 Consider Theorem 7 (iv), and let $\alpha = \frac{3}{4}$ and $\lambda = \frac{-35}{48}$. Then $2(1 - 2\alpha) < \lambda < 0$, so by Theorem 7 (iv), (3) has Hyers–Ulam stability with an HUS constant

$$\frac{1}{1 - \sqrt{1 - 2\lambda + 4\alpha\lambda + \lambda^2}} = \frac{48}{5}$$

on $\mathbb{T}$. Consider $\phi : \mathbb{T} \to \mathbb{R}$ given by

$$\phi(t) := \frac{48\varepsilon}{5}(2^{-t}) + c_1 \left(\frac{4}{9}\right)^t + c_2 \left(\frac{-3}{4}\right)^t, \qquad \mathbb{T} = \mathbb{N}_0 := \{0, 1, 2, 3, \cdots\},$$

which satisfies

$$\Diamond_{3/4}\phi(t) + \frac{35}{48}\phi(t) = 2^{-t}\varepsilon \leq \varepsilon \quad \text{for all} \quad t \in \mathbb{T}_\kappa^\kappa = \mathbb{N}$$

for any given arbitrary constant $\varepsilon > 0$ and for arbitrary constants $c_1, c_2 \in \mathbb{R}$. Clearly $x(t) = c_1 \left(\frac{4}{9}\right)^t + c_2 \left(\frac{-3}{4}\right)^t$ is a solution of (3) for these values of α and λ, yielding

$$|\phi(t) - x(t)| = \frac{48\varepsilon}{5}(2^{-t}) \leq \frac{48\varepsilon}{5}, \quad t \in \mathbb{T}.$$

Thus

$$\frac{1}{1 - \sqrt{1 - 2\lambda + 4\alpha\lambda + \lambda^2}} = \frac{48}{5}$$

is the minimum HUS constant in this case.   ◇

Acknowledgement The second author was supported by JSPS KAKENHI Grant Number JP17K14226.

References

1. C. Alsina, R. Ger, On some inequalities and stability results related to the exponential function. J. Inequal. Appl. **2**, 373–380 (1998)
2. S. András, A. R. Mészáros, Ulam–Hyers stability of dynamic equations on time scales via Picard operators. Appl. Math. Comput. **219**, 4853–4864 (2013)
3. D.R. Anderson, J. Bullock, L. Erbe, A. Peterson, H.N. Tran, Nabla dynamic equations on time scales, PanAmerican Math. J. **13**(1), 1–47 (2003).
4. J. Brzdęk, D. Popa, I. Raşa, B. Xu, *Ulam Stability of Operators*. Mathematical Analysis and Its Applications (Academic Press, London, 2018)
5. M. Eshaghi Gordji, H. Khodaei, On the generalized Hyers–Ulam–Rassias stability of quadratic functional equations. Abstr. Appl. Anal. **2009**, 923476 (2009)

6. M. Eshaghi Gordji, N. Ghobadipour, Generalized Ulam–Hyers stabilities of quartic derivations on Banach algebras. Proyecciones J. Math. **29**(3), 209–226 (2010)
7. D.H. Hyers, On the stability of the linear functional equation. Proc. Nat. Acad. Sci. U.S.A. **27**, 222–224 (1941)
8. S.-M. Jung, Hyers–Ulam stability of linear differential equation of the first order (III). J. Math. Anal. Appl. **311**, 139–146 (2005)
9. S.-M. Jung, Hyers–Ulam stability of linear differential equations of first order (II). Appl. Math. Lett. **19**, 854–858 (2006)
10. S.-M. Jung, Hyers–Ulam stability of a system of first order linear differential equations with constant coefficients. J. Math. Anal. Appl. **320**, 549–561 (2006)
11. S.-M. Jung, Hyers–Ulam stability of linear differential equations of first order (I). Int. J. Appl. Math. Stat. **7**(10), 96–100 (2007)
12. S.-M. Jung, B. Kim, Th. M. Rassias, On the Hyers–Ulam stability of a system of Euler differential equations of first order. Tamsui Oxf. J. Math. Sci. **24**(4), 381–388 (2008)
13. T. Miura, S. Miyajima, S.E. Takahasi, A characterization of Hyers–Ulam stability of first order linear differential operators. J. Math. Anal. Appl. **286**(1), 136–146 (2003)
14. T. Miura, S. Miyajima, and S. E. Takahasi, Hyers–Ulam stability of linear differential operator with constant coefficients. Math. Nachr. **258**, 90–96 (2003)
15. M. Onitsuka, Influence of the stepsize on Hyers–Ulam stability of first-order homogeneous linear difference equations. Int. J. Differ. Equ. **12**(2), 281–302 (2017)
16. M. Onitsuka, T. Shoji, Hyers–Ulam stability of first-order homogeneous linear differential equations with a real-valued coefficient. Appl. Math. Lett. **63**, 102–108 (2017)
17. J.M. Rassias, Solution of a problem of Ulam. J. Approx. Theory **57**, 268–273 (1989)
18. Th. M. Rassias, On the stability of linear mapping in Banach spaces. Proc. Am. Math. Soc. **72**, 297–300 (1978)
19. K. Ravi, M. Arunkumar, J.M. Rassias, Ulam stability for the orthogonally general Euler–Lagrange type functional equation. Int. J. Math. Stat. **3**(A08), 36–46 (2008)
20. J.W. Rogers Jr., Q. Sheng, Notes on the diamond-α dynamic derivative on time scales. J. Math. Anal. Appl. **326**, 228–241 (2007)
21. I.A. Rus, Ulam stability of ordinary differential equations. Stud. Univ. Babeş-Bolyai Math. **54**, 125–134 (2009)
22. B.V. Senthil Kumar, K. Ravi, J. M. Rassias, Solution and generalized Ulam–Hyers stability of a reciprocal type functional equation in non-Archimedean fields. World Sci. News **31**, 71–81 (2016)
23. Y.H. Shen, The Ulam stability of first order linear dynamic equations on time scales. Results Math. **72**(4), 1881-1895 (2017). http://dx.doi.org/10.1007/s00025-017-0725-1
24. S.M. Ulam, *A Collection of Mathematical Problems* (Interscience, New York, 1960)
25. G. Wang, M. Zhou, L. Sun, Hyers–Ulam stability of linear differential equations of first order. Appl. Math. Lett. **21**, 1024–1028 (2008)
26. T. Yoshizawa, *Stability Theory by Liapunov's Second Method* (The Mathematical Society of Japan, Tokyo, 1966)

Hyers–Ulam Stability for a First-Order Linear Proportional Nabla Difference Operator

Douglas R. Anderson

Abstract The Hyers–Ulam stability (HUS) of a certain first-order proportional nabla difference equation with a sign-alternating coefficient is established. For those parameter values for which HUS holds, an HUS constant is found, and in special cases it is shown that this is the minimal such constant possible. A 2-cycle solution and a 4-cycle solution are shown to not have HUS.

1 Introduction

Ulam [23, p. 63] first posed the problem of finding conditions such that a linear mapping near an approximately linear mapping exists. This was solved by Hyers [6] for the case of approximately additive mappings, when he proved that the Cauchy equation is stable in Banach spaces; Rassias [19] subsequently generalized these results. Since then there has been a significant amount of interest in Hyers–Ulam stability (HUS), especially in relation to difference and differential equations; for example, see [1, 3–5, 7–14, 17, 18, 20–22, 24].

We focus on the difference equations case first. Let $\mathbb{T} \subset h\mathbb{Z}$ be nonempty, and define $\mathbb{T}^\kappa := \mathbb{T} \backslash \{\max \mathbb{T}\}$ if $\max \mathbb{T} \in h\mathbb{Z}$ exists, and $\mathbb{T}^\kappa := \mathbb{T}$ otherwise. Then, for $\beta \in \mathbb{R}$, consider the equation

$$\Delta_h x(t) - \beta x(t) = 0, \qquad t \in \mathbb{T}^\kappa, \tag{1}$$

for $h > 0$, where Δ_h is the forward difference operator

$$\Delta_h x(t) = \frac{x(t+h) - x(t)}{h}, \qquad t \in h\mathbb{Z} := \{hk : k \in \mathbb{Z}\}.$$

D. R. Anderson (✉)
Concordia College, Department of Mathematics, Moorhead, MN, USA
e-mail: andersod@cord.edu

© Springer Nature Switzerland AG 2019
G. A. Anastassiou and J. M. Rassias (eds.), *Frontiers in Functional Equations and Analytic Inequalities*, https://doi.org/10.1007/978-3-030-28950-8_15

Recently, Onitsuka [15] investigated the influence of the constant step size $h > 0$ on HUS for (1). In this context, (1) has HUS if and only if there exists a constant $K > 0$ with the following property:

> For arbitrary $\varepsilon > 0$, if a function $\phi : \mathbb{T} \to \mathbb{R}$ satisfies $|\Delta\phi(t) - \beta\phi(t)| \leq \varepsilon$ for all $t \in \mathbb{T}^\kappa$, then there exists a solution $x : \mathbb{T} \to \mathbb{R}$ of (1) such that $|\phi(t) - x(t)| \leq K\varepsilon$ for all $t \in \mathbb{T}$.

Such a constant $K > 0$ is called an HUS constant for (1) on $\mathbb{T}$. Onitsuka [15] was able to classify whether (1) has HUS for each possible value of $\beta \in \mathbb{R}$, and for those values of β for which (1) has HUS, whether there is a minimum value of the HUS constant K, what the corresponding solution x looks like, and whether that solution x is unique; see also [16]. We propose to modify and generalize these discrete results of [15] to a new equation utilizing a proportional nabla difference operator with an alternating coefficient, introduced below. For convenience, we will henceforth take $h = 1$.

2 Hyers–Ulam Stability for a Proportional Nabla Difference Operator

In this section we consider on $\mathscr{S} \subset \mathbb{Z}$ the Hyers–Ulam stability of the first-order linear homogeneous proportional nabla difference equation with alternating coefficient given by

$$D_\alpha x(t) - \beta(-1)^t x(t) = 0, \quad D_\alpha x(t) := \alpha \nabla x(t) + (1 - \alpha)x(t), \quad \alpha \in (0, 1], \tag{2}$$

where $\beta \in \mathbb{R}$ is a parameter. Clearly D_0 is the identity operator, while D_1 is the backward difference operator,

$$D_1 x(t) = \nabla x(t) := x(t) - x(t - 1).$$

Consequently, these results modify those available for forward difference equations, and include results for backward difference equations as a special case when $\alpha = 1$. It is easy to see that the operator D_α satisfies

$$D_\alpha x(t) = x(t) - \alpha x(t - 1),$$

allowing one interpretation of this proportional difference operator to be what proportion of $x(t - 1)$ gets subtracted off, with $\alpha = 1$ being the full amount that yields the traditional ∇ difference operator. To maintain the first-order nature of (2), however, we henceforth consider only

$$\alpha \in (0, 1] \quad \text{and} \quad \beta \neq \pm 1.$$

Proportional derivatives and differences were first introduced in [2].

Define

$$\mathscr{S}_\kappa := \begin{cases} \mathscr{S}\backslash\{\min\mathscr{S}\} & : \text{if the minimum of } \mathscr{S} \text{ exists,} \\ \mathscr{S} & : \text{otherwise.} \end{cases}$$

Note that if a function x exists on $\mathscr{S}$, then $D_\alpha x$ exists on $\mathscr{S}_\kappa$. For the remainder of the paper, we assume that $\mathscr{S}$ and $\mathscr{S}_\kappa$ are nonempty sets in $\mathbb{Z}$.

Definition 1 We say that (2) has Hyers–Ulam stability (HUS) on $\mathscr{S}$ if and only if there exists a constant $K > 0$ with the following property. For arbitrary $\varepsilon > 0$, if a function $\phi : \mathscr{S} \to \mathbb{R}$ satisfies

$$|D_\alpha\phi(t) - \beta(-1)^t\phi(t)| \leq \varepsilon \quad \text{for all} \quad t \in \mathscr{S}_\kappa,$$

then there exists a solution $x : \mathscr{S} \to \mathbb{R}$ of (2) such that $|\phi(t) - x(t)| \leq K\varepsilon$ for all $t \in \mathscr{S}$. Such a constant K is called an HUS constant for (2) on $\mathscr{S}$.

Theorem 1 (No HUS) *Equation* (2) *does not have Hyers–Ulam stability on* $\mathbb{Z}$ *if*

$$\beta^2 = 1 \pm \alpha^2.$$

Proof First let $\beta^2 = 1 - \alpha^2$, say $\beta = \sqrt{1-\alpha^2}$. Given $\varepsilon > 0$, note that the function

$$\phi(t) := \begin{cases} \dfrac{\left(1-\alpha-\sqrt{1-\alpha^2}\right)\varepsilon t/2 + c\alpha^2}{\alpha\left(-1+\sqrt{1-\alpha^2}\right)} & : t = 2n \\[3ex] \dfrac{(t-1)\varepsilon}{2\alpha} + \dfrac{(t+1)\left(-1+\sqrt{1-\alpha^2}\right)\varepsilon}{2\alpha^2} + c & : t = 2n+1 \end{cases}$$

satisfies

$$|D_\alpha\phi(t) - (-1)^t\sqrt{1-\alpha^2}\phi(t)| = |(-1)^t\varepsilon| = \varepsilon$$

for all $t \in \mathbb{Z}$. As the 2-cycle

$$x(t) := c \begin{cases} \dfrac{1+\sqrt{1-\alpha^2}}{\alpha} & : t = 2n \\[2ex] 1 & : t = 2n+1 \end{cases}$$

for any $c \in \mathbb{R}$ and any $n \in \mathbb{Z}$ is the general solution of $D_\alpha x(t) - (-1)^t\sqrt{1-\alpha^2}x(t) = 0$, we see that $|\phi(t) - x(t)| \to \infty$ as $t \to \pm\infty$, so that (2) does not have HUS on $\mathbb{Z}$ when $\beta = \sqrt{1-\alpha^2}$; a similar result holds when $\beta = -\sqrt{1-\alpha^2}$, and thus is omitted. Next, let $\beta^2 = 1 + \alpha^2$, say $\beta = \sqrt{1+\alpha^2}$. Given $\varepsilon > 0$, note that the function

$$\phi(t) := \begin{cases} \dfrac{\left(-1+\alpha+\sqrt{1+\alpha^2}\right)\varepsilon t/2+c\alpha^2}{\alpha\left(-1+\sqrt{1+\alpha^2}\right)} & : t = 4n \\[3mm] \dfrac{(t-1)\varepsilon}{2\alpha} + \dfrac{(t+1)\left(-1+\sqrt{1+\alpha^2}\right)\varepsilon}{2\alpha^2} + c & : t = 4n+1 \\[3mm] \dfrac{\left(-1+\alpha+\sqrt{1+\alpha^2}\right)\varepsilon t/2+c\alpha^2}{\alpha\left(1-\sqrt{1+\alpha^2}\right)} & : t = 4n+2 \\[3mm] \dfrac{(1-t)\varepsilon}{2\alpha} + \dfrac{(t+1)\left(1-\sqrt{1+\alpha^2}\right)\varepsilon}{2\alpha^2} - c & : t = 4n+3 \end{cases}$$

satisfies

$$|D_\alpha \phi(t) - (-1)^t \sqrt{1+\alpha^2}\phi(t)| = \left|(-1)^{\left\lceil \frac{1-t}{2}\right\rceil}\varepsilon\right| = \varepsilon$$

for all $t \in \mathbb{Z}$. As the 4-cycle

$$x(t) := c \begin{cases} \dfrac{1+\sqrt{1+\alpha^2}}{\alpha} & : t = 4n \\[2mm] 1 & : t = 4n+1 \\[2mm] \dfrac{\alpha}{1-\sqrt{1+\alpha^2}} & : t = 4n+2 \\[2mm] -1 & : t = 4n+3 \end{cases}$$

for any $c \in \mathbb{R}$ and any $n \in \mathbb{Z}$ is the general solution of $D_\alpha x(t) - (-1)^t\sqrt{1+\alpha^2}x(t) = 0$, we see that $|\phi(t) - x(t)| \to \infty$ as $t \to \pm\infty$, so that (2) does not have HUS on $\mathbb{Z}$ when $\beta = \sqrt{1+\alpha^2}$ either, as is the case when $\beta = -\sqrt{1+\alpha^2}$. The proof is complete.   □

Remark 1 In the following discussion, take

$$E_1(t) := \frac{\alpha + \beta(-1)^t + 1}{\beta^2 + \alpha^2 - 1} \tag{3}$$

and

$$\psi(t) := \begin{cases} \left(\dfrac{\alpha^2}{1-\beta^2}\right)^{\frac{t}{2}} & : t \text{ even,} \\[3mm] \dfrac{\alpha}{1+\beta}\left(\dfrac{\alpha^2}{1-\beta^2}\right)^{\frac{t-1}{2}} & : t \text{ odd.} \end{cases} \tag{4}$$

Note that ψ is a fundamental solution of (2).

Lemma 1 *Assume $\beta \in (-1, 1)$ with $\beta^2 + \alpha^2 - 1 \neq 0$. Let ψ be the fundamental solution of (2) given in (4), and let $E : \mathscr{S} \to \mathbb{R}$ be a particular solution to the difference equation*

$$D_\alpha E(t) - \beta(-1)^t E(t) = -1. \tag{5}$$

Let $\varepsilon > 0$ be a fixed arbitrary constant, and let ϕ be a real-valued function on $\mathscr{S}$. Then the inequality

$$|D_\alpha\phi(t) - \beta(-1)^t\phi(t)| \leq \varepsilon$$

holds for all $t \in \mathscr{S}_\kappa$ if and only if the nabla inequality

$$0 \leq \nabla\left[\frac{\phi(t) - \varepsilon E(t)}{\psi(t)}\right] \leq \frac{2\varepsilon}{\alpha}\left|\psi^{-1}(t-1)\right|$$

holds for all $t \in \mathscr{S}_\kappa$.

Proof Since $\beta \in (-1, 1)$ with $\beta^2 + \alpha^2 - 1 \neq 0$, the fundamental solution ψ in (4) satisfies $\psi(t) > 0$ for all $t \in \mathscr{S}$. Expanding the following nabla difference for $t \in \mathscr{S}_\kappa$, we see that

$$\alpha\nabla\left[\frac{\phi(t) - \varepsilon E(t)}{\psi(t)}\right]$$

$$= \frac{\alpha\psi(t-1)(\phi(t) - \varepsilon E(t)) - \alpha\psi(t)(\phi(t-1) - \varepsilon E(t-1))}{\psi(t-1)\psi(t)}$$

$$= \frac{\left(1 - \beta(-1)^t\right)(\phi(t) - \varepsilon E(t)) - \alpha\phi(t-1) + \varepsilon\alpha E(t-1)}{\psi(t-1)}$$

$$= \frac{\left(D_\alpha\phi(t) - \beta(-1)^t\phi(t)\right) - \varepsilon\left(D_\alpha E(t) - \beta(-1)^t E(t)\right)}{\psi(t-1)}$$

$$= \psi^{-1}(t-1)\left[\left(D_\alpha\phi(t) - \beta(-1)^t\phi(t)\right) + \varepsilon\right]$$

holds for all $t \in \mathscr{S}_\kappa$; the overall result then follows.   □

Proposition 1 *Assume $\beta \in (-1, 1)$ with $\beta^2 + \alpha^2 - 1 \neq 0$. Let $\varepsilon > 0$ be a given arbitrary constant, and let $\phi : \mathscr{S} \to \mathbb{R}$ satisfy $|D_\alpha\phi(t) - \beta(-1)^t\phi(t)| \leq \varepsilon$ for all $t \in \mathscr{S}_\kappa$. Then there exist a non-decreasing function $u : \mathscr{S} \to \mathbb{R}$ and a non-increasing function $v : \mathscr{S} \to \mathbb{R}$ such that*

$$\phi(t) = u(t)\psi(t) + \varepsilon E_1(t) = v(t)\psi(t) - \varepsilon E_1(t) \tag{6}$$

for E_1 and ψ in (3) and (4), respectively, and one of the following holds.

(i) *If $\beta \in \left(-1, -\sqrt{1 - \alpha^2}\right) \cup \left(\sqrt{1 - \alpha^2}, 1\right)$ and $\max \mathscr{S} =: s^*$ exists, then the inequality*

$$u(t) \leq u(s^*) < v(s^*) \leq v(t) \tag{7}$$

holds for all $t \in \mathscr{S}$.

(ii) If $\beta \in \left(-1, -\sqrt{1-\alpha^2}\right) \cup \left(\sqrt{1-\alpha^2}, 1\right)$ and max $\mathscr{S}$ does not exist, then $\lim_{t\to\infty} u(t)$ and $\lim_{t\to\infty} v(t)$ exist, and

$$u(t) \leq \lim_{t\to\infty} u(t) = \lim_{t\to\infty} v(t) \leq v(t) \tag{8}$$

holds for all $t \in \mathscr{S}$.

(iii) If $\beta \in \left(-\sqrt{1-\alpha^2}, \sqrt{1-\alpha^2}\right)$ and min $\mathscr{S} =: s_*$ exists, then the inequality

$$v(t) \leq v(s_*) < u(s_*) \leq u(t) \tag{9}$$

holds for all $t \in \mathscr{S}$.

(iv) If $\beta \in \left(-\sqrt{1-\alpha^2}, \sqrt{1-\alpha^2}\right)$ and min $\mathscr{S}$ does not exist, then $\lim_{t\to-\infty} u(t)$ and $\lim_{t\to-\infty} v(t)$ exist, and

$$v(t) \leq \lim_{t\to-\infty} v(t) = \lim_{t\to-\infty} u(t) \leq u(t) \tag{10}$$

holds for all $t \in \mathscr{S}$.

Proof Fix the parameter $\beta \in (-1, 1)$ with $\beta^2 + \alpha^2 - 1 \neq 0$. It is straightforward to check that E_1 given in (3) solves (5). Define the functions u and v on $\mathscr{S}$ via

$$u(t) := (\phi(t) - \varepsilon E_1(t)) \frac{1}{\psi(t)} \quad \text{and} \quad v(t) := (\phi(t) + \varepsilon E_1(t)) \frac{1}{\psi(t)}$$

for $t \in \mathscr{S}$. Then (6) holds, and thus for $t \in \mathscr{S}$ we obtain

$$v(t) = u(t) + \frac{2\varepsilon E_1(t)}{\psi(t)}, \quad v(t) \begin{cases} < u(t): & \beta \in \left(-\sqrt{1-\alpha^2}, \sqrt{1-\alpha^2}\right), \\ > u(t): & \beta \in \left(-1, -\sqrt{1-\alpha^2}\right) \cup \left(\sqrt{1-\alpha^2}, 1\right). \end{cases} \tag{11}$$

Using Lemma 1, we have that the inequality

$$0 \leq \nabla u(t) \leq \frac{2\varepsilon}{\alpha\psi(t-1)}$$

holds for all $t \in \mathscr{S}_\kappa$. Since

$$\nabla\left(\frac{2\varepsilon E_1(t)}{\psi(t)}\right) = \frac{-2\varepsilon}{\alpha\psi(t-1)},$$

this together with (11) implies that

$$-\frac{2\varepsilon}{\alpha\psi(t-1)} \le \nabla v(t) \le 0, \quad t \in \mathscr{S}.$$

Consequently, u is non-decreasing and v is non-increasing.

First we consider (i). From the results above and the assumptions in (i), we see that $u(s^*)$ is the maximum of u on $\mathscr{S}$, and $v(s^*)$ is the minimum of v on $\mathscr{S}$. Then (7) follows from (11), and (i) holds.

Next we consider (ii). Let $t_0 \in \mathscr{S}$ be a fixed number. From the inequality in (11) with

$$\beta \in \left(-1, -\sqrt{1-\alpha^2}\right) \cup \left(\sqrt{1-\alpha^2}, 1\right), \tag{12}$$

we have

$$u(t) < v(t_0), \quad t \in \mathscr{S},$$

and thus u is bounded above on $\mathscr{S}$. As $\mathscr{S}$ is unbounded in this case, the $\lim_{t\to\infty} u(t)$ exists. Since (12) holds in this case, we have

$$\lim_{t\to\infty} \frac{2\varepsilon E_1(t)}{\psi(t)} = 0.$$

Consequently, we have from the equality in (11) that $\lim_{t\to\infty} u(t) = \lim_{t\to\infty} v(t)$. As a result, (8) is true for $t \in \mathscr{S}$, since u is non-decreasing and v is non-increasing; thus (ii) holds.

The arguments for (iii) and (iv) are similar to those given above for (i) and (ii), and thus are omitted. This completes the proof. $\qquad\qquad\qquad\qquad\qquad\quad \square$

Theorem 2 *Assume* $\beta \in (-1, 1)$ *with* $\beta^2 + \alpha^2 - 1 \ne 0$. *Let* $\varepsilon > 0$ *be a given arbitrary constant, and let* $\phi : \mathscr{S} \to \mathbb{R}$ *satisfy*

$$|D_\alpha \phi(t) - \beta(-1)^t \phi(t)| \le \varepsilon, \quad t \in \mathscr{S}_\kappa.$$

Then one of the following holds.

(i) If $\beta \in \left(-1, -\sqrt{1-\alpha^2}\right) \cup \left(\sqrt{1-\alpha^2}, 1\right)$ *and* $s^* := \max \mathscr{S}$ *exists, then any solution* x *of (2) with* $|\phi(s^*) - x(s^*)| < \varepsilon E_1(s^*)$ *satisfies*

$$|\phi(t) - x(t)| < \varepsilon \frac{\max\{\alpha \pm \beta + 1\}}{\beta^2 + \alpha^2 - 1}$$

for all $t \in \mathscr{S}$, *where* E_1 *is given in (3).*

(ii) If $\beta \in \left(-1, -\sqrt{1-\alpha^2}\right) \cup \left(\sqrt{1-\alpha^2}, 1\right)$ and $\max \mathscr{S}$ does not exist, then $\lim_{t \to \infty} \phi(t)\psi^{-1}(t)$ exists, and the function

$$x(t) := \left(\lim_{t \to \infty} \phi(t)\psi^{-1}(t) \right) \psi(t)$$

is the unique solution of (2) such that

$$|\phi(t) - x(t)| \leq \varepsilon \frac{\max\{\alpha \pm \beta + 1\}}{\beta^2 + \alpha^2 - 1} \tag{13}$$

for all $t \in \mathscr{S}$, where ψ is given in (4).

(iii) If $\beta \in \left(-\sqrt{1-\alpha^2}, \sqrt{1-\alpha^2}\right)$ and $s_* := \min \mathscr{S}$ exists, then any solution x of (2) with $|\phi(s_*) - x(s_*)| < \varepsilon E_1(s_*)$ satisfies

$$|\phi(t) - x(t)| < \varepsilon \frac{\max\{\alpha \pm \beta + 1\}}{1 - \beta^2 - \alpha^2}$$

for all $t \in \mathscr{S}$.

(iv) If $\beta \in \left(-\sqrt{1-\alpha^2}, \sqrt{1-\alpha^2}\right)$ and $\min \mathscr{S}$ does not exist, then $\lim_{t \to -\infty} \phi(t)\psi^{-1}(t)$ exists, and the function

$$x(t) := \left(\lim_{t \to -\infty} \phi(t)\psi^{-1}(t) \right) \psi(t)$$

is the unique solution of (2) such that

$$|\phi(t) - x(t)| \leq \varepsilon \frac{\max\{\alpha \pm \beta + 1\}}{1 - \beta^2 - \alpha^2}$$

for all $t \in \mathscr{S}$.

Proof From Proposition 1, we can find a non-decreasing function $u : \mathscr{S} \to \mathbb{R}$ and a non-increasing function $v : \mathscr{S} \to \mathbb{R}$ such that (6) holds for all $t \in \mathscr{S}$.

First we consider (i). From Proposition 1 (i), we have (7) for all $t \in \mathscr{S}$. Let x be any solution of (2) with $|\phi(s^*) - x(s^*)| < \varepsilon E_1(s^*)$. Then this solution x can be expressed as

$$x(t) := \frac{x(s^*)\psi(t)}{\psi(s^*)}, \quad t \in \mathscr{S},$$

for ψ given in (4). Using (6) and (7) and $|\phi(s^*) - x(s^*)| < \varepsilon E_1(s^*)$, we obtain

$$u(s^*) < \frac{x(s^*)}{\psi(s^*)} < v(s^*).$$

From this inequality and the results above, we have that

$$\phi(t) - x(t) \le u(s^*)\psi(t) + \varepsilon E_1(t) - \frac{x(s^*)}{\psi(s^*)}\psi(t)$$

$$= \left(u(s^*) - \frac{x(s^*)}{\psi(s^*)}\right)\psi(t) + \varepsilon E_1(t) < \varepsilon \frac{\max\{\alpha \pm \beta + 1\}}{\beta^2 + \alpha^2 - 1}$$

and

$$\phi(t) - x(t) \ge v(s^*)\psi(t) - \varepsilon E_1(t) - \frac{x(s^*)}{\psi(s^*)}\psi(t)$$

$$= \left(v(s^*) - \frac{x(s^*)}{\psi(s^*)}\right)\psi(t) - \varepsilon E_1(t) > -\varepsilon \frac{\max\{\alpha \pm \beta + 1\}}{\beta^2 + \alpha^2 - 1}$$

for all $t \in \mathscr{S}$; thus, (i) holds.

Next consider (ii). From Proposition 1 (ii), we have (8) for all $t \in \mathscr{S}$. Since $\lim_{t\to\infty} u(t)$ exists and $\lim_{t\to\infty} \psi^{-1}(t) = 0$, we see that

$$\lim_{t\to\infty} \phi(t)\psi^{-1}(t) = \lim_{t\to\infty} \left(u(t) + \varepsilon E_1(t)\psi^{-1}(t)\right) = \lim_{t\to\infty} u(t)$$

exists. Now, we consider the function

$$x(t) := \left(\lim_{t\to\infty} \phi(t)\psi^{-1}(t)\right)\psi(t)$$

for all $t \in \mathscr{S}$. This x is a well-defined solution of (2). Using (6) and (8), we obtain

$$\phi(t) - x(t) = \left(u(t) - \lim_{t\to\infty} \phi(t)\psi^{-1}(t)\right)\psi(t) + \varepsilon E_1(t) \le \varepsilon \frac{\max\{\alpha \pm \beta + 1\}}{\beta^2 + \alpha^2 - 1}$$

and

$$\phi(t) - x(t) = \left(v(t) - \lim_{t\to\infty} \phi(t)\psi^{-1}(t)\right)\psi(t) - \varepsilon E_1(t) \ge -\varepsilon \frac{\max\{\alpha \pm \beta + 1\}}{\beta^2 + \alpha^2 - 1}$$

for all $t \in \mathscr{S}$. That is, (13) holds for all $t \in \mathscr{S}$. We next show that this x is the unique solution of (2) such that (13) for all $t \in \mathscr{S}$. Pick any constant $c \ne \lim_{t\to\infty} \phi(t)\psi^{-1}(t)$, and define $y(t) := c\psi(t)$ for all $t \in \mathscr{S}$. This y is some solution of (2) different from x, by the uniqueness of solutions for the initial value problem. Since

$$\lim_{t\to\infty} (u(t) - c) = \lim_{t\to\infty} \phi(t)\psi^{-1}(t) - c \ne 0,$$

we have

$$\lim_{t \to \infty} |\phi(t) - y(t)| = \lim_{t \to \infty} |(u(t) - c)\psi(t) + \varepsilon E_1(t)| = \infty.$$

Consequently, x is the unique solution of (2) such that (13) for all $t \in \mathscr{S}$; thus, (ii) holds.

The arguments for (iii) and (iv) are similar to those given above for (i) and (ii), and thus are omitted. This completes the proof. $\qquad\square$

By Theorem 2, we obtain the following result immediately.

Corollary 1 *If $\beta \in (-1, 1)$ with $\beta^2 + \alpha^2 - 1 \neq 0$, then (2) has Hyers–Ulam stability with an HUS constant $\frac{\max\{\alpha \pm \beta + 1\}}{|\beta^2 + \alpha^2 - 1|}$ on $\mathscr{S}$.*

Corollary 2 *Assume $\beta \in (-1, 1)$ with $\beta^2 + \alpha^2 - 1 > 0$. Then the minimum HUS constant for (2) on all of $\mathbb{Z}$ is*

$$\frac{\max\{\alpha \pm \beta + 1\}}{|\beta^2 + \alpha^2 - 1|}.$$

Proof For $t \in \mathbb{Z}$ and arbitrary constant $C \in \mathbb{R}$, the function

$$\phi(t) := \begin{cases} \left(\frac{(\alpha+\beta+1)\varepsilon}{\beta^2+\alpha^2-1} + \frac{C(1+\beta)}{\alpha} \right) \psi(t) - \frac{(\alpha+\beta+1)\varepsilon}{\beta^2+\alpha^2-1} & : t \text{ even} \\ \left(\frac{(\alpha+\beta+1)\varepsilon}{\beta^2+\alpha^2-1} + \frac{C(1+\beta)}{\alpha} \right) \psi(t) - \frac{(\alpha-\beta+1)\varepsilon}{\beta^2+\alpha^2-1} & : t \text{ odd} \end{cases}$$

satisfies

$$D_\alpha \phi(t) - \beta(-1)^t \phi(t) = \varepsilon \quad \text{for all} \quad t \in \mathbb{Z}.$$

As shown in Theorem 2(ii), using ψ as in (4),

$$\lim_{t \to \infty} \left(\phi(t)\psi^{-1}(t) \right) = \frac{(\alpha + \beta + 1)\varepsilon}{\beta^2 + \alpha^2 - 1} + \frac{C(1 + \beta)}{\alpha}$$

exists, and the function

$$x(t) := \left(\frac{(\alpha + \beta + 1)\varepsilon}{\beta^2 + \alpha^2 - 1} + \frac{C(1 + \beta)}{\alpha} \right) \psi(t)$$

solves (2). Then

$$|\phi(t) - x(t)| = |-\varepsilon E_1(t)| = \varepsilon E_1(t)$$

on $\mathbb{Z}$, for E_1 given in (3). This means that the minimum HUS constant for (2) on $\mathbb{Z}$ is greater than or equal to $\max E_1(t) = \max \frac{\alpha \pm \beta + 1}{\beta^2 + \alpha^2 - 1}$. So, this together with

Corollary 1 implies that the minimum HUS constant for (2) on $\mathbb{Z}$ is max $\frac{\alpha \pm \beta + 1}{\beta^2 + \alpha^2 - 1}$ when $\beta \in (-1, 1)$ with $\beta^2 + \alpha^2 - 1 > 0$. $\qquad\qquad\qquad\qquad\qquad\qquad\qquad\qquad\qquad\qquad\qquad\quad$ $\square$

Lemma 2 *Assume $\beta \neq \pm 1$ and $\beta^2 - \alpha^2 - 1 \neq 0$. Let ψ be the fundamental solution of (2) given in (4), and let $G : \mathscr{S} \to \mathbb{R}$ solve the difference equation*

$$D_\alpha G(t) - \beta(-1)^t G(t) = 1 \operatorname{sgn} \psi(t-1), \tag{14}$$

where sgn is the sign of the expression. Let $\varepsilon > 0$ be a fixed arbitrary constant, and let ϕ be a real-valued function on $\mathscr{S}$. Then the inequality

$$|D_\alpha \phi(t) - \beta(-1)^t \phi(t)| \leq \varepsilon$$

holds for all $t \in \mathscr{S}_\kappa$ if and only if the nabla inequality

$$0 \leq \nabla \left[\frac{\phi(t) + \varepsilon G(t)}{\psi(t)} \right] \leq \frac{2\varepsilon}{\alpha} \left| \psi^{-1}(t-1) \right|$$

holds for all $t \in \mathscr{S}_\kappa$.

Proof Since ψ solves (2), we have $\alpha \psi(t-1) = \left(1 - \beta(-1)^t\right) \psi(t)$. Expanding the following nabla difference for t yields

$$
\begin{aligned}
&\alpha \nabla \left[\frac{\phi(t) + \varepsilon G(t)}{\psi(t)} \right] \\
&= \frac{\alpha \psi(t-1)(\phi(t) + \varepsilon G(t)) - \alpha \psi(t)(\phi(t-1) + \varepsilon G(t-1))}{\psi(t-1)\psi(t)} \\
&= \frac{\left(1 - \beta(-1)^t\right)(\phi(t) + \varepsilon G(t)) - \alpha \phi(t-1) - \varepsilon \alpha G(t-1)}{\psi(t-1)} \\
&= \operatorname{sgn} \psi(t-1) \frac{\left(D_\alpha \phi(t) - \beta(-1)^t \phi(t)\right) + \varepsilon \left(D_\alpha G(t) - \beta(-1)^t G(t)\right)}{|\psi(t-1)|} \\
&= \left| \psi^{-1}(t-1) \right| \left[(\operatorname{sgn} \psi(t-1)) \left(D_\alpha \phi(t) - \beta(-1)^t \phi(t)\right) + \varepsilon \right]
\end{aligned}
$$

holds for all $t \in \mathscr{S}_\kappa$, as G solves (14). The overall result then follows for all $t \in \mathscr{S}_\kappa$, completing the proof. $\qquad\qquad\qquad\qquad\qquad\qquad\qquad\qquad\qquad\qquad\qquad\qquad\quad$ $\square$

Remark 2 For $\beta > 1$ with $\beta^2 - \alpha^2 - 1 \neq 0$, the function $G = G_1$, where G_1 is given by

$$
G_1(t) := \begin{cases} (-1)^{\frac{t}{2}} \left(\frac{\alpha + \beta + 1}{\beta^2 - \alpha^2 - 1} \right) & : t \text{ even} \\ (-1)^{\frac{t-1}{2}} \left(\frac{\alpha + \beta - 1}{\beta^2 - \alpha^2 - 1} \right) & : t \text{ odd}, \end{cases} \tag{15}
$$

solves (14) with sgn $\psi(t-1) = -(-1)^{\lceil \frac{t}{2} \rceil}$. Consequently, Lemma 2 holds with $G = G_1$ given here in (15).

Proposition 2 *Assume* $\beta \in \left(1, \sqrt{1+\alpha^2}\right) \cup \left(\sqrt{1+\alpha^2}, \infty\right)$. *Let* $\varepsilon > 0$ *be a given arbitrary constant, and let* $\phi : \mathscr{S} \to \mathbb{R}$ *satisfy* $|D_\alpha \phi(t) - \beta(-1)^t \phi(t)| \le \varepsilon$ *for all* $t \in \mathscr{S}_\kappa$. *Then there exist a non-decreasing function* $u : \mathscr{S} \to \mathbb{R}$ *and a non-increasing function* $v : \mathscr{S} \to \mathbb{R}$ *such that*

$$\phi(t) = u(t)\psi(t) - \varepsilon G_1(t) = v(t)\psi(t) + \varepsilon G_1(t) \tag{16}$$

for ψ *and* $G_1(t)$ *in (4) and (15), respectively, and one of the following holds.*

(i) *If* $\beta \in \left(1, \sqrt{1+\alpha^2}\right)$ *and* $\max \mathscr{S} =: s^*$ *exists, then the inequality*

$$u(t) \le u(s^*) < v(s^*) \le v(t) \tag{17}$$

holds for all $t \in \mathscr{S}$.

(ii) *If* $\beta \in \left(1, \sqrt{1+\alpha^2}\right)$ *and* $\max \mathscr{S}$ *does not exist, then* $\lim_{t\to\infty} u(t)$ *and* $\lim_{t\to\infty} v(t)$ *exist, and*

$$u(t) \le \lim_{t\to\infty} u(t) = \lim_{t\to\infty} v(t) \le v(t) \tag{18}$$

holds for all $t \in \mathscr{S}$.

(iii) *If* $\beta > \sqrt{1+\alpha^2}$ *and* $\min \mathscr{S} =: s_*$ *exists, then the inequality*

$$v(t) \le v(s_*) < u(s_*) \le u(t) \tag{19}$$

holds for all $t \in \mathscr{S}$.

(iv) *If* $\beta > \sqrt{1+\alpha^2}$ *and* $\min \mathscr{S}$ *does not exist, then* $\lim_{t\to-\infty} u(t)$ *and* $\lim_{t\to-\infty} v(t)$ *exist, and*

$$v(t) \le \lim_{t\to-\infty} v(t) = \lim_{t\to-\infty} u(t) \le u(t) \tag{20}$$

holds for all $t \in \mathscr{S}$.

Proof Fix the parameter $\beta \in \left(1, \sqrt{1+\alpha^2}\right) \cup \left(\sqrt{1+\alpha^2}, \infty\right)$, and define the functions u and v on $\mathscr{S}$ via

$$u(t) := (\phi(t) + \varepsilon G_1(t))\, \psi^{-1}(t) \quad \text{and} \quad v(t) := (\phi(t) - \varepsilon G_1(t))\, \psi^{-1}(t)$$

for $t \in \mathscr{S}$. Then (16) holds, and thus for $t \in \mathscr{S}$ we obtain

$$u(t) = v(t) + 2\varepsilon G_1(t)\psi^{-1}(t), \quad u(t) \begin{cases} < v(t): & 1 < \beta < \sqrt{1+\alpha^2}, \\ > v(t): & \beta > \sqrt{1+\alpha^2}. \end{cases} \tag{21}$$

From the assertion in Lemma 2 and the definition of the function u, we have

$$0 \le \nabla u(t) \le \frac{2\varepsilon}{\alpha} \left| \psi^{-1}(t-1) \right|$$

for all $t \in \mathscr{S}_\kappa$. In addition, using the equality found in (21), we get

$$\nabla u(t) = \nabla v(t) + \frac{2\varepsilon}{\alpha} \left| \psi^{-1}(t-1) \right|,$$

so that

$$-\frac{2\varepsilon}{\alpha} \left| \psi^{-1}(t-1) \right| \le \nabla v(t) \le 0$$

for all $t \in \mathscr{S}_\kappa$. It follows that u is non-decreasing and v is non-increasing.

First we consider (i). Since u is non-decreasing and v is non-increasing, by the inequality in (21) with $1 < \beta < \sqrt{1+\alpha^2}$ and the fact that $s^* := \max \mathscr{S}$ exists, inequality (17) holds for $t \in \mathscr{S}$ and (i) is true.

Consider (ii). Let $t_0 \in \mathscr{S}$ be a fixed number. From the inequality in (21) with $1 < \beta < \sqrt{1+\alpha^2}$, we have

$$u(t) < v(t_0), \quad t \in \mathscr{S},$$

and thus u is bounded above on $\mathscr{S}$. With $\mathscr{S}$ unbounded, we know $\lim_{t\to\infty} u(t)$ exists. Moreover, $1 < \beta^2 < 1 + \alpha^2$ implies that

$$\lim_{t\to\infty} G_1(t)\psi^{-1}(t) = 0.$$

As a result, we have from the equality in (21) that $\lim_{t\to\infty} u(t) = \lim_{t\to\infty} v(t)$. Consequently, (18) is true for $t \in \mathscr{S}$, and (ii) holds.

Using the same arguments in the proofs of (i) and (ii), we can easily see that assertions (iii) and (iv) are also true. The proof is now complete. $\qquad \square$

Theorem 3 *Assume* $\beta \in \left(1, \sqrt{1+\alpha^2}\right) \cup \left(\sqrt{1+\alpha^2}, \infty\right)$. *Let* $\varepsilon > 0$ *be a given arbitrary constant, and let* $\phi : \mathscr{S} \to \mathbb{R}$ *satisfy*

$$|D_\alpha \phi(t) - \beta(-1)^t \phi(t)| \le \varepsilon, \quad t \in \mathscr{S}_\kappa.$$

Then one of the following holds.

(i) *If* $\beta \in \left(1, \sqrt{1+\alpha^2}\right)$ *and* $s^* := \max \mathscr{S}$ *exists, then any solution* x *of* (2) *with*

$$|\phi(s^*) - x(s^*)| < \varepsilon \max |G_1(t)| \quad \text{satisfies} \quad |\phi(t) - x(t)| < \varepsilon \max |G_1(t)|$$

for all $t \in \mathscr{S}$, *where* G_1 *is given in* (15).

(ii) *If* $\beta \in \left(1, \sqrt{1+\alpha^2}\right)$ *and* $\max \mathscr{S}$ *does not exist, then* $\lim_{t \to \infty} \phi(t)\psi^{-1}(t)$ *exists, and the function*

$$x(t) := \left(\lim_{t \to \infty} \phi(t)\psi^{-1}(t)\right) \psi(t)$$

is the unique solution of (2) *such that*

$$|\phi(t) - x(t)| \leq \varepsilon \max |G_1(t)|$$

for all $t \in \mathscr{S}$, *where* ψ *is given in* (4).

(iii) *If* $\beta > \sqrt{1+\alpha^2}$ *and* $s_* := \min \mathscr{S}$ *exists, then any solution* x *of* (2) *with*

$$|\phi(s_*) - x(s_*)| < \varepsilon \max |G_1(t)| \quad \text{satisfies} \quad |\phi(t) - x(t)| < \varepsilon \max |G_1(t)|$$

for all $t \in \mathscr{S}$.

(iv) *If* $\beta > \sqrt{1+\alpha^2}$ *and* $\min \mathscr{S}$ *does not exist, then* $\lim_{t \to -\infty} \phi(t)\psi^{-1}(t)$ *exists, and the function*

$$x(t) := \left(\lim_{t \to -\infty} \phi(t)\psi^{-1}(t)\right) \psi(t)$$

is the unique solution of (2) *such that*

$$|\phi(t) - x(t)| \leq \varepsilon \max |G_1(t)|$$

for all $t \in \mathscr{S}$.

Proof From Proposition 2, we can find a non-decreasing function $u : \mathscr{S} \to \mathbb{R}$ and a non-increasing function $v : \mathscr{S} \to \mathbb{R}$ such that (16) holds for all $t \in \mathscr{S}$.

First we consider (i). Assume $\beta \in \left(1, \sqrt{1+\alpha^2}\right)$ and $s^* := \max \mathscr{S}$ exists. From Proposition 2 (i), we have (17) for all $t \in \mathscr{S}$. Let x be any solution of (2) with

$$|\phi(s^*) - x(s^*)| < \varepsilon \max |G_1(t)|.$$

Then this x can be expressed as

$$x(t) := \frac{x(s^*)\psi(t)}{\psi(s^*)}$$

for all $t \in \mathscr{S}$, and the above inequality yields

$$\phi(s^*) - \varepsilon \max |G_1(t)| < x(s^*) < \phi(s^*) + \varepsilon \max |G_1(t)|. \tag{22}$$

The rest of the proof continues as in the proof of the cases of Theorem 2, and thus the details are omitted. □

By Theorem 3, we obtain the following result immediately.

Corollary 3 *If* $\beta \in \left(1, \sqrt{1 + \alpha^2}\right) \cup \left(\sqrt{1 + \alpha^2}, \infty\right)$, *then* (2) *has Hyers–Ulam stability with an HUS constant* $K = \max |G_1(t)|$ *on* $\mathscr{S}$, *for* G_1 *given in* (15).

Remark 3 Assume $\beta \in \left(1, \sqrt{1 + \alpha^2}\right) \cup \left(\sqrt{1 + \alpha^2}, \infty\right)$. We will prove that the minimum HUS constant for (2) on all of $\mathbb{Z}$ is $K = \max |G_1(t)|$, where G_1 is given in (15). The function

$$\phi(t) := \varepsilon G_1(t) + (\max |G_1(t)|) \, \psi(t), \qquad t \in \mathbb{Z}$$

satisfies $|D_\alpha \phi(t) - \beta \phi(t)| = \varepsilon$ for all $t \in \mathbb{Z}$ since $G1$ solves (14), and the function $x(t) := (\max |G_1(t)|) \, \psi(t)$ solves (2). Clearly $|\phi(t) - x(t)| = \varepsilon |G_1(t)|$ on $\mathbb{Z}$. This means that the minimum HUS constant for (2) on $\mathbb{Z}$ is greater than or equal to $\max |G_1(t)|$. So, this together with Corollary 3 implies that the minimum HUS constant for (2) on $\mathbb{Z}$ is $\max |G_1(t)|$.

Remark 4 For $\beta < -1$ with $\beta^2 - \alpha^2 - 1 \neq 0$, the function $G = G_2$, where G_2 is given by

$$G_2(t) := (-1)^{\lceil \frac{t}{2} \rceil} \left(\frac{\alpha - \beta - (-1)^t}{\beta^2 - \alpha^2 - 1} \right) \tag{23}$$

solves (14) with $\operatorname{sgn} \psi(t - 1) = (-1)^{\left\lceil \frac{t-1}{2} \right\rceil}$. Consequently, Lemma 2 holds with $G = G_2$ given here in (23).

Theorem 4 *Assume* $\beta \in \left(-\infty, -\sqrt{1 + \alpha^2}\right) \cup \left(-\sqrt{1 + \alpha^2}, -1\right)$. *Let* $\varepsilon > 0$ *be a given arbitrary constant, and let* $\phi : \mathscr{S} \to \mathbb{R}$ *satisfy*

$$|D_\alpha \phi(t) - \beta(-1)^t \phi(t)| \leq \varepsilon, \qquad t \in \mathscr{S}_\kappa.$$

Then one of the following holds.

(i) *If* $\beta \in \left(-\sqrt{1 + \alpha^2}, -1\right)$ *and* $s^* := \max \mathscr{S}$ *exists, then any solution* x *of* (2) *with*

$$|\phi(s^*) - x(s^*)| < \varepsilon \max |G_2(t)| \quad \text{satisfies} \quad |\phi(t) - x(t)| < \varepsilon \max |G_2(t)|$$

for all $t \in \mathscr{S}$, *where* G_2 *is given in* (23).

(ii) *If $\beta \in \left(-\sqrt{1+\alpha^2}, -1\right)$ and* $\max \mathscr{S}$ *does not exist, then* $\lim\limits_{t\to\infty} \phi(t)\psi^{-1}(t)$ *exists, and the function*

$$x(t) := \left(\lim_{t\to\infty} \phi(t)\psi^{-1}(t)\right) \psi(t)$$

is the unique solution of (2) such that

$$|\phi(t) - x(t)| \leq \varepsilon \max |G_2(t)|$$

for all $t \in \mathscr{S}$, where ψ is given in (4).

(iii) *If $\beta < -\sqrt{1+\alpha^2}$ and $s_* := \min \mathscr{S}$ exists, then any solution x of (2) with*

$$|\phi(s_*) - x(s_*)| < \varepsilon \max |G_2(t)| \quad \text{satisfies} \quad |\phi(t) - x(t)| < \varepsilon \max |G_2(t)|$$

for all $t \in \mathscr{S}$.

(iv) *If $\beta < -\sqrt{1+\alpha^2}$ and $\min \mathscr{S}$ does not exist, then* $\lim\limits_{t\to-\infty} \phi(t)\psi^{-1}(t)$ *exists, and the function*

$$x(t) := \left(\lim_{t\to-\infty} \phi(t)\psi^{-1}(t)\right) \psi(t)$$

is the unique solution of (2) such that

$$|\phi(t) - x(t)| \leq \varepsilon \max |G_2(t)|$$

for all $t \in \mathscr{S}$.

Remark 5 Similar to Remark 3, for $\beta \in \left(-\infty, -\sqrt{1+\alpha^2}\right) \cup \left(-\sqrt{1+\alpha^2}, -1\right)$ the minimum HUS constant for (2) on all of $\mathbb{Z}$ is $K = \max |G_2(t)|$, where G_2 is given in (23).

3 Extension

With care, one could extend these results to the more general equation

$$D_\alpha x(t) - \beta g(t)x(t) = 0, \qquad t \in \mathscr{S} \subseteq \mathbb{Z}, \qquad \alpha \in (0, 1], \qquad \beta \neq \frac{1}{A}, \frac{1}{B}, \tag{24}$$

where the 2-cycle coefficient function g is given by

$$g(t) := \begin{cases} A : & t \text{ even} \\ B : & t \text{ odd} \end{cases}$$

for any $A, B \in \mathbb{R}$ with $A \neq B$. In this case, the fundamental solution ψ to (24) is

$$\psi(t) := \begin{cases} \left(\dfrac{\alpha^2}{(1-\beta A)(1-\beta B)} \right)^{\frac{t}{2}} & : t \text{ even,} \\ \dfrac{\alpha}{1-\beta B} \left(\dfrac{\alpha^2}{(1-\beta A)(1-\beta B)} \right)^{\frac{t-1}{2}} & : t \text{ odd.} \end{cases}$$

4 Conclusion

In this paper, we investigated the Hyers–Ulam stability (HUS) of the proportional nabla difference equation

$$D_\alpha x(t) - \beta(-1)^t x(t) = 0, \qquad t \in \mathscr{S} \subseteq \mathbb{Z}, \qquad \alpha \in (0, 1], \qquad \beta \neq \pm 1. \tag{25}$$

For the parameter values

$$\beta^2 = 1 - \alpha^2 \quad \text{or} \quad \beta^2 = 1 + \alpha^2,$$

the solution of (2) is a 2-cycle and a 4-cycle, respectively, and there is no HUS.

If $\beta \in (-1, 1)$ with $\beta^2 + \alpha^2 - 1 \neq 0$, then (25) has Hyers–Ulam stability with an HUS constant

$$K = \max |E_1(t)|,$$

and this is the minimal HUS constant if $\mathscr{S} = \mathbb{Z}$, for E_1 given in (3). If $\beta \in (1, \sqrt{1 + \alpha^2}) \cup (\sqrt{1 + \alpha^2}, \infty)$, then (25) has Hyers–Ulam stability with an HUS constant

$$K = \max |G_1(t)|,$$

and this is the minimal HUS constant if $\mathscr{S} = \mathbb{Z}$, for G_1 given in (15). If $\beta \in \left(-\infty, -\sqrt{1 + \alpha^2} \right) \cup (-\sqrt{1 + \alpha^2}, -1)$, then (25) has Hyers–Ulam stability with an HUS constant

$$K = \max |G_2(t)|,$$

and this is the minimal HUS constant if $\mathscr{S} = \mathbb{Z}$, for G_2 given in (23).

References

1. C. Alsina, R. Ger, On some inequalities and stability results related to the exponential function. J. Inequal. Appl. **2**, 373–380 (1998)
2. D.R. Anderson, D. J. Ulness, Newly defined conformable derivatives. Adv. Dyn. Syst. Appl. **10**(2), 109–137 (2015)
3. S. András, A. R. Mészáros, Ulam–Hyers stability of dynamic equations on time scales via Picard operators. Appl. Math. Comput. **219**, 4853–4864 (2013)
4. J. Brzdek, P. Wójcik, On approximate solutions of some difference equations. Bull. Aust. Math. Soc. **95**(3), 476–481 (2017)
5. J. Brzdek, D. Popa, I. Raşa, B. Xu, *Ulam Stability of Operators*. Mathematical Analysis and Its Applications (Academic Press, London, 2018)
6. D.H. Hyers, On the stability of the linear functional equation. Proc. Nat. Acad. Sci. U.S.A. **27**, 222–224 (1941)
7. S.-M. Jung, Hyers–Ulam stability of linear differential equation of the first order (III). J. Math. Anal. Appl. **311**, 139–146 (2005)
8. S.-M. Jung, Hyers–Ulam stability of linear differential equations of first order (II). Appl. Math. Lett. **19**, 854–858 (2006)
9. S.-M. Jung, Hyers–Ulam stability of a system of first order linear differential equations with constant coefficients. J. Math. Anal. Appl. **320**, 549–561 (2006)
10. S.-M. Jung, Hyers–Ulam stability of linear differential equations of first order (I). Int. J. Appl. Math. Stat. **7**, 96–100 (2007)
11. S.-M. Jung, B. Kim, Th.M. Rassias, On the Hyers–Ulam stability of a system of Euler differential equations of first order. Tamsui Oxf. J. Math. Sci. **24**(4), 381–388 (2008)
12. T. Miura, S. Miyajima, S.E. Takahasi, A characterization of Hyers–Ulam stability of first order linear differential operators. J. Math. Anal. Appl. **286**(1), 136–146 (2003)
13. T. Miura, S. Miyajima, S.E. Takahasi, Hyers–Ulam stability of linear differential operator with constant coefficients. Math. Nachr. **258**, 90–96 (2003)
14. Y.W. Nam, X.G. Zhang, Hyers–Ulam stability of elliptic Möbius difference equation. Cogent Math. Stat. **5**, 1–9 (2018)
15. M. Onitsuka, Influence of the stepsize on Hyers–Ulam stability of first-order homogeneous linear difference equations. Int. J. Differ. Equ. **12**(2), 281–302 (2017)
16. M. Onitsuka, Hyers–Ulam stability of first-order nonhomogeneous linear difference equations with a constant stepsize. Appl. Math. Comput. **330**, 143–151 (2018)
17. D. Popa, Hyers–Ulam stability of the linear recurrence with constant coefficients. Adv. Differ. Equ. **2005**, 407076 (2005)
18. D. Popa, Hyers–Ulam–Rassias stability of a linear recurrence. J. Math. Anal. Appl. **309**, 591–597 (2005)
19. Th.M. Rassias, On the stability of linear mapping in Banach spaces. Proc. Am. Math. Soc. **72**, 297–300 (1978)
20. I.A. Rus, Ulam stability of ordinary differential equations. Stud. Univ. Babeş-Bolyai Math. **54**, 125–134 (2009)
21. Y.H. Shen, The Ulam stability of first order linear dynamic equations on time scales. Results Math. **72**(4), 1881–1895 (2017). http://dx.doi.org/10.1007/s00025-017-0725-1
22. Y.H. Shen, Y.J. Li, The z-transform method for the Ulam stability of linear difference equations with constant coefficients. Adv. Differ. Equ. **2018**, 396 (2018)
23. S.M. Ulam, *A Collection of the Mathematical Problems* (Interscience, New York, 1960)
24. G. Wang, M. Zhou, L. Sun, Hyers–Ulam stability of linear differential equations of first order. Appl. Math. Lett. **21**, 1024–1028 (2008)

Solution of Generalized Jensen and Quadratic Functional Equations

A. Charifi, D. Zeglami, and S. Kabbaj

Abstract We obtain in terms of additive and multi-additive functions the general solution $f : S \to H$ of each of the functional equations

$$\sum_{\lambda \in \Phi} f(x + \lambda y + a_\lambda) = Nf(x), \quad x, y \in S,$$

$$\sum_{\lambda \in \Phi} f(x + \lambda y + a_\lambda) = Nf(x) + Nf(y), \quad x, y \in S,$$

where $(S, +)$ is an abelian monoid, Φ is a finite group of automorphisms of S, $N = |\Phi|$ designates the number of its elements, $\{a_\lambda, \lambda \in \Phi\}$ are arbitrary elements of S, and $(H, +)$ is an abelian group. In addition, some applications are given. These equations provide a common generalization of many functional equations (Cauchy's, Jensen's, quadratic, Φ-quadratic equations, ...).

1 Introduction

The Φ-quadratic functional equation

$$\sum_{\lambda \in \Phi} f(x + \lambda y) = Nf(x) + Nf(y), \quad x, y \in S, \tag{1}$$

and the Φ-Jensen functional equation

A. Charifi · S. Kabbaj
Department of Mathematics, Faculty of Sciences, Ibn Tofail University, Kenitra, Morocco

D. Zeglami (✉)
Department of Mathematics, ENSAM, Moulay Ismail University, Meknes, Morocco

© Springer Nature Switzerland AG 2019
G. A. Anastassiou and J. M. Rassias (eds.), *Frontiers in Functional Equations and Analytic Inequalities*, https://doi.org/10.1007/978-3-030-28950-8_16

273

$$\sum_{\lambda \in \Phi} f(x + \lambda y) = Nf(x), \ x, y \in S, \tag{2}$$

where S is an abelian semigroup (a non-empty set equipped with an associative operation), Φ is a finite group of automorphisms of S, H is an abelian group and $f : S \to H$ is the unknown function were studied by Łukasik [9] and appeared in several works by Stetkær by considering S an abelian group, see, for example, [13–15]. Let $id : S \to S$ denote the identity function and $\sigma : S \to S$ denote an additive function of S, such that $\sigma(\sigma(x) = x$, for all $x \in S$ then Eq. (1) reduces to the functional equations

$$f(x + y) = f(x) + f(y), \ x, y \in S, \qquad\qquad \Phi = \{id\} \tag{3}$$

$$f(x + y) + f(x + \sigma(y)) = 2f(x) + 2f(y), \ x, y \in S, \qquad \Phi = \{id, \sigma\} \tag{4}$$

$$\sum_{i=0}^{n-1} f(x + \omega^i y) = nf(x) + nf(y), \ x, y \in S = H = \mathbb{C}, \omega = e^{\frac{2i\pi}{n}},$$

$$\Phi = \left\{ \omega^i, \ 0 \le i \le n - 1 \right\}. \tag{5}$$

If f is a solution of (3), it is said to be additive or satisfies Cauchy's equation. Equation (4) is related to symmetric biadditive functions [8, 11]. It is natural that this equation is called quadratic functional equation. In particular, it is well known that a function f between real vector spaces satisfies the equation

$$f(x + y) + f(x - y) = 2f(x) + 2f(y)$$

if and only if there exists a unique symmetric biadditive function B such that $f(x) = B(x, x)$ for all x. The biadditive function B is given by

$$B(x, y) = \frac{1}{4} \{f(x + y) + f(x - y)\}, \text{ for all } x, y. \tag{6}$$

Some information, applications, and numerous references concerning (4) and its further generalizations can be found, e.g., in [3, 4, 6, 7, 12–15].

Let S be an abelian monoid (that is, a semigroup with identity) and H be an abelian group (satisfying some assumptions). As a continuation of the works by Łukasik [9] and by Charifi et al. [1, 2], the purpose of the present paper is first to give an explicit description of the solutions of the generalized Φ-Jensen functional equation

$$\sum_{\lambda \in \Phi} f(x + \lambda y + a_\lambda) = Nf(x), \ x, y \in S, \tag{7}$$

and the generalized Φ-quadratic functional equation

$$\sum_{\lambda \in \Phi} f(x + \lambda y + a_\lambda) = Nf(x) + Nf(y), \quad x, y \in S, \tag{8}$$

where $f : S \to H$ is an application, Φ is a finite group of automorphisms of S, N designates the number of its elements, and $\{a_\lambda, \lambda \in \Phi\}$ are arbitrary elements of S, and secondly, to illustrate our theory, we give some applications.

These linear functional equations encompass, in addition to (1)–(5) on monoid, the following functional equations as special cases:

$$f(x + y + a) = f(x) + f(y), \quad x, y \in S,$$

$$f(x + y + a) + f(x + \sigma(y) + b) = 2f(x), \quad x, y \in S,$$

$$f(x + y + \sigma(a)) + f(x + \sigma(y) + a) = 2f(x) + 2f(y), \quad x, y \in S,$$

$$f(x + y + a) + f(x + \sigma(y) + b) = 2f(x) + 2f(y), \quad x, y \in S,$$

where $a, b \in S$ and σ is an involution of S, i.e. $\sigma(x + y) = \sigma(y) + \sigma(x)$ and $\sigma(\sigma(x)) = x$ for all $x, y \in S$.

We shall adhere to the following notation.

Notation To formulate our results we introduce the following notation and assumptions that, unless otherwise explicitly stated, will be used throughout the paper:

Let S be an abelian monoid with identity element that we denote 0, let H, N, and Φ given as above, let H^S denote the $\mathbb{Z}$-module consisting of all maps from S into H and let $\{a_\lambda, \lambda \in \Phi\}$ denote arbitrary elements of S.

A function $\mathcal{A} : S \to H$ is additive if $\mathcal{A}(x + y) = \mathcal{A}(x) + \mathcal{A}y$ for all $x, y \in S$, in this case it is easily seen that $\mathcal{A}(rx) = r\mathcal{A}(x)$ for all $x \in S$ and all $r \in \mathbb{N}$.

Let $k \in \mathbb{N}$, a function $\mathcal{A}_k : S^k \to F$ is k-additive if it is additive in each variable, in addition we say that $\mathcal{A}_k$ is symmetric if it satisfies $\mathcal{A}_k(x_{\pi(1)}, x_{\pi(2)}, \ldots, x_{\pi(k)}) = \mathcal{A}_k(x_1, x_2, \ldots, x_k)$ for all $(x_1, x_2, \ldots, x_k) \in S^k$ and all permutations π of k elements.

Let $\mathcal{A}_k : S^k \to H$ be a k-additive and symmetric function and let $\mathcal{A}_k^* : S \to H$ defined by $\mathcal{A}_k^*(x) = \mathcal{A}(x, x, \ldots, x)$ for all $x \in S$. Such a function $\mathcal{A}_k^*$ will be called a monomial function of degree k (if $\mathcal{A}_k^* \neq 0$). We note that $\mathcal{A}_k^*(rx) = r^k \mathcal{A}(x)$ for all $x \in S$ and all $r \in \mathbb{N}$.

A function $\mathcal{P} : S \to H$ is called a GP function (generalized polynomial function) of degree $m \in \mathbb{N}$ iff there exist $\mathcal{A}_0 \in H$ and symmetric k-additive functions $\mathcal{A}_k : S^k \to H$ (for $1 \le k \le m$) such that

$$\mathcal{A}_m^* \neq 0 \text{ and } \mathcal{P}(x) = \mathcal{A}_0 + \sum_{k=1}^{m} \mathcal{A}_k^*(x) \text{ for all } x \in S. \tag{9}$$

For $h \in S$ we define the linear difference operator Δ_h on H^S by

$$\Delta_h(f)(x) = f(x + h) - f(x), \tag{10}$$

for all $f \in H^S$ and $x \in S$. Notice that these difference operators commute ($\Delta_h \Delta_{h'} = \Delta_{h'} \Delta_h$ for all $h, h' \in S$) and if $h \in S$, $n \in \mathbb{N}$ then Δ_h^n the n-th iterate of Δ_h satisfies

$$\Delta_h^n(f)(x) = \sum_{k=0}^{n} (-1)^{n-k} \binom{n}{k} f(x + kh), \text{ for all } x, h \in S \text{ and } f \in H^S. \tag{11}$$

2 Auxiliary Results

In this section, we note some results for later use.

Lemma 1 (Łukasik [9]) *Let $n \in \mathbb{N}^*$. Then we have*

(i)

$$\sum_{k=1}^{n} (-1)^{n-k} \binom{n}{k} k^i = 0, \quad i \in \{1, 2, \ldots, n-1\}, \quad n \neq 1. \tag{12}$$

and

$$\sum_{k=1}^{n} (-1)^{n-k} \binom{n}{k} k^n = n!. \tag{13}$$

(ii) Let $(H, +)$ be an abelian group uniquely divisible by $n!$. If $x_1, x_2, \ldots, x_n \in H$ be such that

$$\sum_{i=1}^{n} k^i x_i = 0, \quad k \in \{0, 1, \ldots, n\}. \tag{14}$$

Then $x_1 = \cdots = x_n = 0$.

The following theorem was proved by Mazur and Orlicz [10] and generalized by Djoković [5]:

Theorem 1 *Let $(S, +)$ be an abelian semigroup, $n \in \mathbb{N}$, $(H, +)$ be an abelian group uniquely divisible by $n!$ and $f \in H^S$. Then the following statements are equivalent*

(i) $\Delta_h^n f(x) = 0$ for all $x, h \in E$.
(ii) $\Delta_{h_1 \ldots h_n} f(x) = 0$ for all $x, h_1, \ldots, h_n \in E$.
(iii) f is a GP function of degree at most $n - 1$.

In the next lemma we recall a general property of functions satisfying (7).

Lemma 2 (Charifi et al. [2]) *Let $(S, +)$ be an abelian monoid, Φ be a finite subgroup of the group of automorphisms of S, $N = card(\Phi)$, $(H, +)$ be an abelian group uniquely divisible by $N!$ and $\{a_\lambda, \lambda \in \Phi\}$ are arbitrary elements of S. Assume that the function $f : S \to G$ satisfies the following equation*

$$\sum_{\lambda \in \Phi} f(x + \lambda y + a_\lambda) = N f(x) \quad and \quad \sum_{\lambda \in \Phi} f(\lambda y) = 0 \qquad (15)$$

for all $x, y \in S$. Then, $\Delta_y^N f(x) = 0$ for $x, y \in S$.

3 Solutions of Eq. (7)

In the following theorem we determine, in terms of GP functions, the solutions of the generalized Φ-Jensen functional equation (7).

Theorem 2 *Let $(S, +)$ be an abelian monoid, Φ be a finite subgroup of the group of automorphisms of S, $N = card(\Phi)$, $(H, +)$ be an abelian group uniquely divisible by $N!$ and $\{a_\lambda, \lambda \in \Phi\}$ are arbitrary elements of S. Then the function $f : S \to H$ is a solution of Eq. (7) if and only if f has the following form*

$$f(x) = \mathcal{A}_0 + \sum_{i=1}^{N-1} \mathcal{A}_i^*(x), \quad x \in S, \qquad (16)$$

where $\mathcal{A}_0 \in H$ and $\mathcal{A}_k : S^k \to H$, $k \in \{1, 2, \ldots, N-1\}$ are k-additive and symmetric functions which satisfy the following conditions

$$\sum_{i=max(k+j,k+1) \leq N-1} \binom{i}{k} \binom{i-k}{j} \sum_{\lambda \in \Phi} \mathcal{A}_i(\underbrace{x, \ldots x}_{k}, a_\lambda, \ldots, a_\lambda, \underbrace{\lambda y, \ldots, \lambda y}_{j}) = 0$$

for all

$$x, y \in S, \ 0 \leq k \leq N-2, \ 0 \leq j \leq N-k-1.$$

Proof It is easy to check, by simple computation, that if f has the form (16) then f satisfies Eq. (7). Indeed,

$$\sum_{\lambda \in \Phi} f(x + \lambda y + a_\lambda)$$

$$= N\mathcal{A}_0 + \sum_{\lambda \in \Phi} \sum_{i=1}^{N-1} \mathcal{A}_i^*(x + \lambda y + a_\lambda)$$

$$= N\mathcal{A}_0 + \sum_{\lambda \in \Phi} \sum_{i=1}^{N-1} \mathcal{A}_i^*(x)$$

$$+ \sum_{\lambda \in \Phi} \sum_{j=0}^{N-1} \sum_{k=0}^{N-2} \sum_{i=\max(k+1,j+k)}^{N-1} \binom{i}{j}\binom{i-j}{k} \mathcal{A}_i(\underbrace{x,\ldots,x}_{}, a_\lambda, \ldots, a_\lambda, \underbrace{\lambda y, \ldots, \lambda y}_{j})$$

$$= N\mathcal{A}_0 + \sum_{\lambda \in \Phi} \sum_{i=1}^{N-1} \mathcal{A}_i^*(x)$$

$$= Nf(x),$$

for all $x, y \in S$.

Conversely, suppose that f is a solution of Eq. (7). By taking in (7), respectively, $y = 0$ and $x = 0$ we get

$$\sum_{\lambda \in \Phi} f(x + a_\lambda) = Nf(x), \ x \in S, \tag{17}$$

and

$$\sum_{\lambda \in \Phi} f(\lambda y + a_\lambda) = Nf(0), \ y \in S, \tag{18}$$

By replacing, in the previous equality, y by μy we obtain

$$N^2 f(0) = \sum_{\mu \in \Phi} \sum_{\lambda \in \Phi} f(\mu \lambda y + a_\lambda)$$

$$= \sum_{\lambda \in \Phi} \sum_{\mu \in \Phi} f(\mu \lambda y + a_\lambda)$$

$$= \sum_{\lambda \in \Phi} \sum_{\nu \in \Phi} f(\nu y + a_\lambda)$$

$$= \sum_{\nu \in \Phi} \sum_{\lambda \in \Phi} f(\nu y + a_\lambda)$$

$$= N \sum_{\nu \in \Phi} f(\nu y), \tag{19}$$

for every $y \in S$. It follows, by taking $g := f - f(0)$ in (19) that

$$\sum_{\nu \in \Phi} g(\nu y) = 0 \text{ for all } y \in S, \tag{20}$$

so we can reformulate Eq. (7) as

$$\sum_{\lambda \in \Phi} g(x + \lambda y + a_\lambda) = Ng(x) + \sum_{\lambda \in \Phi} g(\lambda y), \quad x, y \in S, \tag{21}$$

i.e., g is a solution of the last equation and Eq. (7). In view of Lemma 2 and Theorem 1, we infer that g is a GP function of degree at most $N - 1$. Then by putting $\mathcal{A}_0 = f(0)$ we obtain that there exist $\mathcal{A}_k : S^k \to H$, $k \in \{1, 2, \ldots, N - 1\}$ k-additive and symmetric function such that

$$f(x) = \mathcal{A}_0 + \sum_{i=1}^{N-1} \mathcal{A}_i^*(x), \quad x \in S. \tag{22}$$

To prove the condition:

$$\sum_{i=max(k+j,k+1)} \binom{i}{k}\binom{i-k}{j} \sum_{\lambda \in \Phi} \mathcal{A}_i(\underbrace{x, \ldots x}_{k}, a_\lambda, \ldots, a_\lambda, \underbrace{\lambda y, \ldots, \lambda y}_{j}) = 0, \quad x, y \in S,$$

$$\tag{23}$$

for all $0 \leq k \leq N - 2$, $0 \leq j \leq N - 1 - k$ we define the functions I, J_k, $Q_{(k,j)}$: $S \times S \to H$ by the formulas

$$I(x, y) = \sum_{\lambda \in \Phi} f(x + \lambda y + a_\lambda) - Nf(x), \tag{24}$$

$$J_k(x, y) = \sum_{j=0}^{N-k-1} \sum_{i=max(k+j,k+1)} \binom{i}{k}\binom{i-k}{j} \sum_{\lambda \in \Phi} \mathcal{A}_i(\underbrace{x, \ldots x}_{k}, a_\lambda, \ldots, a_\lambda, \underbrace{\lambda y, \ldots, \lambda y}_{j}),$$

$$\tag{25}$$

and

$$Q_{(k,j)}(x, y) = \sum_{i=max(k+j,k+1)} \binom{i}{j}\binom{i-j}{k} \sum_{\lambda \in \Phi} \mathcal{A}_i(\underbrace{x, \ldots x}_{k}, a_\lambda, \ldots, a_\lambda, \underbrace{\lambda y, \ldots, \lambda y}_{j}),$$

$$\tag{26}$$

for all $x, y \in S$. A direct computation, using the expression of f given in (22) and the condition (iii), shows that we have

$$0 = I(x, y)$$

$$= \sum_{\lambda \in \Phi} f(x + \lambda y + a_\lambda) - Nf(x)$$

$$= \sum_{\lambda \in \Phi} \sum_{i=1}^{N-1} \mathcal{A}_i^*(x + \lambda y + a_\lambda) - N \sum_{i=1}^{N-1} \mathcal{A}_i^*(x)$$

$$= \sum_{k=0}^{N-2} \sum_{j=0}^{N-k-1} \sum_{i=max(k+j,k+1) \leq N-1} \binom{i}{k}\binom{i-k}{j} \sum_{\lambda \in \Phi} \mathcal{A}_i(\underbrace{x, \dots x}_{k}, a_\lambda, \dots, a_\lambda, \underbrace{\lambda y, \dots, \lambda y}_{j})$$

$$= \sum_{k=0}^{N-2} \sum_{j=0}^{N-k-1} Q_{(k,j)}(x, y)$$

$$= \sum_{k=0}^{N-2} J_k(x, y)$$

for all $x, y \in S$. Lemma 1 (ii) applied, respectively, to the functions J_k and $Q_{(k,j)}$ completes the proof of Theorem 2.

By using Theorem 2, we get the following corollaries. First if we take $\{a_\lambda, \lambda \in \Phi\} = \{0\}$ in (7) (i.e., using the notations of Theorem 2, $k = i - j$), we obtain the following result which has been proved by Łukasik ([9], Theorem 5).

Corollary 1 (Łukasik [9]) *Let* $(S, +)$ *be an abelian monoid, let* Φ *be a finite subgroup of the group of automorphisms of* S, $N = card(\Phi)$ *and let* H *be an abelian group uniquely divisible by* $N!$. *Then the general solution* $f : S \to H$ *of the functional equation*

$$\sum_{\lambda \in \Phi} f(x + \lambda y) = N f(x), \quad x, y \in S, \tag{27}$$

is

$$f(x) = \mathcal{A}_0 + \sum_{i=1}^{N-1} \mathcal{A}_i^*(x), \quad x \in S \tag{28}$$

where $\mathcal{A}_0 \in H$, $\mathcal{A}_k : S^k \to H, k \in \{1, 2, \dots, N-1\}$ *are arbitrary k-additive and symmetric functions which satisfy the following conditions*

$$\sum_{\lambda \in \Phi} \mathcal{A}_i(\underbrace{x, \dots x, \lambda y, \dots, \lambda y}_{j}) = 0, \quad x, y \in S, \ 1 \leq j \leq i-1, \ 1 \leq i \leq N-1. \tag{29}$$

Proof Using the notations of Theorem 2, if $\{a_\lambda = 0, \ \lambda \in \Phi\} = \{0\}$, then $i = k + j$, $j \geq 1$ and $k \geq 1$. Thus, by Theorem 2 we get

$$0 = \sum_{i=max(k+j,k+1) \leq N-1} \binom{i}{k}\binom{i-k}{j} \sum_{\lambda \in \Phi} \mathcal{A}_i(\underbrace{x, \dots x}_{k}, \underbrace{\lambda y, \dots, \lambda y}_{j})$$

$$= \sum_{2 \leq i=k+j}^{N-1} \binom{i}{k} \sum_{\lambda \in \Phi} \mathcal{A}_{k+j}(\underbrace{x, \dots x}_{k}, \underbrace{\lambda y, \dots, \lambda y}_{j}),$$

$$= \sum_{k=1}^{N-2} \sum_{j=1}^{N-1-k} \binom{i}{j} \sum_{\lambda \in \Phi} \mathcal{A}_{k+j}(\underbrace{x, \ldots x}_{k}, \underbrace{\lambda y, \ldots, \lambda y}_{j})$$

$$= \sum_{j=1}^{N-2} \sum_{k=1}^{N-1-j} \binom{i}{j} \sum_{\lambda \in \Phi} \mathcal{A}_{k+j}(\underbrace{x, \ldots x}_{k}, \underbrace{\lambda y, \ldots, \lambda y}_{j}), \ x, y \in S.$$

We define, for $1 \leq k \leq N - 1 - j$, $0 \leq j \leq N - 2$, the mappings g_j, $h_{(k,j)}$: $S \times S \to H$ by

$$g_j(x, y) = \sum_{k=1}^{N-j-1} \binom{i}{j} \sum_{\lambda \in \Phi} \mathcal{A}_{k+j}(\underbrace{x, \ldots x}_{k}, \underbrace{\lambda y, \ldots, \lambda y}_{j}),$$

$$h_{(k,j)}(x, y) = \binom{i}{j} \sum_{\lambda \in \Phi} \mathcal{A}_{k+j}(\underbrace{x, \ldots x}_{k}, \underbrace{\lambda y, \ldots, \lambda y}_{j}).$$

First, we note

$$g_j(x, ny) = n^j g_j(x, y), \ x, y \in S, \ n \in \mathbb{N}^*, \ 1 \leq j \leq N - 2,$$

$$h_{(k,j)}(nx, y) = n^k h_{(k,j)}(x, y), \ x, y \in S, \ n \in \mathbb{N}^*, \ 1 \leq k \leq N-1-j, \ 1 \leq j \leq N-2$$

and

$$\sum_{k=1}^{N-1-j} h_{(k,j)}(x, y) = g_j(x, y), \ x, y \in S, \ n \in \mathbb{N}^*, \ 1 \leq j \leq N - 2.$$

Now, from the above equalities we get

$$\sum_{j=1}^{N-1} n^j g_j(x, y) = \sum_{j=1}^{N-1} g_j(x, ny) = 0, \ x, y \in S, \ n \in \mathbb{N}^*$$

and then by Lemma 1,

$$g_j(x, y) = 0, \ x, y \in S, \ 1 \leq j \leq N - 2.$$

So, we obtain

$$\sum_{k=1}^{N-1-j} n^k h_{(k,j)}(x, y) = \sum_{k=1}^{N-1-j} h_{(k,j)}(nx, y) = g_j(nx, y) = 0, \ x, y \in S, \ n \in \mathbb{N}^*,$$

$$1 \le j \le N - 2$$

which gives, according to Lemma 1,

$$h_{(k,j)}(x, y) = 0, \ x, y \in S, \ 1 \le j \le N - 2, \ 1 \le k \le N - 1 - j.$$

This ends the proof.

The second corollary was proved by Sinopoulos ([12], Theorem 2) on a semi-group.

Corollary 2 (Sinopoulos [12]) *Let* $(S, +)$ *be an abelian monoid, and let* H *be an abelian group uniquely divisible by* 2. *Suppose that* σ *is an endomorphism of* S *such that* $\sigma(\sigma(x)) = x$ *for* $x \in S$. *Then, the general solution* $f : S \to H$ *of the functional equation*

$$f(x + y) + f(x + \sigma(y)) = 2f(x), \ x, y \in S, \tag{30}$$

is

$$f(x) = \mathcal{A}_0 + \mathcal{A}_1(x) \tag{31}$$

where $\mathcal{A}_0 \in H$ *is an arbitrary constant and* $\mathcal{A}_1 : S \to H$ *is an arbitrary additive function with* $\mathcal{A}_1(\sigma(x)) = -\mathcal{A}_1(x)$ *for all* $x \in S$.

In the following corollary we solve another special case of Eq. (7) that is, according to our knowledge, not in the literature.

Corollary 3 *Let* $(S, +)$ *be an abelian monoid,* $a, b \in S$ *and let* H *be an abelian group uniquely divisible by* 2. *Suppose that* σ *is an endomorphism of* S *such that* $\sigma(\sigma(x)) = x$ *for* $x \in S$. *Then, the general solution* $f : S \to H$ *of the functional equation*

$$f(x + y + a) + f(x + \sigma(y) + b) = 2f(x), \ x, y \in S, \tag{32}$$

is

$$f(x) = \mathcal{A}_0 + \mathcal{A}_1(x), \ x \in S, \tag{33}$$

where $\mathcal{A}_0 \in H$, $\mathcal{A}_1 : S \to H$ *is an arbitrary additive function with* $\mathcal{A}_1(\sigma(x) + x) = 0$ *and* $\mathcal{A}_1(a + b) = 0$ *for all* $x \in S$.

Corollary 4 *Let* j *be a primitive 3rd root of unity, and let* a *be a complex constant. The continuous solution* $f : \mathbb{C} \to \mathbb{C}$ *of the functional equation*

$$f(x+y+ja)+f(x+jy+j^2a)+f(x+j^2y+a) = 3f(x), \quad x, y \in \mathbb{C}, \quad (34)$$

are the functions of the form

$$f(z) = \alpha_0 + \alpha_1 z + \alpha_2 \bar{z} + \alpha_3 z^2 + \alpha_4 \bar{z}^2, \quad z \in \mathbb{C} \qquad (35)$$

where $\alpha_0, \ldots, \alpha_4$ are arbitrary complex numbers.

Proof In view of Theorem 2, we infer that there exist $\alpha_0 \in \mathbb{C}$, an additive function $\mathcal{A}_1 : \mathbb{C} \to \mathbb{C}$ and a symmetric and bi-additive function $\mathcal{A}_2 : \mathbb{C}^2 \to \mathbb{C}$ such that

$$f(z) = \alpha_0 + \mathcal{A}_1(z) + \mathcal{A}_2(z) \text{ for all } z \in \mathbb{C}. \qquad (36)$$

Since j is a primitive 3rd root of unity we have $1 + j + j^2 = 0$. The continuity of f provides that $\mathcal{A}_1$ and $\mathcal{A}_2$ have the following forms:

$$\mathcal{A}_1(z) = \alpha_1 z + \alpha_2 \bar{z} \text{ and } \mathcal{A}_2(z) = \alpha_3 z^2 + \alpha_4 \bar{z}^2 + \alpha_5 |z|^2, \quad z \in \mathbb{C}. \qquad (37)$$

The condition of Theorem 2 can be satisfied only for $\alpha_5 = 0$. This gives the expression of f. $\qquad \square$

4 Solutions of Eq. (8)

Now we characterize the general solution of the generalized Φ-quadratic equation (8).

Theorem 3 *Let $(S, +)$ be an abelian monoid, Φ be a finite subgroup of the group of automorphisms of S, $N = card(\Phi)$, $(H, +)$ be an abelian group uniquely divisible by $(N+1)!$ and $\{a_\lambda, \lambda \in \Phi\}$ are arbitrary elements of S. Then the function $f : S \to H$ is a solution of Eq. (8) if and only if f has the following form:*

$$f(x) = \mathcal{A}_0 + \sum_{i=1}^{N} \mathcal{A}_i^*(x), \quad x \in S, \qquad (38)$$

where $\mathcal{A}_0 \in H$ and $\mathcal{A}_k : S^k \to H, k \in \{1, 2, \ldots, N\}$ are symmetric and k-additive functions satisfying the three conditions:

(i) $\sum_{\lambda \in \Phi} \sum_{k=1}^{N} \mathcal{A}_k^*(a_\lambda) = N\mathcal{A}_0,$

(ii) $\sum_{i=max(k+j,k+1) \leq N} \binom{i}{k}\binom{i-k}{j} \sum_{\lambda \in \Phi} \mathcal{A}_i(\underbrace{x, \ldots x}_{k}, a_\lambda, \ldots, a_\lambda, \underbrace{\lambda y, \ldots, \lambda y}_{j}) = 0,$

   $x, y \in S,$
   $1 \leq k \leq N - 1, \ 0 \leq j \leq N - k \text{ and}$

(iii) $\sum\limits_{k=i}^{N} \binom{i}{k} \sum\limits_{\lambda \in \Phi} \mathcal{A}_k(\underbrace{\lambda x, \dots, \lambda x}_{i}, a_\lambda, \dots, a_\lambda) = N\mathcal{A}_i^*(x), \ x \in S, \ 1 \le i \le N.$

Proof It is easy to check, by simple computations, that if f satisfies (38) then f is a solution of Eq. (8). Indeed, if the above three conditions (i)–(iii) are satisfied, then we have for all $x, y \in S$

$$\sum\limits_{\lambda \in \Phi} f(x + \lambda y + a_\lambda)$$

$$= N\mathcal{A}_0 + \sum\limits_{\lambda \in \Phi} \sum\limits_{i=1}^{N} \mathcal{A}_i^*(x + \lambda y + a_\lambda)$$

$$= N\mathcal{A}_0 + \sum\limits_{\lambda \in \Phi} \sum\limits_{i=1}^{N} \mathcal{A}_i^*(x) + \sum\limits_{\lambda \in \Phi} \sum\limits_{i=1}^{N} \mathcal{A}_i^*(\lambda y + a_\lambda)$$

$$+ \sum\limits_{k=1}^{N-1} \sum\limits_{j=0}^{N-k} \sum\limits_{i=max(k+j,k+1)} \binom{i}{k}\binom{i-k}{j} \sum\limits_{\lambda \in \Phi} \mathcal{A}_i(\underbrace{x, \dots, x}_{k}, a_\lambda, \dots, a_\lambda, \underbrace{\lambda y, \dots, \lambda y}_{j})$$

$$= Nf(x) + \sum\limits_{i=1}^{N} \sum\limits_{k=0}^{i} \binom{i}{k} \sum\limits_{\lambda \in \Phi} \mathcal{A}_i(\underbrace{\lambda y, \dots, \lambda y}_{k}, a_\lambda, \dots, a_\lambda)$$

$$= Nf(x) + \sum\limits_{i=1}^{N} \sum\limits_{\lambda \in \Phi} \mathcal{A}_i^*(a_\lambda) + \sum\limits_{k=1}^{N} \sum\limits_{i=k}^{N} \binom{i}{k} \sum\limits_{\lambda \in \Phi} \mathcal{A}_i(\underbrace{\lambda y, \dots, \lambda y}_{k}, a_\lambda, \dots, a_\lambda)$$

$$= Nf(x) + N\mathcal{A}_0 + \sum\limits_{k=1}^{N} N\mathcal{A}_k^*(y)$$

$$= Nf(x) + Nf(y).$$

So f is a solution of (8). Conversely, suppose that $f : S \to H$ is a solution of Eq. (8). By taking, respectively, $x = y = 0$, $y = 0$ and $x = 0$ in (8) we obtain

$$\sum\limits_{\lambda \in \Phi} f(a_\lambda) = 2Nf(0), \tag{39}$$

$$\sum\limits_{\lambda \in \Phi} f(x + a_\lambda) = Nf(x) + Nf(0), \tag{40}$$

$$\sum\limits_{\lambda \in \Phi} f(\lambda x + a_\lambda) = Nf(x) + Nf(0), \tag{41}$$

for all $x \in S$. Thus, using these equalities and Eq. (8), we obtain:

$$
\begin{aligned}
N^2 f(x) + N \sum_{\mu \in \Phi} f(\mu y) &= \sum_{\lambda \in \Phi} \sum_{\mu \in \Phi} f(x + \lambda \mu y + a_\lambda) \\
&= \sum_{\lambda \in \Phi} \sum_{\nu \in \Phi} f(x + \nu y + a_\lambda) \\
&= \sum_{\nu \in \Phi} \sum_{\lambda \in \Phi} f(x + \nu y + a_\lambda) \\
&= N^2 f(0) + N \sum_{\nu \in \Phi} f(x + \nu y).
\end{aligned}
$$

With the notation $g := f - f(0)$ we can reformulate (8) to

$$
\sum_{\lambda \in \Phi} g(x + \lambda y) = N g(x) + \sum_{\lambda \in \Phi} g(\lambda y), \quad x, y \in S. \tag{42}
$$

So in view of ([9], Lemma 3) we infer that there exist $\mathcal{A}_0 \in H$ and some k-additive and symmetric functions $\mathcal{A}_k : S^k \to H, k \in \{1, 2, \dots, N\}$ such that

$$
f(x) = \mathcal{A}_0 + \sum_{i=1}^{N} \mathcal{A}_i^*(x), \quad x \in S, \tag{43}
$$

and

$$
\sum_{\lambda \in \Phi} \sum_{k=1}^{N} \mathcal{A}_i^*(a_\lambda) = N \mathcal{A}_0. \tag{44}
$$

In virtue of (40) and (41) we have

$$
\sum_{\lambda \in \Phi} g(y + a_\lambda) = \sum_{\lambda \in \Phi} g(\lambda y + a_\lambda), \tag{45}
$$

and in view of the previous equality, (43) and (44) we obtain

$$
\begin{aligned}
\sum_{\lambda \in \Phi} \sum_{i=1}^{N} \{\mathcal{A}_i^*(y) + \mathcal{A}_i^*(a_\lambda)\} &= \sum_{\lambda \in \Phi} \sum_{i=1}^{N} \mathcal{A}_i^*(y + a_\lambda) \\
&= \sum_{\lambda \in \Phi} g(y + a_\lambda) \\
&= \sum_{\lambda \in \Phi} g(\lambda y + a_\lambda)
\end{aligned}
$$

$$= \sum_{i=1}^{N} \sum_{k=0}^{i} \binom{i}{k} \sum_{\lambda \in \Phi} \mathcal{A}_i(\underbrace{\lambda y, \ldots, \lambda y}_{k}, a_\lambda, \ldots, a_\lambda)$$

$$= N\mathcal{A}_0 + \sum_{k=1}^{N} \sum_{i=k}^{N} \binom{i}{k} \sum_{\lambda \in \Phi} \mathcal{A}_i(\underbrace{\lambda y, \ldots, \lambda y}_{k}, a_\lambda, \ldots, a_\lambda).$$

Thus by putting for $1 \leq k \leq N$

$$g_k(y) := N\mathcal{A}_k^*(y) - \sum_{i=k}^{N} \binom{i}{k} \sum_{\lambda \in \Phi} \mathcal{A}_i(\underbrace{\lambda y, \ldots, \lambda y}_{k}, a_\lambda, \ldots, a_\lambda), \ y \in S, \qquad (46)$$

we get

$$\sum_{i=1}^{N} m^i g_i(y) = \sum_{i=1}^{N} g_i(my) = 0, \qquad (47)$$

for all $y \in S$ and $m \in \mathbb{N}^*$. Then the condition (iii) follows from Lemma 1 (ii). Finally, in order to get the condition (ii) we define, for all $1 \leq k \leq N - 1$ and $0 \leq j \leq N - k$ the maps $I, J_k, Q_{(k,j)} : S \times S \to H$ by the formulas

$$I(x, y) = \sum_{\lambda \in \Phi} f(x + \lambda y + a_\lambda) - Nf(x) - Nf(y), \qquad (48)$$

$$J_k(x, y) = \sum_{j=0}^{N-k} \sum_{i=max(k+j,k+1) \leq N} \binom{i}{k}\binom{i-k}{j} \sum_{\lambda \in \Phi} \mathcal{A}_i(\underbrace{x, \ldots x}_{k}, a_\lambda, \ldots, a_\lambda, \underbrace{\lambda y, \ldots, \lambda y}_{j}), \qquad (49)$$

and

$$Q_{(k,j)}(x, y) = \sum_{i=max(k+j,k+1) \leq N} \binom{i}{j}\binom{i-j}{k} \sum_{\lambda \in \Phi} \mathcal{A}_i(\underbrace{x, \ldots x}_{k}, a_\lambda, \ldots, a_\lambda, \underbrace{\lambda y, \ldots, \lambda y}_{j}), \qquad (50)$$

for all $x, y \in S$. A direct computation, using the expression of f given in (43), the conditions (i) and (iii), shows that we have

$$0 = I(x, y)$$

$$= \sum_{\lambda \in \Phi} f(x + \lambda y + a_\lambda) - Nf(x) - Nf(y)$$

$$= \sum_{\lambda \in \Phi} \sum_{i=1}^{N} \mathcal{A}_i^*(x + \lambda y + a_\lambda) - N \sum_{i=1}^{N} \mathcal{A}_i^*(x) - \sum_{\lambda \in \Phi} \sum_{i=1}^{N} \mathcal{A}_i^*(\lambda y + a_\lambda)$$

$$= \sum_{k=1}^{N-1} \sum_{j=0}^{N-k} \sum_{i=max(k+j,k+1)} \binom{i}{k}\binom{i-k}{j} \sum_{\lambda \in \Phi} \mathcal{A}_i(\underbrace{x, \ldots x}_{k}, a_\lambda, \ldots, a_\lambda, \underbrace{\lambda y, \ldots, \lambda y}_{j})$$

$$= \sum_{j=0}^{N-k} \sum_{k=1}^{N-1} \mathcal{Q}_{(k,j)}(x, y)$$

$$= \sum_{j=0}^{N-k} J_k(x, y)$$

for all $x, y \in S$.

Lemma 1 (ii) applied, respectively, to the functions J_k and $\mathcal{Q}_{(k,j)}$ gives the sought result.

By using Theorem 3 we get the following corollaries. First if we take $\{a_\lambda, \lambda \in \Phi\} = \{0\}$ in (8), we obtain the following result which was proved by Łukasik ([9], Theorem 4).

Corollary 5 (Łukasik [9]) *Let $(S, +)$ be an abelian monoid, let Φ be a finite subgroup of the group of automorphisms of S, $N = card(\Phi)$ and let H be an abelian group uniquely divisible by $(N + 1)!$. Then the general solution $f : S \to H$ of the functional equation*

$$\sum_{\lambda \in \Phi} f(x + \lambda y) = Nf(x) + Nf(y), \ x, y \in S, \tag{51}$$

is

$$f(x) = \sum_{i=1}^{N} \mathcal{A}_i^*(x), \ x \in S, \tag{52}$$

where $\mathcal{A}_k : S^k \to H$, $k \in \{1, 2, \ldots, N\}$ are arbitrary k-additive and symmetric functions which satisfy the following conditions

$$\sum_{\lambda \in \Phi} \mathcal{A}_i(\underbrace{x, \ldots x, \lambda y, \ldots, \lambda y}_{j}) = 0, \ x, y \in S, \ 1 \le j \le i - 1, \ 2 \le i \le N \tag{53}$$

and

$$\mathcal{A}_k^*(\mu x) = \mathcal{A}_k^*(x), \ x \in S, \ \mu \in \Phi, \ 1 \le k \le N. \tag{54}$$

Proof With the notations of previous theorem, if $\{a_\lambda, \ \lambda \in \Phi\} = \{0\}$, then $i = j + k$, $j \ge 1$ and $k \ge 1$. For all $1 \le k \le N - 1$ by Theorem 3 we get that

$$N\mathcal{A}_k^*(x) = \sum_{i=k}^{N} \binom{i}{k} \sum_{\lambda \in \Phi} \mathcal{A}_i^*(\lambda x)$$

$$= \sum_{i=k}^{N} \binom{i}{k} \sum_{\nu \in \Phi} \mathcal{A}_i^*(\nu \mu x), \ \mu \in \Phi$$

$$= N\mathcal{A}_k^*(\mu x), \ x \in S, \ \mu \in \Phi.$$

On the other hand, for all $1 \le k \le N-1, \ 1 \le j \le N-k$, we have

$$0 = \sum_{2 \le i = max(k+j,k+1) \le N} \binom{i}{k}\binom{i-k}{j} \sum_{\lambda \in \Phi} \mathcal{A}_i(\underbrace{x, \ldots x}_{k}, \underbrace{a_\lambda, \ldots, a_\lambda}_{j}, \underbrace{\lambda y, \ldots, \lambda y}_{})$$

$$= \sum_{i=2}^{N} \binom{i}{k} \sum_{\lambda \in \Phi} \mathcal{A}_{i=k+j}(\underbrace{x, \ldots x}_{k}, \underbrace{\lambda y, \ldots, \lambda y}_{j}),$$

$$= \sum_{k=1}^{N-j} \sum_{j=1}^{N-1} \binom{i}{j} \sum_{\lambda \in \Phi} \mathcal{A}_{k+j}(\underbrace{x, \ldots x}_{k}, \underbrace{\lambda y, \ldots, \lambda y}_{j})$$

$$= \sum_{k=1}^{N-1} \sum_{j=1}^{N-k} \binom{i}{j} \sum_{\lambda \in \Phi} \mathcal{A}_{k+j}(\underbrace{x, \ldots x}_{k}, \underbrace{\lambda y, \ldots, \lambda y}_{j}), \ x, y \in S.$$

For $1 \le j \le N-1, \ 1 \le k \le N-1$ such that $k+j \le N$ we define the mappings $g_j, \ h_{(k,j)} : S \times S \to H$ by

$$g_j(x, y) = \sum_{k=1}^{N-j} \binom{i}{j} \sum_{\lambda \in \Phi} \mathcal{A}_{k+j}(\underbrace{x, \ldots x}_{k}, \underbrace{\lambda y, \ldots, \lambda y}_{j}),$$

$$h_{(k,j)}(x, y) = \binom{i}{j} \sum_{\lambda \in \Phi} \mathcal{A}_{k+j}(\underbrace{x, \ldots x}_{k}, \underbrace{\lambda y, \ldots, \lambda y}_{j}).$$

First, we have the following equalities

$$g_j(x, ny) = n^j g_j(x, y), \ x, y \in S, \ n \in \mathbb{N}^*, \ 1 \le j \le N-1,$$

$$h_{(k,j)}(nx, y) = n^k h_{(k,j)}(x, y), \ x, y \in S, \ n \in \mathbb{N}^*, \ 1 \le k \le N-j, \ 1 \le j \le N-1$$

and

$$\sum_{k=1}^{N-j} h_{(k,j)}(x, y) = g_j(x, y), \ x, y \in S, \ n \in \mathbb{N}^*, \ 1 \leq j \leq N - 1.$$

Now, from the above equalities we get

$$\sum_{j=1}^{N-1} n^j g_j(x, y) = \sum_{j=1}^{N-1} g_j(x, ny) = 0, \ x, y \in S, \ n \in \mathbb{N}^*$$

and then by Lemma 1,

$$g_j(x, y) = 0, \ x, y \in S, \ 1 \leq j \leq N - 1.$$

So, we obtain

$$\sum_{k=1}^{N-j} n^k h_{(k,j)}(x, y) = \sum_{k=1}^{N-j} h_{(k,j)}(nx, y) = g_j(nx, y) = 0, \ x, y \in S, \ n \in \mathbb{N}^*,$$

$$1 \leq j \leq N - 1$$

which gives in view of Lemma 1,

$$h_{(k,j)}(x, y) = 0, \ x, y \in S, \ 1 \leq j \leq N - 1, \ 1 \leq k \leq N - j.$$

This ends the proof.

The following corollary was proved by Sinopoulos ([12], Theorem 3) on a semi-group.

Corollary 6 (Sinopoulos [12]) *Let $(S, +)$ be an abelian monoid, and let H be an abelian group, uniquely divisible by 2. Suppose that σ is an endomorphism of S such that $\sigma(\sigma(x)) = x$ for $x \in S$. Then, the general solution $f : S \to H$ of the functional equation*

$$f(x + y) + f(x + \sigma(y)) = 2f(x) + 2f(y), \ x, y \in S, \tag{55}$$

is

$$f(x) = \mathcal{A}_1(x) + \mathcal{A}_2^*(x) \tag{56}$$

where $\mathcal{A}_1 : S \to H$ is an arbitrary additive function with $\mathcal{A}_1(\sigma(x)) = \mathcal{A}_1(x)$ for all $x \in S$, and $\mathcal{A}_2 : S \times S \to G$ is an arbitrary symmetric biadditive function with $\mathcal{A}_2(\sigma(x), y) = -\mathcal{A}_2(x, y)$ for all $x, y \in S$.

In the following two corollaries we solve other special cases of Eq. (8) that are, according to our knowledge, not in the literature.

Corollary 7 *Let $(S, +)$ be an abelian group, $a, b \in S$ and let H be an abelian group, uniquely divisible by 2. Then, the general solution $f : S \to H$ of the functional equation*

$$f(x + y + a) + f(x - y + b) = 2f(x) + 2f(y), \quad x, y \in S, \tag{57}$$

is

$$f(x) = \mathcal{A}_2^*(a) + 2\mathcal{A}_2(x, a) + \mathcal{A}_2^*(x), \; \text{for } x \in S, \tag{58}$$

where $\mathcal{A}_2 : S \times S \to G$ is an arbitrary symmetric biadditive function with $\mathcal{A}_2(x, a + b) = 0$ for all $x \in S$.

Corollary 8 *Let $(S, +)$ be an abelian monoid, $a, b \in S$ and let H be an abelian group, uniquely divisible by 2. Suppose that σ is an endomorphism of S such that $\sigma(\sigma(x)) = x$ for $x \in S$. Then, the general solution $f : S \to H$ of the functional equation*

$$f(x + y + a) + f(x + \sigma(y) + b) = 2f(x) + 2f(y), \quad x, y \in S, \tag{59}$$

is

$$f(x) = \frac{1}{2}\mathcal{A}_1(a + b) + \mathcal{A}_2^*(a) + \mathcal{A}_1(x) + \mathcal{A}_2^*(x) \tag{60}$$

where $\mathcal{A}_2 : S \times S \to G$ is an arbitrary symmetric bi-additive function with

$$\mathcal{A}_2(x, a + b) = 0, \quad \mathcal{A}_2(x, y + \sigma(y)) = 0$$

for all $x \in S$ and $\mathcal{A}_1 : S \to H$ is an arbitrary additive function with

$$\mathcal{A}_1(x) = \mathcal{A}_1(\sigma(x)) + 4\mathcal{A}_2(x, a)$$

for all $x \in S$.

The following corollary is a particular case of Corollary 7.

Corollary 9 *Let $(S, +)$ be a commutative group, $a \in S$ and let H be an abelian group, uniquely divisible by 2. Then, the general solution $f : S \to H$ of the functional equation*

$$f(x + y - a) + f(x - y + a) = 2f(x) + 2f(y), \quad x, y \in S, \tag{61}$$

is

$$f(x) = \mathcal{A}_2^*(a) - 2\mathcal{A}_2^*(x, a) + \mathcal{A}_2^*(x), \; \text{for } x \in S, \tag{62}$$

where $\mathcal{A}_2 : S \times S \to H$ is an arbitrary symmetric bi-additive function.

Corollary 10 *Let $(S, +)$ be a commutative group, let H be an abelian group and $a \in E$. Then, the general solution $f : S \to E$ of the functional equation*

$$f(x + y + a) = f(x) + f(y), \quad x, y \in S, \tag{63}$$

is

$$f(x) = \mathcal{A}(a) + \mathcal{A}(x), \quad x \in S, \tag{64}$$

where $\mathcal{A} : S \to H$ is an arbitrary additive function.

Corollary 11 (Stetkær [14]) *Let w be a primitive Nth root of unity, where $N \geq 2$. The continuous solution $f : \mathbb{C} \to \mathbb{C}$ of the functional equation*

$$\sum_{n=0}^{N-1} f(x + w^n y) = Nf(x) + N(f(y), \quad x, y \in \mathbb{C}, \tag{65}$$

are the functions of the form

$$f(z) = \alpha z^N + \beta \bar{z}^N + \gamma |z|^2, \quad z \in \mathbb{C}, \tag{66}$$

where α, β, γ range over $\mathbb{C}$.

References

1. A. Chahbi, A. Charifi, B. Bouikhalene, S. Kabbaj, Operatorial approach to the non-Archimedean stability of a Pexider K-quadratic functional equation. Arab J. Math. Sci. **21**(1), 67–83 (2015)
2. A. Charifi, M. Almahalebi, S. Kabbaj, A generalization of Drygas functional equation on groups. Proyecciones J. Math. **35**(2), 159–176 (2016)
3. J.K. Chung, B.R. Ebanks, C.T. Ng, P.K. Sahoo, On a quadratic trigonometric functional equation and some applications. Trans. Am. Math. Soc. **347**, 1131–1161 (1995)
4. S. Czerwik, On the stability of the quadratic mapping in normed spaces. Abh. Math. Sem. Univ. Hamburg **62**, 59–64 (1992)
5. D.Ž. Djoković, A representation theorem for $(X_{1-1})(X_{2-1})\cdots(X_{n-1})$ and its applications. Ann. Polon. Math. **22**, 189–198 (1969)
6. B. R. Ebanks, P.L. Kannappan, P.K. Sahoo, A common generalization of functional equations characterizing normed and quasi-inner-product spaces. Can. Math. Bull. **35**, 321–327 (1992)
7. S.-M. Jung, Quadratic functional equations of Pexider type. J. Math. Math. Sci. **24**(5), 351–359 (2000)
8. P. Kannappan, Quadratic functional equation and inner product spaces. Results Math. **27**(3–4), 368–372(1995)
9. R. Łukasik, Some generalization of Cauchys and the quadratic functional equations. Aequat. Math. **83**, 75–86 (2012)
10. S. Mazur, W. Orlicz, Grundlegende Eigenschaften der Polynomischen Operationen. Stud. Math. **5**, 50–68 (1934)

11. Th.M. Rassias, *Inner Product Spaces and Applications*. Pitman Research Notes in Mathematics Series (Addison Wesley Longman, Reading, 1997)
12. P. Sinopoulos, Functional equations on semigroups. Aequat. Math. **59**(3), 255–261 (2000)
13. H. Stetkær, Functional equations on abelian groups with involution. Aequat. Math. **54**, 144–172 (1997)
14. H. Stetkær, Functional equations involving means of functions on the complex plane. Aequat. Math. **55**, 47–62 (1998)
15. H. Stetkær, Functional equations and matrix-valued spherical functions. Aequat. Math. **69**, 271–292 (2005)

Part III
Analytic Inequalities

On Some Functional Equations with Applications in Networks

El-Sayed El-Hady

Abstract Functional equations appear in many applications. They provide a powerful tool for narrowing the models used to describe many phenomena. In particular, some class of functional equations arises recently from many applications, e.g. networks and communication. In this chapter on the one hand, we present some functional equations of the same class of interest. On the other hand, we use boundary value problem theory to investigate the solution of a special functional equation: an equation arising from some queueing model.

1 Introduction

Functional equations (FEs) are well known as the equations where the unknowns are functions not variables [4, 12, 27]. They are relatively old subject of mathematics, but thanks to the pioneer mathematician Aczél [4] their theory has prospered. FEs arise in models of various fields, such as queueing models, see, e.g., [8, 15, 16], digital filtering [25], economics [19], population ethics [6], neural networks [21], decision theory [1], and in inventory control of database systems [14].

In particular, the following general class of two-variable functional equations

$$C_1(x, y) f(x, y) = C_2(x, y) f(x, 0) + C_3(x, y) f(0, y) + C_4(x, y) f(0, 0)$$
$$+ C_5(x, y), \qquad (1)$$

where $C_i(x, y), i = 1, \ldots, 5$ are given polynomials in two complex variables x, y, arises from different communication and network systems. The unknown functions $f(x, y)$, $f(x, 0)$, $f(0, y)$ are defined as follows:

E.-S. El-Hady (✉)
Mathematics Department, College of Science, Jouf University, Sakaka, Kingdom of Saudi Arabia

Basic Science Department, Faculty of Computers and Informatics, Suez Canal University, Ismailia, Egypt
e-mail: elsayed_elhady@ci.suez.edu.eg

© Springer Nature Switzerland AG 2019
G. A. Anastassiou and J. M. Rassias (eds.), *Frontiers in Functional Equations and Analytic Inequalities*, https://doi.org/10.1007/978-3-030-28950-8_17

$$f(x, y) = \sum_{m,n=0}^{\infty} p_{m,n} x^m y^n, \ |x| \leq 1, |y| \leq 1$$

$$f(x, 0) = \sum_{m=0}^{\infty} p_{m,0} x^m, \ |x| \leq 1,$$

and

$$f(0, y) = \sum_{n=0}^{\infty} p_{0,n} y^n, \ |y| \leq 1$$

for some sequences of interest $p_{m,n}$, $p_{m,0}$, $p_{0,n}$, respectively.

Various special cases of (1) appear in the literature, see, e.g., [3, 20, 24], such equations have been solved using the theory of boundary value problems [13]. Some special case of (1) has appeared in [2] in the context of analyzing a multi-programmed computer. The technique used there is that of analytic continuation. Another special case of (1) arises from a network gateway queueing model [23]. A similar equation arises from two parallel queues created by arrivals with two demands, see [18]. One more arises in [28] from two parallel processors with coupled inputs. A survey paper [7] gives an array of similar functional equations and techniques for their solution.

This chapter is mainly concerned with a solution of a two-variable FE arising from a double queue model originally published in [22] using boundary value problem. As a motivation, we first recall some functional equations together with their applications and show the solutions as introduced in the original articles. The chapter is organized as follows: In Sects. 2–6 we present some functional equations coming from different applications, in Sect. 7 we recall the functional equation of interest from the original article [22], in Sect. 8 we investigate the solution of the equation by reduction to Riemann–Hilbert boundary value problem, and finally in Sect. 9 we conclude our work.

2   Equation Arising from Symmetric Two-Node Aloha Network

This equation arises [26] from a packet radio model having two symmetric, interfering queues, as illustrated in Fig. 1. The PGF $f(x, y)$ of the two-dimensional distribution characterizing the system yields the two-place FE

$$f(x, y) = g(x, y) p \frac{(x(y - 1) - p(2xy - x - y))}{xy - g(x, y)((x + y) p \tilde{p} + xy(p^2 + \tilde{p}^2))} f(x, 0)$$

Fig. 1 Symmetric two-node
aloha network

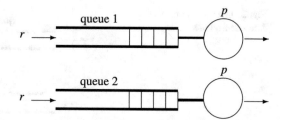

$$+ g(x, y)p\frac{(y(x-1) - p(2xy - x - y))}{xy - g(x, y)((x+y)p\tilde{p} + xy(p^2 + \tilde{p}^2))}f(0, y)$$

$$+ g(x, y)p\frac{p(2xy - x - y)}{xy - g(x, y)((x+y)p\tilde{p} + xy(p^2 + \tilde{p}^2))}f(0, 0), \qquad (2)$$

where

$$g(x, y) = (xr + \tilde{r})(yr + \tilde{r}), \text{ and } \tilde{r} = 1 - r.$$

Up to now there is no solution available to Eq. (2). With some manipulation this FE
can be rewritten in the general form (1) as

$$C_1(x, y)f(x, y) = C_2(x, y)f(x, 0) + C_3(x, y)f(0, y) + C_4(x, y)f(0, 0), \qquad (3)$$

where

$$C_1(x, y) = xy - (xr + \tilde{r})(yr + \tilde{r})((x+y)p\tilde{p} + xy(p^2 + \tilde{p}^2)),$$

$$C_2(x, y) = p(xr + \tilde{r})(yr + \tilde{r})(x(y-1) - p(2xy - x - y)),$$

$$C_3(x, y) = p(xr + \tilde{r})(yr + \tilde{r})(y(x-1) - p(2xy - x - y)),$$

and

$$C_4(x, y) = p^2(xr + \tilde{r})(yr + \tilde{r})(2xy - x - y).$$

3 Equation Arising from Inventory Control of Database Systems

This equation arises [18] from a double queue model, illustrated in Fig. 2. The
resultant two-place FE takes the form

$$Q(x, y)f(x, y) = \beta x(y - 1)f(x, 0) + \alpha y(x - 1)f(0, y), \qquad (4)$$

Fig. 2 Inventory control of database systems

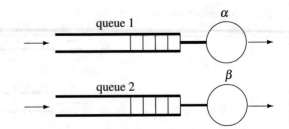

where

$$Q(x, y) = (1 + \alpha + \beta)xy - \alpha y - \beta x - x^2 y^2.$$

This equation is solved first by parameterizing the kernel given by

$$\{(x, y) : Q(x, y) = 0\}$$

by a pair of elliptic functions $x = x(t)$, $y = y(t)$. The functional equation for $f(x, y)$ is converted using the analytic continuation into a set of conditions on $f(x(t), 0)$, $f(0, y(t))$, which in turn lead to the determination of $f(x, y)$ of the form

$$f(x, 0) = \frac{\beta - 1}{\beta} \frac{\Phi(x)}{\Phi(1)}, \qquad |x| \le 1,$$

where

$$\Phi(x) = \frac{\sqrt{a_3 - x} + \sqrt{a_3 - 1}}{[\sqrt{a_3 - x} + \sqrt{a_3 - \alpha/\beta}] \cdot [\sqrt{a_3 - x} - \sqrt{a_3 - \alpha}]},$$

$$f(0, y) = \frac{\alpha - 1}{\alpha} \frac{\Psi(y)}{\Psi(1)}, \qquad |y| \le 1,$$

where

$$\Psi(y) = \frac{\sqrt{b_3 - y} + \sqrt{b_3 - 1}}{[\sqrt{b_3 - y} + \sqrt{b_3 - \beta/\alpha}] \cdot [\sqrt{b_3 - y} + \sqrt{b_3 - \beta}]},$$

for some positive constants a_3, b_3. This FE can be rewritten in the general form (1) as

$$C_1(x, y) f(x, y) = C_2(x, y) f(x, 0) + C_3(x, y) f(0, y), \tag{5}$$

where

$$C_1(x, y) = (1 + \alpha + \beta)xy - \alpha y - \beta x - x^2 y^2,$$

$$C_2(x, y) = \beta x(y - 1),$$

and

$$C_3(x, y) = \alpha y(x - 1).$$

4 Equation Arising from Two Parallel Queues with Batch Server

This equation arises [17] from a model consisting of two parallel $M/M/1$ queues with infinite capacities, illustrated in Fig. 3. It is assumed that the arrivals form two independent Poisson processes with parameters λ_1, λ_2, and that the service times are distributed exponentially with instantaneous service rates S_1 and S_2 depending on the system state in the following manner:

- $S_1 = \mu_1$, $S_2 = \mu_2$ if both queues are nonempty;
- $S_1 = \mu_1^*$, if queue 2 is empty;
- $S_2 = \mu_2^*$, if queue 1 is empty.

The resultant two-place FE is given by

$$T(x, y)f(x, y) = a(x, y)f(x, 0) + b(x, y)f(0, y) + c(x, y)f(0, 0), \qquad (6)$$

where

$$a(x, y) = \mu_1(1 - \frac{1}{x}) + q(1 - \frac{1}{y}),$$

$$b(x, y) = \mu_2(1 - \frac{1}{y}) + p(1 - \frac{1}{x}),$$

$$c(x, y) = p(\frac{1}{x} - 1) + q(\frac{1}{y} - 1),$$

$$T(x, y) = \lambda_1(1 - x) + \mu_1(1 - \frac{1}{x}) + \lambda_2(1 - y) + \mu_2(1 - \frac{1}{y}),$$

$$p = \mu_1 - \mu_1^*, \ q = \mu_2 - \mu_2^*.$$

Fig. 3 Two parallel queues with batch server

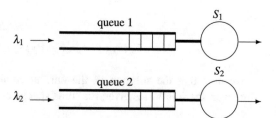

Solution of (6) has been obtained by solving a Dirichlet problem for a circle yielding, for $x = \sqrt{\frac{\mu_1}{\lambda_1}}e^{i\rho}$ (writing x in a polar form with $\sqrt{\frac{\mu_1}{\lambda_1}}$ as its magnitude and ρ as its argument)

$$f(x, 0) = f(\sqrt{\frac{\mu_1}{\lambda_1}}e^{i\rho}, 0)$$

$$= \frac{1}{\pi}\int_0^\pi \frac{z\sin\rho u(\rho)}{1 + z^2 - 2z\cos\rho}d\rho + f(0, 0), \quad |z| < 1,$$

where

$$u(\rho) = \frac{1}{1 - \xi}\Im\frac{\mu_2^*(1 - \frac{1}{h(x)})f(0, 0)}{\mu_1^*(1 - \frac{1}{x}) - \mu_2(1 - \frac{1}{h(x)})}, \quad 0 < \xi < 1$$

where $h(x)$ is one root of the kernel. A similar formula for $f(0, y)$ is given by

$$f(0, y) = f(0, \sqrt{\frac{\mu_2}{\lambda_2}}e^{i\rho})$$

$$= \frac{1}{\pi}\int_0^\pi \frac{z\sin\theta v(\theta)d\theta}{1 + z^2 - 2z\cos\theta} + f(0, 0).$$

The unknown $f(0, 0)$ can be found by using the normalization condition. For the case that $pq \neq \mu_1\mu_2$, the authors determine $f(x, 0)$, $f(0, y)$ by solving a homogenous Riemann–Hilbert problem for a circle. Equation (6) can be rewritten in the general form (1) as

$$C_1(x, y)f(x, y) = C_2(x, y)f(x, 0) + C_3(x, y)f(0, y) + C_4(x, y)f(0, 0), \quad (7)$$

where

$$C_1(x, y) = \lambda_1(xy - x^2y) + \mu_1(xy - y) + \lambda_2(xy - xy^2) + \mu_2(xy - x),$$
$$C_2(x, y) = \mu_1(xy - y) + q(xy - x),$$
$$C_3(x, y) = \mu_2(xy - x) + p(xy - y),$$

and

$$C_4(x, y) = p(y - xy) + q(x - xy).$$

We note that the solution of the current equation is not an explicit solution. Moreover, the solution is given for some special cases.

5 Equation Arising from Two Types of Customers with One Server

This equation arises [9] from a one-server-two-queue model with two types of customers, as illustrated in Fig. 4. The PGF $f(x, y)$ of the two-dimensional distribution characterizing the system yields the two-place FE

$$(1 - (\frac{1}{x} + \frac{1}{y})\epsilon_0\alpha(\chi) - \frac{\epsilon_1}{xy}\alpha^2(\chi))xyf(x, y)$$

$$= (\frac{1}{x}\beta(\chi) - (\frac{1}{x} + \frac{1}{y})\epsilon_0\alpha(\chi) - \frac{\epsilon_1}{xy}\alpha^2(\chi))xyf(x, 0)$$

$$+ (\frac{1}{y}\beta(\chi) - (\frac{1}{x} + \frac{1}{y})\epsilon_0\alpha(\chi) - \frac{\epsilon_1}{xy}\alpha^2(\chi))xyf(0, y)$$

$$+ (\gamma(\chi) - (\frac{1}{x} + \frac{1}{y})\beta(\chi) + (\frac{1}{x} + \frac{1}{y})\epsilon_0\alpha(\chi) + \frac{\epsilon_1}{xy}\alpha^2(\chi))xyf(0, 0), \quad (8)$$

where it is assumed that $\chi(x, y) \triangleq \lambda(1 - \frac{x+y}{2})$. On the other hand, $\alpha = \int_0^\infty t\,dA(t) > 0$, $\beta = \int_0^\infty t\,dB(t) > 0$, and $\gamma = \int_0^\infty t\,dC(t) > 0$ are the first moments, with A, B, and C being the probability distributions. Equation (8) has been solved by reducing it to a Riemann boundary value problem, yielding solutions of the form

$$f(x(z), 0)/f(0, 0) = e^{\Gamma_1(z)}(\Psi(z) + a_1 z + a_0), \quad |z| < 1,$$

$$f(0, y(z))/f(0, 0) = z^{-1}e^{\Gamma_1(z)}(\Psi(z) + a_1 z + a_0), \quad |z| > 1,$$

where for $|z| = 1$

$$\Gamma_1(z) := \frac{1}{2\pi i}\int_{|\zeta|=1}\log \zeta^{-1}G(\zeta)\frac{d\zeta}{\zeta - z},$$

$$\Psi(z) := \frac{1}{2\pi i}\int_{|\zeta|=1}g(\zeta)e^{-\Gamma_1^+(\zeta)}\frac{d\zeta}{\zeta - z},$$

$$G(z) := -\frac{x(z)}{y(z)} \cdot \frac{y(z) - \beta\{\lambda(1 - \frac{x(z)+y(z)}{2})\}}{x(z) - \beta\{\lambda(1 - \frac{x(z)+y(z)}{2})\}},$$

$$g(z) := G(z) - \frac{\gamma\{\lambda\{(1 - \frac{x(z)+y(z)}{2})\} - 1\}x(z)}{\beta\{\lambda(1 - \frac{x(z)+y(z)}{2})\} - x(z)} + 1,$$

Fig. 4 Two types of customers with one server

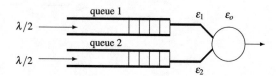

with a_1 and a_0 constants. The unknown $f(0, 0)$ has been obtained by applying the normalization condition $f(1, 1) = 1$. This FE can be rewritten in the general form (1) as

$$C_1(x, y)f(x, y) = C_2(x, y)f(x, 0) + C_3(x, y)f(0, y) + C_4(x, y)f(0, 0), \quad (9)$$

where

$$C_1(x, y) = xy - (y + x)\epsilon_0\alpha(\chi) - \epsilon_1\alpha^2(\chi),$$

$$C_2(x, y) = y\beta(\chi) - (y + x)\epsilon_0\alpha(\chi) - \epsilon_1\alpha^2(\chi),$$

$$C_3(x, y) = x\beta(\chi) - (y + x)\epsilon_0\alpha(\chi) - \epsilon_1\alpha^2(\chi),$$

and

$$C_4(x, y) = \gamma(\chi)xy - (y + x)\beta(\chi) + (y + x)\epsilon_0\alpha(\chi) + \epsilon_1\alpha^2(\chi).$$

6 Equation Arising from Asymmetric Clocked Buffered Switch

This equation arises [10] from a model of an asymmetric 2×2 clocked buffered switch, illustrated in Fig. 5. The PGF $f(x, y)$ of the two-dimensional distribution characterizing the system yields the two-place FE

$$(xy - \phi(x, y))f(x, y) = (x - 1)(y - 1)\phi(x, y)[\frac{f(x, 0)}{(x - 1)} + \frac{f(0, y)}{(y - 1)} + f(0, 0)], \quad (10)$$

where

$$\phi(x, y) = [1 - a_1 + a_1(r_{11}x + r_{12}y)][1 - a_2 + a_2(r_{21}x + r_{22}y)],$$

Fig. 5 Asymmetric 2×2 switch

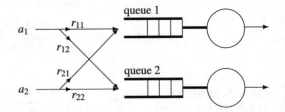

a_1, a_2 are the probabilities that the arrival stream generated at the start of a slot is of stream $1, 2$, respectively, and $r_{i,j}$ is the probability that an i-arrival joins the queue of the j-th service facility, for $i, j = 1, 2$. Equation (10) has been solved by using analytic continuation in locating the zeros and poles of the unknown functions $f(x, 0), f(0, y)$. The unknown functions are given by

$$f(x, 0) = f(1, 0) \frac{P^{(I)}(1) P^{(II)}(1)}{P^{(I)}(x) P^{(II)}(x)} \frac{A^{(I)}(x) A^{(II)}(x)}{A^{(I)}(1) A^{(II)}(1)}, \quad \text{for all } x$$

similarly

$$f(0, y) = f(0, 1) \frac{Q^{(I)}(1) Q^{(II)}(1)}{Q^{(I)}(y) Q^{(II)}(y)} \frac{\Gamma^{(I)}(y) \Gamma^{(II)}(y)}{\Gamma^{(I)}(1) \Gamma^{(II)}(1)}, \quad \text{for all } y$$

where $f(1, 0)$ is given by

$$f(1, 0) = 1 - a_2 r_{22} - a_1 r_{12},$$

and $f(0, 1)$ is given by

$$f(0, 1) = 1 - a_1 r_{11} - a_2 r_{21}.$$

The unknown $f(0, 0)$ is given by

$$f(0, 0) = (1 - a_2 r_{22} - a_1 r_{12}) \frac{Q^{(I)}(1) Q^{(II)}(1)}{\Gamma^{(I)}(1) \Gamma^{(II)}(1)},$$

where the functions $P^{(I)}(.), P^{(II)}(.), Q^{(I)}(.), Q^{(II)}(.), A^{(I)}(.), A^{(II)}(.), \Gamma^{(I)}(.)$, and $\Gamma^{(II)}(.)$ are defined in [10] as infinite products. Equation (10) can be rewritten in the general form (1) as

$$C_1(x, y) f(x, y) = C_2(x, y) f(x, 0) + C_3(x, y) f(0, y) + C_4(x, y) f(0, 0), \quad (11)$$

where

$$C_1(x, y) = xy - [1 - a_1 + a_1(r_{11}x + r_{12}y)][1 - a_2 + a_2(r_{21}x + r_{22}y)],$$

$$C_2(x, y) = (y - 1)[1 - a_1 + a_1(r_{11}x + r_{12}y)][1 - a_2 + a_2(r_{21}x + r_{22}y)],$$

$$C_3(x, y) = (x - 1)[1 - a_1 + a_1(r_{11}x + r_{12}y)][1 - a_2 + a_2(r_{21}x + r_{22}y)],$$

and

$$C_4(x, y) = (x - 1)(y - 1)[1 - a_1 + a_1(r_{11}x + r_{12}y)][1 - a_2 + a_2(r_{21}x + r_{22}y)].$$

We note that the solution of the current equation is not a closed-form one as it is given in terms of an infinite product.

7 The Functional Equation of Interest

In this section, we investigate the analytical solution of a particular case of (1). Such special case is a two-place functional equation arising from a LANE Gateway queueing model depicted in Fig. 6 (see [22]) for more details.

It should be noted that the authors in [23] investigated the analytical solution of such functional equation. A solution has been introduced using only the physical properties of the underlying queueing system. Here, we deal with the equation as a mathematical entity without taking the application into our consideration. Our solution is based on the following assumptions:

1. $\tilde{r}_1 = \tilde{r}_2 = \tilde{r}$,
2. $\tilde{s}_1 = \tilde{s}_2 = \tilde{s}$.
3. $\xi_1 = \xi_2 = 0$.

Using the above assumptions in [22] to get

$$C_1(x, y)f(x, y) = C_2(x, y)f(x, 0) + C_3(x, y)f(0, y) + C_4(x, y)f(0, 0), \qquad (12)$$

where

$$C_1(x, y) = xy - (\tilde{r} + r\tilde{s}y)(\tilde{r} + r_2\tilde{s}x),$$

$$C_2(x, y) = (y - 1)(\tilde{r})(\tilde{r} + r_2\tilde{s}x),$$

$$C_3(x, y) = (x - 1)(\tilde{r})(\tilde{r} + r\tilde{s}y),$$

and

$$C_4(x, y) = (x - 1)(y - 1)(\tilde{r})^2.$$

Fig. 6 Gateway modeled as two back-to-back interfering queues

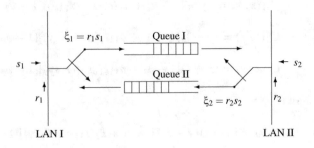

A crucial role in the solution of (12) is played by the kernel, see, e.g., [24], defined by

$$\{(x, y) : C_1(x, y) = 0\}. \tag{13}$$

The main idea in the solution of (12) is based on the analyticity property of the main unknown $f(x, y)$ which means that if $C_1(x, y) = 0$ then

$$C_2(x, y) f(x, 0) + C_3(x, y) f(0, y) + C_4(x, y) f(0, 0) = 0 \tag{14}$$

Now the solution of the main functional equation (12) is reduced to the solution of (14). Equation (14) can be written

$$(y - 1)(\tilde{r})(\tilde{r} + r\tilde{s}x) f(x, 0) + (x - 1)(\tilde{r})(\tilde{r} + r\tilde{s}y) f(0, y)$$
$$+ (x - 1)(y - 1)(\tilde{r})^2 f(0, 0) = 0, \tag{15}$$

which can be divided by $\tilde{r}(x - 1)(y - 1) \neq 0$ to get

$$\frac{(\tilde{r} + r\tilde{s}x)}{(x - 1)} f(x, 0) + \frac{(\tilde{r} + r\tilde{s}y)}{(y - 1)} f(0, y)$$
$$+ \tilde{r} f(0, 0) = 0. \tag{16}$$

Now, it is clear that the solution of (12) is reduced to the solution of (16) on the set defined as

$$G := \{(x, y) : C_1(x, y) = xy - (\tilde{r} + r\tilde{s}y)(\tilde{r} + r_2\tilde{s}x) = 0\}, \tag{17}$$

which is obviously an infinite set of ordered pairs. We can choose a special set from (17). That is the set defined as

$$G_0 := \{(x, \bar{x}) : C_1(x, \bar{x}) = x\bar{x} - (\tilde{r} + r\tilde{s}\bar{x})(\tilde{r} + r_2\tilde{s}x) = 0\} \subset G, \tag{18}$$

where $\bar{x}$ is the complex conjugate of x. The problem now is to solve (16) in the new subset (18). Introducing the function

$$F(x) := \frac{(\tilde{r} + r\tilde{s}x)}{(x - 1)} f(x, 0) + \frac{1}{2} \tilde{r} f(0, 0), \tag{19}$$

in (16) to get

$$F(x) + F(y) = 0, \tag{20}$$

which is satisfied in the subset (18). So that we can rewrite it as

$$F(x) + F(\bar{x}) = 0. \tag{21}$$

The main functional equation is now reduced to the following boundary value problem.

8   Boundary Value Problem

Find a function $F(.)$ satisfying the following conditions:

1. Analytic inside the unit disk
2. Has a single pole at 1
3. $\Re(F) = 0$
4. $\lim_{x \to 1} F(x) = f(1, 0)$

on the set G_0 which is a simply connected domain, see, e.g., [11]. To solve the stated problem we use a conformal mapping which exists by the Riemann conformal mapping theorem. Assume that Φ with inverse Ψ is the conformal mapping of the set G_0 (which is a simply connected domain) to the unit disk with normalization conditions $\Phi(0) = 0$, $\Phi(1) = 1$. This gives rise to the following Riemann–Hilbert boundary value problem on the unit disk: Determine a function $H(.)$ defined as

$$H := F \circ \Phi$$

satisfying the conditions

1. analytic inside the unit disk and continuous on the closure of the disk.
2. $\Re H = 0$
3. $\lim_{w \to 1}(w - 1)H(w) = \frac{f(1,0)}{\Phi_x(1)}$,

where

$$\Phi_x := \frac{d\Phi}{dx}.$$

The problem just stated on the unit disk is a Dirichlet problem with a pole at 1 (see [11] chapter 1). The solution of this problem is given by

$$H(w) = \frac{1}{2} \frac{f(1, 0)}{\Phi_x(1)} \frac{w + 1}{w - 1}, \qquad w \in D,$$

where D is the unit disk. This gives

$$F(x) = H(\Psi(x)) = \frac{1}{2} \frac{f(1, 0)}{\Phi_x(1)} \frac{\Psi(x) + 1}{\Psi(x) - 1},$$

inside the set G_0. Substituting in the original equation to yield

$$f(x, y) = f(1, 0)\Psi_x(1)\frac{(x-1)(y-1)M(x, y)}{(\Psi(x)-1)(\Psi(y)-1)}\frac{\Psi(x)\Psi(y)-1}{xy-M(x, y)}.$$

The obtained expression is a potential solution of the functional equation of interest. Let us also mention that the particular conformal mapping used is explicitly expressed in terms of the Jacobi elliptic function (see, e.g., [5]) for some details.

9 Conclusion

In this chapter, we shed the light on a certain class of interesting functional equations. For such class of equations, no exact form solutions obtained so far. We recall certain special cases of such class of equations with their different applications and solutions. At the end, we investigate the solution of a particular functional equation using the theory of boundary value problems. The solution is given in terms of some conformal mapping. Potential future work could be to use some numerical techniques to investigate the solution of such interesting class of equations.

Funds

This work is funded by Jouf university, Kingdom of Saudi Arabia under the research project number 39/600.

References

1. A. Abbas, J. Aczél, The role of some functional equations in decision analysis. Decis. Anal. **7**, 215–228 (2010)
2. I. Adan, J. Wessels, W. Zijm, Analyzing multiprogramming queues by generating functions. SIAM J. Appl. Math **53**, 1123–1131 (1993)
3. I.-J.-B.-F. Adan, O.-J. Boxma, J.-A. Resing, Queueing models with multiple waiting lines. Queueing Syst. **37**, 65–98 (2001)
4. J. Aczél, *Lectures on Functional Equations and Their Applications* (Academic Press, New York, 1966)
5. J.V. Armitage, W.F. Eberlein, *Elliptic Functions* (Cambridge University Press, Cambridge, 2006), p. 67
6. C. Blackorby, W. Bossert, D. Donaldson, Functional equations and population ethics. Aequat. Math. **58**(3), 272–284 (1999)
7. O. Boxma, G. Koole, Z. Liu, Queueing theoretic solution methods for models of parallel and distributed systems, in *Performance Evaluation of Parallel and Distributed Systems Solution Methods*, ed. by O. Boxma, G. Koole (1994), pp. 1–24
8. J. Brzdęk, El-S. El-hady, W. Förg-Rob, Z. Leśniak, A note on solutions of a functional equation arising in a queuing model for a LAN gateway. Aequat. Math. **90**(4), 671–681 (2016)
9. J.W. Cohen, Boundary value problems in queueing theory. Queueing Syst. **3**, 97–128 (1988)
10. J.W. Cohen, On the asymmetric clocked buffered switch. Queueing Syst. **30**, 985–404 (1998)

11. J.W. Cohen, O.J. Boxma, *Boundary Value Problems in Queueing System Analysis*. Mathematics Studies (Elsevier, Amsterdam, 2000)
12. Z. Daroczy, Z. Pales, *Functional Equations Results and Advances*. Advances in Mathematics, vol. 3 (Springer, Dordecht, 2002)
13. R. Dautray, J.L Lions, *Mathematical Analysis and Numerical Methods for Science and Technology*. Physics Today, vol. 4 (Springer, Berlin, 1991)
14. El-S. El-hady, W. Förg-Rob, M. Mahmoud, On a two-variable functional equation arising from databases. WSEAS Trans. Math. **14**, 265–270 (2015)
15. El-S. El-hady, J. Brzdęk, H. Nassar, On the structure and solutions of functional equations arising from queueing models. Aequat. Math. **91**, 445–477 (2017)
16. El-S. El-hady, J. Brzdęk, W. Förg-Rob, H. Nassar, Remarks on solutions of a functional equation arising from an asymmetric switch, in *Contributions in Mathematics and Engineering* (Springer, Berlin, 2016), pp. 153–163
17. G. Fayoll, R. Iasnogorodski, Two coupled processors: the reduction to a Riemann-Hilbert problem, in *Zeitschrift für Wahrscheinlichkeitstheorie und verwandte Gebiete*, vol. 47 (Springer, Berlin, 1979), pp. 325–351
18. L. Flatto, S. Hahn, Two parallel queues created by arrivals with two demands I. SIAM J. Appl. Math. **44**, 1041–1053 (1984)
19. W. Gehrig, Functional equation methods applied to economic problems: some examples, in *Functional Equations: History, Applications and Theory* (Springer, Berlin, 1985), pp. 33–52
20. F. Guillemin, J.-S.-H. van Leeuwaarden, Rare event asymptotic for a random walk in the quarter plane. Queueing Syst. **67**, 1–32 (2011)
21. L. Kindermann, A. Lewandowski, P. Protzel, A framework for solving functional equations with neural networks. Neural Inf. Process. **2**, 1075–1078 (2001)
22. H. Nassar, Two-dimensional queueing model for a LAN gateway. WSEAS Trans. Commun. **5**, 1585 (2006)
23. H. Nassar, El-S. El-hady, Closed form solution of a LAN gateway queueing model, in *Contributions in Mathematics and Engineering* (Springer, Berlin, 2016), pp. 393–427
24. J. Resing, L. Ormeci, A tandem queueing model with coupled processors. Oper. Res. Lett. **31**, 383–389 (2003)
25. P.-K. Sahoo, L. Székelyhidi, On a functional equation related to digital filtering. Aequat. Math. **62**, 280–285 (2001)
26. M. Sidi, A. Segall, Two interfering queues in packet-radio networks. IEEE Trans. Commun. **31**, 123–129 (1983)
27. C.-G. Small, *Functional Equation and How to Solve Them* (Springer, Waterloo, 2007)
28. P.-E. Wright, Two parallel processors with coupled inputs. Adv. Appl. Probab. **24**, 986–1007 (1992)

Approximate Solutions of an Additive-Quadratic-Quartic (AQQ) Functional Equation

Tianzhou Xu, Yali Ding, and John Michael Rassias

Abstract In this paper, the authors prove some stability and hyperstability results for an (AQQ): additive-quadratic-quartic functional equation of the form

$$f(x + y + z) + f(x + y - z) + f(x - y + z) + f(x - y - z)$$
$$= 2[f(x + y) + f(x - y) + f(y + z) + f(y - z) + f(x + z) + f(x - z)]$$
$$- 4f(x) - 4f(y) - 2[f(z) + f(-z)]$$

by using fixed point theory.

Mathematics Subject Classification Primary 39B82; Secondary 39B52

1 Introduction and Preliminaries

Throughout this paper, $\mathbb{N}$ and $\mathbb{Z}$ stand for the sets of all positive integers and integers, respectively; moreover, $\mathbb{Z}_0 := \mathbb{Z}\backslash\{0\}$. $\mathbb{R}$ and $\mathbb{C}$ stand for the sets of reals and complex numbers, respectively, and $\mathbb{R}_+ := [0, \infty)$.

The study of stability problems for functional equations is related to a question of Ulam [16] concerning the stability of group homomorphisms. A lot of papers on

T. Xu (✉)
School of Mathematics and Statistics, Beijing Institute of Technology, Beijing, People's Republic of China
e-mail: xutianzhou@bit.edu.cn

Y. Ding
School of Arts and Sciences, Shaanxi University of Science and Technology, Xian, People's Republic of China
e-mail: dingyaliding@126.com

J. M. Rassias
National and Kapodistrian University of Athens, Pedagogical Department E. E., Section of Mathematics and Informatics, Athens, Greece
e-mail: jrassias@primedu.uoa.gr; jrass@otenet.gr

© Springer Nature Switzerland AG 2019
G. A. Anastassiou and J. M. Rassias (eds.), *Frontiers in Functional Equations and Analytic Inequalities*, https://doi.org/10.1007/978-3-030-28950-8_18

the stability for various classes of functional equations in various spaces have been published, and there are many interesting results concerning this problem, see, for instance, [1, 2, 5, 10–13, 15, 17] and the references therein. The fixed point method is one of the most effective tools in studying these problems (for more details see, e.g., [4, 6, 9]).

The next two definitions describe roughly the main ideas of such stability and hyperstability notions (A^B denotes the family of all functions mapping a set $B \neq \emptyset$ to a set $A \neq \emptyset$). Definitions 1.1 and 1.2 are actually Definitions 1 and 7 from the survey paper [7].

Definition 1.1 Let $A \neq \emptyset$ be a set, (X, d) be a metric space, $\mathcal{P} \subset \mathcal{C} \subset \mathbb{R}_+^{A^n}$ be nonempty, $\mathcal{T} : \mathcal{C} \to \mathbb{R}_+^A$, and $\mathcal{F}_1, \mathcal{F}_2$ map a nonempty $\mathcal{D} \subset X^A$ into X^{A^n}. The equation

$$\mathcal{F}_1\varphi(x_1, \ldots, x_n) = \mathcal{F}_2\varphi(x_1, \ldots, x_n) \tag{1.1}$$

is stable provided for any $\varepsilon \in \mathcal{P}$ and $\varphi_0 \in \mathcal{D}$ with

$$d(\mathcal{F}_1\varphi_0(x_1, \ldots, x_n), \mathcal{F}_2\varphi_0(x_1, \ldots, x_n)) \leq \varepsilon(x_1, \ldots, x_n), \quad x_1, \ldots, x_n \in A, \tag{1.2}$$

there is a solution $\varphi \in \mathcal{D}$ of (1.1) with

$$d(\varphi(x), \varphi_0(x)) \leq \mathcal{T}\varepsilon(x), \quad x \in A.$$

Definition 1.2 Let A and (X, d) be as before, $\varepsilon \in \mathbb{R}_+^{A^n}$ and $\mathcal{F}_1, \mathcal{F}_2$ map a nonempty $\mathcal{D} \subset X^A$ into X^{A^n}. We say that (1.1) is ε-hyperstable provided every $\varphi_0 \in \mathcal{D}$, satisfying (1.2), fulfills (1.1).

Let $(X, +)$ be a commutative group with the neutral element denoted by 0, Y be a Banach space over a field $\mathbb{F} \in \{\mathbb{R}, \mathbb{C}\}$. Recently, interesting results concerning additive-quadratic-quartic functional equation

$$f(x + y + z) + f(x + y - z) + f(x - y + z) + f(x - y - z)$$
$$= 2[f(x + y) + f(x - y) + f(y + z) + f(y - z) + f(x + z) + f(x - z)]$$
$$- 4f(x) - 4f(y) - 2[f(z) + f(-z)], \quad x, y, z \in X \tag{1.3}$$

have been obtained in [14]. Every solution of the functional equation (1.3) is said to be an additive-quadratic-quartic mapping. Indeed, general solution of the Eq. (1.3) was found in [14]. We say that a function $f : X \to Y$ fulfills the additive-quadratic-quartic functional equation (1.3) on $X_0 := X \setminus \{0\}$ (or is a solution to (1.3) on X_0) provided

$$f(x + y + z) + f(x + y - z) + f(x - y + z) + f(x - y - z)$$
$$= 2[f(x + y) + f(x - y) + f(y + z) + f(y - z) + f(x + z) + f(x - z)]$$
$$- 4f(x) - 4f(y) - 2[f(z) + f(-z)], \quad x, y, z \in X_0. \tag{1.4}$$

In this paper we prove stability and hyperstability results for the Eq. (1.4). The proof of our main result is based on the following fixed point result obtained in [8, Theorem 1] (see also [3, Theorem 2], [4, Theorem 11] and [6, Theorem 19]).

Theorem 1.1 *Let the following three hypotheses be valid:*

(H1) *E is a nonempty set, (V, d) is a complete metric space, $k \in \mathbb{N}$, $f_1, \ldots, f_k :$*
$E \to E$ and $L_1, \ldots, L_k : E \to \mathbb{R}_+$ are given maps;
(H2) *$\mathcal{T} : V^E \to V^E$ is an operator satisfying the inequality*

$$d\left((\mathcal{T}\xi)(x), (\mathcal{T}\mu)(x)\right) \leq \sum_{i=1}^{k} L_i(x) d\left(\xi(f_i(x)), \mu(f_i(x))\right), \qquad \xi, \mu \in V^E, x \in E;$$

(1.5)

(H3) *$\Lambda : \mathbb{R}_+^E \to \mathbb{R}_+^E$ is an operator defined by*

$$(\Lambda\delta)(x) := \sum_{i=1}^{k} L_i(x)\delta(f_i(x)), \qquad \delta \in \mathbb{R}_+^E, x \in E. \qquad (1.6)$$

If functions $\varepsilon : E \to \mathbb{R}_+$ and $\varphi : E \to V$ fulfill the following two conditions :

$$d\left((\mathcal{T}\varphi)(x), \varphi(x)\right) \leq \varepsilon(x), \qquad x \in E, \qquad (1.7)$$

and

$$\varepsilon^*(x) := \sum_{l=0}^{\infty} \left(\Lambda^l \varepsilon\right)(x) < \infty, \qquad x \in E, \qquad (1.8)$$

then there exists a unique fixed point ψ of $\mathcal{T}$ such that

$$d(\varphi(x), \psi(x)) \leq \varepsilon^*(x), \qquad x \in E. \qquad (1.9)$$

Moreover,

$$\psi(x) := \lim_{l\to\infty} \left(\mathcal{T}^l \varphi\right)(x), \qquad x \in E. \qquad (1.10)$$

2   Main Results

Throughout this section, unless otherwise explicitly stated, we will assume that $(X, +)$ be a commutative group, Y be a Banach space over $\mathbb{K} \in \{\mathbb{R}, \mathbb{C}\}$. Given $f : X \to Y$, put

$$(Df)(x, y, z) := f(x + y + z) + f(x + y - z) + f(x - y + z) + f(x - y - z)$$
$$- 2[f(x + y) + f(x - y) + f(y + z) + f(y - z) + f(x + z)$$
$$+ f(x - z)] + 4f(x) + 4f(y) + 2[f(z) + f(-z)]$$

for all $x, y, z \in X$.

Theorem 2.1 *Let* $(X, +)$ *be a commutative group,* $X_0 := X \backslash \{0\}$, Y *be a Banach space over* $\mathbb{K} \in \{\mathbb{R}, \mathbb{C}\}$, *and* $\psi_1, \psi_2, \psi_3 : X_0 \to \mathbb{R}_+$ *be three functions such that*

$$\mathcal{M} := \{m \in \mathbb{Z}_0 : \ \beta_m := 2c(m + 1) + 2c(5m + 1) + 2c(-3m) + c(2m + 1)$$
$$+ 2c(4m + 1) + 5c(3m + 1) + c(6m + 1) + 4c(-2m) + 2c(m) < 1\} \neq \emptyset,$$
$$c_i(m) := \inf\{t \in \mathbb{R}_+ : \ \psi_i(mx) \le t\psi_i(x), \ x \in X\}, \qquad i \in \{1, 2, 3\}, m \in \mathbb{Z}_0,$$
$$c(u) := c_1(u)c_2(u)c_3(u), \qquad u \in \mathbb{Z}_0. \tag{2.1}$$

Suppose that $f : X \to Y$ *satisfies the inequality*

$$\|(Df)(x, y, z)\| \le \psi_1(x)\psi_2(y)\psi_3(z), \qquad x, y, z \in X_0. \tag{2.2}$$

Then there exists a unique solution $F : X_0 \to Y$ *of Eq.* (1.4) *such that*

$$\|f(x) - F(x)\| \le c_0 \psi_1(x)\psi_2(x)\psi_3(x), \qquad x \in X_0, \tag{2.3}$$

where $c_0 := \inf_{m \in \mathcal{M}} \{\frac{c_1(3m+1)c_2(-2m)c_3(-m)}{1 - \beta_m}\}$.

Proof Replacing (x, y, z) by $((3m + 1)x, -2mx, -mx)$ in (2.2) we get

$$\|f(x) + 5f((3m + 1)x) + f((6m + 1)x) - 2f((m + 1)x) - 2f((5m + 1)x)$$
$$- 2f(-3mx) - f((2m + 1)x) - 2f((4m + 1)x) + 4f(-2mx) + 2f(mx)\|$$
$$\le \psi_1((3m + 1)x)\psi_2(-2mx)\psi_3(-mx), \qquad x \in X_0, m \in \mathbb{Z}_0. \tag{2.4}$$

Define

$$\varepsilon_m(x) := \psi_1((3m + 1)x)\psi_2(-2mx)\psi_3(-mx)$$
$$\le c_1(3m + 1)c_2(-2m)c_3(-m)\psi_1(x)\psi_2(x)\psi_3(x), \qquad x \in X_0, m \in \mathbb{Z}_0, \tag{2.5}$$

and

$$(\mathcal{T}_m \xi)(x) := 2\xi((m + 1)x) + 2\xi((5m + 1)x) + 2\xi(-3mx) + \xi((2m + 1)x)$$
$$+ 2\xi((4m + 1)x) - 5\xi((3m + 1)x) - \xi((6m + 1)x)$$
$$- 4\xi(-2mx) - 2\xi(mx), \qquad x \in X_0, \xi \in Y^{X_0}, m \in \mathbb{Z}_0. \tag{2.6}$$

Then, it follows from (2.4) that

$$\|(\mathcal{T}_m f)(x) - f(x)\| \le \varepsilon_m(x), \qquad x \in X_0, m \in \mathbb{Z}_0.$$

We will show that the hypotheses of Theorem 1.1 are fulfilled. For $k = 9$ and $f_1(x) := (m + 1)x$, $f_2(x) := (5m + 1)x$, $f_3(x) := -3mx$, $f_4(x) := (2m + 1)x$, $f_5(x) := (4m + 1)x$, $f_6(x) := (3m + 1)x$, $f_7(x) := (6m + 1)x$, $f_8(x) := -2mx$, $f_9(x) := mx$, $L_1(x) = L_2(x) = L_3(x) = L_5(x) = L_9(x) := 2$, $L_4(x) = L_7(x) := 1$, $L_6(x) := 5$, $L_8(x) := 4$, $x \in X_0$, inequality (1.5) becomes

$$\|(\mathcal{T}_m \xi)(x) - (\mathcal{T}_m \mu)(x)\|$$
$$= \|2\xi((m + 1)x) + 2\xi((5m + 1)x) + 2\xi(-3mx) + \xi((2m + 1)x) + 2\xi((4m + 1)x)$$
$$-5\xi((3m + 1)x) - \xi((6m + 1)x) - 4\xi(-2mx) - 2\xi(mx) - 2\mu((m + 1)x)$$
$$-2\mu((5m + 1)x) - 2\mu(-3mx) - \mu((2m + 1)x) - 2\mu((4m + 1)x)$$
$$+5\mu((3m + 1)x) + \mu((6m + 1)x) + 4\mu(-2mx) + 2\mu(mx)\|$$
$$\le 2\|(\xi - \mu)((m + 1)x)\| + 2\|(\xi - \mu)((5m + 1)x)\| + 2\|(\xi - \mu)(-3mx)\|$$
$$+\|(\xi - \mu)((2m + 1)x)\| + 2\|(\xi - \mu)((4m + 1)x)\| + 5\|(\xi - \mu)((3m + 1)x)\|$$
$$+\|(\xi - \mu)((6m + 1)x)\| + 4\|(\xi - \mu)(-2mx)\| + 2\|(\xi - \mu)(mx)\|,$$
$$x \in X_0, \xi, \mu \in Y^{X_0}, m \in \mathbb{Z}_0,$$

where

$$(\xi - \mu)(u) := \xi(u) - \mu(u), \qquad u \in X_0,$$

so hypothesis (H2) is valid. Next, put $\Lambda_m : \mathbb{R}_+^{X_0} \to \mathbb{R}_+^{X_0}$ for $m \in \mathbb{Z}_0$ by

$$(\Lambda_m \delta)(x) := 2\delta((m + 1)x) + 2\delta((5m + 1)x) + 2\delta(-3mx) + \delta((2m + 1)x)$$
$$+2\delta((4m + 1)x) + 5\delta((3m + 1)x) + \delta((6m + 1)x) + 4\delta(-2mx)$$
$$+2\delta(mx), \quad x \in X_0, \delta \in \mathbb{R}_+^{X_0}.$$

$$(2.7)$$

It is easily seen that Λ_m has the form described in (H3).

Now, using mathematical induction, we will show that for each $x \in X_0$ we have

$$\left(\Lambda_m^l \varepsilon_m\right)(x) \le c_1(3m + 1)c_2(-2m)c_3(-m)\beta_m^n \psi_1(x)\psi_2(x)\psi_3(x), \qquad (2.8)$$

for all $l \in \mathbb{N}_0$ and $m \in \mathcal{M}$. Fix an $m \in \mathcal{M}$, from (2.1) and (2.5), we obtain that the inequality (2.8) holds for $l = 0$. Fix an $x \in X_0$, assume that (2.8) holds for $n = l \in \mathbb{N}_0$. Then we have

$$\left(\Lambda_m^{l+1} \varepsilon_m\right)(x) = \left(\Lambda_m \left(\Lambda_m^l \varepsilon_m\right)\right)(x)$$
$$= 2\left(\Lambda_m^l \varepsilon_m\right)((m + 1)x) + 2\left(\Lambda_m^l \varepsilon_m\right)((5m + 1)x) + 2\left(\Lambda_m^l \varepsilon_m\right)(-3mx)$$
$$+\left(\Lambda_m^l \varepsilon_m\right)((2m + 1)x) + 2\left(\Lambda_m^l \varepsilon_m\right)((4m + 1)x) + 5\left(\Lambda_m^l \varepsilon_m\right)((3m + 1)x)$$
$$+\left(\Lambda_m^l \varepsilon_m\right)((6m + 1)x) + 4\left(\Lambda_m^l \varepsilon_m\right)(-2mx) + 2\left(\Lambda_m^l \varepsilon_m\right)(mx)$$
$$\le c_1(3m + 1)c_2(-2m)c_3(-m)\beta_m^{l+1}\psi_1(x)\psi_2(x)\psi_3(x),$$

and hence (2.8) holds for any $l \in \mathbb{N}_0$, $m \in \mathcal{M}$ and $x \in X_0$. So, we receive the following estimation

$$\varepsilon_m^*(x) = \sum_{n=0}^{\infty} \left(\Lambda_m^n \varepsilon_m \right)(x) \le \frac{c_1(3m+1)c_2(-2m)c_3(-m)\psi_1(x)\psi_2(x)\psi_3(x)}{1-\beta_m}$$

(2.9)

for all $x \in X_0$ and $m \in \mathcal{M}$. By using Theorem 1.1 with $E = X_0$, $V = Y$ and $\varphi = f$, for each $m \in \mathcal{M}$ there exists a function $F_m : X_0 \to Y$ such that

$$
\begin{aligned}
F_m(x) = {} & 2F_m((m+1)x) + 2F_m((5m+1)x) + 2F_m(-3mx) + F_m((2m+1)x) \\
& + 2F_m((4m+1)x) - 5F_m((3m+1)x) - F_m((6m+1)x) - 4F_m(-2mx) \\
& - 2F_m(mx), \quad x \in X_0,
\end{aligned}
$$

(2.10)

and

$$\| f(x) - F_m(x) \| \le \frac{c_1(3m+1)c_2(-2m)c_3(-m)\psi_1(x)\psi_2(x)\psi_3(x)}{1-\beta_m}, \quad x \in X_0.$$

(2.11)

Moreover,

$$F_m(x) = \lim_{n\to\infty} (\mathcal{T}_m^n f)(x), \qquad x \in X_0, m \in \mathcal{M}.$$

(2.12)

Now, we show by mathematical induction that

$$
\begin{aligned}
\| (\mathcal{T}_m^n f)(x+y+z) &+ (\mathcal{T}_m^n f)(x+y-z) + (\mathcal{T}_m^n f)(x-y+z) + (\mathcal{T}_m^n f)(x-y-z) \\
&- 2[(\mathcal{T}_m^n f)(x+y) + (\mathcal{T}_m^n f)(x-y) + (\mathcal{T}_m^n f)(y+z) \\
&+ (\mathcal{T}_m^n f)(y-z) + (\mathcal{T}_m^n f)(x+z) + (\mathcal{T}_m^n f)(x-z)] + 4(\mathcal{T}_m^n f)(x) \\
&+ 4(\mathcal{T}_m^n f)(y) + 2[(\mathcal{T}_m^n f)(z) + (\mathcal{T}_m^n f)(-z)] \| \le \beta_m^n \psi_1(x)\psi_2(y)\psi_3(z)
\end{aligned}
$$

(2.13)

for all $x, y, z \in X_0$, $n \in \mathbb{N}_0$ and $m \in \mathcal{M}$.

Fix $m \in \mathcal{M}$. For $n = 0$, then (2.13) is just (2.2). So, fix $l \in \mathbb{N}_0$ and suppose that (2.13) holds for $n = l$ and every $x, y, z \in X_0$. Then, for every $x, y, z \in X_0$,

$$
\begin{aligned}
\| (\mathcal{T}_m^{l+1} f)(x+y+z) &+ (\mathcal{T}_m^{l+1} f)(x+y-z) + (\mathcal{T}_m^{l+1} f)(x-y+z) \\
&+ (\mathcal{T}_m^{l+1} f)(x-y-z) - 2[(\mathcal{T}_m^{l+1} f)(x+y) + (\mathcal{T}_m^{l+1} f)(x-y) \\
&+ (\mathcal{T}_m^{l+1} f)(y+z) + (\mathcal{T}_m^{l+1} f)(y-z) + (\mathcal{T}_m^{l+1} f)(x+z) + (\mathcal{T}_m^{l+1} f)(x-z)] \\
&+ 4(\mathcal{T}_m^{l+1} f)(x) + 4(\mathcal{T}_m^{l+1} f)(y) + 2[(\mathcal{T}_m^{l+1} f)(z) + (\mathcal{T}_m^{l+1} f)(-z)] \| \\
= \| 2(\mathcal{T}_m^l f)&((m+1)(x+y+z)) + 2(\mathcal{T}_m^l f)((5m+1)(x+y+z)) \\
&+ 2(\mathcal{T}_m^l f)(-3m(x+y+z)) + (\mathcal{T}_m^l f)((2m+1)(x+y+z)) \\
&+ 2(\mathcal{T}_m^l f)((4m+1)(x+y+z)) - 5(\mathcal{T}_m^l f)((3m+1)(x+y+z)) \\
&- (\mathcal{T}_m^l f)((6m+1)(x+y+z)) - 4(\mathcal{T}_m^l f)(-2m(x+y+z)) \\
&- 2(\mathcal{T}_m^l f)(m(x+y+z)) + 2(\mathcal{T}_m^l f)((m+1)(x+y-z)) \\
&+ 2(\mathcal{T}_m^l f)((5m+1)(x+y-z)) + 2(\mathcal{T}_m^l f)(-3m(x+y-z)) \\
&+ (\mathcal{T}_m^l f)((2m+1)(x+y-z)) + 2(\mathcal{T}_m^l f)((4m+1)(x+y-z)) \\
&- 5(\mathcal{T}_m^l f)((3m+1)(x+y-z)) - (\mathcal{T}_m^l f)((6m+1)(x+y-z)) \\
&- 4(\mathcal{T}_m^l f)(-2m(x+y-z)) - 2(\mathcal{T}_m^l f)(m(x+y-z))
\end{aligned}
$$

$$+2(T_m^l f)((m+1)(x-y+z)) + 2(T_m^l f)((5m+1)(x-y+z))$$
$$+2(T_m^l f)(-3m(x-y+z)) + (T_m^l f)((2m+1)(x-y+z))$$
$$+2(T_m^l f)((4m+1)(x-y+z)) - 5(T_m^l f)((3m+1)(x-y+z))$$
$$-(T_m^l f)((6m+1)(x-y+z)) - 4(T_m^l f)(-2m(x-y+z))$$
$$-2(T_m^l f)(m(x-y+z)) + 2(T_m^l f)((m+1)(x-y-z))$$
$$+2(T_m^l f)((5m+1)(x-y-z)) + 2(T_m^l f)(-3m(x-y-z))$$
$$+(T_m^l f)((2m+1)(x-y-z)) + 2(T_m^l f)((4m+1)(x-y-z))$$
$$-5(T_m^l f)((3m+1)(x-y-z)) - (T_m^l f)((6m+1)(x-y-z))$$
$$-4(T_m^l f)(-2m(x-y-z)) - 2(T_m^l f)(m(x-y-z))$$
$$-2[2(T_m^l f)((m+1)(x+y)) + 2(T_m^l f)((5m+1)(x+y))$$
$$+2(T_m^l f)(-3m(x+y)) + (T_m^l f)((2m+1)(x+y))$$
$$+2(T_m^l f)((4m+1)(x+y)) - 5(T_m^l f)((3m+1)(x+y))$$
$$-(T_m^l f)((6m+1)(x+y)) - 4(T_m^l f)(-2m(x+y))$$
$$-2(T_m^l f)(m(x+y)) + 2(T_m^l f)((m+1)(x-y))$$
$$+2(T_m^l f)((5m+1)(x-y)) + 2(T_m^l f)(-3m(x-y))$$
$$+(T_m^l f)((2m+1)(x-y)) + 2(T_m^l f)((4m+1)(x-y))$$
$$-5(T_m^l f)((3m+1)(x-y)) - (T_m^l f)((6m+1)(x-y))$$
$$-4(T_m^l f)(-2m(x-y)) - 2(T_m^l f)(m(x-y))$$
$$+2(T_m^l f)((m+1)(y+z)) + 2(T_m^l f)((5m+1)(y+z))$$
$$+2(T_m^l f)(-3m(y+z)) + (T_m^l f)((2m+1)(y+z))$$
$$+2(T_m^l f)((4m+1)(y+z)) - 5(T_m^l f)((3m+1)(y+z))$$
$$-(T_m^l f)((6m+1)(y+z)) - 4(T_m^l f)(-2m(y+z))$$
$$-2(T_m^l f)(m(y+z)) + 2(T_m^l f)((m+1)(y-z))$$
$$+2(T_m^l f)((5m+1)(y-z)) + 2(T_m^l f)(-3m(y-z))$$
$$+(T_m^l f)((2m+1)(y-z)) + 2(T_m^l f)((4m+1)(y-z))$$
$$-5(T_m^l f)((3m+1)(y-z)) - (T_m^l f)((6m+1)(y-z))$$
$$-4(T_m^l f)(-2m(y-z)) - 2(T_m^l f)(m(y-z))$$
$$+2(T_m^l f)((m+1)(x-z)) + 2(T_m^l f)((5m+1)(x-z))$$
$$+2(T_m^l f)(-3m(x-z)) + (T_m^l f)((2m+1)(x-z))$$
$$+2(T_m^l f)((4m+1)(x-z)) - 5(T_m^l f)((3m+1)(x-z))$$
$$-(T_m^l f)((6m+1)(x-z)) - 4(T_m^l f)(-2m(x-z))$$
$$-2(T_m^l f)(m(x-z)) + 4[2(T_m^l f)((m+1)x)$$
$$+2(T_m^l f)((5m+1)x) + 2(T_m^l f)(-3mx)$$
$$+(T_m^l f)((2m+1)x) + 2(T_m^l f)((4m+1)x)$$
$$-5(T_m^l f)((3m+1)x) - (T_m^l f)((6m+1)x)$$
$$-4(T_m^l f)(-2mx) - 2(T_m^l f)(mx)]$$
$$+4[2(T_m^l f)((m+1)y) + 2(T_m^l f)((5m+1)y)$$
$$+2(T_m^l f)(-3my) + (T_m^l f)((2m+1)y)$$
$$+2(T_m^l f)((4m+1)y) - 5(T_m^l f)((3m+1)y)$$
$$-(T_m^l f)((6m+1)y) - 4(T_m^l f)(-2my)$$
$$-2(T_m^l f)(my)] + 2[2(T_m^l f)((m+1)z)$$

$$+2(\mathcal{T}_m^l f)((5m+1)z) + 2(\mathcal{T}_m^l f)(-3mz)$$
$$+(\mathcal{T}_m^l f)((2m+1)z) + 2(\mathcal{T}_m^l f)((4m+1)z)$$
$$-5(\mathcal{T}_m^l f)((3m+1)z) - (\mathcal{T}_m^l f)((6m+1)z)$$
$$-4(\mathcal{T}_m^l f)(-2mz) - 2(\mathcal{T}_m^l f)(mz)$$
$$+2(\mathcal{T}_m^l f)(-(m+1)z) + 2(\mathcal{T}_m^l f)(-(5m+1)z)$$
$$+2(\mathcal{T}_m^l f)(3mz) + (\mathcal{T}_m^l f)(-(2m+1)z)$$
$$+2(\mathcal{T}_m^l f)(-(4m+1)z) - 5(\mathcal{T}_m^l f)(-(3m+1)z)$$
$$-(\mathcal{T}_m^l f)(-(6m+1)z) - 4(\mathcal{T}_m^l f)(2mz)$$
$$-2(\mathcal{T}_m^l f)(-mz)]\|$$
$$\leq \beta_m^l (2\psi_1((m+1)x)\psi_2((m+1)y)\psi_3((m+1)z)$$
$$+2\psi_1((5m+1)x)\psi_2((5m+1)y)\psi_3((5m+1)z)$$
$$+2\psi_1(-3mx)\psi_2(-3my)\psi_3(-3mz) + \psi_1((2m+1)x)\psi_2((2m+1)y)$$
$$\times\psi_3((2m+1)z) + 2\psi_1((4m+1)x)\psi_2((4m+1)y)\psi_3((4m+1)z)$$
$$+5\psi_1((3m+1)x)\psi_2((3m+1)y)\psi_3((3m+1)z)$$
$$+\psi_1((6m+1)x)\psi_2((6m+1)y)\psi_3((6m+1)z)$$
$$+4\psi_1(-2mx)\psi_2(-2my)\psi_3(-2mz) + 2\psi_1(mx)\psi_2(my)\psi_3(mz))$$
$$\leq \beta_m^{l+1}\psi_1(x)\psi_2(y)\psi_3(z).$$

Thus, by mathematical induction, it results that (2.13) holds for every $x, y, z \in X_0$ and $n \in \mathbb{N}_0$. Letting $n \to \infty$ in (2.13) we get

$$F_m(x+y+z) + F_m(x+y-z) + F_m(x-y+z) + F_m(x-y-z)$$
$$= 2[F_m(x+y) + F_m(x-y) + F_m(y+z) + F_m(y-z) + F_m(x+z)$$
$$+F_m(x-z)] - 4F_m(x) - 4F_m(y) - 2[F_m(z) + F_m(-z)].$$

(2.14)

So, we have proved that, for each $m \in \mathcal{M}$ there exists a function $F_m : X_0 \to Y$ satisfying the Eq. (1.4) for all $x, y, z \in X_0$ and such that

$$\|f(x) - F_m(x)\| \leq \frac{c_1(3m+1)c_2(-2m)c_3(-m)\psi_1(x)\psi_2(x)\psi_3(x)}{1 - \beta_m}, \qquad x \in X_0.$$

(2.15)

Next, we show that $F_m = F_k$ for all $m, k \in \mathcal{M}$. So, fix $m, k \in \mathcal{M}$. Note that F_k satisfies (2.14) with m replaced by k. Thus, replacing (x, y, z) by $((3m+1)x, -2mx, -mx)$ in (2.14), we obtain that $\mathcal{T}_m F_j = F_j$ for $j = m, k$ and

$$\|F_m(x) - F_k(x)\|$$
$$\leq \left(\frac{c_1(3m+1)c_2(-2m)c_3(-m)}{1 - \beta_m} + \frac{c_1(3k+1)c_2(-2k)c_3(-k)}{1 - \beta_k} \right)$$
$$\times \psi_1(x)\psi_2(x)\psi_3(x)$$

$$= \left(c_1(3m+1)c_2(-2m)c_3(-m) + \frac{c_1(3k+1)c_2(-2k)c_3(-k)(1-\beta_m)}{1-\beta_k}\right)$$

$$\times \psi_1(x)\psi_2(x)\psi_3(x)\sum_{n=0}^{\infty}\beta_m^n$$

$$= \mu\psi_1(x)\psi_2(x)\psi_3(x)\sum_{n=0}^{\infty}\beta_m^n, \tag{2.16}$$

where

$$\mu := c_1(3m+1)c_2(-2m)c_3(-m) + \frac{c_1(3k+1)c_2(-2k)c_3(-k)(1-\beta_m)}{1-\beta_k}.$$

By mathematical induction we will show that

$$\|F_m(x) - F_k(x)\| \le \mu\psi_1(x)\psi_2(x)\psi_3(x)\sum_{n=j}^{\infty}\beta_m^n. \tag{2.17}$$

Fix an $x \in X_0$. For $j = 0$ inequality (2.17) is simply (2.16). So, take $l \in \mathbb{N}_0$ and suppose that (2.17) holds for $j = l$. Then, we have

$$\|F_m(x) - F_k(x)\| = \|(\mathcal{T}_m F_m)(x) - (\mathcal{T}_m F_k)(x)\|$$

$$= \|2F_m((m+1)x) + 2F_m((5m+1)x) + 2F_m(-3mx) + F_m((2m+1)x)$$
$$+2F_m((4m+1)x) - 5F_m((3m+1)x) - F_m((6m+1)x)$$
$$-4F_m(-2mx) - 2F_m(mx) - 2F_k((m+1)x) - 2F_k((5m+1)x)$$
$$-2F_k(-3mx) - F_k((2m+1)x) - 2F_k((4m+1)x)$$
$$+5F_k((3m+1)x) + F_k((6m+1)x) + 4F_k(-2mx) + 2F_k(mx)\|$$

$$= 2\|F_m((m+1)x) - F_k((m+1)x)\| + 2\|F_m((5m+1)x) - F_k((5m+1)x)\|$$
$$+2\|F_m(-3mx) - Fk(-3mx)\| + \|F_m((2m+1)x) - F_k((2m+1)x)\|$$
$$+2\|F_m((4m+1)x) - F_k((4m+1)x)\| + 5\|F_m((3m+1)x)$$
$$-F_k((3m+1)x)\| + \|F_m((6m+1)x) - F_k((6m+1)x)\|$$
$$+4\|F_m(-2mx) - F_k(-2mx)\| + 2\|F_m(mx) - F_k(mx)\|$$

$$\le \mu(2\psi_1((m+1)x)\psi_2((m+1)x)\psi_3((m+1)x)$$
$$+2\psi_1((5m+1)x)\psi_2((5m+1)x)\psi_3((5m+1)x)$$
$$+2\psi_1(-3mx)\psi_2(-3mx)\psi_3(-3mx) + \psi_1((2m+1)x)$$
$$\times\psi_2((2m+1)x)\psi_3((2m+1)x) + 2\psi_1((4m+1)x)\psi_2((4m+1)x)$$
$$\times\psi_3((4m+1)x) + 5\psi_1((3m+1)x)\psi_2((3m+1)x)\psi_3((3m+1)x)$$
$$+\psi_1((6m+1)x)\psi_2((6m+1)x)\psi_3((6m+1)x)$$
$$+4\psi_1(-2mx)\psi_2(-2mx)\psi_3(-2mx) + 2\psi_1(mx)\psi_2(mx)\psi_3(mx))\sum_{n=l}^{\infty}\beta_m^n$$

$$\leq \mu(2c(m+1) + 2c(5m+1) + 2c(-3m) + c(2m+1) + 2c(4m+1)$$

$$+5c(3m+1) + c(6m+1) + 4c(-2m) + 2c(m))\psi_1(x)\psi_2(x)\psi_3(x) \sum_{n=l}^{\infty} \beta_m^n$$

$$\leq \mu\psi_1(x)\psi_2(x)\psi_3(x) \sum_{n=l}^{\infty} \beta_m^{n+1} \leq \mu\psi_1(x)\psi_2(x)\psi_3(x) \sum_{n=l+1}^{\infty} \beta_m^n.$$

Hence (2.17) is true for each $j \in \mathbb{N}_0$ and $x \in X_0$. Letting $j \to \infty$ in (2.17) and using the fact that the series $\sum_{n=0}^{\infty} \beta_m^n$ is convergent we obtain that $F_m = F_k =: F$. From (2.14), F is a solution to (1.4), it remains to prove the statement concerning the uniqueness of F. So, assume that $G : X \to Y$ is another function satisfying equation (1.4) and inequality (2.3). Then

$$\|G(x) - F(x)\| \leq 2c_0\psi_1(x)\psi_2(x)\psi_3(x), \qquad x \in X_0. \tag{2.18}$$

Further, $\mathcal{T}_m G = G$ for each $m \in \mathbb{Z}_0$. Next, with a fixed $m \in \mathcal{M}$, we will show by mathematical induction that

$$\|G(x) - F(x)\| \leq 2c_0\beta_m^n\psi_1(x)\psi_2(x)\psi_3(x), \qquad x \in X_0. \tag{2.19}$$

Clearly, if $n = 0$, then (2.19) is simply (2.18). So, fix $l \in \mathbb{N}_0$ and suppose that (2.19) holds for l and $x \in X_0$. Then, for every $x \in X_0$,

$$\|G(x) - F(x)\| = \|(\mathcal{T}_m^{l+1}G)(x) - (\mathcal{T}_m^{l+1}F)(x)\|$$
$$= \|2(\mathcal{T}_m^l G)((m+1)x) + 2(\mathcal{T}_m^l G)((5m+1)x) + 2(\mathcal{T}_m^l G)(-3mx)$$
$$+(\mathcal{T}_m^l G)((2m+1)x) + 2(\mathcal{T}_m^l G)((4m+1)x) - 5(\mathcal{T}_m^l G)((3m+1)x)$$
$$-(\mathcal{T}_m^l G)((6m+1)x) - 4(\mathcal{T}_m^l G)(-2mx) - 2(\mathcal{T}_m^l G)(mx)$$
$$-(2(\mathcal{T}_m^l F)((m+1)x) + 2(\mathcal{T}_m^l F)((5m+1)x) + 2(\mathcal{T}_m^l F)(-3mx)$$
$$+(\mathcal{T}_m^l F)((2m+1)x) + 2(\mathcal{T}_m^l F)((4m+1)x) - 5(\mathcal{T}_m^l F)((3m+1)x)$$
$$-(\mathcal{T}_m^l F)((6m+1)x) - 4(\mathcal{T}_m^l F)(-2mx) - 2(\mathcal{T}_m^l F)(mx))\|$$
$$\leq 2\|(\mathcal{T}_m^l G)((m+1)x) - (\mathcal{T}_m^l F)((m+1)x)\| + 2\|(\mathcal{T}_m^l G)((5m+1)x)$$
$$-(\mathcal{T}_m^l F)((5m+1)x)\| + 2\|(\mathcal{T}_m^l G)(-3mx) - (\mathcal{T}_m^l F)(-3mx)\|$$
$$+\|(\mathcal{T}_m^l G)((2m+1)x) - (\mathcal{T}_m^l F)((2m+1)x)\|$$
$$+2\|(\mathcal{T}_m^l G)((4m+1)x) - (\mathcal{T}_m^l F)((4m+1)x)\| + 5\|(\mathcal{T}_m^l G)((3m+1)x)$$
$$-(\mathcal{T}_m^l F)((3m+1)x)\| + \|(\mathcal{T}_m^l G)((6m+1)x) - (\mathcal{T}_m^l F)((6m+1)x)\|$$
$$+4\|(\mathcal{T}_m^l G)(-2mx) - (\mathcal{T}_m^l F)(-2mx)\|$$
$$+2\|(\mathcal{T}_m^l G)(mx) - (\mathcal{T}_m^l F)(mx)\|$$
$$\leq [2\psi_1((m+1)x)\psi_2((m+1)x)\psi_3((m+1)x) + 2\psi_1((5m+1)x)$$
$$\times\psi_2((5m+1)x)\psi_3((5m+1)x) + 2\psi_1(-3mx)\psi_2(-3mx)\psi_3(-3mx)$$
$$+\psi_1((2m+1)x)\psi_2((2m+1)x)\psi_3((2m+1)x) + 2\psi_1((4m+1)x)$$
$$\times\psi_2((4m+1)x)\psi_3((4m+1)x) + 5\psi_1((3m+1)x)\psi_2((3m+1)x)$$
$$\times\psi_3((3m+1)x) + \psi_1((6m+1)x)\psi_2((6m+1)x)\psi_3((6m+1)x)$$
$$+4\psi_1(-2mx)\psi_2(-2mx)\psi_3(-2mx) + 2\psi_1(mx)\psi_2(mx)\psi_3(mx)]\cdot 2c_0\beta_m^l$$

$$\leq 2c_0\beta_m^l \cdot (2c(m+1) + 2c(5m+1) + 2c(-3m) + c(2m+1)$$
$$+2c(4m+1) + 5c(3m+1) + c(6m+1) + 4c(-2m) + 2c(m))$$
$$\cdot \psi_1(x)\psi_2(x)\psi_3(x)$$
$$= 2c_0\beta_m^{l+1}\psi_1(x)\psi_2(x)\psi_3(x)$$

for all $x \in X_0$, $l \in \mathbb{N}_0$ and $m \in \mathcal{M}$. Thus we have shown (2.19). Now, letting $j \to \infty$ in (2.19), we get $G = F$. This completes the proof. □

The following hyperstability result can be deduced from Theorem 2.1.

Corollary 2.1 *Let $(X, +)$ be a commutative group, Y be a Banach space over $\mathbb{K} \in \{\mathbb{R}, \mathbb{C}\}$. Suppose that $f : X \to Y$ and $c_1, c_2, c_3 : \mathbb{Z}_0 \to \mathbb{R}_+$ satisfy conditions (2.1), (2.2) and*

$$\lim_{k\to\infty} \inf\{c_1(3k+1)c_2(-2k)c_3(-k)\} = 0. \tag{2.20}$$

Then f satisfies (1.4).

Corollary 2.2 *Let X be a normed space and Y be a Banach space over $\mathbb{K} \in \{\mathbb{R}, \mathbb{C}\}$. Suppose that $f : X \to Y$ satisfies the inequality*

$$\|(Df)(x, y, z)\| \leq \alpha \|x\|^p \|y\|^q \|z\|^r, \qquad x, y, z \in X\backslash\{0\}, \tag{2.21}$$

for some $\alpha \geq 0$ and $p, q, r \in \mathbb{R}$ such that $p + q + r < 0$. Then f satisfies (1.4).

Proof Let $\psi_1, \psi_2, \psi_3 : X \to \mathbb{R}_+$ be defined by $\psi_1(x) = \alpha_1\|x\|^p$, $\psi_2(x) = \alpha_2\|x\|^q$ and $\psi_3(x) = \alpha_3\|x\|^r$, where $\alpha_1, \alpha_2, \alpha_3 \in \mathbb{R}_+$ and $p, q, r \in \mathbb{R}$ such that $p + q + r < 0$ and $\alpha = \alpha_1\alpha_2\alpha_3$. For each $k \in \mathbb{N}$,

$$c_1(k) = \inf\{t \in \mathbb{R}_+ : \psi_1(kx) \leq t\psi_1(x), \ x \in X\backslash\{0\}\}$$
$$= \inf\{t \in \mathbb{R}_+ : \alpha_1\|kx\|^p \leq t\alpha_1\|x\|^p, \ x \in X\backslash\{0\}\} = k^p.$$

Also, we have $c_2(k) = k^q$ and $c_3(k) = k^r$ for each $k \in \mathbb{N}$. Then we get

$$\lim_{k\to\infty} (2c(k+1) + 2c(5k+1) + 2c(-3k) + c(2k+1) + 2c(4k+1) + 5c(3k+1)$$
$$+c(6k+1) + 4c(-2k) + 2c(k))$$
$$= \lim_{k\to\infty} (2(k+1)^{p+q+r} + 2(5k+1)^{p+q+r} + 2(3k)^{p+q+r}$$
$$+(2k+1)^{p+q+r} + 2(4k+1)^{p+q+r} + 5(3k+1)^{p+q+r}$$
$$+(6k+1)^{p+q+r} + 4(2k)^{p+q+r} + 2k^{p+q+r})$$
$$= 0.$$

Thus, there is $k_0 \in \mathbb{N}$ such that, for $k \geq k_0$,

$$2c(k+1) + 2c(5k+1) + 2c(-3k) + c(2k+1) + 2c(4k+1) + 5c(3k+1)$$
$$+ c(6k+1) + 4c(-2k) + 2c(k) < 1.$$

Since $p, q, r \in \mathbb{R}$ with $p + q + r < 0$, then

$$
\begin{aligned}
\lim_{k \to \infty} c_1(3k+1)c_2(-2k)c_3(-k) &= \lim_{k \to \infty} k^p(3 + \tfrac{1}{k}) \cdot 2^q k^q \cdot k^r \\
&= \lim_{k \to \infty} k^{p+q+r} \cdot (3 + \tfrac{1}{k}) \cdot 2^q = 0.
\end{aligned}
$$

Thus, the conditions (2.1), (2.2), and (2.20) are fulfilled. Then by Corollary 2.1, we get the desired results.   □

Theorem 2.2 *Let* $(X, +)$ *be a commutative group,* $X_0 := X \backslash \{0\}$, *Y be a Banach space over* $\mathbb{K} \in \{\mathbb{R}, \mathbb{C}\}$ *and* $\psi_1, \psi_2, \psi_3 : X_0 \to \mathbb{R}_+$ *be three functions such that*

$$
\begin{aligned}
\mathcal{M} \quad &:= \{m \in \mathbb{Z}_0 : \ \gamma_m := 2c(m+1) + 2c(5m+1) + 2c(-3m) + c(2m+1) \\
&\quad + 2c(4m+1) + 5c(3m+1) + c(6m+1) + 4c(-2m) + 2c(m) < 1\} \neq \emptyset, \\
c_i(m) &:= \inf\{t \in \mathbb{R}_+ : \ \psi_i(mx) \le t\psi_i(x), \ x \in X\}, \qquad i \in \{1,2,3\}, m \in \mathbb{Z}_0, \\
c(u) \quad &:= \max_{i \in \{1,2,3\}} c_i(u), \qquad u \in \mathbb{Z}_0.
\end{aligned}
$$

$$(2.22)$$

Suppose that $f : X \to Y$ *satisfies the inequality*

$$\|(Df)(x, y, z)\| \le \psi_1(x) + \psi_2(y) + \psi_3(z), \qquad x, y, z \in X_0. \tag{2.23}$$

Then there exists a unique solution $F : X_0 \to Y$ *of Eq.(1.4) such that*

$$\|f(x) - F(x)\| \le c_0(\psi_1(x) + \psi_2(x) + \psi_3(x)), \qquad x \in X_0, \tag{2.24}$$

where $c_0 := \inf_{m \in \mathcal{M}} \{ \frac{\max\{c_1(3m+1), c_2(-2m), c_3(-m)\}}{1 - \gamma_m} \}$.

Proof Replacing (x, y, z) by $((3m+1)x, -2mx, -mx)$ in (2.23) we get

$$
\begin{aligned}
\| f(x) &+ 5f((3m+1)x) + f((6m+1)x) - 2f((m+1)x) - 2f((5m+1)x) \\
&- 2f(-3mx) - f((2m+1)x) - 2f((4m+1)x) + 4f(-2mx) + 2f(mx)\| \\
&\le \psi_1((3m+1)x) + \psi_2(-2mx) + \psi_3(-mx), \qquad x \in X_0, m \in \mathbb{Z}_0.
\end{aligned}
$$

$$(2.25)$$

Define

$$
\begin{aligned}
\varepsilon_m(x) &:= \psi_1((3m+1)x) + \psi_2(-2mx) + \psi_3(-mx) \\
&\le c_1(3m+1)\psi_1(x) + c_2(-2m)\psi_2(x) + c_3(-m)\psi_3(x) \\
&\le \max\{c_1(3m+1), c_2(-2m), c_3(-m)\}(\psi_1(x) + \psi_2(x) \\
&\quad + \psi_3(x)), \qquad x \in X_0, m \in \mathbb{Z}_0.
\end{aligned}
$$

$$(2.26)$$

$\mathcal{T}_m$ is defined as in (2.6) and $\Lambda_m : \mathbb{R}_+^X \to \mathbb{R}_+^X$ for $m \in \mathbb{Z}_0$ is defined as in (2.7). Then, it follows from (2.25) that

$$\|(\mathcal{T}_m f)(x) - f(x)\| \le \varepsilon_m(x), \qquad x \in X_0, m \in \mathbb{Z}_0. \tag{2.27}$$

We will show that the hypotheses of Theorem 1.1 are fulfilled. For $k = 9$ and $f_1(x) := (m + 1)x$, $f_2(x) := (5m + 1)x$, $f_3(x) := -3mx$, $f_4(x) := (2m + 1)x$, $f_5(x) := (4m + 1)x$, $f_6(x) := (3m + 1)x$, $f_7(x) := (6m + 1)x$, $f_8(x) := -2mx$, $f_9(x) := mx$, $L_1(x) = L_2(x) = L_3(x) = L_5(x) = L_9(x) := 2$, $L_4(x) = L_7(x) := 1$, $L_6(x) := 5$, $L_8(x) := 4$, $x \in X_0$, inequality (1.5) becomes

$$
\begin{aligned}
&\|(\mathcal{T}_m\xi)(x) - (\mathcal{T}_m\mu)(x)\| \\
&= \|2\xi((m + 1)x) + 2\xi((5m + 1)x) + 2\xi(-3mx) + \xi((2m + 1)x) \\
&\quad + 2\xi((4m + 1)x) - 5\xi((3m + 1)x) - \xi((6m + 1)x) - 4\xi(-2mx) \\
&\quad - 2\xi(mx) - 2\mu((m + 1)x) - 2\mu((5m + 1)x) - 2\mu(-3mx) \\
&\quad - \mu((2m + 1)x) - 2\mu((4m + 1)x) + 5\mu((3m + 1)x) + \mu((6m + 1)x) \\
&\quad + 4\mu(-2mx) + 2\mu(mx)\| \\
&\leq 2\|(\xi - \mu)((m + 1)x)\| + 2\|(\xi - \mu)((5m + 1)x)\| \\
&\quad + 2\|(\xi - \mu)(-3mx)\| + \|(\xi - \mu)((2m + 1)x)\| + 2\|(\xi - \mu)((4m + 1)x)\| \\
&\quad + 5\|(\xi - \mu)((3m + 1)x)\| + \|(\xi - \mu)((6m + 1)x)\| + 4\|(\xi - \mu)(-2mx)\| \\
&\quad + 2\|(\xi - \mu)(mx)\|, \qquad x \in X_0, \xi, \mu \in Y^{X_0}, m \in \mathbb{Z}_0,
\end{aligned}
$$

where

$$
(\xi - \mu)(u) := \xi(u) - \mu(u), \qquad u \in X_0,
$$

so hypothesis (H2) is valid. Next, put $\Lambda_m : \mathbb{R}_+^{X_0} \to \mathbb{R}_+^{X_0}$ for $m \in \mathbb{Z}_0$ by

$$
\begin{aligned}
(\Lambda_m\delta)(x) := {}&2\delta((m + 1)x) + 2\delta((5m + 1)x) + 2\delta(-3mx) + \delta((2m + 1)x) \\
&+ 2\delta((4m + 1)x) + 5\delta((3m + 1)x) + \delta((6m + 1)x) \\
&+ 4\delta(-2mx) + 2\delta(mx), \qquad x \in X, \delta \in \mathbb{R}_+^{X_0}.
\end{aligned}
$$

Then it is easily seen that Λ_m has the form described in (H3).

Next, it is easily seen that, by mathematical induction on l, from (2.26) we get

$$
\left(\Lambda_m^l\varepsilon_m\right)(x) \leq \max\{c_1(3m + 1), c_2(-2m), c_3(-m)\}\gamma_m^n(\psi_1(x) + \psi_2(x) + \psi_3(x)),
\tag{2.28}
$$

for all $l \in \mathbb{N}_0$ and $m \in \mathcal{M}$. Thus, by (2.28), we obtain that

$$
\begin{aligned}
\varepsilon_m^*(x) &= \sum_{n=0}^{\infty} \left(\Lambda_m^n\varepsilon_m\right)(x) \\
&\leq \frac{\max\{c_1(3m + 1), c_2(-2m), c_3(-m)\}(\psi_1(x) + \psi_2(x) + \psi_3(x))}{1 - \gamma_m}
\end{aligned}
\tag{2.29}
$$

for all $x \in X_0$ and $m \in \mathcal{M}$. By using Theorem 1.1 with $E = X_0$, $V = Y$ and $\varphi = f$, for each $m \in \mathcal{M}$ there exists a function $F_m : X_0 \to Y$, given by $F_m(x) =$

$\lim_{n\to\infty}(\mathcal{T}_m^n f)(x)$ for $x \in X_0$, is a unique fixed point of $\mathcal{T}_m$, i.e.,

$$
\begin{aligned}
F_m(x) &= 2F_m((m+1)x) + 2F_m((5m+1)x) + 2F_m(-3mx) + F_m((2m+1)x) \\
&\quad + 2F_m((4m+1)x) - 5F_m((3m+1)x) - F_m((6m+1)x) \\
&\quad - 4F_m(-2mx) - 2F_m(mx), \qquad x \in X_0,
\end{aligned}
$$
(2.30)

and

$$
\begin{aligned}
&\|f(x) - F_m(x)\| \\
&\le \frac{\max\{c_1(3m+1), c_2(-2m), c_3(-m)\}(\psi_1(x) + \psi_2(x) + \psi_3(x))}{1 - \gamma_m}, \qquad x \in X_0.
\end{aligned}
$$
(2.31)

Similarly as in the proof of Theorem 2.1 we can show that

$$
\begin{aligned}
&\|(\mathcal{T}_m^n f)(x+y+z) + (\mathcal{T}_m^n f)(x+y-z) + (\mathcal{T}_m^n f)(x-y+z) \\
&+ (\mathcal{T}_m^n f)(x-y-z) - 2[(\mathcal{T}_m^n f)(x+y) + (\mathcal{T}_m^n f)(x-y) \\
&+ (\mathcal{T}_m^n f)(y+z) + (\mathcal{T}_m^n f)(y-z) + (\mathcal{T}_m^n f)(x+z) \\
&+ (\mathcal{T}_m^n f)(x-z)] + 4(\mathcal{T}_m^n f)(x) + 4(\mathcal{T}_m^n f)(y) \\
&+ 2[(\mathcal{T}_m^n f)(z) + (\mathcal{T}_m^n f)(-z)]\| \le \gamma_m^n(\psi_1(x) + \psi_2(y) + \psi_3(z))
\end{aligned}
$$
(2.32)

for all $x, y, z \in X_0$, $n \in \mathbb{N}_0$ and $m \in \mathcal{M}$. Letting $n \to \infty$ in (2.32) we get

$$
\begin{aligned}
&F_m(x+y+z) + F_m(x+y-z) + F_m(x-y+z) + F_m(x-y-z) \\
&= 2[F_m(x+y) + F_m(x-y) + F_m(y+z) + F_m(y-z) + F_m(x+z) \\
&\quad + F_m(x-z)] - 4F_m(x) - 4F_m(y) - 2[F_m(z) + F_m(-z)].
\end{aligned}
$$
(2.33)

So, we have proved that, for each $m \in \mathcal{M}$ there exists a function $F_m : X_0 \to Y$ satisfying the Eq. (1.4) for all $x, y, z \in X_0$ and such that

$$
\begin{aligned}
&\|f(x) - F_m(x)\| \\
&\le \frac{\max\{c_1(3m+1), c_2(-2m), c_3(-m)\}(\psi_1(x) + \psi_2(x) + \psi_3(x))}{1 - \gamma_m}, \qquad x \in X_0.
\end{aligned}
$$
(2.34)

The rest of the proof is similar to the proof of Theorem 2.1. $\qquad\square$

One can prove the following hyperstability results.

Corollary 2.3 *Let $(X, +)$ be a commutative group, Y be a Banach space over $\mathbb{K} \in \{\mathbb{R}, \mathbb{C}\}$. Suppose that $f : X \to Y$ and $c_1, c_2, c_3 : \mathbb{Z}_0 \to \mathbb{R}_+$ satisfy conditions (2.22), (2.23) and*

$$\lim_{k \to \infty} \max\{c_1(3k + 1), c_2(-2k), c_3(-k)\} = 0. \tag{2.35}$$

Then f satisfies (1.4).

Corollary 2.4 *Let X be a normed space and Y be a Banach space over* $\mathbb{K} \in \{\mathbb{R}, \mathbb{C}\}$. *Suppose that* $f : X \to Y$ *satisfies the inequality*

$$\|(Df)(x, y, z)\| \leq \alpha \left(\|x\|^p + \|y\|^p + \|z\|^p\right), \qquad x, y, z \in X \backslash \{0\}, \tag{2.36}$$

for some $\alpha \geq 0$ *and* $p \in \mathbb{R}$ *such that* $p < 0$. *Then f satisfies* (1.4).

References

1. A. Bahyrycz, K. Ciepliński, On an equation characterizing multi-Jensen-quadratic mappings and its Hyers-Ulam stability via a fixed point method. J. Fixed Point Theory Appl. **18**, 737–751 (2016)
2. A. Bahyrycz, J. Brzdęk, M. Piszczek, J. Sikorska, Hyperstability of the Fréchet equation and a characterization of inner product spaces. J. Funct. Spaces **2013**, 496361, 7pp. (2013)
3. A. Bahyrycz, K. Ciepliński, J. Olko, On an equation characterizing multi-additive-quadratic mappings and its Hyers-Ulam stability. Appl. Math. Comput. **265**, 448–455 (2015)
4. A. Bahyrycz, J. Brzdęk, E. Jabłońska, R. Malejki, Ulam's stability of a generalization of the Fréchet functional equation. J. Math. Anal. Appl. **442**, 537–553 (2016)
5. A. Bahyrycz, K. Ciepliński, J. Olko, On Hyers-Ulam stability of two functional equations in non-Archimedean spaces. J. Fixed Point Theory Appl. **18**, 433–444 (2016)
6. J. Brzdęk, L. Cădariu, Stability for a family of equations generalizing the equation of p-Wright affine functions. Appl. Math. Comput. **276**, 158–171 (2016)
7. J. Brzdęk, K. Ciepliński, Hyperstability and superstability. Abstr. Appl. Anal. **2013**, 401756, 13pp. (2013)
8. J. Brzdęk, J. Chudziak, Z. Páles, A fixed point approach to stability of functional equations. Nonlinear Anal. **74**, 6728–6732 (2011)
9. J. Brzdęk, L. Cadariu, K. Ciepliński, Fixed point theory and the Ulam stability. J. Funct. Spaces **2014**, 829419, 16pp. (2014)
10. J. Brzdęk, D. Popa, I. Raşa, B. Xu, *Ulam Stability of Operators, Mathematical Analysis and its Applications* (Academic Press, Cambridge, 2018)
11. K. Ciepliński, A. Surowczyk, On the Hyers-Ulam stability of an equation characterizing multi-quadratic mappings. Acta Math. Sci. **35**, 690–702 (2015)
12. M. Eshaghi Gordji, M.B. Savadkouhi, Stability of mixed type cubic and quartic functional equations in random normed spaces. J. Inequal. Appl. **2009**, 527462, 9pp. (2009)
13. P. Găvruţa, A generalization of the Hyers-Ulam-Rassias stability of approximately additive mappings. J. Math. Anal. Appl. **184**, 431–436 (1994)
14. J.M. Rassias, M. Arunkumar, E. Sathya, N.M. Kumar, Generalized Ulam-Hyers stability of on (AQQ): additive-quadratic-quartic functional equation. Malaya J. Mat. **5**, 122–142 (2017)
15. R. Saadati, S.M. Vaezpour, C. Park, The stability of the cubic functional equation in various spaces. Math. Commun. **16**, 131–145 (2011)
16. S.M. Ulam, *A Collection of the Mathematical Problems* (Interscience, New York, 1960)
17. T.Z. Xu, Stability of multi-Jensen mappings in non-Archimedean normed spaces. J. Math. Phys. **53**, 023507, 9pp. (2012)

Ostrowski Type Inequalities Involving Sublinear Integrals

George A. Anastassiou

Abstract The topic here is estimation of a function from its average. This is done by the Ostrowski type inequalities. Here we define a very general sublinear integral which generalizes the well-known Choquet and Shilkret integrals. Then we produce a series of Ostrowski type inequalities at all levels: univariate, multivariate, and fractional, acting to all possible cases. At the end we give a fractional Polya–Choquet inequality. Our upper bounds are simple, very tight, and accurate.

1 Introduction

The famous Ostrowski [1, 12] inequality motivates this work and has as follows:

$$\left| \frac{1}{b-a} \int_a^b f(y)\, dy - f(x) \right| \le \left(\frac{1}{4} + \frac{\left(x - \frac{a+b}{2}\right)^2}{(b-a)^2} \right)(b-a) \left\| f' \right\|_\infty,$$

where $f \in C^1([a,b])$, $x \in [a,b]$, and it is a sharp inequality.

Another motivation comes from author's [3, pp. 507–508], see also [2]:

Let $f \in C^1\left(\prod_{i=1}^k [a_i, b_i] \right)$, where $a_i < b_i$; $a_i, b_i \in \mathbb{R}$, $i = 1, \ldots, k$, and let $x_0 := (x_{01}, \ldots, x_{0k}) \in \prod_{i=1}^k [a_i, b_i]$ be fixed. Then

$$\left| \frac{1}{\prod_{i=1}^k (b_i - a_i)} \int_{a_1}^{b_1} \cdots \int_{a_i}^{b_i} \cdots \int_{a_k}^{b_k} f(z_1, \ldots, z_k)\, dz_1 \ldots dz_k - f(x_0) \right| \le$$

G. A. Anastassiou (✉)
Department of Mathematical Sciences, University of Memphis, Memphis, TN, USA
e-mail: ganastss@memphis.edu

© Springer Nature Switzerland AG 2019
G. A. Anastassiou and J. M. Rassias (eds.), *Frontiers in Functional Equations and Analytic Inequalities*, https://doi.org/10.1007/978-3-030-28950-8_19

$$\sum_{i=1}^{k} \left(\frac{(x_{0i} - a_i)^2 + (b_i - x_{0i})^2}{2(b_i - a_i)} \right) \left\| \frac{\partial f}{\partial z_i} \right\|_{\infty}.$$

The last inequality is sharp, the optimal function is

$$f^*(z_1, \ldots, z_k) := \sum_{i=1}^{k} |z_i - x_{0i}|^{\alpha_i}, \quad \alpha_i > 1.$$

A further inspiration is author's next fractional result, see [5, p. 44]:

Let $[a, b] \subset \mathbb{R}$, $\alpha > 0$, $m = \lceil \alpha \rceil$ ($\lceil \cdot \rceil$ ceiling of the number), $f \in AC^m([a, b])$ (i.e., $f^{(m-1)}$ is absolutely continuous), and $\left\| D_{x_0-}^{\alpha} f \right\|_{\infty, [a, x_0]}$, $\left\| D_{*x_0}^{\alpha} f \right\|_{\infty, [x_0, b]} < \infty$ (where $D_{x_0-}^{\alpha} f$, $D_{*x_0}^{\alpha} f$ are the right and left Caputo fractional derivatives of f of order α, respectively), $x_0 \in [a, b]$. Assume $f^{(k)}(x_0) = 0$, $k = 1, \ldots, m-1$. Then

$$\left| \frac{1}{b-a} \int_a^b f(x) \, dx - f(x_0) \right| \leq \frac{1}{(b-a) \Gamma(\alpha+2)} \cdot$$

$$\left\{ \left\| D_{x_0-}^{\alpha} f \right\|_{\infty, [a, x_0]} (x_0 - a)^{\alpha+1} + \left\| D_{*x_0}^{\alpha} f \right\|_{\infty, [x_0, b]} (b - x_0)^{\alpha+1} \right\} \leq$$

$$\frac{1}{\Gamma(\alpha+2)} \max \left\{ \left\| D_{x_0-}^{\alpha} f \right\|_{\infty, [a, x_0]}, \left\| D_{*x_0}^{\alpha} f \right\|_{\infty, [x_0, b]} \right\} (b-a)^{\alpha}.$$

Another great source for inspiration is [7].

In this work we define a general sublinear integral, which has Choquet and Shilkret integrals as special cases, and we derive a rich set of Ostrowski type inequalities at the univariate, multivariate, and fractional levels, acting to all possible directions. We finish with a fractional Polya–Choquet inequality.

2 Background-I

Consider $\Omega \neq \emptyset$ and let $\mathscr{F}$ be a σ-algebra in Ω. Here μ is a set function $\mu : \mathscr{F} \to [0, +\infty)$ which is monotone, i.e. for $A, B \in \Omega : A \subset B$ we have $\mu(A) \leq \mu(B)$, furthermore it holds $\mu(\emptyset) = 0$.

Here $f, g : \Omega \to \mathbb{R}_+ = [0, +\infty)$ are $\mathscr{F}$-measurable, we write it as $f, g \in M(\Omega, \mathbb{R}_+)$.

We consider a functional denoted by the integral symbol $(SL) \int_A f d\mu$, $\forall A \in \mathscr{F}$, which is positive, i.e. $\int_A f d\mu \geq 0$.

We assume the following properties:

(1) (positive homogeneous)

$$(SL) \int_A \alpha f d\mu = \alpha (SL) \int_A f d\mu, \quad \forall \alpha \geq 0, \forall f \in M(\Omega, \mathbb{R}_+).$$

(2) (Monotonicity) if $f, g \in M(\Omega, \mathbb{R}_+)$ satisfy $f \leq g$, then $(SL) \int_A f d\mu \leq (SL) \int_A g d\mu, \forall A \in \mathcal{F}$.
And

(3) (Subadditivity)

$$(SL) \int_A (f + g) \, d\mu \leq (SL) \int_A f d\mu + (SL) \int_A g d\mu, \ \forall \, A \in \mathcal{F}.$$

(4)

$$(SL) \int_A 1 d\mu = \mu(A), \ \forall \, A \in \mathcal{F}.$$

(5) If $\Omega = \mathbb{R}^d$, $d \in \mathbb{N}$, we assume that μ is strictly positive, i.e. $\mu(A) > 0$, for any A compact subset of $\mathbb{R}^d$. Here $\mathcal{F} = \mathcal{B}$ the Borel σ-algebra.

We call $(SL) \int_A f d\mu$ a sublinear integral.
We notice the following:

$$f(x) = f(x) - g(x) + g(x) \leq |f(x) - g(x)| + g(x),$$

hence

$$(SL) \int_A f(x) \, d\mu(x) \leq (SL) \int_A (|f(x) - g(x)| + g(x)) \, d\mu(x) \leq$$

$$(SL) \int_A |f(x) - g(x)| \, d\mu(x) + (SL) \int_A g(x) \, d\mu(x),$$

i.e.,

$$(SL) \int_A f(x) \, d\mu(x) - (SL) \int_A g(x) \, d\mu(x) \leq (SL) \int_A |f(x) - g(x)| \, d\mu(x).$$

Similarly, we get that

$$(SL) \int_A g(x) \, d\mu(x) - (SL) \int_A f(x) \, d\mu(x) \leq (SL) \int_A |f(x) - g(x)| \, d\mu(x).$$

In conclusion, it holds

$$\left| (SL) \int_A f(x) \, d\mu(x) - (SL) \int_A g(x) \, d\mu(x) \right| \leq (SL) \int_A |f(x) - g(x)| \, d\mu(x),$$

$$\tag{1}$$

$\forall \, A \in \mathcal{F}$ and $\forall \, f, g \in M(\Omega, \mathbb{R}_+)$.

3 Background-II

About the Choquet integral:
We make

Definition 1 Consider $\Omega \neq \emptyset$ and let $\mathscr{C}$ be a σ-algebra of subsets in Ω.

(i) (see, e.g., [14, p. 63]) The set function $\mu : \mathscr{C} \to [0, +\infty]$ is called a monotone set function (or capacity) if $\mu(\emptyset) = 0$ and $\mu(A) \leq \mu(B)$ for all $A, B \in \mathscr{C}$, with $A \subset B$. Also, μ is called submodular if

$$\mu(A \cup B) + \mu(A \cap B) \leq \mu(A) + \mu(B), \text{ for all } A, B \in \mathscr{C}. \tag{2}$$

μ is called bounded if $\mu(\Omega) < +\infty$ and normalized if $\mu(\Omega) = 1$.

(ii) (see, e.g., [14, p. 233], or [9]) If μ is a monotone set function on $\mathscr{C}$ and if $f : \Omega \to \mathbb{R}$ is $\mathscr{C}$-measurable (that is, for any Borel subset $B \subset \mathbb{R}$ it follows $f^{-1}(B) \in \mathscr{C}$), then for any $A \in \mathscr{C}$, the Choquet integral is defined by

$$(C) \int_A f d\mu = \int_0^{+\infty} \mu\left(F_\beta(f) \cap A\right) d\beta + \int_{-\infty}^0 \left[\mu\left(F_\beta(f) \cap A\right) - \mu(A)\right] d\beta, \tag{3}$$

where we used the notation $F_\beta(f) = \{\omega \in \Omega : f(\omega) \geq \beta\}$. Notice that if $f \geq 0$ on A, then in the above formula we get $\int_{-\infty}^0 = 0$.

The integrals on the right-hand side are the usual Riemann integral.
The function f will be called Choquet integrable on A if $(C) \int_A f d\mu \in \mathbb{R}$.

Next we list some well-known properties of the Choquet integral.

Remark 1 If $\mu : \mathscr{C} \to [0, +\infty]$ is a monotone set function, then the following properties hold:

(i) For all $a \geq 0$ we have $(C) \int_A a f d\mu = a \cdot (C) \int_A f d\mu$ (if $f \geq 0$, then see, e.g., [14], Theorem 11.2, (5), p. 228 and if f is arbitrary sign, then see, e.g., [10, p. 64], Proposition 5.1, (ii)).

(ii) For all $c \in \mathbb{R}$ and f of arbitrary sign, we have (see, e.g., [14, pp. 232–233], or [10, p. 65]) $(C) \int_A (f + c) d\mu = (C) \int_A f d\mu + c \cdot \mu(A)$.

If μ is submodular too, then for all f, g of arbitrary sign and lower bounded, we have (see, e.g., [10, p. 75], Theorem 6.3)

$$(C) \int_A (f + g) d\mu \leq (C) \int_A f d\mu + (C) \int_A g d\mu. \tag{4}$$

(iii) If $f \leq g$ on A, then $(C) \int_A f d\mu \leq (C) \int_A g d\mu$ (see, e.g., [14, p. 228], Theorem 11.2, (3) if $f, g \geq 0$ and p. 232 if f, g are of arbitrary sign).

(iv) Let $f \geq 0$. If $A \subset B$, then $(C) \int_A f d\mu \leq (C) \int_B f d\mu$. In addition, if μ is finitely subadditive, then

$$(C) \int_{A \cup B} f d\mu \leq (C) \int_A f d\mu + (C) \int_B f d\mu. \tag{5}$$

(v) It is immediate that $(C) \int_A 1 \cdot d\mu (t) = \mu (A)$.
(vi) If μ is a countably additive bounded measure, then the Choquet integral $(C) \int_A f d\mu$ reduces to the usual Lebesgue type integral (see, e.g., [10, p. 62], or [13, p. 226]).
(vii) If $\Omega = \mathbb{R}^d$, $d \in \mathbb{N}$, we assume μ is strictly positive, i.e. $\mu (A) > 0$, for every A compact subset of $\mathbb{R}^d$. Here $\mathscr{C} = \mathscr{B}$ the Borel σ-algebra.

Clearly here, for μ being submodular, we get

$$\left| (C) \int_A f(x) d\mu (x) - (C) \int_A g(x) d\mu (x) \right| \leq (C) \int_A |f(x) - g(x)| d\mu (x), \tag{6}$$

$\forall A \in \mathscr{C}$ and $\forall f, g \in M(\Omega, \mathbb{R}_+)$ (f, g are measurable with respect to $\mathscr{C}$ σ-algebra).

From now on in this article we assume that $\mu : \mathscr{C} \to [0, +\infty)$ and is submodular.

4 Background-III

Here we follow [13].

Let $\mathscr{F}$ be a σ-field of subsets of an arbitrary set Ω. An extended non-negative real valued function μ on $\mathscr{F}$ is called maxitive if $\mu (\varnothing) = 0$ and

$$\mu (\cup_{i \in I} E_i) = \sup_{i \in I} \mu (E_i), \tag{7}$$

where the set I is of cardinality at most countable, where $\{E_i\}_{i \in I}$ is a disjoint collection of sets from $\mathscr{F}$. We notice that μ is monotone and (7) is true even $\{E_i\}_{i \in I}$ are not disjoint. For more properties of μ see [13]. We also call μ a maxitive measure. Here f stands for a non-negative measurable function on Ω. In [13], Niel Shilkret developed his non-additive integral defined as follows:

$$(N^*) \int_D f d\mu := \sup_{y \in Y} \{y \cdot \mu (D \cap \{f \geq y\})\}, \tag{8}$$

where $Y = [0, m]$ or $Y = [0, m)$ with $0 < m \leq \infty$, and $D \in \mathscr{F}$. Here we take $Y = [0, \infty)$.

It is easily proved that

$$
(N^*) \int_D f d\mu = \sup_{y > 0} \{y \cdot \mu (D \cap \{f > y\})\}. \tag{9}
$$

The Shilkret integral takes values in $[0, \infty]$.

The Shilkret integral [13] has the following properties:

$$
(N^*) \int_\Omega \chi_E d\mu = \mu(E), \tag{10}
$$

where χ_E is the indicator function on $E \in \mathscr{F}$,

$$
(N^*) \int_D cf d\mu = c (N^*) \int_D f d\mu, \quad c \geq 0, \tag{11}
$$

$$
(N^*) \int_D \sup_{n \in \mathbb{N}} f_n d\mu = \sup_{n \in \mathbb{N}} (N^*) \int_D f_n d\mu, \tag{12}
$$

where f_n, $n \in \mathbb{N}$, is an increasing sequence of elementary (countably valued) functions converging uniformly to f. Furthermore we have

$$
(N^*) \int_D f d\mu \geq 0, \tag{13}
$$

$$
f \geq g \text{ implies } (N^*) \int_D f d\mu \geq (N^*) \int_D g d\mu, \tag{14}
$$

where $f, g : \Omega \to [0, \infty]$ are measurable.

Let $a \leq f(\omega) \leq b$ for almost every $\omega \in E$, then

$$
a\mu(E) \leq (N^*) \int_E f d\mu \leq b\mu(E); \tag{15}
$$

$$
(N^*) \int_E 1 d\mu = \mu(E); \tag{16}
$$

$f > 0$ almost everywhere and $(N^*) \int_E f d\mu = 0$ imply $\mu(E) = 0$;
$(N^*) \int_\Omega f d\mu = 0$ if and only $f = 0$ almost everywhere;
$(N^*) \int_\Omega f d\mu < \infty$ implies that

$$
\overline{N}(f) := \{\omega \in \Omega | f(\omega) \neq 0\} \text{ has } \sigma\text{-finite measure;}
$$

$$
(N^*) \int_D (f + g) d\mu \leq (N^*) \int_D f d\mu + (N^*) \int_D g d\mu; \tag{17}
$$

and

$$\left| (N^*) \int_D f d\mu - (N^*) \int_D g d\mu \right| \le (N^*) \int_D |f - g| d\mu. \qquad (18)$$

From now on in this article we assume that $\mu : \mathscr{F} \to [0, +\infty)$.

If $\Omega = \mathbb{R}^d$, $d \in \mathbb{N}$, we assume μ is strictly positive, i.e. $\mu(A) > 0$, for every A compact subset of $\mathbb{R}^d$. Here $\mathscr{F} = \mathscr{B}$ the Borel σ-algebra.

Conclusion 1 *We observe that the Choquet integral* $(C) \int_A f d\mu$ *and Shilkret integral* $(N^*) \int_A f d\mu$ *are perfect examples of the sublinear integral* $(SL) \int_A f d\mu$ *of Sect. 2, fulfilling all properties and they have great applications in many areas of pure and applied mathematics and mathematical economics.*

Therefore, all the results presented in this article which are for the general integral $(SL) \int_A f d\mu$ *are of course valid for the Choquet and Shilkret integrals.*

5   Univariate Ostrowski (*SL*)-Integral Inequalities

From now on we work in the setting of Sect. 2: Background-I.

We make

Remark 2 Let $f \in C^1([a, b], \mathbb{R}_+)$, and $\mu([a, b]) > 0$, $x \in [a, b]$. We will estimate

$$E := \left| \frac{(SL) \int_{[a,b]} f(t) d\mu(t)}{\mu([a, b])} - f(x) \right| = \qquad (19)$$

$$\left| \frac{1}{\mu([a, b])} (SL) \int_{[a,b]} f(t) d\mu(t) - \frac{\mu([a, b]) f(x)}{\mu([a, b])} \right| =$$

$$\left| \frac{1}{\mu([a, b])} \left((SL) \int_{[a,b]} f(t) d\mu(t) - (SL) \int_{[a,b]} f(x) d\mu(t) \right) \right| \le$$

$$\frac{1}{\mu([a, b])} (SL) \int_{[a,b]} |f(t) - f(x)| d\mu(t) \le$$

$$\frac{\|f'\|_\infty}{\mu([a, b])} (SL) \int_{[a,b]} |t - x| d\mu(t). \qquad (20)$$

If $f : [a, b] \to \mathbb{R}_+$ is a Lipschitz function of order $0 < \alpha \le 1$, i.e. $|f(x) - f(y)| \le K |x - y|^\alpha$, $\forall\ x, y \in [a, b]$, where $K > 0$, denoted by $f \in Lip_{\alpha, K}([a, b], \mathbb{R}_+)$, then we get

$$E \leq \frac{1}{\mu\left([a,b]\right)} \left(SL\right) \int_{[a,b]} |f\left(t\right) - f\left(x\right)| \, d\mu\left(t\right) \leq$$

$$\frac{K}{\mu\left([a,b]\right)} \left(SL\right) \int_{[a,b]} |t - x|^{\alpha} \, d\mu\left(t\right). \tag{21}$$

We have proved the following Ostrowski type inequalities:

Theorem 1 *Here* $\mu\left([a,b]\right) > 0$ *and* $x \in [a,b]$.

(1) Let $f \in C^1\left([a,b], \mathbb{R}_+\right)$, *then*

$$\left| \frac{1}{\mu\left([a,b]\right)} \left(SL\right) \int_{[a,b]} f\left(t\right) d\mu\left(t\right) - f\left(x\right) \right| \leq$$

$$\frac{\|f'\|_{\infty}}{\mu\left([a,b]\right)} \left(SL\right) \int_{[a,b]} |t - x| \, d\mu\left(t\right). \tag{22}$$

(2) Let $f \in Lip_{\alpha,K}\left([a,b], \mathbb{R}_+\right), 0 < \alpha \leq 1$, *then*

$$\left| \frac{1}{\mu\left([a,b]\right)} \left(SL\right) \int_{[a,b]} f\left(t\right) d\mu\left(t\right) - f\left(x\right) \right| \leq$$

$$\frac{K}{\mu\left([a,b]\right)} \left(SL\right) \int_{[a,b]} |t - x|^{\alpha} \, d\mu\left(t\right). \tag{23}$$

We make

Remark 3 Let $f \in C^1\left([a,b], \mathbb{R}_+\right)$ and $g \in C^1\left([a,b]\right)$, by Cauchy's mean value theorem we get that

$$\left(f\left(t\right) - f\left(x\right)\right) g'\left(c\right) = \left(g\left(t\right) - g\left(x\right)\right) f'\left(c\right),$$

for some c between t and x; for any $t, x \in [a,b]$.

If $g'\left(c\right) \neq 0$, we have

$$\left(f\left(t\right) - f\left(x\right)\right) = \left(\frac{f'\left(c\right)}{g'\left(c\right)}\right) \left(g\left(t\right) - g\left(x\right)\right).$$

Here we assume that $g'\left(t\right) \neq 0, \forall \, t \in [a,b]$. Hence it holds

$$|f\left(t\right) - f\left(x\right)| \leq \left\| \frac{f'}{g'} \right\|_{\infty} |g\left(t\right) - g\left(x\right)|, \tag{24}$$

$\forall \, t, x \in [a,b]$.

We have again as before

$$E \leq \frac{1}{\mu\left([a,b]\right)} (SL) \int_{[a,b]} |f\left(t\right) - f\left(x\right)| \, d\mu\left(t\right) \leq$$

$$\frac{1}{\mu\left([a,b]\right)} (SL) \int_{[a,b]} \left\| \frac{f'}{g'} \right\|_{\infty} |g\left(t\right) - g\left(x\right)| \, d\mu\left(t\right) \leq$$

$$\frac{1}{\mu\left([a,b]\right)} \left\| \frac{f'}{g'} \right\|_{\infty} (SL) \int_{[a,b]} |g\left(t\right) - g\left(x\right)| \, d\mu\left(t\right). \tag{25}$$

We have established the following general Ostrowski type inequality:

Theorem 2 *Here μ is such that $\mu\left([a,b]\right) > 0$, $x \in [a,b]$. Let $f \in C^1\left([a,b], \mathbb{R}_+\right)$ and $g \in C^1\left([a,b]\right)$ with $g'\left(t\right) \neq 0$, $\forall\, t \in [a,b]$. Then*

$$\left| \frac{1}{\mu\left([a,b]\right)} (SL) \int_{[a,b]} f\left(t\right) \, d\mu\left(t\right) - f\left(x\right) \right| \leq$$

$$\frac{\left\| \frac{f'}{g'} \right\|_{\infty}}{\mu\left([a,b]\right)} (SL) \int_{[a,b]} |g\left(t\right) - g\left(x\right)| \, d\mu\left(t\right). \tag{26}$$

We give for $g\left(t\right) = e^t$ the next result

Corollary 1 *Here μ is such that $\mu\left([a,b]\right) > 0$, $x \in [a,b]$. Let $f \in C^1\left([a,b], \mathbb{R}_+\right)$, then*

$$\left| \frac{1}{\mu\left([a,b]\right)} (SL) \int_{[a,b]} f\left(t\right) \, d\mu\left(t\right) - f\left(x\right) \right| \leq$$

$$\frac{\left\| \frac{f'}{e^t} \right\|_{\infty}}{\mu\left([a,b]\right)} (SL) \int_{[a,b]} |e^t - e^x| \, d\mu\left(t\right). \tag{27}$$

When $g\left(t\right) = \ln t$ we get

Corollary 2 *Here μ is such that $\mu\left([a,b]\right) > 0$, $x \in [a,b]$ and $a > 0$. Let $f \in C^1\left([a,b], \mathbb{R}_+\right)$. Then*

$$\left| \frac{1}{\mu\left([a,b]\right)} (SL) \int_{[a,b]} f\left(t\right) \, d\mu\left(t\right) - f\left(x\right) \right| \leq$$

$$\frac{\left\| t f'\left(t\right) \right\|_{\infty}}{\mu\left([a,b]\right)} (SL) \int_{[a,b]} \left| \ln \frac{t}{x} \right| \, d\mu\left(t\right). \tag{28}$$

Many other applications of Theorem 2 could follow but we stop it here.
We make

Remark 4 Let $f \in [C([a, b], \mathbb{R}_+) \cap C^{n+1}([a, b])]$, $n \in \mathbb{N}$, $x \in [a, b]$. Then by Taylor's theorem we get

$$f(y) - f(x) = \sum_{k=1}^{n} \frac{f^{(k)}(x)}{k!}(y - x)^k + R_n(x, y), \tag{29}$$

where the remainder

$$R_n(x, y) := \int_x^y \left(f^{(n)}(t) - f^{(n)}(x) \right) \frac{(y - t)^{n-1}}{(n - 1)!} dt; \tag{30}$$

here y can be $\geq x$ or $\leq x$.

By [3] we get that

$$|R_n(x, y)| \leq \frac{\left\| f^{(n+1)} \right\|_\infty}{(n + 1)!} |y - x|^{n+1}, \quad \text{for all } x, y \in [a, b]. \tag{31}$$

Here we assume $f^{(k)}(x) = 0$, for all $k = 1, \ldots, n$.

Therefore it holds

$$|f(t) - f(x)| \leq \frac{\left\| f^{(n+1)} \right\|_\infty}{(n + 1)!} |t - x|^{n+1}, \quad \text{for all } t, x \in [a, b]. \tag{32}$$

Here we have again

$$E \leq \frac{1}{\mu([a, b])} (SL) \int_{[a,b]} |f(t) - f(x)| \, d\mu(t) \leq$$

$$\frac{1}{\mu([a, b])} (SL) \int_{[a,b]} \frac{\left\| f^{(n+1)} \right\|_\infty}{(n + 1)!} |t - x|^{n+1} \, d\mu(t) =$$

$$\frac{\left\| f^{(n+1)} \right\|_\infty}{\mu([a, b])(n + 1)!} (SL) \int_{[a,b]} |t - x|^{n+1} \, d\mu(t). \tag{33}$$

We have derived the following high order Ostrowski-type inequality:

Theorem 3 *Let* $f \in [C([a, b], \mathbb{R}_+) \cap C^{n+1}([a, b])]$, $n \in \mathbb{N}$, $x \in [a, b]$. *We assume that* $f^{(k)}(x) = 0$, *all* $k = 1, \ldots, n$. *Here* μ *is such that* $\mu([a, b]) > 0$. *Then*

$$\left| \frac{1}{\mu\left([a,b]\right)} (SL) \int_{[a,b]} f(t) \, d\mu(t) - f(x) \right| \le$$

$$\frac{\left\| f^{(n+1)} \right\|_\infty}{(n+1)! \mu\left([a,b]\right)} (SL) \int_{[a,b]} |t - x|^{n+1} \, d\mu(t), \tag{34}$$

which generalizes (22).

When $x = \frac{a+b}{2}$ we get

Corollary 3 *Let* $f \in \left[C\left([a,b], \mathbb{R}_+\right) \cap C^{n+1}\left([a,b]\right) \right]$, $n \in \mathbb{N}$. *Assume that* $f^{(k)}\left(\frac{a+b}{2}\right) = 0$, $k = 1, \dots, n$. *Here* μ *is such that* $\mu\left([a,b]\right) > 0$. *Then*

$$\left| \frac{1}{\mu\left([a,b]\right)} (SL) \int_{[a,b]} f(t) \, d\mu(t) - f\left(\frac{a+b}{2}\right) \right| \le$$

$$\frac{\left\| f^{(n+1)} \right\|_\infty}{(n+1)! \mu\left([a,b]\right)} (SL) \int_{[a,b]} \left| t - \frac{a+b}{2} \right|^{n+1} d\mu(t). \tag{35}$$

6 Multivariate Ostrowski (SL)-Integral Inequalities

Here Q is a compact and convex subset of $\mathbb{R}^k$, $k \ge 1$.

We make

Remark 5 Let $f \in C\left(Q, \mathbb{R}_+\right)$ and μ is such that $\mu(Q) > 0$, $x \in Q$. We will estimate

$$\left| \frac{(SL) \int_Q f(t) \, d\mu(t)}{\mu(Q)} - f(x) \right| =$$

$$\left| \frac{1}{\mu(Q)} (SL) \int_Q f(t) \, d\mu(t) - \frac{\mu(Q)}{\mu(Q)} f(x) \right| =$$

$$\frac{1}{\mu(Q)} \left| (SL) \int_Q f(t) \, d\mu(t) - (SL) \int_Q f(x) \, d\mu(t) \right| \le$$

$$\frac{1}{\mu(Q)} (SL) \int_Q |f(t) - f(x)| \, d\mu(t), \tag{36}$$

where $t = (t_1, \dots, t_k)$, $x = (x_1, \dots, x_k)$.

That is,

$$\left| \frac{1}{\mu(Q)} (SL) \int_Q f(t) \, d\mu(t) - f(x) \right| \le$$

$$\frac{1}{\mu(Q)} (SL) \int_Q |f(t) - f(x)| \, d\mu(t). \tag{37}$$

We make

Remark 6 Here $Q := \prod_{i=1}^{k} [a_i, b_i]$, where $a_i < b_i$; $a_i, b_i \in \mathbb{R}$, $i = 1, \ldots, k$; $x = (x_1, \ldots, x_k) \in \prod_{i=1}^{k} [a_i, b_i]$ is fixed, and $f \in C^1 \left(\prod_{i=1}^{k} [a_i, b_i], \mathbb{R}_+ \right)$. Consider $g_t(r) := f(x + r(t - x))$, $r \ge 0$. Note that $g_t(0) = f(x)$, $g_t(1) = f(t)$. Thus

$$f(t) - f(x) = g_t(1) - g_t(0) = g_t'(\xi)(1 - 0) = g_t'(\xi), \tag{38}$$

where $\xi \in (0, 1)$.

That is,

$$f(t) - f(x) = \sum_{i=1}^{k} (t_i - x_i) \frac{\partial f}{\partial t_i} (x + \xi(t - x)). \tag{39}$$

Hence

$$|f(t) - f(x)| \le \sum_{i=1}^{k} |t_i - x_i| \left| \frac{\partial f}{\partial t_i} (x + \xi(t - x)) \right|$$

$$\le \sum_{i=1}^{k} |t_i - x_i| \left\| \frac{\partial f}{\partial t_i} \right\|_{\infty}. \tag{40}$$

By (37) we get

$$\left| \frac{1}{\mu \left(\prod_{i=1}^{k} [a_i, b_i] \right)} (SL) \int_{\prod_{i=1}^{k} [a_i, b_i]} f(t) \, d\mu(t) - f(x) \right| \le$$

$$\frac{1}{\mu \left(\prod_{i=1}^{k} [a_i, b_i] \right)} (SL) \int_{\prod_{i=1}^{k} [a_i, b_i]} |f(t) - f(x)| \, d\mu(t) \le$$

$$\frac{1}{\mu \left(\prod_{i=1}^{k} [a_i, b_i] \right)} (SL) \int_{\prod_{i=1}^{k} [a_i, b_i]} \left(\sum_{i=1}^{k} |t_i - x_i| \left\| \frac{\partial f}{\partial t_i} \right\|_{\infty} \right) d\mu(t) \le$$

$$\frac{1}{\mu \left(\prod\limits_{i=1}^{k} [a_i, b_i] \right)} \left(\sum_{i=1}^{k} (SL) \int_{\prod\limits_{i=1}^{k} [a_i,b_i]} |t_i - x_i| \left\| \frac{\partial f}{\partial t_i} \right\|_{\infty} d\mu(t) \right) =$$

$$\frac{1}{\mu \left(\prod\limits_{i=1}^{k} [a_i, b_i] \right)} \left(\sum_{i=1}^{k} \left\| \frac{\partial f}{\partial t_i} \right\|_{\infty} (SL) \int_{\prod\limits_{i=1}^{k} [a_i,b_i]} |t_i - x_i| d\mu(t) \right). \tag{41}$$

Here μ is such that $\mu \left(\prod\limits_{i=1}^{k} [a_i, b_i] \right) > 0$.

Therefore we get

$$\left| \frac{1}{\mu \left(\prod\limits_{i=1}^{k} [a_i, b_i] \right)} (SL) \int_{\prod\limits_{i=1}^{k} [a_i,b_i]} f(t) d\mu(t) - f(x) \right| \leq \tag{42}$$

$$\sum_{i=1}^{k} \left(\frac{\left\| \frac{\partial f}{\partial t_i} \right\|_{\infty}}{\mu \left(\prod\limits_{i=1}^{k} [a_i, b_i] \right)} \right) \left((SL) \int_{\prod\limits_{i=1}^{k} [a_i,b_i]} |t_i - x_i| d\mu(t) \right).$$

If $f : \prod\limits_{i=1}^{k} [a_i, b_i] \to \mathbb{R}_+$ is a Lipschitz function of order $0 < \alpha \leq 1$, i.e.

$|f(x) - f(y)| \leq K \|x - y\|_{l_1}^{\alpha}, \forall\, x, y \in \prod\limits_{i=1}^{k} [a_i, b_i], K > 0$, where $\|x - y\|_{l_1} :=$

$\sum\limits_{i=1}^{k} |x_i - y_i|$, denoted by $f \in Lip_{\alpha, K} \left(\prod\limits_{i=1}^{k} [a_i, b_i], \mathbb{R}_+ \right)$, then by (37) we get

$$\left| \frac{1}{\mu \left(\prod\limits_{i=1}^{k} [a_i, b_i] \right)} (SL) \int_{\prod\limits_{i=1}^{k} [a_i,b_i]} f(t) d\mu(t) - f(x) \right| \leq \tag{43}$$

$$\frac{1}{\mu \left(\prod\limits_{i=1}^{k} [a_i, b_i] \right)} (SL) \int_{\prod\limits_{i=1}^{k} [a_i,b_i]} |f(t) - f(x)| d\mu(t) \leq$$

$$\frac{1}{\mu\left(\prod_{i=1}^{k}[a_i,b_i]\right)} (SL) \int_{\prod_{i=1}^{k}[a_i,b_i]} K \|t-x\|_{l_1}^{\alpha} \, d\mu(t) =$$

$$\frac{K}{\mu\left(\prod_{i=1}^{k}[a_i,b_i]\right)} (SL) \int_{\prod_{i=1}^{k}[a_i,b_i]} \|t-x\|_{l_1}^{\alpha} \, d\mu(t).$$

We have proved

$$\left| \frac{1}{\mu\left(\prod_{i=1}^{k}[a_i,b_i]\right)} (SL) \int_{\prod_{i=1}^{k}[a_i,b_i]} f(t) \, d\mu(t) - f(x) \right| \leq \qquad (44)$$

$$\frac{K}{\mu\left(\prod_{i=1}^{k}[a_i,b_i]\right)} (SL) \int_{\prod_{i=1}^{k}[a_i,b_i]} \|t-x\|_{l_1}^{\alpha} \, d\mu(t).$$

We have established the following multivariate Ostrowski (SL)-inequalities.

Theorem 4 *Here μ is such that $\mu\left(\prod_{i=1}^{k}[a_i,b_i]\right) > 0$, $x \in \prod_{i=1}^{k}[a_i,b_i]$.*

(1) Let $f \in C^1\left(\prod_{i=1}^{k}[a_i,b_i], \mathbb{R}_+\right)$, then

$$\left| \frac{1}{\mu\left(\prod_{i=1}^{k}[a_i,b_i]\right)} (SL) \int_{\prod_{i=1}^{k}[a_i,b_i]} f(t) \, d\mu(t) - f(x) \right| \leq \qquad (45)$$

$$\sum_{i=1}^{k} \left(\frac{\left\|\frac{\partial f}{\partial t_i}\right\|_{\infty}}{\mu\left(\prod_{i=1}^{k}[a_i,b_i]\right)} \right) \left((SL) \int_{\prod_{i=1}^{k}[a_i,b_i]} |t_i - x_i| \, d\mu(t) \right).$$

(2) Let $f \in Lip_{\alpha,K}\left(\prod_{i=1}^{k}[a_i,b_i], \mathbb{R}_+\right)$, $0 < \alpha \leq 1$, then

$$\left| \frac{1}{\mu \left(\prod_{i=1}^{k} [a_i, b_i] \right)} (SL) \int_{\prod_{i=1}^{k} [a_i, b_i]} f(t) \, d\mu(t) - f(x) \right| \leq \tag{46}$$

$$\frac{K}{\mu \left(\prod_{i=1}^{k} [a_i, b_i] \right)} (SL) \int_{\prod_{i=1}^{k} [a_i, b_i]} \|t - x\|_{l_1}^{\alpha} \, d\mu(t).$$

We make

Remark 7 Let Q be a compact and convex subset of $\mathbb{R}^k$, $k \geq 1$. Let $f \in \left(C(Q, \mathbb{R}_+) \cap C^{n+1}(Q) \right)$, $n \in \mathbb{N}$ and $x \in Q$ is fixed such that all partial derivatives $f_\alpha := \frac{\partial^\alpha f}{\partial t^\alpha}$, where $\alpha = (\alpha_1, \ldots, \alpha_k)$, $\alpha_i \in \mathbb{Z}^+$, $i = 1, \ldots, k$, $|\alpha| = \sum_{i=1}^{k} \alpha_i = j$, $j = 1, \ldots, n$ fulfill $f_\alpha(x) = 0$.

By [3, p. 513], we get that

$$|f(t) - f(x)| \leq \frac{\left[\left(\sum_{i=1}^{k} |t_i - x_i| \left\| \frac{\partial}{\partial t_i} \right\|_\infty \right)^{n+1} f \right]}{(n+1)!}, \quad \forall \, t \in Q. \tag{47}$$

Call

$$D_{n+1}(f) := \max_{\alpha: |\alpha| = n+1} \|f_\alpha\|_\infty. \tag{48}$$

For example, when $k = 2$ and $n = 1$, we get that

$$\left[\left(\sum_{i=1}^{2} |t_i - x_i| \left\| \frac{\partial}{\partial t_i} \right\|_\infty \right)^2 f \right] =$$

$$(t_1 - x_1)^2 \left\| \frac{\partial^2 f}{\partial t_1^2} \right\|_\infty + 2 |t_1 - x_1| |t_2 - x_2| \left\| \frac{\partial^2 f}{\partial t_1 \partial t_2} \right\|_\infty + (t_2 - x_2)^2 \left\| \frac{\partial^2 f}{\partial t_2^2} \right\|_\infty \tag{49}$$

and

$$D_2(f) = \max_{\alpha: |\alpha| = 2} \|f_\alpha\|_\infty. \tag{50}$$

Clearly, it holds

$$\left[\left(\sum_{i=1}^{2} |t_i - x_i| \left\| \frac{\partial}{\partial t_i} \right\|_\infty \right)^2 f \right] \leq D_2(f) (|t_1 - x_1| + |t_2 - x_2|)^2. \tag{51}$$

Consequently, we derive that

$$\left[\left(\sum_{i=1}^{k} |t_i - x_i| \left\|\frac{\partial}{\partial t_i}\right\|_\infty\right)^{n+1} f\right] \le D_{n+1}(f) \|t - x\|_{l_1}^{n+1}, \quad \forall\, t \in Q. \tag{52}$$

By (37) we get

$$\left|\frac{1}{\mu(Q)}(SL)\int_Q f(t)\,d\mu(t) - f(x)\right| \le$$

$$\frac{1}{\mu(Q)}(SL)\int_Q |f(t) - f(x)|\,d\mu(t) \le \tag{53}$$

$$\frac{1}{\mu(Q)}(SL)\int_Q \frac{\left[\left(\sum_{i=1}^{k} |t_i - x_i| \left\|\frac{\partial}{\partial t_i}\right\|_\infty\right)^{n+1} f\right]}{(n+1)!}\,d\mu(t) \le$$

$$\frac{1}{\mu(Q)}(SL)\int_Q \frac{D_{n+1}(f)\|t - x\|_{l_1}^{n+1}}{(n+1)!}\,d\mu(t) = \tag{54}$$

$$\frac{1}{\mu(Q)}\frac{D_{n+1}(f)}{(n+1)!}(SL)\int_Q \|t - x\|_{l_1}^{n+1}\,d\mu(t).$$

Here μ is such that $\mu(Q) > 0$.

By (53) and (54) we obtain

$$\left|\frac{1}{\mu(Q)}(SL)\int_Q f(t)\,d\mu(t) - f(x)\right| \le$$

$$\frac{D_{n+1}(f)}{(n+1)!\mu(Q)}(SL)\int_Q \|t - x\|_{l_1}^{n+1}\,d\mu(t). \tag{55}$$

We have established the following multivariate Ostrowski general (SL)-inequality:

Theorem 5 *Let Q be a compact and convex subset of $\mathbb{R}^k$, $k \ge 1$. Let $f \in \left(C(Q, \mathbb{R}_+) \cap C^{n+1}(Q)\right)$, $n \in \mathbb{N}$, $x \in Q$ be fixed: $f_\alpha(x) = 0$, all $\alpha : |\alpha| = j$, $j = 1, \ldots, n$. Here μ is such that $\mu(Q) > 0$. Then*

$$\left|\frac{1}{\mu(Q)}(SL)\int_Q f(t)\,d\mu(t) - f(x)\right| \le$$

$$\frac{D_{n+1}(f)}{(n+1)!\mu(Q)}(SL)\int_Q \|t-x\|_{l_1}^{n+1}\,d\mu(t).\tag{56}$$

Corollary 4 *All as in Theorem 5. Then*

$$\left|\frac{1}{\mu(Q)}(SL)\int_Q f(t)\,d\mu(t)-f(x)\right|\le$$

$$\frac{1}{(n+1)!\mu(Q)}(SL)\int_Q\left[\left(\sum_{i=1}^k|t_i-x_i|\left\|\frac{\partial}{\partial x_i}\right\|_\infty\right)^{n+1}f\right]d\mu(t).\tag{57}$$

Next we take again $Q:=\prod_{i=1}^k[a_i,b_i]$, we set $a:=(a_1,\ldots,a_k)$, $b:=(b_1,\ldots,b_k)$, and $\frac{a+b}{2}=\left(\frac{a_1+b_1}{2},\ldots,\frac{a_k+b_k}{2}\right)\in\prod_{i=1}^k[a_i,b_i]$.

Corollary 5 *Let* $f\in\left(C\left(\prod_{i=1}^k[a_i,b_i]\mathbb{R}_+\right)\cap C^{n+1}\left(\prod_{i=1}^k[a_i,b_i]\right)\right)$, $n\in\mathbb{N}$, *such that* $f_\alpha\left(\frac{a+b}{2}\right)=0$, *all* $\alpha:|\alpha|=j$, $j=1,\ldots,n$. *Here* μ *is such that* $\mu\left(\prod_{i=1}^k[a_i,b_i]\right)>0$. *Then*

$$\left|\frac{1}{\mu\left(\prod_{i=1}^k[a_i,b_i]\right)}(SL)\int_{\prod_{i=1}^k[a_i,b_i]}f(t)\,d\mu(t)-f\left(\frac{a+b}{2}\right)\right|\le$$

$$\frac{D_{n+1}(f)}{(n+1)!\mu\left(\prod_{i=1}^k[a_i,b_i]\right)}(SL)\int_{\prod_{i=1}^k[a_i,b_i]}\left\|t-\frac{a+b}{2}\right\|_{l_1}^{n+1}d\mu(t).\tag{58}$$

Proof By Theorem 5.

We make

Remark 8 By multinomial theorem we have that

$$\|t-x\|_{l_1}^{n+1}=\left(\sum_{i=1}^k|t_i-x_i|\right)^{n+1}=$$

$$\sum_{r_1+r_2+\ldots+r_k=n+1}\binom{n+1}{r_1,r_2,\ldots,r_k}|t_1-x_1|^{r_1}|t_2-x_2|^{r_2}\ldots|t_k-x_k|^{r_k},\tag{59}$$

where

$$\binom{n+1}{r_1, r_2, \ldots, r_k} = \frac{(n+1)!}{r_1! r_2! \ldots r_k!}. \tag{60}$$

By (59), (60), (54) we get

$$\left| \frac{1}{\mu(Q)} (SL) \int_Q f(t) \, d\mu(t) - f(x) \right| \le$$

$$\frac{1}{\mu(Q)} (SL) \int_Q \frac{D_{n+1}(f)}{(n+1)!} \|t - x\|_{l_1}^{n+1} \, d\mu(t) =$$

$$\frac{1}{\mu(Q)} (SL) \int_Q \left[\sum_{r_1 + r_2 + \ldots + r_k = n+1} \left(\frac{D_{n+1}(f)}{r_1! r_2! \ldots r_k!} \right) \left(\prod_{i=1}^{k} |t_i - x_i|^{r_i} \right) \right] d\mu(t) \le$$

$$\frac{1}{\mu(Q)} \left(\sum_{r_1 + r_2 + \ldots + r_k = n+1} (SL) \int_Q \left(\frac{D_{n+1}(f)}{r_1! r_2! \ldots r_k!} \right) \left(\prod_{i=1}^{k} |t_i - x_i|^{r_i} \right) d\mu(t) \right) =$$

$$\frac{1}{\mu(Q)} \left(\sum_{r_1 + r_2 + \ldots + r_k = n+1} \left(\frac{D_{n+1}(f)}{r_1! r_2! \ldots r_k!} \right) (SL) \int_Q \left(\prod_{i=1}^{k} |t_i - x_i|^{r_i} \right) d\mu(t) \right). \tag{61}$$

We have proved the following multivariate Ostrowski general SL-inequality:

Theorem 6 *Here all as in Theorem 5. Then*

$$\left| \frac{1}{\mu(Q)} (SL) \int_Q f(t) \, d\mu(t) - f(x) \right| \le$$

$$\frac{D_{n+1}(f)}{\mu(Q)} \left[\sum_{r_1 + r_2 + \ldots + r_k = n+1} \left(\frac{1}{r_1! r_2! \ldots r_k!} \right) (SL) \int_Q \left(\prod_{i=1}^{k} |t_i - x_i|^{r_i} \right) d\mu(t) \right]. \tag{62}$$

We make

Remark 9 In case $k = 2, n = 1$, by (53), (54) we get

$$\left| \frac{1}{\mu(Q)} (SL) \int_Q f(t) \, d\mu(t) - f(x) \right| \le$$

$$\frac{1}{\mu(Q)} (SL) \int_Q \frac{D_2(f)}{2} \|t - x\|_{l_1}^2 \, d\mu(t) =$$

$$\frac{1}{\mu(Q)}(SL)\int_Q \frac{D_2(f)}{2}\left[(t_1-x_1)^2+2|t_1-x_1||t_2-x_2|+(t_2-x_2)^2\right]d\mu(t)\leq$$

(63)

$$\frac{1}{\mu(Q)}\left[(SL)\int_Q \frac{D_2(f)}{2}(t_1-x_1)^2\,d\mu(t)+\right.$$

$$(SL)\int_Q D_2(f)|t_1-x_1||t_2-x_2|\,d\mu(t)+(SL)\int_Q \frac{D_2(f)}{2}(t_2-x_2)^2\,d\mu(t)\right]=$$

$$\frac{1}{\mu(Q)}\left[\frac{D_2(f)}{2}(SL)\int_Q (t_1-x_1)^2\,d\mu(t)+\right.$$

$$D_2(f)(SL)\int_Q |t_1-x_1||t_2-x_2|\,d\mu(t)+\frac{D_2(f)}{2}(SL)\int_Q (t_2-x_2)^2\,d\mu(t)\right].$$

We have proved.

Corollary 6 *Let Q be a compact and convex subset of $\mathbb{R}^2$. Let $f \in (C(Q,\mathbb{R}_+)\cap C^2(Q))$, $x = (x_1,x_2) \in Q$ be fixed: $\frac{\partial f}{\partial t_1}(x_1,x_2) = \frac{\partial f}{\partial t_2}(x_1,x_2) = 0$. Here μ is such that $\mu(Q) > 0$. Then*

$$\left|\frac{1}{\mu(Q)}(SL)\int_Q f(t)\,d\mu(t)-f(x)\right|\leq$$

$$\frac{D_2(f)}{2\mu(Q)}(SL)\int_Q (t_1-x_1)^2\,d\mu(t)+\frac{D_2(f)}{\mu(Q)}(SL)\int_Q |t_1-x_1||t_2-x_2|\,d\mu(t)$$

(64)

$$+\frac{D_2(f)}{2\mu(Q)}(SL)\int_Q (t_2-x_2)^2\,d\mu(t).$$

7 Canavati Fractional Background

We need

Remark 10 Here $[\cdot]$ denotes the integral part of the number. Let $\alpha > 0$, $m = [\alpha]$, $\beta = \alpha - m$, $0 < \beta < 1$, $f \in C([a,b])$, $[a,b] \subset \mathbb{R}$, $x \in [a,b]$. The gamma function Γ is given by $\Gamma(\alpha) = \int_0^\infty e^{-t}t^{\alpha-1}dt$. We define the left Riemann–Liouville integral

$$\left(J_\alpha^{a+}f\right)(x)=\frac{1}{\Gamma(\alpha)}\int_a^x (x-t)^{\alpha-1}f(t)\,dt,$$

(65)

$a \leq x \leq b$. We define the subspace $C_{\alpha+}^\alpha([a,b])$ of $C^m([a,b])$:

$$C_{a+}^{\alpha} ([a, b]) = \{ f \in C^m ([a, b]) : J_{1-\beta}^{a+} f^{(m)} \in C^1 ([a, b]) \}. \tag{66}$$

For $f \in C_{a+}^{\alpha} ([a, b])$, we define the left generalized α-fractional derivative of f over $[a, b]$ as

$$D_{a+}^{\alpha} f := \left(J_{1-\beta}^{a+} f^{(m)} \right)', \tag{67}$$

see [4, p. 24]. Canavati first in [8] introduced the above over $[0, 1]$. Notice that $D_{a+}^{\alpha} f \in C ([a, b])$.

We need the following left fractional Taylor's formula, see [4, pp. 8–10], and in [8] the same over $[0, 1]$ that appeared first.

Let $f \in C_{a+}^{\alpha} ([a, b])$:

(i) If $\alpha \geq 1$, then

$$f(x) = f(a) + f'(a)(x-a) + f''(a)\frac{(x-a)^2}{2} + \ldots + f^{(m-1)}(a)\frac{(x-a)^{m-1}}{(m-1)!} +$$

$$\frac{1}{\Gamma(\alpha)} \int_a^x (x-t)^{\alpha-1} \left(D_{a+}^{\alpha} f \right)(t)\, dt, \tag{68}$$

all $x \in [a, b]$.

(ii) If $0 < \alpha < 1$, we have

$$f(x) = \frac{1}{\Gamma(\alpha)} \int_a^x (x-t)^{\alpha-1} \left(D_{a+}^{\alpha} f \right)(t)\, dt, \tag{69}$$

all $x \in [a, b]$.

Notice that

$$\int_a^x (x-t)^{\alpha-1} \left(D_{a+}^{\alpha} f \right)(t)\, dt = \int_a^x \left(D_{a+}^{\alpha} f \right)(t)\, d\left(\frac{(x-t)^{\alpha}}{-\alpha} \right) \tag{70}$$

$$= \left(D_{a+}^{\alpha} f \right)(\xi_x) \frac{(x-a)^{\alpha}}{\alpha}, \quad \text{where } \xi_x \in [a, x],$$

by first integral mean value theorem.

Hence, when $\alpha \geq 1$ and $f^{(i)}(a) = 0$, $i = 0, 1, \ldots, m-1$ or when $0 < \alpha < 1$, we get

$$f(x) = \left(D_{a+}^{\alpha} f \right)(\xi_x) \frac{(x-a)^{\alpha}}{\Gamma(\alpha+1)}, \quad \text{all } x \in [a, b]. \tag{71}$$

Furthermore we need:

Let again $\alpha > 0$, $m = [\alpha]$, $\beta = \alpha - m$, $f \in C([a, b])$, call the right Riemann–Liouville fractional integral operator by

$$\left(J_{b-}^{\alpha} f\right)(x) := \frac{1}{\Gamma(\alpha)} \int_{x}^{b} (t - x)^{\alpha-1} f(t) \, dt, \tag{72}$$

$x \in [a, b]$, see also [6, pp. 333, 345].

Define the subspace of functions

$$C_{b-}^{\alpha}([a, b]) = \{f \in C^{m}([a, b]) : J_{b-}^{1-\beta} f^{(m)} \in C^{1}([a, b])\}. \tag{73}$$

Define the right generalized α-fractional derivative of f over $[a, b]$ as

$$D_{b-}^{\alpha} f = (-1)^{m-1} \left(J_{b-}^{1-\beta} f^{(m)}\right)', \tag{74}$$

see [6, p. 345]. We set $D_{b-}^{0} f = f$.

Notice that $D_{b-}^{\alpha} f \in C([a, b])$.

From [6, p. 348], we need the following right Taylor fractional formula:
Let $f \in C_{b-}^{\alpha}([a, b])$, $\alpha > 0$, $m = [\alpha]$. Then

(i) If $\alpha \geq 1$, we get

$$f(x) = \sum_{k=0}^{m-1} \frac{f^{(k)}(b)}{k!} (x - b)^{k} + \left(J_{b-}^{\alpha} D_{b-}^{\alpha} f\right)(x), \tag{75}$$

all $x \in [a, b]$.

(ii) If $0 < \alpha < 1$, we get

$$f(x) = J_{b-}^{\alpha} D_{b-}^{\alpha} f(x) = \frac{1}{\Gamma(\alpha)} \int_{x}^{b} (t - x)^{\alpha-1} \left(D_{b-}^{\alpha} f\right)(t) \, dt, \tag{76}$$

all $x \in [a, b]$.

Notice that

$$\int_{x}^{b} (t - x)^{\alpha-1} \left(D_{b-}^{\alpha} f\right)(t) \, dt = \int_{x}^{b} \left(D_{b-}^{\alpha} f\right)(t) \, d\left(\frac{(t - x)^{\alpha}}{\alpha}\right)$$

$$= \left(D_{b-}^{\alpha} f\right)(\eta_{x}) \frac{(b - x)^{\alpha}}{\alpha}, \quad \text{where } \eta_{x} \in [x, b], \tag{77}$$

by first integral mean value theorem.

Hence, when $\alpha \geq 1$ and $f^{(k)}(b) = 0$, $k = 0, 1, \ldots, m-1$ or $0 < \alpha < 1$, we obtain

$$f(x) = \left(D_{b-}^\alpha f\right)(\eta_x) \frac{(b-x)^\alpha}{\Gamma(\alpha+1)}, \text{ all } x \in [a,b]. \tag{78}$$

Let $f \in C_{a+}^\alpha([a,b])$, $\alpha \geq 1$, and $f^{(i)}(a) = 0$, $i = 1, \ldots, m-1$, then

$$|f(x) - f(a)| \leq \left\|D_{a+}^\alpha f\right\|_\infty \frac{(x-a)^\alpha}{\Gamma(\alpha+1)}, \tag{79}$$

all $x \in [a,b]$, by (68).

Again let $f \in C_{a+}^\alpha([a,b])$, $\alpha \geq 1$, and $f^{(i)}(a) = 0$, $i = 1, \ldots, m-1$, then by (68) we have

$$f(x) - f(a) = \frac{1}{\Gamma(\alpha)} \int_a^x (x-t)^{\alpha-1} \left(D_{a+}^\alpha f\right)(t)\, dt, \tag{80}$$

hence

$$|f(x) - f(a)| \leq \frac{(x-a)^{\alpha-1}}{\Gamma(\alpha)} \left\|D_{a+}^\alpha f\right\|_{L_1([a,b])}, \tag{81}$$

all $x \in [a,b]$.

Let $p, q > 1 : \frac{1}{p} + \frac{1}{q} = 1$, continuing from (80), $\alpha \geq 1$, we get

$$|f(x) - f(a)| \leq \frac{1}{\Gamma(\alpha)} \left(\int_a^x (x-t)^{p(\alpha-1)}\, dt\right)^{\frac{1}{p}} \left\|D_{a+}^\alpha f\right\|_{L_q([a,b])} =$$

$$\frac{(x-a)^{\frac{(p(\alpha-1)+1)}{p}}}{\Gamma(\alpha)(p(\alpha-1)+1)^{\frac{1}{p}}} \left\|D_{a+}^\alpha f\right\|_{L_q([a,b])}, \tag{82}$$

$\forall\, x \in [a,b]$.

Let $f \in C_{b-}^\alpha([a,b])$, $\alpha \geq 1$, $m = [\alpha]$, $f^{(k)}(b) = 0$, $k = 1, \ldots, m-1$, then by (75) we get:

$$f(x) - f(b) = \frac{1}{\Gamma(\alpha)} \int_x^b (t-x)^{\alpha-1} \left(D_{b-}^\alpha f\right)(t)\, dt. \tag{83}$$

We derive the following estimates:

(1)

$$|f(x) - f(b)| \leq \frac{(b-x)^{\alpha-1}}{\Gamma(\alpha+1)} \left\|D_{b-}^\alpha f\right\|_\infty, \tag{84}$$

(2)

$$|f(x) - f(b)| \leq \frac{(b-x)^{\alpha-1}}{\Gamma(\alpha)} \left\| D_{b-}^{\alpha} f \right\|_{L_1([a,b])}, \tag{85}$$

(3) let $p, q > 1 : \frac{1}{p} + \frac{1}{q} = 1$, then

$$|f(x) - f(b)| \leq \frac{(b-x)^{\frac{(p(\alpha-1)+1)}{p}}}{\Gamma(\alpha)(p(\alpha-1)+1)^{\frac{1}{p}}} \left\| D_{b-}^{\alpha} f \right\|_{L_q([a,b])}, \tag{86}$$

$\forall x \in [a, b]$.

8 Caputo Fractional Background

We need

Remark 11 Let $v > 0$, $n := \lceil v \rceil$, $\lceil \cdot \rceil$ is the ceiling of the number, $f \in AC^n([a, b])$ (i.e., $f^{(n-1)}$ is absolutely continuous on $[a, b]$). We call the left Caputo fractional derivative ([11])

$$D_{*a}^{v} f(x) := \frac{1}{\Gamma(n-v)} \int_a^x (x-t)^{n-v-1} f^{(n)}(t) \, dt, \tag{87}$$

$\forall x \in [a, b]$.

The above function $D_{*a}^{v} f(x)$ exists almost everywhere for $x \in [a, b]$.

If $v \in \mathbb{N}$, then $D_{*a}^{v} f = f^{(v)}$ the ordinary derivative, it is also $D_{*a}^0 f = f$.

We have the left fractional Taylor formula for left Caputo fractional derivatives [11, p. 40].

Assume $v > 0$, $n = \lceil v \rceil$, and $f \in AC^n([a, b])$. Then

$$f(x) = \sum_{k=0}^{n-1} \frac{f^{(k)}(a)}{k!} (x-a)^k + \frac{1}{\Gamma(v)} \int_a^x (x-t)^{v-1} D_{*a}^{v} f(t) \, dt, \tag{88}$$

$\forall x \in [a, b]$.

Additionally assume that

$$f^{(k)}(a) = 0, \ k = 1, \dots, n-1;$$

then

$$f(x) - f(a) = \frac{1}{\Gamma(v)} \int_a^x (x-t)^{v-1} D_{*a}^{v} f(t) \, dt, \tag{89}$$

$\forall x \in [a, b]$.

We get the following estimates:

(1) if $D_{*a}^{v} f \in L_{\infty}([a, b])$, then

$$|f(x) - f(a)| \leq \frac{\|D_{*a}^{v} f\|_{\infty}}{\Gamma(v+1)} (x-a)^{v}, \tag{90}$$

$\forall x \in [a, b]$, see [4, p. 619];

(2) if $v \geq 1$, and $D_{*a}^{v} f \in L_{1}([a, b])$, then

$$|f(x) - f(a)| \leq \frac{\|D_{*a}^{v} f\|_{L_{1}([a,b])}}{\Gamma(v)} (x-a)^{v-1}, \tag{91}$$

$\forall x \in [a, b]$, see [4, p. 620];

(3) let $p, q > 1 : \frac{1}{p} + \frac{1}{q} = 1$, and $v > \frac{1}{q}$, and $D_{*a}^{v} f \in L_{q}([a, b])$, then

$$|f(x) - f(a)| \leq \frac{\|D_{*a}^{v} f\|_{L_{q}([a,b])}}{\Gamma(v)(p(v-1)+1)^{\frac{1}{p}}} (x-a)^{v-\frac{1}{q}}, \tag{92}$$

$\forall x \in [a, b]$, see [4, p. 621].

Furthermore we need:

Let $f \in AC^{m}([a, b])$ ($f^{(m-1)}$ is absolutely continuous on $[a, b]$), $m \in \mathbb{N}$, $m = \lceil \alpha \rceil$, $\alpha > 0$. We define the right Caputo fractional derivative of order $\alpha > 0$ by

$$D_{cb-}^{\alpha} f(x) = \frac{(-1)^{m}}{\Gamma(m-\alpha)} \int_{x}^{b} (J-x)^{m-\alpha-1} f^{(m)}(J) dJ, \tag{93}$$

$\forall x \in [a, b]$, see [6, p. 336].

If $\alpha = m \in \mathbb{N}$, then

$$D_{cb-}^{\alpha} f(x) = (-1)^{m} f^{(m)}(x), \quad \forall x \in [a, b]. \tag{94}$$

If $x > b$ we define $D_{cb-}^{\alpha} f(x) = 0$.

We also need:

Let $f \in AC^{m}([a, b])$, $\alpha > 0$, $m = \lceil \alpha \rceil$. Then

$$f(x) = \sum_{k=0}^{m-1} \frac{f^{(k)}(b)}{k!} (x-b)^{k} + \frac{1}{\Gamma(\alpha)} \int_{x}^{b} (J-x)^{\alpha-1} D_{cb-}^{\alpha} f(J) dJ, \tag{95}$$

$\forall x \in [a, b]$, the right Caputo fractional Taylor formula with integral remainder, see [6, p. 338].

Additionally assume that

$$f^{(k)}(b) = 0, \ k = 1, \ldots, m-1,$$

then

$$f(x) - f(b) = \frac{1}{\Gamma(\alpha)} \int_x^b (J - x)^{\alpha-1} D_{cb-}^\alpha f(J) \, dJ, \tag{96}$$

$\forall \, x \in [a, b]$.

Following (96) we get the following estimates:

(1) if $D_{cb-}^\alpha f \in L_\infty([a, b])$, then

$$|f(x) - f(b)| \le \frac{(b-x)^\alpha}{\Gamma(\alpha+1)} \left\| D_{cb-}^\alpha f \right\|_\infty, \tag{97}$$

$\forall \, x \in [a, b]$, see [5, p. 23];

(2) if $D_{cb-}^\alpha f \in L_1([a, b])$, $\alpha \ge 1$, then

$$|f(x) - f(b)| \le \frac{\left\| D_{cb-}^\alpha f \right\|_{L_1([a,b])}}{\Gamma(\alpha)} (b-x)^{\alpha-1}, \tag{98}$$

$\forall \, x \in [a, b]$, see [5, p. 24];

(3) let $p, q > 1 : \frac{1}{p} + \frac{1}{q} = 1$, and $\alpha > \frac{1}{q}$, $m = \lceil \alpha \rceil$, $D_{cb-}^\alpha f \in L_q([a, b])$, then

$$|f(x) - f(b)| \le \frac{\left\| D_{cb-}^\alpha f \right\|_{L_q([a,b])}}{\Gamma(\alpha)(p(\alpha-1)+1)^{\frac{1}{p}}} (b-x)^{\alpha-\frac{1}{q}}, \tag{99}$$

$\forall \, x \in [a, b]$, see [5, p. 25].

9 Main Fractional Results

We make

Remark 12 Here μ is such that $\mu([a, b]) > 0$. Let $\alpha \ge 1$, $f \in C_{a+}^\alpha([a, b])$ and $f^{(i)}(a) = 0$, $i = 1, \ldots, m-1$; $m = \lceil \alpha \rceil$. By (19), (20), for $x = a$, we get that

$$\left| \frac{1}{\mu([a, b])} (SL) \int_{[a,b]} f(t) \, d\mu(t) - f(a) \right| \le \tag{100}$$

$$\frac{1}{\mu([a, b])} (SL) \int_{[a,b]} |f(t) - f(a)| \, d\mu(t) =: \Delta(a).$$

By (79) we obtain

$$\Delta\left(a\right) \le \frac{1}{\mu\left(\left[a,b\right]\right)}\left(SL\right)\int_{\left[a,b\right]}\frac{\left\|D_{a+}^{\alpha}f\right\|_{\infty}}{\Gamma\left(\alpha+1\right)}\left(t-a\right)^{\alpha}d\mu\left(t\right) =$$

$$\frac{1}{\mu\left(\left[a,b\right]\right)}\frac{\left\|D_{a+}^{\alpha}f\right\|_{\infty}}{\Gamma\left(\alpha+1\right)}\left(SL\right)\int_{\left[a,b\right]}\left(t-a\right)^{\alpha}d\mu\left(t\right). \tag{101}$$

By (81) we obtain

$$\Delta\left(a\right) \le \frac{1}{\mu\left(\left[a,b\right]\right)}\left(SL\right)\int_{\left[a,b\right]}\frac{\left\|D_{a+}^{\alpha}f\right\|_{L_1\left(\left[a,b\right]\right)}}{\Gamma\left(\alpha\right)}\left(t-a\right)^{\alpha-1}d\mu\left(t\right) =$$

$$\frac{1}{\mu\left(\left[a,b\right]\right)}\frac{\left\|D_{a+}^{\alpha}f\right\|_{L_1\left(\left[a,b\right]\right)}}{\Gamma\left(\alpha\right)}\left(SL\right)\int_{\left[a,b\right]}\left(t-a\right)^{\alpha-1}d\mu\left(t\right). \tag{102}$$

And by (82) $(p,q>1:\frac{1}{p}+\frac{1}{q}=1)$ we derive

$$\Delta\left(a\right) \le \frac{1}{\mu\left(\left[a,b\right]\right)}\left(SL\right)\int_{\left[a,b\right]}\frac{\left\|D_{a+}^{\alpha}f\right\|_{L_q\left(\left[a,b\right]\right)}}{\Gamma\left(\alpha\right)\left(p\left(\alpha-1\right)+1\right)^{\frac{1}{p}}}\left(t-a\right)^{\frac{\left(p\left(\alpha-1\right)+1\right)}{p}}d\mu\left(t\right) =$$

$$\frac{1}{\mu\left(\left[a,b\right]\right)}\frac{\left\|D_{a+}^{\alpha}f\right\|_{L_q\left(\left[a,b\right]\right)}}{\Gamma\left(\alpha\right)\left(p\left(\alpha-1\right)+1\right)^{\frac{1}{p}}}\left(SL\right)\int_{\left[a,b\right]}\left(t-a\right)^{\left(\alpha-1+\frac{1}{p}\right)}d\mu\left(t\right). \tag{103}$$

We have proved that

$$\left|\frac{1}{\mu\left(\left[a,b\right]\right)}\left(SL\right)\int_{\left[a,b\right]}f\left(t\right)d\mu\left(t\right)-f\left(a\right)\right| \le$$

$$\frac{1}{\mu\left(\left[a,b\right]\right)}\min\left\{\frac{\left\|D_{a+}^{\alpha}f\right\|_{\infty}}{\Gamma\left(\alpha+1\right)}\left(SL\right)\int_{\left[a,b\right]}\left(t-a\right)^{\alpha}d\mu\left(t\right),\right.$$

$$\frac{\left\|D_{a+}^{\alpha}f\right\|_{L_1\left(\left[a,b\right]\right)}}{\Gamma\left(\alpha\right)}\left(SL\right)\int_{\left[a,b\right]}\left(t-a\right)^{\alpha-1}d\mu\left(t\right),$$

$$\left.\frac{\left\|D_{a+}^{\alpha}f\right\|_{L_q\left(\left[a,b\right]\right)}}{\Gamma\left(\alpha\right)\left(p\left(\alpha-1\right)+1\right)^{\frac{1}{p}}}\left(SL\right)\int_{\left[a,b\right]}\left(t-a\right)^{\alpha-\frac{1}{q}}d\mu\left(t\right)\right\}. \tag{104}$$

We have established the following left generalized fractional Ostrowski SL-integral inequality.

Theorem 7 *Here μ is such that $\mu\left([a,b]\right) > 0$. Let $p, q > 1 : \frac{1}{p} + \frac{1}{q} = 1$; $\alpha \geq 1$. Let $f \in C^{\alpha}_{a+}\left([a,b]\right)$ with $f^{(i)}\left(a\right) = 0$, $i = 1, \ldots, m-1$; $m = [\alpha]$; and f is $\mathbb{R}_+$-valued. Then*

$$\left| \frac{1}{\mu\left([a,b]\right)} \, (SL) \int_{[a,b]} f\left(t\right) d\mu\left(t\right) - f\left(a\right) \right| \leq$$

$$\frac{1}{\mu\left([a,b]\right)} \min \left\{ \frac{\left\| D^{\alpha}_{a+} f \right\|_{\infty}}{\Gamma\left(\alpha+1\right)} \, (SL) \int_{[a,b]} \left(t-a\right)^{\alpha} d\mu\left(t\right), \right.$$

$$\frac{\left\| D^{\alpha}_{a+} f \right\|_{L_1\left([a,b]\right)}}{\Gamma\left(\alpha\right)} \, (SL) \int_{[a,b]} \left(t-a\right)^{\alpha-1} d\mu\left(t\right),$$

$$\left. \frac{\left\| D^{\alpha}_{a+} f \right\|_{L_q\left([a,b]\right)}}{\Gamma\left(\alpha\right)\left(p\left(\alpha-1\right)+1\right)^{\frac{1}{p}}} \, (SL) \int_{[a,b]} \left(t-a\right)^{\alpha-\frac{1}{q}} d\mu\left(t\right) \right\}. \tag{105}$$

Similarly (as in Remark 12), we get the right generalized fractional Ostrowski *SL*-integral inequality (use of (19), (20) for $x = b$, and (84)–(86)).

Theorem 8 *Here μ is such that $\mu\left([a,b]\right) > 0$. Let $p, q > 1 : \frac{1}{p} + \frac{1}{q} = 1$; $\alpha \geq 1$. Let $f \in C^{\alpha}_{b-}\left([a,b]\right)$ with $f^{(k)}\left(b\right) = 0$, $k = 1, \ldots, m-1$; $[\alpha] = m$; and f is $\mathbb{R}_+$-valued. Then*

$$\left| \frac{1}{\mu\left([a,b]\right)} \, (SL) \int_{[a,b]} f\left(t\right) d\mu\left(t\right) - f\left(b\right) \right| \leq$$

$$\frac{1}{\mu\left([a,b]\right)} \min \left\{ \frac{\left\| D^{\alpha}_{b-} f \right\|_{\infty}}{\Gamma\left(\alpha+1\right)} \, (SL) \int_{[a,b]} \left(b-t\right)^{\alpha} d\mu\left(t\right), \right.$$

$$\frac{\left\| D^{\alpha}_{b-} f \right\|_{L_1\left([a,b]\right)}}{\Gamma\left(\alpha\right)} \, (SL) \int_{[a,b]} \left(b-t\right)^{\alpha-1} d\mu\left(t\right),$$

$$\left. \frac{\left\| D^{\alpha}_{b-} f \right\|_{L_q\left([a,b]\right)}}{\Gamma\left(\alpha\right)\left(p\left(\alpha-1\right)+1\right)^{\frac{1}{p}}} \, (SL) \int_{[a,b]} \left(b-t\right)^{\alpha-\frac{1}{q}} d\mu\left(t\right) \right\}. \tag{106}$$

We present the following left Caputo fractional Ostrowski *SL*-integral inequalities:

Theorem 9 *Here μ is such that $\mu\left([a,b]\right) > 0$. Let $f : [a,b] \to \mathbb{R}_+$ such that $f \in AC^n\left([a,b]\right)$, where $n = \lceil v \rceil$, $v > 0$. Assume $f^{(k)}\left(a\right) = 0$, $k = 1, \ldots, n-1$. We have*

(1) if $D_{*a}^{v} f \in L_{\infty}([a, b])$, then

$$\left| \frac{1}{\mu([a,b])} (SL) \int_{[a,b]} f(t) \, d\mu(t) - f(a) \right| \leq$$

$$\frac{\|D_{*a}^{v} f\|_{\infty}}{\mu([a,b]) \, \Gamma(v+1)} (SL) \int_{[a,b]} (t-a)^{v} \, d\mu(t), \tag{107}$$

(2) if $v \geq 1$, and $D_{*a}^{v} f \in L_{1}([a, b])$, then

$$\left| \frac{1}{\mu([a,b])} (SL) \int_{[a,b]} f(t) \, d\mu(t) - f(a) \right| \leq$$

$$\frac{\|D_{*a}^{v} f\|_{L_{1}([a,b])}}{\mu([a,b]) \, \Gamma(v)} (SL) \int_{[a,b]} (t-a)^{v-1} \, d\mu(t), \tag{108}$$

and
(3) let $p, q > 1 : \frac{1}{p} + \frac{1}{q} = 1$, and $v > \frac{1}{q}$, and $D_{*a}^{v} f \in L_{q}([a, b])$, then

$$\left| \frac{1}{\mu([a,b])} (SL) \int_{[a,b]} f(t) \, d\mu(t) - f(a) \right| \leq$$

$$\frac{\|D_{*a}^{v} f\|_{L_{q}([a,b])}}{\mu([a,b]) \, \Gamma(v) \, (p(v-1)+1)^{\frac{1}{p}}} (SL) \int_{[a,b]} (t-a)^{v-\frac{1}{q}} \, d\mu(t). \tag{109}$$

Proof By the use of (90)–(92), acting as in the proof of Theorem 7.

Next, we give the following right Caputo fractional Ostrowski SL-integral inequalities:

Theorem 10 *Here μ is such that $\mu([a, b]) > 0$. Let $f : [a, b] \to \mathbb{R}_{+}$ such that $f \in AC^{m}([a, b])$, $m \in \mathbb{N}$, $m = \lceil \alpha \rceil$, $\alpha > 0$. Assume $f^{(k)}(b) = 0$, $k = 1, \ldots, m-1$. We have*

(1) if $D_{cb-}^{\alpha} f \in L_{\infty}([a, b])$, then

$$\left| \frac{1}{\mu([a,b])} (SL) \int_{[a,b]} f(t) \, d\mu(t) - f(b) \right| \leq$$

$$\frac{\|D_{cb-}^{\alpha} f\|_{\infty}}{\mu([a,b]) \, \Gamma(\alpha+1)} (SL) \int_{[a,b]} (b-t)^{\alpha} \, d\mu(t), \tag{110}$$

(2) if $\alpha \geq 1$, and $D_{cb-}^{\alpha} f \in L_{1}([a, b])$, then

$$\left| \frac{1}{\mu\left([a,b]\right)} \, (SL) \int_{[a,b]} f\left(t\right) d\mu\left(t\right) - f\left(b\right) \right| \le$$

$$\frac{\left\| D_{cb-}^{\alpha} f \right\|_{L_1([a,b])}}{\mu\left([a,b]\right) \Gamma\left(\alpha\right)} \, (SL) \int_{[a,b]} (b-t)^{\alpha-1} \, d\mu\left(t\right), \tag{111}$$

(3) if $p, q > 1 : \frac{1}{p} + \frac{1}{q} = 1$, $\alpha > \frac{1}{q}$, $D_{cb-}^{\alpha} f \in L_q\left([a,b]\right)$, then

$$\left| \frac{1}{\mu\left([a,b]\right)} \, (SL) \int_{[a,b]} f\left(t\right) d\mu\left(t\right) - f\left(b\right) \right| \le$$

$$\frac{\left\| D_{cb-}^{\alpha} f \right\|_{L_q([a,b])}}{\mu\left([a,b]\right) \Gamma\left(\alpha\right) \left(p\left(\alpha-1\right)+1\right)^{\frac{1}{p}}} \, (SL) \int_{[a,b]} (b-t)^{\alpha-\frac{1}{q}} \, d\mu\left(t\right). \tag{112}$$

Proof Use of (97)–(99), acting again as in the proof of Theorem 7.

We make

Remark 13 Let $x_0 \in [a, b]$. Of interest will be to fractionally estimate the quantity

$$\left| \frac{(SL) \int_{[a,x_0]} f\left(t\right) d\mu\left(t\right)}{\mu\left([a,x_0]\right)} + \frac{(SL) \int_{[x_0,b]} f\left(t\right) d\mu\left(t\right)}{\mu\left([x_0,b]\right)} - 2f\left(x_0\right) \right| \le$$

$$\left| \frac{1}{\mu\left([a,x_0]\right)} \, (SL) \int_{[a,x_0]} f\left(t\right) d\mu\left(t\right) - f\left(x_0\right) \right| + \tag{113}$$

$$\left| \frac{1}{\mu\left([x_0,b]\right)} \, (SL) \int_{[x_0,b]} f\left(t\right) d\mu\left(t\right) - f\left(x_0\right) \right|,$$

above we have $\mu\left([a,x_0]\right), \mu\left([x_0,b]\right) > 0$.
An important case is when $x_0 = \frac{a+b}{2}$.

The above can be done with the use of our earlier fractional results.
We make

Remark 14 Here μ is positive on any non-empty closed subsets of $[a, b]$ and is finitely subadditive.
Let $f : [a, b] \to \mathbb{R}_+$ be continuous, such that $f \in C_{a+}^{\alpha}\left([a, \frac{a+b}{2}]\right)$, $0 < \alpha < 1$. By (71) we get

$$f\left(x\right) \le \left\| D_{a+}^{\alpha} f \right\|_{\infty, \left[a, \frac{a+b}{2}\right]} \frac{(x-a)^{\alpha}}{\Gamma\left(\alpha+1\right)}, \text{ all } x \in \left[a, \frac{a+b}{2}\right]. \tag{114}$$

Assume also $f \in C_{b-}^{\alpha}\left(\left[\frac{a+b}{2}, b\right]\right)$, $0 < \alpha < 1$.

By (78) we obtain

$$f(x) \le \left\| D_{b-}^{\alpha} f \right\|_{\infty, \left[\frac{a+b}{2}, b \right]} \frac{(b-x)^{\alpha}}{\Gamma(\alpha+1)}, \forall x \in \left[\frac{a+b}{2}, b \right]. \tag{115}$$

We notice that the Choquet integral

$$(C) \int_{[a,b]} f(t) \, d\mu(t) = C \int_{\left[a, \frac{a+b}{2}\right] \cup \left[\frac{a+b}{2}, b\right]} f(t) \, d\mu(t) \overset{(5)}{\le}$$

$$(C) \int_{\left[a, \frac{a+b}{2}\right]} f(t) \, d\mu(t) + (C) \int_{\left[\frac{a+b}{2}, b\right]} f(t) \, d\mu(t) \overset{\text{(by (114), (115))}}{\le}$$

$$\frac{\left\| D_{a+}^{\alpha} f \right\|_{\infty, \left[a, \frac{a+b}{2}\right]}}{\Gamma(\alpha+1)} (C) \int_{\left[a, \frac{a+b}{2}\right]} (t-a)^{\alpha} \, d\mu(t) + \tag{116}$$

$$\frac{\left\| D_{b-}^{\alpha} f \right\|_{\infty, \left[\frac{a+b}{2}, b\right]}}{\Gamma(\alpha+1)} (C) \int_{\left[\frac{a+b}{2}, b\right]} (b-t)^{\alpha} \, d\mu(t).$$

We have proved the generalized α-fractional Polya–Choquet type inequality:

$$(C) \int_{[a,b]} f(t) \, d\mu(t) \le$$

$$\frac{1}{\Gamma(\alpha+1)} \left[\left\| D_{a+}^{\alpha} f \right\|_{\infty, \left[a, \frac{a+b}{2}\right]} (C) \int_{\left[a, \frac{a+b}{2}\right]} (t-a)^{\alpha} \, d\mu(t) + \right.$$

$$\left. \left\| D_{b-}^{\alpha} f \right\|_{\infty, \left[\frac{a+b}{2}, b\right]} (C) \int_{\left[\frac{a+b}{2}, b\right]} (b-t)^{\alpha} \, d\mu(t) \right]. \tag{117}$$

References

1. G.A. Anastassiou, Ostrowski type inequalities. Proc. Am. Math. Soc. **123**(12), 3775–3781 (1995)
2. G.A. Anastassiou, Multivariate Ostrowski type inequalities. Acta Math. Hung. **76**(4), 267–278 (1997)
3. G.A. Anastassiou, *Quantitative Approximations* (Chapman & Hall/CRC, Boca Raton, 2001)
4. G.A. Anastassiou, *Fractional Differentiation Inequalities* (Springer, New York, 2009)
5. G.A. Anastassiou, *Advances on Fractional Inequalities* (Springer, New York, 2011)

6. G.A. Anastassiou, *Intelligent Mathematics: Computational Analysis* (Springer, Heidelberg, 2011)
7. G.A. Anastassiou, *Intelligent Comparisons: Analytic Inequalities* (Springer, Heidelberg, 2016)
8. J.A. Canavati, The Riemann-Liouville integral. Nieuw Archief Voor Wiskunde **5**(1), 53–75 (1987)
9. G. Choquet, Theory of capacities. Ann. Inst. Fourier (Grenoble) **5**, 131–295 (1954)
10. D. Denneberg, *Non-additive Measure and Integral* (Kluwer, Dordrecht, 1994)
11. K. Diethelm, *The Analysis of Fractional Differential Equations*. Lecture Notes in Mathematics, vol. 2004, 1st edn. (Springer, New York, 2010)
12. A. Ostrowski, Über die Absolutabweichung einer differentiebaren Funktion von ihrem Integralmittelwert. Comment. Math. Helv. **10**, 226–227 (1938)
13. N. Shilkret, Maxitive measure and integration. Indag. Math. **33**, 109–116 (1971)
14. Z. Wang, G.J. Klir, *Fuzzy Measure Theory* (Plenum, New York, 1992)

Inequalities for Special Strong Differential Superordinations Using a Generalized Sălăgean Operator and Ruscheweyh Derivative

Alina Alb Lupaş

Abstract In the present paper we establish several inequalities for strong differential superordinations regarding the extended new operator $RD_{\lambda,\alpha}^m$ defined by using the extended Sălăgean operator and the extended Ruscheweyh derivative, $RD_{\lambda,\alpha}^m$: $\mathscr{A}_{n\zeta}^* \to \mathscr{A}_{n\zeta}^*$, $RD_{\lambda,\alpha}^m f(z,\zeta) = (1-\alpha)R^m f(z,\zeta) + \alpha D_\lambda^m f(z,\zeta)$, $z \in U$, $\zeta \in \overline{U}$, where $R^m f(z,\zeta)$ denote the extended Ruscheweyh derivative, $D_\lambda^m f(z,\zeta)$ is the extended generalized Sălăgean operator, and $\mathscr{A}_{n\zeta}^* = \{f \in \mathscr{H}(U \times \overline{U}),\ f(z,\zeta) = z + a_{n+1}(\zeta) z^{n+1} + \ldots,\ z \in U,\ \zeta \in \overline{U}\}$ is the class of normalized analytic functions.

1 Introduction

Denote by U the unit disc of the complex plane $U = \{z \in \mathbb{C} : |z| < 1\}$, $\overline{U} = \{z \in \mathbb{C} : |z| \leq 1\}$ the closed unit disc of the complex plane, and $\mathscr{H}(U \times \overline{U})$ the class of analytic functions in $U \times \overline{U}$.

Let

$$\mathscr{A}_{n\zeta}^* = \{f \in \mathscr{H}(U \times \overline{U}),\ f(z,\zeta) = z + a_{n+1}(\zeta) z^{n+1} + \ldots,\ z \in U,\ \zeta \in \overline{U}\},$$

where $a_k(\zeta)$ are holomorphic functions in $\overline{U}$ for $k \geq 2$, and

$$\mathscr{H}^*[a,n,\zeta] = \{f \in \mathscr{H}(U \times \overline{U}),\ f(z,\zeta) = a + a_n(\zeta) z^n + a_{n+1}(\zeta) z^{n+1} + \ldots,\ z \in U,\ \zeta \in \overline{U}\},$$

for $a \in \mathbb{C}$, $n \in \mathbb{N}$, $a_k(\zeta)$ are holomorphic functions in $\overline{U}$ for $k \geq n$.

We extend the generalized Sălăgean differential operator [6] and Ruscheweyh derivative [9] to the new class of analytic functions $\mathscr{A}_{n\zeta}^*$ introduced in [8].

A. A. Lupaş (✉)
Department of Mathematics and Computer Science, University of Oradea, Oradea, Romania
e-mail: dalb@uoradea.ro

© Springer Nature Switzerland AG 2019
G. A. Anastassiou and J. M. Rassias (eds.), *Frontiers in Functional Equations and Analytic Inequalities*, https://doi.org/10.1007/978-3-030-28950-8_20

Definition 1 ([5]) For $f \in \mathcal{A}_{n\zeta}^*$, $n, m \in \mathbb{N}$, the operator R^m is defined by R^m: $\mathcal{A}_{n\zeta}^* \to \mathcal{A}_{n\zeta}^*$,

$$R^0 f(z, \zeta) = f(z, \zeta),$$

$$R^1 f(z, \zeta) = z f_z'(z, \zeta), \dots,$$

$$(m+1) R^{m+1} f(z, \zeta) = z \left(R^m f(z, \zeta) \right)_z' + m R^m f(z, \zeta), \quad z \in U, \; \zeta \in \overline{U}.$$

Remark 1 ([5]) If $f \in \mathcal{A}_{n\zeta}^*$, $f(z, \zeta) = z + \sum_{j=n+1}^{\infty} a_j(\zeta) z^j$, then
$R^m f(z, \zeta) = z + \sum_{j=n+1}^{\infty} C_{m+j-1}^m a_j(\zeta) z^j, z \in U, \zeta \in \overline{U}.$

Definition 2 ([3]) For $f \in \mathcal{A}_{n\zeta}^*$, $\lambda \geq 0$ and $n, m \in \mathbb{N}$, the operator D_λ^m is defined by $D_\lambda^m : \mathcal{A}_{n\zeta}^* \to \mathcal{A}_{n\zeta}^*$,

$$D_\lambda^0 f(z, \zeta) = f(z, \zeta)$$

$$D_\lambda^1 f(z, \zeta) = (1 - \lambda) f(z, \zeta) + \lambda z f_z'(z, \zeta) = D_\lambda f(z, \zeta), \dots,$$

$$D_\lambda^{m+1} f(z, \zeta) = (1 - \lambda) D_\lambda^m f(z, \zeta) + \lambda z \left(D_\lambda^m f(z, \zeta) \right)_z'$$

$$= D_\lambda \left(D_\lambda^m f(z, \zeta) \right), \; z \in U, \zeta \in \overline{U}.$$

Remark 2 ([3]) If $f \in \mathcal{A}_{n\zeta}^*$ and $f(z) = z + \sum_{j=n+1}^{\infty} a_j(\zeta) z^j$, then
$D_\lambda^m f(z, \zeta) = z + \sum_{j=n+1}^{\infty} [1 + (j-1)\lambda]^m a_j(\zeta) z^j$, for $z \in U, \zeta \in \overline{U}.$

As a dual notion of strong differential subordination G.I. Oros has introduced and developed the notion of strong differential superordinations in [7].

Definition 3 ([7]) Let $f(z, \zeta)$, $H(z, \zeta)$ be analytic in $U \times \overline{U}$. The function $f(z, \zeta)$ is said to be strongly superordinate to $H(z, \zeta)$ if there exists a function w analytic in U, with $w(0) = 0$ and $|w(z)| < 1$, such that $H(z, \zeta) = f(w(z), \zeta)$, for all $\zeta \in \overline{U}$. In such a case we write $H(z, \zeta) \prec\prec f(z, \zeta), z \in U, \zeta \in \overline{U}.$

Remark 3 ([7])

(i) Since $f(z, \zeta)$ is analytic in $U \times \overline{U}$, for all $\zeta \in \overline{U}$, and univalent in U, for all $\zeta \in \overline{U}$, Definition 3 is equivalent to $H(0, \zeta) = f(0, \zeta)$, for all $\zeta \in \overline{U}$, and $H(U \times \overline{U}) \subset f(U \times \overline{U}).$

(ii) If $H(z, \zeta) \equiv H(z)$ and $f(z, \zeta) \equiv f(z)$, the strong superordination becomes the usual notion of superordination.

Definition 4 ([4]) We denote by Q^* the set of functions that are analytic and injective on $\overline{U} \times \overline{U} \backslash E(f, \zeta)$, where $E(f, \zeta) = \{ y \in \partial U : \lim_{z \to y} f(z, \zeta) = \infty \}$, and are such that $f_z'(y, \zeta) \neq 0$ for $y \in \partial U \times \overline{U} \backslash E(f, \zeta)$. The subclass of Q^* for which $f(0, \zeta) = a$ is denoted by $Q^*(a).$

We need the following lemmas to obtain some inequalities for the strong differential superordinations.

Lemma 1 ([4]) *Let $h(z, \zeta)$ be a convex function with $h(0, \zeta) = a$ and let $\gamma \in \mathbb{C}^*$ be a complex number with $\mathrm{Re}\,\gamma \geq 0$. If $p \in \mathscr{H}^*[a, n, \zeta] \cap Q^*$, $p(z, \zeta) + \frac{1}{\gamma} z p'_z(z, \zeta)$ is univalent in $U \times \overline{U}$ and*

$$h(z, \zeta) \prec\prec p(z, \zeta) + \frac{1}{\gamma} z p'_z(z, \zeta), \quad z \in U, \ \zeta \in \overline{U},$$

then

$$q(z, \zeta) \prec\prec p(z, \zeta), \quad z \in U, \ \zeta \in \overline{U},$$

where $q(z, \zeta) = \frac{\gamma}{n z^{\frac{\gamma}{n}}} \int_0^z h(t, \zeta) t^{\frac{\gamma}{n} - 1} dt$, $z \in U$, $\zeta \in \overline{U}$. The function q is convex and is the best subordinant.

Lemma 2 ([4]) *Let $q(z, \zeta)$ be a convex function in $U \times \overline{U}$ and let $h(z, \zeta) = q(z, \zeta) + \frac{1}{\gamma} z q'_z(z, \zeta)$, $z \in U$, $\zeta \in \overline{U}$, where $\mathrm{Re}\,\gamma \geq 0$.*

If $p \in \mathscr{H}^*[a, n, \zeta] \cap Q^*$, $p(z, \zeta) + \frac{1}{\gamma} z p'_z(z, \zeta)$ is univalent in $U \times \overline{U}$ and

$$q(z, \zeta) + \frac{1}{\gamma} z q'_z(z, \zeta) \prec\prec p(z, \zeta) + \frac{1}{\gamma} z p'_z(z, \zeta), \quad z \in U, \ \zeta \in \overline{U},$$

then

$$q(z, \zeta) \prec\prec p(z, \zeta), \quad z \in U, \ \zeta \in \overline{U},$$

where $q(z, \zeta) = \frac{\gamma}{n z^{\frac{\gamma}{n}}} \int_0^z h(t, \zeta) t^{\frac{\gamma}{n} - 1} dt$, $z \in U$, $\zeta \in \overline{U}$. The function q is the best subordinant.

We extend the differential operator studied in [1, 2] to the new class of analytic functions $\mathscr{A}^*_{n\zeta}$.

Definition 5 ([3]) Let $\alpha, \lambda \geq 0$, $n, m \in \mathbb{N}$. Denote by $RD^m_{\lambda, \alpha}$ the extended operator given by $RD^m_{\lambda, \alpha} : \mathscr{A}^*_{n\zeta} \to \mathscr{A}^*_{n\zeta}$,

$$RD^m_{\lambda, \alpha} f(z, \zeta) = (1 - \alpha) R^m f(z, \zeta) + \alpha D^m_\lambda f(z, \zeta), \quad z \in U, \ \zeta \in \overline{U}.$$

Remark 4 ([3]) If $f \in \mathscr{A}^*_{n\zeta}$, $f(z) = z + \sum_{j=n+1}^\infty a_j(\zeta) z^j$, then

$$RD^m_{\lambda, \alpha} f(z, \zeta) = z + \sum_{j=n+1}^\infty \left\{ \alpha [1 + (j - 1)\lambda]^m + (1 - \alpha) C^m_{m+j-1} \right\} a_j(\zeta) z^j,$$

$z \in U, \zeta \in \overline{U}$.

Remark 5 For $\alpha = 0$, $RD^m_{\lambda, 0} f(z, \zeta) = R^m f(z, \zeta)$, where $z \in U$, $\zeta \in \overline{U}$, and for $\alpha = 1$, $RD^m_{\lambda, 1} f(z, \zeta) = D^m_\lambda f(z, \zeta)$, where $z \in U$, $\zeta \in \overline{U}$.

For $\lambda = 1$, we obtain $RD^m_{1, \alpha} f(z, \zeta) = L^m_\alpha f(z, \zeta)$ which was studied in [4, 5].

For $m = 0$, $RD_{\lambda,\alpha}^0 f(z,z) = (1-\alpha) R^0 f(z,\zeta) + \alpha D_\lambda^0 f(z,\zeta) = f(z,\zeta) = R^0 f(z,\zeta) = D_\lambda^0 f(z,\zeta)$, where $z \in U$, $\zeta \in \overline{U}$.

2   Main Results

Theorem 1 *Let* $h(z,\zeta)$ *be a convex function in* $U \times \overline{U}$ *with* $h(0,\zeta) = 1$. *Let* $m \in \mathbb{N}$, $\lambda, \alpha \geq 0$, $f(z,\zeta) \in \mathscr{A}_{n\zeta}^*$, $F(z,\zeta) = I_c(f)(z,\zeta) = \frac{c+2}{z^{c+1}} \int_0^z t^c f(t,\zeta) dt$, $z \in U$, $\zeta \in \overline{U}$, $Rec > -2$, *and suppose that* $\left(RD_{\lambda,\alpha}^m f(z,\zeta)\right)_z'$ *is univalent in* $U \times \overline{U}$, $\left(RD_{\lambda,\alpha}^m F(z,\zeta)\right)_z' \in \mathscr{H}^*[1,n,\zeta] \cap Q^*$ *and*

$$h(z,\zeta) \prec\prec \left(RD_{\lambda,\alpha}^m f(z,\zeta)\right)_z', \quad z \in U, \ \zeta \in \overline{U}, \tag{1}$$

then

$$q(z,\zeta) \prec\prec \left(RD_{\lambda,\alpha}^m F(z,\zeta)\right)_z', \quad z \in U, \ \zeta \in \overline{U},$$

where $q(z,\zeta) = \frac{c+2}{nz^{\frac{c+2}{n}}} \int_0^z h(t,\zeta) t^{\frac{c+2}{n}-1} dt$. *The function* q *is convex and it is the best subordinant.*

Proof We have

$$z^{c+1} F(z,\zeta) = (c+2) \int_0^z t^c f(t,\zeta) dt$$

and differentiating it, with respect to z, we obtain $(c+1) F(z,\zeta) + zF_z'(z,\zeta) = (c+2) f(z,\zeta)$ and

$$(c+1) RD_{\lambda,\alpha}^m F(z,\zeta) + z\left(RD_{\lambda,\alpha}^m F(z,\zeta)\right)_z' = (c+2) RD_{\lambda,\alpha}^m f(z,\zeta), \quad z \in U, \ \zeta \in \overline{U}.$$

Differentiating the last relation with respect to z we have

$$\left(RD_{\lambda,\alpha}^m F(z,\zeta)\right)_z' + \frac{1}{c+2} z\left(RD_{\lambda,\alpha}^m F(z,\zeta)\right)_{z^2}'' = \left(RD_{\lambda,\alpha}^m f(z,\zeta)\right)_z', \quad z \in U, \ \zeta \in \overline{U}. \tag{2}$$

Using (2), the strong differential superordination (1) becomes

$$h(z,\zeta) \prec\prec \left(RD_{\lambda,\alpha}^m F(z,\zeta)\right)_z' + \frac{1}{c+2} z\left(RD_{\lambda,\alpha}^m F(z,\zeta)\right)_{z^2}''. \tag{3}$$

Denote

$$p(z,\zeta) = \left(RD_{\lambda,\alpha}^m F(z,\zeta)\right)_z', \quad z \in U, \ \zeta \in \overline{U}. \tag{4}$$

Replacing (4) in (3) we obtain

$$h(z, \zeta) \prec\prec p(z, \zeta) + \frac{1}{c+2} z p'_z(z, \zeta), \quad z \in U, \ \zeta \in \overline{U}.$$

Using Lemma 1 for $\gamma = c + 2$, we have

$$q(z, \zeta) \prec\prec p(z, \zeta), \ z \in U, \ \zeta \in \overline{U}, \ \text{i.e.} \ q(z, \zeta) \prec\prec \left(RD^m_{\lambda, \alpha} F(z, \zeta)\right)'_z, z \in U, \ \zeta \in \overline{U},$$

where $q(z, \zeta) = \frac{c+2}{n z^{\frac{c+2}{n}}} \int_0^z h(t, \zeta) t^{\frac{c+2}{n}-1} dt$. The function q is convex and it is the best subordinant.

Corollary 1 Let $h(z, \zeta) = \frac{\zeta + (2\beta - \zeta) z}{1+z}$, where $\beta \in [0, 1)$. Let $m \in \mathbb{N}, \lambda, \alpha \geq 0, \ f(z, \zeta) \in \mathscr{A}^*_{n\zeta}, \ F(z, \zeta) = I_c(f)(z, \zeta) = \frac{c+2}{z^{c+1}} \int_0^z t^c f(t, \zeta) dt, \ z \in U, \ \zeta \in \overline{U}, \ Rec > -2$, and suppose that $\left(RD^m_{\lambda, \alpha} f(z, \zeta)\right)'_z$ is univalent in $U \times \overline{U}$, $\left(RD^m_{\lambda, \alpha} F(z, \zeta)\right)'_z \in \mathscr{H}^*[1, n, \zeta] \cap Q^*$ and

$$h(z, \zeta) \prec\prec \left(RD^m_{\lambda, \alpha} f(z, \zeta)\right)'_z, \ z \in U, \ \zeta \in \overline{U}, \tag{5}$$

then

$$q(z, \zeta) \prec\prec \left(RD^m_{\lambda, \alpha} F(z, \zeta)\right)'_z, \quad z \in U, \ \zeta \in \overline{U},$$

*where q is given by $q(z, \zeta) = 2\beta - \zeta + \frac{2(c+2)(\zeta - \beta)}{n z^{\frac{c+2}{n}}} \int_0^z \frac{t^{\frac{c+2}{n}-1}}{t+1} dt, z \in U, \zeta \in \overline{U}$.
The function q is convex and it is the best subordinant.*

Proof Following the same steps as in the proof of Theorem 1 and considering $p(z, \zeta) = \left(RD^m_{\lambda, \alpha} F(z, \zeta)\right)'_z$, the strong differential superordination (5) becomes

$$h(z, \zeta) = \frac{\zeta + (2\beta - \zeta) z}{1+z} \prec\prec p(z, \zeta) + \frac{1}{c+2} z p'_z(z, \zeta), \quad z \in U, \ \zeta \in \overline{U}.$$

By using Lemma 1 for $\gamma = c + 2$, we have $q(z, \zeta) \prec\prec p(z, \zeta)$, i.e.

$$q(z, \zeta) = \frac{c+2}{n z^{\frac{c+2}{n}}} \int_0^z h(t, \zeta) t^{\frac{c+2}{n}-1} dt = \frac{c+2}{n z^{\frac{c+2}{n}}} \int_0^z \frac{\zeta + (2\beta - \zeta) t}{1+t} t^{\frac{c+2}{n}-1} dt$$

$$= 2\beta - \zeta + \frac{2(c+2)(\zeta - \beta)}{n z^{\frac{c+2}{n}}} \int_0^z \frac{t^{\frac{c+2}{n}-1}}{t+1} dt \prec\prec \left(RD^m_{\lambda, \alpha} F(z, \zeta)\right)'_z, \quad z \in U, \ \zeta \in \overline{U}.$$

The function q is convex and it is the best subordinant.

Theorem 2 *Let $q(z, \zeta)$ be a convex function in $U \times \overline{U}$ and let $h(z, \zeta) = q(z, \zeta) + \frac{1}{c+2} z q'_z(z, \zeta)$, where $z \in U, \zeta \in \overline{U}, Rec > -2$.*

Let $m \in \mathbb{N}, \lambda, \alpha \geq 0, f(z, \zeta) \in \mathscr{A}^*_{n\zeta}, F(z, \zeta) = I_c(f)(z, \zeta) = \frac{c+2}{z^{c+1}} \int_0^z t^c f(t, \zeta) dt, z \in U, \zeta \in \overline{U}$, and suppose that $\left(RD^m_{\lambda, \alpha} f(z, \zeta) \right)'_z$ is univalent in $U \times \overline{U}, \left(RD^m_{\lambda, \alpha} F(z, \zeta) \right)'_z \in \mathscr{H}^*[1, n, \zeta] \cap Q^*$ and

$$h(z, \zeta) \prec\prec \left(RD^m_{\lambda, \alpha} f(z, \zeta) \right)'_z, \quad z \in U, \zeta \in \overline{U}, \tag{6}$$

then

$$q(z, \zeta) \prec\prec \left(RD^m_{\lambda, \alpha} F(z, \zeta) \right)'_z, \quad z \in U, \zeta \in \overline{U},$$

where $q(z, \zeta) = \frac{c+2}{nz^{\frac{c+2}{n}}} \int_0^z h(t, \zeta) t^{\frac{c+2}{n}-1} dt$. The function q is the best subordinant.

Proof Following the same steps as in the proof of Theorem 1 and considering $p(z, \zeta) = \left(RD^m_{\lambda, \alpha} F(z, \zeta) \right)'_z, z \in U, \zeta \in \overline{U}$, the strong differential superordination (6) becomes

$$h(z, \zeta) = q(z, \zeta) + \frac{1}{c+2} z q'_z(z, \zeta) \prec\prec p(z, \zeta) + \frac{1}{c+2} z p'_z(z, \zeta), \quad z \in U, \zeta \in \overline{U}.$$

Using Lemma 2 for $\gamma = c + 2$, we have

$$q(z, \zeta) \prec\prec p(z, \zeta), z \in U, \zeta \in \overline{U}, \text{ i.e. } q(z, \zeta) \prec\prec \left(RD^m_{\lambda, \alpha} F(z, \zeta) \right)'_z, z \in U, \zeta \in \overline{U},$$

where $q(z, \zeta) = \frac{c+2}{nz^{\frac{c+2}{n}}} \int_0^z h(t, \zeta) t^{\frac{c+2}{n}-1} dt$. The function q is the best subordinant.

Theorem 3 *Let $h(z, \zeta)$ be a convex function, $h(0, \zeta) = 1$. Let $m \in \mathbb{N}, \lambda, \alpha \geq 0, f(z, \zeta) \in \mathscr{A}^*_{n\zeta}$ and suppose that $\left(RD^m_{\lambda, \alpha} f(z, \zeta) \right)'_z$ is univalent and $\frac{RD^m_{\lambda, \alpha} f(z, \zeta)}{z} \in \mathscr{H}^*[1, n, \zeta] \cap Q^*$. If*

$$h(z, \zeta) \prec\prec \left(RD^m_{\lambda, \alpha} f(z, \zeta) \right)'_z, \quad z \in U, \zeta \in \overline{U}, \tag{7}$$

then

$$q(z, \zeta) \prec\prec \frac{RD^m_{\lambda, \alpha} f(z, \zeta)}{z}, \quad z \in U, \zeta \in \overline{U},$$

where $q(z, \zeta) = \frac{1}{nz^{\frac{1}{n}}} \int_0^z h(t, \zeta) t^{\frac{1}{n}-1} dt$. The function q is convex and it is the best subordinant.

Proof By using the properties of operator $RD_{\lambda,\alpha}^m$, we have for $z \in U$, $\zeta \in \overline{U}$,

$$RD_{\lambda,\alpha}^m f(z, \zeta) = z + \sum_{j=n+1}^{\infty} \left\{ \alpha \left[1 + (j-1)\lambda\right]^m + (1-\alpha) C_{m+j-1}^m \right\} a_j(\zeta) z^j.$$

Consider $p(z, \zeta) = \dfrac{RD_{\lambda,\alpha}^m f(z,\zeta)}{z} = \dfrac{z + \sum_{j=n+1}^{\infty} \left\{ \alpha[1+(j-1)\lambda]^m + (1-\alpha) C_{m+j-1}^m \right\} a_j(\zeta) z^j}{z} =$

$1 + p_n(\zeta) z^n + p_{n+1}(\zeta) z^{n+1} + \ldots$, $z \in U, \zeta \in \overline{U}$.

We deduce that $p \in \mathscr{H}^*[1, n, \zeta]$.

Let $RD_{\lambda,\alpha}^m f(z, \zeta) = zp(z, \zeta)$, $z \in U$, $\zeta \in \overline{U}$. Differentiating with respect to z we obtain $\left(RD_{\lambda,\alpha}^m f(z, \zeta)\right)_z' = p(z, \zeta) + zp_z'(z, \zeta)$, $z \in U$, $\zeta \in \overline{U}$.

Then (7) becomes

$$h(z, \zeta) \prec\prec p(z, \zeta) + zp_z'(z, \zeta), \qquad z \in U, \zeta \in \overline{U}.$$

By using Lemma 1 for $\gamma = 1$, we have

$$q(z, \zeta) \prec\prec p(z, \zeta), \ z \in U, \ \zeta \in \overline{U} \text{ i.e. } q(z, \zeta) \prec\prec \frac{RD_{\lambda,\alpha}^m f(z, \zeta)}{z}, \ z \in U, \ \zeta \in \overline{U},$$

where $q(z, \zeta) = \frac{1}{nz^{\frac{1}{n}}} \int_0^z h(t, \zeta) t^{\frac{1}{n}-1} dt$. The function q is convex and it is the best subordinant.

Corollary 2 *Let* $h(z, \zeta) = \frac{\zeta+(2\beta-\zeta)z}{1+z}$ *be a convex function in* $U \times \overline{U}$, *where* $0 \le \beta < 1$. *Let* $m \in \mathbb{N}$, $\lambda, \alpha \ge 0$, $f(z, \zeta) \in \mathscr{A}_{n\zeta}^*$ *and suppose that* $\left(RD_{\lambda,\alpha}^m f(z, \zeta)\right)_z'$ *is univalent and* $\frac{RD_{\lambda,\alpha}^m f(z,\zeta)}{z} \in \mathscr{H}^*[1, n, \zeta] \cap Q^*$. *If*

$$h(z, \zeta) \prec\prec \left(RD_{\lambda,\alpha}^m f(z, \zeta)\right)_z', \qquad z \in U, \ \zeta \in \overline{U}, \tag{8}$$

then

$$q(z, \zeta) \prec\prec \frac{RD_{\lambda,\alpha}^m f(z, \zeta)}{z}, \qquad z \in U, \ \zeta \in \overline{U},$$

where q is given by $q(z, \zeta) = 2\beta - \zeta + \frac{2(\zeta-\beta)}{nz^{\frac{1}{n}}} \int_0^z \frac{t^{\frac{1}{n}-1}}{t+1} dt$, $z \in U, \zeta \in \overline{U}$. *The function q is convex and it is the best subordinant.*

Proof Following the same steps as in the proof of Theorem 3 and considering $p(z, \zeta) = \frac{RD_{\lambda,\alpha}^m f(z,\zeta)}{z}$, the strong differential superordination (8) becomes

$$h(z, \zeta) = \frac{\zeta + (2\beta - \zeta)z}{1 + z} \prec\prec p(z, \zeta) + zp_z'(z, \zeta), \qquad z \in U, \zeta \in \overline{U}.$$

By using Lemma 1 for $\gamma = 1$, we have $q(z, \zeta) \prec\prec p(z, \zeta)$, i.e.

$$q(z, \zeta) = \frac{1}{nz^{\frac{1}{n}}} \int_0^z h(t, \zeta) t^{\frac{1}{n}-1} dt = \frac{1}{nz^{\frac{1}{n}}} \int_0^z \frac{\zeta + (2\beta - \zeta)t}{1+t} t^{\frac{1}{n}-1} dt$$

$$= 2\beta - \zeta + \frac{2(\zeta - \beta)}{nz^{\frac{1}{n}}} \int_0^z \frac{t^{\frac{1}{n}-1}}{t+1} dt \prec\prec \frac{RD_{\lambda,\alpha}^m f(z, \zeta)}{z}, \quad z \in U, \; \zeta \in \overline{U}.$$

The function q is convex and it is the best subordinant.

Theorem 4 *Let $q(z, \zeta)$ be convex in $U \times \overline{U}$ and let h be defined by $h(z, \zeta) = q(z, \zeta) + zq'_z(z, \zeta)$. If $m \in \mathbb{N}$, $\lambda, \alpha \geq 0$, $f(z, \zeta) \in \mathscr{A}_{n\zeta}^*$, suppose that $\left(RD_{\lambda,\alpha}^m f(z, \zeta) \right)'_z$ is univalent and $\frac{RD_{\lambda,\alpha}^m f(z,\zeta)}{z} \in \mathscr{H}^*[1, n, \zeta] \cap Q^*$ and satisfies the differential superordination*

$$h(z, \zeta) = q(z, \zeta) + zq'_z(z, \zeta) \prec\prec \left(RD_{\lambda,\alpha}^m f(z, \zeta) \right)'_z, \quad z \in U, \; \zeta \in \overline{U}, \quad (9)$$

then

$$q(z, \zeta) \prec\prec \frac{RD_{\lambda,\alpha}^m f(z, \zeta)}{z}, \quad z \in U, \; \zeta \in \overline{U},$$

where $q(z, \zeta) = \frac{1}{nz^{\frac{1}{n}}} \int_0^z h(t, \zeta) t^{\frac{1}{n}-1} dt$. The function q is the best subordinant.

Proof Following the same steps as in the proof of Theorem 3 and considering $p(z, \zeta) = \frac{RD_{\lambda,\alpha}^m f(z,\zeta)}{z}$, the strong differential superordination (9) becomes

$$q(z, \zeta) + zq'_z(z, \zeta) \prec\prec p(z, \zeta) + zp'_z(z, \zeta), \quad z \in U, \; \zeta \in \overline{U}.$$

Using Lemma 2 for $\gamma = 1$, we have

$$q(z, \zeta) \prec\prec p(z, \zeta), \quad z \in U, \; \zeta \in \overline{U}, \quad \text{i.e.}$$

$$q(z, \zeta) = \frac{1}{nz^{\frac{1}{n}}} \int_0^z h(t, \zeta) t^{\frac{1}{n}-1} dt \prec\prec \frac{RD_{\lambda,\alpha}^m f(z, \zeta)}{z}, \quad z \in U, \; \zeta \in \overline{U},$$

and q is the best subordinant.

Theorem 5 *Let $h(z, \zeta)$ be a convex function, $h(0, \zeta) = 1$. Let $m \in \mathbb{N}$, $\lambda, \alpha \geq 0$, $f(z, \zeta) \in \mathscr{A}_{n\zeta}^*$ and suppose that $\left(\frac{zRD_{\lambda,\alpha}^{m+1} f(z,\zeta)}{RD_{\lambda,\alpha}^m f(z,\zeta)} \right)'_z$ is univalent and $\frac{RD_{\lambda,\alpha}^{m+1} f(z,\zeta)}{RD_{\lambda,\alpha}^m f(z,\zeta)} \in \mathscr{H}^*[1, n, \zeta] \cap Q^*$. If*

$$h(z, \zeta) \prec\prec \left(\frac{zRD_{\lambda,\alpha}^{m+1} f(z, \zeta)}{RD_{\lambda,\alpha}^m f(z, \zeta)} \right)'_z, \quad z \in U, \; \zeta \in \overline{U}, \quad (10)$$

then

$$q(z, \zeta) \prec\prec \frac{RD_{\lambda,\alpha}^{m+1} f(z, \zeta)}{RD_{\lambda,\alpha}^{m} f(z, \zeta)}, \qquad z \in U, \ \zeta \in \overline{U},$$

where $q(z, \zeta) = \frac{1}{nz^{\frac{1}{n}}} \int_0^z h(t, \zeta) t^{\frac{1}{n}-1} dt$. *The function q is convex and it is the best subordinant.*

Proof For $f \in \mathscr{A}_{n\zeta}^*$, $f(z) = z + \sum_{j=n+1}^{\infty} a_j(\zeta) z^j$ we have

$RD_{\lambda,\alpha}^{m} f(z, \zeta) = z + \sum_{j=n+1}^{\infty} \left\{ \alpha [1 + (j-1)\lambda]^m + (1-\alpha) C_{m+j-1}^m \right\} a_j(\zeta) z^j,$
$z \in U, \ \zeta \in \overline{U}.$

Consider

$$p(z, \zeta) = \frac{RD_{\lambda,\alpha}^{m+1} f(z, \zeta)}{RD_{\lambda,\alpha}^{m} f(z, \zeta)}$$

$$= \frac{z + \sum_{j=n+1}^{\infty} \left\{ \alpha [1 + (j-1)\lambda]^{m+1} + (1-\alpha) C_{m+j}^{m+1} \right\} a_j(\zeta) z^j}{z + \sum_{j=n+1}^{\infty} \left\{ \alpha [1 + (j-1)\lambda]^m + (1-\alpha) C_{m+j-1}^m \right\} a_j(\zeta) z^j}.$$

We have $p_z'(z, \zeta) = \frac{\left(RD_{\lambda,\alpha}^{m+1} f(z,\zeta)\right)_z'}{RD_{\lambda,\alpha}^{m} f(z,\zeta)} - p(z, \zeta) \cdot \frac{\left(RD_{\lambda,\alpha}^{m} f(z,\zeta)\right)_z'}{RD_{\lambda,\alpha}^{m} f(z,\zeta)}$ and we obtain

$$p(z, \zeta) + z \cdot p_z'(z, \zeta) = \left(\frac{z RD_{\lambda,\alpha}^{m+1} f(z,\zeta)}{RD_{\lambda,\alpha}^{m} f(z,\zeta)} \right)_z'.$$

Relation (10) becomes

$$h(z, \zeta) \prec\prec p(z, \zeta) + z p_z'(z, \zeta), \qquad z \in U, \ \zeta \in \overline{U}.$$

By using Lemma 1 for $\gamma = 1$, we have

$$q(z, \zeta) \prec\prec p(z, \zeta), \ z \in U, \ \zeta \in \overline{U}, \ \text{i.e.} \ q(z, \zeta) \prec\prec \frac{RD_{\lambda,\alpha}^{m+1} f(z, \zeta)}{RD_{\lambda,\alpha}^{m} f(z, \zeta)}, \ z \in U, \ \zeta \in \overline{U},$$

where $q(z, \zeta) = \frac{1}{nz^{\frac{1}{n}}} \int_0^z h(t, \zeta) t^{\frac{1}{n}-1} dt$. The function q is convex and it is the best subordinant.

Corollary 3 *Let* $h(z, \zeta) = \frac{\zeta + (2\beta - \zeta)z}{1+z}$ *be a convex function in* $U \times \overline{U}$, *where* $0 \le \beta < 1$. *Let* $\lambda, \alpha \ge 0$, $m \in \mathbb{N}$, $f(z, \zeta) \in \mathscr{A}_{n\zeta}^*$ *and suppose that* $\left(\frac{z RD_{\lambda,\alpha}^{m+1} f(z,\zeta)}{RD_{\lambda,\alpha}^{m} f(z,\zeta z)} \right)_z'$ *is univalent and* $\frac{RD_{\lambda,\alpha}^{m+1} f(z,\zeta)}{RD_{\lambda,\alpha}^{m} f(z,\zeta)} \in \mathscr{H}^*[1, n, \zeta] \cap Q^*$. *If*

$$h(z, \zeta) \prec\prec \left(\frac{z RD_{\lambda,\alpha}^{m+1} f(z, \zeta)}{RD_{\lambda,\alpha}^m f(z, \zeta)} \right)_z', \qquad z \in U, \ \zeta \in \overline{U}, \tag{11}$$

then

$$q(z, \zeta) \prec\prec \frac{RD_{\lambda,\alpha}^{m+1} f(z, \zeta)}{RD_{\lambda,\alpha}^m f(z, \zeta)}, \qquad z \in U, \ \zeta \in \overline{U},$$

where q is given by $q(z, \zeta) = 2\beta - \zeta + \frac{2(\zeta - \beta)}{nz^{\frac{1}{n}}} \int_0^z \frac{t^{\frac{1}{n}-1}}{t+1} dt, \ z \in U, \ \zeta \in \overline{U}. \ The$
function q is convex and it is the best subordinant.

Proof Following the same steps as in the proof of Theorem 5 and considering
$p(z, \zeta) = \frac{RD_{\lambda,\alpha}^{m+1} f(z,\zeta)}{RD_{\lambda,\alpha}^m f(z,\zeta)}$, the strong differential superordination (11) becomes

$$h(z, \zeta) = \frac{\zeta + (2\beta - \zeta)z}{1 + z} \prec\prec p(z, \zeta) + z p_z'(z, \zeta), \qquad z \in U, \ \zeta \in \overline{U}.$$

By using Lemma 1 for $\gamma = 1$, we have $q(z, \zeta) \prec\prec p(z, \zeta)$, i.e.

$$q(z, \zeta) = \frac{1}{nz^{\frac{1}{n}}} \int_0^z h(t, \zeta) t^{\frac{1}{n}-1} = \frac{1}{nz^{\frac{1}{n}}} \int_0^z \frac{\zeta + (2\beta - \zeta)t}{1 + t} t^{\frac{1}{n}-1} dt$$

$$= 2\beta - \zeta + \frac{2(\zeta - \beta)}{nz^{\frac{1}{n}}} \int_0^z \frac{t^{\frac{1}{n}-1}}{t+1} dt \prec\prec \frac{RD_{\lambda,\alpha}^{m+1} f(z, \zeta)}{RD_{\lambda,\alpha}^m f(z, \zeta)}, \qquad z \in U, \ \zeta \in \overline{U}.$$

The function q is convex and it is the best subordinant.

Theorem 6 *Let* $q(z, \zeta)$ *be a convex function and h be defined by* $h(z, \zeta) = q(z, \zeta) + z q_z'(z, \zeta)$. *Let* $\lambda, \alpha \geq 0, \ m \in \mathbb{N}, \ f(z, \zeta) \in \mathscr{A}_{n\zeta}^*$ *and suppose that*
$\left(\frac{z RD_{\lambda,\alpha}^{m+1} f(z,\zeta)}{RD_{\lambda,\alpha}^m f(z,\zeta)} \right)_z'$ *is univalent and* $\frac{RD_{\lambda,\alpha}^{m+1} f(z,\zeta)}{RD_{\lambda,\alpha}^m f(z,\zeta)} \in \mathscr{H}^*[1, n, \zeta] \cap Q^*$. *If*

$$h(z, \zeta) = q(z, \zeta) + z q_z'(z, \zeta) \prec\prec \left(\frac{z RD_{\lambda,\alpha}^{m+1} f(z, \zeta)}{RD_{\lambda,\alpha}^m f(z, \zeta)} \right)_z', \qquad z \in U, \ \zeta \in \overline{U},$$

$$\tag{12}$$

then

$$q(z, \zeta) \prec \frac{RD_{\lambda,\alpha}^{m+1} f(z, \zeta)}{RD_{\lambda,\alpha}^m f(z, \zeta)}, \qquad z \in U, \ \zeta \in \overline{U},$$

where $q(z, \zeta) = \frac{1}{nz^{\frac{1}{n}}} \int_0^z h(t, \zeta) t^{\frac{1}{n}-1} dt$. *The function q is the best subordinant.*

Proof Following the same steps as in the proof of Theorem 5 and considering $p(z, \zeta) = \frac{RD_{\lambda,\alpha}^{m+1} f(z,\zeta)}{RD_{\lambda,\alpha}^m f(z,\zeta)}$, the strong differential superordination (12) becomes

$$h(z, \zeta) = q(z, \zeta) + zq_z'(z, \zeta) \prec\prec p(z, \zeta) + zp_z'(z, \zeta), \quad z \in U, \ \zeta \in \overline{U}.$$

By using Lemma 2 for $\gamma = 1$, we have

$$q(z, \zeta) \prec\prec p(z, \zeta), \quad z \in U, \ \zeta \in \overline{U}, \quad \text{i.e.}$$

$$q(z, \zeta) = \frac{1}{nz^{\frac{1}{n}}} \int_0^z h(t, \zeta) t^{\frac{1}{n}-1} dt \prec\prec \frac{RD_{\lambda,\alpha}^{m+1} f(z,\zeta)}{RD_{\lambda,\alpha}^m f(z,\zeta)}, \quad z \in U, \ \zeta \in \overline{U},$$

and q is the best subordinant.

Theorem 7 *Let $h(z, \zeta)$ be a convex function in $U \times \overline{U}$ with $h(0, \zeta) = 1$ and let $\lambda, \alpha \geq 0$, $m \in \mathbb{N}$, $f(z, \zeta) \in \mathscr{A}_{n\zeta}^*$, $\frac{(m+1)(m+2)}{z} RD_{\lambda,\alpha}^{m+2} f(z, \zeta) - \frac{(m+1)(2m+1)}{z} RD_{\lambda,\alpha}^{m+1} f(z, \zeta) + \frac{m^2}{z} RD_{\lambda,\alpha}^m f(z, \zeta) - \frac{\alpha\left[(m+1)(m+2)-\frac{1}{\lambda^2}\right]}{z} D_\lambda^{m+2} f(z, \zeta) + \frac{\alpha\left[(m+1)(2m+1)-\frac{2(1-\lambda)}{\lambda^2}\right]}{z} D_\lambda^{m+1} f(z, \zeta) - \frac{\alpha\left[m^2 - \frac{(1-\lambda)^2}{\lambda^2}\right]}{z} D_\lambda^m f(z, \zeta)$ is univalent and $[RD_{\lambda,\alpha}^m f(z,\zeta)]_z' \in \mathscr{H}^*[1, n, \zeta] \cap Q^*$. If*

$$h(z, \zeta) \prec\prec \frac{(m+1)(m+2)}{z} RD_{\lambda,\alpha}^{m+2} f(z, \zeta) - \frac{(m+1)(2m+1)}{z} RD_{\lambda,\alpha}^{m+1} f(z, \zeta) +$$

$$\tag{13}$$

$$\frac{m^2}{z} RD_{\lambda,\alpha}^m f(z, \zeta) - \frac{\alpha\left[(m+1)(m+2)-\frac{1}{\lambda^2}\right]}{z} D_\lambda^{m+2} f(z, \zeta) +$$

$$\frac{\alpha\left[(m+1)(2m+1)-\frac{2(1-\lambda)}{\lambda^2}\right]}{z} D_\lambda^{m+1} f(z, \zeta) - \frac{\alpha\left[m^2 - \frac{(1-\lambda)^2}{\lambda^2}\right]}{z} D_\lambda^m f(z, \zeta),$$

$z \in U, \ \zeta \in \overline{U}$, holds, then

$$q(z, \zeta) \prec\prec [RD_{\lambda,\alpha}^m f(z, \zeta)]_z', \quad z \in U, \ \zeta \in \overline{U},$$

where $q(z, \zeta) = \frac{1}{nz^{\frac{1}{n}}} \int_0^z h(t, \zeta) t^{\frac{1}{n}-1}$. The function q is convex and it is the best subordinant.

Proof For $f \in \mathscr{A}_{n\zeta}^*$, $f(z) = z + \sum_{j=n+1}^\infty a_j(\zeta) z^j$ we have
$$RD_{\lambda,\alpha}^m f(z, \zeta) = z + \sum_{j=n+1}^\infty \left\{\alpha[1 + (j-1)\lambda]^m + (1-\alpha) C_{m+j-1}^m\right\} a_j(\zeta) z^j,$$
$z \in U, \ \zeta \in \overline{U}$.

Let

$$p(z, \zeta) = \left(RD_{\lambda,\alpha}^m f(z, \zeta)\right)'_z \tag{14}$$

$$= 1 + \sum_{j=n+1}^{\infty} \left\{ \alpha \left[1 + (j-1)\lambda\right]m + (1-\alpha)C_{m+j-1}^m \right\} ja_j(\zeta) z^{j-1}$$

$$= 1 + p_n(\zeta) z^n + p_{n+1}(\zeta) z^{n+1} + \dots.$$

By using the properties of operators $RD_{\lambda,\alpha}^m$, R^m and D_λ^m, after a short calculation, we obtain

$$p(z, \zeta) + zp_z'(z, \zeta) = \frac{(m+1)(m+2)}{z} RD_{\lambda,\alpha}^{m+2} f(z, \zeta) - \frac{(m+1)(2m+1)}{z} RD_{\lambda,\alpha}^{m+1} f(z, \zeta)$$

$$+ \frac{m^2}{z} RD_{\lambda,\alpha}^m f(z, \zeta) - \frac{\alpha\left[(m+1)(m+2) - \frac{1}{\lambda^2}\right]}{z} D_\lambda^{m+2} f(z, \zeta) + \frac{\alpha\left[(m+1)(2m+1) - \frac{2(1-\lambda)}{\lambda^2}\right]}{z}$$

$$D_\lambda^{m+1} f(z, \zeta) - \frac{\alpha\left[m^2 - \frac{(1-\lambda)^2}{\lambda^2}\right]}{z} D_\lambda^m f(z, \zeta).$$

Using the notation in (14), the strong differential superordination becomes

$$h(z, \zeta) \prec\prec p(z, \zeta) + zp_z'(z, \zeta).$$

By using Lemma 1 for $\gamma = 1$, we have

$$q(z, \zeta) \prec\prec p(z, \zeta), \ z \in U, \ \zeta \in \overline{U}, \ \text{i.e.} \ q(z, \zeta) \prec\prec \left(RD_{\lambda,\alpha}^m f(z, \zeta)\right)'_z, \ z \in U, \ \zeta \in \overline{U},$$

where $q(z, \zeta) = \frac{1}{nz^{\frac{1}{n}}} \int_0^z h(t, \zeta) t^{\frac{1}{n}-1}$. The function q is convex and it is the best subordinant.

Corollary 4 *Let* $h(z, \zeta) = \frac{\zeta + (2\beta - \zeta)z}{1+z}$ *be a convex function in* $U \times \overline{U}$, *where* $0 \leq \beta < 1$. *Let* $\lambda, \alpha \geq 0$, $m \in \mathbb{N}$, $f(z, \zeta) \in \mathscr{A}_{n\zeta}^*$, *suppose that*

$$\frac{(m+1)(m+2)}{z} RD_\lambda^{m+2} f(z, \zeta) - \frac{(m+1)(2m+1)}{z} RD_{\lambda,\alpha}^{m+1} f(z, \zeta) + \frac{m^2}{z} RD_{\lambda,\alpha}^m f(z, \zeta)$$

$$- \frac{\alpha\left[(m+1)(m+2) - \frac{1}{\lambda^2}\right]}{z} D_\lambda^{m+2} f(z, \zeta) + \frac{\alpha\left[(m+1)(2m+1) - \frac{2(1-\lambda)}{\lambda^2}\right]}{z} D_\lambda^{m+1} f(z, \zeta)$$

$$- \frac{\alpha\left[m^2 - \frac{(1-\lambda)^2}{\lambda^2}\right]}{z} D_\lambda^m f(z, \zeta) \ \text{is univalent in} \ U \times \overline{U} \ \text{and} \ [RD_{\lambda,\alpha}^m f(z, \zeta)]'_z \in \mathscr{H}^*[1, n, \zeta] \cap Q^*. \ \text{If}$$

$$h(z, \zeta) \prec \frac{(m+1)(m+2)}{z} RD_{\lambda,\alpha}^{m+2} f(z, \zeta) - \frac{(m+1)(2m+1)}{z} RD_{\lambda,\alpha}^{m+1} f(z, \zeta) + \tag{15}$$

$$\frac{m^2}{z} RD_{\lambda,\alpha}^m f(z, \zeta) - \frac{\alpha\left[(m+1)(m+2) - \frac{1}{\lambda^2}\right]}{z} D_\lambda^{m+2} f(z, \zeta) +$$

$$\frac{\alpha\left[(m+1)(2m+1) - \frac{2(1-\lambda)}{\lambda^2}\right]}{z} D_\lambda^{m+1} f(z, \zeta) - \frac{\alpha\left[m^2 - \frac{(1-\lambda)^2}{\lambda^2}\right]}{z} D_\lambda^m f(z, \zeta),$$

$z \in U, \zeta \in \overline{U}$, then

$$q(z, \zeta) \prec\prec \left(RD_{\lambda,\alpha}^m f(z, \zeta)\right)_z', \quad z \in U, \ \zeta \in \overline{U},$$

where q is given by $q(z, \zeta) = 2\beta - \zeta + \frac{2(\zeta - \beta)}{nz^{\frac{1}{n}}} \int_0^z \frac{t^{\frac{1}{n}-1}}{t+1} dt$, $z \in U, \ \zeta \in \overline{U}$. The function q is convex and it is the best subordinant.

Proof Following the same steps as in the proof of Theorem 7 and considering $p(z, \zeta) = \left(RD_{\lambda,\alpha}^m f(z, \zeta)\right)_z'$, the strong differential superordination (15) becomes

$$h(z, \zeta) = \frac{\zeta + (2\beta - \zeta)z}{1+z} \prec\prec p(z, \zeta) + zp_z'(z, \zeta), \quad z \in U, \ \zeta \in \overline{U}.$$

By using Lemma 1 for $\gamma = 1$, we have $q(z, \zeta) \prec\prec p(z, \zeta)$, i.e.

$$q(z, \zeta) = \frac{1}{nz^{\frac{1}{n}}} \int_0^z h(t, \zeta) t^{\frac{1}{n}-1} = \frac{1}{nz^{\frac{1}{n}}} \int_0^z \frac{\zeta + (2\beta - \zeta)t}{1+t} t^{\frac{1}{n}-1} dt$$

$$= 2\beta - \zeta + \frac{2(\zeta - \beta)}{nz^{\frac{1}{n}}} \int_0^z \frac{t^{\frac{1}{n}-1}}{t+1} dt \prec\prec \left(RD_{\lambda,\alpha}^m f(z, \zeta)\right)_z', \quad z \in U, \ \zeta \in \overline{U}.$$

The function q is convex and it is the best subordinant.

Theorem 8 Let $q(z, \zeta)$ be a convex function in $U \times \overline{U}$ and $h(z, \zeta) = q(z, \zeta) + zq_z'(z, \zeta)$. Let $\lambda, \alpha \geq 0$, $m \in \mathbb{N}$, $f(z, \zeta) \in \mathcal{A}_{n\zeta}^*$, suppose that $\frac{(m+1)(m+2)}{z} RD_{\lambda,\alpha}^{m+2} f(z, \zeta) - \frac{(m+1)(2m+1)}{z} RD_{\lambda,\alpha}^{m+1} f(z, \zeta) + \frac{m^2}{z} RD_{\lambda,\alpha}^m$ $f(z, \zeta) - \frac{\alpha\left[(m+1)(m+2)-\frac{1}{\lambda^2}\right]}{z} D_\lambda^{m+2} f(z, \zeta) + \frac{\alpha\left[(m+1)(2m+1)-\frac{2(1-\lambda)}{\lambda^2}\right]}{z} D_\lambda^{m+1} f(z, \zeta) - \frac{\alpha\left[m^2 - \frac{(1-\lambda)^2}{\lambda^2}\right]}{z} D_\lambda^m f(z, \zeta)$ is univalent in $U \times \overline{U}$ and $[RD_{\lambda,\alpha}^m f(z, \zeta)]_z' \in \mathcal{H}^*[1, n, \zeta] \cap Q^*$. If

$$h(z, \zeta) = q(z, \zeta) + zq_z'(z, \zeta) \prec\prec \frac{(m+1)(m+2)}{z} RD_{\lambda,\alpha}^{m+2} f(z, \zeta) - \quad (16)$$

$$\frac{(m+1)(2m+1)}{z} RD_{\lambda,\alpha}^{m+1} f(z, \zeta) + \frac{m^2}{z} RD_{\lambda,\alpha}^m f(z, \zeta) -$$

$$\frac{\alpha\left[(m+1)(m+2)-\frac{1}{\lambda^2}\right]}{z} D_\lambda^{m+2} f(z, \zeta) + \frac{\alpha\left[(m+1)(2m+1)-\frac{2(1-\lambda)}{\lambda^2}\right]}{z} D_\lambda^{m+1} f(z, \zeta)$$

$$- \frac{\alpha\left[m^2 - \frac{(1-\lambda)^2}{\lambda^2}\right]}{z} D_\lambda^m f(z, \zeta), \quad z \in U, \ \zeta \in \overline{U},$$

then

$$q(z, \zeta) \prec\prec \left(RD_{\lambda,\alpha}^m f(z, \zeta)\right)_z', \quad z \in U, \ \zeta \in \overline{U},$$

where $q(z, \zeta) = \frac{1}{nz^{\frac{1}{n}}} \int_0^z h(t, \zeta) t^{\frac{1}{n}-1}$. *The function q is the best subordinant.*

Proof Following the same steps as in the proof of Theorem 7 and considering $p(z, \zeta) = \left(RD_{\lambda,\alpha}^m f(z, \zeta)\right)_z'$, the strong differential superordination (16) becomes

$$h(z, \zeta) = q(z, \zeta) + zq_z'(z, \zeta) \prec\prec p(z, \zeta) + zp_z'(z, \zeta), \quad z \in U, .$$

By using Lemma 2 for $\gamma = 1$, we have $q(z, \zeta) \prec\prec p(z, \zeta)$, i.e.

$$q(z, \zeta) = \frac{1}{nz^{\frac{1}{n}}} \int_0^z h(t, \zeta) t^{\frac{1}{n}-1} \prec\prec \left(RD_{\lambda,\alpha}^m f(z, \zeta)\right)_z', \quad z \in U, \ \zeta \in \overline{U}.$$

The function q is the best subordinant.

References

1. A. Alb Lupaş, On special differential superordinations using a generalized Sălăgean operator and Ruscheweyh derivative. Comput. Math. Appl. **61**, 1048–1058 (2011). https://doi.org/10.1016/j.camwa.2010.12.055
2. A. Alb Lupaş, On special differential subordinations using a generalized Sălăgean operator and Ruscheweyh derivative. J. Comput. Anal. Appl. **13**(1), 98–107 (2011)
3. A. Alb Lupaş, On special strong differential subordinations using a generalized Sălăgean operator and Ruscheweyh derivative. J. Concr. Appl. Math. **10**(1–2), 17–23 (2012)
4. A. Alb Lupaş, On special strong differential superordinations using Sălăgean and Ruscheweyh operators. J. Adv. Appl. Comput. Math. **1**(1), 1–7 (2014)
5. A. Alb Lupaş, G.I. Oros, Gh. Oros, On special strong differential subordinations using Sălăgean and Ruscheweyh operators. J. Comput. Anal. Appl. **14**(2), 266–270 (2012)
6. F.M. Al-Oboudi, On univalent functions defined by a generalized Sălăgean operator. Ind. J. Math. Math. Sci. (25–28), 1429–1436 (2004)
7. G.I. Oros, Strong differential superordination. Acta Univ. Apulensis (19), 101–106 (2009)
8. G.I. Oros, On a new strong differential subordination. Acta Univ. Apulensis (32), 243–250 (2012)
9. St. Ruscheweyh, New criteria for univalent functions. Proc. Amet. Math. Soc. **49**, 109–115 (1975)

Conformable Fractional Inequalities

George A. Anastassiou

Abstract This is a long journey in the modern realm of Conformable fractional differentiation. In that setting the author presents the following types of analytic inequalities: Landau, Hilbert–Pachpatte, Ostrowski, Opial, Poincare, and Sobolev inequalities. We present uniform and L_p results, involving left and right conformable fractional derivatives, as well engaging several functions. We discuss many interesting special cases.

1 Introduction

Our motivations to write this work follow. The first inspiration comes next.

Let $p \in [1, \infty]$, $I = \mathbb{R}_+$ or $I = \mathbb{R}$ and $f : I \to \mathbb{R}$ is twice differentiable with $f, f'' \in L_p(I)$, then $f' \in L_p(I)$.

Moreover, there exists a constant $C_p(I) > 0$ independent of f, such that

$$\left\| f' \right\|_{p,I} \leq C_p(I) \left\| f \right\|_{p,I}^{\frac{1}{2}} \left\| f'' \right\|_{p,I}^{\frac{1}{2}},$$

where $\|\cdot\|_{p,I}$ is the p-norm on the interval I, see [3, 14].

The research on these inequalities started by Landau [20] in 1914. For the case of $p = \infty$ he proved that

$$C_\infty(\mathbb{R}_+) = 2 \text{ and } C_\infty(\mathbb{R}) = \sqrt{2},$$

are the best constants above.

G. A. Anastassiou (✉)
Department of Mathematical Sciences, University of Memphis, Memphis, TN, USA
e-mail: ganastss@memphis.edu

© Springer Nature Switzerland AG 2019   371
G. A. Anastassiou and J. M. Rassias (eds.), *Frontiers in Functional Equations and Analytic Inequalities*, https://doi.org/10.1007/978-3-030-28950-8_21

In 1932, Hardy and Littlewood [17] proved the above inequality for $p = 2$, with the best constants

$$C_2\left(\mathbb{R}_+\right) = \sqrt{2} \text{ and } C_2\left(\mathbb{R}\right) = 1.$$

In 1935, Hardy et al. [16] showed that the best constant $C_p\left(\mathbb{R}_+\right)$ above satisfies the estimate

$$C_p\left(\mathbb{R}_+\right) \le 2 \text{ for } p \in [1, \infty),$$

which yields $C_p\left(\mathbb{R}\right) \le 2$ for $p \in [1, \infty)$.

In fact in [15] and [18], it was shown that $C_p\left(\mathbb{R}\right) \le \sqrt{2}$.

The author in [9] studied extensively fractional Landau type inequalities involving right and left Caputo fractional derivatives.

The famous Ostrowski ([21]) inequality motivates this work and has as follows:

$$\left| \frac{1}{b-a} \int_a^b f\left(y\right) dy - f\left(x\right) \right| \le \left(\frac{1}{4} + \frac{\left(x - \frac{a+b}{2}\right)^2}{\left(b-a\right)^2} \right) \left(b-a\right) \left\| f' \right\|_\infty,$$

where $f \in C^1\left([a, b]\right)$, $x \in [a, b]$, and it is a sharp inequality.

Another motivation is author's next Ostrowski type fractional result, see [9], p. 44:

Let $[a, b] \subset \mathbb{R}$, $\alpha > 0$, $m = \lceil \alpha \rceil$ ($\lceil \cdot \rceil$ ceiling of the number), $f \in AC^m\left([a, b]\right)$ (i.e., $f^{(m-1)}$ is absolutely continuous), and $\left\| D_{x_0-}^\alpha f \right\|_{\infty, [a, x_0]}$, $\left\| D_{*x_0}^\alpha f \right\|_{\infty, [x_0, b]} < \infty$ (where $D_{x_0-}^\alpha f$, $D_{*x_0}^\alpha f$ are the right and left Caputo fractional derivatives of f of order α, respectively), $x_0 \in [a, b]$. Assume $f^{(k)}\left(x_0\right) = 0$, $k = 1, \ldots, m - 1$.

Then

$$\left| \frac{1}{b-a} \int_a^b f\left(x\right) dx - f\left(x_0\right) \right| \le \frac{1}{\left(b-a\right) \Gamma\left(\alpha + 2\right)} \cdot$$

$$\left\{ \left\| D_{x_0-}^\alpha f \right\|_{\infty, [a, x_0]} \left(x_0 - a\right)^{\alpha+1} + \left\| D_{*x_0}^\alpha f \right\|_{\infty, [x_0, b]} \left(b - x_0\right)^{\alpha+1} \right\} \le$$

$$\frac{1}{\Gamma\left(\alpha + 2\right)} \max \left\{ \left\| D_{x_0-}^\alpha f \right\|_{\infty, [a, x_0]}, \left\| D_{*x_0}^\alpha f \right\|_{\infty, [x_0, b]} \right\} \left(b - a\right)^\alpha.$$

The author's monographs [4–10] motivate and support greatly this work too.

From the point of view of Conformable fractional differentiation the author scans the broad area of analytic inequalities and reveals a great variety of well-known inequalities in the Conformable fractional environment to all possible directions.

2 Main Results—I

We need

Definition 1 ([2, 19]) Let $f : [0, \infty) \to \mathbb{R}$. The conformable α-fractional derivative for $\alpha \in (0, 1]$ is given by

$$D_\alpha f(t) := \lim_{\varepsilon \to 0} \frac{f\left(t + \varepsilon t^{1-\alpha}\right) - f(t)}{\varepsilon}, \tag{1}$$

$$D_\alpha f(0) = \lim_{t \to 0+} D_\alpha f(t). \tag{2}$$

If f is differentiable, then

$$D_\alpha f(t) = t^{1-\alpha} f'(t), \tag{3}$$

where f' is the usual derivative.

We define

$$D_\alpha^n f = D_\alpha^{n-1}(D_\alpha f). \tag{4}$$

If $f : [0, \infty) \to \mathbb{R}$ is α-differentiable at $t_0 > 0$, $\alpha \in (0, 1]$, then f is continuous at t_0, see [19].

Definition 2 ([13]) Let $\alpha \in (0, 1]$ and $0 \le a < b$. A function $f : [a, b] \to \mathbb{R}$ is α-fractional integrable on $[a, b]$ if the integral

$$I_\alpha^a f(b) := \int_a^b f(t) \, d_\alpha t := \int_a^b f(t) \, t^{\alpha-1} dt, \tag{5}$$

exists and is finite.

We need

Theorem 1 ([13]) (Ostrowski Type Inequality) *Let $a, b, t \in \mathbb{R}_+$ with $0 \le a < b$, and let $f : [a, b] \to \mathbb{R}$ be α-fractional differentiable for $\alpha \in (0, 1]$. Then*

$$\left| \frac{\alpha}{b^\alpha - a^\alpha} \int_a^b f(t) \, d_\alpha t - f(t) \right| \le \frac{M_1}{2\alpha (b^\alpha - a^\alpha)} \left[(t^\alpha - a^\alpha)^2 + (b^\alpha - t^\alpha)^2 \right], \tag{6}$$

where

$$M_1 := \sup_{t \in (a,b)} |D_\alpha f(t)|. \tag{7}$$

Inequality (6) is sharp.

Corollary 1 (to Theorem 1) *Let $a, b \in \mathbb{R}_+$ with $0 \leq a < b$, and let $f : [a, b] \to \mathbb{R}$ be α-fractional differentiable for $\alpha \in (0, 1]$. Then*

$$\left| \frac{\alpha}{b^\alpha - a^\alpha} \int_a^b f(t) \, d_\alpha t - f(a) \right| \leq \frac{M_1}{2\alpha} \left(b^\alpha - a^\alpha \right), \tag{8}$$

where

$$M_1 := \sup_{t \in (a,b)} |D_\alpha f(t)|.$$

We need

Theorem 2 ([11]) *Let $\alpha \in (0, 1]$, and $f : [a, b] \to \mathbb{R}$, $a \geq 0$, be α-fractional differentiable on $[a, b]$. Assume that $D_\alpha f$ is continuous on $[a, b]$. Then*

$$I_\alpha^a D_\alpha f(t) = f(t) - f(a), \quad \forall \, t \in [a, b]. \tag{9}$$

We make

Remark 1 Let $\alpha \in (0, 1]$, and any $a, b \in \mathbb{R}_+ : 0 \leq a < b$, and $D_\alpha f$ is α-fractional differentiable and continuous on every $[a, b] \subset \mathbb{R}_+$. By Corollary 1 we get

$$\left| \frac{\alpha}{b^\alpha - a^\alpha} \int_a^b D_\alpha f(t) \, d_\alpha t - D_\alpha f(a) \right| \leq \frac{M_2}{2\alpha} \left(b^\alpha - a^\alpha \right), \tag{10}$$

where

$$M_2 := \sup_{t \in (a,b)} \left| D_\alpha^2 f(t) \right|.$$

By Theorem 2, equivalently we have

$$\left| \frac{\alpha}{b^\alpha - a^\alpha} \left(f(b) - f(a) \right) - D_\alpha f(a) \right| \leq \frac{M_2}{2\alpha} \left(b^\alpha - a^\alpha \right). \tag{11}$$

Hence it holds

$$|D_\alpha f(a)| - \frac{\alpha}{b^\alpha - a^\alpha} |f(b) - f(a)| \leq \frac{M_2}{2\alpha} \left(b^\alpha - a^\alpha \right). \tag{12}$$

Equivalently, we can write

$$|D_\alpha f(a)| \leq \frac{\alpha}{b^\alpha - a^\alpha} |f(b) - f(a)| + \frac{M_2}{2\alpha} \left(b^\alpha - a^\alpha \right) \leq \tag{13}$$

$$\left(\frac{\alpha}{b^\alpha - a^\alpha} \right) \left(2 \, \|f\|_{\infty, [0, +\infty)} \right) + \left(\frac{b^\alpha - a^\alpha}{2\alpha} \right) \left\| D_\alpha^2 f \right\|_{\infty, [0, +\infty)},$$

$\forall \, a, b \in \mathbb{R}_+ : a < b$.

Notice that the right-hand side of (13) depends only on $b^\alpha - a^\alpha$. Therefore it holds

$$\|D_\alpha f\|_{\infty,[0,+\infty)} \le \left(\frac{2\alpha}{b^\alpha - a^\alpha}\right) \|f\|_{\infty,[0,+\infty)} + \left(\frac{\|D_\alpha^2 f\|_{\infty,[0,+\infty)}}{2\alpha}\right)(b^\alpha - a^\alpha).$$

(14)

Set $t := b^\alpha - a^\alpha > 0$. Thus

$$\|D_\alpha f\|_{\infty,[0,+\infty)} \le \left(\frac{2\alpha}{t}\right) \|f\|_{\infty,[0,+\infty)} + \left(\frac{\|D_\alpha^2 f\|_{\infty,[0,+\infty)}}{2\alpha}\right) t, \ \forall \, t > 0.$$

(15)

Call

$$\mu := 2\alpha \|f\|_{\infty,[0,+\infty)},$$

$$and$$

(16)

$$\theta := \left(\frac{\|D_\alpha^2 f\|_{\infty,[0,+\infty)}}{2\alpha}\right),$$

both are greater than zero.

That is, we have

$$\|D_\alpha f\|_{\infty,[0,+\infty)} \le \frac{\mu}{t} + \theta \cdot t, \forall \, t > 0.$$

(17)

Consider the function

$$y(t) := \frac{\mu}{t} + \theta \cdot t, t > 0.$$

(18)

As in [9], pp. 80–82, y has a global minimum at

$$t_0 = \left(\frac{\mu}{\theta}\right)^{\frac{1}{2}},$$

(19)

which is

$$y(t_0) = 2\sqrt{\theta \mu}.$$

(20)

Consequently we derive

$$y(t_0) = 2\sqrt{\|f\|_{\infty,[0,+\infty)} \|D_\alpha^2 f\|_{\infty,[0,+\infty)}}.$$

(21)

We have proved that

$$\|D_\alpha f\|_{\infty,[0,+\infty)} \le 2\sqrt{\|f\|_{\infty,[0,+\infty)} \|D_\alpha^2 f\|_{\infty,[0,+\infty)}}. \qquad (22)$$

We have established the following conformable fractional Landau type inequality:

Theorem 3 *Let* $f : \mathbb{R}_+ \to \mathbb{R}$ *be* α-*fractional differentiable,* $\alpha \in (0, 1]$. *And* $D_\alpha f$ *is also* α-*fractional differentiable and continuous on* $\mathbb{R}_+$. *Assume that* $\|f\|_{\infty,\mathbb{R}_+}$, $\|D_\alpha^2 f\|_{\infty,\mathbb{R}_+} < \infty$. *Then*

$$\|D_\alpha f\|_{\infty,\mathbb{R}_+} \le 2 \|f\|_{\infty,\mathbb{R}_+}^{\frac{1}{2}} \|D_\alpha^2 f\|_{\infty,\mathbb{R}_+}^{\frac{1}{2}}, \qquad (23)$$

that is, $\|D_\alpha f\|_{\infty,\mathbb{R}_+} < \infty$.

Note 1 If f is differentiable then $D_\alpha f(t) = t^{1-\alpha} f'(t)$, $t > 0$, $\alpha \in (0, 1]$. When $t > 0$, $t^{1-\alpha}$ is differentiable. If f is twice differentiable and $t > 0$, then we have

$$D_\alpha^2 f(t) = D_\alpha (D_\alpha f(t)) = D_\alpha \left(t^{1-\alpha} f'(t)\right) = t^{1-\alpha} \left(t^{1-\alpha} f'(t)\right)'$$

$$= t^{1-\alpha} \left((1-\alpha) t^{-\alpha} f'(t) + t^{1-\alpha} f''(t)\right).$$

That is an interesting formula:

$$D_\alpha^2 f(t) = (1-\alpha) t^{1-2\alpha} f'(t) + t^{2(1-\alpha)} f''(t), t > 0. \qquad (24)$$

We need

Definition 3 Let $\alpha \in (0, 1]$. We define the spaces of functions:

$$L_\alpha^p([a, b]) := \left\{ f : [a, b] \subset \mathbb{R}_+ \to \mathbb{R} : \int_a^b |f(t)|^p \, d_\alpha t < +\infty, p \ge 1 \right\},$$

and

$$L_\alpha^p(\mathbb{R}_+) := \left\{ f : \mathbb{R}_+ \to \mathbb{R} : \int_{\mathbb{R}_+} |f(x)|^p \, d_\alpha x := \int_{\mathbb{R}_+} |f(x)|^p x^{\alpha-1} dx < +\infty, p \ge 1 \right\}.$$

We need the conformable fractional L_p Ostrowski type inequality:

Theorem 4 ([22]) *Let* $a \ge 0$, $f : [a, b] \to \mathbb{R}$ *be an* α-*fractional differentiable function for* $\alpha \in (0, 1]$, $D_\alpha(f) \in L_\alpha^p([a, b])$; $p, q > 1 : \frac{1}{p} + \frac{1}{q} = 1$. *Then for all* $x \in [a, b]$, *we have the inequality:*

$$\left| \frac{\alpha}{b^\alpha - a^\alpha} \int_a^b f(t) \, d_\alpha t - f(x) \right| \le A_\alpha(x, q) \, \|D_\alpha(f)\|_p, \qquad (25)$$

where

$$A_\alpha(x, q) = \frac{1}{(b^\alpha - a^\alpha)} \left(\frac{1}{\alpha(q+1)} \left(\frac{b^\alpha - a^\alpha}{2} \right)^{\alpha(q+1)} \right)^{\frac{1}{q}} +$$

$$\left| \frac{1}{\alpha} \left(x^\alpha - \frac{a^\alpha + b^\alpha}{2} \right) \right|^{\frac{1}{q}}. \qquad (26)$$

When $x = a$, we get:

$$\left| \frac{\alpha}{b^\alpha - a^\alpha} \int_a^b f(t) \, d_\alpha t - f(a) \right| \le A_\alpha(a, q) \, \|D_\alpha(f)\|_p, \qquad (27)$$

where

$$A_\alpha(a, q) = \frac{1}{b^\alpha - a^\alpha} \left(\frac{1}{\alpha(q+1)} \left(\frac{b^\alpha - a^\alpha}{2} \right)^{\alpha(q+1)} \right)^{\frac{1}{q}}$$

$$+ \left[\frac{1}{\alpha} \left(\frac{b^\alpha - a^\alpha}{2} \right) \right]^{\frac{1}{q}}. \qquad (28)$$

We need

Corollary 2 ([22]) *Let $a \ge 0$, $f : [a, b] \to \mathbb{R}$ be an α-fractional differentiable function for $\alpha \in (0, 1]$, $D_\alpha(f) \in L_\alpha^p([a, b])$, $p, q > 1 : \frac{1}{p} + \frac{1}{q} = 1$. Then*

$$\left| \frac{\alpha}{b^\alpha - a^\alpha} \int_a^b f(t) \, d_\alpha t - f\left(\left(\frac{a^\alpha + b^\alpha}{2} \right)^{\frac{1}{\alpha}} \right) \right| \le \qquad (29)$$

$$\frac{1}{2} \left(\frac{1}{\alpha(q+1)} \right)^{\frac{1}{2}} \left(\frac{b^\alpha - a^\alpha}{2} \right)^{\alpha\left(1 + \frac{1}{q}\right) - 1} \|D_\alpha(f)\|_{p,[a,b]}.$$

We make

Remark 2 Here $f : \mathbb{R}_+ \to \mathbb{R}$ be α-fractional differentiable and $0 < \alpha \le 1$, and $D_\alpha f$ is also α-fractional differentiable and continuous function on $\mathbb{R}_+$, and $D_\alpha^2(f) \in L_\alpha^p(\mathbb{R}_+)$, $p, q > 1 : \frac{1}{p} + \frac{1}{q} = 1$. Here $[a, b] \subset \mathbb{R}_+$.

Then, by (29), we get:

$$\left| D_\alpha f\left(\left(\frac{a^\alpha + b^\alpha}{2}\right)^{\frac{1}{\alpha}}\right) - \frac{\alpha}{b^\alpha - a^\alpha} \int_a^b D_\alpha f(t)\, d_\alpha t \right| \le$$

$$\frac{1}{2}\left(\frac{1}{\alpha(q+1)}\right)^{\frac{1}{2}}\left(\frac{b^\alpha - a^\alpha}{2}\right)^{\alpha\left(1+\frac{1}{q}\right)-1}\left\| D_\alpha^2(f) \right\|_{p,[a,b]}, \tag{30}$$

equivalently it holds

$$\left| D_\alpha f\left(\left(\frac{a^\alpha + b^\alpha}{2}\right)^{\frac{1}{\alpha}}\right) - \frac{\alpha}{b^\alpha - a^\alpha}(f(b) - f(a)) \right| \le$$

$$\frac{1}{2}\left(\frac{1}{\alpha(q+1)}\right)^{\frac{1}{2}}\left(\frac{b^\alpha - a^\alpha}{2}\right)^{\alpha\left(1+\frac{1}{q}\right)-1}\left\| D_\alpha^2(f) \right\|_{p,[a,b]}. \tag{31}$$

Hence it follows

$$\left| D_\alpha f\left(\left(\frac{a^\alpha + b^\alpha}{2}\right)^{\frac{1}{\alpha}}\right) \right| - \frac{\alpha}{b^\alpha - a^\alpha}|f(b) - f(a)| \le$$

$$\frac{1}{2}\left(\frac{1}{\alpha(q+1)}\right)^{\frac{1}{2}}\left\| D_\alpha^2(f) \right\|_{p,[a,b]}\left(\frac{b^\alpha - a^\alpha}{2}\right)^{\alpha\left(1+\frac{1}{q}\right)-1}. \tag{32}$$

Thus, it holds

$$\left| D_\alpha f\left(\left(\frac{a^\alpha + b^\alpha}{2}\right)^{\frac{1}{\alpha}}\right) \right| \le \frac{\left(2\alpha \|f\|_{\infty,\mathbb{R}_+}\right)}{b^\alpha - a^\alpha} +$$

$$\left[\frac{1}{2^{\alpha\left(1+\frac{1}{q}\right)}}\left(\frac{1}{\alpha(q+1)}\right)^{\frac{1}{q}}\left\| D_\alpha^2(f) \right\|_{p,\mathbb{R}_+}\right](b^\alpha - a^\alpha)^{\alpha\left(1+\frac{1}{q}\right)-1} \tag{33}$$

true, $\forall\, a, b \in \mathbb{R}_+, a < b$.

The right-hand side of (33) depends only on $b^\alpha - a^\alpha$.

We have $a^\alpha \le \frac{a^\alpha + b^\alpha}{2} \le b^\alpha$, iff $a \le \left(\frac{a^\alpha + b^\alpha}{2}\right)^{\frac{1}{\alpha}} \le b$.

From now on we assume that $|D_\alpha f|$ is increasing (or decreasing) then

$$|D_\alpha f(a)| \le \left| D_\alpha f\left(\left(\frac{a^\alpha + b^\alpha}{2}\right)^{\frac{1}{\alpha}}\right) \right| \tag{34}$$

(or $|D_\alpha f(b)| \le \left| D_\alpha f\left(\left(\frac{a^\alpha + b^\alpha}{2}\right)^{\frac{1}{\alpha}}\right) \right|$).

Therefore it holds

$$\|D_\alpha f\|_{\infty,\mathbb{R}_+} \leq \frac{\left(2\alpha \|f\|_{\infty,\mathbb{R}_+}\right)}{b^\alpha - a^\alpha} + \tag{35}$$

$$\left(\frac{1}{2^{\alpha\left(1+\frac{1}{q}\right)}} \left(\frac{1}{\alpha\,(q+1)}\right)^{\frac{1}{q}} \left\|D_\alpha^2\,(f)\right\|_{p,\mathbb{R}_+}\right) \left(b^\alpha - a^\alpha\right)^{\alpha\left(1+\frac{1}{q}\right)-1}.$$

Set $t := b^\alpha - a^\alpha > 0$, so that

$$\|D_\alpha f\|_{\infty,\mathbb{R}_+} \leq \frac{\left(2\alpha \|f\|_{\infty,\mathbb{R}_+}\right)}{t} + \tag{36}$$

$$\left(\frac{1}{2^{\alpha\left(1+\frac{1}{q}\right)}} \left(\frac{1}{\alpha\,(q+1)}\right)^{\frac{1}{q}} \left\|D_\alpha^2\,(f)\right\|_{p,\mathbb{R}_+}\right) t^{\alpha\left(1+\frac{1}{q}\right)-1}, \ \forall\, t > 0.$$

Call

$$\widetilde{\mu} := 2\alpha \|f\|_{\infty,\mathbb{R}_+},$$

$$and \tag{37}$$

$$\widetilde{\theta} := \left(\frac{1}{2^{\alpha\left(1+\frac{1}{q}\right)}} \left(\frac{1}{\alpha\,(q+1)}\right)^{\frac{1}{q}} \left\|D_\alpha^2\,(f)\right\|_{p,\mathbb{R}_+}\right),$$

both are greater than 0.

From now on we consider $\alpha \in (0, 1)$, i.e. $0 < \alpha < 1$, thus $\frac{1}{\alpha} > 1$.
We would like to have

$$0 < \alpha\left(1 + \frac{1}{q}\right) - 1 < 1 \Leftrightarrow$$

$$(0 <)\frac{1-\alpha}{\alpha} < \frac{1}{q} < \frac{2-\alpha}{\alpha}; \tag{38}$$

where $0 < \frac{1}{q} < 1$.

By $\alpha < 1$ we get $\frac{2-\alpha}{\alpha} > 1$. Therefore $\frac{1}{q} < \frac{2-\alpha}{\alpha}$, always correct.
Inequalities (38) are written, equivalently, as

$$\frac{\alpha}{2-\alpha} < q < \frac{\alpha}{1-\alpha}. \tag{39}$$

Notice that $\frac{1}{2} < \alpha < 1$ is equivalently to $\frac{\alpha}{1-\alpha} > 1$.

From now on we assume that

$$\frac{1}{2} < \alpha < 1 \text{ and } 1 < q < \frac{\alpha}{1-\alpha}, \tag{40}$$

and it holds

$$0 < \alpha\left(1 + \frac{1}{q}\right) - 1 < 1. \tag{41}$$

Next, we call

$$\widetilde{v} := \alpha\left(1 + \frac{1}{q}\right) - 1, \ \widetilde{v} \in (0, 1). \tag{42}$$

We consider the function

$$\widetilde{y}(t) = \frac{\widetilde{\mu}}{t} + \widetilde{\theta} t^{\widetilde{v}}, \ t \in (0, \infty). \tag{43}$$

Next we act as in [9], pp. 80–82.

The only critical number here is

$$\widetilde{t_0} = \left(\frac{\widetilde{\mu}}{\widetilde{v}\widetilde{\theta}}\right)^{\frac{1}{\widetilde{v}+1}}, \tag{44}$$

and $\widetilde{y}$ has a global minimum at $\widetilde{t_0}$, which is

$$\widetilde{y}\left(\widetilde{t_0}\right) = \left(\widetilde{\theta}\widetilde{\mu}^{\widetilde{v}}\right)^{\frac{1}{\widetilde{v}+1}} (\widetilde{v} + 1) \widetilde{v}^{-\left(\frac{\widetilde{v}}{\widetilde{v}+1}\right)}. \tag{45}$$

Thus, we have proved

$$\|D_\alpha f\|_{\infty, \mathbb{R}_+} \leq \tag{46}$$

$$\left[\left(\frac{1}{2^{\alpha\left(1+\frac{1}{q}\right)}}\right)\left(\frac{1}{\alpha(q+1)}\right)^{\frac{1}{q}} \left\|D_\alpha^2(f)\right\|_{p,\mathbb{R}_+}\right) \left(2\alpha \|f\|_{\infty,\mathbb{R}_+}\right)^{a\left(1+\frac{1}{q}\right)-1}\right]^{\frac{1}{\alpha\left(1+\frac{1}{q}\right)}}$$

$$\left(\alpha\left(1 + \frac{1}{q}\right)\right)\left(\alpha\left(1 + \frac{1}{q}\right) - 1\right)^{-\frac{\left(\alpha\left(1+\frac{1}{q}\right)-1\right)}{\alpha\left(1+\frac{1}{q}\right)}}.$$

We have established the following L_p conformable fractional Landau type inequality:

Theorem 5 *Here* $f : \mathbb{R}_+ \to \mathbb{R}$ *be* α-*fractional differentiable with* $\frac{1}{2} < \alpha < 1$, *and* $D_\alpha f$ *is also* α-*fractional differentiable and continuous function on* $\mathbb{R}_+$, *and* $D_\alpha^2 (f) \in L_\alpha^p (\mathbb{R}_+)$, *where* $p, q > 1 : \frac{1}{p} + \frac{1}{q} = 1, 1 < q < \frac{\alpha}{1-\alpha}$. *Assume* $|D_\alpha f|$ *is monotone and* $\|f\|_{\infty, \mathbb{R}_+} < \infty$. *Then*

$$\|D_\alpha f\|_{\infty, \mathbb{R}_+} \le \left[\|f\|_{\infty, \mathbb{R}_+}^{\left(\frac{\left(\alpha \left(1 + \frac{1}{q}\right) - 1 \right)}{\alpha \left(1 + \frac{1}{q}\right)} \right)} \left\| D_\alpha^2 (f) \right\|_{p, \mathbb{R}_+}^{\frac{1}{\alpha \left(1 + \frac{1}{q}\right)}} \right] .$$

$$\left[\frac{(2\alpha)^{\alpha \left(1 + \frac{1}{q}\right) - 1}}{2^{\alpha \left(1 + \frac{1}{q}\right)} \left(\alpha \left(q + 1 \right) \right)^{\frac{1}{q}}} \right]^{\frac{1}{\alpha \left(1 + \frac{1}{q}\right)}} \left(\alpha \left(1 + \frac{1}{q} \right) \right) \left(\alpha \left(1 + \frac{1}{q} \right) - 1 \right)^{- \frac{\left(\alpha \left(1 + \frac{1}{q}\right) - 1 \right)}{\alpha \left(1 + \frac{1}{q}\right)}} .$$

$$(47)$$

That is, $\|D_\alpha f\|_{\infty, \mathbb{R}_+} < +\infty$.

3 Main Results—II

In this section we use generalized Conformable fractional calculus.

Here we follow [1] for the basics of generalized Conformable fractional calculus, see also [19].

We need

Definition 4 ([1]) Let $a, b \in \mathbb{R}$. The left conformable fractional derivative starting from a of a function $f : [a, \infty) \to \mathbb{R}$ of order $0 < \alpha \le 1$ is defined by

$$\left(T_\alpha^a f \right) (t) = \lim_{\varepsilon \to 0} \frac{f \left(t + \varepsilon \left(t - a \right)^{1-\alpha} \right) - f(t)}{\varepsilon}. \tag{48}$$

If $\left(T_\alpha^a f \right) (t)$ exists on (a, b), then

$$\left(T_\alpha^a f \right) (a) = \lim_{t \to a+} \left(T_\alpha^a f \right) (t). \tag{49}$$

The right conformable fractional derivative of order $0 < \alpha \le 1$ terminating at b of $f : (-\infty, b] \to \mathbb{R}$ is defined by

$$\left({}_\alpha^b T f \right) (t) = - \lim_{\varepsilon \to 0} \frac{f \left(t + \varepsilon \left(b - t \right)^{1-\alpha} \right) - f(t)}{\varepsilon}. \tag{50}$$

If $\left({}_\alpha^b T f \right) (t)$ exists on (a, b), then

$$\left({}_\alpha^b T f \right) (b) = \lim_{t \to b-} \left({}_\alpha^b T f \right) (t). \tag{51}$$

Note that if f is differentiable then

$$\left(T_\alpha^a f\right)(t) = (t-a)^{1-\alpha} f'(t),\tag{52}$$

and

$$\left(_\alpha^b Tf\right)(t) = -(b-t)^{1-\alpha} f'(t).\tag{53}$$

Denote by

$$\left(I_\alpha^a f\right)(t) = \int_a^t (x-a)^{\alpha-1} f(x)\,dx,\tag{54}$$

and

$$\left(^b I_\alpha f\right)(t) = \int_t^b (b-x)^{\alpha-1} f(x)\,dx,\tag{55}$$

these are the left and right conformable fractional integrals of order $0 < \alpha \le 1$.

In the higher order case we can generalize things as follows:

Definition 5 ([1]) Let $\alpha \in (n, n+1]$, and set $\beta = \alpha - n$. Then, the left conformable fractional derivative starting from a of a function $f : [a, \infty) \to \mathbb{R}$ of order α, where $f^{(n)}(t)$ exists, is defined by

$$\left(\mathbf{T}_\alpha^a f\right)(t) = \left(T_\beta^a f^{(n)}\right)(t),\tag{56}$$

The right conformable fractional derivative of order α terminating at b of $f : (-\infty, b] \to \mathbb{R}$, where $f^{(n)}(t)$ exists, is defined by

$$\left(_\alpha^b \mathbf{T} f\right)(t) = (-1)^{n+1} \left(_\beta^b T f^{(n)}\right)(t).\tag{57}$$

If $\alpha = n+1$, then $\beta = 1$ and $\mathbf{T}_{n+1}^a f = f^{(n+1)}$.

If n is odd, then $_{n+1}^b \mathbf{T} f = -f^{(n+1)}$, and if n is even, then $_{n+1}^b \mathbf{T} f = f^{(n+1)}$.

When $n = 0$ (or $\alpha \in (0, 1]$), then $\beta = \alpha$, and (56), (57) collapse to $\{(48)-(51)\}$, respectively.

Lemma 1 ([1]) *Let $f : (a, b) \to \mathbb{R}$ be continuously differentiable and $0 < \alpha \le 1$. Then, for all $t > a$ we have*

$$I_\alpha^a T_\alpha^a (f)(t) = f(t) - f(a).\tag{58}$$

We need

Definition 6 (See Also [1]) If $\alpha \in (n, n+1]$, then the left fractional integral of order α starting at a is defined by

$$\left(\mathbf{I}_\alpha^a f\right)(t) = \frac{1}{n!} \int_a^t (t-x)^n (x-a)^{\beta-1} f(x)\, dx. \tag{59}$$

Similarly, (author's definition, see [12]) the right fractional integral of order α terminating at b is defined by

$$\left({}^b\mathbf{I}_\alpha f\right)(t) = \frac{1}{n!} \int_t^b (x-t)^n (b-x)^{\beta-1} f(x)\, dx. \tag{60}$$

We need

Proposition 1 ([1]) *Let* $\alpha \in (n, n+1]$ *and* $f : [a, \infty) \to \mathbb{R}$ *be* $(n+1)$ *times continuously differentiable for* $t > a$. *Then, for all* $t > a$ *we have*

$$\mathbf{I}_\alpha^a \mathbf{T}_\alpha^a (f)(t) = f(t) - \sum_{k=0}^n \frac{f^{(k)}(a)(t-a)^k}{k!}. \tag{61}$$

We also have

Proposition 2 ([12]) *Let* $\alpha \in (n, n+1]$ *and* $f : (-\infty, b] \to \mathbb{R}$ *be* $(n+1)$ *times continuously differentiable for* $t < b$. *Then, for all* $t < b$ *we have*

$$-{}^b \mathbf{I}_\alpha \, {}_\alpha^b \mathbf{T} (f)(t) = f(t) - \sum_{k=0}^n \frac{f^{(k)}(b)(t-b)^k}{k!}. \tag{62}$$

If $n = 0$ *or* $0 < \alpha \le 1$, *then (see also [1])*

$$^b I_\alpha \, {}_\alpha^b T (f)(t) = f(t) - f(b). \tag{63}$$

In conclusion we derive

Theorem 6 ([12]) *Let* $\alpha \in (n, n+1]$ *and* $f \in C^{n+1}([a, b])$, $n \in \mathbb{N}$. *Then*

(1)

$$f(t) - \sum_{k=0}^n \frac{f^{(k)}(a)(t-a)^k}{k!} = \frac{1}{n!} \int_a^t (t-x)^n (x-a)^{\beta-1} \left(\mathbf{T}_\alpha^a (f)\right)(x)\, dx, \tag{64}$$

and

(2)

$$f(t) - \sum_{k=0}^{n} \frac{f^{(k)}(b)(t-b)^k}{k!} = -\frac{1}{n!} \int_t^b (b-x)^{\beta-1}(x-t)^n \left({}_{\alpha}^{b}\mathbf{T}(f)\right)(x)\,dx,$$

(65)

$\forall\, t \in [a, b]$.

We need

Remark 3 ([12]) We notice the following: let $\alpha \in (n, n+1]$ and $f \in C^{n+1}([a, b])$, $n \in \mathbb{N}$. Then ($\beta := \alpha - n$, $0 < \beta \leq 1$)

$$\left(\mathbf{T}_{\alpha}^{a}(f)\right)(x) = \left(T_{\beta}^{\alpha} f^{(n)}\right)(x) = (x-a)^{1-\beta}\, f^{(n+1)}(x),$$

(66)

and

$$\left({}_{\alpha}^{b}\mathbf{T}(f)\right)(x) = (-1)^{n+1}\left({}_{\beta}^{b}T f^{(n)}\right)(x) =$$

$$(-1)^{n+1}(-1)(b-x)^{1-\beta}\, f^{(n+1)}(x) = (-1)^n (b-x)^{1-\beta}\, f^{(n+1)}(x).$$

(67)

Consequently we get that

$$\left(\mathbf{T}_{\alpha}^{a}(f)\right)(x),\ \left({}_{\alpha}^{b}\mathbf{T}(f)\right)(x) \in C([a, b]).$$

Furthermore it is obvious that

$$\left(\mathbf{T}_{\alpha}^{a}(f)\right)(a) = \left({}_{\alpha}^{b}\mathbf{T}(f)\right)(b) = 0,$$

(68)

when $0 < \beta < 1$, i.e. when $\alpha \in (n, n+1)$.

If $f^{(k)}(a) = 0$, $k = 1, \ldots, n$, then

$$f(t) - f(a) = \frac{1}{n!} \int_a^t (t-x)^n (x-a)^{\beta-1} \left(\mathbf{T}_{\alpha}^{a}(f)\right)(x)\,dx,$$

(69)

$\forall\, t \in [a, b]$.

If $f^{(k)}(b) = 0$, $k = 1, \ldots, n$, then

$$f(t) - f(b) = -\frac{1}{n!} \int_t^b (b-x)^{\beta-1}(x-t)^n \left({}_{\alpha}^{b}\mathbf{T}(f)\right)(x)\,dx,$$

(70)

$\forall\, t \in [a, b]$.

We make

Remark 4 Here let $\alpha_i \in (n_i, n_i + 1]$, $f_i \in C^{n_i+1}([a_i, b_i])$, $n_i \in \mathbb{Z}_+$; $\beta_i := \alpha_i - n_i$ $(0 < \beta_i \leq 1)$, where $i = 1, 2$.

By definition we have

$$\left(T_{\alpha_i}^{a_i}\left(f_i\right)\right)(t_i) = \left(T_{\beta_i}^{\alpha_i}\left(f_i^{(n_i)}\right)\right)(t_i), \quad i = 1, 2.$$

Assume that $f_i^{(k_i)}(a_i) = 0, k_i = 0, 1, \ldots, n_i; i = 1, 2.$
Then (by (69))

$$f_i(t_i) = \frac{1}{n_i!} \int_{a_i}^{t_i} (t_i - x_i)^{n_i} (x_i - a_i)^{\beta_i - 1} \left(T_{\alpha_i}^{a_i}(f_i)\right)(x_i) \, dx_i, \tag{71}$$

$\forall\, t_i \in [a_i, b_i]; i = 1, 2.$
Let $p, q > 1 : \frac{1}{p} + \frac{1}{q} = 1$, then

$$|f_i(t_i)| \le \frac{(b_i - a_i)^{n_i}}{n_i!} \int_{a_i}^{t_i} (x_i - a_i)^{\beta_i - 1} \left|\left(T_{\alpha_i}^{a_i}(f_i)\right)(x_i)\right| dx_i, \tag{72}$$

$i = 1, 2.$
Therefore we get

$$|f_1(t_1)| \le \frac{(b_1 - a_1)^{n_1}}{n_1!} \int_{a_1}^{t_1} (x_1 - a_1)^{\beta_1 - 1} \left|T_{\alpha_1}^{a_1}(f_1)(x_1)\right| dx_1 \le$$

$$\frac{(b_1 - a_1)^{n_1}}{n_1!} \left(\int_{a_1}^{t_1} (x_1 - a_1)^{p(\beta_1 - 1)} dx_1\right)^{\frac{1}{p}} \left(\int_{a_1}^{t_1} \left|\left(T_{\alpha_1}^{a_1}(f_1)\right)(x_1)\right|^q dx_1\right)^{\frac{1}{q}} \le$$

$$\frac{(b_1 - a_1)^{n_1}}{n_1!} \left(\frac{(t_1 - a_1)^{p(\beta_1 - 1) + 1}}{p(\beta_1 - 1) + 1}\right)^{\frac{1}{p}} \left\|T_{\alpha_1}^{a_1}(f_1)\right\|_{L_q([a_1, b_1])}, \tag{73}$$

under the assumption $\beta_1 > \frac{1}{q} \Leftrightarrow p(\beta_1 - 1) + 1 > 0.$
We have proved that

$$|f_1(t_1)| \le \frac{(b_1 - a_1)^{n_1}}{n_1!} \left(\frac{(t_1 - a_1)^{p(\beta_1 - 1) + 1}}{p(\beta_1 - 1) + 1}\right)^{\frac{1}{p}} \left\|T_{\alpha_1}^{a_1}(f_1)\right\|_{L_q([a_1, b_1])}, \tag{74}$$

$\forall\, t_1 \in [a_1, b_1]$, where $\beta_1 > \frac{1}{q}$.
Similarly, by assuming $\beta_2 > \frac{1}{p}$, we get

$$|f_2(t_2)| \le \frac{(b_2 - a_2)^{n_2}}{n_2!} \left(\frac{(t_2 - a_2)^{q(\beta_2 - 1) + 1}}{q(\beta_2 - 1) + 1}\right)^{\frac{1}{q}} \left\|T_{\alpha_2}^{a_2}(f_2)\right\|_{L_p([a_2, b_2])}, \tag{75}$$

$\forall\, t_2 \in [a_2, b_2].$

Hence we have (by (74) and (75) multiplication)

$$|f_1(t_1)||f_2(t_2)| \leq \left[\frac{(b_1-a_1)^{n_1}}{n_1!} \cdot \frac{(b_2-a_2)^{n_2}}{n_2!}\right]$$

$$\frac{1}{(p(\beta_1-1)+1)^{\frac{1}{p}}(q(\beta_2-1)+1)^{\frac{1}{q}}}(t_1-a_1)^{\frac{p(\beta_1-1)+1}{p}}(t_2-a_2)^{\frac{q(\beta_2-1)+1}{q}}$$

$$\left\|T_{\alpha_1}^{a_1}(f_1)\right\|_{L_q([a_1,b_1])}\left\|T_{\alpha_2}^{a_2}(f_2)\right\|_{L_p([a_2,b_2])} \leq \tag{76}$$

(using Young's inequality for $a, b \geq 0$, $a^{\frac{1}{p}}b^{\frac{1}{q}} \leq \frac{a}{p}+\frac{b}{q}$)

$$\left(\frac{(b_1-a_1)^{n_1}(b_2-a_2)^{n_2}}{n_1!n_2!}\right)\frac{1}{(p(\beta_1-1)+1)^{\frac{1}{p}}(q(\beta_2-1)+1)^{\frac{1}{q}}}$$

$$\left[\frac{(t_1-a_1)^{p(\beta_1-1)+1}}{p}+\frac{(t_2-a_2)^{q(\beta_2-1)+1}}{q}\right]$$

$$\left\|T_{\alpha_1}^{a_1}(f_1)\right\|_{L_q([a_1,b_1])}\left\|T_{\alpha_2}^{a_2}(f_2)\right\|_{L_p([a_2,b_2])}, \tag{77}$$

$\forall\, t_i \in [a_i, b_i]; i = 1, 2.$

Therefore we can write

$$\frac{|f_1(t_1)||f_2(t_2)|}{\left[\frac{(t_1-a_1)^{p(\beta_1-1)+1}}{p}+\frac{(t_2-a_2)^{q(\beta_2-1)+1}}{q}\right]} \leq$$

$$\frac{(b_1-a_1)^{n_1}(b_2-a_2)^{n_2}}{n_1!n_2!(p(\beta_1-1)+1)^{\frac{1}{p}}(q(\beta_2-1)+1)^{\frac{1}{q}}} \tag{78}$$

$$\left\|T_{\alpha_1}^{a_1}(f_1)\right\|_{L_q([a_1,b_1])}\left\|T_{\alpha_2}^{a_2}(f_2)\right\|_{L_p([a_2,b_2])},$$

$\forall\, t_i \in [a_i, b_i]; i = 1, 2.$

The denominator of the left-hand side of (78) can be zero only when both $t_1 = a_1$ and $t_2 = a_2$.

Therefore it holds

$$\int_{a_1}^{b_1}\int_{a_2}^{b_2}\frac{|f_1(t_1)||f_2(t_2)|\,dt_1dt_2}{\left[\frac{(t_1-a_1)^{p(\beta_1-1)+1}}{p}+\frac{(t_2-a_2)^{q(\beta_2-1)+1}}{q}\right]} \leq$$

$$\frac{(b_1 - a_1)^{n_1+1} (b_2 - a_2)^{n_2+1} \left\| T_{\alpha_1}^{a_1} (f_1) \right\|_{L_q([a_1,b_1])} \left\| T_{\alpha_2}^{a_2} (f_2) \right\|_{L_p([a_2,b_2])}}{n_1! n_2! \left(p (\beta_1 - 1) + 1 \right)^{\frac{1}{p}} \left(q (\beta_2 - 1) + 1 \right)^{\frac{1}{q}}}. \tag{79}$$

Notice here that $T_{\alpha_i}^{a_i} (f_i) \in C ([a_i, b_i])$.

We have proved the left Conformable fractional Hilbert–Pachpatte inequality:

Theorem 7 *Let* $\alpha_i \in (n_i, n_i + 1]$, $f_i \in C^{n_i+1} ([a_i, b_i])$, $[a_i, b_i] \subset \mathbb{R}$, $n_i \in \mathbb{Z}_+$; $\beta_i := \alpha_i - n_i$, $i = 1, 2$; $p, q > 1 : \frac{1}{p} + \frac{1}{q} = 1$. *Assume that* $f_i^{(k_i)} (a_i) = 0$, $k_i = 0, 1, \ldots, n_i$; $i = 1, 2$. *Suppose that* $\beta_1 > \frac{1}{q}$ *and* $\beta_2 > \frac{1}{p}$. *Then*

$$\int_{a_1}^{b_1} \int_{a_2}^{b_2} \frac{|f_1 (t_1)| \, |f_2 (t_2)| \, dt_1 dt_2}{\left[\frac{(t_1 - a_1)^{p(\beta_1-1)+1}}{p} + \frac{(t_2 - a_2)^{q(\beta_2-1)+1}}{q} \right]} \leq \tag{80}$$

$$\frac{(b_1 - a_1)^{n_1+1} (b_2 - a_2)^{n_2+1} \left\| T_{\alpha_1}^{a_1} (f_1) \right\|_{L_q([a_1,b_1])} \left\| T_{\alpha_2}^{a_2} (f_2) \right\|_{L_p([a_2,b_2])}}{n_1! n_2! \left(p (\beta_1 - 1) + 1 \right)^{\frac{1}{p}} \left(q (\beta_2 - 1) + 1 \right)^{\frac{1}{q}}}.$$

We make

Remark 5 Here let $\alpha_i \in (n_i, n_i + 1]$, $f_i \in C^{n_i+1} ([a_i, b_i])$, $n_i \in \mathbb{Z}_+$; $\beta_i := \alpha_i - n_i$ $(0 < \beta_i \leq 1)$, where $i = 1, 2$.

By definition we have

$$\left({}_{\alpha_i}^{b_i} T (f_i) \right) (t_i) = (-1)^{n_i+1} \left({}_{\beta_i}^{b_i} T \left(f_i^{(n_i)} \right) \right) (t_i), \quad i = 1, 2.$$

Assume that $f_i^{(k_i)} (b_i) = 0$, $k_i = 0, 1, \ldots, n_i$; $i = 1, 2$.

Then (by (70))

$$f_i (t_i) = -\frac{1}{n_i!} \int_{t_i}^{b_i} (b_i - x_i)^{\beta_i - 1} (x_i - t_i)^{n_i} \left({}_{\alpha_i}^{b_i} T (f_i) \right) (x_i) \, dx_i, \tag{81}$$

$\forall \, t_i \in [a_i, b_i]$; $i = 1, 2$ $(\beta_i := \alpha_i - n_i, 0 < \beta_i < 1$ when $\alpha_i \in (n_i, n_i + 1))$.

Let $p, q > 1 : \frac{1}{p} + \frac{1}{q} = 1$, then

$$|f_i (t_i)| \leq \frac{(b_i - a_i)^{n_i}}{n_i!} \int_{t_i}^{b_i} (b_i - x_i)^{\beta_i - 1} \left| \left({}_{\alpha_i}^{b_i} T (f_i) \right) (x_i) \right| dx_i, \tag{82}$$

$i = 1, 2$.

We have

$$|f_1(t_1)| \le \frac{(b_1 - a_1)^{n_1}}{n_1!} \int_{t_1}^{b_1} (b_1 - x_1)^{\beta_1 - 1} \left| \left({}_{\alpha_1}^{b_1} T(f_1) \right)(x_1) \right| dx_1 \le$$

$$\frac{(b_1 - a_1)^{n_1}}{n_1!} \left(\int_{t_1}^{b_1} (b_1 - x_1)^{p(\beta_1 - 1)} dx_1 \right)^{\frac{1}{p}} \left\| {}_{\alpha_1}^{b_1} T(f_1) \right\|_{L_q([a_1, b_1])} = \qquad (83)$$

$$\frac{(b_1 - a_1)^{n_1}}{n_1!} \left(\frac{(b_1 - t_1)^{p(\beta_1 - 1) + 1}}{p(\beta_1 - 1) + 1} \right)^{\frac{1}{p}} \left\| {}_{\alpha_1}^{b_1} T(f_1) \right\|_{L_q([a_1, b_1])}.$$

We assume $\beta_1 > \frac{1}{q}$ and we have proved

$$|f_1(t_1)| \le \frac{(b_1 - a_1)^{n_1}}{n_1!} \left(\frac{(b_1 - t_1)^{p(\beta_1 - 1) + 1}}{p(\beta_1 - 1) + 1} \right)^{\frac{1}{p}} \left\| {}_{\alpha_1}^{b_1} T(f_1) \right\|_{L_q([a_1, b_1])}, \qquad (84)$$

$\forall\, t_1 \in [a_1, b_1]$.

Similarly, by assuming $\beta_2 > \frac{1}{p}$, we get

$$|f_2(t_2)| \le \frac{(b_2 - a_2)^{n_2}}{n_2!} \left(\frac{(b_2 - t_2)^{q(\beta_2 - 1) + 1}}{q(\beta_2 - 1) + 1} \right)^{\frac{1}{q}} \left\| {}_{\alpha_2}^{b_2} T(f_2) \right\|_{L_p([a_2, b_2])}, \qquad (85)$$

$\forall\, t_2 \in [a_2, b_2]$.

Hence it holds

$$|f_1(t_1)|\, |f_2(t_2)| \le \left[\frac{(b_1 - a_1)^{n_1}}{n_1!} \frac{(b_2 - a_2)^{n_2}}{n_2!} \right]$$

$$\frac{1}{(p(\beta_1 - 1) + 1)^{\frac{1}{p}} (q(\beta_2 - 1) + 1)^{\frac{1}{q}}} (b_1 - t_1)^{\frac{p(\beta_1 - 1) + 1}{p}} (b_2 - t_2)^{\frac{q(\beta_2 - 1) + 1}{q}}$$

$$\qquad (86)$$

$$\left\| {}_{\alpha_1}^{b_1} T(f_1) \right\|_{L_q([a_1, b_1])} \left\| {}_{\alpha_2}^{b_2} T(f_2) \right\|_{L_p([a_2, b_2])} \le$$

(using Young's inequality for $a, b \ge 0$, $a^{\frac{1}{p}} b^{\frac{1}{q}} \le \frac{a}{p} + \frac{b}{q}$)

$$\left(\frac{(b_1 - a_1)^{n_1} (b_2 - a_2)^{n_2}}{n_1! n_2!} \right) \frac{1}{(p(\beta_1 - 1) + 1)^{\frac{1}{p}} (q(\beta_2 - 1) + 1)^{\frac{1}{q}}}$$

$$\left[\frac{(b_1 - t_1)^{p(\beta_1 - 1) + 1}}{p} + \frac{(b_2 - t_2)^{q(\beta_2 - 1) + 1}}{q} \right] \tag{87}$$

$$\left\| {}_{\alpha_1}^{b_1} T(f_1) \right\|_{L_q([a_1, b_1])} \left\| {}_{\alpha_2}^{b_2} T(f_2) \right\|_{L_p([a_2, b_2])},$$

$\forall \, t_i \in [a_i, b_i]; i = 1, 2.$

Therefore we can write

$$\frac{|f_1(t_1)| \, |f_2(t_2)|}{\left[\frac{(b_1 - t_1)^{p(\beta_1 - 1) + 1}}{p} + \frac{(b_2 - t_2)^{q(\beta_2 - 1) + 1}}{q} \right]} \leq$$

$$\frac{(b_1 - a_1)^{n_1} (b_2 - a_2)^{n_2}}{n_1! n_2! \left(p(\beta_1 - 1) + 1 \right)^{\frac{1}{p}} \left(q(\beta_2 - 1) + 1 \right)^{\frac{1}{q}}} \tag{88}$$

$$\left\| {}_{\alpha_1}^{b_1} T(f_1) \right\|_{L_q([a_1, b_1])} \left\| {}_{\alpha_2}^{b_2} T(f_2) \right\|_{L_p([a_2, b_2])},$$

$\forall \, t_i \in [a_i, b_i]; i = 1, 2.$

The denominator of the left-hand side of (88) equals 0 only when both $t_1 = b_1$ and $t_2 = b_2$.

Therefore it holds

$$\int_{a_1}^{b_1} \int_{a_2}^{b_2} \frac{|f_1(t_1)| \, |f_2(t_2)| \, dt_1 dt_2}{\left[\frac{(b_1 - t_1)^{p(\beta_1 - 1) + 1}}{p} + \frac{(b_2 - t_2)^{q(\beta_2 - 1) + 1}}{q} \right]} \leq$$

$$\frac{(b_1 - a_1)^{n_1 + 1} (b_2 - a_2)^{n_2 + 1} \left\| {}_{\alpha_1}^{b_1} T(f_1) \right\|_{L_q([a_1, b_1])} \left\| {}_{\alpha_2}^{b_2} T(f_2) \right\|_{L_p([a_2, b_2])}}{n_1! n_2! \left(p(\beta_1 - 1) + 1 \right)^{\frac{1}{p}} \left(q(\beta_2 - 1) + 1 \right)^{\frac{1}{q}}}. \tag{89}$$

Notice here that ${}_{\alpha_i}^{b_i} T(f_i) \in C([a_i, b_i])$.

We have proved the right conformable fractional Hilbert–Pachpatte inequality:

Theorem 8 *Let $\alpha_i \in (n_i, n_i + 1]$, $f_i \in C^{n_i + 1}([a_i, b_i])$, $[a_i, b_i] \subset \mathbb{R}$, $n_i \in \mathbb{Z}_+$; $\beta_i := \alpha_i - n_i, i = 1, 2$. Assume that $f_i^{(k_i)}(b_i) = 0, k_i = 0, 1, \ldots, n_i; i = 1, 2$. Let $p, q > 1 : \frac{1}{p} + \frac{1}{q} = 1$, with $\beta_1 > \frac{1}{q}, \beta_2 > \frac{1}{p}$. Then*

$$\int_{a_1}^{b_1} \int_{a_2}^{b_2} \frac{|f_1(t_1)| \, |f_2(t_2)| \, dt_1 dt_2}{\left[\frac{(b_1 - t_1)^{p(\beta_1 - 1) + 1}}{p} + \frac{(b_2 - t_2)^{q(\beta_2 - 1) + 1}}{q} \right]} \leq$$

$$\frac{(b_1 - a_1)^{n_1 + 1} (b_2 - a_2)^{n_2 + 1} \left\| {}_{\alpha_1}^{b_1} T(f_1) \right\|_{L_q([a_1, b_1])} \left\| {}_{\alpha_2}^{b_2} T(f_2) \right\|_{L_p([a_2, b_2])}}{n_1! n_2! \left(p(\beta_1 - 1) + 1 \right)^{\frac{1}{p}} \left(q(\beta_2 - 1) + 1 \right)^{\frac{1}{q}}}. \tag{90}$$

Next we present Conformable fractional Ostrowski type inequalities:

Theorem 9 *Let $\alpha \in (n, n+1]$, $n \in \mathbb{Z}_+$, $f \in C^{n+1}([a, b])$, $\beta := \alpha - n$; $x_0 \in [a, b]$ be fixed. Assume $f^{(k)}(x_0) = 0$, $k = 1, \ldots, n$. Then*

$$\left| \frac{1}{b-a} \int_a^b f(t)\, dt - f(x_0) \right| \leq \frac{\Gamma(\beta)}{\Gamma(\alpha+2)(b-a)}$$

$$\left\{ (x_0 - a)^{\alpha+1} \left\| {}_\alpha^{x_0} T(f) \right\|_{\infty, [a, x_0]} + (b - x_0)^{\alpha+1} \left\| T_\alpha^{x_0}(f) \right\|_{\infty, [x_0, b]} \right\}. \tag{91}$$

Proof We have (by (69))

$$f(t) - f(x_0) = \frac{1}{n!} \int_{x_0}^t (t-x)^n (x-x_0)^{\beta-1} \left(T_\alpha^{x_0}(f) \right)(x)\, dx, \tag{92}$$

$\forall\, t \in [x_0, b]$, and (by (70))

$$f(t) - f(x_0) = -\frac{1}{n!} \int_t^{x_0} (x_0 - x)^{\beta-1} (x-t)^n \left({}_\alpha^{x_0} T(f) \right)(x)\, dx, \tag{93}$$

$\forall\, t \in [a, x_0]$.

We observe that

$$|f(t) - f(x_0)| \leq \frac{1}{n!} \int_{x_0}^t (t-x)^n (x-x_0)^{\beta-1} \left| T_\alpha^{x_0}(f)(x) \right| dx \leq$$

$$\frac{\left\| T_\alpha^{x_0}(f) \right\|_{\infty, [x_0, b]}}{n!} \int_{x_0}^t (t-x)^{(n+1)-1} (x-x_0)^{\beta-1}\, dx =$$

$$\frac{\left\| T_\alpha^{x_0}(f) \right\|_{\infty, [x_0, b]}}{n!} \frac{\Gamma(n+1)\,\Gamma(\beta)}{\Gamma(n+\beta+1)} (t-x_0)^{n+\beta} =$$

$$\frac{\left\| T_\alpha^{x_0}(f) \right\|_{\infty, [x_0, b]} \Gamma(\beta)}{\Gamma(n+\beta+1)} (t-x_0)^{n+\beta}. \tag{94}$$

That is,

$$|f(t) - f(x_0)| \leq \frac{\left\| T_\alpha^{x_0}(f) \right\|_{\infty, [x_0, b]} \Gamma(\beta)}{\Gamma(n+\beta+1)} (t-x_0)^{n+\beta}, \ \forall\, t \in [x_0, b]. \tag{95}$$

Similarly, it holds

$$|f(t) - f(x_0)| \leq \frac{1}{n!} \left\| {}_\alpha^{x_0} T(f) \right\|_{\infty, [a, x_0]} \int_t^{x_0} (x_0 - x)^{\beta-1} (x-t)^{(n+1)-1}\, dx =$$

$$\frac{\left\|_\alpha^{x_0} T(f)\right\|_{\infty,[a,x_0]} \Gamma(\beta) \Gamma(n+1)}{n!} (x_0 - t)^{\beta+n} =$$

$$\frac{\left\|_\alpha^{x_0} T(f)\right\|_{\infty,[a,x_0]} \Gamma(\beta)}{\Gamma(\beta+n+1)} (x_0 - t)^{\beta+n}. \tag{96}$$

That is,

$$|f(t) - f(x_0)| \le \frac{\left\|_\alpha^{x_0} T(f)\right\|_{\infty,[a,x_0]} \Gamma(\beta)}{\Gamma(\beta+n+1)} (x_0 - t)^{\beta+n}, \ \forall \, t \in [a, x_0]. \tag{97}$$

Hence, we can write

$$\left| \frac{1}{b-a} \int_a^b f(t)\, dt - f(x_0) \right| \le$$

$$\frac{1}{b-a} \left\{ \int_a^{x_0} |f(t) - f(x_0)|\, dt + \int_{x_0}^b |f(t) - f(x_0)|\, dt \right\} \le$$

$$\frac{\Gamma(\beta)}{\Gamma(n+\beta+1)(b-a)} \left\{ \left(\int_a^{x_0} (x_0 - t)^{\beta+n}\, dt \right) \left\|_\alpha^{x_0} T(f)\right\|_{\infty,[a,x_0]} + \right.$$

$$\left. \left(\int_{x_0}^b (t - x_0)^{n+\beta}\, dt \right) \left\| T_\alpha^{x_0}(f)\right\|_{\infty,[x_0,b]} \right\} =$$

$$\frac{\Gamma(\beta)}{\Gamma(n+\beta+1)(b-a)} \left\{ (x_0 - a)^{\beta+n+1} \left\|_\alpha^{x_0} T(f)\right\|_{\infty,[a,x_0]} + \right.$$

$$\left. (b - x_0)^{n+\beta+1} \left\| T_\alpha^{x_0}(f)\right\|_{\infty,[x_0,b]} \right\}, \tag{98}$$

proving (91).

Theorem 10 *Here all as in Theorem 9. Let $p_1, p_2, p_3 > 1 : \frac{1}{p_1} + \frac{1}{p_2} + \frac{1}{p_3} = 1$, with $\beta > \frac{1}{p_1} + \frac{1}{p_3}$. Then*

$$\left| \frac{1}{b-a} \int_a^b f(t)\, dt - f(x_0) \right| \le$$

$$\frac{1}{(b-a)\, n!\, (p_1 n + 1)^{\frac{1}{p_1}} (p_2(\beta-1)+1)^{\frac{1}{p_2}} \left(\alpha + \frac{1}{p_2} + \frac{1}{p_1} \right)}$$

$$\left\{ (b - x_0)^{\alpha + \frac{1}{p_2} + \frac{1}{p_1}} \left\| T_\alpha^{x_0}(f)\right\|_{p_3,[x_0,b]} + (x_0 - a)^{\alpha + \frac{1}{p_2} + \frac{1}{p_1}} \left\|_\alpha^{x_0} T(f)\right\|_{p_3,[a,x_0]} \right\}. \tag{99}$$

Proof By (92) we get

$$|f(t) - f(x_0)| \leq \frac{1}{n!} \int_{x_0}^{t} (t-x)^n (x-x_0)^{\beta-1} \left| T_{\alpha}^{x_0}(f)(x) \right| dx \leq$$

$$\frac{1}{n!} \left(\int_{x_0}^{t} (t-x)^{p_1 n} dx \right)^{\frac{1}{p_1}} \left(\int_{x_0}^{t} (x-x_0)^{p_2(\beta-1)} dx \right)^{\frac{1}{p_2}} \left\| T_{\alpha}^{x_0}(f) \right\|_{p_3,[x_0,b]} =$$

$$\frac{\left\| T_{\alpha}^{x_0}(f) \right\|_{p_3,[x_0,b]}}{n!} \left(\frac{(t-x_0)^{p_1 n+1}}{p_1 n+1} \right)^{\frac{1}{p_1}} \left(\frac{(t-x_0)^{p_2(\beta-1)+1}}{p_2(\beta-1)+1} \right)^{\frac{1}{p_2}} =$$

$$\frac{\left\| T_{\alpha}^{x_0}(f) \right\|_{p_3,[x_0,b]} (t-x_0)^{n+\frac{1}{p_1}+\beta-1+\frac{1}{p_2}}}{n! (p_1 n+1)^{\frac{1}{p_1}} (p_2(\beta-1)+1)^{\frac{1}{p_2}}}. \tag{100}$$

Notice that $p_2(\beta-1)+1 > 0$, iff $\beta > \frac{1}{p_1} + \frac{1}{p_3}$.

We have proved

$$|f(t) - f(x_0)| \leq \frac{\left\| T_{\alpha}^{x_0}(f) \right\|_{p_3,[x_0,b]} (t-x_0)^{n+\beta-\frac{1}{p_3}}}{n! (p_1 n+1)^{\frac{1}{p_1}} (p_2(\beta-1)+1)^{\frac{1}{p_2}}}, \tag{101}$$

$\forall\, t \in [x_0, b]$.

Similarly, we have (by (93))

$$|f(t) - f(x_0)| \leq$$

$$\frac{1}{n!} \left(\int_{t}^{x_0} (x_0-x)^{p_2(\beta-1)} dx \right)^{\frac{1}{p_2}} \left(\int_{t}^{x_0} (x-t)^{p_1 n} dx \right)^{\frac{1}{p_1}} \left\| {}_{\alpha}^{x_0}T(f) \right\|_{p_3,[a,x_0]} =$$

$$\frac{1}{n!} \left(\frac{(x_0-t)^{p_2(\beta-1)+1}}{p_2(\beta-1)+1} \right)^{\frac{1}{p_2}} \left(\frac{(x_0-t)^{p_1 n+1}}{p_1 n+1} \right)^{\frac{1}{p_1}} \left\| {}_{\alpha}^{x_0}T(f) \right\|_{p_3,[a,x_0]} =$$

$$\frac{\left\| {}_{\alpha}^{x_0}T(f) \right\|_{p_3,[a,x_0]} (x_0-t)^{\beta+n-\frac{1}{p_3}}}{n! (p_2(\beta-1)+1)^{\frac{1}{p_2}} (p_1 n+1)^{\frac{1}{p_1}}}. \tag{102}$$

We have proved that

$$|f(t) - f(x_0)| \leq \frac{\left\| {}_{\alpha}^{x_0}T(f) \right\|_{p_3,[a,x_0]} (x_0-t)^{\beta+n-\frac{1}{p_3}}}{n! (p_2(\beta-1)+1)^{\frac{1}{p_2}} (p_1 n+1)^{\frac{1}{p_1}}}, \tag{103}$$

$\forall\, t \in [a, x_0]$, where $\beta > \frac{1}{p_1} + \frac{1}{p_3}$.

Therefore, we derive

$$\left| \frac{1}{b-a} \int_a^b f(t)\,dt - f(x_0) \right| \leq$$

$$\frac{1}{b-a} \left\{ \int_a^{x_0} |f(t) - f(x_0)|\,dt + \int_{x_0}^b |f(t) - f(x_0)|\,dt \right\} \leq$$

$$\frac{1}{n!\,(p_1 n + 1)^{\frac{1}{p_1}}\,(p_2\,(\beta-1)+1)^{\frac{1}{p_2}}\,(b-a)}$$

$$\left\{ \left(\int_a^{x_0} (x_0 - t)^{\beta+n-\frac{1}{p_3}}\,dt \right) \left\| {}_\alpha^{x_0} T(f) \right\|_{p_3,[a,x_0]} \right.$$

$$\left. + \left(\int_{x_0}^b (t - x_0)^{n+\beta-\frac{1}{p_3}}\,dt \right) \left\| T_\alpha^{x_0}(f) \right\|_{p_3,[x_0,b]} \right\} = \tag{104}$$

$$\frac{1}{n!\,(p_1 n + 1)^{\frac{1}{p_1}}\,(p_2\,(\beta-1)+1)^{\frac{1}{p_2}}\,(b-a)} \left\{ \frac{(x_0 - a)^{\beta+n+\frac{1}{p_2}+\frac{1}{p_1}}}{\left(\beta + n + \frac{1}{p_2} + \frac{1}{p_1}\right)} \left\| {}_\alpha^{x_0} T(f) \right\|_{p_3,[a,x_0]} \right.$$

$$\left. + \frac{(b - x_0)^{n+\beta+\frac{1}{p_2}+\frac{1}{p_1}}}{\left(\beta + n + \frac{1}{p_2} + \frac{1}{p_1}\right)} \left\| T_\alpha^{x_0}(f) \right\|_{p_3,[x_0,b]} \right\}, \tag{105}$$

proving (99).

We make

Remark 6 Here we will discuss about generalized conformable fractional Ostrowski and Grüss type inequalities involving several functions.

Let $\alpha \in (n, n+1]$, $n \in \mathbb{Z}_+$, $f_i \in C^{n+1}([a, b])$, $i = 1, \ldots, r \in \mathbb{N}$, $[a, b] \subset \mathbb{R}$, $\beta := \alpha - n$, $x_0 \in [a, b]$, and $f_i^{(k)}(x_0) = 0$, $k = 1, \ldots, n$; $i = 1, \ldots, r$.

If $n = 0$, initial conditions are void, i.e. $0 < \alpha \leq 1$.

By (69) and (70) we get that

$$f_i(t) - f_i(x_0) = \frac{1}{n!} \int_{x_0}^t (t - x)^n (x - x_0)^{\beta-1} \left(T_\alpha^{x_0}(f_i) \right)(x)\,dx, \tag{106}$$

$\forall\, t \in [x_0, b]$, all $i = 1, \ldots, r$,
and

$$f_i(t) - f_i(x_0) = -\frac{1}{n!} \int_t^{x_0} (x_0 - x)^{\beta-1} (x - t)^n \left({}_\alpha^{x_0} T(f_i) \right)(x)\,dx, \tag{107}$$

$\forall\, t \in [a, x_0]$, all $i = 1, \ldots, r$.

Multiply (106), (107) by $\prod_{\substack{j=1 \\ j \neq i}}^{r} f_j(t)$ to get

$$\prod_{k=1}^{r} f_k(t) - \left(\prod_{\substack{j=1 \\ j \neq i}}^{r} f_j(t) \right) f_i(x_0) =$$

$$\frac{\prod_{\substack{j=1 \\ j \neq i}}^{r} f_j(t)}{n!} \int_{x_0}^{t} (t-x)^n (x-x_0)^{\beta-1} \left(T_\alpha^{x_0}(f_i) \right)(x) \, dx, \tag{108}$$

and

$$\prod_{k=1}^{r} f_k(t) - \left(\prod_{\substack{j=1 \\ j \neq i}}^{r} f_j(t) \right) f_i(x_0) =$$

$$-\frac{\prod_{\substack{j=1 \\ j \neq i}}^{r} f_j(t)}{n!} \int_{t}^{x_0} (x_0-x)^{\beta-1} (x-t)^n \left({}_{\alpha}^{x_0}T(f_i) \right)(x) \, dx, \tag{109}$$

$\forall i = 1, \ldots, r$.

Adding (108), (109) per set, we obtain

$$r \left(\prod_{k=1}^{r} f_k(t) \right) - \sum_{i=1}^{r} \left[\left(\prod_{\substack{j=1 \\ j \neq i}}^{r} f_j(t) \right) f_i(x_0) \right] =$$

$$\frac{1}{n!} \sum_{i=1}^{r} \left[\prod_{\substack{j=1 \\ j \neq i}}^{r} f_j(t) \int_{x_0}^{t} (t-x)^n (x-x_0)^{\beta-1} \left(T_\alpha^{x_0}(f_i) \right)(x) \, dx \right], \tag{110}$$

$\forall t \in [x_0, b]$, and

$$r \left(\prod_{k=1}^{r} f_k(t) \right) - \sum_{i=1}^{r} \left[\left(\prod_{\substack{j=1 \\ j \neq i}}^{r} f_j(t) \right) f_i(x_0) \right] =$$

$$
- \frac{1}{n!} \sum_{i=1}^{r} \left[\prod_{\substack{j=1 \\ j \neq i}}^{r} f_j(t) \int_t^{x_0} (x_0 - x)^{\beta-1} (x - t)^n \left({}_{\alpha}^{x_0} T(f_i) \right) (x) \, dx \right], \qquad (111)
$$

$\forall \, t \in [a, x_0]$.

Next we integrate (110), (111) with respect to $t \in [a, b]$. We have

$$
r \int_{x_0}^{b} \left(\prod_{k=1}^{r} f_k(t) \right) dt - \sum_{i=1}^{r} \left[f_i(x_0) \int_{x_0}^{b} \left(\prod_{\substack{j=1 \\ j \neq i}}^{r} f_j(t) \right) dt \right] =
$$

$$
\frac{1}{n!} \sum_{i=1}^{r} \left[\int_{x_0}^{b} \left(\prod_{\substack{j=1 \\ j \neq i}}^{r} f_j(t) \right) \left(\int_{x_0}^{t} (t - x)^n (x - x_0)^{\beta-1} \left(T_{\alpha}^{x_0}(f_i) \right) (x) \, dx \right) dt \right],
$$

$$(112)$$

and

$$
r \int_{a}^{x_0} \left(\prod_{k=1}^{r} f_k(t) \right) dt - \sum_{i=1}^{r} \left[f_i(x_0) \int_{a}^{x_0} \left(\prod_{\substack{j=1 \\ j \neq i}}^{r} f_j(t) \right) dt \right] =
$$

$$
- \frac{1}{n!} \sum_{i=1}^{r} \left[\int_{a}^{x_0} \left(\prod_{\substack{j=1 \\ j \neq i}}^{r} f_j(t) \right) \left(\int_{t}^{x_0} (x_0 - x)^{\beta-1} (x - t)^n \left({}_{\alpha}^{x_0} T(f_i) \right) (x) \, dx \right) dt \right].
$$

$$(113)$$

Adding (112), (113) we obtain

$$
\theta(f_1, \ldots, f_r)(x_0) := r \int_{a}^{b} \left(\prod_{k=1}^{r} f_k(t) \right) dt - \sum_{i=1}^{r} \left[f_i(x_0) \int_{a}^{b} \left(\prod_{\substack{j=1 \\ j \neq i}}^{r} f_j(t) \right) dt \right]
$$

$$
= \frac{1}{n!} \sum_{i=1}^{r} \left[\left[\int_{x_0}^{b} \left(\prod_{\substack{j=1 \\ j \neq i}}^{r} f_j(t) \right) \left(\int_{x_0}^{t} (t - x)^n (x - x_0)^{\beta-1} \left(T_{\alpha}^{x_0}(f_i) \right) (x) \, dx \right) dt \right] \right.
$$

$$-\left[\int_a^{x_0}\left(\prod_{\substack{j=1\\j\neq i}}^r f_j\left(t\right)\right)\left(\int_t^{x_0}\left(x_0-x\right)^{\beta-1}\left(x-t\right)^n\left(_\alpha^{x_0}T\left(f_i\right)\right)\left(x\right)dx\right)dt\right]\right].$$

$$(114)$$

Hence, it holds

$$\left|\theta\left(f_1,\ldots,f_r\right)\left(x_0\right)\right|\leq$$

$$\frac{1}{n!}\sum_{i=1}^r\left[\left[\int_{x_0}^b\left(\prod_{\substack{j=1\\j\neq i}}^r\left|f_j\left(t\right)\right|\right)\left(\int_{x_0}^t\left(t-x\right)^n\left(x-x_0\right)^{\beta-1}\left|T_\alpha^{x_0}\left(f_i\right)\left(x\right)\right|dx\right)dt\right]$$

$$+\left[\int_a^{x_0}\left(\prod_{\substack{j=1\\j\neq i}}^r\left|f_j\left(t\right)\right|\right)\left(\int_t^{x_0}\left(x_0-x\right)^{\beta-1}\left(x-t\right)^n\left|_\alpha^{x_0}T\left(f_i\right)\left(x\right)\right|dx\right)dt\right]\right]=:(*).$$

$$(115)$$

We notice that

$$(*)\leq\sum_{i=1}^r\left[\left[\int_{x_0}^b\left(\prod_{\substack{j=1\\j\neq i}}^r\left|f_j\left(t\right)\right|\right)\frac{\left\|T_\alpha^{x_0}\left(f_i\right)\right\|_{\infty,[x_0,b]}\Gamma\left(\beta\right)}{\Gamma\left(n+\beta+1\right)}\left(t-x_0\right)^{n+\beta}dt\right]+\right.$$

$$\left.+\left[\int_a^{x_0}\left(\prod_{\substack{j=1\\j\neq i}}^r\left|f_j\left(t\right)\right|\right)\frac{\left\|_\alpha^{x_0}T\left(f_i\right)\right\|_{\infty,[a,x_0]}\Gamma\left(\beta\right)}{\Gamma\left(\beta+n+1\right)}\left(x_0-t\right)^{\beta+n}dt\right]\right].\quad(116)$$

Thus we have proved so far

$$\left|\theta\left(f_1,\ldots,f_r\right)\left(x_0\right)\right|\leq\frac{\Gamma\left(\beta\right)}{\Gamma\left(\beta+n+1\right)}$$

$$\sum_{i=1}^r\left[\left[\left\|T_\alpha^{x_0}\left(f_i\right)\right\|_{\infty,[x_0,b]}\int_{x_0}^b\left(t-x_0\right)^{n+\beta}\left(\prod_{\substack{j=1\\j\neq i}}^r\left|f_j\left(t\right)\right|\right)dt\right]+\right.$$

$$\left[\left\| {}_\alpha^{x_0} T\left(f_i\right) \right\|_{\infty,[a,x_0]} \int_a^{x_0} (x_0 - t)^{\beta+n} \left(\prod_{\substack{j=1 \\ j \neq i}}^{r} \left| f_j\left(t\right) \right| \right) dt \right] \right]. \tag{117}$$

We further notice that

$$\left| \theta\left(f_1, \ldots, f_r\right)\left(x_0\right) \right| \leq \frac{\Gamma\left(\beta\right)}{\Gamma\left(\beta + n + 2\right)}$$

$$\sum_{i=1}^{r} \left[\left[\left\| T_\alpha^{x_0}\left(f_i\right) \right\|_{\infty,[x_0,b]} \left(\prod_{\substack{j=1 \\ j \neq i}}^{r} \left\| f_j \right\|_{\infty,[x_0,b]} \right) (b - x_0)^{n+\beta+1} \right] + \right.$$

$$\left. \left[\left\| {}_\alpha^{x_0} T\left(f_i\right) \right\|_{\infty,[a,x_0]} \left(\prod_{\substack{j=1 \\ j \neq i}}^{r} \left\| f_j \right\|_{\infty,[a,x_0]} \right) (x_0 - a)^{n+\beta+1} \right] \right], \tag{118}$$

which is an ∞-Ostrowski type inequality.

Next let $p_1, p_2, p_3 > 1 : \frac{1}{p_1} + \frac{1}{p_2} + \frac{1}{p_3} = 1$, such that $\beta > \frac{1}{p_1} + \frac{1}{p_3}$.
Hence we can write

$$(*) \leq \frac{1}{n!} \sum_{i=1}^{r} \left[\left[\frac{\int_{x_0}^{b} \left(\prod_{\substack{j=1 \\ j \neq i}}^{r} \left| f_j\left(t\right) \right| \right) \left\| T_\alpha^{x_0}\left(f_i\right) \right\|_{p_3,[x_0,b]} (t - x_0)^{n+\beta-\frac{1}{p_3}}}{(p_1 n + 1)^{\frac{1}{p_1}} (p_2 (\beta - 1) + 1)^{\frac{1}{p_2}}} dt \right] \right. \tag{119}$$

$$\left. + \left[\frac{\int_a^{x_0} \left(\prod_{\substack{j=1 \\ j \neq i}}^{r} \left| f_j\left(t\right) \right| \right) \left\| {}_\alpha^{x_0} T\left(f_i\right) \right\|_{p_3,[a,x_0]} (x_0 - t)^{\beta+n-\frac{1}{p_3}}}{(p_1 n + 1)^{\frac{1}{p_1}} (p_2 (\beta - 1) + 1)^{\frac{1}{p_2}}} dt \right] \right] =$$

$$\frac{1}{n! \, (p_1 n + 1)^{\frac{1}{p_1}} (p_2 (\beta - 1) + 1)^{\frac{1}{p_2}}}$$

$$\sum_{i=1}^{r} \left[\left[\left(\int_{x_0}^{b} (t - x_0)^{n+\beta-\frac{1}{p_3}} \left(\prod_{\substack{j=1 \\ j \neq i}}^{r} |f_j(t)| \right) dt \right) \left\| T_\alpha^{x_0}(f_i) \right\|_{p_3,[x_0,b]} \right] \right.$$

$$\left. + \left[\left(\int_{a}^{x_0} (x_0 - t)^{\beta+n-\frac{1}{p_3}} \left(\prod_{\substack{j=1 \\ j \neq i}}^{r} |f_j(t)| \right) dt \right) \left\| {}_\alpha^{x_0} T(f_i) \right\|_{p_3,[a,x_0]} \right] \right] \leq \tag{120}$$

$$\frac{1}{n! \, (p_1 n + 1)^{\frac{1}{p_1}} \, (p_2(\beta - 1) + 1)^{\frac{1}{p_2}}}$$

$$\sum_{i=1}^{r} \left[\left[\frac{(b - x_0)^{n+\beta+\frac{1}{p_1}+\frac{1}{p_2}}}{\left(n + \beta + \frac{1}{p_1} + \frac{1}{p_2} \right)} \left(\prod_{\substack{j=1 \\ j \neq i}}^{r} \left\| f_j \right\|_{\infty,[x_0,b]} \right) \left\| T_\alpha^{x_0}(f_i) \right\|_{p_3,[x_0,b]} \right] \right. \tag{121}$$

$$\left. + \left[\frac{(x_0 - a)^{n+\beta+\frac{1}{p_1}+\frac{1}{p_2}}}{\left(n + \beta + \frac{1}{p_1} + \frac{1}{p_2} \right)} \left(\prod_{\substack{j=1 \\ j \neq i}}^{r} \left\| f_j \right\|_{\infty,[a,x_0]} \right) \left\| {}_\alpha^{x_0} T(f_i) \right\|_{p_3,[a,x_0]} \right] \right].$$

We have proved the L_p-Ostrowski type inequality:

$$|\theta(f_1, \ldots, f_r)(x_0)| \leq$$

$$\frac{1}{n! \, (p_1 n + 1)^{\frac{1}{p_1}} \, (p_2(\beta - 1) + 1)^{\frac{1}{p_2}} \left(n + \beta + \frac{1}{p_1} + \frac{1}{p_2} \right)}$$

$$\sum_{i=1}^{r} \left[\left[(b - x_0)^{n+\beta+\frac{1}{p_1}+\frac{1}{p_2}} \left(\prod_{\substack{j=1 \\ j \neq i}}^{r} \left\| f_j \right\|_{\infty,[x_0,b]} \right) \left\| T_\alpha^{x_0}(f_i) \right\|_{p_3,[x_0,b]} \right] \right.$$

$$\left. + \left[(x_0 - a)^{n+\beta+\frac{1}{p_1}+\frac{1}{p_2}} \left(\prod_{\substack{j=1 \\ j \neq i}}^{r} \left\| f_j \right\|_{\infty,[a,x_0]} \right) \left\| {}_\alpha^{x_0} T(f_i) \right\|_{p_3,[a,x_0]} \right] \right]. \tag{122}$$

From now on we assume $0 < \alpha \leq 1$, i.e. $n = 0$. So no initial conditions are needed.

Notice that

$$\Delta (f_1, \ldots, f_r) := \int_a^b \theta (f_1, \ldots, f_r) (x) \, dx =$$

$$r (b - a) \left(\int_a^b \left(\prod_{k=1}^r f_k (x) \, dx \right) \right) -$$

$$\sum_{i=1}^r \left[\left(\int_a^b f_i (x) \, dx \right) \left(\int_a^b \left(\prod_{\substack{j=1 \\ j \neq i}}^r f_j (x) \right) dx \right) \right], \tag{123}$$

and it holds

$$|\Delta (f_1, \ldots, f_r)| \leq \int_a^b |\theta (f_1, \ldots, f_r) (x)| \, dx. \tag{124}$$

By (124) and (118) we get the ∞-Gruss type inequality (here $\alpha = \beta$):

$$|\Delta (f_1, \ldots, f_r)| \leq \frac{\Gamma (\alpha) (b - a)^{a+2}}{\Gamma (\alpha + 3)}$$

$$\sum_{i=1}^r \left[\left(\sup_{x_0 \in [a,b]} \| T_\alpha^{x_0} (f_i) \|_{\infty, [x_0, b]} \right) \left(\prod_{\substack{j=1 \\ j \neq i}}^r \sup_{x_0 \in [a,b]} \| f_j \|_{\infty, [x_0, b]} \right) \tag{125}$$

$$+ \left(\sup_{x_0 \in [a,b]} \| {}_\alpha^{x_0} T (f_i) \|_{\infty, [a, x_0]} \right) \left(\prod_{\substack{j=1 \\ j \neq i}}^r \sup_{x_0 \in [a,b]} \| f_j \|_{\infty, [a, x_0]} \right) \right].$$

We have proved that

$$|\Delta (f_1, \ldots, f_r)| \leq \frac{\Gamma (\alpha) (b - a)^{a+2}}{\Gamma (\alpha + 3)}$$

$$\sum_{i=1}^r \left[\left(\sup_{x_0 \in [a,b]} \| {}_\alpha^{x_0} T (f_i) \|_{\infty, [a, x_0]} + \sup_{x_0 \in [a,b]} \| T_\alpha^{x_0} (f_i) \|_{\infty, [x_0, b]} \right) \left(\prod_{\substack{j=1 \\ j \neq i}}^r \| f_j \|_{\infty, [a, b]} \right) \right]. \tag{126}$$

Next by (122) we get the L_p-Gruss inequality:

$$|\Delta (f_1, \ldots, f_r)| \leq$$

$$\frac{(b-a)^{\alpha + \frac{1}{p_1} + \frac{1}{p_2} + 1}}{n! \, (p_1 n + 1)^{\frac{1}{p_1}} \, (p_2 (\alpha - 1) + 1)^{\frac{1}{p_2}} \left(\alpha + \frac{1}{p_1} + \frac{1}{p_2}\right) \left(\alpha + \frac{1}{p_1} + \frac{1}{p_2} + 1\right)}$$

$$\sum_{i=1}^{r} \left[\left(\sup_{x_0 \in [a,b]} \left\| T_\alpha^{x_0} (f_i) \right\|_{p_3, [x_0, b]} + \sup_{x_0 \in [a,b]} \left\| {}_\alpha^{x_0} T (f_i) \right\|_{p_3, [a, x_0]} \right) \left(\prod_{\substack{j=1 \\ j \neq i}}^{r} \left\| f_j \right\|_{\infty, [a,b]} \right) \right].$$

$$(127)$$

We have proved the following results:

An ∞-Ostrowski type Conformable fractional inequality for several functions follows:

Theorem 11 *Let* $\alpha \in (n, n+1]$, $n \in \mathbb{Z}_+$, $f_i \in C^{n+1} ([a, b])$, $i = 1, \ldots, r \in \mathbb{N}$, $[a, b] \subset \mathbb{R}$, $\beta := \alpha - n$, $x_0 \in [a, b]$, *and* $f_i^{(k)} (x_0) = 0$, $k = 1, \ldots, n$; $i = 1, \ldots, r$. *Call*

$$\theta (f_1, \ldots, f_r) (x_0) := r \int_a^b \left(\prod_{k=1}^{r} f_k (t) \right) dt - \sum_{i=1}^{r} \left[f_i (x_0) \int_a^b \left(\prod_{\substack{j=1 \\ j \neq i}}^{r} f_j (t) \right) dt \right].$$

$$(128)$$

Then

$$|\theta (f_1, \ldots, f_r) (x_0)| \leq \frac{\Gamma (\beta)}{\Gamma (\alpha + 2)}$$

$$\sum_{i=1}^{r} \left[\left[\left\| T_\alpha^{x_0} (f_i) \right\|_{\infty, [x_0, b]} \left(\prod_{\substack{j=1 \\ j \neq i}}^{r} \left\| f_j \right\|_{\infty, [x_0, b]} \right) (b - x_0)^{\alpha + 1} \right] + \right.$$

$$\left. \left[\left\| {}_\alpha^{x_0} T (f_i) \right\|_{\infty, [a, x_0]} \left(\prod_{\substack{j=1 \\ j \neq i}}^{r} \left\| f_j \right\|_{\infty, [a, x_0]} \right) (x_0 - a)^{\alpha + 1} \right] \right]. \qquad (129)$$

Next follows the corresponding L_p-Ostrowski inequality for several functions.

Theorem 12 *All as in Theorem 11. Let $p_1, p_2, p_3 > 1 : \frac{1}{p_1} + \frac{1}{p_2} + \frac{1}{p_3} = 1$ such that $\beta > \frac{1}{p_1} + \frac{1}{p_3}$. Then*

$$|\theta (f_1, \ldots, f_r) (x_0)| \le \tag{130}$$

$$\frac{1}{n! (p_1 n + 1)^{\frac{1}{p_1}} (p_2 (\beta - 1) + 1)^{\frac{1}{p_2}} \left(\alpha + \frac{1}{p_1} + \frac{1}{p_2} \right)}$$

$$\sum_{i=1}^{r} \left[\left[(b - x_0)^{\alpha + \frac{1}{p_1} + \frac{1}{p_2}} \left(\prod_{\substack{j=1 \\ j \ne i}}^{r} \|f_j\|_{\infty, [x_0, b]} \right) \|T_\alpha^{x_0} (f_i)\|_{p_3, [x_0, b]} \right] \right.$$

$$\left. + \left[(x_0 - a)^{\alpha + \frac{1}{p_1} + \frac{1}{p_2}} \left(\prod_{\substack{j=1 \\ j \ne i}}^{r} \|f_j\|_{\infty, [a, x_0]} \right) \|{}_\alpha^{x_0} T (f_i)\|_{p_3, [a, x_0]} \right] \right].$$

The corresponding Gruss type inequalities follow:

Theorem 13 *Let all as in Theorem 11, with $0 < \alpha \le 1$. We denote*

$$\Delta (f_1, \ldots, f_r) := \int_a^b \theta (f_1, \ldots, f_r) (x) \, dx =$$

$$r (b - a) \left(\int_a^b \left(\prod_{k=1}^{r} f_k (x) \, dx \right) \right) - \tag{131}$$

$$\sum_{i=1}^{r} \left[\left(\int_a^b f_i (x) \, dx \right) \left(\int_a^b \left(\prod_{\substack{j=1 \\ j \ne i}}^{r} f_j (x) \right) dx \right) \right].$$

Then

$$|\Delta (f_1, \ldots, f_r)| \le \frac{\Gamma (\alpha) (b - a)^{a+2}}{\Gamma (\alpha + 3)}$$

$$\sum_{i=1}^{r} \left[\left(\sup_{x_0 \in [a,b]} \|{}_\alpha^{x_0} T (f_i)\|_{\infty, [a, x_0]} + \sup_{x_0 \in [a,b]} \|T_\alpha^{x_0} (f_i)\|_{\infty, [x_0, b]} \right) \left(\prod_{\substack{j=1 \\ j \ne i}}^{r} \|f_j\|_{\infty, [a,b]} \right) \right].$$

$$\tag{132}$$

Theorem 14 *Here all as in Theorems 11 and 13. Let $p_1, p_2, p_3 > 1 : \frac{1}{p_1} + \frac{1}{p_2} + \frac{1}{p_3} = 1$, with $0 < \alpha \le 1$, such that $\beta > \frac{1}{p_1} + \frac{1}{p_3}$. Then*

$$|\Delta (f_1, \ldots, f_r)| \le$$

$$\frac{(b-a)^{\alpha + \frac{1}{p_1} + \frac{1}{p_2} + 1}}{n! \, (p_1 n + 1)^{\frac{1}{p_1}} \, (p_2 (\alpha - 1) + 1)^{\frac{1}{p_2}} \left(\alpha + \frac{1}{p_1} + \frac{1}{p_2}\right) \left(\alpha + \frac{1}{p_1} + \frac{1}{p_2} + 1\right)}$$

$$\sum_{i=1}^{r} \left[\left(\sup_{x_0 \in [a,b]} \left\| T_\alpha^{x_0} (f_i) \right\|_{p_3,[x_0,b]} + \sup_{x_0 \in [a,b]} \left\| {}_{\alpha}^{x_0} T (f_i) \right\|_{p_3,[a,x_0]} \right) \left(\prod_{\substack{j=1 \\ j \ne i}}^{r} \left\| f_j \right\|_{\infty,[a,b]} \right) \right].$$

$$(133)$$

We make

Remark 7 Here we discuss about Conformable fractional left Opial inequality.
Let $\alpha \in (n, n+1], n \in \mathbb{Z}_+, f \in C^{n+1} ([a, b]) (\beta := \alpha - n, 0 < \beta \le 1)$. Assume $f^{(k)} (a) = 0, k = 0, 1, \ldots, n$, then (by (69))

$$f(t) = \frac{1}{n!} \int_a^t (t - x)^n (x - a)^{\beta - 1} \left(T_\alpha^a (f) \right) (x) \, dx, \qquad (134)$$

$\forall \, t \in [a, b]$.
Let $a \le w \le t$, then we have

$$f(w) = \frac{1}{n!} \int_a^w (w - x)^n (x - a)^{\beta - 1} \left(T_\alpha^a (f) \right) (x) \, dx. \qquad (135)$$

Then

$$|f(w)| \le \frac{(b-a)^n}{n!} \int_a^w (x - a)^{\beta - 1} \left| T_\alpha^a (f) (x) \right| dx \le$$

$$\frac{(b-a)^n}{n!} \left(\int_a^w (x - a)^{(\beta - 1)p} \, dx \right)^{\frac{1}{p}} \left(\int_a^w \left| T_\alpha^a (f) (x) \right|^q dx \right)^{\frac{1}{q}} =$$

$$\frac{(b-a)^n}{n!} \left(\frac{(w - a)^{p(\beta - 1) + 1}}{p (\beta - 1) + 1} \right)^{\frac{1}{p}} (z (w))^{\frac{1}{q}}, \qquad (136)$$

where

$$z(w) := \int_a^w \left| T_\alpha^a (f) (x) \right|^q dx, \quad \text{all } a \le w \le t, \qquad (137)$$

[we need $p(\beta - 1) + 1 > 0 \Leftrightarrow p(\beta - 1) > -1 \Leftrightarrow \beta - 1 > -\frac{1}{p} \Leftrightarrow \beta > 1 - \frac{1}{p} = \frac{1}{q}$, so we assume that $\beta > \frac{1}{q}$]
   and

$$z(a) = 0. \tag{138}$$

Thus

$$z'(w) = \left| T_\alpha^a(f)(w) \right|^q, \text{ and } \left| T_\alpha^a f(w) \right| = (z'(w))^{\frac{1}{q}}. \tag{139}$$

Therefore we obtain

$$|f(w)| \left| T_\alpha^a f(w) \right| \le \frac{(b-a)^n}{n!} \frac{(w-a)^{\frac{p(\beta-1)+1}{p}}}{(p(\beta-1)+1)^{\frac{1}{p}}} (z(w) z'(w))^{\frac{1}{q}}. \tag{140}$$

Integrating the last inequality we get

$$\int_a^t |f(w)| \left| T_\alpha^a f(w) \right| dw \le$$

$$\frac{(b-a)^n}{n!(p(\beta-1)+1)^{\frac{1}{p}}} \int_a^t (w-a)^{\frac{p(\beta-1)+1}{p}} (z(w) z'(w))^{\frac{1}{q}} dw \le$$

$$\frac{(b-a)^n}{n!(p(\beta-1)+1)^{\frac{1}{p}}} \left(\int_a^t (w-a)^{p(\beta-1)+1} dw \right)^{\frac{1}{p}} \left(\int_a^t z(w) z'(w) dw \right)^{\frac{1}{q}} =$$

$$\tag{141}$$

$$\frac{(b-a)^n}{n!(p(\beta-1)+1)^{\frac{1}{p}}} \left(\frac{(t-a)^{p(\beta-1)+2}}{p(\beta-1)+2} \right)^{\frac{1}{p}} \left(\int_a^t z(w) \, dz(w) \right)^{\frac{1}{q}} =$$

$$\frac{(b-a)^n}{n!(p(\beta-1)+1)^{\frac{1}{p}}} \frac{(t-a)^{(\beta-1)+\frac{2}{p}}}{(p(\beta-1)+2)^{\frac{1}{p}}} \left(\frac{z^2(t)}{2} \right)^{\frac{1}{q}} =$$

$$\frac{(b-a)^n (t-a)^{\beta-1+\frac{2}{p}}}{n! 2^{\frac{1}{q}} [(p(\beta-1)+1)(p(\beta-1)+2)]^{\frac{1}{p}}} \left(\int_a^t \left| T_\alpha^a(f)(x) \right|^q dx \right)^{\frac{2}{q}}. $$

$$\tag{142}$$

We have proved the conformable left fractional Opial inequality:

Theorem 15 *Let* $\alpha \in (n, n+1]$, $n \in \mathbb{Z}_+$, $f \in C^{n+1}([a, b])$, $\beta := \alpha - n$. *Assume* $f^{(k)}(a) = 0$, $k = 0, 1, \ldots, n$. *Let* $p, q > 1 : \frac{1}{p} + \frac{1}{q} = 1$, $\beta > \frac{1}{q}$. *Then*

$$\int_a^t |f(w)| \left| T_\alpha^a f(w) \right| dw \leq$$

$$\frac{(b-a)^n (t-a)^{\beta - 1 + \frac{2}{p}}}{n! 2^{\frac{1}{q}} \left[(p(\beta - 1) + 1)(p(\beta - 1) + 2) \right]^{\frac{1}{p}}} \left(\int_a^t \left| T_\alpha^a (f)(x) \right|^q dx \right)^{\frac{2}{q}}, \qquad (143)$$

$\forall\, t \in [a, b]$.

We make

Remark 8 Here we discuss the Conformable right fractional Opial inequality.

Let $\alpha \in (n, n+1]$, $n \in \mathbb{Z}_+$, $f \in C^{n+1}([a, b])$ $(\beta := \alpha - n, 0 < \beta \leq 1)$. Assume that $f^{(k)}(b) = 0$, $k = 0, 1, \ldots, n$, then (by (70))

$$f(t) = -\frac{1}{n!} \int_t^b (b-x)^{\beta - 1} (x-t)^n \left({}_\alpha^b T(f) \right)(x)\, dx, \qquad (144)$$

$\forall\, t \in [a, b]$.

Let $t \leq w \leq b$, then we have

$$f(w) = -\frac{1}{n!} \int_w^b (b-x)^{\beta - 1} (x-w)^n \left({}_\alpha^b T(f) \right)(x)\, dx. \qquad (145)$$

Then

$$|f(w)| \leq \frac{(b-a)^n}{n!} \int_w^b (b-x)^{\beta - 1} \left| {}_\alpha^b T(f)(x) \right| dx \leq$$

$$\frac{(b-a)^n}{n!} \left(\int_w^b (b-x)^{p(\beta - 1)} dx \right)^{\frac{1}{p}} \left(\int_w^b \left| {}_\alpha^b T(f)(x) \right|^q dx \right)^{\frac{1}{q}} =$$

$$\frac{(b-a)^n}{n!} \frac{(b-w)^{\frac{p(\beta - 1)+1}{p}}}{(p(\beta - 1) + 1)^{\frac{1}{p}}} (z(w))^{\frac{1}{q}}, \qquad (146)$$

where

$$z(w) := \int_w^b \left| {}_\alpha^b T(f)(x) \right|^q dx, \qquad (147)$$

$t \leq w \leq b$, $z(b) = 0$. Thus

$$-z(w) := \int_b^w \left| {}_\alpha^b T(f)(x) \right|^q dx, \tag{148}$$

and

$$(-z(w))' = \left| {}_\alpha^b T(f)(x) \right|^q \geq 0, \tag{149}$$

and

$$\left| {}_\alpha^b T(f)(x) \right| = \left((-z(w))' \right)^{\frac{1}{q}} = \left(-z'(w) \right)^{\frac{1}{q}}. \tag{150}$$

(want $p(\beta - 1) + 1 > 0 \Leftrightarrow p(\beta - 1) > -1 \Leftrightarrow \beta - 1 > -\frac{1}{p} \Leftrightarrow \beta > 1 - \frac{1}{p} = \frac{1}{q}$,
so we assume $\beta > \frac{1}{q}$).

Therefore we obtain

$$|f(w)| \left| {}_\alpha^b T(f)(w) \right| \leq \frac{(b-a)^n}{n!} \frac{(b-w)^{\frac{p(\beta-1)+1}{p}}}{(p(\beta-1)+1)^{\frac{1}{p}}} \left(z(w) \left(-z'(w) \right) \right)^{\frac{1}{q}}, \tag{151}$$

all $t \leq w \leq b$.

Hence it holds

$$\int_t^b |f(w)| \left| {}_\alpha^b T(f)(w) \right| dw \leq$$

$$\frac{(b-a)^n}{n!(p(\beta-1)+1)^{\frac{1}{p}}} \int_t^b (b-w)^{\frac{p(\beta-1)+1}{p}} \left(z(w) \left(-z'(w) \right) \right)^{\frac{1}{q}} dw \leq$$

$$\frac{(b-a)^n}{n!(p(\beta-1)+1)^{\frac{1}{p}}} \left(\int_t^b (b-w)^{p(\beta-1)+1} dw \right)^{\frac{1}{p}} \left(\int_t^b z(w) \left(-z'(w) \right) dw \right)^{\frac{1}{q}} = \tag{152}$$

$$\frac{(b-a)^n}{n!(p(\beta-1)+1)^{\frac{1}{p}}} \frac{(b-t)^{(\beta-1)+\frac{2}{p}}}{(p(\beta-1)+2)^{\frac{1}{p}}} \frac{(z(t))^{\frac{2}{q}}}{2^{\frac{1}{q}}} =$$

$$\frac{(b-a)^n (b-t)^{\beta-1+\frac{2}{p}}}{n! 2^{\frac{1}{q}} [(p(\beta-1)+1)(p(\beta-1)+2)]^{\frac{1}{p}}} \left(\int_t^b \left| {}_\alpha^b T(f)(x) \right|^q dx \right)^{\frac{2}{q}}. \tag{153}$$

We have proved the Conformable right fractional Opial type inequality:

Theorem 16 *Let $\alpha \in (n, n+1]$, $n \in \mathbb{Z}_+$, $\beta := \alpha - n$, $f \in C^{n+1}([a,b])$. Assume $f^{(k)}(b) = 0$, $k = 0, 1, \ldots, n$. Let $p, q > 1 : \frac{1}{p} + \frac{1}{q} = 1$ such that $\beta > \frac{1}{q}$. Then*

$$\int_t^b |f(w)| \left| {}_\alpha^b T(f)(w) \right| dw \leq$$

$$\frac{(b-a)^n (b-t)^{\beta-1+\frac{2}{p}}}{2^{\frac{1}{q}} n! \left[(p(\beta-1)+1)(p(\beta-1)+2) \right]^{\frac{1}{p}}} \left(\int_t^b \left| {}_\alpha^b T(f)(x) \right|^q dx \right)^{\frac{2}{q}}, \qquad (154)$$

$\forall t \in [a,b]$.

Next we give a left conformable fractional Poincare type inequality:

Theorem 17 *Let $\alpha \in (n, n+1]$, $n \in \mathbb{Z}_+$, $f \in C^{n+1}([a,b])$, $\beta := \alpha - n$. Assume $f^{(k)}(a) = 0$, $k = 0, 1, \ldots, n$. Let $p_1, p_2, p_3 > 1 : \frac{1}{p_1} + \frac{1}{p_2} + \frac{1}{p_3} = 1$, such that $\beta > \frac{1}{p_1} + \frac{1}{p_3}$. Then*

$$\|f\|_{p_3,[a,b]} \leq \frac{(b-a)^\alpha \left\| T_\alpha^a f \right\|_{p_3,[a,b]}}{n! (p_1 n + 1)^{\frac{1}{p_1}} (p_2(\beta-1)+1)^{\frac{1}{p_2}} (\alpha p_3)^{\frac{1}{p_3}}}. \qquad (155)$$

Proof Since $f^{(k)}(a) = 0$, $k = 0, 1, \ldots, n$, then (by (69))

$$f(t) = \frac{1}{n!} \int_a^t (t-x)^n (x-a)^{\beta-1} \left(T_\alpha^a(f) \right)(x) \, dx, \qquad (156)$$

$\forall t \in [a,b]$.

Let $p_1, p_2, p_3 > 1 : \frac{1}{p_1} + \frac{1}{p_2} + \frac{1}{p_3} = 1$. Then

$$|f(t)| \leq \frac{1}{n!} \int_a^t (t-x)^n (x-a)^{\beta-1} \left| T_\alpha^a(f)(x) \right| dx \leq$$

$$\frac{1}{n!} \left(\int_a^t (t-x)^{p_1 n} dx \right)^{\frac{1}{p_1}} \left(\int_a^t (x-a)^{p_2(\beta-1)} dx \right)^{\frac{1}{p_2}} \left\| T_\alpha^a f \right\|_{p_3,[a,b]} = \qquad (157)$$

$$\frac{1}{n!} \frac{(t-a)^{\frac{p_1 n+1}{p_1}}}{(p_1 n + 1)^{\frac{1}{p_1}}} \frac{(t-a)^{\frac{p_2(\beta-1)+1}{p_2}}}{(p_2(\beta-1)+1)^{\frac{1}{p_2}}} \left\| T_\alpha^a f \right\|_{p_3,[a,b]} =$$

$$\frac{(t-a)^{n+\frac{1}{p_1}+\beta-1+\frac{1}{p_2}}}{n! (p_1 n + 1)^{\frac{1}{p_1}} (p_2(\beta-1)+1)^{\frac{1}{p_2}}} \left\| T_\alpha^a f \right\|_{p_3,[a,b]} = \qquad (158)$$

$$\frac{(t-a)^{n+\beta-\frac{1}{p_3}}}{n! (p_1 n + 1)^{\frac{1}{p_1}} (p_2(\beta-1)+1)^{\frac{1}{p_2}}} \left\| T_\alpha^a f \right\|_{p_3,[a,b]}.$$

We have proved

$$|f(t)| \leq \frac{(t-a)^{n+\beta-\frac{1}{p_3}}}{n! \, (p_1 n + 1)^{\frac{1}{p_1}} \, (p_2 (\beta - 1) + 1)^{\frac{1}{p_2}}} \left\| T_\alpha^a f \right\|_{p_3,[a,b]}, \tag{159}$$

$\forall \, t \in [a, b]$.

Then

$$|f(t)|^{p_3} \leq \frac{(t-a)^{p_3(n+\beta)-1}}{(n!)^{p_3} \, (p_1 n + 1)^{\frac{p_3}{p_1}} \, (p_2 (\beta - 1) + 1)^{\frac{p_3}{p_2}}} \left\| T_\alpha^a f \right\|_{p_3,[a,b]}^{p_3}. \tag{160}$$

Therefore, it holds

$$\int_a^b |f(t)|^{p_3} \, dt \leq \frac{\int_a^b (t-a)^{p_3(n+\beta)-1} \, dt}{(n!)^{p_3} \, (p_1 n + 1)^{\frac{p_3}{p_1}} \, (p_2 (\beta - 1) + 1)^{\frac{p_3}{p_2}}} \left\| T_\alpha^a f \right\|_{p_3,[a,b]}^{p_3} =$$

$$\frac{(b-a)^{p_3(n+\beta)} \left\| T_\alpha^a f \right\|_{p_3,[a,b]}^{p_3}}{(n!)^{p_3} \, (p_1 n + 1)^{\frac{p_3}{p_1}} \, (p_2 (\beta - 1) + 1)^{\frac{p_3}{p_2}} \, p_3 (n + \beta)}. \tag{161}$$

Consequently, we get

$$\|f\|_{p_3,[a,b]} \leq \frac{(b-a)^{(n+\beta)} \left\| T_\alpha^a f \right\|_{p_3,[a,b]}}{n! \, (p_1 n + 1)^{\frac{1}{p_1}} \, (p_2 (\beta - 1) + 1)^{\frac{1}{p_2}} \, (p_3 (n + \beta))^{\frac{1}{p_3}}}. \tag{162}$$

(we want $p_2 (\beta - 1) + 1 > 0 \Leftrightarrow p_2 (\beta - 1) > -1 \Leftrightarrow \beta - 1 > -\frac{1}{p_2} \Leftrightarrow \beta > 1 - \frac{1}{p_2} \Leftrightarrow \beta > \frac{1}{p_1} + \frac{1}{p_3}$, by assumption).

It follows the right conformable fractional Poincare type inequality:

Theorem 18 Let $\alpha \in (n, n+1]$, $n \in \mathbb{Z}_+$, $f \in C^{n+1}([a, b])$, $\beta := \alpha - n$, $f^{(k)}(b) = 0$, $k = 0, 1, \ldots, n$. Let $p_1, p_2, p_3 > 1 : \frac{1}{p_1} + \frac{1}{p_2} + \frac{1}{p_3} = 1$, such that $\beta > \frac{1}{p_1} + \frac{1}{p_3}$. Then

$$\|f\|_{p_3,[a,b]} \leq \frac{(b-a)^\alpha \left\| {}_\alpha^b T (f) \right\|_{p_3,[a,b]}}{n! \, (p_1 n + 1)^{\frac{1}{p_1}} \, (p_2 (\beta - 1) + 1)^{\frac{1}{p_2}} \, (\alpha p_3)^{\frac{1}{p_3}}}. \tag{163}$$

Proof By (70) we get ($\forall \, t \in [a, b]$)

$$f(t) = -\frac{1}{n!} \int_t^b (b-x)^{\beta-1} (x-t)^n \left({}_\alpha^b T (f) \right)(x) \, dx. \tag{164}$$

Hence

$$\left| f\left(t\right) \right| \leq \frac{1}{n!} \int_{t}^{b} \left(b-x\right)^{\beta-1} \left(x-t\right)^{n} \left| {}_{\alpha}^{b}T\left(f\right)\left(x\right) \right| dx \leq$$

$$\frac{1}{n!} \left(\int_{t}^{b} \left(x-t\right)^{p_1 n} dx \right)^{\frac{1}{p_1}} \left(\int_{t}^{b} \left(b-x\right)^{p_2\left(\beta-1\right)} dx \right)^{\frac{1}{p_2}} \left\| {}_{\alpha}^{b}T\left(f\right) \right\|_{p_3,[a,b]} = \quad (165)$$

$$\frac{1}{n!} \frac{\left(b-t\right)^{\frac{p_1 n+1}{p_1}}}{\left(p_1 n+1\right)^{\frac{1}{p_1}}} \frac{\left(b-t\right)^{\frac{p_2\left(\beta-1\right)+1}{p_2}}}{\left(p_2\left(\beta-1\right)+1\right)^{\frac{1}{p_2}}} \left\| {}_{\alpha}^{b}T\left(f\right) \right\|_{p_3,[a,b]} =$$

$$\frac{\left(b-t\right)^{n+\frac{1}{p_1}+\beta-1+\frac{1}{p_2}}}{n!\left(p_1 n+1\right)^{\frac{1}{p_1}}\left(p_2\left(\beta-1\right)+1\right)^{\frac{1}{p_2}}} \left\| {}_{\alpha}^{b}T\left(f\right) \right\|_{p_3,[a,b]} =$$

$$\frac{\left(b-t\right)^{n+\beta-\frac{1}{p_3}}}{n!\left(p_1 n+1\right)^{\frac{1}{p_1}}\left(p_2\left(\beta-1\right)+1\right)^{\frac{1}{p_2}}} \left\| {}_{\alpha}^{b}T\left(f\right) \right\|_{p_3,[a,b]}.$$

We have proved

$$\left| f\left(t\right) \right| \leq \frac{\left(b-t\right)^{n+\beta-\frac{1}{p_3}}}{n!\left(p_1 n+1\right)^{\frac{1}{p_1}}\left(p_2\left(\beta-1\right)+1\right)^{\frac{1}{p_2}}} \left\| {}_{\alpha}^{b}T\left(f\right) \right\|_{p_3,[a,b]}, \quad (166)$$

$\forall\, t \in [a,b]$.

Then, it holds

$$\left| f\left(t\right) \right|^{p_3} \leq \frac{\left(b-t\right)^{p_3\left(n+\beta\right)-1} \left\| {}_{\alpha}^{b}T\left(f\right) \right\|_{p_3,[a,b]}^{p_3}}{\left(n!\right)^{p_3}\left(p_1 n+1\right)^{\frac{p_3}{p_1}}\left(p_2\left(\beta-1\right)+1\right)^{\frac{p_3}{p_2}}}, \quad (167)$$

$\forall\, t \in [a,b]$. Hence, we derive

$$\int_{a}^{b} \left| f\left(t\right) \right|^{p_3} dt \leq \frac{\left(b-a\right)^{p_3\left(n+\beta\right)} \left\| {}_{\alpha}^{b}T\left(f\right) \right\|_{p_3,[a,b]}^{p_3}}{\left(n!\right)^{p_3}\left(p_1 n+1\right)^{\frac{p_3}{p_1}}\left(p_2\left(\beta-1\right)+1\right)^{\frac{p_3}{p_2}} p_3\left(n+\beta\right)}. \quad (168)$$

Then, raise (168) to the power $\frac{1}{p_3}$, and we are done.

Next we give a left conformable fractional Sobolev type inequality:

Theorem 19 *All assumptions as in Theorem 17 and $r > 0$. Then*

$$\left\| f \right\|_{r,[a,b]} \leq \frac{\left(b-a\right)^{\left(\alpha-\frac{1}{p_3}+\frac{1}{r}\right)} \left\| T_{\alpha}^{a}f \right\|_{p_3,[a,b]}}{n!\left(p_1 n+1\right)^{\frac{1}{p_1}}\left(p_2\left(\beta-1\right)+1\right)^{\frac{1}{p_2}} \left[r\left(\alpha-\frac{1}{p_3}\right)+1\right]^{\frac{1}{r}}}. \quad (169)$$

Proof We use (159). Hence it holds

$$|f(t)|^r \leq \frac{(t-a)^{r\left(n+\beta-\frac{1}{p_3}\right)} \left\|T_\alpha^a f\right\|_{p_3,[a,b]}^r}{(n!)^r (p_1 n + 1)^{\frac{r}{p_1}} (p_2 (\beta - 1) + 1)^{\frac{r}{p_2}}}, \tag{170}$$

$\forall t \in [a, b]$.

Consequently we obtain

$$\int_a^b |f(t)|^r \, dt \leq \frac{(b-a)^{r\left(n+\beta-\frac{1}{p_3}\right)+1} \left\|T_\alpha^a f\right\|_{p_3,[a,b]}^r}{(n!)^r (p_1 n + 1)^{\frac{r}{p_1}} (p_2 (\beta - 1) + 1)^{\frac{r}{p_2}} \left[r\left(n+\beta-\frac{1}{p_3}\right)+1\right]}. \tag{171}$$

We have proved that

$$\|f\|_{r,[a,b]} \leq \frac{(b-a)^{\left(n+\beta-\frac{1}{p_3}+\frac{1}{r}\right)} \left\|T_\alpha^a f\right\|_{p_3,[a,b]}}{n! (p_1 n + 1)^{\frac{1}{p_1}} (p_2 (\beta - 1) + 1)^{\frac{1}{p_2}} \left[r\left(n+\beta-\frac{1}{p_3}\right)+1\right]^{\frac{1}{r}}}. \tag{172}$$

We have established (169). ∎

It follows the right conformable fractional Sobolev type inequality:

Theorem 20 *All assumptions as in Theorem 18, and $r > 0$. Then*

$$\|f\|_{r,[a,b]} \leq \frac{(b-a)^{\left(\alpha-\frac{1}{p_3}+\frac{1}{r}\right)} \left\|{}_\alpha^b T (f)\right\|_{p_3,[a,b]}}{n! (p_1 n + 1)^{\frac{1}{p_1}} (p_2 (\beta - 1) + 1)^{\frac{1}{p_2}} \left[r\left(\alpha-\frac{1}{p_3}\right)+1\right]^{\frac{1}{r}}}. \tag{173}$$

Proof We use (166). We get that

$$|f(t)|^r \leq \frac{(b-t)^{r\left(n+\beta-\frac{1}{p_3}\right)} \left\|{}_\alpha^b T (f)\right\|_{p_3,[a,b]}^r}{(n!)^r (p_1 n + 1)^{\frac{r}{p_1}} (p_2 (\beta - 1) + 1)^{\frac{r}{p_2}}}, \tag{174}$$

and

$$\int_a^b |f(t)|^r \, dt \leq \frac{(b-a)^{r\left(n+\beta-\frac{1}{p_3}\right)+1} \left\|{}_\alpha^b T (f)\right\|_{p_3,[a,b]}^r}{(n!)^r (p_1 n + 1)^{\frac{r}{p_1}} (p_2 (\beta - 1) + 1)^{\frac{r}{p_2}} \left[r\left(n+\beta-\frac{1}{p_3}\right)+1\right]}. \tag{175}$$

Finally, we derive

$$\|f\|_{r,[a,b]} \leq \frac{(b-a)^{\left(n+\beta-\frac{1}{p_3}+\frac{1}{r}\right)} \|{}_\alpha^b T (f)\|_{p_3,[a,b]}}{n! (p_1 n + 1)^{\frac{1}{p_1}} (p_2 (\beta - 1) + 1)^{\frac{1}{p_2}} \left[r \left(n + \beta - \frac{1}{p_3}\right) + 1\right]^{\frac{1}{r}}},$$

(176)

proving the claim.

We need

Corollary 3 (of Theorem 9) *Let* $\alpha \in (0, 1]$, $f \in C^1 ([a, b])$, $[a, b] \subset \mathbb{R}$. *Then*

$$\left|\frac{1}{b-a} \int_a^b f (t) dt - f (a)\right| \leq \frac{(b-a)^\alpha}{\alpha (\alpha + 1)} \|T_\alpha^a (f)\|_{\infty,[a,b]},$$

(177)

and

$$\left|\frac{1}{b-a} \int_a^b f (t) dt - f (b)\right| \leq \frac{(b-a)^\alpha}{\alpha (\alpha + 1)} \|{}_\alpha^b T (f)\|_{\infty,[a,b]}.$$

(178)

We need

Corollary 4 *Let* $\alpha \in (0, 1]$, *any* $[a, b] \subset \mathbb{R}_+$, $f \in C^1 (\mathbb{R}_+)$ *with* $\|T_\alpha^0 (f)\|_{\infty,\mathbb{R}_+} < +\infty$. *Then*

$$\left|\frac{1}{b-a} \int_a^b f (t) dt - f (a)\right| \leq \frac{(b-a)^\alpha}{\alpha (\alpha + 1)} \|T_\alpha^0 (f)\|_{\infty,\mathbb{R}_+}.$$

(179)

Proof It comes from (177), and the following:
Here

$$T_\alpha^a (f) (x) = (x - a)^{1-\alpha} f' (x),$$

all $x \in [a, b], 0 \leq a < b$.
Then

$$\left|T_\alpha^a (f) (x)\right| = (x - a)^{1-\alpha} \left|f' (x)\right| \leq x^{1-\alpha} \left|f' (x)\right| = \left|T_\alpha^0 (f) (x)\right|,$$

$\forall x \in [a, b]$.
Therefore it holds

$$\|T_\alpha^a (f)\|_{\infty,[a,b]} \leq \|T_\alpha^0 (f)\|_{\infty,\mathbb{R}_+}.$$

(180)

Corollary 5 *Let* $\alpha \in (0, 1]$, *any* $[a, b] \subset \mathbb{R}_-$, $f \in C^1 (\mathbb{R}_-)$ *with* $\left\| {}_\alpha^0 T (f) \right\|_{\infty, \mathbb{R}_-} < +\infty$. *Then*

$$\left| \frac{1}{b - a} \int_a^b f (t) \, dt - f (b) \right| \leq \frac{(b - a)^\alpha}{\alpha (\alpha + 1)} \left\| {}_\alpha^0 T (f) \right\|_{\infty, \mathbb{R}_-}. \tag{181}$$

Proof It comes from (178), and the following:
Here

$$- \left({}_\alpha^b T (f) \right) (x) = (b - x)^{1 - \alpha} f' (x),$$

all $x \in [a, b]$, $a < b \leq 0$.
Then

$$\left| {}_\alpha^b T (f) (x) \right| = (b - x)^{1 - \alpha} \left| f' (x) \right| \leq (-x)^{1 - \alpha} \left| f' (x) \right| = \left| {}_\alpha^0 T (f) (x) \right|,$$

$\forall \, x \in [a, b]$.
Therefore it holds

$$\left\| {}_\alpha^b T (f) \right\|_{\infty, [a, b]} \leq \left\| {}_\alpha^0 T (f) \right\|_{\infty, \mathbb{R}_-}. \tag{182}$$

We need

Corollary 6 (to Theorem 10) *Let* $\alpha \in (0, 1]$, $f \in C^1 ([a, b])$, $[a, b] \subset \mathbb{R}$. *Let* $p_1, p_2, p_3 > 1 : \frac{1}{p_1} + \frac{1}{p_2} + \frac{1}{p_3} = 1$, *with* $\alpha > \frac{1}{p_1} + \frac{1}{p_3}$. *Then*

$$\left| \frac{1}{b - a} \int_a^b f (t) \, dt - f (a) \right| \leq$$

$$\frac{1}{(p_1 + 1)^{\frac{1}{p_1}} (p_2 (\alpha - 1) + 1)^{\frac{1}{p_2}} \left(\alpha + \frac{1}{p_2} + \frac{1}{p_3} \right)} (b - a)^{\alpha - \frac{1}{p_1}} \left\| T_\alpha^a (f) \right\|_{p_3, [a, b]}, \tag{183}$$

and

$$\left| \frac{1}{b - a} \int_a^b f (t) \, dt - f (b) \right| \leq$$

$$\frac{1}{(p_1 + 1)^{\frac{1}{p_1}} (p_2 (\alpha - 1) + 1)^{\frac{1}{p_2}} \left(\alpha + \frac{1}{p_2} + \frac{1}{p_3} \right)} (b - a)^{\alpha - \frac{1}{p_1}} \left\| {}_\alpha^b T (f) \right\|_{p_3, [a, b]}. \tag{184}$$

We need

Corollary 7 *Let* $\alpha \in (0, 1]$, $f \in C^1(\mathbb{R}_+)$, *any* $[a, b] \subset \mathbb{R}_+$. *Let* $p_1, p_2, p_3 > 1$:
$\frac{1}{p_1} + \frac{1}{p_2} + \frac{1}{p_3} = 1$, *with* $\alpha > \frac{1}{p_1} + \frac{1}{p_3}$. *We assume that* $\left\| T_\alpha^0(f) \right\|_{p_3, \mathbb{R}_+} < +\infty$. *Then*

$$\left| \frac{1}{b-a} \int_a^b f(t)\, dt - f(a) \right| \le$$

$$\frac{1}{(p_1 + 1)^{\frac{1}{p_1}} (p_2(\alpha - 1) + 1)^{\frac{1}{p_2}} \left(\alpha + \frac{1}{p_2} + \frac{1}{p_3} \right)} (b - a)^{\alpha - \frac{1}{p_1}} \left\| T_\alpha^0(f) \right\|_{p_3, \mathbb{R}_+}.$$

$$(185)$$

Proof As in the proof of Corollary 4 we have that

$$\left| T_\alpha^a(f)(x) \right| \le \left| T_\alpha^0(f)(x) \right|,$$

$\forall\, x \in [a, b]$.

Clearly then

$$\left\| T_\alpha^a(f) \right\|_{p_3, [a,b]} \le \left\| T_\alpha^0(f) \right\|_{p_3, [a,b]} \le \left\| T_\alpha^0(f) \right\|_{p_3, \mathbb{R}_+}. \qquad (186)$$

Corollary 8 *Let* $\alpha \in (0, 1]$, $f \in C^1(\mathbb{R}_-)$, *any* $[a, b] \subset \mathbb{R}_-$. *Let* $p_1, p_2, p_3 > 1$:
$\frac{1}{p_1} + \frac{1}{p_2} + \frac{1}{p_3} = 1$, *with* $\alpha > \frac{1}{p_1} + \frac{1}{p_3}$. *We assume that* $\left\| {}_\alpha^0 T(f) \right\|_{p_3, \mathbb{R}_-} < +\infty$. *Then*

$$\left| \frac{1}{b-a} \int_a^b f(t)\, dt - f(b) \right| \le$$

$$\frac{1}{(p_1 + 1)^{\frac{1}{p_1}} (p_2(\alpha - 1) + 1)^{\frac{1}{p_2}} \left(\alpha + \frac{1}{p_2} + \frac{1}{p_3} \right)} (b - a)^{\alpha - \frac{1}{p_1}} \left\| {}_\alpha^0 T(f) \right\|_{p_3, \mathbb{R}_-}.$$

$$(187)$$

We make

Proof As in the proof of Corollary 5 we have that

$$\left| {}_\alpha^b T(f)(x) \right| \le \left| {}_\alpha^0 T(f)(x) \right|,$$

$\forall\, x \in [a, b]$.

Clearly then

$$\left\| {}_\alpha^b T(f) \right\|_{p_3, [a,b]} \le \left\| {}_\alpha^0 T(f) \right\|_{p_3, [a,b]} \le \left\| {}_\alpha^0 T(f) \right\|_{p_3, \mathbb{R}_-}. \qquad (188)$$

We make

Remark 9 Let $\alpha \in (0, 1]$, any $[a, b] \subset \mathbb{R}_+$, $f \in C^2(\mathbb{R}_+)$, with $\left\| T_\alpha^0 (f') \right\|_{\infty, \mathbb{R}_+} < +\infty$. Then (by (179))

$$\left| \frac{1}{b - a} \int_a^b f'(t) \, dt - f'(a) \right| \leq \frac{(b - a)^\alpha}{\alpha (\alpha + 1)} \left\| T_\alpha^0 (f') \right\|_{\infty, \mathbb{R}_+}. \qquad (189)$$

That is,

$$\left| \frac{1}{b - a} (f(b) - f(a)) - f'(a) \right| \leq \frac{(b - a)^\alpha}{\alpha (\alpha + 1)} \left\| T_\alpha^0 (f') \right\|_{\infty, \mathbb{R}_+}. \qquad (190)$$

Hence it holds

$$\left| f'(a) \right| - \frac{1}{b - a} |f(b) - f(a)| \leq \frac{(b - a)^\alpha}{\alpha (\alpha + 1)} \left\| T_\alpha^0 (f') \right\|_{\infty, \mathbb{R}_+}. \qquad (191)$$

Equivalently, we can write

$$\left| f'(a) \right| \leq \frac{1}{b - a} |f(b) - f(a)| + \frac{(b - a)^\alpha}{\alpha (\alpha + 1)} \left\| T_\alpha^0 (f') \right\|_{\infty, \mathbb{R}_+} \leq \qquad (192)$$

$$\frac{2 \|f\|_{\infty, \mathbb{R}_+}}{b - a} + \frac{(b - a)^\alpha}{\alpha (\alpha + 1)} \left\| T_\alpha^0 (f') \right\|_{\infty, \mathbb{R}_+},$$

$\forall\, a, b \in \mathbb{R}_+ : a < b$.

The last right-hand side of (192) depends only on $(b - a)$.

Therefore it holds

$$\left\| f' \right\|_{\infty, \mathbb{R}_+} \leq \frac{2 \|f\|_{\infty, \mathbb{R}_+}}{b - a} + \frac{(b - a)^\alpha}{\alpha (\alpha + 1)} \left\| T_\alpha^0 (f') \right\|_{\infty, \mathbb{R}_+}. \qquad (193)$$

Set $t := b - a > 0$. Thus

$$\left\| f' \right\|_{\infty, \mathbb{R}_+} \leq \frac{2 \|f\|_{\infty, \mathbb{R}_+}}{t} + \frac{t^\alpha}{\alpha (\alpha + 1)} \left\| T_\alpha^0 (f') \right\|_{\infty, \mathbb{R}_+}, \qquad (194)$$

$\forall\, t > 0$.

Call

$$\mu := 2 \|f\|_{\infty, \mathbb{R}_+},$$

$$and \qquad (195)$$

$$\theta := \frac{\left\| T_\alpha^0 (f') \right\|_{\infty, \mathbb{R}_+}}{\alpha (\alpha + 1)}, \quad \alpha \in (0, 1],$$

both μ, θ are greater than zero.

That is, we have

$$\|f'\|_{\infty,\mathbb{R}_+} \le \frac{\mu}{t} + \theta t^{\alpha}, \ \forall \, t > 0. \tag{196}$$

Consider the function

$$y(t) := \frac{\mu}{t} + \theta t^{\alpha}, \ t > 0, \ \alpha \in (0,1]. \tag{197}$$

Next we act as in [9], pp. 80–82. The only critical number here is

$$t_0 = \left(\frac{\mu}{\alpha\theta}\right)^{\frac{1}{\alpha+1}}, \tag{198}$$

and y has a global minimum at t_0, which is

$$y(t_0) = \left(\theta\mu^{\alpha}\right)^{\frac{1}{\alpha+1}} (\alpha+1) \, \alpha^{-\left(\frac{\alpha}{\alpha+1}\right)}. \tag{199}$$

Thus, we have proved:

$$\|f'\|_{\infty,\mathbb{R}_+} \le \left[\frac{\left\|T_\alpha^0(f')\right\|_{\infty,\mathbb{R}_+}}{\alpha(\alpha+1)} \left(2\|f\|_{\infty,\mathbb{R}_+}\right)^{\alpha}\right]^{\frac{1}{\alpha+1}} (\alpha+1) \, \alpha^{-\left(\frac{\alpha}{\alpha+1}\right)}, \tag{200}$$

under the assumption $\|f\|_{\infty,\mathbb{R}_+} < +\infty$.

We have established the following ∞-conformable left fractional alternative Landau type inequality:

Theorem 21 *Let* $\alpha \in (0,1]$, $f \in C^2(\mathbb{R}_+)$; $\|f\|_{\infty,\mathbb{R}_+}, \left\|T_\alpha^0(f')\right\|_{\infty,\mathbb{R}_+} < +\infty$. *Then*

$$\|f'\|_{\infty,\mathbb{R}_+} \le \left(\frac{\alpha+1}{\alpha}\right) \left(\frac{2^{\alpha}}{\alpha+1}\right)^{\frac{1}{\alpha+1}} \|f\|_{\infty,\mathbb{R}_+}^{\frac{\alpha}{\alpha+1}} \left\|T_\alpha^0(f')\right\|_{\infty,\mathbb{R}_+}^{\frac{1}{\alpha+1}}. \tag{201}$$

That is, $\|f'\|_{\infty,\mathbb{R}_+} < +\infty$.

We make

Remark 10 Let $\alpha \in (0,1]$, any $[a,b] \subset \mathbb{R}_-$, $f \in C^2(\mathbb{R}_-)$, with $\left\|{}_\alpha^0 T(f')\right\|_{\infty,\mathbb{R}_-} < +\infty$. Assume also $\|f\|_{\infty,\mathbb{R}_-} < +\infty$. Then (by (181)) we get

$$\left|\frac{1}{b-a}\int_a^b f'(t)\,dt - f'(b)\right| \le \frac{(b-a)^{\alpha}}{\alpha(\alpha+1)} \left\|{}_\alpha^0 T(f')\right\|_{\infty,\mathbb{R}_-}. \tag{202}$$

That is,

$$\left| \frac{1}{b-a} \left(f(b) - f(a) \right) - f'(b) \right| \leq \frac{(b-a)^\alpha}{\alpha(\alpha+1)} \left\| {}^0_\alpha T(f') \right\|_{\infty, \mathbb{R}_-}. \tag{203}$$

Hence it holds

$$\left| f'(b) \right| - \frac{1}{b-a} \left| f(b) - f(a) \right| \leq \frac{(b-a)^\alpha}{\alpha(\alpha+1)} \left\| T^0_\alpha(f') \right\|_{\infty, \mathbb{R}_-}. \tag{204}$$

Equivalently, we can write

$$\left| f'(b) \right| \leq \frac{1}{b-a} \left| f(b) - f(a) \right| + \frac{(b-a)^\alpha}{\alpha(\alpha+1)} \left\| {}^0_\alpha T(f') \right\|_{\infty, \mathbb{R}_-} \leq$$

$$\frac{2 \left\| f \right\|_{\infty, \mathbb{R}_-}}{b-a} + \frac{(b-a)^\alpha}{\alpha(\alpha+1)} \left\| {}^0_\alpha T(f') \right\|_{\infty, \mathbb{R}_-}, \tag{205}$$

$\forall\, a, b \in \mathbb{R}_- : a < b.$

The last right-hand side of (205) depends only on $(b-a)$.

Therefore it holds

$$\left\| f' \right\|_{\infty, \mathbb{R}_-} \leq \frac{2 \left\| f \right\|_{\infty, \mathbb{R}_-}}{b-a} + \frac{(b-a)^\alpha}{\alpha(\alpha+1)} \left\| {}^0_\alpha T(f') \right\|_{\infty, \mathbb{R}_-}. \tag{206}$$

Set $t := b - a > 0$. Thus

$$\left\| f' \right\|_{\infty, \mathbb{R}_-} \leq \frac{2 \left\| f \right\|_{\infty, \mathbb{R}_-}}{t} + \frac{t^\alpha}{\alpha(\alpha+1)} \left\| {}^0_\alpha T(f') \right\|_{\infty, \mathbb{R}_-}, \tag{207}$$

$\forall\, t > 0.$

Call

$$\overline{\mu} := 2 \left\| f \right\|_{\infty, \mathbb{R}_-},$$

$$and \tag{208}$$

$$\overline{\theta} := \frac{\left\| {}^0_\alpha T(f') \right\|_{\infty, \mathbb{R}_-}}{\alpha(\alpha+1)}, \quad \alpha \in (0, 1],$$

both $\overline{\mu}, \overline{\theta} > 0$.

That is, we have

$$\left\| f' \right\|_{\infty, \mathbb{R}_-} \leq \frac{\overline{\mu}}{t} + \overline{\theta} t^\alpha, \ \forall\, t > 0. \tag{209}$$

Consider the function

$$\overline{y}(t) := \frac{\overline{\mu}}{t} + \overline{\theta}t^{\alpha}, \ t > 0, \ \alpha \in (0, 1]. \tag{210}$$

Next we act as in [9], pp. 80–82. The only critical number here is

$$\overline{t_0} = \left(\frac{\overline{\mu}}{\alpha\overline{\theta}}\right)^{\frac{1}{\alpha+1}}, \tag{211}$$

and $\overline{y}$ has a global minimum at $\overline{t_0}$, which is

$$\overline{y}(\overline{t_0}) = \left(\overline{\theta}\,\overline{\mu}^{\alpha}\right)^{\frac{1}{\alpha+1}} (\alpha + 1) \, \alpha^{-\left(\frac{\alpha}{\alpha+1}\right)}. \tag{212}$$

Thus, we have proved:

$$\|f'\|_{\infty,\mathbb{R}_-} \leq \left[\frac{\left\|{}_{\alpha}^{0}T(f')\right\|_{\infty,\mathbb{R}_-}}{\alpha(\alpha+1)} \left(2\|f\|_{\infty,\mathbb{R}_-}\right)^{\alpha}\right]^{\frac{1}{\alpha+1}} (\alpha + 1) \, \alpha^{-\left(\frac{\alpha}{\alpha+1}\right)}. \tag{213}$$

We have established the following ∞-conformable right fractional alternative Landau type inequality:

Theorem 22 *Let* $\alpha \in (0, 1]$, $f \in C^2(\mathbb{R}_-)$; $\|f\|_{\infty,\mathbb{R}_-}$, $\left\|{}_{\alpha}^{0}T(f')\right\|_{\infty,\mathbb{R}_-} < +\infty$. *Then*

$$\|f'\|_{\infty,\mathbb{R}_-} \leq \left(\frac{\alpha+1}{\alpha}\right) \left(\frac{2^{\alpha}}{\alpha+1}\right)^{\frac{1}{\alpha+1}} \|f\|_{\infty,\mathbb{R}_-}^{\frac{\alpha}{\alpha+1}} \left\|{}_{\alpha}^{0}T(f')\right\|_{\infty,\mathbb{R}_-}^{\frac{1}{\alpha+1}}. \tag{214}$$

That is, $\|f'\|_{\infty,\mathbb{R}_-} < +\infty$.

We make

Remark 11 Let $\alpha \in (0, 1]$, any $[a, b] \subset \mathbb{R}_+$, $f \in C^2(\mathbb{R}_+)$, $\|f\|_{\infty,\mathbb{R}_+} < +\infty$. Let $p_1, p_2, p_3 > 1 : \frac{1}{p_1} + \frac{1}{p_2} + \frac{1}{p_3} = 1$, with $\alpha > \frac{1}{p_1} + \frac{1}{p_3}$. We assume that $\left\|T_{\alpha}^{0}(f')\right\|_{p_3,\mathbb{R}_+} < +\infty$. Then (by (185))

$$\left|\frac{1}{b-a}\int_a^b f'(t)\,dt - f'(a)\right| \leq \gamma(b-a)^{\delta} \left\|T_{\alpha}^{0}(f')\right\|_{p_3,\mathbb{R}_+}, \tag{215}$$

where

$$\gamma := \frac{1}{(p_1+1)^{\frac{1}{p_1}} (p_2(\alpha-1)+1)^{\frac{1}{p_2}} \left(\alpha + \frac{1}{p_2} + \frac{1}{p_3}\right)}, \tag{216}$$

and

$$\delta := \alpha - \frac{1}{p_1}. \tag{217}$$

Since $\alpha < 1 + \frac{1}{p_1}$, then $\alpha - \frac{1}{p_1} < 1$. Since $\alpha > \frac{1}{p_1} + \frac{1}{p_3} > \frac{1}{p_1}$, then $\alpha - \frac{1}{p_1} > 0$. Hence $0 < \delta < 1$. That is,

$$\left| \frac{1}{b-a} (f(b) - f(a)) - f'(a) \right| \leq \gamma (b-a)^\delta \left\| T_\alpha^0 (f') \right\|_{p_3, \mathbb{R}_+}. \tag{218}$$

Hence it holds

$$\left| f'(a) \right| - \frac{1}{b-a} |f(b) - f(a)| \leq \gamma (b-a)^\delta \left\| T_\alpha^0 (f') \right\|_{p_3, \mathbb{R}_+}. \tag{219}$$

Equivalently, we can write

$$\left| f'(a) \right| \leq \frac{1}{b-a} |f(b) - f(a)| + \gamma (b-a)^\delta \left\| T_\alpha^0 (f') \right\|_{p_3, \mathbb{R}_+} \leq \tag{220}$$

$$\frac{2 \|f\|_{\infty, \mathbb{R}_+}}{b-a} + (b-a)^\delta \gamma \left\| T_\alpha^0 (f') \right\|_{p_3, \mathbb{R}_+},$$

$\forall\, a, b \in \mathbb{R}_+ : a < b$.

The last right-hand side of (220) depends only on $(b-a)$.

Therefore it holds

$$\|f'\|_{\infty, \mathbb{R}_+} \leq \frac{2 \|f\|_{\infty, \mathbb{R}_+}}{b-a} + (b-a)^\delta \gamma \left\| T_\alpha^0 (f') \right\|_{p_3, \mathbb{R}_+}. \tag{221}$$

Set $t := b - a > 0$. Thus

$$\|f'\|_{\infty, \mathbb{R}_+} \leq \frac{2 \|f\|_{\infty, \mathbb{R}_+}}{t} + t^\delta \gamma \left\| T_\alpha^0 (f') \right\|_{p_3, \mathbb{R}_+}, \tag{222}$$

$\forall\, t > 0$.

Call

$$\mu := 2 \|f\|_{\infty, \mathbb{R}_+},$$

$$and \tag{223}$$

$$\theta := \gamma \left\| T_\alpha^0 (f') \right\|_{p_3, \mathbb{R}_+}, \quad \alpha \in (0, 1],$$

both $\mu, \theta > 0$.

That is, we have

$$\|f'\|_{\infty,\mathbb{R}_+} \le \frac{\mu}{t} + \theta t^\delta, \ \forall\, t > 0. \tag{224}$$

Consider the function

$$y(t) := \frac{\mu}{t} + \theta t^\delta, \ \forall\, t > 0, \ 0 < \delta < 1. \tag{225}$$

Next we act as in [9], pp. 80–82. The only critical number here is

$$t_0 = \left(\frac{\mu}{\delta\theta}\right)^{\frac{1}{\delta+1}}, \tag{226}$$

and y has a global minimum at t_0, which is

$$y(t_0) = \left(\theta\mu^\delta\right)^{\frac{1}{\delta+1}} (\delta + 1)\, \delta^{-\left(\frac{\delta}{\delta+1}\right)}. \tag{227}$$

Thus, we have proved:

$$\|f'\|_{\infty,\mathbb{R}_+} \le \left[\left(\gamma\,\left\|T_\alpha^0(f')\right\|_{p_3,\mathbb{R}_+}\right)\left(2\,\|f\|_{\infty,\mathbb{R}_+}\right)^\delta\right]^{\frac{1}{\delta+1}} (\delta + 1)\, \delta^{-\left(\frac{\delta}{\delta+1}\right)}. \tag{228}$$

We have established the following L_p-conformable left fractional alternative Landau type inequality:

Theorem 23 *Let $\alpha \in (0, 1]$, $f \in C^2(\mathbb{R}_+)$, $\|f\|_{\infty,\mathbb{R}_+} < +\infty$. Let $p_1, p_2, p_3 > 1$: $\frac{1}{p_1} + \frac{1}{p_2} + \frac{1}{p_3} = 1$, with $\alpha > \frac{1}{p_1} + \frac{1}{p_3}$. We assume that $\left\|T_\alpha^0(f')\right\|_{p_3,\mathbb{R}_+} < +\infty$. Set*

$$\gamma := \frac{1}{(p_1 + 1)^{\frac{1}{p_1}} (p_2(\alpha - 1) + 1)^{\frac{1}{p_2}} \left(\alpha + \frac{1}{p_2} + \frac{1}{p_3}\right)},$$

$$and \tag{229}$$

$$\delta := \alpha - \frac{1}{p_1}.$$

Then

$$\|f'\|_{\infty,\mathbb{R}_+} \le \|f\|_{\infty,\mathbb{R}_+}^{\frac{\delta}{\delta+1}} \left\|T_\alpha^0(f')\right\|_{p_3,\mathbb{R}_+}^{\frac{1}{\delta+1}} \gamma^{\frac{1}{\delta+1}} 2^{\frac{\delta}{\delta+1}} (\delta + 1)\, \delta^{-\left(\frac{\delta}{\delta+1}\right)}. \tag{230}$$

That is, $\|f'\|_{\infty,\mathbb{R}_+} < +\infty$.

We make

Remark 12 Let $\alpha \in (0, 1]$, any $[a, b] \subset \mathbb{R}_-$, $f \in C^2(\mathbb{R}_-)$, $\|f\|_{\infty, \mathbb{R}_-} < +\infty$. Let $p_1, p_2, p_3 > 1 : \frac{1}{p_1} + \frac{1}{p_2} + \frac{1}{p_3} = 1$, with $\alpha > \frac{1}{p_1} + \frac{1}{p_3}$. We assume that $\left\|{}_{\alpha}^{0}T(f')\right\|_{p_3, \mathbb{R}_-} < +\infty$. Then (by (187))

$$\left| \frac{1}{b-a} \int_a^b f'(t)\, dt - f'(b) \right| \le \gamma (b-a)^{\delta} \left\|{}_{\alpha}^{0}T(f')\right\|_{p_3, \mathbb{R}_-}, \tag{231}$$

where

$$\gamma := \frac{1}{(p_1 + 1)^{\frac{1}{p_1}} (p_2(\alpha - 1) + 1)^{\frac{1}{p_2}} \left(\alpha + \frac{1}{p_2} + \frac{1}{p_3} \right)}, \tag{232}$$

and

$$\delta := \alpha - \frac{1}{p_1}. \tag{233}$$

It holds $0 < \delta < 1$. That is,

$$\left| \frac{1}{b-a} (f(b) - f(a)) - f'(b) \right| \le \gamma (b-a)^{\delta} \left\|{}_{\alpha}^{0}T(f')\right\|_{p_3, \mathbb{R}_-}. \tag{234}$$

Hence it holds

$$\left| f'(b) \right| - \frac{1}{b-a} \left| f(b) - f(a) \right| \le \gamma (b-a)^{\delta} \left\|{}_{\alpha}^{0}T(f')\right\|_{p_3, \mathbb{R}_-}. \tag{235}$$

Equivalently, we can write

$$\left| f'(b) \right| \le \frac{1}{b-a} \left| f(b) - f(a) \right| + \gamma (b-a)^{\delta} \left\|{}_{\alpha}^{0}T(f')\right\|_{p_3, \mathbb{R}_-} \le \tag{236}$$

$$\frac{2 \|f\|_{\infty, \mathbb{R}_-}}{b-a} + (b-a)^{\delta} \gamma \left\|{}_{\alpha}^{0}T(f')\right\|_{p_3, \mathbb{R}_-},$$

$\forall a, b \in \mathbb{R}_- : a < b$.

The last right-hand side of (236) depends only on $(b-a)$.

Therefore it holds

$$\|f'\|_{\infty, \mathbb{R}_-} \le \frac{2 \|f\|_{\infty, \mathbb{R}_-}}{b-a} + (b-a)^{\delta} \gamma \left\|{}_{\alpha}^{0}T(f')\right\|_{p_3, \mathbb{R}_-}. \tag{237}$$

Set $t := b - a > 0$. Thus

$$\left\| f' \right\|_{\infty, \mathbb{R}_-} \leq \frac{2 \left\| f \right\|_{\infty, \mathbb{R}_-}}{t} + t^\delta \gamma \left\| {}_\alpha^0 T \left(f' \right) \right\|_{p3, \mathbb{R}_-}, \tag{238}$$

$\forall\, t > 0$.

Call

$$\overline{\mu} := 2 \left\| f \right\|_{\infty, \mathbb{R}_-},$$

$$and \tag{239}$$

$$\overline{\theta} := \gamma \left\| {}_\alpha^0 T \left(f' \right) \right\|_{p3, \mathbb{R}_-}, \quad \alpha \in (0, 1],$$

both $\overline{\mu}, \overline{\theta} > 0$.

That is, we have

$$\left\| f' \right\|_{\infty, \mathbb{R}_-} \leq \frac{\overline{\mu}}{t} + \overline{\theta} t^\delta, \quad \forall\, t > 0. \tag{240}$$

Consider the function

$$\overline{y}(t) := \frac{\overline{\mu}}{t} + \overline{\theta} t^\delta, \quad \forall\, t > 0,\ 0 < \delta < 1. \tag{241}$$

Next we act as in [9], pp. 80–82. The only critical number here is

$$\overline{t_0} = \left(\frac{\overline{\mu}}{\delta \overline{\theta}} \right)^{\frac{1}{\delta+1}}, \tag{242}$$

and $\overline{y}$ has a global minimum at $\overline{t_0}$, which is

$$\overline{y}\left(\overline{t_0} \right) = \left(\overline{\theta} \, \overline{\mu}^\delta \right)^{\frac{1}{\delta+1}} (\delta + 1) \, \delta^{-\left(\frac{\delta}{\delta+1} \right)}. \tag{243}$$

Thus, we have proved:

$$\left\| f' \right\|_{\infty, \mathbb{R}_-} \leq \left[\left(\gamma \left\| {}_\alpha^0 T \left(f' \right) \right\|_{p3, \mathbb{R}_-} \right) \left(2 \left\| f \right\|_{\infty, \mathbb{R}_-} \right)^\delta \right]^{\frac{1}{\delta+1}} (\delta + 1) \, \delta^{-\left(\frac{\delta}{\delta+1} \right)}. \tag{244}$$

We have established the following L_p-conformable right fractional alternative Landau type inequality:

Theorem 24 *Let* $\alpha \in (0, 1]$, $f \in C^2(\mathbb{R}_-)$, $\|f\|_{\infty, \mathbb{R}_-} < +\infty$. *Let* $p_1, p_2, p_3 > 1$: $\frac{1}{p_1} + \frac{1}{p_2} + \frac{1}{p_3} = 1$, *with* $\alpha > \frac{1}{p_1} + \frac{1}{p_3}$. *We assume that* $\left\|T_\alpha^0(f')\right\|_{p_3, \mathbb{R}_-} < +\infty$. *Set*

$$\gamma := \frac{1}{(p_1 + 1)^{\frac{1}{p_1}}(p_2(\alpha - 1) + 1)^{\frac{1}{p_2}}\left(\alpha + \frac{1}{p_2} + \frac{1}{p_3}\right)},$$

and (245)

$$\delta := \alpha - \frac{1}{p_1}.$$

Then

$$\|f'\|_{\infty, \mathbb{R}_-} \le \|f\|_{\infty, \mathbb{R}_-}^{\frac{\delta}{\delta+1}} \left\|{}^0_\alpha T(f')\right\|_{p_3, \mathbb{R}_-}^{\frac{1}{\delta+1}} \gamma^{\frac{1}{\delta+1}} 2^{\frac{\delta}{\delta+1}} (\delta + 1) \delta^{-\left(\frac{\delta}{\delta+1}\right)}. \quad (246)$$

That is, $\|f'\|_{\infty, \mathbb{R}_-} < +\infty$.

Comment 3 *Let* $f, g \ge 0$ *be functions. Then it is well-known that*

$$\sup(fg) \le (\sup f)(\sup g). \quad (247)$$

Property (247) strongly supports our investigations throughout Section 3.

References

1. T. Abdeljawad, On conformable fractional calculus. J. Comput. Appl. Math. **279**, 57–66 (2015)
2. M. Abu Hammad, R. Khalil, Abel's formula and Wronskian for conformable fractional differentiation equations. Int. J. Differ. Equ. Appl. **13**(3), 177–183 (2014)
3. A. Aglic Aljinović, Lj. Marangunić, J. Pečarić, On Landau type inequalities via Ostrowski inequalities. Nonlinear Funct. Anal. Appl. **10**(4), 565–579 (2005)
4. G.A. Anastassiou, *Quantitative Approximations* (CRC Press, Boca Raton, 2001)
5. G.A. Anastassiou, *Fractional Differentiation Inequalities* (Springer, Heidelberg, 2009)
6. G.A. Anastassiou, *Probabilistic Inequalities* (World Scientific, Singapore, 2010)
7. G.A. Anastassiou, *Advances Inequalities* (World Scientific, Singapore, 2010)
8. G.A. Anastassiou, *Intelligent Mathematics: Computational Analysis* (Springer, Heidelberg, 2011)
9. G.A. Anastassiou, *Advances on Fractional Inequalities* (Springer, Heidelberg, 2011)
10. G.A. Anastassiou, *Intelligent Comparisons: Analytic Inequalities* (Springer, Heidelberg, 2016)
11. G.A. Anastassiou, Conformable fractional approximation by max-product operators. Stud. Univ. Babes-Bolyai Math. **63**(1), 3–22 (2018)
12. G.A. Anastassiou, Mixed conformable fractional approximation by sublinear operators. Indian J. Math. **60**(1), 107–140 (2018)
13. D. Anderson, Taylor's formula and integral inequalities for conformable fractional derivatives, in *Contributions in Mathematics and Engineering, in Honor of Constantin Carathéodory* (Springer, Berlin, 2016), pp. 25–43

14. N.S. Barnett, S.S. Dragomir, Some Landau type inequalities for functions whose derivatives are of locally bounded variation. Tamkang J. Math. **37**(4), 301–308 (2006)
15. Z. Ditzian, Remarks, questions and conjectures on Landau-Kolmogorov-type inequalities. Math. Inequal. Appl. **3**, 15–24 (2000)
16. G.H. Hardy, E. Landau, J.E. Littlewood, Some inequalities satisfied by the integrals or derivatives of real or analytic functions. Math. Z. **39**, 677–695 (1935)
17. G.H. Hardy, J.E. Littlewood, Some integral inequalities connected with the calculus of variations. Q. J. Math. Oxford Ser. **3**, 241–252 (1932)
18. R.R. Kallman, G.C. Rota, On the inequality $\left\| f' \right\|^2 \leq 4 \left\| f \right\| \left\| f'' \right\|$, in *"Inequalities"*, vol. II, ed. by O. Shisha (Academic Press, New York, 1970), pp. 187–192
19. R. Khalil, M. Al. Horani, A. Yousef, M. Sababheh, A new definition of fractional derivative. J. Comput. Appl. Math. **264**, 65–70 (2014)
20. E. Landau, Einige Ungleichungen für zweimal differentzierban Funktionnen. Proc. Lond. Math. Soc. **13**, 43–49 (1913)
21. A. Ostrowski, Über die Absolutabweichung einer differentiebaren Funktion von ihrem Integralmittelwert. Comment. Math. Helv. **10**, 226–227 (1938)
22. F. Usta, H. Budak, T. Tunc, M.Z. Sarikaya, New bounds for the Ostrowski-type inequalities via conformable fractional calculus. Arab. J. Math. **7**(4), 317–328 (2018)

New Inequalities for η-Quasiconvex Functions

Eze R. Nwaeze and Delfim F. M. Torres

Abstract The class of η-quasiconvex functions was introduced in 2016. Here we establish novel inequalities of Ostrowski type for functions whose second derivative, in absolute value raised to the power $q \geq 1$, is η-quasiconvex. Several interesting inequalities are deduced as special cases. Furthermore, we apply our results to the arithmetic, geometric, Harmonic, logarithmic, generalized log and identric means, getting new relations amongst them.

2010 MSC: 26D15, 26E60 (Primary); 26A51 (Secondary)

1 Introduction

A function $G : I \to \mathbb{R}$ is said to be convex on the interval $I \subset \mathbb{R}$ if

$$G(xu + (1 - x)v) \leq xG(u) + (1 - x)G(v)$$

holds for all $u, v \in I$ and $x \in [0, 1]$. Many interesting inequalities have been established for convex functions. Worthy of mention is the following result proved in 2011 by Sarikaya and Aktan [8].

Theorem 1 (See [8]) *Let* $I \subset \mathbb{R}$ *be an open interval,* $\alpha, \beta \in \mathbb{R}$ *with* $\alpha < \beta$, $\lambda \in [0, 1]$, *and* $G : I \to \mathbb{R}$ *be a twice differentiable mapping such that* G'' *is integrable. If* $|G''|$ *is a convex function on* $[\alpha, \beta]$, *then*

E. R. Nwaeze
Department of Mathematics and Computer Science, Alabama State University,
Montgomery, AL, USA
e-mail: enwaeze@alasu.edu

D. F. M. Torres (✉)
Center for Research and Development in Mathematics and Applications (CIDMA),
Department of Mathematics, University of Aveiro, Aveiro, Portugal
e-mail: delfim@ua.pt

© Springer Nature Switzerland AG 2019
G. A. Anastassiou and J. M. Rassias (eds.), *Frontiers in Functional Equations and Analytic Inequalities*, https://doi.org/10.1007/978-3-030-28950-8_22

$$\left| (\lambda - 1) G \left(\frac{\alpha + \beta}{2} \right) - \lambda \frac{G(\alpha) + G(\beta)}{2} + \frac{1}{\beta - \alpha} \int_\alpha^\beta G(x) \, dx \right|$$

$$\leq \begin{cases} \frac{(\beta - \alpha)^2}{12} \left[\left(\lambda^4 + (1 + \lambda)(1 - \lambda)^3 + \frac{5\lambda - 3}{4} \right) |G''(\alpha)| \right. \\ \qquad \left. + \left(\lambda^4 + (2 - \lambda)\lambda^3 + \frac{1 - 3\lambda}{4} \right) |G''(\beta)| \right], & \text{if } 0 \leq \lambda \leq \frac{1}{2}, \\ \frac{(\beta - \alpha)^2 (3\lambda - 1)}{48} \left[|G''(\alpha)| + |G''(\beta)| \right], & \text{if } \frac{1}{2} \leq \lambda \leq 1. \end{cases}$$

In 2015, Liu obtained a related inequality for s-convex functions [7]. The notion of s-convexity was introduced in 1994 by Hudzik and Maligranda [5]. Let us recall it here.

Definition 2 (See [5]) A function $G : [0, \infty) \to \mathbb{R}$ is said to be s-convex if

$$G(ux + (1 - x)v) \leq x^s G(u) + (1 - x)^s G(v)$$

holds for all $u, v \in I$, $x \in [0, 1]$ and for some fixed $s \in (0, 1]$.

Evidently, the notion of s-convexity given in Definition 2 generalizes the classical concept of convexity. For this class of functions, Liu [7], among other things, established the following result:

Theorem 3 (See [7]) *Let $I \subset [0, \infty)$, $G : I \to \mathbb{R}$ be a twice differentiable function on I° such that $G'' \in L_1[\alpha, \beta]$, where $\alpha, \beta \in I$ with $\alpha < \beta$. If $|G''|^q$ is s-convex on $[\alpha, \beta]$ for some fixed $s \in (0, 1]$ and $q \geq 1$, then*

$$\left| \frac{1}{\beta - \alpha} \int_\alpha^\beta G(x) \, dx - (1 - \lambda) G\left(\frac{\alpha + \beta}{2} \right) - \lambda \frac{G(\alpha) + G(\beta)}{2} \right|$$

$$\leq \frac{(\beta - \alpha)^2}{16} \left(\frac{8\lambda^3 - 3\lambda + 1}{3} \right)^{1 - \frac{1}{q}} \left\{ \left[\frac{2(2\lambda)^{s+3} - 2(s + 3)\lambda + s + 2}{(s + 2)(s + 3)} \left| G''\left(\frac{\alpha + \beta}{2} \right) \right|^q \right. \right.$$

$$+ \left. \frac{4(1 - 2\lambda)^{s+2} [(s + 1)\lambda + 1] + 2(s + 3)\lambda - 2}{(s + 1)(s + 2)(s + 3)} |G''(\alpha)|^q \right]^{\frac{1}{q}}$$

$$+ \left[\frac{2(2\lambda)^{s+3} - 2(s + 3)\lambda + s + 2}{(s + 2)(s + 3)} \left| G''\left(\frac{\alpha + \beta}{2} \right) \right|^q \right.$$

$$+ \left. \left. \frac{4(1 - 2\lambda)^{s+2} [(s + 1)\lambda + 1] + 2(s + 3)\lambda - 2}{(s + 1)(s + 2)(s + 3)} |G''(\beta)|^q \right]^{\frac{1}{q}} \right\}$$

for $0 \leq \lambda \leq \frac{1}{2}$ and

$$\left| \frac{1}{\beta - \alpha} \int_\alpha^\beta G(x) \, dx - (1 - \lambda) G\left(\frac{\alpha + \beta}{2} \right) - \lambda \frac{G(\alpha) + G(\beta)}{2} \right|$$

$$\leq \frac{(\beta - \alpha)^2}{16} \left(\lambda - \frac{1}{3} \right)^{1 - \frac{1}{q}} \left\{ \left[\frac{2(s+3)\lambda - s - 2}{(s+2)(s+3)} \left| G'' \left(\frac{\alpha + \beta}{2} \right) \right|^q \right. \right.$$

$$\left. + \frac{2(s+3)\lambda - 2}{(s+1)(s+2)(s+3)} |G''(\alpha)|^q \right]^{\frac{1}{q}} + \left[\frac{2(s+3)\lambda - s - 2}{(s+2)(s+3)} \left| G'' \left(\frac{\alpha + \beta}{2} \right) \right|^q \right.$$

$$\left. \left. + \frac{2(s+3)\lambda - 2}{(s+1)(s+2)(s+3)} |G''(\beta)|^q \right]^{\frac{1}{q}} \right\}$$

for $\frac{1}{2} \leq \lambda \leq 1$.

In 2016, Eshaghi Gordji et al. [4] proposed a larger class of functions called η-quasiconvex.

Definition 4 (See [4]) A function $G : I \subset \mathbb{R} \to \mathbb{R}$ is said to be an η-quasiconvex function with respect to $\eta : \mathbb{R} \times \mathbb{R} \to \mathbb{R}$, if

$$G(xu + (1 - x)v) \leq \max \{G(v), G(v) + \eta(G(u), G(v))\}$$

for all $u, v \in I$ and $x \in [0, 1]$.

An η-quasiconvex function $G : [\alpha, \beta] \to \mathbb{R}$ is integrable if η is bounded from above on $G([\alpha, \beta]) \times G([\alpha, \beta])$ (see [2, Remark 4]). By taking $\eta(x, y) = x - y$ in Definition 4, one recovers the classical definition of quasiconvexity. It is also important to note that any convex function is η-quasiconvex with respect to $\eta(x, y) = x - y$. For some results around this recent class of functions, we invite the interested reader to see [1, 3, 6] and the references therein.

Motivated by the above results, it is our purpose to generalize Theorems 1 and 3 for the class of η-quasiconvex functions. To the best of our knowledge, the results we prove here (see Theorems 7 and 10) are novel and provide an interesting contribution to the literature of Ostrowski type results. In addition, we apply our results to some special known means of positive real numbers.

The paper is organized as follows. We begin by recalling in Sect. 2 two results, needed in the sequel. In Sect. 3, we formulate and prove our main results, that is, Theorems 7 and 10, followed by several interesting corollaries. Section 4 contains applications of our results to special means, in particular to the arithmetic, geometric, harmonic, logarithmic, the generalized log-mean, and identric means (see Propositions 14, 15 and 16). We end with Sect. 5 of conclusion.

2 Preliminaries

In this section, we recall two results that will be needed in the proof of our main results.

Lemma 5 (See [7]) *Let $I \subset \mathbb{R}$ and $G : I \to \mathbb{R}$ be a twice differentiable function on I° such that $G'' \in L^1[\alpha, \beta]$, where $\alpha, \beta \in I$ with $\alpha < \beta$. Then,*

$$\frac{1}{\beta - \alpha} \int_\alpha^\beta G(x)\,dx - (1 - \lambda)G\left(\frac{\alpha + \beta}{2}\right) - \lambda\frac{G(\alpha) + G(\beta)}{2}$$

$$= \frac{(\beta - \alpha)^2}{16}\left[\int_0^1 (x^2 - 2\lambda x)G''\left(x\frac{\alpha + \beta}{2} + (1 - x)\alpha\right)dx\right.$$

$$\left. + \int_0^1 (x^2 - 2\lambda x)G''\left(x\frac{\alpha + \beta}{2} + (1 - x)\beta\right)dx\right]$$

holds for any $\lambda \in [0, 1]$.

Lemma 6 (See [8]) *Let $I \subset \mathbb{R}$ and $G : I \to \mathbb{R}$ be a twice differentiable function on I° such that $G'' \in L^1[\alpha, \beta]$, where $\alpha, \beta \in I$ with $\alpha < \beta$. Then,*

$$\frac{1}{\beta - \alpha} \int_\alpha^\beta G(x)\,dx - (1 - \lambda)G\left(\frac{\alpha + \beta}{2}\right) - \lambda\frac{G(\alpha) + G(\beta)}{2}$$

$$= (\beta - \alpha)^2 \int_0^1 p(x)G''(x\alpha + (1 - x)\beta)\,dx$$

holds for any $\lambda \in [0, 1]$, where

$$p(x) = \begin{cases} \frac{1}{2}x(x - \lambda), & 0 \leq x \leq \frac{1}{2}, \\ \frac{1}{2}(1 - x)(1 - \lambda - x), & \frac{1}{2} \leq x \leq 1. \end{cases} \tag{1}$$

3   Main Results

We now state and prove our first main result.

Theorem 7 *Let $I \subset [0, \infty)$ and $G : [\alpha, \beta] \subset I \to \mathbb{R}$ be a twice differentiable function on (α, β) with $\alpha < \beta$. If $|G''|^q$, $q \geq 1$, is η-quasiconvex on $[\alpha, \beta]$ and η-bounded from above on $|G''|^q([\alpha, \beta]) \times |G''|^q([\alpha, \beta])$, then*

$$\left|\frac{1}{\beta - \alpha} \int_\alpha^\beta G(x)\,dx - (1 - \lambda)G\left(\frac{\alpha + \beta}{2}\right) - \lambda\frac{G(\alpha) + G(\beta)}{2}\right|$$

$$\leq \begin{cases} \frac{(\beta - \alpha)^2}{16}\left(\frac{8\lambda^3 - 3\lambda + 1}{3}\right)\left(\mathcal{N}_{q,\eta}^{\frac{1}{q}} + \mathcal{M}_{q,\eta}^{\frac{1}{q}}\right), & \text{if } 0 \leq \lambda \leq \frac{1}{2}, \\ \frac{(\beta - \alpha)^2}{16}\left(\lambda - \frac{1}{3}\right)\left(\mathcal{N}_{q,\eta}^{\frac{1}{q}} + \mathcal{M}_{q,\eta}^{\frac{1}{q}}\right), & \text{if } \frac{1}{2} \leq \lambda \leq 1, \end{cases}$$

holds, where

$$\mathcal{M}_{q,\eta} := \max\left\{\left|G''(\alpha)\right|^q, \left|G''(\alpha)\right|^q + \eta\left(\left|G''\left(\frac{\alpha+\beta}{2}\right)\right|^q, \left|G''(\alpha)\right|^q\right)\right\}$$

and

$$\mathcal{N}_{q,\eta} := \max\left\{\left|G''(\beta)\right|^q, \left|G''(\beta)\right|^q + \eta\left(\left|G''\left(\frac{\alpha+\beta}{2}\right)\right|^q, \left|G''(\beta)\right|^q\right)\right\}.$$

Proof The hypothesis that function $|G''|^q$, $q \geq 1$, is η-quasiconvex on $[\alpha, \beta]$, implies that for $x \in [0, 1]$ we have

$$\left|G''\left(x\frac{\alpha+\beta}{2} + (1-x)\alpha\right)\right|^q \leq \mathcal{M}_{q,\eta} \tag{2}$$

and

$$\left|G''\left(x\frac{\alpha+\beta}{2} + (1-x)\beta\right)\right|^q \leq \mathcal{N}_{q,\eta}. \tag{3}$$

Using Lemma 5, (2), (3) and Hölder's inequality, one obtains that

$$\left|\frac{1}{\beta-\alpha}\int_\alpha^\beta G(x)\,dx - (1-\lambda)G\left(\frac{\alpha+\beta}{2}\right) - \lambda\frac{G(\alpha)+G(\beta)}{2}\right|$$

$$\leq \frac{(\beta-\alpha)^2}{16}\left[\int_0^1 |x^2 - 2\lambda x|\left|G''\left(x\frac{\alpha+\beta}{2} + (1-x)\alpha\right)\right|dx\right.$$

$$\left. + \int_0^1 |x^2 - 2\lambda x|\left|G''\left(x\frac{\alpha+\beta}{2} + (1-x)\beta\right)\right|dx\right]$$

$$\leq \frac{(\beta-\alpha)^2}{16}\left[\left(\int_0^1 |x^2 - 2\lambda x|\,dx\right)^{1-\frac{1}{q}}\right.$$

$$\times \left(\int_0^1 |x^2 - 2\lambda x|\left|G''\left(x\frac{\alpha+\beta}{2} + (1-x)\alpha\right)\right|^q dx\right)^{\frac{1}{q}}$$

$$\left. + \left(\int_0^1 |x^2 - 2\lambda x|\,dx\right)^{1-\frac{1}{q}}\left(\int_0^1 |x^2 - 2\lambda x|\left|G''\left(x\frac{\alpha+\beta}{2} + (1-x)\beta\right)\right|^q dx\right)^{\frac{1}{q}}\right]$$

$$\leq \frac{(\beta-\alpha)^2}{16}\left[\left(\int_0^1 |x^2 - 2\lambda x|\,dx\right)^{1-\frac{1}{q}}\left(\int_0^1 |x^2 - 2\lambda x|\mathcal{M}_{q,\eta}\,dx\right)^{\frac{1}{q}}\right.$$

$$+ \left(\int_0^1 |x^2 - 2\lambda x| \, dx \right)^{1-\frac{1}{q}} \left(\int_0^1 |x^2 - 2\lambda x| \mathcal{N}_{q,\eta} \, dx \right)^{\frac{1}{q}} \right]$$

$$\leq \frac{(\beta - \alpha)^2}{16} \left[\left(\mathcal{N}_{q,\eta}^{\frac{1}{q}} + \mathcal{M}_{q,\eta}^{\frac{1}{q}} \right) \int_0^1 |x^2 - 2\lambda x| \, dx \right]. \tag{4}$$

To finish the proof, we need to evaluate $\int_0^1 |x^2 - 2\lambda x| \, dx$. For this, we consider two cases.

Case I: $0 \leq \lambda \leq \frac{1}{2}$ We get $0 \leq 2\lambda \leq 1$ and

$$\int_0^1 |x^2 - 2\lambda x| \, dx = \int_0^{2\lambda} |x^2 - 2\lambda x| \, dx + \int_{2\lambda}^1 |x^2 - 2\lambda x| \, dx$$

$$= \int_0^{2\lambda} (2\lambda x - x^2) \, dx + \int_{2\lambda}^1 (x^2 - 2\lambda x) \, dx \tag{5}$$

$$= \frac{8\lambda^3 - 3\lambda + 1}{3}.$$

Case II: $\frac{1}{2} \leq \lambda \leq 1$ We get $2\lambda \geq 1$ and $x^2 \leq 2\lambda x^2 \leq 2\lambda x$, because $x \in [0, 1]$. It follows that

$$\int_0^1 |x^2 - 2\lambda x| \, dx = \int_0^1 (2\lambda x - x^2) \, dx$$

$$= \lambda - \frac{1}{3}. \tag{6}$$

The desired inequalities are obtained by using (5) and (6) in inequality (4).

Corollary 8 *Let $I \subset [0, \infty)$ and $G : [\alpha, \beta] \subset I \to \mathbb{R}$ be a twice differentiable function on (α, β) with $\alpha < \beta$. If $|G''|$ is η-quasiconvex on $[\alpha, \beta]$ and η bounded from above on $|G''|([\alpha, \beta]) \times |G''|([\alpha, \beta])$, then the inequality*

$$\left| \frac{1}{\beta - \alpha} \int_\alpha^\beta G(x) \, dx - (1 - \lambda) G \left(\frac{\alpha + \beta}{2} \right) - \lambda \frac{G(\alpha) + G(\beta)}{2} \right|$$

$$\leq \begin{cases} \frac{(\beta - \alpha)^2}{16} \left(\frac{8\lambda^3 - 3\lambda + 1}{3} \right) \left(\mathcal{N}_\eta + \mathcal{M}_\eta \right), & \text{if } 0 \leq \lambda \leq \frac{1}{2}, \\ \frac{(\beta - \alpha)^2}{16} \left(\lambda - \frac{1}{3} \right) \left(\mathcal{N}_\eta + \mathcal{M}_\eta \right), & \text{if } \frac{1}{2} \leq \lambda \leq 1, \end{cases}$$

holds, where

$$\mathcal{M}_\eta := \max \left\{ |G''(\alpha)|, |G''(\alpha)| + \eta \left(\left| G'' \left(\frac{\alpha + \beta}{2} \right) \right|, |G''(\alpha)| \right) \right\},$$

and

$$\mathcal{N}_\eta := \max\left\{ \left|G''(\beta)\right|, \left|G''(\beta)\right| + \eta\left(\left|G''\left(\frac{\alpha+\beta}{2}\right)\right|, \left|G''(\beta)\right|\right)\right\}.$$

Proof The proof follows by setting $q = 1$ in Theorem 7.

Remark 9 By choosing different values of $\lambda \in [0, 1]$ in the inequality of Corollary 8, we obtain different results for $\eta-$quasiconvex functions. For example,

1. for $\lambda = 0$, we get a midpoint type inequality:

$$\left|\frac{1}{\beta-\alpha}\int_\alpha^\beta G(x)\,dx - G\left(\frac{\alpha+\beta}{2}\right)\right| \leq \frac{(\beta-\alpha)^2}{48}\left(\mathcal{N}_\eta + \mathcal{M}_\eta\right); \qquad (7)$$

2. for $\lambda = \frac{1}{3}$, we get a Simpson type inequality:

$$\left|\frac{1}{\beta-\alpha}\int_\alpha^\beta G(x)\,dx - \frac{2}{3}G\left(\frac{\alpha+\beta}{2}\right) - \frac{G(\alpha)+G(\beta)}{6}\right| \leq \frac{(\beta-\alpha)^2}{162}\left(\mathcal{N}_\eta + \mathcal{M}_\eta\right); \qquad (8)$$

3. for $\lambda = \frac{1}{2}$, we obtain a midpoint-trapezoid type inequality:

$$\left|\frac{1}{\beta-\alpha}\int_\alpha^\beta G(x)\,dx - \frac{1}{2}G\left(\frac{\alpha+\beta}{2}\right) - \frac{G(\alpha)+G(\beta)}{4}\right| \leq \frac{(\beta-\alpha)^2}{96}\left(\mathcal{N}_\eta + \mathcal{M}_\eta\right); \qquad (9)$$

4. for $\lambda = 1$, we have a trapezoid type inequality:

$$\left|\frac{1}{\beta-\alpha}\int_\alpha^\beta G(x)\,dx - \frac{G(\alpha)+G(\beta)}{2}\right| \leq \frac{(\beta-\alpha)^2}{24}\left(\mathcal{N}_\eta + \mathcal{M}_\eta\right). \qquad (10)$$

Follows the second main result of our paper.

Theorem 10 *Let $I \subset [0, \infty)$ and $G : [\alpha, \beta] \subset I \to \mathbb{R}$ be a twice differentiable function on (α, β) with $\alpha < \beta$. If $|G''|^q$, $q \geq 1$, is η-quasiconvex on $[\alpha, \beta]$ and η bounded from above on $|G''|^q([\alpha, \beta]) \times |G''|^q([\alpha, \beta])$, then the inequality*

$$\left|\frac{1}{\beta-\alpha}\int_\alpha^\beta G(x)\,dx - (1-\lambda)G\left(\frac{\alpha+\beta}{2}\right) - \lambda\frac{G(\alpha)+G(\beta)}{2}\right|$$

$$\leq \begin{cases} \frac{(\beta-\alpha)^2}{8}\left(\frac{8\lambda^3-3\lambda+1}{3}\right)\mathscr{U}_{q,\eta}^{\frac{1}{q}}, & \text{if } 0 \leq \lambda \leq \frac{1}{2}, \\[2mm] \frac{(\beta-\alpha)^2}{8}\left(\frac{3\lambda-1}{3}\right)\mathscr{U}_{q,\eta}^{\frac{1}{q}}, & \text{if } \frac{1}{2} \leq \lambda \leq 1, \end{cases}$$

holds, where

$$\mathscr{U}_{q,\eta} := \max \left\{ \left| G''(\beta) \right|^q, \left| G''(\beta) \right|^q + \eta \left(\left| G''(\alpha) \right|^q, \left| G''(\beta) \right|^q \right) \right\}.$$

Proof Since $|G''|^q$ is η-quasiconvex on $[\alpha, \beta]$, the inequality

$$\left| G''(x\alpha + (1-x)\beta) \right|^q \leq \mathscr{U}_{q,\eta} \tag{11}$$

holds for $x \in [0, 1]$. From the definition of $p(x)$ given by (1), we observe that for $0 \leq \lambda \leq \frac{1}{2}$ one has

$$
\begin{aligned}
\int_0^1 |p(x)|\, dx &= \int_0^{\frac{1}{2}} \left| \frac{1}{2} x(x-\lambda) \right| dx + \int_{\frac{1}{2}}^1 \left| \frac{1}{2}(1-x)(1-\lambda-x) \right| dx \\
&= \frac{1}{2} \left[\int_0^\lambda x(\lambda-x)\, dx + \int_\lambda^{\frac{1}{2}} x(x-\lambda)\, dx \right. \\
&\quad \left. + \int_{\frac{1}{2}}^{1-\lambda} (1-x)(1-\lambda-x)\, dx + \int_{1-\lambda}^1 (1-x)(x-1+\lambda)\, dx \right] \\
&= \frac{8\lambda^3 - 3\lambda + 1}{24}.
\end{aligned}
\tag{12}
$$

Also, for $\frac{1}{2} \leq \lambda \leq 1$, we get

$$
\begin{aligned}
\int_0^1 |p(x)|\, dx &= \int_0^{\frac{1}{2}} \left| \frac{1}{2} x(x-\lambda) \right| dx + \int_{\frac{1}{2}}^1 \left| \frac{1}{2}(1-x)(1-\lambda-x) \right| dx \\
&= \frac{1}{2} \left[\int_0^{\frac{1}{2}} x(\lambda-x)\, dx + \int_{\frac{1}{2}}^1 (1-x)(\lambda+x-1)\, dx \right] \\
&= \frac{3\lambda - 1}{24}.
\end{aligned}
\tag{13}
$$

Now, using Lemma 6, the Hölder inequality and (11), we obtain that

$$
\begin{aligned}
&\left| \frac{1}{\beta-\alpha} \int_\alpha^\beta G(x)\, dx - (1-\lambda)G\left(\frac{\alpha+\beta}{2} \right) - \lambda \frac{G(\alpha)+G(\beta)}{2} \right| \\
&\leq (\beta-\alpha)^2 \int_0^1 |p(x)| \left| G''(x\alpha + (1-x)\beta) \right| dx \\
&\leq (\beta-\alpha)^2 \left(\int_0^1 |p(x)|\, dx \right)^{1-\frac{1}{q}} \left(\int_0^1 |p(x)| \left| G''(x\alpha + (1-x)\beta) \right|^q dx \right)^{\frac{1}{q}} \\
&\leq (\beta-\alpha)^2 \mathscr{U}_{q,\eta}^{\frac{1}{q}} \int_0^1 |p(x)|\, dx.
\end{aligned}
$$

We get the intended result by using (12) and (13).

Corollary 11 *Let $I \subset [0, \infty)$ and $G : [\alpha, \beta] \subset I \to \mathbb{R}$ be a twice differentiable function on (α, β) with $\alpha < \beta$. If $|G''|$ is η-quasiconvex on $[\alpha, \beta]$ and η bounded from above on $|G''|([\alpha, \beta]) \times |G''|([\alpha, \beta])$, then the inequality*

$$\left| \frac{1}{\beta - \alpha} \int_{\alpha}^{\beta} G(x)\, dx - (1 - \lambda) G\left(\frac{\alpha + \beta}{2} \right) - \lambda \frac{G(\alpha) + G(\beta)}{2} \right|$$

$$\leq \begin{cases} \frac{(\beta - \alpha)^2}{8} \left(\frac{8\lambda^3 - 3\lambda + 1}{3} \right) \mathscr{U}_\eta, & \text{if } 0 \leq \lambda \leq \frac{1}{2}, \\ \frac{(\beta - \alpha)^2}{8} \left(\frac{3\lambda - 1}{3} \right) \mathscr{U}_\eta, & \text{if } \frac{1}{2} \leq \lambda \leq 1, \end{cases}$$

holds, where

$$\mathscr{U}_\eta := \max \left\{ \left| G''(\beta) \right|, \left| G''(\beta) \right| + \eta \left(\left| G''(\alpha) \right|, \left| G''(\beta) \right| \right) \right\}.$$

Proof Let $q = 1$ in Theorem 10.

Remark 12 Choosing different values of $\lambda \in [0, 1]$, we obtain, from Corollary 11, the succeeding results:

1. for $\lambda = 0$, we get

$$\left| \frac{1}{\beta - \alpha} \int_{\alpha}^{\beta} G(x)\, dx - G\left(\frac{\alpha + \beta}{2} \right) \right| \leq \frac{(\beta - \alpha)^2}{24} \mathscr{U}_\eta; \tag{14}$$

2. for $\lambda = \frac{1}{3}$, we obtain

$$\left| \frac{1}{\beta - \alpha} \int_{\alpha}^{\beta} G(x)\, dx - \frac{2}{3} G\left(\frac{\alpha + \beta}{2} \right) - \frac{G(\alpha) + G(\beta)}{6} \right| \leq \frac{(\beta - \alpha)^2}{81} \mathscr{U}_\eta; \tag{15}$$

3. for $\lambda = \frac{1}{2}$, we have

$$\left| \frac{1}{\beta - \alpha} \int_{\alpha}^{\beta} G(x)\, dx - \frac{1}{2} G\left(\frac{\alpha + \beta}{2} \right) - \frac{G(\alpha) + G(\beta)}{4} \right| \leq \frac{(\beta - \alpha)^2}{48} \mathscr{U}_\eta; \tag{16}$$

4. for $\lambda = 1$, we get

$$\left| \frac{1}{\beta - \alpha} \int_{\alpha}^{\beta} G(x)\, dx - \frac{G(\alpha) + G(\beta)}{2} \right| \leq \frac{(\beta - \alpha)^2}{12} \mathscr{U}_\eta. \tag{17}$$

Remark 13 Let $0 < \alpha < \beta$. By setting $G(x) = \ln x$ with $x \in [\alpha, \beta]$ and $\eta(x, y) = x - y$ in inequalities (14)–(17), one gets [7, Proposition 3].

4   Application to Special Means

In this section, we apply our results to the following special means of arbitrary positive numbers μ and ν with $\mu \neq \nu$:

1. the arithmetic mean

$$A(\mu, \nu) = \frac{\mu + \nu}{2};$$

2. the geometric mean

$$G(\mu, \nu) = \sqrt{\mu\nu};$$

3. the harmonic mean

$$H(\mu, \nu) = \frac{2\mu\nu}{\mu + \nu};$$

4. the logarithmic mean

$$L(\mu, \nu) = \frac{\nu - \mu}{\ln \nu - \ln \mu};$$

5. the generalized log-mean

$$L_p(\mu, \nu) = \left[\frac{\nu^{p+1} - \mu^{p+1}}{(p+1)(\nu - \mu)} \right]^{\frac{1}{p}}, \quad p \neq -1, 0;$$

6. the identric mean

$$I(\mu, \nu) = \frac{1}{e} \left(\frac{\nu^\nu}{\mu^\mu} \right)^{\frac{1}{\nu - \mu}}.$$

We now state our findings in the following propositions.

Proposition 14 *Let μ and ν be two positive numbers, $\mu < \nu$. The following inequalities hold:*

1. $\left| L_2^2(\mu, \nu) - A^2(\mu, \nu) \right| \leq \dfrac{(\nu - \mu)^2}{12};$

2. $\left| L_2^2(\mu, \nu) - \dfrac{2A^2(\mu,\nu) + A(\mu^2,\nu^2)}{3} \right| \leq \dfrac{2(\nu - \mu)^2}{81};$

3. $\left| L_2^2(\mu, \nu) - \dfrac{A^2(\mu,\nu) + A(\mu^2,\nu^2)}{2} \right| \leq \dfrac{(\nu - \mu)^2}{24};$

4. $\left| L_2^2(\mu, \nu) - A(\mu^2, \nu^2) \right| \leq \dfrac{(\nu - \mu)^2}{6}.$

Proof The desired inequalities follow by employing (7)–(10) to function $G(x) = x^2$ defined on the interval $[\mu, v]$. In this case, $|G''(x)| = 2$. By taking $\eta(x, y) = x - y$, we easily see that $|G''(x)|$ is η−quasiconvex. Moreover, $\mathcal{M}_\eta = \mathcal{N}_\eta = 2$.

Proposition 15 *Let μ and v be two positive numbers, $\mu < v$. The following inequalities hold:*

1. $\left| A^{-1}(\mu, v) - L^{-1}(\mu, v) \right| \leq \frac{(v-\mu)^2}{48} \left[\max \left\{ \frac{2}{\mu^3}, \frac{16}{(\mu+v)^3} \right\} + \max \left\{ \frac{2}{v^3}, \frac{16}{(\mu+v)^3} \right\} \right];$

2. $\left| \frac{2A^{-1}(\mu,v)+H^{-1}(\mu,v)}{3} - L^{-1}(\mu, v) \right| \leq \frac{(v-\mu)^2}{162} \left[\max \left\{ \frac{2}{\mu^3}, \frac{16}{(\mu+v)^3} \right\} + \max \left\{ \frac{2}{v^3}, \frac{16}{(\mu+v)^3} \right\} \right];$

3. $\left| \frac{A^{-1}(\mu,v)+H^{-1}(\mu,v)}{2} - L^{-1}(\mu, v) \right| \leq \frac{(v-\mu)^2}{96} \left[\max \left\{ \frac{2}{\mu^3}, \frac{16}{(\mu+v)^3} \right\} + \max \left\{ \frac{2}{v^3}, \frac{16}{(\mu+v)^3} \right\} \right];$

4. $\left| H^{-1}(\mu, v) - L^{-1}(\mu, v) \right| \leq \frac{(v-\mu)^2}{24} \left[\max \left\{ \frac{2}{\mu^3}, \frac{16}{(\mu+v)^3} \right\} + \max \left\{ \frac{2}{v^3}, \frac{16}{(\mu+v)^3} \right\} \right].$

Proof We apply inequalities (7)–(10) to the function $G : [\mu, v] \to \mathbb{R}$ defined by $G(x) = \frac{1}{x}$. For this, we observe that $|G''(x)| = \frac{2}{x^3}$ is convex on $[\mu, v]$ and so η−quasiconvex with respect to $\eta(x, y) = x - y$.

We end with more four new inequalities.

Proposition 16 *Let μ and v be two positive numbers with $\mu < v$. Then the following inequalities hold:*

1. $|\ln A(\mu, v) - \ln I(\mu, v)| \leq \frac{(v-\mu)^2}{48} \left[\max \left\{ \frac{1}{\mu^2}, \frac{4}{(\mu+v)^2} \right\} + \max \left\{ \frac{1}{v^2}, \frac{4}{(\mu+v)^2} \right\} \right];$

2. $\left| \frac{2\ln A(\mu,v)+\ln G(\mu,v)}{3} - \ln I(\mu, v) \right| \leq \frac{(v-\mu)^2}{162} \left[\max \left\{ \frac{1}{\mu^2}, \frac{4}{(\mu+v)^2} \right\} + \max \left\{ \frac{1}{v^2}, \frac{4}{(\mu+v)^2} \right\} \right];$

3. $\left| \frac{\ln A(\mu,v)+\ln G(\mu,v)}{2} - \ln I(\mu, v) \right| \leq \frac{(v-\mu)^2}{96} \left[\max \left\{ \frac{1}{\mu^2}, \frac{4}{(\mu+v)^2} \right\} + \max \left\{ \frac{1}{v^2}, \frac{4}{(\mu+v)^2} \right\} \right];$

4. $|\ln G(\mu, v) - \ln I(\mu, v)| \leq \frac{(v-\mu)^2}{24} \left[\max \left\{ \frac{1}{\mu^2}, \frac{4}{(\mu+v)^2} \right\} + \max \left\{ \frac{1}{v^2}, \frac{4}{(\mu+v)^2} \right\} \right].$

Proof Result follows by applying (7)–(10) to the function $G(x) = \ln x$, $x \in [\mu, v]$, taking $\eta(x, y) = x - y$ and noting that $|G''(x)| = \frac{1}{x^2}$ is η−quasiconvex.

5 Conclusion

We proved two main theorems that establish Ostrowski type inequalities in terms of a parameter $\lambda \in [0, 1]$. By choosing $\lambda = 0, 1/3, 1/2, 1$, we deduced midpoint, Simpson, midpoint-trapezoid, and trapezoid type inequalities, respectively. Thereafter, we illustrated the importance of our results by applying them to special means of positive real numbers.

Acknowledgements This research was partially supported by the Portuguese Foundation for Science and Technology (FCT) through CIDMA, project UID/MAT/04106/2019.

References

1. M.U. Awan, M.A. Noor, K.I. Noor, F. Safdar, On strongly generalized convex functions. Filomat **31**(18), 5783–5790 (2017)
2. M.R. Delavar, M. De La Sen, Some generalizations of Hermite–Hadamard type inequalities. SpringerPlus **5**, 1661, 9pp. (2016)
3. M.R. Delavar, S.S. Dragomir, On η-convexity. Math. Inequal. Appl. **20**(1), 203–216 (2017)
4. M. Eshaghi Gordji, M. Rostamian Delavar, M. De La Sen, On ϕ-convex functions. J. Math. Inequal. **10**(1), 173–183 (2016)
5. H. Hudzik, L. Maligranda, Some remarks on s-convex functions. Aequationes Math. **48**(1), 100–111 (1994)
6. M.A. Khan, Y. Khurshid, T. Ali, Hermite-Hadamard inequality for fractional integrals via η-convex functions. Acta Math. Univ. Comenian. **86**(1), 153–164 (2017)
7. Z. Liu, Remarks on some inequalities for s-convex functions and applications. J. Inequal. Appl. **2015**, 333, 17pp. (2015)
8. M.Z. Sarikaya, N. Aktan, On the generalization of some integral inequalities and their applications. Math. Comput. Modell. **54**(9–10), 2175–2182 (2011)

Local Fractional Inequalities

George A. Anastassiou

Abstract This research is about inequalities in a local fractional setting. The author presents the following types of analytic local fractional inequalities: Opial, Hilbert-Pachpatte, Ostrowski, comparison of means, Poincare, Sobolev, Landau, and Polya–Ostrowski. The results are with respect to uniform and L_p norms, involving left and right Riemann–Liouville fractional derivatives. We derive also several interesting special cases.

1 Introduction

Several sources motivate us to write this work. The first one comes next. It is the famous Opial inequality [13]:

$$\int_0^a \left| y'(x) \, y(x) \right| dx \le \frac{a}{2} \int_0^a \left| y'(x) \right|^2 dx,$$

where $y(x)$ is absolutely continuous function and $y(0) = 0$. The above inequality is proved sharp.

The well-known Ostrowski [14] inequality also motivates this work and has as follows:

$$\left| \frac{1}{b-a} \int_a^b f(y) \, dy - f(x) \right| \le \left(\frac{1}{4} + \frac{\left(x - \frac{a+b}{2}\right)^2}{(b-a)^2} \right) (b-a) \left\| f' \right\|_\infty,$$

where $f \in C^1([a, b])$, $x \in [a, b]$, and it is a sharp inequality.

Next $D_{*a}^\rho f$ indicates the left Caputo fractional derivative of order $\rho > 0$, anchored at $a \in \mathbb{R}$, see [10, p. 50].

G. A. Anastassiou (✉)
Department of Mathematical Sciences, University of Memphis, Memphis, TN, USA
e-mail: ganastss@memphis.edu

© Springer Nature Switzerland AG 2019
G. A. Anastassiou and J. M. Rassias (eds.), *Frontiers in Functional Equations and Analytic Inequalities*, https://doi.org/10.1007/978-3-030-28950-8_23

The author in [7, pp. 82–83], proved the following left Caputo fractional Landau inequality: Let $0 < \nu \leq 1$, $f \in AC^2([0, b])$ (i.e. $f' \in AC([0, b])$, absolutely continuous functions), $\forall\, b > 0$. Suppose $\|f\|_{\infty,\mathbb{R}_+} < +\infty$, $D_{*0}^{\nu+1} f \in L_\infty(\mathbb{R}_+)$, and

$$\left\| D_{*a}^{\nu+1} f \right\|_{\infty,[a,+\infty)} \leq \left\| D_{*0}^{\nu+1} f \right\|_{\infty,\mathbb{R}_+}, \quad \forall\, a \geq 0.$$

Then

$$\|f'\|_{\infty,\mathbb{R}_+} \leq (\nu+1) \left(\frac{2}{\nu}\right)^{\frac{\nu}{\nu+1}} (\Gamma(\nu+2))^{-\frac{1}{\nu+1}} \left(\|f\|_{\infty,\mathbb{R}_+}\right)^{\frac{\nu}{\nu+1}} \left(\left\| D_{*0}^{\nu+1} f \right\|_{\infty,\mathbb{R}_+}\right)^{\frac{1}{\nu+1}},$$

that is $\|f'\|_{\infty,\mathbb{R}_+}$ is finite.

The last inequality is another inspiration.

The author's monographs [2–6, 8] motivate and support greatly this work too.

Under the point of view of local fractional differentiation the author examines the broad area of analytic inequalities and produces a great variety of well-known inequalities in the local fractional environment to all possible directions.

2 Background

We mention

Definition 1 ([11]) Let $x, x' \in [a, b]$, $f \in C([a, b])$. The Riemann–Liouville (R-L) fractional derivative of a function f of order q $(0 < q < 1)$ is defined as

$$D_x^q f(x') = \left\{ \begin{array}{ll} D_{x+}^q f(x'), & x' > x, \\ D_{x-}^q f(x'), & x' < x \end{array} \right\} =$$

$$\frac{1}{\Gamma(1-q)} \left\{ \begin{array}{ll} \frac{d}{dx'} \int_x^{x'} (x'-t)^{-q} f(t)\, dt, & x' > x, \\ -\frac{d}{dx'} \int_{x'}^x (t-x')^{-q} f(t)\, dt, & x' < x, \end{array} \right. \tag{1}$$

the left and right R-L fractional derivatives, respectively.

We need

Definition 2 ([11, 12]) The local fractional derivative of order q $(0 < q < 1)$ of a function $f \in C([a, b])$ is defined as

$$D^q f(x) = \lim_{x' \to x} D_x^q \left(f(x') - f(x) \right). \tag{2}$$

More generally we define

Definition 3 ([9]) Let $N \in \mathbb{Z}_+$, $0 < q < 1$, the local fractional derivative of order $(N+q)$ of a function $f \in C^N([a,b])$ is defined by

$$D^{N+q} f(x) = \lim_{x' \to x} D_x^q \left(f(x') - \sum_{n=0}^{N} \frac{f^{(n)}(x)}{n!} (x'-x)^n \right). \tag{3}$$

If $N = 0$, then Definition 3 collapses to Definition 2.

We need

Definition 4 (Related to Definition 3) Let $f \in C^N([a,b])$, $N \in \mathbb{Z}_+$. Set

$$F(x, x'-x; q, N) := D_x^q \left(f(x') - \sum_{n=0}^{N} \frac{f^{(n)}(x)}{n!} (x'-x)^n \right). \tag{4}$$

Let $x' - x := t$, then $x' = x + t$, and

$$F(x, t; q, N) = D_x^q \left(f(x+t) - \sum_{n=0}^{N} \frac{f^{(n)}(x)}{n!} t^n \right). \tag{5}$$

We make

Remark 1 Here $x', x \in [a,b]$, and $a \le x + t \le b$, equivalently $a - x \le t \le b - x$. From $a \le x \le b$, we get $a - x \le 0 \le b - x$. We assume here that $F(x, \cdot; q, N) \in C^1([a-x, b-x])$. Clearly, then it holds

$$D^{N+q} f(x) = F(x, 0; q, N), \tag{6}$$

and $D^{N+q} f(x)$ exists in $\mathbb{R}$.

We would need:

Theorem 1 ([9]) Let $f \in C^N([a,b])$, $N \in \mathbb{Z}_+$. Here $x, x' \in [a,b]$, and $F(x, \cdot; q, N) \in C^1([a-x, b-x])$. Then

$$f(x') = \sum_{n=0}^{N} \frac{f^{(n)}(x)}{n!} (x'-x)^n + \frac{D^{N+q} f(x)}{\Gamma(q+1)} |x'-x|^q + \tag{7}$$

$$\frac{1}{\Gamma(q+1)} \int_0^{x'-x} \frac{dF(x,t;q,N)}{dt} |(x'-x) - t|^q \, dt.$$

Corollary 1 (To Theorem 1, $N = 0$) Let $f \in C([a,b])$, $x, x' \in [a,b]$, and $F(x, \cdot; q, 0) \in C^1([a-x, b-x])$. Then

$$f\left(x'\right) = f\left(x\right) + \frac{D^q f\left(x\right)}{\Gamma\left(q+1\right)} \left|x' - x\right|^q + \tag{8}$$

$$\frac{1}{\Gamma\left(q+1\right)} \int_0^{x'-x} \frac{dF\left(x, t; q, 0\right)}{dt} \left|\left(x'-x\right) - t\right|^q dt.$$

3 Main Results

We make

Remark 2 Let $f \in C^N\left([0, a]\right)$, $N \in \mathbb{Z}_+$, $a > 0$, $x \in [0, a]$; $F\left(0, \cdot; q, N\right) \in C^1\left([0, a]\right)$, $0 < q < 1$. Then, by (7), we have

$$f\left(x\right) = \sum_{n=0}^N \frac{f^{(n)}\left(0\right)}{n!} x^n + \frac{D^{N+q} f\left(0\right)}{\Gamma\left(q+1\right)} x^q + \tag{9}$$

$$\frac{1}{\Gamma\left(q+1\right)} \int_0^x \frac{dF\left(0, t; q, N\right)}{dt}\left(x-t\right)^q dt.$$

Assume that $f^{(n)}\left(0\right) = 0$, $n = 0, 1, \ldots, N$, and $D^{N+q} f\left(0\right) = 0 (= F\left(0, 0; q, N\right) = D_0^q f\left(0\right))$.
   Then

$$f\left(x\right) = \frac{1}{\Gamma\left(q+1\right)} \int_0^x \frac{dF\left(0, t; q, N\right)}{dt}\left(x-t\right)^q dt, \tag{10}$$

$\forall\, x \in [0, a]$.
   Here it is

$$F\left(0, t; q, N\right) = D_0^q\left(f\left(t\right)\right) \in C^1\left([0, a]\right),$$

where D_0^q is the left Riemann–Liouville fractional derivative.
   We have that

$$\left|f\left(x\right)\right| \leq \frac{1}{\Gamma\left(q+1\right)} \int_0^x \left|\frac{dF\left(0, t; q, N\right)}{dt}\right|\left(x-t\right)^q dt \tag{11}$$

$$\leq \frac{a^q}{\Gamma\left(q+1\right)} \int_0^x \left|\frac{dF\left(0, t; q, N\right)}{dt}\right| dt,$$

$\forall\, x \in [0, a]$.

Consider the function

$$z(x) := \frac{a^q}{\Gamma(q+1)} \int_0^x \left| \frac{dF(0, t; q, N)}{dt} \right| dt, \tag{12}$$

$\forall x \in [0, a]$; $z(0) = 0$. Then

$$z'(x) = \frac{a^q}{\Gamma(q+1)} \left| \frac{dF(0, x; q, N)}{dx} \right|. \tag{13}$$

Notice that $|f(x)| \leq z(x)$, $\forall x \in [0, a]$, and we have:

$$2 \int_0^a \frac{a^q}{\Gamma(q+1)} \left| \frac{dF(0, x; q, N)}{dx} \right| |f(x)| dx \leq$$

$$2 \int_0^a z'(x) z(x) dx = z^2(a). \tag{14}$$

By Cauchy–Schwarz inequality we get:

$$z^2(a) = \left(\int_0^a z'(x) dx \right)^2 \leq \left(\int_0^a 1 dx \right) \left(\int_0^a (z'(x))^2 dx \right)$$

$$= a \int_0^a \left(\frac{a^q}{\Gamma(q+1)} \right)^2 \left(\frac{dF(0, x; q, N)}{dx} \right)^2 dx. \tag{15}$$

That is, it holds

$$\frac{2a^q}{\Gamma(q+1)} \int_0^a \left| \frac{dF(0, x; q, N)}{dx} \right| |f(x)| dx \leq$$

$$a \left(\frac{a^q}{\Gamma(q+1)} \right)^2 \int_0^a \left(\frac{dF(0, x; q, N)}{dx} \right)^2 dx. \tag{16}$$

We have proved that

$$\int_0^a |f(x)| \left| \frac{dF(0, x; q, N)}{dx} \right| dx \leq \frac{a^{q+1}}{2\Gamma(q+1)} \int_0^a \left(\frac{dF(0, x; q, N)}{dx} \right)^2 dx, \tag{17}$$

an Opial type inequality.

Equivalently we have proved:

Theorem 2 *Let $f \in C^N([0, a])$, $N \in \mathbb{Z}_+$, $a > 0$, $0 < q < 1$, $D_0^q f \in C^1([0, a])$, and $f^{(n)}(0) = 0$, $n = 0, 1, \ldots, N$, and $D^{N+q} f(0) = 0$. Then*

$$\int_0^a |f(x)| \left| \frac{d}{dx} \left(D_0^q f(x) \right) \right| dx \le \frac{a^{q+1}}{2\Gamma(q+1)} \int_0^a \left(\frac{d}{dx} \left(D_0^q f(x) \right) \right)^2 dx,$$

(18)

above D_0^q is the left Riemann–Liouville fractional derivative.

Inequality (18) is a local fractional Opial type inequality.

Corollary 2 *Let $f \in C([0, a])$, $a > 0$, $0 < q < 1$, $D_0^q f \in C^1([0, a])$, and $f(0) = D^q f(0) = 0$. Then*

$$\int_0^a |f(x)| \left| \frac{d}{dx} \left(D_0^q f(x) \right) \right| dx \le \frac{a^{q+1}}{2\Gamma(q+1)} \int_0^a \left(\frac{d}{dx} \left(D_0^q f(x) \right) \right)^2 dx. \quad (19)$$

Proof Similar to Theorem 2.

We make

Remark 3 Here all as in Remark 2. Let $p_1, q_1 > 1 : \frac{1}{p_1} + \frac{1}{q_1} = 1$. Then it holds

$$f(x) = \frac{1}{\Gamma(q+1)} \int_0^x \frac{dF(0, t; q, N)}{dt} (x - t)^q \, dt,$$

(20)

$\forall x \in [0, a]$.

Let $0 \le w \le x$, then we have

$$f(w) = \frac{1}{\Gamma(q+1)} \int_0^w \frac{dF(0, t; q, N)}{dt} (w - t)^q \, dt.$$

(21)

Hence we derive

$$|f(w)| \le \frac{1}{\Gamma(q+1)} \int_0^w \left| \frac{dF(0, t; q, N)}{dt} \right| (w - t)^q \, dt$$

(22)

$$\le \frac{1}{\Gamma(q+1)} \left(\int_0^w (w - t)^{qp_1} \, dt \right)^{\frac{1}{p_1}} \left(\int_0^w \left| \frac{dF(0, t; q, N)}{dt} \right|^{q_1} dt \right)^{\frac{1}{q_1}}$$

$$= \frac{1}{\Gamma(q+1)} \left(\frac{w^{qp_1+1}}{qp_1 + 1} \right)^{\frac{1}{p_1}} \left(\int_0^w \left| \frac{dF(0, t; q, N)}{dt} \right|^{q_1} dt \right)^{\frac{1}{q_1}}$$

$$= \frac{1}{\Gamma(q+1)} \left(\frac{w^{q+\frac{1}{p_1}}}{(qp_1 + 1)^{\frac{1}{p_1}}} \right) (z(w))^{\frac{1}{q_1}},$$

(23)

where

$$z(w) := \int_0^w \left| \frac{dF(0, t; q, N)}{dt} \right|^{q_1} dt, \tag{24}$$

all $0 \le w \le x$, $z(0) = 0$.

Thus

$$z'(w) = \left| \frac{dF(0, w; q, N)}{dw} \right|^{q_1}, \tag{25}$$

and

$$\left| \frac{dF(0, w; q, N)}{dw} \right| = (z'(w))^{\frac{1}{q_1}}. \tag{26}$$

Therefore we obtain

$$|f(w)| \left| \frac{dF(0, w; q, N)}{dw} \right| \le \frac{w^{q + \frac{1}{p_1}}}{\Gamma(q+1)(qp_1+1)^{\frac{1}{p_1}}} (z(w) z'(w))^{\frac{1}{q_1}}. \tag{27}$$

Hence it holds

$$\int_0^x |f(w)| \left| \frac{dF(0, w; q, N)}{dw} \right| dw \le$$

$$\frac{1}{\Gamma(q+1)(qp_1+1)^{\frac{1}{p_1}}} \int_0^x w^{\frac{qp_1+1}{p_1}} (z(w) z'(w))^{\frac{1}{q_1}} dw \le \tag{28}$$

$$\frac{1}{\Gamma(q+1)(qp_1+1)^{\frac{1}{p_1}}} \left(\int_0^x w^{qp_1+1} dw \right)^{\frac{1}{p_1}} \left(\int_0^x z(w) z'(w) dw \right)^{\frac{1}{q_1}} =$$

$$\frac{1}{\Gamma(q+1)(qp_1+1)^{\frac{1}{p_1}}} \left(\frac{x^{qp_1+2}}{qp_1+2} \right)^{\frac{1}{p_1}} \left(\int_0^x z(w) \, dz(w) \right)^{\frac{1}{q_1}} =$$

$$\frac{x^{q+\frac{2}{p_1}}}{\Gamma(q+1)[(qp_1+1)(qp_1+2)]^{\frac{1}{p_1}}} \left(\frac{z^2(x)}{2} \right)^{\frac{1}{q_1}} =$$

$$\frac{x^{q+\frac{2}{p_1}}}{2^{\frac{1}{q_1}} \Gamma(q+1)[(qp_1+1)(qp_1+2)]^{\frac{1}{p_1}}} \left(\int_0^x \left| \frac{dF(0, t; q, N)}{dt} \right|^{q_1} dt \right)^{\frac{2}{q_1}}. \tag{29}$$

We have proved an L_p-Opial type local fractional inequality:

Theorem 3 *Let* $p_1, q_1 > 1 : \frac{1}{p_1} + \frac{1}{q_1} = 1;$ $f \in C^N([0, a]),$ $N \in \mathbb{Z}_+,$ $a > 0;$ $F(0, \cdot; q, N) \in C^1([0, a]),$ $0 < q < 1.$ *Assume that* $f^{(n)}(0) = 0,$ $n = 0, 1, \ldots, N,$ *and* $D^{N+q} f(0) = 0$ *(i.e.* $F(0, 0; q, N) = D_0^q f(0) = 0$).

[Here it is $F(0, t; q, N) = D_0^q(f(t)) \in C^1([0, a]),$ *where* D_0^q *is the left Riemann–Liouville fractional derivative.]*

Then

$$\int_0^x |f(w)| \left| \frac{dF(0, w; q, N)}{dw} \right| dw \leq \tag{30}$$

$$\frac{x^{q + \frac{2}{p_1}}}{2^{\frac{1}{q_1}} \Gamma(q+1) [(qp_1 + 1)(qp_1 + 2)]^{\frac{1}{p_1}}} \left(\int_0^x \left| \frac{dF(0, w; q, N)}{dw} \right|^{q_1} dw \right)^{\frac{2}{q_1}},$$

$\forall \, x \in [0, a].$

$\Leftrightarrow$ *it holds*

$$\int_0^x |f(w)| \left| \frac{dD_0^q f(w)}{dw} \right| dw \leq$$

$$\frac{x^{q + \frac{2}{p_1}}}{2^{\frac{1}{q_1}} \Gamma(q+1) [(qp_1 + 1)(qp_1 + 2)]^{\frac{1}{p_1}}} \left(\int_0^x \left| \frac{dD_0^q f(w)}{dw} \right|^{q_1} dw \right)^{\frac{2}{q_1}}, \tag{31}$$

$\forall \, x \in [0, a].$

We make

Remark 4 Here all as in Remark 2 for f_1, f_2. Let $i = 1, 2$, then

$$f_i(x_i) = \frac{1}{\Gamma(q+1)} \int_0^{x_i} \frac{dF_i(0, t_i; q, N)}{dt_i} (x_i - t_i)^q \, dt_i, \tag{32}$$

$\forall \, x_i \in [0, a_i].$

Hence it holds

$$|f_i(x_i)| \leq \frac{1}{\Gamma(q+1)} \int_0^{x_i} \left| \frac{dF_i(0, t_i; q, N)}{dt_i} \right| (x_i - t_i)^q \, dt_i. \tag{33}$$

Let $p_1, q_1 > 1 : \frac{1}{p_1} + \frac{1}{q_1} = 1.$ We get by Hölder's inequality:

$$|f_1(x_1)| \leq \frac{1}{\Gamma(q+1)} \left(\int_0^{x_1} (x_1 - t_1)^{qp_1} \, dt_1 \right)^{\frac{1}{p_1}}.$$

$$\left(\int_0^{x_1} \left| \frac{dF_1(0, t_1; q, N)}{dt_1} \right|^{q_1} dt_1 \right)^{\frac{1}{q_1}} \leq \tag{34}$$

$$\frac{1}{\Gamma(q+1)} \frac{x_1^{\frac{qp_1+1}{p_1}}}{(qp_1+1)^{\frac{1}{p_1}}} \left\| \frac{dF_1(0, t_1; q, N)}{dt_1} \right\|_{q_1, [0, a_1]},$$

$\forall \, x_1 \in [0, a_1]$.

Similarly, we obtain

$$|f_2(x_2)| \leq \frac{1}{\Gamma(q+1)} \frac{x_2^{\frac{qq_1+1}{q_1}}}{(qq_1+1)^{\frac{1}{q_1}}} \left\| \frac{dF_2(0, t_2; q, N)}{dt_2} \right\|_{p_1, [0, a_2]}. \tag{35}$$

Therefore we have

$$|f_1(x_1)| \, |f_2(x_2)| \leq \frac{1}{(\Gamma(q+1))^2 (qp_1+1)^{\frac{1}{p_1}} (qq_1+1)^{\frac{1}{q_1}}} \cdot$$

$$x_1^{\frac{qp_1+1}{p_1}} x_2^{\frac{qq_1+1}{q_1}} \left\| \frac{dF_1(0, t_1; q, N)}{dt_1} \right\|_{q_1, [0, a_1]} \left\| \frac{dF_2(0, t_2; q, N)}{dt_2} \right\|_{p_1, [0, a_2]} \leq \tag{36}$$

(using Young's inequality for $a, b \geq 0$, $a^{\frac{1}{p_1}} b^{\frac{1}{q_1}} \leq \frac{a}{p_1} + \frac{b}{q_1}$)

$$\frac{1}{(\Gamma(q+1))^2 (qp_1+1)^{\frac{1}{p_1}} (qq_1+1)^{\frac{1}{q_1}}} \left[\frac{x_1^{qp_1+1}}{p_1} + \frac{x_2^{qq_1+1}}{q_1} \right] \cdot$$

$$\left\| \frac{dF_1(0, t_1; q, N)}{dt_1} \right\|_{q_1, [0, a_1]} \left\| \frac{dF_2(0, t_2; q, N)}{dt_2} \right\|_{p_1, [0, a_2]},$$

$\forall \, x_i \in [0, a_i]; \, i = 1, 2$.

So far we have established

$$\frac{|f_1(x_1)| \, |f_2(x_2)|}{\left[\frac{x_1^{qp_1+1}}{p_1} + \frac{x_2^{qq_1+1}}{q_1} \right]} \leq \frac{1}{(\Gamma(q+1))^2 (qp_1+1)^{\frac{1}{p_1}} (qq_1+1)^{\frac{1}{q_1}}} \cdot \tag{37}$$

$$\left\| \frac{dF_1(0, t_1; q, N)}{dt_1} \right\|_{q_1, [0, a_1]} \left\| \frac{dF_2(0, t_2; q, N)}{dt_2} \right\|_{p_1, [0, a_2]},$$

$\forall \, x_i \in [0, a_i]; \, i = 1, 2$.

The denominator of left-hand side (37) can be zero only when $x_1 = 0$ and $x_2 = 0$. By integrating (37) over $[0, a_1] \times [0, a_2]$ we get

$$\int_0^{a_1} \int_0^{a_2} \frac{|f_1(x_1)| \, |f_2(x_2)| \, dx_1 dx_2}{\left[\frac{x_1^{qp_1+1}}{p_1} + \frac{x_2^{qq_1+1}}{q_1} \right]} \leq \frac{a_1 a_2}{(\Gamma(q+1))^2 (qp_1+1)^{\frac{1}{p_1}} (qq_1+1)^{\frac{1}{q_1}}} \cdot$$

$$\left\| \frac{dF_1(0, t_1; q, N)}{dt_1} \right\|_{q_1,[0,a_1]} \left\| \frac{dF_2(0, t_2; q, N)}{dt_2} \right\|_{p_1,[0,a_2]} . \tag{38}$$

We have proved the following local fractional Hilbert-Pachpatte inequality:

Theorem 4 *Let* $p_1, q_1 > 1 : \frac{1}{p_1} + \frac{1}{q_1} = 1; i = 1, 2$ *for* $f_i \in C^N([0, a_i])$, $N \in \mathbb{Z}_+$, $a_i > 0$; $F_i(0, \cdot; q, N) \in C^1([0, a_i])$, $0 < q < 1$. *Assume that* $f_i^{(n)}(0) = 0$, $n = 0, 1, \ldots, N$, *and* $D^{N+q} f_i(0) = 0$, $i = 1, 2$ *(i.e.* $F_i(0, 0; q, N) = D_0^q f_i(0) = 0$).

[Here it is $F_i(0, t_i; q, N) = D_0^q(f_i(t_i)) \in C^1([0, a_i])$, *where* D_0^q *is the left Riemann–Liouville fractional derivative]*

Then

$$\int_0^{a_1} \int_0^{a_2} \frac{|f_1(x_1)| \, |f_2(x_2)| \, dx_1 dx_2}{\left[\frac{x_1^{qp_1+1}}{p_1} + \frac{x_2^{qq_1+1}}{q_1} \right]} \leq \frac{a_1 a_2}{(\Gamma(q+1))^2 (qp_1+1)^{\frac{1}{p_1}} (qq_1+1)^{\frac{1}{q_1}}} \cdot$$

$$\left\| \frac{dF_1(0, t_1; q, N)}{dt_1} \right\|_{q_1,[0,a_1]} \left\| \frac{dF_2(0, t_2; q, N)}{dt_2} \right\|_{p_1,[0,a_2]} . \tag{39}$$

$\Leftrightarrow$ *it holds*

$$\int_0^{a_1} \int_0^{a_2} \frac{|f_1(x_1)| \, |f_2(x_2)| \, dx_1 dx_2}{\left[\frac{x_1^{qp_1+1}}{p_1} + \frac{x_2^{qq_1+1}}{q_1} \right]} \leq \frac{a_1 a_2}{(\Gamma(q+1))^2 (qp_1+1)^{\frac{1}{p_1}} (qq_1+1)^{\frac{1}{q_1}}} \cdot$$

$$\left\| \frac{d}{dt_1} D_0^q(f_1(t_1)) \right\|_{q_1,[0,a_1]} \left\| \frac{d}{dt_2} D_0^q(f_2(t_2)) \right\|_{p_1,[0,a_2]} . \tag{40}$$

We need

Remark 5 *Here* $f \in C^N([a, b])$, $x, x' \in [a, b]$, $0 < q < 1$, $N \in \mathbb{Z}_+$, *and*

$$D_{x\pm}^q \left(f(\cdot) - \sum_{n=0}^N \frac{f^{(n)}(x)}{n!} (\cdot - x)^n \right) \in L_\infty(a, b).$$

By [1] we obtain:

(i) if $x' > x$, then

$$f(x') - \sum_{n=0}^N \frac{f^{(n)}(x)}{n!} (x' - x)^n = \tag{41}$$

$$\frac{1}{\Gamma(q)} \int_x^{x'} (x'-z)^{q-1} D_{x+}^q \left(f(z) - \sum_{n=0}^N \frac{f^{(n)}(x)}{n!} (z-x)^n \right) dz,$$

(ii) if $x' < x$, then

$$f(x') - \sum_{n=0}^N \frac{f^{(n)}(x)}{n!} (x'-x)^n = \tag{42}$$

$$\frac{1}{\Gamma(q)} \int_{x'}^x (z-x')^{q-1} D_{x-}^q \left(f(z) - \sum_{n=0}^N \frac{f^{(n)}(x)}{n!} (z-x)^n \right) dz.$$

By Proposition 15.114, [3, p. 388], indeed we have that the functions in x' of the right-hand sides of (41) and (42) are continuous.

When $N = 0$, we derive

(i) if $x' > x$, then

$$f(x') - f(x) = \frac{1}{\Gamma(q)} \int_x^{x'} (x'-z)^{q-1} D_{x+}^q (f(z) - f(x)) dz, \tag{43}$$

(ii) if $x' < x$, then

$$f(x') - f(x) = \frac{1}{\Gamma(q)} \int_{x'}^x (z-x')^{q-1} D_{x-}^q (f(z) - f(x)) dz. \tag{44}$$

For $x' > x$ we get

$$\left| f(x') - f(x) \right| \le \left\| D_{x+}^q (f(\cdot) - f(x)) \right\|_{\infty,[x,b]} \frac{(x'-x)^q}{\Gamma(q+1)}, \tag{45}$$

and for $x' < x$ we find that

$$\left| f(x') - f(x) \right| \le \left\| D_{x-}^q (f(\cdot) - f(x)) \right\|_{\infty,[a,x]} \frac{(x-x')^q}{\Gamma(q+1)}. \tag{46}$$

We have that

$$\frac{1}{b-a} \int_a^b f(x') dx' - f(x) = \frac{1}{b-a} \int_a^b (f(x') - f(x)) dx' =$$

$$\frac{1}{b-a} \left[\int_a^x (f(x') - f(x)) dx' + \int_x^b (f(x') - f(x)) dx' \right]. \tag{47}$$

Hence it holds

$$\left| \frac{1}{b-a} \int_a^b f\left(x'\right) dx' - f\left(x\right) \right| \le$$

$$\frac{1}{b-a} \left[\int_a^x \left| f\left(x'\right) - f\left(x\right) \right| dx' + \int_x^b \left| f\left(x'\right) - f\left(x\right) \right| dx' \right] \le$$

(by (45) and (46))

$$\frac{1}{b-a} \left[\frac{\left\| D_{x-}^q \left(f\left(\cdot\right) - f\left(x\right) \right) \right\|_{\infty,[a,x]}}{\Gamma\left(q+1\right)} \int_a^x \left(x-x'\right)^q dx' + \right.$$

$$\left. \frac{\left\| D_{x+}^q \left(f\left(\cdot\right) - f\left(x\right) \right) \right\|_{\infty,[x,b]}}{\Gamma\left(q+1\right)} \int_x^b \left(x'-x\right)^q dx' \right] = \qquad (48)$$

$$\frac{1}{(b-a)\,\Gamma\left(q+2\right)} \left[(x-a)^{q+1} \left\| D_{x-}^q \left(f\left(\cdot\right) - f\left(x\right) \right) \right\|_{\infty,[a,x]} + \right.$$

$$\left. (b-x)^{q+1} \left\| D_{x+}^q \left(f\left(\cdot\right) - f\left(x\right) \right) \right\|_{\infty,[x,b]} \right].$$

We have proved a local fractional ∞-Ostrowski inequality:

Theorem 5 *Let* $f \in C\left([a,b]\right)$, $x \in [a,b]$, $0 < q < 1$, *and* $D_{x+}^q \left(f\left(\cdot\right) - f\left(x\right) \right) \in L_\infty \left([x,b]\right)$, $D_{x-}^q \left(f\left(\cdot\right) - f\left(x\right) \right) \in L_\infty \left([a,x]\right)$. *Then*

$$\left| \frac{1}{b-a} \int_a^b f\left(x'\right) dx' - f\left(x\right) \right| \le$$

$$\frac{1}{(b-a)\,\Gamma\left(q+2\right)} \left[(b-x)^{q+1} \left\| D_{x+}^q \left(f\left(\cdot\right) - f\left(x\right) \right) \right\|_{\infty,[x,b]} + \right.$$

$$\left. (x-a)^{q+1} \left\| D_{x-}^q \left(f\left(\cdot\right) - f\left(x\right) \right) \right\|_{\infty,[a,x]} \right]. \qquad (49)$$

We make

Remark 6 Here $p_1, q_1 > 1 : \frac{1}{p_1} + \frac{1}{q_1} = 1$, with $\frac{1}{q_1} < q < 1$, $f \in C\left([a,b]\right)$, $x \in [a,b]$, $D_{x+}^q \left(f\left(\cdot\right) - f\left(x\right) \right) \in L_{q_1} \left([x,b]\right)$, $D_{x-}^q \left(f\left(\cdot\right) - f\left(x\right) \right) \in L_{q_1} \left([a,x]\right)$.
Notice here that $q > \frac{1}{q_1} \Leftrightarrow (q-1)\, p_1 + 1 > 0$.
If $x' > x$, then

$$\left| f\left(x'\right) - f\left(x\right) \right| \le \frac{1}{\Gamma\left(q\right)} \int_x^{x'} \left(x'-z\right)^{q-1} \left| D_{x+}^q \left(f\left(z\right) - f\left(x\right) \right) \right| dz \le$$

$$\frac{1}{\Gamma(q)} \left(\int_x^{x'} (x'-z)^{(q-1)p_1} dz \right)^{\frac{1}{p_1}} \left\| D_{x+}^q \left(f\left(\cdot\right) - f\left(x\right) \right) \right\|_{q_1,[x,b]} =$$

$$\frac{1}{\Gamma(q)} \frac{(x'-x)^{\frac{(q-1)p_1+1}{p_1}}}{((q-1)p_1+1)^{\frac{1}{p_1}}} \left\| D_{x+}^q \left(f\left(\cdot\right) - f\left(x\right) \right) \right\|_{q_1,[x,b]}. \tag{50}$$

Thus

$$\left| f\left(x'\right) - f\left(x\right) \right| \le \frac{(x'-x)^{\frac{(q-1)p_1+1}{p_1}}}{\Gamma(q)\left((q-1)p_1+1\right)^{\frac{1}{p_1}}} \left\| D_{x+}^q \left(f\left(\cdot\right) - f\left(x\right) \right) \right\|_{q_1,[x,b]}, \tag{51}$$

$\forall\, x' \in [x,b]$, and $\frac{1}{q_1} < q < 1$.

Similarly, if $x' < x$, then

$$\left| f\left(x'\right) - f\left(x\right) \right| \le \frac{1}{\Gamma(q)} \int_{x'}^x (z-x')^{q-1} \left| D_{x-}^q \left(f\left(z\right) - f\left(x\right) \right) \right| dz \le$$

$$\frac{1}{\Gamma(q)} \left(\int_{x'}^x (z-x')^{(q-1)p_1} dz \right)^{\frac{1}{p_1}} \left\| D_{x-}^q \left(f\left(\cdot\right) - f\left(x\right) \right) \right\|_{q_1,[a,x]} =$$

$$\frac{1}{\Gamma(q)} \frac{(x-x')^{\frac{(q-1)p_1+1}{p_1}}}{((q-1)p_1+1)^{\frac{1}{p_1}}} \left\| D_{x-}^q \left(f\left(\cdot\right) - f\left(x\right) \right) \right\|_{q_1,[a,x]}. \tag{52}$$

Thus it holds

$$\left| f\left(x'\right) - f\left(x\right) \right| \le \frac{(x-x')^{\frac{(q-1)p_1+1}{p_1}}}{\Gamma(q)\left((q-1)p_1+1\right)^{\frac{1}{p_1}}} \left\| D_{x-}^q \left(f\left(\cdot\right) - f\left(x\right) \right) \right\|_{q_1,[a,x]}, \tag{53}$$

$\forall\, x' \in [a,x]$, and $\frac{1}{q_1} < q < 1$.

Consequently, we derive

$$\left| \frac{1}{b-a} \int_a^b f\left(x'\right) dx' - f\left(x\right) \right| \le$$

$$\frac{1}{b-a} \left[\int_a^x \left| f\left(x'\right) - f\left(x\right) \right| dx' + \int_x^b \left| f\left(x'\right) - f\left(x\right) \right| dx' \right] \le$$

$$\frac{1}{(b-a)} \left[\frac{\left\| D_{x-}^q \left(f\left(\cdot\right) - f\left(x\right) \right) \right\|_{q_1,[a,x]}}{\Gamma(q)\left((q-1)p_1+1\right)^{\frac{1}{p_1}}} \left(\int_a^x (x-x')^{(q-1)+\frac{1}{p_1}} dx' \right) + \tag{54} \right.$$

$$\frac{\left\| D_{x+}^q \left(f \left(\cdot \right) - f \left(x \right) \right) \right\|_{q_1,[x,b]}}{\Gamma \left(q \right) \left(\left(q - 1 \right) p_1 + 1 \right)^{\frac{1}{p_1}}} \left(\int_x^b \left(x' - x \right)^{\left(q - 1 \right) + \frac{1}{p_1}} dx' \right) \right] =$$

$$\frac{1}{\left(b - a \right) \Gamma \left(q \right) \left(\left(q - 1 \right) p_1 + 1 \right)^{\frac{1}{p_1}}} \cdot$$

$$\left[\left\| D_{x-}^q \left(f \left(\cdot \right) - f \left(x \right) \right) \right\|_{q_1,[a,x]} \frac{\left(x - a \right)^{q + \frac{1}{p_1}}}{\left(q + \frac{1}{p_1} \right)} + \left\| D_{x+}^q \left(f \left(\cdot \right) - f \left(x \right) \right) \right\|_{q_1,[x,b]} \frac{\left(b - x \right)^{q + \frac{1}{p_1}}}{\left(q + \frac{1}{p_1} \right)} \right].$$

We have proved the following L_p-Ostrowski type local fractional inequality:

Theorem 6 *Let $p_1, q_1 > 1 : \frac{1}{p_1} + \frac{1}{q_1} = 1$, with $\frac{1}{q_1} < q < 1$, $f \in C \left([a, b] \right)$, $x \in [a, b]$, $D_{x+}^q \left(f \left(\cdot \right) - f \left(x \right) \right) \in L_{q_1} \left([x, b] \right)$, $D_{x-}^q \left(f \left(\cdot \right) - f \left(x \right) \right) \in L_{q_1} \left([a, x] \right)$. Then*

$$\left| \frac{1}{b - a} \int_a^b f \left(x' \right) dx' - f \left(x \right) \right| \leq \frac{1}{\left(b - a \right) \Gamma \left(q \right) \left(\left(q - 1 \right) p_1 + 1 \right)^{\frac{1}{p_1}} \left(q + \frac{1}{p_1} \right)} \cdot$$
$$\tag{55}$$
$$\left[\left(b - x \right)^{q + \frac{1}{p_1}} \left\| D_{x+}^q \left(f \left(\cdot \right) - f \left(x \right) \right) \right\|_{q_1,[x,b]} + \left(x - a \right)^{q + \frac{1}{p_1}} \left\| D_{x-}^q \left(f \left(\cdot \right) - f \left(x \right) \right) \right\|_{q_1,[a,x]} \right].$$

Corollary 3 *Here $f \in C \left([a, b] \right)$, $x \in [a, b]$, $D_{x+}^q \left(f \left(\cdot \right) - f \left(x \right) \right) \in L_2 \left([x, b] \right)$, $D_{x-}^q \left(f \left(\cdot \right) - f \left(x \right) \right) \in L_2 \left([a, x] \right)$, $\frac{1}{2} < q < 1$. Then*

$$\left| \frac{1}{b - a} \int_a^b f \left(x' \right) dx' - f \left(x \right) \right| \leq \frac{1}{\left(b - a \right) \Gamma \left(q \right) \sqrt{\left(\left(q - 1 \right) 2 + 1 \right)} \left(q + \frac{1}{2} \right)} \cdot$$
$$\tag{56}$$
$$\left[\left(b - x \right)^{q + \frac{1}{2}} \left\| D_{x+}^q \left(f \left(\cdot \right) - f \left(x \right) \right) \right\|_{2,[x,b]} + \left(x - a \right)^{q + \frac{1}{2}} \left\| D_{x-}^q \left(f \left(\cdot \right) - f \left(x \right) \right) \right\|_{2,[a,x]} \right].$$

We make

Remark 7 Here we will discuss the related comparison of means.

Let $f \in C^N \left([0, a] \right)$, $N \in \mathbb{Z}_+$, $a > 0$, $x \in [0, a]$; $F \left(0, \cdot; q, N \right) \in C^1 \left([0, a] \right)$, $0 < q < 1$. Then

$$f \left(x \right) = \sum_{n=0}^N \frac{f^{(n)} \left(0 \right)}{n!} x^n + \frac{D^{N+q} f \left(0 \right)}{\Gamma \left(q + 1 \right)} x^q + \tag{57}$$

$$\frac{1}{\Gamma \left(q + 1 \right)} \int_0^x \frac{dF \left(0, t; q, N \right)}{dt} \left(x - t \right)^q dt.$$

Assume that $f^{(n)}(0) = 0$, $n = 0, 1, \ldots, N$. Here $D^{N+q} f(0) = F(0, 0; q, N) = \left(D_0^q f\right)(0)$, where D_0^q is the left Riemann–Liouville fractional derivative.

So far we have

$$f(x) = \frac{\left(D_0^q f\right)(0)}{\Gamma(q+1)} x^q + R(x), \tag{58}$$

where

$$R(x) := \frac{1}{\Gamma(q+1)} \int_0^x \left(\frac{d}{dt} D_0^q f(t)\right) (x-t)^q \, dt. \tag{59}$$

We also assume that $D_0^q f \in C^1([0, a])$.

We notice that

$$|R(x)| \le \frac{1}{\Gamma(q+1)} \left\| \frac{d}{dt} D_0^q f \right\|_{\infty, [0,a]} \int_0^x (x-t)^q \, dt$$

$$= \frac{x^{q+1}}{\Gamma(q+2)} \left\| \frac{d}{dt} D_0^q f \right\|_{\infty, [0,a]}. \tag{60}$$

That is,

$$|R(x)| \le \frac{x^{q+1}}{\Gamma(q+2)} \left\| \frac{d}{dt} D_0^q f \right\|_{\infty, [0,a]}, \tag{61}$$

$\forall \, x \in [0, a]$.

Hence, it holds

$$\int_0^a f(x) \, dx - \frac{\left(D_0^q f\right)(0)}{\Gamma(q+1)} \int_0^a x^q \, dx = \int_0^a R(x) \, dx, \tag{62}$$

equivalently,

$$\int_0^a f(x) \, dx - \frac{\left(D_0^q f\right)(0)}{\Gamma(q+1)} \left(\frac{x^{q+1}}{q+1}\Big|_0^a\right) = \int_0^a R(x) \, dx, \tag{63}$$

equivalently,

$$\int_0^a f(x) \, dx - \frac{\left(D_0^q f\right)(0)}{\Gamma(q+2)} a^{q+1} = \int_0^a R(x) \, dx. \tag{64}$$

Therefore, we find

$$\left| \int_0^a f(x)\, dx - \frac{\left(D_0^q f \right)(0)}{\Gamma(q+2)} a^{q+1} \right| \le \frac{\left\| \frac{d}{dt} D_0^q f \right\|_{\infty,[0,a]}}{\Gamma(q+2)} \int_0^a x^{q+1} dx =$$

$$\frac{\left\| \frac{d}{dt} D_0^q f \right\|_{\infty,[0,a]}}{\Gamma(q+2)} \frac{a^{q+2}}{q+2} = \frac{\left\| \frac{d}{dt} D_0^q f \right\|_{\infty,[0,a]}}{\Gamma(q+3)} a^{q+2}. \tag{65}$$

We have proved the following local fractional comparison of means result:

Theorem 7 *Let $f \in C^N([0,a])$, $N \in \mathbb{Z}_+$, $a > 0$, $F(0, \cdot; q, N) \in C^1([0,a])$, $0 < q < 1$. Assume that $f^{(n)}(0) = 0$, $n = 0, 1, \ldots, N$; and that $D_0^q f \in C^1([0,a])$. Then*

$$\left| \int_0^a f(x)\, dx - \frac{\left(D_0^q f \right)(0)}{\Gamma(q+2)} a^{q+1} \right| \le \frac{a^{q+2}}{\Gamma(q+3)} \left\| \frac{d}{dt} D_0^q f \right\|_{\infty,[0,a]}, \tag{66}$$

$\Leftrightarrow$

$$\left| \frac{1}{a} \int_0^a f(x)\, dx - \frac{\left(D_0^q f \right)(0)}{\Gamma(q+2)} a^q \right| \le \frac{a^{q+1}}{\Gamma(q+3)} \left\| \frac{d}{dt} D_0^q f \right\|_{\infty,[0,a]}. \tag{67}$$

We make

Remark 8 Here we discuss the related Poincare inequalities. All assumptions are as in Theorem 7, plus $D^{N+q} f(0) = 0$.

Hence we have

$$f(x) = \frac{1}{\Gamma(q+1)} \int_0^x \frac{dF(0, t; q, N)}{dt} (x - t)^q \, dt, \tag{68}$$

$\forall\, x \in [0, a]$.

Let $p_1, q_1 > 1 : \frac{1}{p_1} + \frac{1}{q_1} = 1$. Then

$$|f(x)| \le \frac{1}{\Gamma(q+1)} \int_0^x \left| \frac{dF(0, t; q, N)}{dt} \right| (x - t)^q \, dt \le \tag{69}$$

$$\frac{1}{\Gamma(q+1)} \left(\int_0^x (x - t)^{qp_1} \, dt \right)^{\frac{1}{p_1}} \left(\int_0^x \left| \frac{dF(0, t; q, N)}{dt} \right|^{q_1} dt \right)^{\frac{1}{q_1}} \le$$

$$\frac{1}{\Gamma(q+1)} \frac{x^{\frac{qp_1+1}{p_1}}}{(qp_1 + 1)^{\frac{1}{p_1}}} \left\| \frac{dF(0, t; q, N)}{dt} \right\|_{q_1,[0,a]}. \tag{70}$$

That is,

$$|f(x)| \leq \frac{x^{\frac{qp_1+1}{p_1}}}{\Gamma(q+1)(qp_1+1)^{\frac{1}{p_1}}} \left\| \frac{dF(0,t;q,N)}{dt} \right\|_{q_1,[0,a]},$$

$\forall x \in [0,a]$.

Thus

$$|f(x)|^{q_1} \leq \frac{x^{q_1(q+1)-1}}{\left(\Gamma(q+1)(qp_1+1)^{\frac{1}{p_1}}\right)^{q_1}} \left\| \frac{dF(0,t;q,N)}{dt} \right\|_{q_1,[0,a]}^{q_1}. \tag{71}$$

Therefore we have

$$\int_0^a |f(x)|^{q_1} dx \leq \frac{a^{q_1(q+1)}}{\left(\Gamma(q+1)(qp_1+1)^{\frac{1}{p_1}}\right)^{q_1} q_1(q+1)} \left\| \frac{dF(0,t;q,N)}{dt} \right\|_{q_1,[0,a]}^{q_1}. \tag{72}$$

That is,

$$\|f\|_{q_1,[0,a]} \leq \frac{a^{(q+1)}}{\Gamma(q+1)(qp_1+1)^{\frac{1}{p_1}}(q_1(q+1))^{\frac{1}{q_1}}} \left\| \frac{dF(0,t;q,N)}{dt} \right\|_{q_1,[0,a]}. \tag{73}$$

We have proved the following local fractional Poincare inequality:

Theorem 8 *Let* $f \in C^N([0,a])$, $N \in \mathbb{Z}_+$, $a > 0$, $F(0,\cdot;q,N) \in C^1([0,a])$, $0 < q < 1$; $f^{(n)}(0) = 0$, $n = 0,1,\ldots,N$; $D_0^{N+q}f(0) = 0$, $D_0^q f \in C^1([0,a])$; $p_1, q_1 > 1 : \frac{1}{p_1} + \frac{1}{q_1} = 1$.

[we have $F(0,t;q,N) = D_0^q f(t)$, *where* D_0^q *is the left Riemann–Liouville fractional derivative.]*

Then

$$\|f\|_{q_1,[0,a]} \leq \frac{a^{q+1}}{\Gamma(q+1)(qp_1+1)^{\frac{1}{p_1}}(q_1(q+1))^{\frac{1}{q_1}}} \left\| (D_0^q f)' \right\|_{q_1,[0,a]}. \tag{74}$$

Next comes the related local fractional Sobolev type inequality:

Theorem 9 *All as in Theorem 8, plus* $r > 0$. *Then*

$$\|f\|_{r,[0,a]} \leq \frac{a^{q+\frac{1}{p_1}+\frac{1}{r}} \left\| (D_0^q f)' \right\|_{q_1,[0,a]}}{\Gamma(q+1)(qp_1+1)^{\frac{1}{p_1}}\left(r\left(q+\frac{1}{p_1}\right)+1\right)^{\frac{1}{r}}}. \tag{75}$$

Proof By (70) we get:

$$|f(x)| \leq \frac{x^{q+\frac{1}{p_1}}}{\Gamma(q+1)(qp_1+1)^{\frac{1}{p_1}}} \left\|\left(D_0^q f\right)'\right\|_{q_1,[0,a]}, \tag{76}$$

$\forall\, x \in [0,a]$.

Hence

$$|f(x)|^r \leq \frac{x^{r\left(q+\frac{1}{p_1}\right)}}{\left(\Gamma(q+1)(qp_1+1)^{\frac{1}{p_1}}\right)^r} \left\|\left(D_0^q f\right)'\right\|_{q_1,[0,a]}^r. \tag{77}$$

Consequently, it holds

$$\int_0^a |f(x)|^r\, dx \leq \frac{a^{r\left(q+\frac{1}{p_1}\right)+1}}{\left(\Gamma(q+1)(qp_1+1)^{\frac{1}{p_1}}\right)^r \left(r\left(q+\frac{1}{p_1}\right)+1\right)} \left\|\left(D_0^q f\right)'\right\|_{q_1,[0,a]}^r. \tag{78}$$

We make

Remark 9 Let $f \in C^1([a,b])$, $0 < q < 1$, $D_{a+}^q\left(f'(\cdot) - f'(a)\right) \in L_\infty([a,b])$. Then, by Theorem 5, we get

$$\left|\frac{1}{b-a}\int_a^b f'(x')\, dx' - f'(a)\right| \leq \frac{(b-a)^q}{\Gamma(q+2)} \left\|D_{a+}^q\left(f'(\cdot) - f'(a)\right)\right\|_{\infty,[a,b]}, \tag{79}$$

$\Leftrightarrow$

$$\left|\frac{1}{b-a}\left(f(b) - f(a)\right) - f'(a)\right| \leq \frac{(b-a)^q}{\Gamma(q+2)} \left\|D_{a+}^q\left(f'(\cdot) - f'(a)\right)\right\|_{\infty,[a,b]}. \tag{80}$$

Hence it holds

$$\left|f'(a)\right| - \frac{1}{b-a}\left|f(b) - f(a)\right| \leq \frac{(b-a)^q}{\Gamma(q+2)} \left\|D_{a+}^q\left(f'(\cdot) - f'(a)\right)\right\|_{\infty,[a,b]}, \tag{81}$$

$\Leftrightarrow$

$$\left|f'(a)\right| \leq \frac{|f(b) - f(a)|}{b-a} + \frac{(b-a)^q}{\Gamma(q+2)} \left\|D_{a+}^q\left(f'(\cdot) - f'(a)\right)\right\|_{\infty,[a,b]}. \tag{82}$$

Therefore

$$\left| f'(a) \right| \le \frac{2 \left\| f \right\|_{\infty,[a,b]}}{b-a} + \frac{(b-a)^q}{\Gamma(q+2)} \left\| D_{a+}^q \left(f'(\cdot) - f'(a) \right) \right\|_{\infty,[a,b]}. \tag{83}$$

Complete assumptions follow:

Assume here that $f \in C^1 ([A_0, +\infty))$, where $A_0 \in \mathbb{R}$ is fixed, $0 < q < 1$, and for every $[a, b] \subset [A_0, +\infty)$ we have that $D_{a+}^q \left(f'(\cdot) - f'(a) \right) \in L_\infty ([a, b])$, and that

$$\left\| D_{a+}^q \left(f'(\cdot) - f'(a) \right) \right\|_{\infty,[a,b]} \le \left\| D_{A_0+}^q \left(f'(\cdot) - f'(A_0) \right) \right\|_{\infty,[A_0,+\infty)} < \infty, \tag{84}$$

$\forall\, a \ge A_0$,

along with

$$\left\| f \right\|_{\infty,[A_0,+\infty)} < \infty. \tag{85}$$

Therefore (by (83)) we have

$$\left| f'(a) \right| \le \frac{2 \left\| f \right\|_{\infty,[A_0,+\infty)}}{b-a} + \frac{(b-a)^q}{\Gamma(q+2)} \left\| D_{A_0+}^q \left(f'(\cdot) - f'(A_0) \right) \right\|_{\infty,[A_0,+\infty)}, \tag{86}$$

where the right-hand side of (86) depends only on $(b - a)$.

Calling $t = b - a > 0$; we can write

$$\left\| f' \right\|_{\infty,[A_0,+\infty)} \le \frac{2 \left\| f \right\|_{\infty,[A_0,+\infty)}}{t} + \frac{t^q}{\Gamma(q+2)} \left\| D_{A_0+}^q \left(f'(\cdot) - f'(A_0) \right) \right\|_{\infty,[A_0,+\infty)}, \tag{87}$$

$\forall\, t > 0$.

Set

$$\mu := 2 \left\| f \right\|_{\infty,[A_0,+\infty)}$$

$$and \tag{88}$$

$$\theta := \frac{\left\| D_{A_0+}^q \left(f'(\cdot) - f'(A_0) \right) \right\|_{\infty,[A_0,+\infty)}}{\Gamma(q+2)},$$

both are greater than zero.

We consider the function

$$y(t) = \frac{\mu}{t} + \theta t^q, \ 0 < q < 1, t > 0. \tag{89}$$

As in [7, pp. 81–82], y has a global minimum at

$$t_0 = \left(\frac{\mu}{q\theta}\right)^{\frac{1}{q+1}}, \tag{90}$$

which is

$$y(t_0) = \left(\theta\mu^q\right)^{\frac{1}{q+1}} (q+1) q^{-\frac{q}{q+1}}. \tag{91}$$

Consequently it is

$$y(t_0) = \left[\frac{\left\|D_{A_0+}^q \left(f'(\cdot) - f'(A_0)\right)\right\|_{\infty,[A_0,+\infty)}}{\Gamma(q+2)} \left(2\|f\|_{\infty,[A_0,+\infty)}\right)^q\right]^{\frac{1}{q+1}} (q+1) q^{-\frac{q}{q+1}}. \tag{92}$$

We have proved that

$$\|f'\|_{\infty,[A_0,+\infty)} \leq (q+1) \left(\frac{2}{q}\right)^{\frac{q}{q+1}} (\Gamma(q+2))^{-\frac{1}{q+1}}. \tag{93}$$

$$\left(\|f\|_{\infty,[A_0,+\infty)}\right)^{\frac{q}{q+1}} \left(\left\|D_{A_0+}^q \left(f'(\cdot) - f'(A_0)\right)\right\|_{\infty,[A_0,+\infty)}\right)^{\frac{1}{q+1}}.$$

We have established a local left fractional Riemann–Liouville ∞-Landau inequality:

Theorem 10 *Assume here that $f \in C^1([A_0, +\infty))$, where $A_0 \in \mathbb{R}$ is fixed, $0 < q < 1$, and for every $[a, b] \subset [A_0, +\infty)$ we have that $D_{a+}^q \left(f'(\cdot) - f'(a)\right) \in L_\infty([a, b])$, and that*

$$\left\|D_{a+}^q \left(f'(\cdot) - f'(a)\right)\right\|_{\infty,[a,b]} \leq \left\|D_{A_0+}^q \left(f'(\cdot) - f'(A_0)\right)\right\|_{\infty,[A_0,+\infty)} < \infty, \tag{94}$$

$\forall\, a \geq A_0$,
along with

$$\|f\|_{\infty,[A_0,+\infty)} < \infty. \tag{95}$$

Then

$$\|f'\|_{\infty,[A_0,+\infty)} \leq (q+1) \left(\frac{2}{q}\right)^{\frac{q}{q+1}} (\Gamma(q+2))^{-\frac{1}{q+1}}.$$

$$\|f\|_{\infty,[A_0,+\infty)}^{\frac{q}{q+1}} \left\|D_{A_0+}^q \left(f'(\cdot) - f'(A_0)\right)\right\|_{\infty,[A_0,+\infty)}^{\frac{1}{q+1}}. \tag{96}$$

If $A_0 = 0$, we have

Corollary 4 *Assume here that $f \in C^1(\mathbb{R}_+)$, $0 < q < 1$, and for any $[a,b] \subset \mathbb{R}_+$ we have that $D_{a+}^q \left(f'(\cdot) - f'(a)\right) \in L_\infty([a,b])$, and that*

$$\left\|D_{a+}^q \left(f'(\cdot) - f'(a)\right)\right\|_{\infty,[a,b]} \leq \left\|D_{0+}^q \left(f'(\cdot) - f'(0)\right)\right\|_{\infty,\mathbb{R}_+} < \infty, \tag{97}$$

$\forall \, a \geq 0$,
  along with

$$\|f\|_{\infty,\mathbb{R}_+} < \infty. \tag{98}$$

Then

$$\left\|f'\right\|_{\infty,\mathbb{R}_+} \leq (q+1) \left(\frac{2}{q}\right)^{\frac{q}{q+1}} (\Gamma(q+2))^{-\frac{1}{q+1}}.$$

$$\|f\|_{\infty,\mathbb{R}_+}^{\frac{q}{q+1}} \left\|D_{0+}^q \left(f'(\cdot) - f'(0)\right)\right\|_{\infty,\mathbb{R}_+}^{\frac{1}{q+1}}. \tag{99}$$

That is,

$$\left\|f'\right\|_{\infty,\mathbb{R}_+} < \infty.$$

It follows a local left fractional Riemann–Liouville L_p-Landau inequality:

Theorem 11 *Let $p_1, q_1 > 1 : \frac{1}{p_1} + \frac{1}{q_1} = 1$, $\frac{1}{q_1} < q < 1$; $f \in C^1([A_0, +\infty))$, $A_0 \in \mathbb{R}$ is fixed, $\|f\|_{\infty,[A_0,+\infty)} < \infty$; $D_{a+}^q \left(f'(\cdot) - f'(a)\right) \in L_{q_1}([a,b])$, $\forall$ $[a,b] \subseteq [A_0, +\infty)$; and $\forall \, a \geq A_0$ it holds*

$$\left\|D_{a+}^q \left(f'(\cdot) - f'(a)\right)\right\|_{q_1,[a,b]} \leq \left\|D_{A_0+}^q \left(f'(\cdot) - f'(A_0)\right)\right\|_{q_1,[A_0,+\infty)} < +\infty. \tag{100}$$

Then

$$\left\|f'\right\|_{\infty,[A_0,+\infty)} \leq \left(\frac{2\left(q + \frac{1}{p_1}\right)}{q - \frac{1}{q_1}}\right)^{\frac{q - \frac{1}{q_1}}{q + \frac{1}{p_1}}} \frac{1}{(\Gamma(q))^{\frac{1}{q+\frac{1}{p_1}}} ((q-1)p_1 + 1)^{\frac{1}{qp_1+1}}} \tag{101}$$

$$\left\|D_{A_0+}^q \left(f'(\cdot) - f'(A_0)\right)\right\|_{q_1,[A_0,+\infty)}^{\frac{1}{q+\frac{1}{p_1}}} \left(\|f\|_{\infty,[A_0,+\infty)}\right)^{\frac{q - \frac{1}{q_1}}{q + \frac{1}{p_1}}}.$$

That is,

$$\|f'\|_{\infty,[A_0,+\infty)} < +\infty.$$

The case $A_0 = 0$ follows:

Corollary 5 *Let* $p_1, q_1 > 1 : \frac{1}{p_1} + \frac{1}{q_1} = 1, \frac{1}{q_1} < q < 1; f \in C^1(\mathbb{R}_+), \|f\|_{\infty,\mathbb{R}_+} <$
$+\infty; D_{a+}^q(f'(\cdot) - f'(a)) \in L_{q_1}([a,b]), \forall [a,b] \subseteq \mathbb{R}_+;$ *and* $\forall a \geq 0$ *it holds*

$$\left\|D_{a+}^q(f'(\cdot) - f'(a))\right\|_{q_1,[a,b]} \leq \left\|D_{0+}^q(f'(\cdot) - f'(0))\right\|_{q_1,\mathbb{R}_+} < +\infty. \quad (102)$$

Then

$$\|f'\|_{\infty,\mathbb{R}_+} \leq \left(\frac{2\left(q + \frac{1}{p_1}\right)}{q - \frac{1}{q_1}}\right)^{\frac{q - \frac{1}{q_1}}{q + \frac{1}{p_1}}} \frac{1}{(\Gamma(q))^{\frac{1}{q + \frac{1}{p_1}}} ((q-1)p_1 + 1)^{\frac{1}{qp_1+1}}} \cdot \quad (103)$$

$$\left\|D_{0+}^q(f'(\cdot) - f'(0))\right\|_{q_1,\mathbb{R}_+}^{\frac{1}{q + \frac{1}{p_1}}} \left(\|f\|_{\infty,\mathbb{R}_+}\right)^{\frac{q - \frac{1}{q_1}}{q + \frac{1}{p_1}}}.$$

That is,

$$\|f'\|_{\infty,\mathbb{R}_+} < +\infty.$$

Proof of Theorem 11 By Theorem 6 we get:

$$\left|\frac{1}{b-a}\int_a^b f'(x')\,dx' - f'(a)\right| \leq \quad (104)$$

$$\frac{(b-a)^{q - \frac{1}{q_1}}}{\Gamma(q)((q-1)p_1 + 1)^{\frac{1}{p_1}}\left(q + \frac{1}{p_1}\right)}\left\|D_{a+}^q(f'(\cdot) - f'(a))\right\|_{q_1,[a,b]} =: \rho.$$

That is,

$$\left|\frac{f(b) - f(a)}{b - a} - f'(a)\right| \leq \rho. \quad (105)$$

Hence it holds

$$\left|f'(a)\right| - \frac{1}{b-a}|f(b) - f(a)| \leq \rho, \quad (106)$$

equivalently,

$$\left| f'(a) \right| \leq \frac{\left| f(b) - f(a) \right|}{b - a} + \rho \leq \frac{2 \, \| f \|_{\infty, [A_0, +\infty)}}{b - a} + \rho. \tag{107}$$

So far we have derived

$$\left| f'(a) \right| \leq \frac{2 \, \| f \|_{\infty, [A_0, +\infty)}}{b - a} + \tag{108}$$

$$\frac{(b - a)^{q - \frac{1}{q_1}}}{\Gamma(q) \left((q - 1) p_1 + 1 \right)^{\frac{1}{p_1}} \left(q + \frac{1}{p_1} \right)} \left\| D^q_{A_0 +} \left(f'(\cdot) - f'(A_0) \right) \right\|_{q_1, [A_0, +\infty)},$$

and the right-hand side of (108) depends only on $b - a =: t > 0$.

Therefore it holds

$$\left\| f' \right\|_{\infty, [A_0, +\infty)} \leq \frac{2 \, \| f \|_{\infty, [A_0, +\infty)}}{t} + \tag{109}$$

$$\frac{t^{q - \frac{1}{q_1}}}{\Gamma(q) \left((q - 1) p_1 + 1 \right)^{\frac{1}{p_1}} \left(q + \frac{1}{p_1} \right)} \left\| D^q_{A_0 +} \left(f'(\cdot) - f'(A_0) \right) \right\|_{q_1, [A_0, +\infty)},$$

$\forall \, t \in (0, \infty)$.

Notice that $0 < q - \frac{1}{q_1} < 1$. Call

$$\tilde{\mu} := 2 \, \| f \|_{\infty, [A_0, +\infty)},$$

$$\text{and} \tag{110}$$

$$\tilde{\theta} := \frac{\left\| D^q_{A_0 +} \left(f'(\cdot) - f'(A_0) \right) \right\|_{q_1, [A_0, +\infty)}}{\Gamma(q) \left((q - 1) p_1 + 1 \right)^{\frac{1}{p_1}} \left(q + \frac{1}{p_1} \right)},$$

both are positive,

and

$$\tilde{\nu} := q - \frac{1}{q_1} \in (0, 1). \tag{111}$$

We consider the function

$$\tilde{y}(t) = \frac{\tilde{\mu}}{t} + \tilde{\theta} t^{\tilde{\nu}}, \; t \in (0, \infty). \tag{112}$$

The only critical number is

$$\widetilde{t}_0 = \left(\frac{\widetilde{\mu}}{\widetilde{\nu}\theta}\right)^{\frac{1}{\widetilde{\nu}+1}}, \tag{113}$$

and $\widetilde{y}$ has a global minimum at $\widetilde{t}_0$, which is

$$\widetilde{y}\left(\widetilde{t}_0\right) = \left(\theta\widetilde{\mu}^{\widetilde{\nu}}\right)^{\frac{1}{\widetilde{\nu}+1}} (\widetilde{\nu}+1)\,\widetilde{\nu}^{-\frac{\widetilde{\nu}}{\widetilde{\nu}+1}}. \tag{114}$$

Consequently, we derive

$$\widetilde{y}\left(\widetilde{t}_0\right) = \left(\frac{\left\|D_{A_0+}^q\left(f'\left(\cdot\right)-f'\left(A_0\right)\right)\right\|_{q_1,[A_0,+\infty)}}{\Gamma\left(q\right)\left(\left(q-1\right)p_1+1\right)^{\frac{1}{p_1}}\left(q+\frac{1}{p_1}\right)}\right)^{\frac{p_1}{qp_1+1}}.$$

$$2^{\frac{q-\frac{1}{q_1}}{q+\frac{1}{p_1}}}\left(\|f\|_{\infty,[A_0,+\infty)}\right)^{\frac{q-\frac{1}{q_1}}{q+\frac{1}{p_1}}}\left(q+\frac{1}{p_1}\right)\left(q-\frac{1}{q_1}\right)^{-\frac{q-\frac{1}{q_1}}{q+\frac{1}{p_1}}}. \tag{115}$$

We have proved that

$$\|f'\|_{\infty,[A_0,+\infty)} \le \left(\frac{2\left(q+\frac{1}{p_1}\right)}{q-\frac{1}{q_1}}\right)^{\frac{q-\frac{1}{q_1}}{q+\frac{1}{p_1}}} \frac{1}{\left(\Gamma\left(q\right)\right)^{\frac{1}{q+\frac{1}{p_1}}}\left(\left(q-1\right)p_1+1\right)^{\frac{1}{qp_1+1}}}.$$

$$\left\|D_{A_0+}^q\left(f'\left(\cdot\right)-f'\left(A_0\right)\right)\right\|_{q_1,[A_0,+\infty)}^{\frac{1}{q+\frac{1}{p_1}}}\left(\|f\|_{\infty,[A_0,+\infty)}\right)^{\frac{q-\frac{1}{q_1}}{q+\frac{1}{p_1}}}. \tag{116}$$

Next comes a local fractional Polya–Ostrowski type inequality:

Theorem 12 *Let* $f \in C\left([a,b]\right)$, $0 < q < 1$, *and assume that*
$D^q_{\left(\frac{a+b}{2}\right)+}\left(f\left(\cdot\right)-f\left(\frac{a+b}{2}\right)\right) \in L_\infty\left(\left[\frac{a+b}{2},b\right]\right)$, $D^q_{\left(\frac{a+b}{2}\right)-}\left(f\left(\cdot\right)-f\left(\frac{a+b}{2}\right)\right) \in$
$L_\infty\left(\left[a,\frac{a+b}{2}\right]\right)$.

Set $M\left(f\right) := \max\left\{\left\|D^q_{\left(\frac{a+b}{2}\right)-}\left(f\left(\cdot\right)-f\left(\frac{a+b}{2}\right)\right)\right\|_{\infty,\left[a,\frac{a+b}{2}\right]},\right.$

$$\left.\left\|D^q_{\left(\frac{a+b}{2}\right)+}\left(f\left(\cdot\right)-f\left(\frac{a+b}{2}\right)\right)\right\|_{\infty,\left[\frac{a+b}{2},b\right]}\right\}.$$

Then

$$\left| \int_a^b f(x')\, dx' - f\left(\frac{a+b}{2}\right)(b-a) \right| \le$$

$$\int_a^b \left| f(x') - f\left(\frac{a+b}{2}\right) \right| dx' \le M(f) \frac{(b-a)^{q+1}}{\Gamma(q+2)\, 2^q}, \tag{117}$$

$\Leftrightarrow$

$$\left| \frac{1}{b-a} \int_a^b f(x')\, dx' - f\left(\frac{a+b}{2}\right) \right| \le$$

$$\frac{1}{b-a} \int_a^b \left| f(x') - f\left(\frac{a+b}{2}\right) \right| dx' \le M(f) \frac{(b-a)^q}{\Gamma(q+2)\, 2^q}. \tag{118}$$

Corollary 6 *All as in Theorem 12. Additionally assume that* $f\left(\frac{a+b}{2}\right) = 0$. *Then*

$$\int_a^b |f(x')|\, dx' \le M^*(f) \frac{(b-a)^{q+1}}{\Gamma(q+2)\, 2^q}, \tag{119}$$

where $M^*(f) := \max \left\{ \left\| D^q_{\left(\frac{a+b}{2}\right)-} f \right\|_{\infty, \left[a, \frac{a+b}{2}\right]}, \left\| D^q_{\left(\frac{a+b}{2}\right)+} f \right\|_{\infty, \left[\frac{a+b}{2}, b\right]} \right\}.$

Proof of Theorem 12 We have

$$f(x') - f\left(\frac{a+b}{2}\right) =$$

$$\frac{1}{\Gamma(q)} \int_{\frac{a+b}{2}}^{x'} (x'-z)^{q-1} D^q_{\left(\frac{a+b}{2}\right)+} \left(f(z) - f\left(\frac{a+b}{2}\right) \right) dz, \tag{120}$$

all $\frac{a+b}{2} \le x' \le b$,
and

$$f(x') - f\left(\frac{a+b}{2}\right) =$$

$$\frac{1}{\Gamma(q)} \int_{x'}^{\frac{a+b}{2}} (z-x')^{q-1} D^q_{\left(\frac{a+b}{2}\right)-} \left(f(z) - f\left(\frac{a+b}{2}\right) \right) dz, \tag{121}$$

all $a \le x' \le \frac{a+b}{2}$.

Then it holds

$$\left| \int_a^b \left(f(x') - f\left(\frac{a+b}{2}\right) \right) dx' \right| \leq \int_a^b \left| f(x') - f\left(\frac{a+b}{2}\right) \right| dx' = \qquad (122)$$

$$\int_a^{\frac{a+b}{2}} \left| f(x') - f\left(\frac{a+b}{2}\right) \right| dx' + \int_{\frac{a+b}{2}}^b \left| f(x') - f\left(\frac{a+b}{2}\right) \right| dx' =: (*).$$

Notice that

$$\left| f(x') - f\left(\frac{a+b}{2}\right) \right| \leq$$

$$\frac{1}{\Gamma(q+1)} \left\| D^q_{\left(\frac{a+b}{2}\right)-} \left(f(\cdot) - f\left(\frac{a+b}{2}\right) \right) \right\|_{\infty, \left[a, \frac{a+b}{2} \right]} \left(\frac{a+b}{2} - x' \right)^q,$$

$$\qquad (123)$$

all $a \leq x' \leq \frac{a+b}{2}$,

and

$$\left| f(x') - f\left(\frac{a+b}{2}\right) \right| \leq$$

$$\frac{1}{\Gamma(q+1)} \left\| D^q_{\left(\frac{a+b}{2}\right)+} \left(f(\cdot) - f\left(\frac{a+b}{2}\right) \right) \right\|_{\infty, \left[\frac{a+b}{2}, b \right]} \left(x' - \frac{a+b}{2} \right)^q,$$

$$\qquad (124)$$

all $\frac{a+b}{2} \leq x' \leq b$.

Therefore we have

$$(*) \leq \frac{1}{\Gamma(q+1)} \left[\left(\int_a^{\frac{a+b}{2}} \left(\frac{a+b}{2} - x' \right)^q dx' \right) \left\| D^q_{\left(\frac{a+b}{2}\right)-} \left(f(\cdot) - f\left(\frac{a+b}{2}\right) \right) \right\|_{\infty, \left[a, \frac{a+b}{2} \right]} \right.$$

$$\qquad (125)$$

$$\left. + \left(\int_{\frac{a+b}{2}}^b \left(x' - \frac{a+b}{2} \right)^q dx' \right) \left\| D^q_{\left(\frac{a+b}{2}\right)+} \left(f(\cdot) - f\left(\frac{a+b}{2}\right) \right) \right\|_{\infty, \left[\frac{a+b}{2}, b \right]} \right] \leq$$

$$\frac{M(f)}{\Gamma(q+1)} \left[\frac{\left(\frac{a+b}{2} - a \right)^{q+1}}{q+1} + \frac{\left(b - \frac{a+b}{2} \right)^{q+1}}{q+1} \right] =$$

$$\frac{M(f)}{\Gamma(q+2)}\left[\left(\frac{b-a}{2}\right)^{q+1}+\left(\frac{b-a}{2}\right)^{q+1}\right]=$$

$$\frac{M(f)}{2^q\,\Gamma(q+2)}(b-a)^{q+1}=M(f)\frac{(b-a)^{q+1}}{\Gamma(q+2)\,2^q}.$$

We make

Remark 10 Let $f\in C^N([a,b])$, $N\in\mathbb{Z}_+$. Here $x,x'\in[a,b]:x'<x$, and $F(x,\cdot;q,N)\in C^1([a-x,b-x])$, $0<q<1$. By Theorem 1 we get

$$f(x')=\sum_{n=0}^{N}\frac{f^{(n)}(x)}{n!}(x'-x)^n+\frac{D^{N+q}f(x)}{\Gamma(q+1)}(x-x')^q-$$

$$\frac{1}{\Gamma(q+1)}\int_{x'-x}^{0}\frac{dF(x,t;q,N)}{dt}(t-x'+x)^q\,dt. \tag{127}$$

Clearly then we get:

Let $f\in C^N([a,0])$, $a<0$, $N\in\mathbb{Z}_+$, $F(0,\cdot;q,N)\in C^1([a,0])$, $0<q<1$. Then, for any $x\in[a,0]$, we derive

$$f(x)=\sum_{n=0}^{N}\frac{f^{(n)}(0)}{n!}x^n+\frac{D^{N+q}f(0)}{\Gamma(q+1)}(-x)^q-$$

$$\frac{1}{\Gamma(q+1)}\int_{x}^{0}\frac{dF(0,t;q,N)}{dt}(t-x)^q\,dt. \tag{128}$$

One can go and act right fractionally, by using (128), and establish all corresponding results of this work to this right fractional direction. We omit this task.

We make

Remark 11 Let $f\in C^1([a,b])$, $0<q<1$, $D_{b-}^q\left(f'(\cdot)-f'(b)\right)\in L_\infty([a,b])$. Then, by Theorem 5, we get

$$\left|\frac{1}{b-a}\int_{a}^{b}f'(x')\,dx'-f'(b)\right|\le\frac{(b-a)^q}{\Gamma(q+2)}\left\|D_{b-}^q\left(f'(\cdot)-f'(b)\right)\right\|_{\infty,[a,b]}, \tag{129}$$

$\Leftrightarrow$

$$\left|\frac{1}{b-a}\left(f(b)-f(a)\right)-f'(b)\right|\le\frac{(b-a)^q}{\Gamma(q+2)}\left\|D_{b-}^q\left(f'(\cdot)-f'(b)\right)\right\|_{\infty,[a,b]}. \tag{130}$$

Hence it holds

$$\left| f'(b) \right| - \frac{1}{b-a} \left| f(b) - f(a) \right| \leq \frac{(b-a)^q}{\Gamma(q+2)} \left\| D_{b-}^q \left(f'(\cdot) - f'(b) \right) \right\|_{\infty,[a,b]},$$

(131)

$\Leftrightarrow$

$$\left| f'(b) \right| \leq \frac{\left| f(b) - f(a) \right|}{b-a} + \frac{(b-a)^q}{\Gamma(q+2)} \left\| D_{b-}^q \left(f'(\cdot) - f'(b) \right) \right\|_{\infty,[a,b]}.$$

(132)

Therefore

$$\left| f'(b) \right| \leq \frac{2 \left\| f \right\|_{\infty,[a,b]}}{b-a} + \frac{(b-a)^q}{\Gamma(q+2)} \left\| D_{b-}^q \left(f'(\cdot) - f'(b) \right) \right\|_{\infty,[a,b]}.$$

(133)

Complete assumptions follow:

Assume here that $f \in C^1 \left((-\infty, B_0] \right)$, where $B_0 \in \mathbb{R}$ is fixed, $0 < q < 1$, and for every $[a, b] \subset (-\infty, B_0]$ we have that $D_{b-}^q \left(f'(\cdot) - f'(b) \right) \in L_\infty \left([a, b] \right)$, and that

$$\left\| D_{b-}^q \left(f'(\cdot) - f'(b) \right) \right\|_{\infty,[a,b]} \leq \left\| D_{B_0-}^q \left(f'(\cdot) - f'(B_0) \right) \right\|_{\infty,(-\infty,B_0]} < \infty,$$

(134)

$\forall \, b \leq B_0$,

along with

$$\left\| f \right\|_{\infty,(-\infty,B_0]} < \infty.$$

(135)

Therefore (by (133)) we have

$$\left| f'(b) \right| \leq \frac{2 \left\| f \right\|_{\infty,(-\infty,B_0]}}{b-a} + \frac{(b-a)^q}{\Gamma(q+2)} \left\| D_{B_0-}^q \left(f'(\cdot) - f'(B_0) \right) \right\|_{\infty,(-\infty,B_0]},$$

(136)

where the right-hand side of (136) depends only on $(b - a)$.

Calling $t = b - a > 0$; we can write

$$\left\| f' \right\|_{\infty,(-\infty,B_0]} \leq \frac{2 \left\| f \right\|_{\infty,(-\infty,B_0]}}{t} + \frac{t^q}{\Gamma(q+2)} \left\| D_{B_0-}^q f'(\cdot) - f'(B_0) \right\|_{\infty,(-\infty,B_0]},$$

(137)

$\forall \, t > 0$.

Set

$$\mu := 2 \left\| f \right\|_{\infty,(-\infty,B_0]}$$

$$and$$

(138)

$$\theta := \frac{\left\| D_{B_0-}^q \left(f'(\cdot) - f'(B_0) \right) \right\|_{\infty,(-\infty,B_0]}}{\Gamma(q+2)},$$

both are greater than zero.

We consider the function

$$y(t) = \frac{\mu}{t} + \theta t^q, \ 0 < q < 1, t > 0. \tag{139}$$

As in [7, pp. 81–82], y has a global minimum at

$$t_0 = \left(\frac{\mu}{q\theta} \right)^{\frac{1}{q+1}}, \tag{140}$$

which is

$$y(t_0) = \left(\theta \mu^q \right)^{\frac{1}{q+1}} (q+1) q^{-\frac{q}{q+1}}. \tag{141}$$

Consequently it is

$$y(t_0) = \left[\frac{\left\| D_{B_0-}^q \left(f'(\cdot) - f'(B_0) \right) \right\|_{\infty,(-\infty,B_0]}}{\Gamma(q+2)} \left(2 \| f \|_{\infty,(-\infty,B_0]} \right)^q \right]^{\frac{1}{q+1}} (q+1) q^{-\frac{q}{q+1}}. \tag{142}$$

We have proved that

$$\| f' \|_{\infty,(-\infty,B_0]} \le (q+1) \left(\frac{2}{q} \right)^{\frac{q}{q+1}} \left(\Gamma(q+2) \right)^{-\frac{1}{q+1}}. \tag{143}$$

$$\left(\| f \|_{\infty,(-\infty,B_0]} \right)^{\frac{q}{q+1}} \left(\left\| D_{B_0-}^q \left(f'(\cdot) - f'(B_0) \right) \right\|_{\infty,(-\infty,B_0]} \right)^{\frac{1}{q+1}}.$$

We have established a local right fractional Riemann–Liouville ∞-Landau inequality:

Theorem 13 *Assume here that $f \in C^1((-\infty, B_0])$, where $B_0 \in \mathbb{R}$ is fixed, $0 < q < 1$, and for every $[a, b] \subset (-\infty, B_0]$ we have that $D_{b-}^q \left(f'(\cdot) - f'(b) \right) \in L_\infty([a, b])$, and that*

$$\left\| D_{b-}^q \left(f'(\cdot) - f'(b) \right) \right\|_{\infty,[a,b]} \le \left\| D_{B_0-}^q \left(f'(\cdot) - f'(B_0) \right) \right\|_{\infty,(-\infty,B_0]} < \infty, \tag{144}$$

$\forall \, b \le B_0,$

along with

$$\|f\|_{\infty,(-\infty,B_0]} < \infty. \tag{145}$$

Then

$$\|f'\|_{\infty,(-\infty,B_0]} \leq (q+1)\left(\frac{2}{q}\right)^{\frac{q}{q+1}} \left(\Gamma\left(q+2\right)\right)^{-\frac{1}{q+1}}.$$

$$\|f\|_{\infty,(-\infty,B_0]}^{\frac{q}{q+1}} \left\|D_{B_0-}^q\left(f'\left(\cdot\right) - f'\left(B_0\right)\right)\right\|_{\infty,(-\infty,B_0]}^{\frac{1}{q+1}}. \tag{146}$$

If $B_0 = 0$, we have

Corollary 7 *Assume here that* $f \in C^1(\mathbb{R}_-)$, $0 < q < 1$, *and for every* $[a, b] \subset \mathbb{R}_-$ *we have that* $D_{b-}^q\left(f'\left(\cdot\right) - f'\left(b\right)\right) \in L_\infty\left([a, b]\right)$, *and that*

$$\left\|D_{b-}^q\left(f'\left(\cdot\right) - f'\left(b\right)\right)\right\|_{\infty,[a,b]} \leq \left\|D_{0-}^q\left(f'\left(\cdot\right) - f'\left(0\right)\right)\right\|_{\infty,\mathbb{R}_-} < \infty, \tag{147}$$

$\forall\, b \leq 0$,
along with

$$\|f\|_{\infty,\mathbb{R}_-} < \infty. \tag{148}$$

Then

$$\|f'\|_{\infty,\mathbb{R}_-} \leq (q+1)\left(\frac{2}{q}\right)^{\frac{q}{q+1}} \left(\Gamma\left(q+2\right)\right)^{-\frac{1}{q+1}}.$$

$$\|f\|_{\infty,\mathbb{R}_-}^{\frac{q}{q+1}} \left\|D_{0-}^q\left(f'\left(\cdot\right) - f'\left(0\right)\right)\right\|_{\infty,\mathbb{R}_-}^{\frac{1}{q+1}}. \tag{149}$$

That is,

$$\|f'\|_{\infty,\mathbb{R}_-} < \infty.$$

It follows a local right fractional Riemann–Liouville L_p-Landau inequality:

Theorem 14 *Let* $p_1, q_1 > 1 : \frac{1}{p_1} + \frac{1}{q_1} = 1$, $\frac{1}{q_1} < q < 1$; $f \in C^1\left((-\infty, B_0]\right)$, $B_0 \in \mathbb{R}$ *is fixed,* $\|f\|_{\infty,(-\infty,B_0]} < \infty$; $D_{b-}^q\left(f'\left(\cdot\right) - f'\left(b\right)\right) \in L_{q_1}\left([a, b]\right)$, $\forall\, [a, b] \subseteq (-\infty, B_0]$; *and* $\forall\, b \leq B_0$ *it holds*

$$\left\|D_{b-}^q\left(f'\left(\cdot\right) - f'\left(b\right)\right)\right\|_{q_1,[a,b]} \leq \left\|D_{B_0-}^q\left(f'\left(\cdot\right) - f'\left(B_0\right)\right)\right\|_{q_1,(-\infty,B_0]} < +\infty. \tag{150}$$

Then

$$\left\| f' \right\|_{\infty,(-\infty,B_0]} \leq \left(\frac{2\left(q + \frac{1}{p_1}\right)}{q - \frac{1}{q_1}} \right)^{\frac{q - \frac{1}{q_1}}{q + \frac{1}{p_1}}} \frac{1}{\left(\Gamma(q)\right)^{\frac{1}{q + \frac{1}{p_1}}} \left((q-1)p_1 + 1\right)^{\frac{1}{qp_1 + 1}}} \cdot$$
(151)

$$\left\| D_{B_0-}^q \left(f'(\cdot) - f'(B_0)\right) \right\|_{q_1,(-\infty,B_0]}^{\frac{1}{q + \frac{1}{p_1}}} \left(\left\| f \right\|_{\infty,(-\infty,B_0]} \right)^{\frac{q - \frac{1}{q_1}}{q + \frac{1}{p_1}}}.$$

That is,

$$\left\| f' \right\|_{\infty,(-\infty,B_0]} < +\infty.$$

The case $B_0 = 0$ follows:

Corollary 8 *Let* $p_1, q_1 > 1 : \frac{1}{p_1} + \frac{1}{q_1} = 1, \frac{1}{q_1} < q < 1; f \in C^1(\mathbb{R}_-), \left\| f \right\|_{\infty,\mathbb{R}_-} < +\infty; D_{b-}^q \left(f'(\cdot) - f'(b)\right) \in L_{q_1}([a,b]), \forall [a,b] \subseteq \mathbb{R}_-; \text{ and } \forall b \leq 0 \text{ it holds}$

$$\left\| D_{b-}^q \left(f'(\cdot) - f'(b)\right) \right\|_{q_1,[a,b]} \leq \left\| D_{0-}^q \left(f'(\cdot) - f'(0)\right) \right\|_{q_1,\mathbb{R}_-} < +\infty. \quad (152)$$

Then

$$\left\| f' \right\|_{\infty,\mathbb{R}_-} \leq \left(\frac{2\left(q + \frac{1}{p_1}\right)}{q - \frac{1}{q_1}} \right)^{\frac{q - \frac{1}{q_1}}{q + \frac{1}{p_1}}} \frac{1}{\left(\Gamma(q)\right)^{\frac{1}{q + \frac{1}{p_1}}} \left((q-1)p_1 + 1\right)^{\frac{1}{qp_1 + 1}}} \cdot \quad (153)$$

$$\left\| D_{0-}^q \left(f'(\cdot) - f'(0)\right) \right\|_{q_1,\mathbb{R}_-}^{\frac{1}{q + \frac{1}{p_1}}} \left(\left\| f \right\|_{\infty,\mathbb{R}_-} \right)^{\frac{q - \frac{1}{q_1}}{q + \frac{1}{p_1}}}.$$

That is,

$$\left\| f' \right\|_{\infty,\mathbb{R}_-} < +\infty.$$

Proof of Theorem 14 By Theorem 6 we get:

$$\left| \frac{1}{b-a} \int_a^b f'(x') \, dx' - f'(b) \right| \leq \quad (154)$$

$$\frac{(b-a)^{q - \frac{1}{q_1}}}{\Gamma(q) \left((q-1)p_1 + 1\right)^{\frac{1}{p_1}} \left(q + \frac{1}{p_1}\right)} \left\| D_{b-}^q \left(f'(\cdot) - f'(b)\right) \right\|_{q_1,[a,b]} =: \overline{\rho}.$$

That is,

$$\left| \frac{f(b) - f(a)}{b - a} - f'(b) \right| \le \overline{\rho}. \tag{155}$$

Hence it holds

$$\left| f'(b) \right| - \frac{1}{b - a} |f(b) - f(a)| \le \overline{\rho}, \tag{156}$$

equivalently,

$$\left| f'(b) \right| \le \frac{|f(b) - f(a)|}{b - a} + \overline{\rho} \le \frac{2\|f\|_{\infty,(-\infty,B_0]}}{b - a} + \overline{\rho}. \tag{157}$$

So far we have derived

$$\left| f'(b) \right| \le \frac{2\|f\|_{\infty,(-\infty,B_0]}}{b - a} + \tag{158}$$

$$\frac{(b-a)^{q-\frac{1}{q_1}}}{\Gamma(q)\,((q-1)\,p_1 + 1)^{\frac{1}{p_1}}\left(q + \frac{1}{p_1}\right)} \left\| D^q_{B_0-}\left(f'(\cdot) - f'(B_0)\right) \right\|_{q_1,(-\infty,B_0]},$$

and the right-hand side of (158) depends only on $b - a =: t > 0$.
Therefore it holds

$$\left\| f' \right\|_{\infty,(-\infty,B_0]} \le \frac{2\|f\|_{\infty,(-\infty,B_0]}}{t} + \tag{159}$$

$$\frac{t^{q-\frac{1}{q_1}}}{\Gamma(q)\,((q-1)\,p_1 + 1)^{\frac{1}{p_1}}\left(q + \frac{1}{p_1}\right)} \left\| D^q_{B_0-} f'(\cdot) - f'(B_0) \right\|_{q_1,(-\infty,B_0]},$$

$\forall\, t \in (0, \infty)$.
Notice that $0 < q - \frac{1}{q_1} < 1$. Call

$$\widetilde{\mu} := 2\|f\|_{\infty,(-\infty,B_0]},$$

$$and \tag{160}$$

$$\widetilde{\theta} := \frac{\left\| D^q_{B_0-} f'(\cdot) - f'(B_0) \right\|_{q_1,(-\infty,B_0]}}{\Gamma(q)\,((q-1)\,p_1 + 1)^{\frac{1}{p_1}}\left(q + \frac{1}{p_1}\right)},$$

both are positive,

and

$$\widetilde{\nu} := q - \frac{1}{q_1} \in (0, 1).$$

(161)

We consider the function

$$\widetilde{y}(t) = \frac{\widetilde{\mu}}{t} + \widetilde{\theta} t^{\widetilde{\nu}}, \ t \in (0, \infty).$$

(162)

The only critical number is

$$\widetilde{t_0} = \left(\frac{\widetilde{\mu}}{\widetilde{\nu}\widetilde{\theta}}\right)^{\frac{1}{\widetilde{\nu}+1}},$$

(163)

and $\widetilde{y}$ has a global minimum at $\widetilde{t_0}$, which is

$$\widetilde{y}(\widetilde{t_0}) = \left(\widetilde{\theta}\widetilde{\mu}^{\widetilde{\nu}}\right)^{\frac{1}{\widetilde{\nu}+1}} (\widetilde{\nu} + 1) \, \widetilde{\nu}^{-\frac{\widetilde{\nu}}{\widetilde{\nu}+1}}.$$

(164)

Consequently, we derive

$$\widetilde{y}(\widetilde{t_0}) = \left(\frac{\left\|D_{B_0-}^q \left(f'(\cdot) - f'(B_0)\right)\right\|_{q_1,(-\infty,B_0]}}{\Gamma(q)\left((q-1)p_1+1\right)^{\frac{1}{p_1}}\left(q+\frac{1}{p_1}\right)}\right)^{\frac{p_1}{qp_1+1}}.$$

$$2^{\frac{q-\frac{1}{q_1}}{q+\frac{1}{p_1}}} \left(\|f\|_{\infty,(-\infty,B_0]}\right)^{\frac{q-\frac{1}{q_1}}{q+\frac{1}{p_1}}} \left(q+\frac{1}{p_1}\right) \left(q-\frac{1}{q_1}\right)^{-\frac{q-\frac{1}{q_1}}{q+\frac{1}{p_1}}}.$$

(165)

We have proved that

$$\|f'\|_{\infty,(-\infty,B_0]} \le \left(\frac{2\left(q+\frac{1}{p_1}\right)}{q-\frac{1}{q_1}}\right)^{\frac{q-\frac{1}{q_1}}{q+\frac{1}{p_1}}} \frac{1}{(\Gamma(q))^{\frac{1}{q+\frac{1}{p_1}}}\left((q-1)p_1+1\right)^{\frac{1}{qp_1+1}}}$$

$$\left\|D_{B_0-}^q \left(f'(\cdot) - f'(B_0)\right)\right\|_{q_1,(-\infty,B_0]}^{\frac{1}{q+\frac{1}{p_1}}} \left(\|f\|_{\infty,(-\infty,B_0]}\right)^{\frac{q-\frac{1}{q_1}}{q+\frac{1}{p_1}}}.$$

(166)

References

1. F.B. Adda, J. Cresson, Fractional differentiation equations and the Schrödinger equation. Appl. Math. Comput. **161**, 323–345 (2005)
2. G.A. Anastassiou, *Quantitative Approximations* (CRC Press, Boca Raton, 2001)
3. G.A. Anastassiou, *Fractional Differentiation Inequalities* (Springer, Heidelberg, 2009)
4. G.A. Anastassiou, *Probabilistic Inequalities* (World Scientific, Singapore, 2010)
5. G.A. Anastassiou, *Advanced Inequalities* (World Scientific, Singapore, 2010)
6. G.A. Anastassiou, *Intelligent Mathematics: Computational Analysis* (Springer, Heidelberg, 2011)
7. G.A. Anastassiou, *Advances on Fractional Inequalities* (Springer, Heidelberg, 2011)
8. G.A. Anastassiou, *Intelligent Comparisons: Analytic Inequalities* (Springer, Heidelberg, 2016)
9. G.A. Anastassiou, Local fractional Taylor formula. J. Comput. Anal. Appl. **28**, 709–713 (2020)
10. K. Diethelm, *The Analysis of Fractional Differential Equations* (Springer, Heidelberg, 2010)
11. K.M. Kolwankar, Local fractional calculus: a review. arXiv: 1307:0739v1 [nlin.CD] (2013)
12. K.M. Kolwankar, A.D. Gangal, *Local Fractional Calculus: A Calculus for Fractal Space-Time*. Fractals: Theory and Applications in Engineering (Springer, New York, 1999), pp. 171–181
13. Z. Opial, Sur une inégalité. Ann. Polon. Math. **8**, 29–32 (1960)
14. A. Ostrowski, Über die Absolutabweichung einer differentiebaren Funktion von ihrem Integralmittelwert. Comment. Math. Helv. **10**, 226–227 (1938)

Some New Hermite–Hadamard Type Integral Inequalities for Twice Differentiable Generalized $((h_1, h_2); (\eta_1, \eta_2))$-Convex Mappings and Their Applications

Artion Kashuri and Rozana Liko

Abstract In this article, we first introduced a new class of generalized $((h_1, h_2); (\eta_1, \eta_2))$-convex mappings and an interesting lemma regarding Hermite–Hadamard type integral inequalities. By using the notion of generalized $((h_1, h_2); (\eta_1, \eta_2))$-convexity and lemma as an auxiliary result, some new estimates difference between the left and middle part in Hermite–Hadamard type integral inequality associated with twice differentiable generalized $((h_1, h_2); (\eta_1, \eta_2))$-convex mappings are established. It is pointed out that some new special cases can be deduced from main results. At the end, some applications to special means for different positive real numbers are provided.

1 Introduction

The following notations are used throughout this paper. We use I to denote an interval on the real line $\mathbb{R} = (-\infty, +\infty)$. For any subset $K \subseteq \mathbb{R}^n$, where $\mathbb{R}^n$ is used to denote an n-dimensional vector space, K° is the interior of K. The set of integrable functions on the interval $[a, b]$ is denoted by $L[a, b]$.

The following inequality, named Hermite–Hadamard inequality, is one of the most famous inequalities in the literature for convex functions.

Theorem 1 *Let $f : I \subseteq \mathbb{R} \longrightarrow \mathbb{R}$ be a convex function on I and $a, b \in I$ with $a < b$. Then the following inequality holds:*

$$f\left(\frac{a+b}{2}\right) \leq \frac{1}{b-a} \int_a^b f(x)dx \leq \frac{f(a) + f(b)}{2}. \tag{1}$$

This inequality (1) is also known as trapezium inequality.

A. Kashuri (✉) · R. Liko
Department of Mathematics, Faculty of Technical Science, University Ismail Qemali of Vlora, Vlora, Albania

© Springer Nature Switzerland AG 2019
G. A. Anastassiou and J. M. Rassias (eds.), *Frontiers in Functional Equations and Analytic Inequalities*, https://doi.org/10.1007/978-3-030-28950-8_24

The trapezium type inequality has remained an area of great interest due to its wide applications in the field of mathematical analysis. For other recent results which generalize, improve, and extend the inequality (1) through various classes of convex functions interested readers are referred to [1–30, 32, 34, 35, 37–41, 45, 47, 48]. Let us recall some special functions and evoke some basic definitions as follows.

Definition 1 The incomplete beta function is defined for $a, b > 0$ as

$$\beta_x(a, b) = \int_0^x t^{a-1}(1 - t)^{b-1}dt, \quad 0 < x \leq 1. \tag{2}$$

Definition 2 ([46]) A set $S \subseteq \mathbb{R}^n$ is said to be invex set with respect to the mapping $\eta : S \times S \longrightarrow \mathbb{R}^n$, if $x + t\eta(y, x) \in S$ for every $x, y \in S$ and $t \in [0, 1]$.

The invex set S is also termed an η-connected set.

Definition 3 ([31]) Let $h : [0, 1] \longrightarrow \mathbb{R}$ be a non-negative function and $h \neq 0$. The function f on the invex set K is said to be h-preinvex with respect to η, if

$$f\big(x + t\eta(y, x)\big) \leq h(1 - t)f(x) + h(t)f(y) \tag{3}$$

for each $x, y \in K$ and $t \in [0, 1]$ where $f(\cdot) > 0$.

Clearly, when putting $h(t) = t$ in Definition 3, f becomes a preinvex function [36]. If the mapping $\eta(y, x) = y - x$ in Definition 3, then the non-negative function f reduces to h-convex mappings [43].

Definition 4 ([44]) Let $S \subseteq \mathbb{R}^n$ be an invex set with respect to $\eta : S \times S \longrightarrow \mathbb{R}^n$. A function $f : S \longrightarrow [0, +\infty)$ is said to be s-preinvex (or s-Breckner-preinvex) with respect to η and $s \in (0, 1]$, if for every $x, y \in S$ and $t \in [0, 1]$,

$$f\big(x + t\eta(y, x)\big) \leq (1 - t)^s f(x) + t^s f(y). \tag{4}$$

Definition 5 ([33]) A function $f : K \longrightarrow \mathbb{R}$ is said to be s-Godunova–Levin–Dragomir-preinvex of second kind, if

$$f\big(x + t\eta(y, x)\big) \leq (1 - t)^{-s} f(x) + t^{-s} f(y), \tag{5}$$

for each $x, y \in K, t \in (0, 1)$ and $s \in (0, 1]$.

Definition 6 ([42]) A non-negative function $f : K \subseteq \mathbb{R} \longrightarrow \mathbb{R}$ is said to be tgs-convex on K if the inequality

$$f\big((1 - t)x + ty\big) \leq t(1 - t)[f(x) + f(y)] \tag{6}$$

grips for all $x, y \in K$ and $t \in (0, 1)$.

Definition 7 ([28]) A function $f : I \subseteq \mathbb{R} \longrightarrow \mathbb{R}$ is said to be MT-convex functions, if it is non-negative and $\forall\, x, y \in I$ and $t \in (0, 1)$ satisfies the subsequent inequality

$$f(tx + (1 - t)y) \leq \frac{\sqrt{t}}{2\sqrt{1 - t}} f(x) + \frac{\sqrt{1 - t}}{2\sqrt{t}} f(y). \tag{7}$$

The concept of η-convex functions (at the beginning was named by φ-convex functions), considered in [13], has been introduced as the following.

Definition 8 Consider a convex set $I \subseteq \mathbb{R}$ and a bifunction $\eta : f(I) \times f(I) \longrightarrow \mathbb{R}$. A function $f : I \longrightarrow \mathbb{R}$ is called convex with respect to η (briefly η-convex), if

$$f\big(\lambda x + (1 - \lambda)y\big) \leq f(y) + \lambda \eta(f(x), f(y)), \tag{8}$$

is valid for all $x, y \in I$ and $\lambda \in [0, 1]$.

Geometrically it says that if a function is η-convex on I, then for any $x, y \in I$, its graph is on or under the path starting from $(y, f(y))$ and ending at $(x, f(y) + \eta(f(x), f(y)))$. If $f(x)$ should be the end point of the path for every $x, y \in I$, then we have $\eta(x, y) = x - y$ and the function reduces to a convex one. For more results about η-convex functions, see [7, 8, 12, 13].

Definition 9 ([1]) Let $I \subseteq \mathbb{R}$ be an invex set with respect to $\eta_1 : I \times I \longrightarrow \mathbb{R}$. Consider $f : I \longrightarrow \mathbb{R}$ and $\eta_2 : f(I) \times f(I) \longrightarrow \mathbb{R}$. The function f is said to be (η_1, η_2)-convex if

$$f\big(x + \lambda \eta_1(y, x)\big) \leq f(x) + \lambda \eta_2(f(y), f(x)), \tag{9}$$

is valid for all $x, y \in I$ and $\lambda \in [0, 1]$.

Motivated by the above works and references therein, the main objective of this article is to apply a new class of generalized $((h_1, h_2); (\eta_1, \eta_2))$-convex mappings and an interesting lemma to establish some new estimates difference between the left and middle part in Hermite–Hadamard type integral inequality associated with twice differentiable generalized $((h_1, h_2); (\eta_1, \eta_2))$-convex mappings. Also, some new special cases will be deduced. At the end, some applications to special means for different positive real numbers will be given as well.

2 Main Results

The following definitions will be used in this section.

Definition 10 ([10]) A set $K \subseteq \mathbb{R}^n$ is named as m-invex with respect to the mapping $\eta : K \times K \longrightarrow \mathbb{R}^n$ for some fixed $m \in (0, 1]$, if $mx + t\eta(y, mx) \in K$ grips for each $x, y \in K$ and any $t \in [0, 1]$.

Remark 1 In Definition 10, under certain conditions, the mapping $\eta(y, mx)$ could reduce to $\eta(y, x)$. When $m = 1$, we get Definition 2.

We next introduce the concept of generalized $((h_1, h_2); (\eta_1, \eta_2))$-convex mappings.

Definition 11 Let $K \subseteq \mathbb{R}$ be an open m-invex set with respect to the mapping $\eta_1 : K \times K \longrightarrow \mathbb{R}$. Suppose $h_1, h_2 : [0, 1] \longrightarrow [0, +\infty)$ and $\varphi : I \longrightarrow K$ are continuous. Consider $f : K \longrightarrow (0, +\infty)$ and $\eta_2 : f(K) \times f(K) \longrightarrow \mathbb{R}$. The mapping f is said to be generalized $((h_1, h_2); (\eta_1, \eta_2))$-convex if

$$
f\big(m\varphi(x) + t\eta_1(\varphi(y), m\varphi(x))\big) \leq \big[mh_1(t)f^r(x) + h_2(t)\eta_2(f^r(y), f^r(x))\big]^{\frac{1}{r}},
$$
(10)

holds for all $x, y \in I$, $r \neq 0$, $t \in [0, 1]$ and some fixed $m \in (0, 1]$.

Remark 2 In Definition 11, if we choose $m = r = 1$, $h_1(t) = 1$, $h_2(t) = t$, $\eta_1(\varphi(y), m\varphi(x)) = \varphi(y) - m\varphi(x)$, $\eta_2(f^r(y), f^r(x)) = \eta(f^r(y), f^r(x))$ and $\varphi(x) = x$, $\forall x \in I$, then we get Definition 8. Also, in Definition 11, if we choose $m = r = 1$, $h_1(t) = 1$, $h_2(t) = t$ and $\varphi(x) = x$, $\forall x \in I$, then we get Definition 9. Under some suitable choices as we have done above, we can also get Definitions 4 and 5.

Remark 3 For $r = 1$, let us discuss some special cases in Definition 11 as follows.

(I) If taking $h_1(t) = h(1 - t)$, $h_2(t) = h(t)$, then we get generalized $((m, h); (\eta_1, \eta_2))$-convex mappings.
(II) If taking $h_1(t) = (1 - t)^s$, $h_2(t) = t^s$ for $s \in (0, 1]$, then we get generalized $((m, s); (\eta_1, \eta_2))$-Breckner-convex mappings.
(III) If taking $h_1(t) = (1-t)^{-s}$, $h_2(t) = t^{-s}$ for $s \in (0, 1]$, then we get generalized $((m, s); (\eta_1, \eta_2))$-Godunova–Levin–Dragomir-convex mappings.
(IV) If taking $h_1(t) = h_2(t) = t(1 - t)$, then we get generalized $((m, tgs); (\eta_1, \eta_2))$-convex mappings.
(V) If taking $h_1(t) = \dfrac{\sqrt{1 - t}}{2\sqrt{t}}$, $h_2(t) = \dfrac{\sqrt{t}}{2\sqrt{1 - t}}$, then we get generalized $(m; (\eta_1, \eta_2))$-MT-convex mappings.

It is worth to mention here that to the best of our knowledge all the special cases discussed above are new in the literature.

Let us see the following example of a generalized $((h_1, h_2); (\eta_1, \eta_2))$-convex mapping which is not convex.

Example Let us take $m = r = 1$, $h_1(t) = 1$, $h_2(t) = t$ and φ an identity function. Consider the function $f : [0, +\infty) \longrightarrow [0, +\infty)$ by

$$f(x) = \begin{cases} x, & 0 \le x \le 2; \\ 2, & x > 2. \end{cases}$$

Define two bifunctions $\eta_1 : [0, +\infty) \times [0, +\infty) \longrightarrow \mathbb{R}$ and $\eta_2 : [0, +\infty) \times [0, +\infty) \longrightarrow [0, +\infty)$ by

$$\eta_1(x, y) = \begin{cases} -y, & 0 \le y \le 2; \\ x + y, & y > 2, \end{cases}$$

and

$$\eta_2(x, y) = \begin{cases} x + y, & x \le y; \\ 4(x + y), & x > y. \end{cases}$$

Then f is generalized $((1, t); (\eta_1, \eta_2))$-convex mapping. But f is not preinvex with respect to η_1 and also it is not convex (consider $x = 0$, $y = 3$ and $t \in (0, 1]$).

For establishing our main results regarding some new estimates difference between the left and middle part in Hermite–Hadamard type integral inequality associated with generalized $((h_1, h_2); (\eta_1, \eta_2))$-convexity, we need the following lemma.

Lemma 1 *Let $\varphi : I \longrightarrow K$ be a continuous function. Suppose $K \subseteq \mathbb{R}$ be an open m-invex subset with respect to $\eta : K \times K \longrightarrow \mathbb{R}$ for some fixed $m \in (0, 1]$, where $\eta(\varphi(x), m\varphi(y)) \ne 0$ and $\eta(\varphi(y), m\varphi(x)) \ne 0$. If $f : K \longrightarrow \mathbb{R}$ is a twice differentiable mapping on K° such that $f'' \in L(K)$, then the following identity holds:*

$$-\frac{2}{\eta^2(\varphi(y), m\varphi(x))} f\left(m\varphi(x) + \frac{\eta(\varphi(y), m\varphi(x))}{2}\right)$$

$$-\frac{2}{\eta^2(\varphi(x), m\varphi(y))} f\left(m\varphi(y) + \frac{\eta(\varphi(x), m\varphi(y))}{2}\right)$$

$$+\frac{2}{\eta^3(\varphi(y), m\varphi(x))} \int_{m\varphi(x)}^{m\varphi(x)+\eta(\varphi(y), m\varphi(x))} f(t)\,dt$$

$$+\frac{2}{\eta^3(\varphi(x), m\varphi(y))} \int_{m\varphi(y)}^{m\varphi(y)+\eta(\varphi(x), m\varphi(y))} f(t)\,dt$$

$$= \int_0^{\frac{1}{2}} t^2 \big[f'' \left(m\varphi(x) + t\eta(\varphi(y), m\varphi(x)) \right) + f'' \left(m\varphi(y) + t\eta(\varphi(x), m\varphi(y)) \right) \big] dt \tag{11}$$

$$+ \int_{\frac{1}{2}}^1 (1-t)^2 \big[f'' \left(m\varphi(x) + t\eta(\varphi(y), m\varphi(x)) \right) + f'' \left(m\varphi(y) + t\eta(\varphi(x), m\varphi(y)) \right) \big] dt.$$

We denote

$$T_f(\eta, \varphi; x, y, m)$$

$$:= \int_0^{\frac{1}{2}} t^2 \big[f'' \left(m\varphi(x) + t\eta(\varphi(y), m\varphi(x)) \right) + f'' \left(m\varphi(y) + t\eta(\varphi(x), m\varphi(y)) \right) \big] dt \tag{12}$$

$$+ \int_{\frac{1}{2}}^1 (1-t)^2 \big[f'' \left(m\varphi(x) + t\eta(\varphi(y), m\varphi(x)) \right) + f'' \left(m\varphi(y) + t\eta(\varphi(x), m\varphi(y)) \right) \big] dt.$$

Proof

$$T_f(\eta, \varphi; x, y, m) = T_{11} + T_{12} + T_{21} + T_{22},$$

where

$$T_{11} = \int_0^{\frac{1}{2}} t^2 f'' \left(m\varphi(y) + t\eta(\varphi(x), m\varphi(y)) \right) dt;$$

$$T_{12} = \int_0^{\frac{1}{2}} t^2 f'' \left(m\varphi(x) + t\eta(\varphi(y), m\varphi(x)) \right) dt;$$

$$T_{21} = \int_{\frac{1}{2}}^1 (1-t)^2 f'' \left(m\varphi(y) + t\eta(\varphi(x), m\varphi(y)) \right) dt;$$

$$T_{22} = \int_{\frac{1}{2}}^1 (1-t)^2 f'' \left(m\varphi(x) + t\eta(\varphi(y), m\varphi(x)) \right) dt.$$

Now, using twice integration by parts, we have

$$T_{11} = \frac{t^2 f' \left(m\varphi(y) + t\eta(\varphi(x), m\varphi(y)) \right)}{\eta(\varphi(x), m\varphi(y))} \Bigg|_0^{\frac{1}{2}}$$

$$- \frac{2}{\eta(\varphi(x), m\varphi(y))} \int_0^{\frac{1}{2}} t f' \left(m\varphi(y) + t\eta(\varphi(x), m\varphi(y)) \right) dt$$

$$
= \frac{f'\left(m\varphi(y) + \frac{\eta(\varphi(x),m\varphi(y))}{2}\right)}{4\eta(\varphi(x), m\varphi(y))} - \frac{2}{\eta(\varphi(x), m\varphi(y))}
$$

$$
\times \left\{ \left. \frac{tf\left(m\varphi(y) + t\eta(\varphi(x), m\varphi(y))\right)}{\eta(\varphi(x), m\varphi(y))} \right|_0^{\frac{1}{2}} \right.
$$

$$
\left. - \frac{1}{\eta(\varphi(x), m\varphi(y))} \int_0^{\frac{1}{2}} f\left(m\varphi(y) + t\eta(\varphi(x), m\varphi(y))\right) dt \right\}
$$

$$
= \frac{f'\left(m\varphi(y) + \frac{\eta(\varphi(x),m\varphi(y))}{2}\right)}{4\eta(\varphi(x), m\varphi(y))} - \frac{2}{\eta(\varphi(x), m\varphi(y))} \tag{13}
$$

$$
\times \left\{ \frac{f\left(m\varphi(y) + \frac{\eta(\varphi(x),m\varphi(y))}{2}\right)}{2\eta(\varphi(x), m\varphi(y))} - \frac{1}{\eta^2(\varphi(x), m\varphi(y))} \int_{m\varphi(y)}^{m\varphi(y)+\frac{\eta(\varphi(x),m\varphi(y))}{2}} f(t)dt \right\}.
$$

In a similar way, we find

$$
T_{12} = \frac{f'\left(m\varphi(x) + \frac{\eta(\varphi(y),m\varphi(x))}{2}\right)}{4\eta(\varphi(y), m\varphi(x))} - \frac{2}{\eta(\varphi(y), m\varphi(x))} \tag{14}
$$

$$
\times \left\{ \frac{f\left(m\varphi(x) + \frac{\eta(\varphi(y),m\varphi(x))}{2}\right)}{2\eta(\varphi(y), m\varphi(x))} - \frac{1}{\eta^2(\varphi(y), m\varphi(x))} \int_{m\varphi(x)}^{m\varphi(x)+\frac{\eta(\varphi(y),m\varphi(x))}{2}} f(t)dt \right\}.
$$

$$
T_{21} = -\frac{f'\left(m\varphi(y) + \frac{\eta(\varphi(x),m\varphi(y))}{2}\right)}{4\eta(\varphi(x), m\varphi(y))} + \frac{2}{\eta(\varphi(x), m\varphi(y))} \tag{15}
$$

$$
\times \left\{ -\frac{f\left(m\varphi(y) + \frac{\eta(\varphi(x),m\varphi(y))}{2}\right)}{2\eta(\varphi(x), m\varphi(y))} + \frac{1}{\eta^2(\varphi(x), m\varphi(y))} \int_{m\varphi(y)+\frac{\eta(\varphi(x),m\varphi(y))}{2}}^{m\varphi(y)+\eta(\varphi(x),m\varphi(y))} f(t)dt \right\}.
$$

$$
T_{22} = -\frac{f'\left(m\varphi(x) + \frac{\eta(\varphi(y),m\varphi(x))}{2}\right)}{4\eta(\varphi(y), m\varphi(x))} + \frac{2}{\eta(\varphi(y), m\varphi(x))} \tag{16}
$$

$$
\times \left\{ -\frac{f\left(m\varphi(x) + \frac{\eta(\varphi(y),m\varphi(x))}{2}\right)}{2\eta(\varphi(y), m\varphi(x))} + \frac{1}{\eta^2(\varphi(y), m\varphi(x))} \int_{m\varphi(x)+\frac{\eta(\varphi(y),m\varphi(x))}{2}}^{m\varphi(x)+\eta(\varphi(y),m\varphi(x))} f(t)dt \right\}.
$$

Adding Eqs. (13)–(16), we get our lemma. $\qquad\qquad\square$

Remark 4 In Lemma 1, if we take $m = 1$, $a < b$, $x = \mu a + (1 - \mu)b$, $y = \mu b + (1 - \mu)a$, where $\mu \in [0, 1]\setminus\{\frac{1}{2}\}$ and $\eta(\varphi(x), m\varphi(y)) = \varphi(x) - m\varphi(y)$, $\eta(\varphi(y), m\varphi(x)) = \varphi(y) - m\varphi(x)$, where $\varphi(x) = x$ for all $x \in I$, in identity (11), then it becomes identity of Lemma 2.1 in [37].

Theorem 2 *Let* $h_1, h_2 : [0, 1] \longrightarrow [0, +\infty)$ *and* $\varphi : I \longrightarrow K$ *are continuous functions. Suppose* $K \subseteq \mathbb{R}$ *be an open m-invex subset with respect to* $\eta_1 : K \times K \longrightarrow \mathbb{R}$ *for some fixed* $m \in (0, 1]$*, where* $\eta_1(\varphi(x), m\varphi(y)) \neq 0$ *and* $\eta_1(\varphi(y), m\varphi(x)) \neq 0$*. Assume that* $f : K \longrightarrow (0, +\infty)$ *is a twice differentiable mapping on* K° *such that* $f'' \in L(K)$ *and* $\eta_2 : f(K) \times f(K) \longrightarrow \mathbb{R}$*. If* f''^q *is generalized* $((h_1, h_2); (\eta_1, \eta_2))$*-convex mapping,* $0 < r \leq 1$*,* $q > 1$*,* $p^{-1} + q^{-1} = 1$*, then the following inequality holds:*

$$\left| T_f(\eta_1, \varphi; x, y, m) \right| \leq \left(\frac{1}{(2p+1)2^{2p+1}} \right)^{\frac{1}{p}} \tag{17}$$

$$\times \left\{ \left[m \left(f''(x) \right)^{rq} I^r(h_1(t); r) + \eta_2 \left(\left(f''(y) \right)^{rq}, \left(f''(x) \right)^{rq} \right) I^r(h_2(t); r) \right]^{\frac{1}{rq}} \right.$$

$$+ \left[m \left(f''(y) \right)^{rq} I^r(h_1(t); r) + \eta_2 \left(\left(f''(x) \right)^{rq}, \left(f''(y) \right)^{rq} \right) I^r(h_2(t); r) \right]^{\frac{1}{rq}}$$

$$+ \left[m \left(f''(x) \right)^{rq} J^r(h_1(t); r) + \eta_2 \left(\left(f''(y) \right)^{rq}, \left(f''(x) \right)^{rq} \right) J^r(h_2(t); r) \right]^{\frac{1}{rq}}$$

$$+ \left. \left[m \left(f''(y) \right)^{rq} J^r(h_1(t); r) + \eta_2 \left(\left(f''(x) \right)^{rq}, \left(f''(y) \right)^{rq} \right) J^r(h_2(t); r) \right]^{\frac{1}{rq}} \right\},$$

where

$$I(h_i(t); r) := \int_0^{\frac{1}{2}} h_i^{\frac{1}{r}}(t)dt, \quad J(h_i(t); r) := \int_{\frac{1}{2}}^1 h_i^{\frac{1}{r}}(t)dt, \quad \forall i = 1, 2.$$

Proof From Lemma 1, generalized $((h_1, h_2); (\eta_1, \eta_2))$-convexity of f''^q, Hölder inequality, Minkowski inequality, and properties of the modulus, we have

$$\left| T_f(\eta_1, \varphi; x, y, m) \right|$$

$$\leq \int_0^{\frac{1}{2}} t^2 \left[|f'' \left(m\varphi(x) + t\eta_1(\varphi(y), m\varphi(x)) \right)| + |f'' \left(m\varphi(y) + t\eta_1(\varphi(x), m\varphi(y)) \right)| \right]dt$$

$$+ \int_{\frac{1}{2}}^1 (1-t)^2 \left[|f'' \left(m\varphi(x) + t\eta_1(\varphi(y), m\varphi(x)) \right)| + |f'' \left(m\varphi(y) + t\eta_1(\varphi(x), m\varphi(y)) \right)| \right]dt$$

$$
\leq \left(\int_0^{\frac{1}{2}} t^{2p} dt \right)^{\frac{1}{p}} \times \Bigg\{ \left(\int_0^{\frac{1}{2}} \left(f''(m\varphi(x) + t\eta_1(\varphi(y), m\varphi(x))) \right)^q dt \right)^{\frac{1}{q}}
$$

$$
+ \left(\int_0^{\frac{1}{2}} \left(f''(m\varphi(y) + t\eta_1(\varphi(x), m\varphi(y))) \right)^q dt \right)^{\frac{1}{q}} \Bigg\}
$$

$$
+ \left(\int_{\frac{1}{2}}^{1} (1-t)^{2p} dt \right)^{\frac{1}{p}} \times \Bigg\{ \left(\int_{\frac{1}{2}}^{1} \left(f''(m\varphi(x) + t\eta_1(\varphi(y), m\varphi(x))) \right)^q dt \right)^{\frac{1}{q}}
$$

$$
+ \left(\int_{\frac{1}{2}}^{1} \left(f''(m\varphi(y) + t\eta_1(\varphi(x), m\varphi(y))) \right)^q dt \right)^{\frac{1}{q}} \Bigg\}
$$

$$
\leq \left(\frac{1}{(2p+1)2^{2p+1}} \right)^{\frac{1}{p}}
$$

$$
\times \Bigg\{ \left(\int_0^{\frac{1}{2}} \left[mh_1(t) \left(f''(x) \right)^{rq} + h_2(t)\eta_2 \left(\left(f''(y) \right)^{rq}, \left(f''(x) \right)^{rq} \right) \right]^{\frac{1}{r}} dt \right)^{\frac{1}{q}}
$$

$$
+ \left(\int_0^{\frac{1}{2}} \left[mh_1(t) \left(f''(y) \right)^{rq} + h_2(t)\eta_2 \left(\left(f''(x) \right)^{rq}, \left(f''(y) \right)^{rq} \right) \right]^{\frac{1}{r}} dt \right)^{\frac{1}{q}}
$$

$$
+ \left(\int_{\frac{1}{2}}^{1} \left[mh_1(t) \left(f''(x) \right)^{rq} + h_2(t)\eta_2 \left(\left(f''(y) \right)^{rq}, \left(f''(x) \right)^{rq} \right) \right]^{\frac{1}{r}} dt \right)^{\frac{1}{q}}
$$

$$
+ \left(\int_{\frac{1}{2}}^{1} \left[mh_1(t) \left(f''(y) \right)^{rq} + h_2(t)\eta_2 \left(\left(f''(x) \right)^{rq}, \left(f''(y) \right)^{rq} \right) \right]^{\frac{1}{r}} dt \right)^{\frac{1}{q}} \Bigg\}
$$

$$
\leq \left(\frac{1}{(2p+1)2^{2p+1}} \right)^{\frac{1}{p}}
$$

$$
\times \Bigg\{ \left[\left(\int_0^{\frac{1}{2}} m^{\frac{1}{r}} \left(f''(x) \right)^q h_1^{\frac{1}{r}}(t) dt \right)^r + \left(\int_0^{\frac{1}{2}} \eta_2^{\frac{1}{r}} \left(\left(f''(y) \right)^{rq}, \left(f''(x) \right)^{rq} \right) h_2^{\frac{1}{r}}(t) dt \right)^r \right]^{\frac{1}{rq}}
$$

$$
+ \left[\left(\int_0^{\frac{1}{2}} m^{\frac{1}{r}} \left(f''(y) \right)^q h_1^{\frac{1}{r}}(t) dt \right)^r + \left(\int_0^{\frac{1}{2}} \eta_2^{\frac{1}{r}} \left(\left(f''(x) \right)^{rq}, \left(f''(y) \right)^{rq} \right) h_2^{\frac{1}{r}}(t) dt \right)^r \right]^{\frac{1}{rq}}
$$

$$
+\left[\left(\int_{\frac{1}{2}}^{1} m^{\frac{1}{r}} \left(f''(x)\right)^q h_1^{\frac{1}{r}}(t)dt\right)^r + \left(\int_{\frac{1}{2}}^{1} \eta_2^{\frac{1}{r}} \left(\left(f''(y)\right)^{rq}, \left(f''(x)\right)^{rq}\right) h_2^{\frac{1}{r}}(t)dt\right)^r\right]^{\frac{1}{rq}}
$$

$$
+\left[\left(\int_{\frac{1}{2}}^{1} m^{\frac{1}{r}} \left(f''(y)\right)^q h_1^{\frac{1}{r}}(t)dt\right)^r + \left(\int_{\frac{1}{2}}^{1} \eta_2^{\frac{1}{r}} \left(\left(f''(x)\right)^{rq}, \left(f''(y)\right)^{rq}\right) h_2^{\frac{1}{r}}(t)dt\right)^r\right]^{\frac{1}{rq}}\right\}
$$

$$
= \left(\frac{1}{(2p+1)2^{2p+1}}\right)^{\frac{1}{p}}
$$

$$
\times\left\{\left[m \left(f''(x)\right)^{rq} I^r(h_1(t); r) + \eta_2 \left(\left(f''(y)\right)^{rq}, \left(f''(x)\right)^{rq}\right) I^r(h_2(t); r)\right]^{\frac{1}{rq}}\right.
$$

$$
+\left[m \left(f''(y)\right)^{rq} I^r(h_1(t); r) + \eta_2 \left(\left(f''(x)\right)^{rq}, \left(f''(y)\right)^{rq}\right) I^r(h_2(t); r)\right]^{\frac{1}{rq}}
$$

$$
+\left[m \left(f''(x)\right)^{rq} J^r(h_1(t); r) + \eta_2 \left(\left(f''(y)\right)^{rq}, \left(f''(x)\right)^{rq}\right) J^r(h_2(t); r)\right]^{\frac{1}{rq}}
$$

$$
+\left.\left[m \left(f''(y)\right)^{rq} J^r(h_1(t); r) + \eta_2 \left(\left(f''(x)\right)^{rq}, \left(f''(y)\right)^{rq}\right) J^r(h_2(t); r)\right]^{\frac{1}{rq}}\right\}.
$$

So, the proof of this theorem is completed. □

We point out some special cases of Theorem 2.

Corollary 1 *In Theorem 2 for $h_1(t) = h(1-t)$, $h_2(t) = h(t)$ and $f''(x) \leq L$, $\forall x \in I$, we get the following Hermite–Hadamard type inequality for generalized $((m, h); (\eta_1, \eta_2))$-convex mappings*

$$
\left|T_f(\eta_1, \varphi; x, y, m)\right| \leq 2 \left(\frac{1}{(2p+1)2^{2p+1}}\right)^{\frac{1}{p}} \tag{18}
$$

$$
\times\left\{\left[mL^{rq} I^r(h(t); r) + \eta_2 \left(L^{rq}, L^{rq}\right) I^r(h(1-t); r)\right]^{\frac{1}{rq}}\right.
$$

$$
+\left.\left[mL^{rq} I^r(h(1-t); r) + \eta_2 \left(L^{rq}, L^{rq}\right) I^r(h(t); r)\right]^{\frac{1}{rq}}\right\}.
$$

Corollary 2 *In Corollary 1 for $h_1(t) = (1-t)^s$ and $h_2(t) = t^s$, we get the following Hermite–Hadamard type inequality for generalized $((m, s); (\eta_1, \eta_2))$-Breckner-convex mappings*

$$|T_f(\eta_1, \varphi; x, y, m)| \le 2 \left(\frac{1}{(2p+1)2^{2p+1}} \right)^{\frac{1}{p}} \left(\frac{r}{(s+r)2^{\frac{s}{r}+1}} \right)^{\frac{1}{q}} \tag{19}$$

$$\times \left\{ \left[mL^{rq} + \eta_2 \left(L^{rq}, L^{rq} \right) \left(2^{\frac{s}{r}+1} - 1 \right)^r \right]^{\frac{1}{rq}} \right.$$

$$\left. + \left[mL^{rq} \left(2^{\frac{s}{r}+1} - 1 \right)^r + \eta_2 \left(L^{rq}, L^{rq} \right) \right]^{\frac{1}{rq}} \right\}.$$

Corollary 3 *In Corollary 1 for $h_1(t) = (1-t)^{-s}$ and $h_2(t) = t^{-s}$ and $0 < s < r$, we get the following Hermite–Hadamard type inequality for generalized $((m, s); (\eta_1, \eta_2))$-Godunova–Levin–Dragomir-convex mappings*

$$|T_f(\eta_1, \varphi; x, y, m)| \le 2 \left(\frac{1}{(2p+1)2^{2p+1}} \right)^{\frac{1}{p}} \left(\frac{r}{(r-s)2^{1-\frac{s}{r}}} \right)^{\frac{1}{q}} \tag{20}$$

$$\times \left\{ \left[mL^{rq} + \eta_2 \left(L^{rq}, L^{rq} \right) \left(2^{1-\frac{s}{r}} - 1 \right)^r \right]^{\frac{1}{rq}} \right.$$

$$\left. + \left[mL^{rq} \left(2^{1-\frac{s}{r}} - 1 \right)^r + \eta_2 \left(L^{rq}, L^{rq} \right) \right]^{\frac{1}{rq}} \right\}.$$

Corollary 4 *In Theorem 2 for $h_1(t) = h_2(t) = t(1-t)$ and $f''(x) \le L, \forall x \in I$, we get the following Hermite–Hadamard type inequality for generalized $((m, tgs); (\eta_1, \eta_2))$-convex mappings*

$$|T_f(\eta_1, \varphi; x, y, m)| \le 4 \left(\frac{1}{(2p+1)2^{2p+1}} \right)^{\frac{1}{p}} \beta_{1/2}^{\frac{1}{q}} \left(1 + \frac{1}{r}, 1 + \frac{1}{r} \right) \tag{21}$$

$$\times \left[mL^{rq} + \eta_2 \left(L^{rq}, L^{rq} \right) \right]^{\frac{1}{rq}}.$$

Corollary 5 *In Corollary 1 for $h_1(t) = \dfrac{\sqrt{1-t}}{2\sqrt{t}}$, $h_2(t) = \dfrac{\sqrt{t}}{2\sqrt{1-t}}$ and $r \in \left(\frac{1}{2}, 1 \right]$, we get the following Hermite–Hadamard type inequality for generalized $(m; (\eta_1, \eta_2))$-MT-convex mappings*

$$|T_f(\eta_1, \varphi; x, y, m)| \le 2 \left(\frac{1}{(2p+1)2^{2p+1}} \right)^{\frac{1}{p}} \left(\frac{1}{2} \right)^{\frac{1}{rq}} \tag{22}$$

$$\times \left\{ \left[mL^{rq}\beta^r_{1/2}\left(1 - \frac{1}{2r}, 1 + \frac{1}{2r}\right) + \eta_2\left(L^{rq}, L^{rq}\right)\beta^r_{1/2}\left(1 + \frac{1}{2r}, 1 - \frac{1}{2r}\right) \right]^{\frac{1}{rq}} \right.$$

$$\left. + \left[mL^{rq}\beta^r_{1/2}\left(1 + \frac{1}{2r}, 1 - \frac{1}{2r}\right) + \eta_2\left(L^{rq}, L^{rq}\right)\beta^r_{1/2}\left(1 - \frac{1}{2r}, 1 + \frac{1}{2r}\right) \right]^{\frac{1}{rq}} \right\}.$$

Theorem 3 *Let $h_1, h_2 : [0, 1] \longrightarrow [0, +\infty)$ and $\varphi : I \longrightarrow K$ are continuous functions. Suppose $K \subseteq \mathbb{R}$ be an open m-invex subset with respect to $\eta_1 : K \times K \longrightarrow \mathbb{R}$ for some fixed $m \in (0, 1]$, where $\eta_1(\varphi(x), m\varphi(y)) \neq 0$ and $\eta_1(\varphi(y), m\varphi(x)) \neq 0$. Assume that $f : K \longrightarrow (0, +\infty)$ is a twice differentiable mapping on K° such that $f'' \in L(K)$ and $\eta_2 : f(K) \times f(K) \longrightarrow \mathbb{R}$. If f''^q is generalized $((h_1, h_2); (\eta_1, \eta_2))$-convex mapping, $0 < r \leq 1$ and $q \geq 1$, then the following inequality holds:*

$$\left| T_f(\eta_1, \varphi; x, y, m) \right| \leq \left(\frac{1}{24} \right)^{1 - \frac{1}{q}} \tag{23}$$

$$\times \left\{ \left[m\left(f''(x)\right)^{rq} F^r(h_1(t); r) + \eta_2\left(\left(f''(y)\right)^{rq}, \left(f''(x)\right)^{rq}\right) F^r(h_2(t); r) \right]^{\frac{1}{rq}} \right.$$

$$+ \left[m\left(f''(y)\right)^{rq} F^r(h_1(t); r) + \eta_2\left(\left(f''(x)\right)^{rq}, \left(f''(y)\right)^{rq}\right) F^r(h_2(t); r) \right]^{\frac{1}{rq}}$$

$$+ \left[m\left(f''(x)\right)^{rq} G^r(h_1(t); r) + \eta_2\left(\left(f''(y)\right)^{rq}, \left(f''(x)\right)^{rq}\right) G^r(h_2(t); r) \right]^{\frac{1}{rq}}$$

$$\left. + \left[m\left(f''(y)\right)^{rq} G^r(h_1(t); r) + \eta_2\left(\left(f''(x)\right)^{rq}, \left(f''(y)\right)^{rq}\right) G^r(h_2(t); r) \right]^{\frac{1}{rq}} \right\},$$

where

$$F(h_i(t); r) := \int_0^{\frac{1}{2}} t^2 h_i^{\frac{1}{r}}(t)dt, \quad G(h_i(t); r) := \int_{\frac{1}{2}}^1 (1-t)^2 h_i^{\frac{1}{r}}(t)dt, \quad \forall i = 1, 2.$$

Proof From Lemma 1, generalized $((h_1, h_2); (\eta_1, \eta_2))$-convexity of f''^q, the well-known power mean inequality, Minkowski inequality, and properties of the modulus, we have

$$\left| T_f(\eta_1, \varphi; x, y, m) \right|$$

$$\leq \int_0^{\frac{1}{2}} t^2 \left[\left| f''\left(m\varphi(x) + t\eta_1(\varphi(y), m\varphi(x))\right) \right| + \left| f''\left(m\varphi(y) + t\eta_1(\varphi(x), m\varphi(y))\right) \right| \right] dt$$

$$+ \int_{\frac{1}{2}}^{1} (1-t)^2 \left[|f'' \left(m\varphi(x) + t\eta_1(\varphi(y), m\varphi(x)) \right) | + | f'' \left(m\varphi(y) + t\eta_1(\varphi(x), m\varphi(y)) \right) | \right] dt$$

$$\leq \left(\int_0^{\frac{1}{2}} t^2 dt \right)^{1-\frac{1}{q}} \times \left\{ \left(\int_0^{\frac{1}{2}} t^2 \left(f''(m\varphi(x) + t\eta_1(\varphi(y), m\varphi(x))) \right)^q dt \right)^{\frac{1}{q}} \right.$$

$$\left. + \left(\int_0^{\frac{1}{2}} t^2 \left(f''(m\varphi(y) + t\eta_1(\varphi(x), m\varphi(y))) \right)^q dt \right)^{\frac{1}{q}} \right\}$$

$$+ \left(\int_{\frac{1}{2}}^{1} (1-t)^2 dt \right)^{1-\frac{1}{q}} \times \left\{ \left(\int_{\frac{1}{2}}^{1} (1-t)^2 \left(f''(m\varphi(x) + t\eta_1(\varphi(y), m\varphi(x))) \right)^q dt \right)^{\frac{1}{q}} \right.$$

$$\left. + \left(\int_{\frac{1}{2}}^{1} (1-t)^2 \left(f''(m\varphi(y) + t\eta_1(\varphi(x), m\varphi(y))) \right)^q dt \right)^{\frac{1}{q}} \right\}$$

$$\leq \left(\frac{1}{24} \right)^{1-\frac{1}{q}}$$

$$\times \left\{ \left(\int_0^{\frac{1}{2}} t^2 \left[mh_1(t) \left(f''(x) \right)^{rq} + h_2(t)\eta_2 \left(\left(f''(y) \right)^{rq}, \left(f''(x) \right)^{rq} \right) \right]^{\frac{1}{r}} dt \right)^{\frac{1}{q}} \right.$$

$$+ \left(\int_0^{\frac{1}{2}} t^2 \left[mh_1(t) \left(f''(y) \right)^{rq} + h_2(t)\eta_2 \left(\left(f''(x) \right)^{rq}, \left(f''(y) \right)^{rq} \right) \right]^{\frac{1}{r}} dt \right)^{\frac{1}{q}}$$

$$+ \left(\int_{\frac{1}{2}}^{1} (1-t)^2 \left[mh_1(t) \left(f''(x) \right)^{rq} + h_2(t)\eta_2 \left(\left(f''(y) \right)^{rq}, \left(f''(x) \right)^{rq} \right) \right]^{\frac{1}{r}} dt \right)^{\frac{1}{q}}$$

$$\left. + \left(\int_{\frac{1}{2}}^{1} (1-t)^2 \left[mh_1(t) \left(f''(y) \right)^{rq} + h_2(t)\eta_2 \left(\left(f''(x) \right)^{rq}, \left(f''(y) \right)^{rq} \right) \right]^{\frac{1}{r}} dt \right)^{\frac{1}{q}} \right\}$$

$$\leq \left(\frac{1}{24} \right)^{1-\frac{1}{q}}$$

$$\times \left\{ \left[\left(\int_0^{\frac{1}{2}} m^{\frac{1}{r}} \left(f''(x) \right)^q t^2 h_1^{\frac{1}{r}}(t) dt \right)^r + \left(\int_0^{\frac{1}{2}} t^2 \eta_2^{\frac{1}{r}} \left(\left(f''(y) \right)^{rq}, \left(f''(x) \right)^{rq} \right) h_2^{\frac{1}{r}}(t) dt \right)^r \right]^{\frac{1}{rq}} \right.$$

$$+\left[\left(\int_0^{\frac{1}{2}} m^{\frac{1}{r}} \left(f''(y)\right)^q t^2 h_1^{\frac{1}{r}}(t)dt\right)^r + \left(\int_0^{\frac{1}{2}} t^2 \eta_2^{\frac{1}{r}} \left(\left(f''(x)\right)^{rq}, \left(f''(y)\right)^{rq}\right) h_2^{\frac{1}{r}}(t)dt\right)^r\right]^{\frac{1}{rq}}$$

$$+\left[\left(\int_{\frac{1}{2}}^1 m^{\frac{1}{r}} \left(f''(x)\right)^q (1-t)^2 h_1^{\frac{1}{r}}(t)dt\right)^r\right.$$

$$+\left.\left(\int_{\frac{1}{2}}^1 (1-t)^2 \eta_2^{\frac{1}{r}} \left(\left(f''(y)\right)^{rq}, \left(f''(x)\right)^{rq}\right) h_2^{\frac{1}{r}}(t)dt\right)^r\right]^{\frac{1}{rq}}$$

$$+\left[\left(\int_{\frac{1}{2}}^1 m^{\frac{1}{r}} \left(f''(y)\right)^q (1-t)^2 h_1^{\frac{1}{r}}(t)dt\right)^r\right.$$

$$+\left.\left(\int_{\frac{1}{2}}^1 (1-t)^2 \eta_2^{\frac{1}{r}} \left(\left(f''(x)\right)^{rq}, \left(f''(y)\right)^{rq}\right) h_2^{\frac{1}{r}}(t)dt\right)^r\right]^{\frac{1}{rq}}\right\}$$

$$= \left(\frac{1}{24}\right)^{1-\frac{1}{q}}$$

$$\times\left\{\left[m\left(f''(x)\right)^{rq} F^r(h_1(t); r) + \eta_2\left(\left(f''(y)\right)^{rq}, \left(f''(x)\right)^{rq}\right) F^r(h_2(t); r)\right]^{\frac{1}{rq}}\right.$$

$$+\left[m\left(f''(y)\right)^{rq} F^r(h_1(t); r) + \eta_2\left(\left(f''(x)\right)^{rq}, \left(f''(y)\right)^{rq}\right) F^r(h_2(t); r)\right]^{\frac{1}{rq}}$$

$$+\left[m\left(f''(x)\right)^{rq} G^r(h_1(t); r) + \eta_2\left(\left(f''(y)\right)^{rq}, \left(f''(x)\right)^{rq}\right) G^r(h_2(t); r)\right]^{\frac{1}{rq}}$$

$$+\left.\left[m\left(f''(y)\right)^{rq} G^r(h_1(t); r) + \eta_2\left(\left(f''(x)\right)^{rq}, \left(f''(y)\right)^{rq}\right) G^r(h_2(t); r)\right]^{\frac{1}{rq}}\right\}.$$

So, the proof of this theorem is completed. □

We point out some special cases of Theorem 3.

Corollary 6 *In Theorem 3 for* $h_1(t) = h(1-t)$, $h_2(t) = h(t)$ *and* $f''(x) \leq L$, $\forall x \in I$, *we get the following Hermite–Hadamard type inequality for generalized* $((m, h); (\eta_1, \eta_2))$-*convex mappings*

$$\left|T_f(\eta_1, \varphi; x, y, m)\right| \leq 2\left(\frac{1}{24}\right)^{1-\frac{1}{q}} \tag{24}$$

$$\times \left\{ \left[mL^{rq} F^r (h(t); r) + \eta_2 \left(L^{rq}, L^{rq} \right) F^r (h(1-t); r) \right]^{\frac{1}{rq}} \right.$$

$$\left. + \left[mL^{rq} F^r (h(1-t); r) + \eta_2 \left(L^{rq}, L^{rq} \right) F^r (h(t); r) \right]^{\frac{1}{rq}} \right\}.$$

Corollary 7 *In Corollary 6 for $h_1(t) = (1-t)^s$ and $h_2(t) = t^s$, we get the following Hermite–Hadamard type inequality for generalized $((m, s); (\eta_1, \eta_2))$-Breckner-convex mappings*

$$\left| T_f (\eta_1, \varphi; x, y, m) \right| \leq 2 \left(\frac{1}{24} \right)^{1-\frac{1}{q}} \tag{25}$$

$$\times \left\{ \left[mL^{rq} \left(\frac{r}{(s+3r)2^{\frac{s}{r}+3}} \right)^r + \eta_2 \left(L^{rq}, L^{rq} \right) \beta^r_{1/2} \left(3, 1 + \frac{s}{r} \right) \right]^{\frac{1}{rq}} \right.$$

$$\left. + \left[mL^{rq} \beta^r_{1/2} \left(3, 1 + \frac{s}{r} \right) + \eta_2 \left(L^{rq}, L^{rq} \right) \left(\frac{r}{(s+3r)2^{\frac{s}{r}+3}} \right)^r \right]^{\frac{1}{rq}} \right\}.$$

Corollary 8 *In Corollary 6 for $h_1(t) = (1-t)^{-s}$ and $h_2(t) = t^{-s}$ and $0 < s < r$, we get the following Hermite–Hadamard type inequality for generalized $((m, s); (\eta_1, \eta_2))$-Godunova–Levin–Dragomir-convex mappings*

$$\left| T_f (\eta_1, \varphi; x, y, m) \right| \leq 2 \left(\frac{1}{24} \right)^{1-\frac{1}{q}} \tag{26}$$

$$\times \left\{ \left[mL^{rq} \left(\frac{r}{(3r-s)2^{3-\frac{s}{r}}} \right)^r + \eta_2 \left(L^{rq}, L^{rq} \right) \beta^r_{1/2} \left(3, 1 - \frac{s}{r} \right) \right]^{\frac{1}{rq}} \right.$$

$$\left. + \left[mL^{rq} \beta^r_{1/2} \left(3, 1 - \frac{s}{r} \right) + \eta_2 \left(L^{rq}, L^{rq} \right) \left(\frac{r}{(3r-s)2^{3-\frac{s}{r}}} \right)^r \right]^{\frac{1}{rq}} \right\}.$$

Corollary 9 *In Theorem 3 for $h_1(t) = h_2(t) = t(1-t)$ and $f''(x) \leq L, \forall x \in I$, we get the following Hermite–Hadamard type inequality for generalized $((m, tgs); (\eta_1, \eta_2))$-convex mappings*

$$\left| T_f (\eta_1, \varphi; x, y, m) \right| \leq 4 \left(\frac{1}{24} \right)^{1-\frac{1}{q}} \beta^{\frac{1}{q}}_{1/2} \left(3 + \frac{1}{r}, 1 + \frac{1}{r} \right) \tag{27}$$

$$\times \left[mL^{rq} + \eta_2 \left(L^{rq}, L^{rq} \right) \right]^{\frac{1}{rq}}.$$

Corollary 10 *In Corollary 6 for* $h_1(t) = \dfrac{\sqrt{1-t}}{2\sqrt{t}}$, $h_2(t) = \dfrac{\sqrt{t}}{2\sqrt{1-t}}$ *and* $r \in$
$\left(\frac{1}{2}, 1\right]$, *we get the following Hermite–Hadamard type inequality for generalized*
$(m; (\eta_1, \eta_2))$-MT-*convex mappings*

$$\left| T_f(\eta_1, \varphi; x, y, m) \right| \le 2 \left(\frac{1}{24}\right)^{1 - \frac{1}{q}} \left(\frac{1}{2}\right)^{\frac{1}{rq}} \tag{28}$$

$$\times \left\{ \left[m L^{rq} \beta^r_{1/2} \left(3 - \frac{1}{2r}, 1 + \frac{1}{2r}\right) + \eta_2 \left(L^{rq}, L^{rq}\right) \beta^r_{1/2} \left(3 + \frac{1}{2r}, 1 - \frac{1}{2r}\right) \right]^{\frac{1}{rq}} \right.$$

$$\left. + \left[m L^{rq} \beta^r_{1/2} \left(3 + \frac{1}{2r}, 1 - \frac{1}{2r}\right) + \eta_2 \left(L^{rq}, L^{rq}\right) \beta^r_{1/2} \left(3 - \frac{1}{2r}, 1 + \frac{1}{2r}\right) \right]^{\frac{1}{rq}} \right\}.$$

3   Applications to Special Means

Definition 12 A function $M : \mathbb{R}^2_+ \longrightarrow \mathbb{R}_+$ is called a Mean function if it has the
following properties:

1. Homogeneity: $M(ax, ay) = a M(x, y)$, for all $a > 0$,
2. Symmetry: $M(x, y) = M(y, x)$,
3. Reflexivity: $M(x, x) = x$,
4. Monotonicity: If $x \le x'$ and $y \le y'$, then $M(x, y) \le M(x', y')$,
5. Internality: $\min\{x, y\} \le M(x, y) \le \max\{x, y\}$.

We consider some means for different positive real numbers α, β.

1. The arithmetic mean:

$$A := A(\alpha, \beta) = \frac{\alpha + \beta}{2}.$$

2. The geometric mean:

$$G := G(\alpha, \beta) = \sqrt{\alpha \beta}.$$

3. The harmonic mean:

$$H := H(\alpha, \beta) = \frac{2}{\frac{1}{\alpha} + \frac{1}{\beta}}.$$

4. The power mean:

$$P_r := P_r(\alpha, \beta) = \left(\frac{\alpha^r + \beta^r}{2}\right)^{\frac{1}{r}}, \quad r \geq 1.$$

5. The identric mean:

$$I := I(\alpha, \beta) = \begin{cases} \frac{1}{e}\left(\frac{\beta^\beta}{\alpha^\alpha}\right), & \alpha \neq \beta; \\ \alpha, & \alpha = \beta. \end{cases}$$

6. The logarithmic mean:

$$L := L(\alpha, \beta) = \frac{\beta - \alpha}{\ln \beta - \ln \alpha}.$$

7. The generalized log-mean:

$$L_p := L_p(\alpha, \beta) = \left[\frac{\beta^{p+1} - \alpha^{p+1}}{(p+1)(\beta - \alpha)}\right]^{\frac{1}{p}}; \quad p \in \mathbb{R} \setminus \{-1, 0\}.$$

It is well known that L_p is monotonic nondecreasing over $p \in \mathbb{R}$ with $L_{-1} := L$ and $L_0 := I$. In particular, we have the following inequality $H \leq G \leq L \leq I \leq A$. Now, let a and b be positive real numbers such that $a < b$. Let us consider continuous functions $\varphi : I \longrightarrow K$, $\eta_1 : K \times K \longrightarrow \mathbb{R}$ and $M := M(\varphi(a), \varphi(b)) : [\varphi(a), \varphi(a) + \eta_1(\varphi(b), \varphi(a))] \times [\varphi(a), \varphi(a) + \eta_1(\varphi(b), \varphi(a))] \longrightarrow \mathbb{R}_+$, which is one of the above-mentioned means. Therefore one can obtain various inequalities using the results of Sect. 2 for these means as follows. Replace $\eta_1(\varphi(a), \varphi(b)) = \eta_1(\varphi(b), \varphi(a)) = M(\varphi(a), \varphi(b))$, for $m = 1$ in (17) and (23), one can obtain the following interesting inequalities involving means:

$$\left|T_f(M(\cdot, \cdot), \varphi; a, b, 1)\right| \leq \left(\frac{1}{(2p+1)2^{2p+1}}\right)^{\frac{1}{p}} \tag{29}$$

$$\times \left\{\left[\left(f''(a)\right)^{rq} I^r(h_1(t); r) + \eta_2\left(\left(f''(b)\right)^{rq}, \left(f''(a)\right)^{rq}\right) I^r(h_2(t); r)\right]^{\frac{1}{rq}}\right.$$

$$+ \left[\left(f''(b)\right)^{rq} I^r(h_1(t); r) + \eta_2\left(\left(f''(a)\right)^{rq}, \left(f''(b)\right)^{rq}\right) I^r(h_2(t); r)\right]^{\frac{1}{rq}}$$

$$+ \left[\left(f''(a)\right)^{rq} J^r(h_1(t); r) + \eta_2\left(\left(f''(b)\right)^{rq}, \left(f''(a)\right)^{rq}\right) J^r(h_2(t); r)\right]^{\frac{1}{rq}}$$

$$+ \left.\left[\left(f''(b)\right)^{rq} J^r(h_1(t); r) + \eta_2\left(\left(f''(a)\right)^{rq}, \left(f''(b)\right)^{rq}\right) J^r(h_2(t); r)\right]^{\frac{1}{rq}}\right\},$$

$$\left|T_f(M(\cdot,\cdot),\varphi;a,b,1)\right| \le \left(\frac{1}{24}\right)^{1-\frac{1}{q}} \tag{30}$$

$$\times \left\{ \left[\left(f''(a)\right)^{rq} F^r(h_1(t);r) + \eta_2 \left(\left(f''(b)\right)^{rq}, \left(f''(a)\right)^{rq}\right) F^r(h_2(t);r) \right]^{\frac{1}{rq}} \right.$$

$$+ \left[\left(f''(b)\right)^{rq} F^r(h_1(t);r) + \eta_2 \left(\left(f''(a)\right)^{rq}, \left(f''(b)\right)^{rq}\right) F^r(h_2(t);r) \right]^{\frac{1}{rq}}$$

$$+ \left[\left(f''(a)\right)^{rq} G^r(h_1(t);r) + \eta_2 \left(\left(f''(b)\right)^{rq}, \left(f''(a)\right)^{rq}\right) G^r(h_2(t);r) \right]^{\frac{1}{rq}}$$

$$+ \left. \left[\left(f''(b)\right)^{rq} G^r(h_1(t);r) + \eta_2 \left(\left(f''(a)\right)^{rq}, \left(f''(b)\right)^{rq}\right) G^r(h_2(t);r) \right]^{\frac{1}{rq}} \right\}.$$

Letting $M(\varphi(a), \varphi(b)) := A, G, H, P_r, I, L, L_p, M_p$ in (29) and (30), we get inequalities involving means for a particular choices of twice differentiable generalized $((h_1, h_2); (\eta_1, \eta_2))$-convex mapping f at certain powers. The details are left to the interested reader.

4 Conclusion

In this article, we first presented a new integral identity concerning twice differentiable mappings defined on m-invex set. By using the notion of generalized $((h_1, h_2); (\eta_1, \eta_2))$-convexity and lemma as an auxiliary result, some new estimates difference between the left and middle part in Hermite–Hadamard type integral inequality associated with twice differentiable generalized $((h_1, h_2); (\eta_1, \eta_2))$-convex mappings are established. It is pointed out that some new special cases are deduced from main results. At the end, some applications to special means for different positive real numbers are provided. Motivated by this new interesting class we can indeed see to be vital for fellow researchers and scientists working in the same domain. We conclude that our methods considered here may be a stimulant for further investigations concerning Hermite–Hadamard, Ostrowski, and Simpson type integral inequalities for various kinds of convex and preinvex functions involving local fractional integrals, fractional integral operators, Caputo k-fractional derivatives, q-calculus, (p, q)-calculus, time scale calculus, and conformable fractional integrals.

References

1. S.M. Aslani, M.R. Delavar, S.M. Vaezpour, Inequalities of Fejér type related to generalized convex functions with applications. Int. J. Anal. Appl. **16**(1), 38–49 (2018)
2. F. Chen, A note on Hermite-Hadamard inequalities for products of convex functions via Riemann-Liouville fractional integrals. Ital. J. Pure Appl. Math. **33**, 299–306 (2014)
3. Y.-M. Chu, G.D. Wang, X.H. Zhang, Schur convexity and Hadamard's inequality. Math. Inequal. Appl. **13**(4), 725–731 (2010)
4. Y.-M. Chu, M.A. Khan, T.U. Khan, T. Ali, Generalizations of Hermite-Hadamard type inequalities for MT-convex functions. J. Nonlinear Sci. Appl. **9**(5), 4305–4316 (2016)
5. Y.-M. Chu, M.A. Khan, T. Ali, S.S. Dragomir, Inequalities for α-fractional differentiable functions. J. Inequal. Appl. **2017**(93), 12 (2017)
6. Z. Dahmani, On Minkowski and Hermite-Hadamard integral inequalities via fractional integration. Ann. Funct. Anal. **1**(1), 51–58 (2010)
7. M.R. Delavar, M. De La Sen, Some generalizations of Hermite-Hadamard type inequalities. SpringerPlus **5**, 1661 (2016)
8. M.R. Delavar, S.S. Dragomir, On η-convexity. Math. Inequal. Appl. **20**, 203–216 (2017)
9. S.S. Dragomir, J. Pečarić, L.E. Persson, Some inequalities of Hadamard type. Soochow J. Math. **21**, 335–341 (1995)
10. T.S. Du, J.G. Liao, Y.J. Li, Properties and integral inequalities of Hadamard-Simpson type for the generalized (s, m)-preinvex functions. J. Nonlinear Sci. Appl. **9**, 3112–3126 (2016)
11. G. Farid, A.U. Rehman, Generalizations of some integral inequalities for fractional integrals. Ann. Math. Sil. **31**, 14 (2017)
12. M.E. Gordji, S.S. Dragomir, M.R. Delavar, An inequality related to η-convex functions (II). Int. J. Nonlinear Anal. Appl. **6**(2), 26–32 (2016)
13. M.E. Gordji, M.R. Delavar, M. De La Sen, On φ-convex functions. J. Math. Inequal. Wiss **10**(1), 173–183 (2016)
14. A. Iqbal, M.A. Khan, S. Ullah, Y.-M. Chu, A. Kashuri, Hermite-Hadamard type inequalities pertaining conformable fractional integrals and their applications. AIP Adv. **8**(7), 18 (2018)
15. A. Kashuri, R. Liko, On Hermite-Hadamard type inequalities for generalized (s, m, φ)-preinvex functions via k-fractional integrals. Adv. Inequal. Appl. 6, 1–12 (2017)
16. A. Kashuri, R. Liko, Generalizations of Hermite-Hadamard and Ostrowski type inequalities for MT_m-preinvex functions. Proyecciones **36**(1), 45–80 (2017)
17. A. Kashuri, R. Liko, Hermite-Hadamard type fractional integral inequalities for generalized $(r; s, m, \varphi)$-preinvex functions. Eur. J. Pure Appl. Math. **10**(3), 495–505 (2017)
18. A. Kashuri, R. Liko, Hermite-Hadamard type fractional integral inequalities for twice differentiable generalized (s, m, φ)-preinvex functions. Konuralp J. Math. **5**(2), 228–238 (2017)
19. A. Kashuri, R. Liko, Hermite-Hadamard type inequalities for generalized (s, m, φ)-preinvex functions via k-fractional integrals. Tbil. Math. J. **10**(4), 73–82 (2017)
20. A. Kashuri, R. Liko, Hermite-Hadamard type fractional integral inequalities for $MT_{(m,\varphi)}$-preinvex functions. Stud. Univ. Babeş-Bolyai Math. **62**(4), 439–450 (2017)
21. A. Kashuri, R. Liko, Hermite-Hadamard type fractional integral inequalities for twice differentiable generalized beta-preinvex functions. J. Fract. Calc. Appl. **9**(1), 241–252 (2018)
22. M.A. Khan, Y. Khurshid, T. Ali, N. Rehman, Inequalities for three times differentiable functions. J. Math. Punjab Univ. **48**(2), 35–48 (2016)
23. M.A. Khan, T. Ali, S.S. Dragomir, M.Z. Sarikaya, Hermite-Hadamard type inequalities for conformable fractional integrals. Revista de la Real Academia de Ciencias Exactas, Físicas y Naturales. Serie A. Matemáticas (2017). https://doi.org/10.1007/s13398-017-0408-5
24. M.A. Khan, Y.-M. Chu, T.U. Khan, J. Khan, Some new inequalities of Hermite-Hadamard type for s-convex functions with applications. Open Math. **15**, 1414–1430 (2017)
25. M.A. Khan, Y. Khurshid, T. Ali, Hermite-Hadamard inequality for fractional integrals via η-convex functions. Acta Math. Univ. Comenianae **79**(1), 153–164 (2017)

26. M.A. Khan, Y.-M. Chu, A. Kashuri, R. Liko, G. Ali, New Hermite-Hadamard inequalities for conformable fractional integrals. J. Funct. Spaces **2018**, 6928130, 9 (2018)
27. M.A. Khan, Y.-M. Chu, A. Kashuri, R. Liko, Hermite-Hadamard type fractional integral inequalities for $MT_{(r;g,m,\varphi)}$-preinvex functions. J. Comput. Anal. Appl. **26**(8), 1487–1503 (2019)
28. W. Liu, W. Wen, J. Park, Ostrowski type fractional integral inequalities for MT-convex functions. Miskolc Math. Notes **16**(1), 249–256 (2015)
29. W. Liu, W. Wen, J. Park, Hermite-Hadamard type inequalities for MT-convex functions via classical integrals and fractional integrals. J. Nonlinear Sci. Appl. **9**, 766–777 (2016)
30. C. Luo, T.S. Du, M.A. Khan, A. Kashuri, Y. Shen, Some k-fractional integrals inequalities through generalized $\lambda_{\phi m}$-MT-preinvexity. J. Comput. Anal. Appl. **27**(4), 690–705 (2019)
31. M. Matłoka, Inequalities for h-preinvex functions. Appl. Math. Comput. **234**, 52–57 (2014)
32. S. Mubeen, G.M. Habibullah, k-Fractional integrals and applications. Int. J. Contemp. Math. Sci. 7, 89–94 (2012)
33. M.A. Noor, K.I. Noor, M.U. Awan, S. Khan, Hermite-Hadamard inequalities for s-Godunova-Levin preinvex functions. J. Adv. Math. Stud. **7**(2), 12–19 (2014)
34. O. Omotoyinbo, A. Mogbodemu, Some new Hermite-Hadamard integral inequalities for convex functions. Int. J. Sci. Innovation Tech. **1**(1), 1–12 (2014)
35. C. Peng, C. Zhou, T.S. Du, Riemann-Liouville fractional Simpson's inequalities through generalized (m, h_1, h_2)-preinvexity. Ital. J. Pure Appl. Math. **38**, 345–367 (2017)
36. R. Pini, Invexity and generalized convexity. Optimization **22**, 513–525 (1991)
37. E. Set, Some new generalized Hermite-Hadamard type inequalities for twice differentiable functions (2017). https://www.researchgate.net/publication/327601181
38. E. Set, S.S. Karataş, M.A. Khan, Hermite-Hadamard type inequalities obtained via fractional integral for differentiable m-convex and (α, m)-convex functions. Int. J. Anal. **2016**, 4765691, 8 (2016)
39. E. Set, A. Gözpinar, J. Choi, Hermite-Hadamard type inequalities for twice differentiable m-convex functions via conformable fractional integrals. Far East J. Math. Sci. **101**(4), 873–891 (2017)
40. E. Set, M.Z. Sarikaya, A. Gözpinar, Some Hermite-Hadamard type inequalities for convex functions via conformable fractional integrals and related inequalities. Creat. Math. Inform. **26**(2), 221–229 (2017)
41. H.N. Shi, Two Schur-convex functions related to Hadamard-type integral inequalities. Publ. Math. Debrecen **78**(2), 393–403 (2011)
42. M. Tunç, E. Göv, Ü. Şanal, On tgs-convex function and their inequalities. Facta Univ. Ser. Math. Inform. **30**(5), 679–691 (2015)
43. S. Varošanec, On h-convexity. J. Math. Anal. Appl. **326**(1), 303–311 (2007)
44. Y. Wang, S.H. Wang, F. Qi, Simpson type integral inequalities in which the power of the absolute value of the first derivative of the integrand is s-preinvex. Facta Univ. Ser. Math. Inform. **28**(2), 151–159 (2013)
45. H. Wang, T.S. Du, Y. Zhang, k-fractional integral trapezium-like inequalities through (h, m)-convex and (α, m)-convex mappings. J. Inequal. Appl. **2017**(311), 20 (2017)
46. T. Weir, B. Mond, Preinvex functions in multiple objective optimization. J. Math. Anal. Appl. **136**, 29–38 (1988)
47. X.M. Zhang, Y.-M. Chu, X.H. Zhang, The Hermite-Hadamard type inequality of GA-convex functions and its applications. J. Inequal. Appl. **2010**, 507560, 11 (2010)
48. Y. Zhang, T.S. Du, H. Wang, Y.J. Shen, A. Kashuri, Extensions of different type parameterized inequalities for generalized (m, h)-preinvex mappings via k-fractional integrals. J. Inequal. Appl. **2018**(49), 30 (2018)

Hardy's Type Inequalities via Conformable Calculus

S. H. Saker, M. R. Kenawy, and D. Baleanu

Abstract In this chapter, we establish some inequalities of Hardy and Leindler type and their converses via conformable calculus with weighted functions. As applications, we obtain some classical integral inequalities as special cases.

1 Introduction

The classical discrete Hardy inequality is given by

$$\sum_{n=1}^{\infty}\left(\frac{1}{n}\sum_{i=1}^{n}g(i)\right)^{p} \leq \left(\frac{p}{p-1}\right)^{p}\sum_{n=1}^{\infty}g^{p}(n), \quad p > 1. \tag{1}$$

where $g(n)$ is a sequence with nonnegative terms. Some generalizations of the discrete Hardy inequality (1) have been obtained by Leindler [11, 12] with new weighted function. In particular, Leindler in [11] proved that if $p > 1$ and $\lambda(n) > 0$ then the inequality with heads

S. H. Saker (✉)
Samir Atwa Saker, Department of Mathematics, Faculty of Science,
Mansoura University, Mansoura, Egypt
e-mail: shsaker@mans.edu.eg

M. R. Kenawy
Mohammed Ragab Zakie Kenawy, Department of Mathematics,
Faculty of Science, Fayoum University, Fayoum, Egypt
e-mail: mrz00@fayoum.edu.eg

D. Baleanu
Dumitru Baleanu, Department of Mathematics and Computer Science,
Cankaya University, Ankara, Turkey
e-mail: dumitru@cankaya.edu.tr

© Springer Nature Switzerland AG 2019
G. A. Anastassiou and J. M. Rassias (eds.), *Frontiers in Functional Equations and Analytic Inequalities*, https://doi.org/10.1007/978-3-030-28950-8_25

$$\sum_{n=1}^{\infty} \lambda(n) \left(\sum_{k=1}^{n} g(k) \right)^p \le p^p \sum_{n=1}^{\infty} \lambda^{1-p}(n) \left(\sum_{k=n}^{\infty} \lambda(k) \right)^p g^p(n), \tag{2}$$

holds and the inequality with tails

$$\sum_{n=1}^{\infty} \lambda(n) \left(\sum_{k=n}^{\infty} g(k) \right)^p \le p^p \sum_{n=1}^{\infty} \lambda^{1-p}(n) \left(\sum_{k=1}^{n} \lambda(k) \right)^p g^p(n), \tag{3}$$

also holds.

In 1928 Copson [5] proved some new types of discrete inequalities (see also [7, Theorem 344]). In particular one of his inequalities is given by

$$\sum_{n=1}^{\infty} \left(\sum_{k=n}^{\infty} g(k) \right)^p \ge p^p \sum_{n=1}^{\infty} (ng(n))^p, \text{ for } 0 < p < 1, \tag{4}$$

where g_n is a sequence with nonnegative terms. The converses of (2) and (3) are proved by Leindler in [12]. In particular, he proved that if $0 < p \le 1$, then

$$\sum_{n=1}^{\infty} \lambda(n) \left(\sum_{k=1}^{n} g(k) \right)^p \ge p^p \sum_{n=1}^{\infty} \lambda^{1-p}(n) \left(\sum_{k=n}^{\infty} \lambda(k) \right)^p g^p(n), \tag{5}$$

and

$$\sum_{n=1}^{\infty} \lambda(n) \left(\sum_{k=n}^{\infty} g(k) \right)^p \ge p^p \sum_{n=1}^{\infty} \lambda^{1-p}(n) \left(\sum_{k=1}^{n} \lambda(p) \right)^p g^p(n). \tag{6}$$

An interesting variant of the Hardy–Copson inequalities was given by Leindler [13]. In fact Leindler in [13] generalized the above inequalities and proved that if $\sum_{i=n}^{\infty} \lambda(i) < \infty$, $p > 1$ and $0 \le c < 1$, then

$$\sum_{n=1}^{\infty} \frac{\lambda(n)}{(\Omega(n))^c} \left(\sum_{i=1}^{n} \lambda(i) g(i) \right)^p \le \left(\frac{p}{1-c} \right)^p \sum_{n=1}^{\infty} \lambda(n)(\Omega(n))^{p-c} g^p(n), \tag{7}$$

where $\Omega_n = \sum_{i=n}^{\infty} \lambda(i)$, and if $1 < c \le p$, then

$$\sum_{n=1}^{\infty} \frac{\lambda(n)}{(\Omega(n))^c} \left(\sum_{i=n}^{\infty} \lambda(i) g(i) \right)^p \le \left(\frac{p}{c-1} \right)^p \sum_{n=1}^{\infty} \lambda(n)(\Omega(n))^{p-c} g^p(n). \tag{8}$$

In recent years, some authors studied the fractional inequalities by using the fractional Caputo and Riemann–Liouville derivative, we refer to [4, 8], and [18] for the results. In [1] and [9] the authors extended the calculus of fractional order to conformable calculus. Recently, some authors have extended classical inequalities by using conformable calculus such as Opial's inequality [15] and [14],

Hermite–Hadamard's inequality [3, 6, 10] and [17], Chebyshev's inequality [2], and Steffensen's inequality [16].

The following question now arises: Is it possible to prove inequalities of Leindler-type and their converses by using conformable calculus? Our aim in this chapter is to give an affirmative answer to this question and obtain the classical integral inequalities as special cases.

The chapter is organized as follows: In Sect. 2, we will present some preliminaries about the conformable calculus and also the Hölder's inequality for $\alpha-$ differentiable functions which will be needed in the proofs of the main results. In Sect. 3, we will establish some Leindler-type inequalities and some of their generalizations for $\alpha-$ differentiable functions and obtain the classical ones as $\alpha = 1$. In Sect. 4, we will prove some converses of Leindler-type inequalities and some of their generalizations and obtain the classical ones when $\alpha = 1$.

2   Basic Concepts and Lemmas

In this section, we present the concepts of conformable derivative and integral of order $0 < \alpha \leq 1$, that will be used throughout the article. For more details, we refer the reader to [1] and [9] for the recent results on conformable calculus.

Definition 1 Let $f : [0, \infty) \to \mathbb{R}$. Then the conformable derivative of order α of f is defined by

$$D_\alpha f(t) = \lim_{\epsilon \to 0} \frac{f(t + \epsilon t^{1-\alpha}) - f(t)}{\epsilon},$$

for all $t > 0$ and $0 < \alpha \leq 1$, and $D_\alpha f(0) = \lim_{t \to 0^+} D_\alpha f(t)$.

Let $\alpha \in (0, 1]$ and f, g be $\alpha-$differentiable at a point t. Then

$$D_\alpha(fg) = f D_\alpha g + g D_\alpha f. \tag{9}$$

Further, let $\alpha \in (0, 1]$ and f, g be $\alpha-$differentiable at a point t, with $g(t) \neq 0$. Then

$$D_\alpha \left(\frac{f}{g} \right) = \frac{g D_\alpha f - f D_\alpha g}{g^2}. \tag{10}$$

Remark 1 If f is a differentiable function, then

$$D_\alpha f(t) = t^{1-\alpha} \frac{df(t)}{dt}.$$

Definition 2 Let $f : [0, \infty) \to \mathbb{R}$. Then the conformable integral of order α of f is defined by

$$I_\alpha f(t) = \int_0^t f(s)d_\alpha s = \int_0^t s^{\alpha-1}f(s)ds, \tag{11}$$

for all $t > 0$ and $0 < \alpha \leq 1$.

Now, we state an integration by parts (see [1] and [9]).

Lemma 1 *Assume that U, $V : [0, \infty) \to \mathbb{R}$ are two functions such that U, V are differentiable and $0 < \alpha \leq 1$. Then for any $b > 0$,*

$$\int_0^b U(s)D_\alpha V(s)d_\alpha s = U(s)V(s)|_0^b - \int_0^b V(s)D_\alpha U(s)d_\alpha s. \tag{12}$$

Next, we state a Hölder type inequality needed in the next section (of course it is the usual Hölder inequality for the functions considered (i.e., $s^{\frac{(\alpha-1)}{p}}f(s)$ and $s^{\frac{(\alpha-1)}{q}}g(s)$).

Lemma 2 *Let f, $g : [0, \infty) \to \mathbb{R}$ and $0 < \alpha \leq 1$. Then for any $b > 0$,*

$$\int_0^b |f(s)g(s)|\, d_\alpha s \leq \left(\int_0^b |f(s)|^p\, d_\alpha s \right)^{\frac{1}{p}} \left(\int_0^b |g(s)|^q\, d_\alpha s \right)^{\frac{1}{q}}, \tag{13}$$

where $1/p + 1/q = 1$ (provided the integrals exist (and are finite)).

Define the conformable Hardy operator

$$Tg(t) = \int_0^t g(s)d_\alpha s, \tag{14}$$

and its dual

$$T^*g(t) = \int_t^\infty g(s)d_\alpha s. \tag{15}$$

Throughout the chapter, we will assume that the functions are nonnegative locally α−integrable and the integrals throughout are assumed to exist (and are finite, i.e. convergent).

3 Leindler Type Inequalities

In this section, we will establish some Leindler type inequalities and some of their generalizations for α−differentiable functions and obtain the classical ones when $\alpha = 1$.

Theorem 1 *If $p > 1$, then*

$$\int_0^\infty \omega(t)T^p g(t)d_\alpha t \leq p^p \int_0^\infty \omega^{1-p}(t)\Omega^p(t)g^p(t)d_\alpha t, \tag{16}$$

where

$$\Omega(t) := \int_t^\infty \omega(s)d_\alpha s.$$

Proof Integrating the term $\int_0^\infty \omega(t)T^p g(t)d_\alpha t$, by parts formula (12) with

$$U(t) = T^p g(t), \text{ and } V(t) = -\Omega(t),$$

we get that

$$\int_0^\infty \omega(t)T^p g(t)d_\alpha t = -T^p g(t)\Omega(t)\Big|_0^\infty + p\int_0^\infty t^{1-\alpha}T^{p-1}g(t)(t)T' g(t)\Omega(t)d_\alpha t$$

$$= p\int_0^\infty t^{1-\alpha}T^{p-1}g(t)(t)T' g(t)\Omega(t)d_\alpha t, \tag{17}$$

where

$$Tg(0) = 0, \ Tg(\infty) < \infty, \ \Omega(0) < \infty \text{ and } \Omega(\infty) = 0.$$

From the definition of conformable Hardy operator (14), we see that

$$T' g(t) = t^{\alpha-1}g(t).$$

Substituting into (17), we obtain

$$\int_0^\infty \omega(t)T^p g(t)d_\alpha t = p\int_0^\infty \frac{g(t)\Omega(t)}{(\omega(t))^{\frac{p-1}{p}}}(\omega(t))^{\frac{p-1}{p}}T^{p-1}g(t)d_\alpha t. \tag{18}$$

Applying the Hölder inequality (13) on the right-hand side of (18) with indices p and $p/(p-1)$, we see that

$$\int_0^\infty \omega(t)T^p g(t)d_\alpha t \leq p\left(\int_0^\infty \left(\frac{g(t)\Omega(t)}{(\omega(t))^{\frac{p-1}{p}}}\right)^p d_\alpha t\right)^{1/p}$$

$$\times \left(\int_0^\infty \left((\omega(t))^{\frac{p-1}{p}}T^{p-1}g(t)\right)^{\frac{p}{p-1}}d_\alpha t\right)^{\frac{p-1}{p}},$$

then

$$\int_0^\infty \omega(t)T^p g(t)d_\alpha t \leq p \left(\int_0^\infty \frac{(g(t)\Omega(t))^p}{\omega^{p-1}(t)} d_\alpha t \right)^{1/p} \left(\int_0^\infty \omega(t)T^p g(t)d_\alpha t \right)^{1-1/p},$$

since

$$\left(\int_0^\infty \omega(t)T^p g(t)d_\alpha t \right)^{1-1/p} > 0,$$

then

$$\left(\int_0^\infty \omega(t)T^p g(t)d_\alpha t \right)^{\frac{1}{p}} \leq p \left(\int_0^\infty \frac{(g(t)\Omega(t))^p}{\omega^{p-1}(t)} d_\alpha t \right)^{1/p}.$$

This implies that

$$\int_0^\infty \omega(t)T^p g(t)d_\alpha t \leq p^p \int_0^\infty \omega^{1-p}(t)\Omega^p(t)g^p(t)d_\alpha t,$$

which is the desired inequality (16). The proof is complete. □

Remark 2 In Theorem 1 if $\alpha = 1$, then we obtain the inequality

$$\int_0^\infty \omega(t) \left(\int_0^t g(s)ds \right)^p dt \leq p^p \int_0^\infty \omega^{1-p}(t) \left(\int_t^\infty \omega(s)ds \right)^p g^p(t)dt. \tag{19}$$

Remark 3 In Theorem 1 if $\omega(t) = 1/t^p$, then we obtain the inequality

$$\int_0^\infty \left(\frac{1}{t} \int_0^t g(s)d_\alpha s \right)^p d_\alpha t \leq \left(\frac{p}{p-\alpha} \right)^p \int_0^\infty \left(t^{\alpha-1}g(t) \right)^p d_\alpha t. \tag{20}$$

which is the Hardy inequality.

Remark 4 As a consequence if $\alpha = 1$ in (20), we obtain the classical Hardy inequality

$$\int_0^\infty \left(\frac{1}{t} \int_0^t g(s)ds \right)^p dt \leq \left(\frac{p}{p-1} \right)^p \int_0^\infty g^p(t)dt. \tag{21}$$

Theorem 2 *If $p > 1$, then*

$$\int_0^\infty \omega(t)T^{*p} g(t)d_\alpha t \leq p^p \int_0^\infty \omega^{1-p}(t)\Lambda^p(t)g^p(t)d_\alpha t, \tag{22}$$

where

$$\Lambda(t) := \int_0^t \omega(s)d_\alpha s.$$

Proof Integrating the term $\int_0^\infty \omega(t)T^{*^p}g(t)d_\alpha t$, by parts formula (12) with

$$U(t) = T^{*^p}g(t), \text{ and } V(t) = \Lambda(t),$$

we get

$$\int_0^\infty \omega(t)T^{*^p}g(t)d_\alpha t = T^{*^p}g(t)\Lambda(t)\Big|_0^\infty - p\int_0^\infty t^{1-\alpha}T^{*^{p-1}}g(t)T^{*'}g(t)\Lambda(t)d_\alpha t$$

$$= -p\int_0^\infty t^{1-\alpha}T^{*^{p-1}}g(t)T^{*'}g(t)\Lambda(t)d_\alpha t, \tag{23}$$

where

$$T^*g(0) < \infty, \ T^*g(\infty) = 0, \ \Lambda(0) = 0 \text{ and } \Lambda(\infty) < \infty.$$

From the definition of conformable dual Hardy operator (15), we see that

$$T^{*'}g(t) = -t^{\alpha-1}g(t).$$

Substituting into (23), we obtain

$$\int_0^\infty \omega(t)T^{*^p}g(t)d_\alpha t = p\int_0^\infty \frac{g(t)\Lambda(t)}{(\omega(t))^{\frac{p-1}{p}}}(\omega(t))^{\frac{p-1}{p}}T^{*^{p-1}}g(t)d_\alpha t. \tag{24}$$

Applying the Hölder inequality (13) on the right-hand side of (24) with indices p and $p/(p-1)$, we see that

$$\int_0^\infty \omega(t)T^{*^p}g(t)d_\alpha t \leq p\left(\int_0^\infty \left(\frac{g(t)\Lambda(t)}{(\omega(t))^{\frac{p-1}{p}}}\right)^p d_\alpha t\right)^{1/p}$$

$$\times \left(\int_0^\infty \left((\omega(t))^{\frac{p-1}{p}}T^{*^{p-1}}g(t)\right)^{\frac{p}{p-1}} d_\alpha t\right)^{\frac{p-1}{p}}.$$

Then

$$\left(\int_0^\infty \omega(t)T^{*^p}g(t)d_\alpha t\right)^{\frac{1}{p}} \leq p\left(\int_0^\infty \frac{(g(t)\Lambda(t))^p}{\omega^{p-1}(t)}d_\alpha t\right)^{1/p},$$

and so that

$$
\int_0^\infty \omega(t) T^{*^p} g(t) d_\alpha t \le p^p \int_0^\infty \omega^{1-p}(t) \Lambda^p(t) g^p(t) d_\alpha t,
$$

which is the desired inequality (22). The proof is complete. □

Remark 5 In Theorem 2 if $\alpha = 1$, then we obtain the inequality

$$
\int_0^\infty \omega(t) \left(\int_t^\infty g(s) ds \right)^p dt \le p^p \int_0^\infty \omega^{1-p}(t) \left(\int_0^t \omega(s) ds \right)^p g^p(t) dt.
\tag{25}
$$

Remark 6 In Theorem 2 if $\omega(t) = 1$, then we obtain the inequality

$$
\int_0^\infty \left(\int_t^\infty g(s) d_\alpha s \right)^p d_\alpha t \le \left(\frac{p}{\alpha} \right)^p \int_0^\infty \left(t^\alpha g(t) \right)^p d_\alpha t,
\tag{26}
$$

which is the Copson inequality.

Remark 7 As a consequence if $\alpha = 1$ in (26) we obtain the Copson inequality

$$
\int_0^\infty \left(\int_t^\infty g(s) ds \right)^p dt \le p^p \int_0^\infty \left(t g(t) \right)^p dt.
\tag{27}
$$

Theorem 3 *If $p \ge 1$ and $0 \le c < 1$, then*

$$
\int_0^\infty \frac{\omega(t)}{\Omega^c(t)} \Phi^p(t) d_\alpha t \le \left(\frac{p}{1-c} \right)^p \int_0^\infty \frac{\omega(t)}{\Omega^{c-p}(t)} g^p(t) d_\alpha t,
\tag{28}
$$

where

$$
\Omega(t) := \int_t^\infty \omega(s) d_\alpha s, \text{ and } \Phi(t) := \int_0^t \omega(s) g(s) d_\alpha s.
$$

Proof Integrating the term

$$
\int_0^\infty \frac{\omega(t)}{\Omega^c(t)} \Phi^p(t) d_\alpha t,
$$

by parts formula (12) with

$$
U(t) = \Phi^p(t), \text{ and } V(t) = -\frac{\Omega^{1-c}(t)}{1-c},
$$

and

$$D_\alpha U(t) = pt^{1-\alpha}\Phi^{p-1}(t)\Phi'(t) \text{ and } D_\alpha V(t) = \omega(t)\Omega^{-c}(t),$$

we obtain

$$\int_0^\infty \frac{\omega(t)}{\Omega^c(t)}\Phi^p(t)d_\alpha t = -\frac{\Phi^p(t)\Omega^{1-c}(t)}{1-c}\Big|_0^\infty + \int_0^\infty \frac{pt^{1-\alpha}\Phi^{p-1}(t)\Phi'(t)\Omega^{1-c}(t)}{1-c}d_\alpha t.$$

By using

$$\Phi(0) = 0, \ \Phi(\infty) < \infty, \ \Omega(0) < \infty \text{ and } \Omega(\infty) = 0,$$

and since $\Phi'(t) = t^{\alpha-1}\omega(t)g(t)$, we get that

$$\int_0^\infty \frac{\omega(t)}{\Omega^c(t)}\Phi^p(t)d_\alpha t = -\frac{p}{1-c}\int_0^\infty t^{1-\alpha}\Phi^{p-1}(t)t^{\alpha-1}\omega(t)g(t)\Omega^{1-c}(t)d_\alpha t,$$

and hence

$$\int_0^\infty \frac{\omega(t)}{\Omega^c(t)}\Phi^p(t)d_\alpha t = \frac{p}{1-c}\int_0^\infty \frac{\omega(t)\Omega^{1-c}(t)g(t)}{\left(\frac{\omega(t)}{\Omega^c(t)}\right)^{\frac{p-1}{p}}}\left(\frac{\omega(t)}{\Omega^c(t)}\right)^{\frac{p-1}{p}}\Phi^{p-1}(t)d_\alpha t.$$

(29)

Applying the Hölder inequality (13) on the right-hand side of (29) with indices p and $p/p - 1$, we see that

$$\int_0^\infty \frac{\omega(t)}{\Omega^c(t)}\Phi^p(t)d_\alpha t$$

$$\leq \frac{p}{1-c}\left(\int_0^\infty \left(\omega(t)\Omega^{1-c}(t)g(t)\left(\frac{\omega(t)}{\Omega^c(t)}\right)^{-\frac{p-1}{p}}\right)^p d_\alpha t\right)^{\frac{1}{p}}$$

$$\times \left(\int_0^\infty \left(\left(\frac{\omega(t)}{\Omega^c(t)}\right)^{\frac{p-1}{p}}\Phi^{p-1}(t)\right)^{\frac{p}{p-1}}d_\alpha t\right)^{\frac{p-1}{p}}$$

$$= \frac{p}{1-c}\left(\int_0^\infty \omega(t)\Omega^{p-c}(t)g^p(t)d_\alpha t\right)^{\frac{1}{p}}\left(\int_0^\infty \frac{\omega(t)}{\Omega^c(t)}\Phi^p(t)d_\alpha t\right)^{\frac{p-1}{p}}.$$

Thus

$$\left(\int_0^\infty \frac{\omega(t)}{\Omega^c(t)}\Phi^p(t)d_\alpha t\right)^{\frac{1}{p}} \leq \frac{p}{1-c}\left(\int_0^\infty \omega(t)\Omega^{p-c}(t)g^p(t)d_\alpha t\right)^{\frac{1}{p}},$$

and hence

$$\int_0^\infty \frac{\omega(t)}{\Omega^c(t)} \Phi^p(t) d_\alpha t \leq \left(\frac{p}{1-c}\right)^p \int_0^\infty \omega(t) \Omega^{p-c}(t) g^p(t) d_\alpha t,$$

which is the desired inequality (28). The proof is complete. □

Remark 8 In Theorem 3, if $\alpha = 1$, then we obtain the inequality

$$\int_0^\infty \frac{\omega(t)}{\Omega^c(t)} \left(\int_0^t \omega(s) g(s) ds\right)^p dt \leq \left(\frac{p}{1-c}\right)^p \int_0^\infty \frac{\omega(t)}{\Omega^{c-p}(t)} g^p(t) dt, \qquad (30)$$

where $\Omega(t) = \int_t^\infty \omega(s) ds$, $p > 1$ and $0 \leq c < 1$.

Theorem 4 *If $1 < c \leq p$ and $p > c - 1$, then*

$$\int_0^\infty \frac{\omega(t)}{\Omega^c(t)} \Psi^p(t) d_\alpha t \leq \left(\frac{p}{c-1}\right)^p \int_0^\infty \frac{\omega(t)}{\Omega^{c-p}(t)} g^p(t) d_\alpha t, \qquad (31)$$

where

$$\Omega(t) := \int_t^\infty \omega(s) d_\alpha s \text{ and } \Psi(t) := \int_t^\infty \omega(s) g(s) d_\alpha s.$$

Proof Integrating the term

$$\int_0^\infty \frac{\omega(t)}{\Omega^c(t)} \Psi^p(t) d_\alpha t,$$

by parts formula (12) with

$$U(t) = \Psi^p(t), \text{ and } V(t) = -\frac{\Omega^{1-c}(t)}{1-c},$$

and

$$D_\alpha U(t) = p t^{1-\alpha} \Psi^{p-1}(t) \Psi'(t) \text{ and } D_\alpha V(t) = \omega(t) \Omega^{-c}(t),$$

we obtain

$$\int_0^\infty \frac{\omega(t)}{\Omega^c(t)} \Psi^p(t) d_\alpha t = -\frac{\Psi^p(t) \Omega^{1-c}(t)}{1-c} \Big|_0^\infty + \int_0^\infty \frac{p t^{1-\alpha} \Psi^{p-1}(t) \Psi'(t) \Omega^{1-c}(t)}{1-c} d_\alpha t.$$

By using $\Psi'(t) = -t^{\alpha-1} \omega(t) g(t)$,

$$\Psi(0) < \infty, \ \Psi(\infty) = 0, \ \Omega(0) < \infty, \ \Omega(\infty) = 0 \text{ and } c > 1,$$

and noting

$$\lim_{t\to\infty} \Omega^{\frac{1-c}{p}}(t)\Psi(t) = \lim_{t\to\infty} \frac{\int_t^\infty s^{\alpha-1}\omega(s)g(s)\,ds}{(\Omega(t))^{\frac{c-1}{p}}}$$

$$= \lim_{t\to\infty} \frac{-t^{\alpha-1}\omega(t)g(t)}{-\frac{c-1}{p}(\Omega(t))^{\frac{c-1}{p}-1}t^{\alpha-1}\omega(t)}$$

$$= \lim_{t\to\infty} \frac{p(\Omega(t))^{1-\frac{c-1}{p}}g(t)}{c-1} = 0,$$

we get that

$$\int_0^\infty \frac{\omega(t)}{\Omega^c(t)}\Psi^p(t)d_\alpha t \le \frac{p}{c-1}\int_0^\infty t^{1-\alpha}\Psi^{p-1}(t)t^{\alpha-1}\omega(t)g(t)\Omega^{1-c}(t)d_\alpha t,$$

then

$$\int_0^\infty \frac{\omega(t)}{\Omega^c(t)}\Psi^p(t)d_\alpha t \le \frac{p}{c-1}\int_0^\infty \frac{\omega(t)\Omega^{1-c}(t)g(t)}{\left(\frac{\omega(t)}{\Omega^c(t)}\right)^{\frac{p-1}{p}}}\left(\frac{\omega(t)}{\Omega^c(t)}\right)^{\frac{p-1}{p}}\Psi^{p-1}(t)d_\alpha t.$$

$$(32)$$

Applying the Hölder inequality (13) on the right-hand side of (32) with indices p and $p/(p-1)$, we see that

$$\int_0^\infty \frac{\omega(t)}{\Omega^c(t)}\Psi^p(t)d_\alpha t$$

$$\le \frac{p}{c-1}\left(\int_0^\infty \left(\omega(t)\Omega^{1-c}(t)g(t)\left(\frac{\omega(t)}{\Omega^c(t)}\right)^{-\frac{p-1}{p}}\right)^p d_\alpha t\right)^{\frac{1}{p}}$$

$$\times \left(\int_0^\infty \left(\left(\frac{\omega(t)}{\Omega^c(t)}\right)^{\frac{p-1}{p}}\Psi^{p-1}(t)\right)^{\frac{p}{p-1}} d_\alpha t\right)^{\frac{p-1}{p}}$$

$$= \frac{p}{c-1}\left(\int_0^\infty \omega(t)\Omega^{p-c}(t)g^p(t)d_\alpha t\right)^{\frac{1}{p}}\left(\int_0^\infty \frac{\omega(t)}{\Omega^c(t)}\Psi^p(t)d_\alpha t\right)^{\frac{p-1}{p}}.$$

Thus

$$\left(\int_0^\infty \frac{\omega(t)}{\Omega^c(t)}\Psi^p(t)d_\alpha t\right)^{\frac{1}{p}} \le \frac{p}{c-1}\left(\int_0^\infty \omega(t)\Omega^{p-c}(t)g^p(t)d_\alpha t\right)^{\frac{1}{p}},$$

and hence

$$\int_0^\infty \frac{\omega(t)}{\Omega^c(t)} \Psi^p(t)d_\alpha t \le \left(\frac{p}{c-1}\right)^p \int_0^\infty \omega(t)\Omega^{p-c}(t)g^p(t)d_\alpha t,$$

which is the desired inequality (31). The proof is complete   □

Remark 9 From the proof of Theorem 4 we see that if $t^{\alpha-1}\omega(t)g(t)$ and $t^{\alpha-1}\omega(t)$ is continuous on $[0, \infty)$ replaced either by

(i) $t^{\alpha-1}\omega(t)g(t)$, $t^{\alpha-1}\omega(t)$ is continuous on $(0, \infty)$ and $\lim_{t\to\infty} (\Omega(t))^{1-\frac{c-1}{p}} g(t)$
     $= 0$, or
(ii) $\lim_{t\to\infty} \Omega^{1-c}(t)\Psi^p(t) = 0$,
     then (31) is again true.

Remark 10 In Theorem 4 at $\alpha = 1$, then we obtain the inequality

$$\int_0^\infty \frac{\omega(t)}{\Omega^c(t)} \left(\int_t^\infty \omega(s)g(s)ds\right)^p dt \le \left(\frac{p}{c-1}\right)^p \int_0^\infty \frac{\omega(t)}{\Omega^{c-p}(t)} g^p(t)dt, \qquad (33)$$

where $\Omega(t) = \int_t^\infty \omega(s)ds$, and $1 < c \le p$.

4   Reversed Inequalities

In this section, we will prove some converses of inequalities and some generalizations and obtain the classical ones when $\alpha = 1$.

Theorem 5 *If $0 < p \le 1$, then*

$$\int_0^\infty \omega(t)T^p g(t)d_\alpha t \ge p^p \int_0^\infty \omega^{1-p}(t)\Omega^p(t)g^p(t)d_\alpha t, \qquad (34)$$

where

$$\Omega(t) := \int_t^\infty \omega(s)d_\alpha s.$$

Proof Integrating the term $\int_0^\infty \omega(t)T^p g(t)d_\alpha t$ by parts formula (12) with

$$U(t) = T^p g(t) \text{ and } V(t) = -\Omega(t),$$

and we obtain

$$\int_0^\infty \omega(t)T^p g(t)d_\alpha t = -T^p g(t)\Omega(t)\Big|_0^\infty + \int_0^\infty pt^{1-\alpha}T^{p-1}g(t)T'g(t)\Omega(t)d_\alpha t.$$

By using

$$Tg(0) = 0, \ Tg(\infty) < \infty, \ \Omega(0) < \infty \text{ and } \Omega(\infty) = 0,$$

and from (14), we have

$$T'g(t) = t^{\alpha-1}g(t),$$

and so we have

$$\int_0^\infty \omega(t)T^P g(t)d_\alpha t = p \int_0^\infty t^{1-\alpha}T^{P-1}g(t)t^{\alpha-1}g(t)\Omega(t)d_\alpha t,$$

then

$$\int_0^\infty \omega(t)T^P g(t)d_\alpha t = p \int_0^\infty \left(\frac{\Omega^P(t)g^P(t)}{T^{P(1-p)}g(t)} \right)^{\frac{1}{p}} d_\alpha t,$$

which can be rewritten in the form

$$\left(\int_0^\infty \omega(t)T^P g(t)d_\alpha t \right)^P = p^P \left(\int_0^\infty \left(\frac{\Omega^P(t)g^P(t)}{T^{P(1-p)}g(t)} \right)^{\frac{1}{p}} d_\alpha t \right)^P. \tag{35}$$

Applying Hölder's inequality

$$\int_0^\infty F(t)G(t)d_\alpha t \leq \left(\int_0^\infty F^u(t)d_\alpha t \right)^{\frac{1}{u}} \left(\int_0^\infty G^v(t)d_\alpha t \right)^{\frac{1}{v}},$$

with indices $u = 1/p$ and $v = 1/(1-p)$, and

$$F(t) = \frac{\Omega^P(t)g^P(t)}{T^{P(1-p)}g(t)} \text{ and } G(t) = \omega^{1-P}(t)T^{P(1-p)}g(t),$$

we see that

$$\left(\int_0^\infty F^{\frac{1}{p}}(t)d_\alpha t \right)^P = \left(\int_0^\infty \left(\frac{\Omega^P(t)g^P(t)}{T^{P(1-p)}g(t)} \right)^{\frac{1}{p}} d_\alpha t \right)^P$$

$$\geq \frac{\int_0^\infty |F(t)G(t)| \, d_\alpha t}{\left(\int_0^\infty G^{\frac{1}{1-p}}(t)d_\alpha t \right)^{1-p}}$$

$$= \int_0^\infty \frac{\Omega^P(t)g^P(t)}{T^{P(1-P)}g(t)}\omega^{1-P}(t)T^{P(1-P)}g(t)d_\alpha t$$

$$\times \left(\int_0^\infty \left(\omega^{1-P}(t)T^{P(1-P)}g(t) \right)^{\frac{1}{1-P}} d_\alpha t \right)^{P-1},$$

and then

$$\left(\int_0^\infty \left(\frac{\Omega^P(t)g^P(t)}{T^{P(1-P)}g(t)} \right)^{\frac{1}{P}} d_\alpha t \right)^P \geq \left(\int_0^\infty \Omega^P(t)g^P(t)\omega^{1-P}(t)d_\alpha t \right)$$

$$\times \left(\int_0^\infty \left(\omega(t)T^P g(t) \right) d_\alpha t \right)^{P-1}. \qquad (36)$$

Substituting (36) into (35), we have

$$\left(\int_0^\infty \omega(t)T^P g(t)d_\alpha t \right)^P \geq p^P \frac{\int_0^\infty \Omega^P(t)g^P(t)\omega^{1-P}(t)d_\alpha t}{\left(\int_0^\infty \left(\omega(t)T^P g(t) \right) d_\alpha t \right)^{1-P}}.$$

This implies that

$$\int_0^\infty \omega(t)T^P g(t)d_\alpha t \geq p^P \int_0^\infty \omega^{1-P}(t)\Omega^P(t)g^P(t)d_\alpha t,$$

which is the desired inequality (34). The proof is complete.   □

Remark 11 In Theorem 5 if $\alpha = 1$, then we obtain the inequality

$$\int_0^\infty \omega(t)\left(\int_0^t g(s)ds \right)^P dt \geq p^P \int_0^\infty \omega^{1-P}(t)\left(\int_t^\infty \omega(s)ds \right)^P g^P(t)dt, \qquad (37)$$

Remark 12 In Theorem 5 if $\omega(t) = 1/t^P$ and $p > \alpha$, then we obtain the inequality

$$\int_0^\infty \left(\frac{\int_0^t g(s)d_\alpha s}{t} \right)^P d_\alpha t \geq \left(\frac{p}{p-\alpha} \right)^P \int_0^\infty \left(t^{\alpha-1}g(t) \right)^P d_\alpha t. \qquad (38)$$

which is the reversed Hardy inequality.

Remark 13 If $\alpha = 1$ in (38) we obtain the reversed Hardy inequality for $0 < p \leq 1$

$$\int_0^\infty \left(\frac{\int_0^t g(s)ds}{t} \right)^P dt \geq \left(\frac{p}{p-1} \right)^P \int_0^\infty g^P(t)dt. \qquad (39)$$

Theorem 6 *If* $0 < p \le 1$, *then*

$$\int_0^\infty \omega(t) T^{*^p} g(t) d_\alpha t \ge p^p \int_0^\infty \omega^{1-p}(t) \Lambda^p(t) g^p(t) d_\alpha t, \tag{40}$$

where

$$\Lambda(t) := \int_0^t \omega(s) d_\alpha s.$$

Proof Integrating the term $\int_0^\infty \omega(t) T^{*^p} g(t) d_\alpha t$ by parts formula (12) with

$$U(t) = T^{*^p} g(t), \text{ and } V(t) = \Lambda(t),$$

and we obtain

$$\int_0^\infty \omega(t) T^{*^p} g(t) d_\alpha t = T^{*^p} g(t) \Lambda(t) \Big|_0^\infty - \int_0^\infty p t^{1-\alpha} T^{*^{p-1}} g(t) T^{*'} g(t) \Lambda(t) d_\alpha t.$$

By using

$$T^* g(0) < \infty, \ T^* g(\infty) = 0, \ \Lambda(0) = 0 \text{ and } \Lambda(\infty) < \infty,$$

and from (15), we get that

$$T^{*'} g(t) = -t^{\alpha-1} g(t),$$

so we have

$$\int_0^\infty \omega(t) T^{*^p} g(t) d_\alpha t = p \int_0^\infty t^{1-\alpha} T^{*^{p-1}} g(t) t^{\alpha-1} g(t) \Lambda(t) d_\alpha t,$$

then

$$\int_0^\infty \omega(t) T^{*^p} g(t) d_\alpha t = p \int_0^\infty \left(\frac{\Lambda^p(t) g^p(t)}{T^{*p(1-p)} g(t)} \right)^{\frac{1}{p}} d_\alpha t,$$

which can be rewritten in the form

$$\left(\int_0^\infty \omega(t) T^{*^p} g(t) d_\alpha t \right)^p = p^p \left(\int_0^\infty \left(\frac{\Lambda^p(t) g^p(t)}{T^{*p(1-p)} g(t)} \right)^{\frac{1}{p}} d_\alpha t \right)^p. \tag{41}$$

Proceeding as in the proof of Theorem 5, we have that

$$\int_0^\infty \omega(t) T^{*^p} g(t) d_\alpha t \geq p^p \int_0^\infty \omega^{1-p}(t) \Lambda^p(t) g^p(t) d_\alpha t,$$

which is the desired inequality (40). The proof is complete. □

Remark 14 In Theorem 6, if $\alpha = 1$, then we obtain the inequality

$$\int_0^\infty \omega(t) \left(\int_t^\infty g(s) ds \right)^p dt \geq p^p \int_0^\infty \omega^{1-p}(t) \left(\int_0^t \omega(s) ds \right)^p g^p(t) dt. \tag{42}$$

Remark 15 In Theorem 6 if $\omega(t) = 1$ and $p \geq \alpha$, then we obtain the inequality

$$\int_0^\infty \left(\int_t^\infty g(s) d_\alpha s \right)^p d_\alpha t \geq \left(\frac{p}{\alpha} \right)^p \int_0^\infty \left(t^\alpha g(t) \right)^p d_\alpha t,$$

since $(p/\alpha)^p > 1$, we have

$$\int_0^\infty \left(\int_t^\infty g(s) d_\alpha s \right)^p d_\alpha t \geq \int_0^\infty \left(t^\alpha g(t) \right)^p d_\alpha t, \tag{43}$$

which is the reversed Copson, inequality.

Remark 16 If $\alpha = 1$ in (43), we obtain the reversed Copson inequality

$$\int_0^\infty \left(\int_t^\infty g(s) ds \right)^p dt \geq \int_0^\infty (t g(t))^p dt. \tag{44}$$

Theorem 7 *If $c \leq 0 < p < 1$, then*

$$\int_0^\infty \frac{\omega(t)}{\Lambda^c(t)} \Phi^p(t) d_\alpha t \geq \left(\frac{p}{1-c} \right)^p \int_0^\infty \frac{\omega(t)}{\Lambda^{c-p}(t)} g^p(t) d_\alpha t, \tag{45}$$

where

$$\Omega(t) := \int_t^\infty \omega(s) d_\alpha s \ and \ \Phi(t) := \int_0^t \omega(s) g(s) d_\alpha s.$$

Proof Integrate by parts the term

$$\int_0^\infty \frac{\omega(t)}{\Omega^c(t)} \Phi^p(t) d_\alpha t,$$

with

$$U(t) = \Phi^p(t)\Omega^{-c}(t), \text{ and } V(t) = -\Omega(t),$$

and using

$$D_\alpha U(t) = t^{1-\alpha}\left(p\Phi^{p-1}(t)\Phi'(t)\Omega^{-c}(t) - c\Phi^p(t)\Omega^{-c-1}(t)\Omega'(t)\right),$$

we obtain

$$\int_0^\infty \frac{\omega(t)}{\Omega^c(t)}\Phi^p(t)d_\alpha t = -\Omega^{1-c}(t)\Phi^p(t)\Big|_0^\infty$$

$$+ \int_0^\infty t^{1-\alpha}\left(p\Phi^{p-1}(t)\Phi'(t)\Omega^{-c}(t) - c\Phi^p(t)\Omega^{-c-1}(t)\Omega'(t)\right)\Omega(t)d_\alpha t.$$

By using

$$\Phi(0) = 0, \ \Phi(\infty) < \infty, \ \Omega(\infty) = 0 \text{ and } \Omega(0) < \infty,$$

and since $\Phi'(t) = t^{\alpha-1}\omega(t)g(t)$, and $\Omega'(t) = -t^{\alpha-1}\omega(t)$, we get that

$$\int_0^\infty \frac{\omega(t)}{\Omega^c(t)}\Phi^p(t)d_\alpha t = p\int_0^\infty \omega(t)g(t)\Omega^{1-c}(t)\Phi^{p-1}(t)d_\alpha t$$

$$+ c\int_0^\infty \frac{\Phi^p(t)\omega(t)}{\Omega^c(t)}d_\alpha t.$$

Then

$$\int_0^\infty \frac{\omega(t)}{\Omega^c(t)}\Phi^p(t)d_\alpha t = \frac{p}{1-c}\int_0^\infty \omega(t)g(t)\Omega^{1-c}(t)\Phi^{p-1}(t)d_\alpha t,$$

which can be rewritten in the form

$$\left(\int_0^\infty \frac{\omega(t)}{\Omega^c(t)}\Phi^p(t)d_\alpha t\right)^p = \left(\frac{p}{1-c}\right)^p\left(\int_0^\infty \left(\frac{(\omega(t)g(t))^p}{\Omega^{p(c-1)}(t)\Phi^{p(1-p)}(t)}\right)^{\frac{1}{p}}d_\alpha t\right)^p.$$

Applying Hölder's inequality

$$\int_0^\infty F(t)G(t)d_\alpha t \leq \left(\int_0^\infty F^u(t)d_\alpha t\right)^{\frac{1}{u}}\left(\int_0^\infty G^v(t)d_\alpha t\right)^{\frac{1}{v}},$$

with indices $u = 1/p$ and $v = 1/(1-p)$, where

$$F(t) = \frac{(\omega(t)g(t))^p}{\Omega^{p(c-1)}(t)\Phi^{p(1-p)}(t)} \text{ and } G(t) = \left(\frac{\omega(t)}{\Omega^c(t)}\right)^{1-p}\Phi^{p(1-p)}(t),$$

we get that

$$\left(\int_0^\infty F^{\frac{1}{p}}(t)d_\alpha t\right)^p = \left(\int_0^\infty \left(\frac{(\omega(t)g(t))^p}{\Omega^{p(c-1)}(t)\Phi^{p(1-p)}(t)}\right)^{\frac{1}{p}} d_\alpha t,\right)^p$$

$$\geq \frac{\int_0^\infty F(t)G(t)d_\alpha t}{\left(\int_0^\infty G^{\frac{1}{1-p}}(t)d_\alpha t\right)^{1-p}}$$

$$= \int_0^\infty \frac{(\omega(t)g(t))^p}{\Omega^{p(c-1)}(t)\Phi^{p(1-p)}(t)} \left(\frac{\omega(t)}{\Omega^c(t)}\right)^{1-p} \Phi^{p(1-p)}(t)d_\alpha t$$

$$\times \left(\int_0^\infty \left(\left(\frac{\omega(t)}{\Omega^c(t)}\right)^{1-p} \Phi^{p(1-p)}(t)\right)^{\frac{1}{1-p}} d_\alpha t\right)^{p-1},$$

so

$$\left(\int_0^\infty \left(\frac{(\omega(t)g(t))^p}{\Omega^{p(c-1)}(t)\Phi^{p(1-p)}(t)}\right)^{\frac{1}{p}} d_\alpha t\right)^p \geq \left(\int_0^\infty \frac{\omega(t)g^p(t)}{\Omega^{c-p}(t)}d_\alpha t\right)$$

$$\times \left(\int_0^\infty \left(\frac{\omega(t)}{\Omega^c(t)}\right) F^p(t)d_\alpha t\right)^{p-1},$$

then

$$\left(\int_0^\infty \frac{\omega(t)}{\Omega^c(t)}\Phi^p(t)d_\alpha t\right)^p \geq \left(\frac{p}{1-c}\right)^p \left(\int_0^\infty \frac{\omega(t)g^p(t)}{\Omega^{c-p}(t)}d_\alpha t\right)$$

$$\times \left(\int_0^\infty \left(\frac{\omega(t)}{\Omega^c(t)}\right) F^p(t)d_\alpha t\right)^{p-1}.$$

Hence

$$\int_0^\infty \frac{\omega(t)}{\Omega^c(t)}\Phi^p(t)d_\alpha t \geq \left(\frac{p}{1-c}\right)^p \int_0^\infty \frac{\omega(t)g^p(t)}{\Omega^{c-p}(t)}d_\alpha t,$$

which is the desired inequality (45). The proof is complete. $\square$

Remark 17 In Theorem 7 if $\alpha = 1$, then we obtain the inequality

$$\int_0^\infty \frac{\omega(t)}{\Omega^c(t)} \left(\int_0^t \omega(s)g(s)ds\right)^p dt \geq \left(\frac{p}{1-c}\right)^p \int_0^\infty \frac{\omega(t)}{\Omega^{c-p}(t)}g^p(t)dt, \qquad (46)$$

where $\Omega(t) = \int_t^\infty \omega(s)ds$ and $c \leq 0 < p < 1$.

Theorem 8 *If $0 < p < 1 < c$ and $p < c - 1$, then*

$$\int_0^\infty \frac{\omega(t)}{\Lambda^c(t)} \Psi^p(t) d_\alpha t \geq \left(\frac{p}{c-1}\right)^p \int_0^\infty \frac{\omega(t)}{\Lambda^{c-p}(t)} g^p(t) d_\alpha t, \tag{47}$$

where

$$\Omega(t) := \int_t^\infty \omega(s) d_\alpha s \text{ and } \Psi(t) := \int_t^\infty \omega(s) g(s) d_\alpha s.$$

Proof Integrating the term

$$\int_0^\infty \frac{\omega(t)}{\Omega^c(t)} \Psi^p(t) d_\alpha t,$$

by parts formula (12) with

$$U(t) = \Psi^p(t) \text{ and } V(t) = -\frac{\Omega^{1-c}(t)}{1-c},$$

and

$$D_\alpha U(t) = pt^{1-\alpha} \Psi^{p-1}(t) \Psi'(t) \text{ and } D_\alpha V(t) = \omega(t) \Omega^{-c}(t),$$

we obtain

$$\int_0^\infty \frac{\omega(t)}{\Omega^c(t)} \Psi^p(t) d_\alpha t = -\frac{\Psi^p(t) \Omega^{1-c}(t)}{1-c} \Big|_0^\infty$$

$$+ \int_0^\infty \frac{pt^{1-\alpha} \Psi^{p-1}(t) \Psi'(t) \Omega^{1-c}(t)}{1-c} d_\alpha t.$$

By using $\Psi'(t) = -t^{\alpha-1} \omega(t) g(t)$,

$$\Psi(0) < \infty, \quad \Psi(\infty) = 0, \quad \Omega(0) < \infty, \quad \Omega(\infty) = 0 \text{ and } c > 1,$$

and noting that

$$\lim_{t \to \infty} \Omega(t) \Psi^{\frac{p}{1-c}}(t) = \lim_{t \to \infty} \frac{\int_t^\infty s^{\alpha-1} \omega(s) \, ds}{(\Psi(t))^{\frac{p}{c-1}}}$$

$$= \lim_{t \to \infty} \frac{-t^{\alpha-1} \omega(t)}{-\frac{p}{c-1} (\Psi(t))^{\frac{p}{c-1}-1} t^{\alpha-1} \omega(t) g(t)}$$

$$= \lim_{t \to \infty} \frac{(c-1)\left(\Psi(t)\right)^{1-\frac{p}{c-1}}}{pg(t)} = 0.$$

we get that

$$\int_0^\infty \frac{\omega(t)}{\Omega^c(t)} \Psi^p(t) d_\alpha t \le \frac{p}{c-1} \int_0^\infty t^{1-\alpha} \Psi^{p-1}(t) t^{\alpha-1} \omega(t) g(t) \Omega^{1-c}(t) d_\alpha t.$$

Then

$$\int_0^\infty \frac{\omega(t)}{\Omega^c(t)} \Psi^p(t) d_\alpha t \le \frac{p}{c-1} \int_0^\infty \omega(t) g(t) \Omega^{1-c}(t) \Psi^{p-1}(t) d_\alpha t,$$

which can be rewritten in the form

$$\left(\int_0^\infty \frac{\omega(t)}{\Omega^c(t)} \Psi^p(t) d_\alpha t\right)^p = \left(\frac{p}{c-1}\right)^p \left(\int_0^\infty \left(\frac{(\omega(t)g(t))^p}{\Omega^{p(c-1)}(t)\Psi^{p(1-p)}(t)}\right)^{\frac{1}{p}} d_\alpha t\right)^p.$$

Applying Hölder's inequality

$$\int_0^\infty F(t)G(t) d_\alpha t \le \left(\int_0^\infty F^u(t) d_\alpha t\right)^{\frac{1}{u}} \left(\int_0^\infty G^v(t) d_\alpha t\right)^{\frac{1}{v}}$$

with indices $u = 1/p$ and $v = 1/(1-p)$, where

$$F(t) = \frac{(\omega(t)g(t))^p}{\Omega^{p(c-1)}(t)\Psi^{p(1-p)}(t)} \quad \text{and} \quad G(t) = \left(\frac{\omega(t)}{\Omega^c(t)}\right)^{1-p} \Psi^{p(1-p)}(t),$$

we get that

$$\left(\int_0^\infty F^{\frac{1}{p}}(t) d_\alpha t\right)^p = \left(\int_0^\infty \left(\frac{(\omega(t)g(t))^p}{\Omega^{p(c-1)}(t)\Psi^{p(1-p)}(t)}\right)^{\frac{1}{p}} d_\alpha t,\right)^p$$

$$\ge \frac{\int_0^\infty F(t)G(t) d_\alpha t}{\left(\int_0^\infty G^{\frac{1}{1-p}}(t) d_\alpha t\right)^{1-p}}$$

$$= \int_0^\infty \frac{(\omega(t)g(t))^p}{\Omega^{p(c-1)}(t)\Psi^{p(1-p)}(t)} \left(\frac{\omega(t)}{\Omega^c(t)}\right)^{1-p} \Psi^{p(1-p)}(t) d_\alpha t$$

$$\times \left(\int_0^\infty \left(\left(\frac{\omega(t)}{\Omega^c(t)}\right)^{1-p} \Psi^{p(1-p)}(t)\right)^{\frac{1}{1-p}} d_\alpha t\right)^{p-1},$$

this implies that

$$\left(\int_0^\infty \left(\frac{(\omega(t)g(t))^p}{\Omega^{p(c-1)}(t)\Psi^{p(1-p)}(t)}\right)^{\frac{1}{p}} d_\alpha t,\right)^p \geq \left(\int_0^\infty \frac{\omega(t)g^p(t)}{\Omega^{c-p}(t)} d_\alpha t\right)$$
$$\times \left(\int_0^\infty \left(\frac{\omega(t)}{\Omega^c(t)}\right) F^p(t) d_\alpha t\right)^{p-1},$$

since we have

$$\left(\int_0^\infty \frac{\omega(t)}{\Omega^c(t)}\Psi^p(t) d_\alpha t\right)^p = \left(\frac{p}{c-1}\right)^p \left(\int_0^\infty \left(\frac{(\omega(t)g(t))^p}{\Omega^{p(c-1)}(t)\Psi^{p(1-p)}(t)}\right)^{\frac{1}{p}} d_\alpha t,\right)^p,$$

then

$$\left(\int_0^\infty \frac{\omega(t)}{\Omega^c(t)}\Psi^p(t) d_\alpha t\right)^p \geq \left(\frac{p}{c-1}\right)^p \left(\int_0^\infty \frac{\omega(t)g^p(t)}{\Omega^{c-p}(t)} d_\alpha t\right)$$
$$\times \left(\int_0^\infty \left(\frac{\omega(t)}{\Omega^c(t)}\right) F^p(t) d_\alpha t\right)^{p-1}.$$

Hence

$$\int_0^\infty \frac{\omega(t)}{\Omega^c(t)}\Psi^p(t) d_\alpha t \geq \left(\frac{p}{c-1}\right)^p \int_0^\infty \frac{\omega(t)g^p(t)}{\Omega^{c-p}(t)} d_\alpha t,$$

which is the desired inequality (47). The proof is complete. □

Remark 18 From the proof of Theorem 8 we see that if $t^{\alpha-1}\omega(t)g(t)$ and $t^{\alpha-1}\omega(t)$ is continuous on $[0, \infty)$ replaced either by

(i) $t^{\alpha-1}\omega(t)g(t)$, $t^{\alpha-1}\omega(t)$ is continuous on $(0, \infty)$ and $\lim_{t\to\infty} \frac{(\Psi(t))^{1-\frac{p}{c-1}}}{g(t)} = 0$,
   or
(ii) $\lim_{t\to\infty} \Omega^{1-c}(t)\Psi^p(t) = 0$,
   then (47) is again true.

Remark 19 In Theorem 8, if $\alpha = 1$, then we obtain the inequality

$$\int_0^\infty \frac{\omega(t)}{\Omega^c(t)} \left(\int_t^\infty \omega(s)g(s)ds\right)^p dt \geq \left(\frac{p}{c-1}\right)^p \int_0^\infty \frac{\omega(t)}{\Omega^{c-p}(t)} g^p(t)dt, \quad (48)$$

where $\Omega(t) = \int_t^\infty \omega(s)ds$ and $0 < p < 1 < c$.

References

1. T. Abdeljawad, On conformable fractional calculus. J. Comput. Appl. Math. **279**, 57–66 (2015)
2. A. Akkurt, M.E. Yildirim, H. Yildirim, On some integral inequalities for conformable fractional integrals. RGMIA Res. Rep. Collection **19**, 107, 8pp. (2016)
3. M.U. Awan, M.A. Noor, T.S. Du, K.I. Noor, New refinements of fractional Hermite–Hadamard inequality. Revista de la Real Academia de Ciencias Exactas, Físicas y Naturales. Serie A. Matemáticas. https://doi.org/10.1007/s13398-017-0448-x
4. K. Bogdan, B. Dyda, The best constant in a fractional Hardy inequality. Math. Nach. **284**, 629–638 (2011)
5. E.T. Copson, Note on series of positive terms. J. Lond. Math. Soc. **3**, 49–51 (1928)
6. Y.M. Chu, M.A. Khan, T. Ali, S.S. Dragomir, Inequalities for α-fractional differentiable functions. J. Inequal Appl. **1**, 93 (2017)
7. G.H. Hardy, J.E. Littlewood, G. Polya, *Inequalities*, 2nd edn. (Cambridge University Press, Cambridge, 1952)
8. M. Jleli, B. Samet, Lyapunov-type inequalities for a fractional differential equation with mixed boundary conditions. Math. Inequal. Appl. **18**, 443–451 (2015)
9. R. Khalil, M. Al Horani, A. Yousef, M. Sababheh, A new definition of fractional derivative. J. Comput. Appl. Math. **264**, 65–70 (2014)
10. M.A. Khan, T. Ali, S.S. Dragomir, M.Z. Sarikaya, Hermite-Hadamard type inequalities for conformable fractional integrals. Revista de la Real Academia de Ciencias Exactas, Físicas y Naturales. Serie A. Matemáticas **112**(4), 1033–1048
11. L. Leindler, Generalization of inequalities of Hardy and Littlewood. Acta Sci. Math. **31**, 297–285 (1970)
12. L. Leindler, Further sharpening of inequalities of Hardy and Littlewood. Acta Sci. Math. **54**, 285–289 (1990)
13. L. Leindler, Some inequalities pertaining to Bennett's results. Acta Sci. Math. **58**, 261–279 (1993)
14. M.Z. Sarikaya, H. Budak, Opial type inequalities for conformable fractional integrals. RGMIA Res. Rep. Collection **19**, 93, 11pp. (2016)
15. M.Z. Sarikaya, H. Budak, New inequalities of Opial type for conformable fractional integrals. Turkish J. Math. **41**(5), 1164–1173 (2017)
16. M.Z. Sarikaya, H. Yaldiz, H. Budak, Steffensen's integral inequality for conformable fractional integrals. Int. J. Anal. Appl. **15**(1), 23–30 (2017)
17. E. Set, A. Gözpınar, A. Ekinci, Hermite- Hadamard type inequalities via conformable fractional integrals. Acta Math. Univ. Comenian **86**(2), 309–320 (2017)
18. C. Yildiz, M.E. Ozdemir, H.K. Onelan, Fractional integral inequalities for different functions. New Trends Math. Sci. **3**, 110–117 (2015)

Inequalities for Symmetrized or Anti-Symmetrized Inner Products of Complex-Valued Functions Defined on an Interval

Silvestru Sever Dragomir

Abstract For a function $f : [a, b] \to \mathbb{C}$ we consider the *symmetrical transform of* f on the interval $[a, b]$, denoted by $\check{f}$, and defined by

$$\check{f}(t) := \frac{1}{2}[f(t) + f(a + b - t)], \ t \in [a, b]$$

and the *anti-symmetrical transform of* f on the interval $[a, b]$ denoted by $\tilde{f}$ and defined by

$$\tilde{f} := \frac{1}{2}[f(t) - f(a + b - t)], t \in [a, b].$$

We consider in this paper the inner products

$$\langle f, g \rangle_{\smile} := \int_a^b \check{f}(t)\overline{\check{g}(t)}dt \text{ and } \langle f, g \rangle_{\sim} := \int_a^b \tilde{f}(t)\overline{\tilde{g}(t)}dt,$$

the corresponding norms and establish their fundamental properties. Some Schwarz and Grüss' type inequalities are also provided.

1 Introduction

For a function $f : [a, b] \to \mathbb{C}$ we consider the *symmetrical transform of* f on the interval $[a, b]$, denoted by $\check{f}_{[a,b]}$ or simply $\check{f}$, when the interval $[a, b]$ is implicit, as defined by

---

S. S. Dragomir (✉)
Mathematics, College of Engineering & Science Victoria University, Melbourne City, MC, Australia

DST-NRF Centre of Excellence in the Mathematical and Statistical Sciences, School of Computer Science & Applied Mathematics, University of the Witwatersrand, Johannesburg, South Africa
e-mail: sever.dragomir@vu.edu.au

© Springer Nature Switzerland AG 2019
G. A. Anastassiou and J. M. Rassias (eds.), *Frontiers in Functional Equations and Analytic Inequalities*, https://doi.org/10.1007/978-3-030-28950-8_26

$$\check{f}(t) := \frac{1}{2}[f(t) + f(a+b-t)], \ t \in [a, b]. \tag{1}$$

The *anti-symmetrical transform of* f on the interval $[a, b]$ is denoted by $\tilde{f}_{[a,b]}$, or simply $\tilde{f}$ and is defined by

$$\tilde{f} := \frac{1}{2}[f(t) - f(a+b-t)], \ t \in [a, b].$$

It is obvious that for any function f we have $\check{f} + \tilde{f} = f$. We observe that the symmetrical and anti-symmetrical transforms are *linear transforms*, namely

$$(\alpha f + \beta g)^{\vee} = \alpha \check{f} + \beta \check{g}$$

and

$$(\alpha f + \beta g)^{\sim} = \alpha \tilde{f} + \beta \tilde{g}$$

for any functions f, g and any scalars α, $\beta \in \mathbb{C}$.

We say that the function is *symmetrical a.e.* on the interval $[a, b]$ if

$$f(t) = f(a+b-t) \text{ for almost every } t \in [a, b]$$

and *anti-symmetrical a.e.* on the interval $[a, b]$ if

$$f(t) = -f(a+b-t) \text{ for almost every } t \in [a, b].$$

We observe that if the function is (Lebesgue) integrable on $[a, b]$, then by the change of variable $s = a + b - t$, $t \in [a, b]$ we have

$$\int_a^b \check{f}(t)\, dt = \frac{1}{2}\left[\int_a^b f(t)\, dt + \int_a^b f(a+b-s)\, ds\right] = \int_a^b f(t)\, dt$$

and

$$\int_a^b \tilde{f}(t)\, dt = \frac{1}{2}\left[\int_a^b f(t)\, dt - \int_a^b f(a+b-s)\, ds\right] = 0.$$

Assume that all functions below are measurable and the integrals involved are finite, then by considering the functionals

$$\langle f, g \rangle_{\smile} := \int_a^b \check{f}(t)\, \overline{\check{g}(t)}\, dt \text{ and } \langle f, g \rangle_{\sim} := \int_a^b \tilde{f}(t)\, \overline{\tilde{g}(t)}\, dt$$

we have

$$\langle \alpha f + \beta h, g \rangle_{\smile} = \alpha \langle f, g \rangle_{\smile} + \beta \langle h, g \rangle_{\smile}, \quad \langle g, f \rangle_{\smile} = \overline{\langle f, g \rangle_{\smile}}$$

for any scalars α, β and

$$\langle f, f \rangle_{\smile} \geq 0,$$

and the similar relations for the functional $\langle \cdot, \cdot \rangle_{\sim}$.

These show that the functionals $\langle \cdot, \cdot \rangle_{\smile}$ and $\langle \cdot, \cdot \rangle_{\sim}$ are nonnegative Hermitian forms. We also observe that if $\check{f} \in L_2[a, b]$, the Hilbert space of Lebesgue square-integrable functions on $[a, b]$ and $\langle f, f \rangle_{\smile} = 0$, then f must be *anti-symmetrical a.e.* on the interval $[a, b]$. Also, if $\check{f} \in L_2[a, b]$ and $\langle f, f \rangle_{\sim} = 0$, then f must be *symmetrical a.e.* on the interval $[a, b]$.

We can define the equivalence relation " $\smile$ " by $f \smile g \Leftrightarrow f - g$ is *anti-symmetrical a.e.* on the interval $[a, b]$. Similarly, we have the equivalence relation " $\sim$ " by $f \sim g \Leftrightarrow f - g$ is *symmetrical a.e.* on the interval $[a, b]$.

We define the linear space of measurable functions $L_2^{\smile}[a, b]$ as the collections of all " $\smile$ "-classes of measurable functions for which $\int_a^b \left| \check{f}(t) \right|^2 dt < \infty$, and in a similar way the space $L_2^{\sim}[a, b]$. In this situation $\langle \cdot, \cdot \rangle_{\smile}$ becomes a proper inner product on $L_2^{\smile}[a, b]$ and $\langle \cdot, \cdot \rangle_{\sim}$ a proper inner product on $L_2^{\sim}[a, b]$. Therefore $\|\cdot\|_{\smile} := \langle \cdot, \cdot \rangle_{\smile}^{1/2}$ and $\|\cdot\|_{\sim} := \langle \cdot, \cdot \rangle_{\sim}^{1/2}$ are norms on $L_2^{\smile}[a, b]$ and $L_2^{\sim}[a, b]$, respectively.

In what follows we establish some fundamental properties for these inner products. Some Schwarz and Grüss' type inequalities are also provided. For recent results in connection to Grüss' inequality, see [1–12, 14–18, 20–27] and the references therein.

2 Some Fundamental Properties

We have

Theorem 1 *If $f, g \in L_2[a, b]$ then $f, g \in L_2^{\smile}[a, b]$, we have the representations*

$$\langle f, g \rangle_{\smile} = \frac{1}{2} \left[\int_a^b f(t) \overline{g(t)} + \int_a^b f(a + b - t) \overline{g(t)} dt \right] \tag{2}$$

$$= \frac{1}{2} \left[\int_a^b f(t) \overline{g(t)} + \int_a^b f(t) \overline{g(a + b - t)} dt \right]$$

$$= \int_a^b f(t) \overline{\check{g}(t)} dt = \int_a^b \check{f}(t) \overline{g(t)} dt,$$

$$\|f\|_{\smile}^2 = \frac{1}{2}\left[\int_a^b |f(t)|^2 + \int_a^b f(t)\,\overline{f(a+b-t)}dt\right] \tag{3}$$

$$= \int_a^b f(t)\,\overline{\check{f}(t)}dt = \int_a^b \check{f}(t)\,\overline{f(t)}dt$$

and the inequalities

$$\left(\frac{1}{b-a}\int_a^b |f(t)|\,dt\right)^2 \tag{4}$$

$$\leq \frac{1}{2}\left[\frac{1}{b-a}\int_a^b |f(t)|^2 + \frac{1}{b-a}\int_a^b f(t)\,\overline{f(a+b-t)}dt\right]$$

$$\leq \frac{1}{b-a}\int_a^b |f(t)|^2,$$

$$\left|\int_a^b f(t)\,\overline{g(t)} + \int_a^b f(a+b-t)\,\overline{g(t)}dt\right|^2 \tag{5}$$

$$\leq \left[\int_a^b |f(t)|^2 + \int_a^b f(t)\,\overline{f(a+b-t)}dt\right]$$

$$\times \left[\int_a^b |g(t)|^2 + \int_a^b g(t)\,\overline{g(a+b-t)}dt\right].$$

Proof We have by the definition of $\langle \cdot, \cdot \rangle_{\smile}$ that

$$\langle f, g \rangle_{\smile} = \frac{1}{4}\int_a^b [f(t) + f(a+b-t)]\overline{[g(t) + g(a+b-t)]}dt \tag{6}$$

$$= \frac{1}{4}\int_a^b \Big[f(t)\,\overline{g(t)} + f(a+b-t)\,\overline{g(t)}$$

$$+ f(t)\,\overline{g(a+b-t)} + f(a+b-t)\,\overline{g(a+b-t)}\Big]dt$$

$$= \frac{1}{4}\left[\int_a^b f(t)\,\overline{g(t)}dt + \int_a^b f(a+b-t)\,\overline{g(t)}dt\right.$$

$$\left. + \int_a^b f(t)\,\overline{g(a+b-t)}dt + \int_a^b f(a+b-t)\,\overline{g(a+b-t)}dt\right],$$

for any $f, g \in L_2[a, b]$.

Using the change of variable $s = a + b - t$, $t \in [a, b]$, we have

$$\int_a^b f(t) \overline{g(a+b-t)} dt = \int_a^b f(a+b-t) \overline{g(t)} dt$$

and

$$\int_a^b f(a+b-t) \overline{g(a+b-t)} dt = \int_a^b f(t) \overline{g(t)} dt$$

and by (6) we get the first equality in (2). The rest is obvious.

The equality (3) follows by (2) for $g = f$. Also, from (3) we observe that $\int_a^b f(t) \overline{f(a+b-t)} dt$ is a real number for any $f \in L_2[a, b]$.

If $f \in L_2[a, b]$, then by Cauchy–Bunyakovsky–Schwarz inequality we have

$$\|f\|_{\smile}^2 = \frac{1}{4} \int_a^b |f(t) + f(a+b-t)|^2 \, dt$$

$$\geq \frac{1}{4(b-a)} \left| \int_a^b [f(t) + f(a+b-t)] \, dt \right|^2$$

$$= \frac{1}{4(b-a)} \left| \int_a^b f(t) \, dt + \int_a^b f(a+b-t) \, dt \right|^2$$

$$= \frac{1}{b-a} \left| \int_a^b f(t) \, dt \right|^2,$$

which proves the first inequality in (4).

If $f \in L_2[a, b]$, then by Cauchy–Bunyakovsky–Schwarz inequality we also have

$$\|f\|_{\smile}^2 = \int_a^b \check{f}(t) \overline{f(t)} dt \leq \left(\int_a^b \left| \check{f}(t) \right|^2 dt \right)^{1/2} \left(\int_a^b |f(t)|^2 \, dt \right)^{1/2}$$

$$= \|f\|_{\smile} \|f\|_2,$$

which implies that $\|f\|_{\smile} \leq \|f\|_2$ that is equivalent to the second inequality in (4).

By the Schwarz inequality for the inner product $\langle \cdot, \cdot \rangle_{\smile}$, namely

$$|\langle f, g \rangle_{\smile}|^2 \leq \|f\|_{\smile}^2 \|g\|_{\smile}^2,$$

and by employing (2) and (3) we obtain the desired result (5).

We have the corresponding result for $L_2^{\sim}[a, b]$.

Theorem 2 *If* $f, g \in L_2[a, b]$ *then* $f, g \in L_2^{\sim}[a, b]$, *we have the representations*

$$\langle f, g \rangle_{\sim} = \frac{1}{2} \left[\int_a^b f(t) \overline{g(t)} - \int_a^b f(a+b-t) \overline{g(t)} dt \right] \tag{7}$$

$$= \frac{1}{2} \left[\int_a^b f(t) \overline{g(t)} - \int_a^b f(t) \overline{g(a+b-t)} dt \right]$$

$$= \int_a^b f(t) \overline{\tilde{g}(t)} dt = \int_a^b \tilde{f}(t) \overline{g(t)} dt,$$

$$\|f\|_{\sim}^2 = \frac{1}{2} \left[\int_a^b |f(t)|^2 - \int_a^b f(t) \overline{f(a+b-t)} dt \right] \tag{8}$$

$$= \int_a^b f(t) \overline{\tilde{f}(t)} dt = \int_a^b \tilde{f}(t) \overline{f(t)} dt$$

and the inequalities

$$0 \leq \frac{1}{2} \left[\frac{1}{b-a} \int_a^b |f(t)|^2 - \frac{1}{b-a} \int_a^b f(t) \overline{f(a+b-t)} dt \right] \tag{9}$$

$$\leq \frac{1}{b-a} \int_a^b |f(t)|^2,$$

$$\left| \int_a^b f(t) \overline{g(t)} - \int_a^b f(a+b-t) \overline{g(t)} dt \right|^2 \tag{10}$$

$$\leq \left[\int_a^b |f(t)|^2 - \int_a^b f(t) \overline{f(a+b-t)} dt \right]$$

$$\times \left[\int_a^b |g(t)|^2 - \int_a^b g(t) \overline{g(a+b-t)} dt \right].$$

Proof If $f \in L_2[a, b]$ then $f \in L_2^{\sim}[a, b]$, we have the representations

$$\langle f, g \rangle_{\sim} = \frac{1}{4} \int_a^b [f(t) - f(a+b-t)] \overline{[g(t) - g(a+b-t)]} dt$$

$$= \frac{1}{4} \int_a^b \left[f(t) \overline{g(t)} - f(a+b-t) \overline{g(t)} \right.$$

$$\left. - f(t) \overline{g(a+b-t)} + f(a+b-t) \overline{g(a+b-t)} \right] dt$$

$$= \frac{1}{4} \left[\int_a^b f(t) \overline{g(t)} dt - \int_a^b f(a+b-t) \overline{g(t)} dt \right.$$

$$\left. - \int_a^b f(t) \overline{g(a+b-t)} dt + \int_a^b f(a+b-t) \overline{g(a+b-t)} dt \right]$$

$$= \frac{1}{2} \left[\int_a^b f(t) \overline{g(t)} - \int_a^b f(a+b-t) \overline{g(t)} dt \right]$$

for any $f, g \in L_2[a, b]$.

The rest of the equality (7) and (8) follow from this equality.

As above, we observe that the integral $\int_a^b f(t) \overline{f(a+b-t)} dt$ is a real number for any $f \in L_2[a, b]$.

By Cauchy–Bunyakovsky–Schwarz integral inequality we have for $f \in L_2[a, b]$ that

$$\left| \int_a^b f(t) \overline{f(a+b-t)} dt \right| \leq \left(\int_a^b |f(t)|^2 dt \right)^{1/2} \left(\int_a^b \left| \overline{f(a+b-t)} \right|^2 dt \right)^{1/2}$$

$$= \int_a^b |f(t)|^2 dt,$$

namely, since $\int_a^b f(t) \overline{f(a+b-t)} dt$ is real,

$$- \int_a^b |f(t)|^2 dt \leq \int_a^b f(t) \overline{f(a+b-t)} dt \leq \int_a^b |f(t)|^2 dt,$$

which is equivalent to (9).

By the Schwarz inequality for the inner product $\langle \cdot, \cdot \rangle_\sim$, namely

$$\left| \langle f, g \rangle_\sim \right|^2 \leq \|f\|_\sim^2 \|g\|_\sim^2,$$

for any $f, g \in L_2[a, b]$ and the equalities (7) and (8) we get the desired result (10).

3 Inequalities for Bounded Functions

Now, for $\phi, \Phi \in \mathbb{C}$ and $[a, b]$ an interval of real numbers, define the sets of complex-valued functions (see for instance [19])

$$\bar{U}_{[a,b]}(\phi, \Phi)$$

$$:= \left\{ g : [a, b] \to \mathbb{C} \,\middle|\, \mathrm{Re} \left[(\Phi - g(t)) \left(\overline{g(t)} - \bar{\phi} \right) \right] \geq 0 \text{ for almost every } t \in [a, b] \right\}$$

and

$$\bar{\Delta}_{[a,b]}(\phi, \Phi) := \left\{ g : [a, b] \to \mathbb{C} \Big| \left| g(t) - \frac{\phi + \Phi}{2} \right| \le \frac{1}{2} |\Phi - \phi| \text{ for a.e. } t \in [a, b] \right\}.$$

The following representation result may be stated.

Proposition 1 *For any* ϕ, $\Phi \in \mathbb{C}$, $\phi \ne \Phi$, *we have that* $\bar{U}_{[a,b]}(\phi, \Phi)$ *and* $\bar{\Delta}_{[a,b]}(\phi, \Phi)$ *are nonempty, convex, and closed sets and*

$$\bar{U}_{[a,b]}(\phi, \Phi) = \bar{\Delta}_{[a,b]}(\phi, \Phi). \tag{11}$$

Proof We observe that for any $z \in \mathbb{C}$ we have the equivalence

$$\left| z - \frac{\phi + \Phi}{2} \right| \le \frac{1}{2} |\Phi - \phi|$$

if and only if

$$\operatorname{Re}\left[(\Phi - z)(\bar{z} - \bar{\phi}) \right] \ge 0.$$

This follows by the equality

$$\frac{1}{4} |\Phi - \phi|^2 - \left| z - \frac{\phi + \Phi}{2} \right|^2 = \operatorname{Re}\left[(\Phi - z)(\bar{z} - \bar{\phi}) \right]$$

that holds for any $z \in \mathbb{C}$.

The equality (11) is thus a simple consequence of this fact.

On making use of the complex numbers field properties we can also state that:

Corollary 1 *For any* ϕ, $\Phi \in \mathbb{C}$, $\phi \ne \Phi$, *we have that*

$$\bar{U}_{[a,b]}(\phi, \Phi) = \{ g : [a, b] \to \mathbb{C} \mid (\operatorname{Re}\Phi - \operatorname{Re}g(t))(\operatorname{Re}g(t) - \operatorname{Re}\phi) \tag{12}$$

$$+ (\operatorname{Im}\Phi - \operatorname{Im}g(t))(\operatorname{Im}g(t) - \operatorname{Im}\phi) \ge 0 \text{ for a.e. } t \in [a, b] \}.$$

Now, if we assume that $\operatorname{Re}(\Phi) \ge \operatorname{Re}(\phi)$ and $\operatorname{Im}(\Phi) \ge \operatorname{Im}(\phi)$, then we can define the following set of functions as well:

$$\bar{S}_{[a,b]}(\phi, \Phi) := \{ g : [a, b] \to \mathbb{C} \mid \operatorname{Re}(\Phi) \ge \operatorname{Re}g(t) \ge \operatorname{Re}(\phi) \tag{13}$$

$$\text{and } \operatorname{Im}(\Phi) \ge \operatorname{Im}g(t) \ge \operatorname{Im}(\phi) \text{ for a.e. } t \in [a, b] \}.$$

One can easily observe that $\bar{S}_{[a,b]}(\phi, \Phi)$ is closed, convex, and

$$\emptyset \ne \bar{S}_{[a,b]}(\phi, \Phi) \subseteq \bar{U}_{[a,b]}(\phi, \Phi). \tag{14}$$

We have the following Grüss' type inequalities:

Theorem 3 *Let $\phi, \Phi \in \mathbb{C}, \phi \neq \Phi$ and $f \in \bar{\Delta}_{[a,b]}(\phi, \Phi), g \in L_2[a,b]$. Then*

$$\left| \langle f, g \rangle_\smile - \frac{\phi + \Phi}{2} \int_a^b \overline{g(t)} dt \right| \leq \frac{1}{2} |\Phi - \phi| \int_a^b |\breve{g}(t)| \, dt \tag{15}$$

$$\leq \frac{1}{2} |\Phi - \phi| \int_a^b |g(t)| \, dt$$

and

$$\left| \langle f, g \rangle_\sim \right| \leq \frac{1}{2} |\Phi - \phi| \int_a^b |\breve{g}(t)| \, dt \leq \frac{1}{2} |\Phi - \phi| \int_a^b |g(t)| \, dt. \tag{16}$$

We also have

$$\left| \langle f, g \rangle_\smile - \frac{1}{b-a} \int_a^b \overline{g(s)} ds \int_a^b f(t) \, dt \right| \tag{17}$$

$$\leq \frac{1}{2} |\Phi - \phi| \int_a^b \left| \breve{g}(t) - \frac{1}{b-a} \int_a^b g(s) \, ds \right| dt$$

$$\leq \frac{1}{2} |\Phi - \phi| \int_a^b \left| g(t) - \frac{1}{b-a} \int_a^b g(s) \, ds \right| dt.$$

Proof We have by (2) that

$$\int_a^b \left(f(t) - \frac{\phi + \Phi}{2} \right) \overline{\breve{g}(t)} dt = \int_a^b f(t) \overline{\breve{g}(t)} dt - \frac{\phi + \Phi}{2} \int_a^b \overline{\breve{g}(t)} dt \tag{18}$$

$$= \langle f, g \rangle_\smile - \frac{\phi + \Phi}{2} \int_a^b \overline{g(t)} dt.$$

Taking the modulus in this equality, we have

$$\left| \langle f, g \rangle_\smile - \frac{\phi + \Phi}{2} \int_a^b \overline{g(t)} dt \right| \leq \int_a^b \left| f(t) - \frac{\phi + \Phi}{2} \right| |\breve{g}(t)| \, dt$$

$$\leq \frac{1}{2} |\Phi - \phi| \int_a^b |\breve{g}(t)| \, dt$$

$$= \frac{1}{4} |\Phi - \phi| \int_a^b |[g(t) + g(a+b-t)]| \, dt$$

$$\leq \frac{1}{4} |\Phi - \phi| \int_a^b [|g(t)| + |g(a+b-t)|] \, dt$$

$$= \frac{1}{2} |\Phi - \phi| \int_a^b |g(t)| \, dt$$

and the inequality (15) is proved.

We have by (7) that

$$\int_a^b \left(f(t) - \frac{\phi + \Phi}{2} \right) \overline{\breve{g}(t)} dt = \int_a^b f(t) \overline{\breve{g}(t)} dt - \frac{\phi + \Phi}{2} \int_a^b \overline{\breve{g}(t)} dt \qquad (19)$$

$$= \int_a^b f(t) \overline{\breve{g}(t)} dt = \langle f, g \rangle_\sim .$$

Taking the modulus in this equality we have

$$\left| \int_a^b f(t) \overline{\breve{g}(t)} dt \right| \le \int_a^b \left| f(t) - \frac{\phi + \Phi}{2} \right| |\breve{g}(t)| \, dt$$

$$\le \frac{1}{2} |\Phi - \phi| \int_a^b |\breve{g}(t)| \, dt$$

$$= \frac{1}{4} |\Phi - \phi| \int_a^b |[g(t) - g(a + b - t)]| \, dt$$

$$\le \frac{1}{4} |\Phi - \phi| \int_a^b [|g(t)| + |g(a + b - t)|] \, dt$$

$$= \frac{1}{2} |\Phi - \phi| \int_a^b |g(t)| \, dt$$

and the inequality (16) is obtained.
We also have

$$\int_a^b \left(f(t) - \frac{\phi + \Phi}{2} \right) \left(\overline{\breve{g}(t)} - \frac{1}{b - a} \int_a^b \overline{g(s)} ds \right) dt$$

$$= \int_a^b f(t) \left(\overline{\breve{g}(t)} - \frac{1}{b - a} \int_a^b \overline{g(s)} ds \right) dt$$

$$- \frac{\phi + \Phi}{2} \int_a^b \left(\overline{\breve{g}(t)} - \frac{1}{b - a} \int_a^b \overline{g(s)} ds \right) dt$$

$$= \int_a^b f(t) \overline{\breve{g}(t)} dt - \frac{1}{b - a} \int_a^b \overline{g(s)} ds \int_a^b f(t) \, dt$$

$$- \frac{\phi + \Phi}{2} \int_a^b \left(\overline{\breve{g}(t)} - \frac{1}{b - a} \int_a^b \overline{g(s)} ds \right) dt$$

$$= \int_a^b f(t) \overline{\breve{g}(t)} dt - \frac{1}{b - a} \int_a^b \overline{g(s)} ds \int_a^b f(t) \, dt$$

$$= \langle f, g \rangle_\sim - \frac{1}{b - a} \int_a^b \overline{g(s)} ds \int_a^b f(t) \, dt,$$

which gives, by taking the modulus,

$$\left| \langle f, g \rangle_\smile - \frac{1}{b-a} \int_a^b \overline{g(s)} ds \int_a^b f(t)\, dt \right|$$

$$\leq \int_a^b \left| f(t) - \frac{\phi + \Phi}{2} \right| \left| \overline{\check{g}(t)} - \frac{1}{b-a} \int_a^b \overline{g(s)} ds \right| dt$$

$$\leq \frac{1}{2} |\Phi - \phi| \int_a^b \left| \check{g}(t) - \frac{1}{b-a} \int_a^b g(s)\, ds \right| dt$$

$$= \frac{1}{2} |\Phi - \phi| \int_a^b \left| \frac{g(t) + g(a+b-t)}{2} - \frac{1}{b-a} \int_a^b g(s)\, ds \right| dt$$

$$\leq \frac{1}{2} |\Phi - \phi| \int_a^b \left| g(t) - \frac{1}{b-a} \int_a^b g(s)\, ds \right| dt$$

and the last inequality (17) is proved.

We have

Theorem 4 Let $\phi, \Phi \in \mathbb{C}$, $\phi \neq \Phi$ and $f \in \bar{\Delta}_{[a,b]}(\phi, \Phi)$. If $\check{\psi}, \check{\Psi} \in \mathbb{C}$, $\check{\psi} \neq \check{\Psi}$ and $\check{g} \in \bar{\Delta}_{[a,b]}\left(\check{\psi}, \check{\Psi} \right)$, then

$$\left| \langle f, g \rangle_\smile - \frac{\overline{\check{\psi}} + \overline{\check{\Psi}}}{2} \int_a^b f(t)\, dt - \frac{\phi + \Phi}{2} \int_a^b \overline{g(t)} dt \right.$$

$$\left. + \left(\frac{\phi + \Phi}{2} \right) \left(\frac{\overline{\check{\psi}} + \overline{\check{\Psi}}}{2} \right) (b-a) \right|$$

$$\leq \frac{1}{4} |\Phi - \phi| \left| \check{\Psi} - \check{\psi} \right| (b-a) \qquad (20)$$

and

$$\left| \langle f, g \rangle_\smile - \frac{1}{b-a} \int_a^b \overline{g(s)} ds \int_a^b f(t)\, dt \right| \qquad (21)$$

$$\leq \frac{1}{2} |\Phi - \phi| (b-a) \left(\frac{1}{b-a} \int_a^b |\check{g}(t)|^2 - \left| \frac{1}{b-a} \int_a^b g(t)\, dt \right|^2 \right)^{1/2}$$

$$\leq \frac{1}{4} |\Phi - \phi| \left| \check{\Psi} - \check{\psi} \right| (b-a).$$

If $\tilde{\psi}, \check{\Psi} \in \mathbb{C}, \tilde{\psi} \neq \check{\Psi}$ and $\check{g} \in \bar{\Delta}_{[a,b]} \left(\tilde{\psi}, \check{\Psi} \right)$, then

$$\left| \langle f, g \rangle_{\sim} - \frac{\overline{\tilde{\psi}} + \overline{\check{\Psi}}}{2} \int_a^b f(t)\, dt \right| \leq \frac{1}{4} |\Phi - \phi| \left| \check{\Psi} - \tilde{\psi} \right| (b - a). \tag{22}$$

Proof We have by (18) that

$$\int_a^b \left(f(t) - \frac{\phi + \Phi}{2} \right) \overline{\left(\check{g}(t) - \frac{\tilde{\psi} + \check{\psi}}{2} \right)} dt$$

$$= \int_a^b f(t) \overline{\left(\check{g}(t) - \frac{\tilde{\psi} + \check{\psi}}{2} \right)} dt - \frac{\phi + \Phi}{2} \int_a^b \overline{\left(\check{g}(t) - \frac{\tilde{\psi} + \check{\psi}}{2} \right)} dt$$

$$= \int_a^b f(t) \overline{\check{g}(t)} dt - \frac{\overline{\tilde{\psi}} + \overline{\check{\psi}}}{2} \int_a^b f(t)\, dt$$

$$\quad - \frac{\phi + \Phi}{2} \int_a^b \overline{\check{g}(t)} dt + \left(\frac{\phi + \Phi}{2} \right) \left(\frac{\overline{\tilde{\psi}} + \overline{\check{\psi}}}{2} \right)$$

$$= \langle f, g \rangle_{\smile} - \frac{\overline{\tilde{\psi}} + \overline{\check{\psi}}}{2} \int_a^b f(t)\, dt - \frac{\phi + \Phi}{2} \int_a^b \overline{g(t)} dt$$

$$\quad + \left(\frac{\phi + \Phi}{2} \right) \left(\frac{\overline{\tilde{\psi}} + \overline{\check{\psi}}}{2} \right) (b - a).$$

Taking the modulus in this equality, we have

$$\left| \langle f, g \rangle_{\smile} - \frac{\overline{\tilde{\psi}} + \overline{\check{\psi}}}{2} \int_a^b f(t)\, dt - \frac{\phi + \Phi}{2} \int_a^b \overline{g(t)} dt + \left(\frac{\phi + \Phi}{2} \right) \left(\frac{\overline{\tilde{\psi}} + \overline{\check{\psi}}}{2} \right) \right|$$

$$\leq \int_a^b \left| f(t) - \frac{\phi + \Phi}{2} \right| \left| \check{g}(t) - \frac{\tilde{\psi} + \check{\psi}}{2} \right| dt \leq \frac{1}{4} |\Phi - \phi| \left| \check{\Psi} - \tilde{\psi} \right| (b - a),$$

which proves (20).

By the Schwarz and Grüss' inequalities, see, for instance, [13], we have

$$\frac{1}{b - a} \int_a^b \left| \check{g}(t) - \frac{1}{b - a} \int_a^b g(s)\, ds \right| dt$$

$$= \frac{1}{b - a} \int_a^b \left| \check{g}(t) - \frac{1}{b - a} \int_a^b \check{g}(s)\, ds \right| dt$$

$$\leq \left(\frac{1}{b-a} \int_a^b \left| \check{g}(t) - \frac{1}{b-a} \int_a^b \check{g}(s)\, ds \right|^2 dt \right)^{1/2}$$

$$= \left(\frac{1}{b-a} \int_a^b |\check{g}(t)|^2 - \left| \frac{1}{b-a} \int_a^b g(t)\, dt \right|^2 \right)^{1/2} \leq \frac{1}{2} \left| \check{\Psi} - \check{\psi} \right|$$

and by (17) we get (21).

By (19) we have

$$\int_a^b \left(f(t) - \frac{\phi + \Phi}{2} \right) \overline{\left(\tilde{g}(t) - \frac{\tilde{\psi} + \tilde{\Psi}}{2} \right)} dt$$

$$= \int_a^b f(t) \overline{\left(\tilde{g}(t) - \frac{\tilde{\psi} + \tilde{\Psi}}{2} \right)} dt = \langle f, g \rangle_\sim - \frac{\overline{\tilde{\psi} + \tilde{\Psi}}}{2} \int_a^b f(t).$$

By taking the modulus in this equality, we have

$$\left| \langle f, g \rangle_\sim - \frac{\overline{\tilde{\psi} + \tilde{\Psi}}}{2} \int_a^b f(t)\, dt \right| \leq \int_a^b \left| f(t) - \frac{\phi + \Phi}{2} \right| \left| \tilde{g}(t) - \frac{\tilde{\psi} + \tilde{\Psi}}{2} \right| dt$$

$$\leq \frac{1}{4} |\Phi - \phi| \left| \tilde{\Psi} - \tilde{\psi} \right| (b-a).$$

and the inequality (22) is proved.

Remark 1 We observe that if $\phi, \Phi \in \mathbb{R}$, $\phi < \Phi$ and f is real-valued function, then $f \in \bar{\Delta}_{[a,b]}(\phi, \Phi)$ is equivalent to

$$\phi \leq f(t) \leq \Phi \text{ for a.e. } t \in [a, b].$$

If $\check{\psi}, \check{\Psi} \in \mathbb{R}$, $\check{\psi} < \check{\Psi}$ and g is real-valued function, then $\check{g} \in \bar{\Delta}_{[a,b]}\left(\check{\psi}, \check{\psi} \right)$ is equivalent to

$$\check{\psi} \leq \frac{1}{2} [g(t) + g(a + b - t)] \leq \check{\Psi} \text{ for a.e. } t \in [a, b]. \tag{23}$$

If ψ, Ψ are real numbers so that $\psi \leq g(t) \leq \Psi$ for a.e. $t \in [a, b]$, then

$$\psi \leq \frac{1}{2} [g(t) + g(a + b - t)] \leq \Psi \text{ for a.e. } t \in [a, b]. \tag{24}$$

One can find examples of functions for which the bounds provided by (23) are better than (24). For instance, if we consider the function $f : [a, b] \subset (0, \infty) \to \mathbb{R}$ given by $g(t) = \ln t$, then we have

$$\breve{g}(t) = \frac{1}{2} [\ln t + \ln (a + b - t)],$$

$$(\breve{g}(t))' = \frac{1}{2} \left(\frac{1}{t} - \frac{1}{a+b-t} \right) = \frac{\frac{a+b}{2} - t}{t(a+b-t)}, \ t \in (a, b)$$

and

$$(\breve{g}(t))'' = -\frac{1}{2} \left(\frac{1}{t^2} + \frac{1}{(a+b-t)^2} \right), \ t \in (a, b).$$

These show that $\breve{f}$ is strictly increasing on $\left(a, \frac{a+b}{2} \right)$, strictly decreasing on $\left(\frac{a+b}{2}, b \right)$, and strictly concave on (a, b). Therefore

$$\breve{\psi} := \ln G(a, b) \le \breve{g}(t) \le \ln A(a, b) =: \breve{\Psi} \ \text{for any} \ t \in [a, b], \tag{25}$$

where $G(a, b) := \sqrt{ab}$ is the *geometric mean* and $A(a, b) := \frac{1}{2}(a + b)$ is the *arithmetic mean* of positive numbers a, b.

Since $\psi := \ln a \le \ln t \le \ln b =: \Psi$, then by (24) we get

$$\psi \le \breve{g}(t) \le \Psi \ \text{for any} \ t \in [a, b]. \tag{26}$$

We observe that the bounds provided by (25) for $\breve{g}$ are better than (26).

4 The Case of One Function of Bounded Variation

For a function of bounded variation $f : [a, b] \to \mathbb{C}$ we denote by $\bigvee_a^b (f)$ its total variation on $[a, b]$.

Theorem 5 *Assume that* $f : [a, b] \to \mathbb{C}$ *is of bounded variation g is integrable on* $[a, b]$. *Then we have*

$$\left| \langle f, g \rangle_{\smile} - \frac{f(a) + f(b)}{2} \int_a^b \overline{g(t)} dt \right| \le \frac{1}{2} \bigvee_a^b (f) \int_a^b |\breve{g}(t)| \, dt \tag{27}$$

$$\le \frac{1}{2} \bigvee_a^b (f) \int_a^b |g(t)| \, dt$$

and

$$|\langle f, g \rangle_{\sim}| \le \frac{1}{2} \bigvee_a^b (f) \int_a^b |\tilde{g}(t)| \, dt \le \frac{1}{2} \bigvee_a^b (f) \int_a^b |g(t)| \, dt. \tag{28}$$

Proof We have by (2) that

$$\int_a^b \left(f(t) - \frac{f(a) + f(b)}{2} \right) \overline{\breve{g}(t)} dt = \int_a^b f(t) \overline{\breve{g}(t)} dt - \frac{f(a) + f(b)}{2} \int_a^b \overline{\breve{g}(t)} dt$$

$$= \langle f, g \rangle_\smile - \frac{f(a) + f(b)}{2} \int_a^b \overline{g(t)} dt.$$

Taking the modulus in this equality, we get

$$\left| \langle f, g \rangle_\smile - \frac{f(a) + f(b)}{2} \int_a^b \overline{g(t)} dt \right| \leq \int_a^b \left| f(t) - \frac{f(a) + f(b)}{2} \right| |\breve{g}(t)| \, dt.$$

$$(29)$$

Observe that, for any $t \in [a, b]$ we have

$$\left| f(t) - \frac{f(a) + f(b)}{2} \right| = \left| \frac{f(t) - f(a) + f(t) - f(b)}{2} \right|$$

$$\leq \frac{1}{2} [|f(t) - f(a)| + |f(b) - f(t)|] \leq \frac{1}{2} \bigvee_a^b (f)$$

and by (29) we get the first inequality in (27).

Since

$$\int_a^b |\breve{g}(t)| \, dt = \frac{1}{2} \int_a^b |g(t) + g(a + b - t)| \, dt$$

$$\leq \frac{1}{2} \int_a^b [|g(t)| + |g(a + b - t)|] \, dt = \int_a^b |g(t)| \, dt,$$

the last part of (27) also holds.

We have by (7) that

$$\int_a^b \left(f(t) - \frac{f(a) + f(b)}{2} \right) \overline{\tilde{g}(t)} dt = \int_a^b f(t) \overline{\tilde{g}(t)} dt - \frac{f(a) + f(b)}{2} \int_a^b \overline{\tilde{g}(t)} dt$$

$$= \int_a^b f(t) \overline{\tilde{g}(t)} dt = \langle f, g \rangle_\sim .$$

Taking the modulus in this equality, we get

$$|\langle f, g \rangle_\sim| \leq \int_a^b \left| f(t) - \frac{f(a) + f(b)}{2} \right| |\tilde{g}(t)| \, dt \leq \frac{1}{2} \bigvee_a^b (f) \int_a^b |\tilde{g}(t)| \, dt$$

$$= \frac{1}{4} \bigvee_a^b (f) \int_a^b |g(t) - g(a+b-t)| \, dt$$

$$\leq \frac{1}{4} \bigvee_a^b (f) \int_a^b [|g(t)| + |g(a+b-t)|] \, dt = \frac{1}{2} \bigvee_a^b (f) \int_a^b |g(t)| \, dt$$

and the inequality (28) is proved.

We say that the function $h : [a, b] \to \mathbb{R}$ is *H-r-Hölder continuous* with the constant $H > 0$ and power $r \in (0, 1]$ if

$$|h(t) - h(s)| \leq H |t - s|^r \tag{30}$$

for any $t, s \in [a, b]$. If $r = 1$ we call that h is *L-Lipschitzian* when $H = L > 0$.

Corollary 2 *Assume that $f : [a, b] \to \mathbb{C}$ is of bounded variation and g is H-r-Hölder continuous with the constant $H > 0$ and power $r \in (0, 1]$. Then*

$$|\langle f, g \rangle_\sim| \leq \frac{1}{4(r+1)} H \bigvee_a^b (f) (b-a)^{r+1} . \tag{31}$$

In particular, if L-Lipschitzian with $L > 0$, then

$$|\langle f, g \rangle_\sim| \leq \frac{1}{8} L \bigvee_a^b (f) (b-a)^2 . \tag{32}$$

Proof Since g is H-r-Hölder continuous with the constant $H > 0$ and power $r \in (0, 1]$, then

$$|\tilde{g}(t)| = \frac{1}{2} |g(t) - g(a+b-t)| \leq \frac{1}{2} H |2t - a - b|^r$$

$$= \frac{1}{2} 2^r H \left| t - \frac{a+b}{2} \right|^r = \frac{1}{2^{1-r}} H \left| t - \frac{a+b}{2} \right|^r ,$$

which implies that

$$\int_a^b |\tilde{g}(t)| \, dt \leq \frac{1}{2^{1-r}} H \int_a^b \left| t - \frac{a+b}{2} \right|^r \, dt = \frac{1}{2^{1-r}} H \frac{(b-a)^{r+1}}{2^r (r+1)}$$

$$= \frac{1}{2(r+1)} H (b-a)^{r+1}$$

and the inequality (31) is proved.

5 The Case of One Hölder Continuous Function

We say that the function $h : [a, b] \to \mathbb{C}$ is *K-p-Hölder continuous in the middle* with the constant $K > 0$ and power $p > 0$ if

$$\left| h(t) - h\left(\frac{a+b}{2}\right) \right| \le K \left| t - \frac{a+b}{2} \right|^p \tag{33}$$

for any $t \in [a, b]$. We observe that if $h : [a, b] \to \mathbb{C}$ is *H-r-Hölder continuous* with the constant $H > 0$ and power $r \in (0, 1]$, then it is Hölder continuous in the middle with the same constants H and r.

We define the following Lebesgue norms for a measurable function $h : [a, b] \to \mathbb{C}$

$$\|h\|_\infty := \operatorname*{essup}_{t \in [a,b]} |h(t)| < \infty \text{ if } h \in L_\infty [a, b]$$

and, for $\beta \ge 1$,

$$\|h\|_\beta := \left(\int_a^b |h(t)|^\beta \, dt \right)^{1/\beta} < \infty \text{ if } h \in L_\beta [a, b].$$

Theorem 6 *Assume that $f : [a, b] \to \mathbb{C}$ is K-p-Hölder continuous in the middle with the constant $K > 0$ and power $p > 0$, and g is integrable on $[a, b]$. Then we have*

$$\left| \langle f, g \rangle_\smile - f\left(\frac{a+b}{2}\right) \int_a^b \overline{g(t)} dt \right| \le K \int_a^b \left| t - \frac{a+b}{2} \right|^p |\breve{g}(t)| \, dt \tag{34}$$

$$\le K \begin{cases} \frac{1}{2^p} (b-a)^p \|\breve{g}\|_1, \\[2mm] \frac{1}{2^p (p\alpha+1)^{1/\alpha}} (b-a)^{p+1/\alpha} \|\breve{g}\|_\beta \\ \text{where } \alpha, \beta > 1 \text{ with } \frac{1}{\alpha} + \frac{1}{\beta} = 1, \\[2mm] \frac{1}{2^p (p+1)} (b-a)^{p+1} \|\breve{g}\|_\infty, \end{cases}$$

$$\le K \begin{cases} \frac{1}{2^p} (b-a)^p \|g\|_1, \\[2mm] \frac{1}{2^p (p\alpha+1)^{1/\alpha}} (b-a)^{p+1/\alpha} \|g\|_\beta \\ \text{where } \alpha, \beta > 1 \text{ with } \frac{1}{\alpha} + \frac{1}{\beta} = 1, \\[2mm] \frac{1}{2^p (p+1)} (b-a)^{p+1} \|g\|_\infty, \end{cases}$$

and

$$\left|\langle f, g\rangle_\sim\right| \le K \int_a^b \left|t - \frac{a+b}{2}\right|^P |\breve{g}(t)| \, dt \tag{35}$$

$$\le K \begin{cases} \frac{1}{2^p} (b-a)^P \|\breve{g}\|_1, \\[2ex] \frac{1}{2^p (p\alpha+1)^{1/\alpha}} (b-a)^{p+1/\alpha} \|\breve{g}\|_\beta \\ \text{where } \alpha, \beta > 1 \text{ with } \frac{1}{\alpha} + \frac{1}{\beta} = 1, \\[2ex] \frac{1}{2^p (p+1)} (b-a)^{p+1} \|\breve{g}\|_\infty, \end{cases}$$

$$\le K \begin{cases} \frac{1}{2^p} (b-a)^P \|g\|_1, \\[2ex] \frac{1}{2^p (p\alpha+1)^{1/\alpha}} (b-a)^{p+1/\alpha} \|g\|_\beta \\ \text{where } \alpha, \beta > 1 \text{ with } \frac{1}{\alpha} + \frac{1}{\beta} = 1, \\[2ex] \frac{1}{2^p (p+1)} (b-a)^{p+1} \|g\|_\infty. \end{cases}$$

Proof We have by (2) that

$$\int_a^b \left(f(t) - f\left(\frac{a+b}{2}\right)\right) \overline{\breve{g}(t)} \, dt = \int_a^b f(t) \overline{\breve{g}(t)} \, dt - f\left(\frac{a+b}{2}\right) \int_a^b \overline{\breve{g}(t)} \, dt$$

$$= \langle f, g\rangle_\smile - f\left(\frac{a+b}{2}\right) \int_a^b \overline{g(t)} \, dt.$$

Taking the modulus in this equality, we get

$$\left|\langle f, g\rangle_\smile - f\left(\frac{a+b}{2}\right) \int_a^b \overline{g(t)} \, dt\right| \le \int_a^b \left|f(t) - f\left(\frac{a+b}{2}\right)\right| |\breve{g}(t)| \, dt$$

$$\le K \int_a^b \left|t - \frac{a+b}{2}\right|^P |\breve{g}(t)| \, dt.$$

By the Hölder's integral inequality we have

$$\int_a^b \left|t - \frac{a+b}{2}\right|^P |\breve{g}(t)| \, dt \le \begin{cases} \max_{t \in [a,b]} \left|t - \frac{a+b}{2}\right|^P \int_a^b |\breve{g}(t)| \, dt, \\[2ex] \left(\int_a^b \left|t - \frac{a+b}{2}\right|^{p\alpha} dt\right)^{1/\alpha} \left(\int_a^b |\breve{g}(t)|^\beta \, dt\right)^{1/\beta} \\ \text{where } \alpha, \beta > 1 \text{ with } \frac{1}{\alpha} + \frac{1}{\beta} = 1, \\[2ex] \int_a^b \left|t - \frac{a+b}{2}\right|^P dt \, \mathrm{essup}_{t \in [a,b]} |\breve{g}(t)| \end{cases}$$

$$= \begin{cases} \frac{1}{2^p} (b-a)^p \|\check{g}\|_1, \\\\ \frac{1}{2^p (p\alpha+1)^{1/\alpha}} (b-a)^{p+1/\alpha} \|\check{g}\|_\beta \\ \text{where } \alpha, \beta > 1 \text{ with } \frac{1}{\alpha} + \frac{1}{\beta} = 1, \\\\ \frac{1}{2^p (p+1)} (b-a)^{p+1} \|\check{g}\|_\infty, \end{cases}$$

which proves the second inequality in (34).

By the triangle inequality for the Lebesgue norms we have

$$\|\check{g}\|_\beta = \frac{1}{2} \|g + g(a+b-\cdot)\|_\beta \le \frac{1}{2} \left[\|g\|_\beta + \|g(a+b-\cdot)\|_\beta \right] = \|g\|_\beta,$$

which proves the last part of (34).

We have by (7) that

$$\int_a^b \left(f(t) - f\left(\frac{a+b}{2}\right) \right) \overline{\tilde{g}(t)} dt = \int_a^b f(t) \overline{\tilde{g}(t)} dt - f\left(\frac{a+b}{2}\right) \int_a^b \overline{\tilde{g}(t)} dt$$

$$= \int_a^b f(t) \overline{\tilde{g}(t)} dt = \langle f, g \rangle_\sim.$$

Taking the modulus in this equality, we get

$$|\langle f, g \rangle_\sim| \le \int_a^b \left| f(t) - f\left(\frac{a+b}{2}\right) \right| |\tilde{g}(t)| \, dt \le K \int_a^b \left| t - \frac{a+b}{2} \right|^p |\tilde{g}(t)| \, dt,$$

which proves the second inequality in (35).

The rest follows in a similar manner and the details are omitted.

Corollary 3 *Assume that* $f : [a, b] \to \mathbb{C}$ *is* K-p-*Hölder continuous in the middle with the constant* $K > 0$ *and power* $p > 0$, *and* g *is* H-r-*Hölder continuous with the constant* $H > 0$ *and power* $r \in (0, 1]$. *Then*

$$|\langle f, g \rangle_\sim| \le \frac{1}{2^{p+1} (p+r+1)} HK (b-a)^{p+r+1}. \tag{36}$$

In particular, if L-*Lipschitzian with* $L > 0$, *then*

$$|\langle f, g \rangle_\sim| \le \frac{1}{2^{p+1} (p+2)} LK (b-a)^{p+2}. \tag{37}$$

Proof From the first inequality in (35) we have

$$
|\langle f, g \rangle_\sim| \le K \int_a^b \left| t - \frac{a+b}{2} \right|^p |\tilde{g}(t)| \, dt \le \frac{1}{2^{1-r}} HK \int_a^b \left| t - \frac{a+b}{2} \right|^{p+r} dt
$$

$$
= \frac{1}{2^{1-r}} HK \frac{(b-a)^{p+r+1}}{2^{p+r}(p+r+1)} = \frac{1}{2^{p+1}(p+r+1)} HK (b-a)^{p+r+1},
$$

which proves (36).

References

1. A.M. Acu, Improvement of Grüss and Ostrowski type inequalities. Filomat **29**(9), 2027–2035 (2015)
2. A.M. Acu, H. Gonska, I. Raşa, Grüss-type and Ostrowski-type inequalities in approximation theory. Ukr. Math. J. **63**(6), 843–864 (2011)
3. A.M. Acu, F. Sofonea, C.V. Muraru, Grüss and Ostrowski type inequalities and their applications. Sci. Stud. Res. Ser. Math. Inform. **23**(1), 5–14 (2013)
4. M.W. Alomari, Some Grüss type inequalities for Riemann-Stieltjes integral and applications. Acta Math. Univ. Comenian. (N.S.) **81**(2), 211–220 (2012)
5. M.W. Alomari, New Grüss type inequalities for double integrals. Appl. Math. Comput. **228** , 102–107 (2014)
6. M.W. Alomari, New inequalities of Grüss-Lupaş type and applications for selfadjoint operators. Armen. J. Math. **8**(1), 25–37 (2016)
7. G.A. Anastassiou, Basic and s-convexity Ostrowski and Grüss type inequalities involving several functions. Commun. Appl. Anal. **17**(2), 189–212 (2013)
8. G.A. Anastassiou, General Grüss and Ostrowski type inequalities involving S-convexity. Bull. Allahabad Math. Soc. **28**(1), 101–129.880 (2013)
9. G.A. Anastassiou, Fractional Ostrowski and Grüss type inequalities involving several functions. PanAmer. Math. J. **24**(3), 1–14 (2014)
10. G.A. Anastassiou, Further interpretation of some fractional Ostrowski and Grüss type inequalities. J. Appl. Funct. Anal. **9**(3–4), 392–403 (2014)
11. P. Cerone, S.S. Dragomir, Some new Ostrowski-type bounds for the Čebyšev functional and applications. J. Math. Inequal. **8**(1), 159–170 (2014)
12. X.-K. Chai, Y. Miao, Note on the Grüss inequality. Gen. Math. **19**(3), 93–99 (2011)
13. S.S. Dragomir, A generalization of Grüss's inequality in inner product spaces and applications. J. Math. Anal. Appl. **237**(1), 74–82 (1999)
14. S.S. Dragomir, New Grüss' type inequalities for functions of bounded variation and applications. Appl. Math. Lett. **25**(10), 1475–1479 (2012)
15. S.S. Dragomir, Bounds for convex functions of Čebyšev functional via Sonin's identity with applications. Commun. Math. **22**(2), 107–132 (2014)
16. S.S. Dragomir, Some Grüss-type results via Pompeiu's-like inequalities. Arab. J. Math. **4**(3), 159–170 (2015)
17. S.S. Dragomir, Bounding the Čebyšev functional for a differentiable function whose derivative is *h* or λ-convex in absolute value and applications. Matematiche (Catania) **71**(1), 173–202 (2016)

18. S.S. Dragomir, Bounding the Čebyšev functional for a function that is convex in absolute value and applications. Facta Univ. Ser. Math. Inform. **31**(1), 33–54 (2016)
19. S.S. Dragomir, M.S. Moslehian, Y.J. Cho, Some reverses of the Cauchy-Schwarz inequality for complex functions of self-adjoint operators in Hilbert spaces. Math. Inequal. Appl. **17**(4), 1365–1373 (2014). Preprint RGMIA Res. Rep. Coll. **14**, Article 84. (2011). http://rgmia.org/papers/v14/v14a84.pdf
20. Q. Feng, F. Meng, Some generalized Ostrowski-Grüss type integral inequalities. Comput. Math. Appl. **63**(3), 652–659 (2012)
21. B. Gavrea, Improvement of some inequalities of Chebysev-Grüss type. Comput. Math. Appl. **64**(6), 2003–2010 (2012)
22. A.G. Ghazanfari, A Grüss type inequality for vector-valued functions in Hilbert C^*-modules. J. Inequal. Appl. **2014**(16), 10 (2014)
23. S. Hussain, A. Qayyum, A generalized Ostrowski-Grüss type inequality for bounded differentiable mappings and its applications. J. Inequal. Appl. **2013**(1), 7 (2013)
24. N. Minculete and L. Ciurdariu, A generalized form of Grüss type inequality and other integral inequalities. J. Inequal. Appl. **2014**(119), 18 (2014)
25. A. Qayyum, S. Hussain, A new generalized Ostrowski Grüss type inequality and applications. Appl. Math. Lett. **25**(11), 1875–1880 (2012)
26. M.-D. Rusu, On Grüss-type inequalities for positive linear operators. Stud. Univ. Babeş-Bolyai Math. **56**(2), 551–565 (2011)
27. M.Z. Sarikaya, H. Budak, An inequality of Grüss like via variant of Pompeiu's mean value theorem. Konuralp J. Math. **3**(1), 29–35 (2015)

Generalized Finite Hilbert Transform and Some Basic Inequalities

Silvestru Sever Dragomir

Abstract In this paper we consider a generalized finite Hilbert transform of complex valued functions and establish some basic inequalities for several particular classes of interest. Applications for some particular instances of finite Hilbert transforms are given as well.

1 Introduction

Finite Hilbert transform on the open interval (a, b) is defined by

$$(Tf)(a, b; t) := \frac{1}{\pi} PV \int_a^b \frac{f(\tau)}{\tau - t} d\tau := \lim_{\varepsilon \to 0+} \left[\int_a^{t-\varepsilon} + \int_{t+\varepsilon}^b \right] \frac{f(\tau)}{\pi(\tau - t)} d\tau$$

(1)

for $t \in (a, b)$ and for various classes of functions f for which the above Cauchy Principal Value integral exists, see [13, Section 3.2] or [17, Lemma II.1.1].

We say that the function $f : [a, b] \to \mathbb{R}$ is α-H-*Hölder continuous* on (a, b), if

$$|f(t) - f(s)| \le H |t - s|^\alpha \text{ for all } t, s \in (a, b),$$

where $\alpha \in (0, 1]$, $H > 0$.

The following theorem holds.

Theorem 1 (Dragomir et al., [1]) *If* $f : [a, b] \to \mathbb{R}$ *is* α-H-*Hölder continuous on* (a, b), *then we have the estimate*

S. S. Dragomir (✉)
Mathematics, College of Engineering & Science, Victoria University, Melbourne City, MC, Australia

DST-NRF Centre of Excellence in the Mathematical and Statistical Sciences, School of Computer Science & Applied Mathematics, University of the Witwatersrand, Johannesburg, South Africa
e-mail: sever.dragomir@vu.edu.au

© Springer Nature Switzerland AG 2019
G. A. Anastassiou and J. M. Rassias (eds.), *Frontiers in Functional Equations and Analytic Inequalities*, https://doi.org/10.1007/978-3-030-28950-8_27

$$\left| (Tf)(a,b;t) - \frac{f(t)}{\pi} \ln\left(\frac{b-t}{t-a}\right) \right| \leq \frac{H}{\alpha\pi} \left[(t-a)^{\alpha} + (b-t)^{\alpha} \right]$$

for all $t \in (a,b)$.

The following two corollaries are natural.

Corollary 1 *Let $f : [a,b] \to \mathbb{R}$ be an L-Lipschitzian mapping on $[a,b]$, i.e. f satisfies the condition*

$$|f(t) - f(s)| \leq L|t-s| \quad \text{for all } t, s \in [a,b], \ (L > 0).$$

Then we have the inequality

$$\left| (Tf)(a,b;t) - \frac{f(t)}{\pi} \ln\left(\frac{b-t}{t-a}\right) \right| \leq \frac{L(b-a)}{\pi}$$

for all $t \in (a,b)$.

Corollary 2 *Let $f : [a,b] \to \mathbb{R}$ be an absolutely continuous mapping on $[a,b]$. If $f' \in L_{\infty}[a,b]$, then, for all $t \in (a,b)$, we have*

$$\left| (Tf)(a,b;t) - \frac{f(t)}{\pi} \ln\left(\frac{b-t}{t-a}\right) \right| \leq \frac{\|f'\|_{\infty}(b-a)}{\pi},$$

where $\|f'\|_{\infty} = \mathrm{essup}_{t \in (a,b)} |f'(t)| < \infty$.

We also have

Theorem 2 (Dragomir et al., [1]) *Let $f : [a,b] \to \mathbb{R}$ be a monotonic nondecreasing (nonincreasing) function on $[a,b]$. If the finite Hilbert transform $(Tf)(a,b,\cdot)$ exists in every $t \in (a,b)$, then*

$$(Tf)(a,b;t) \geq (\leq) \frac{1}{\pi} f(t) \ln\left(\frac{b-t}{t-a}\right)$$

for all $t \in (a,b)$.

The following result can be useful in practice.

Corollary 3 *Let $f : [a,b] \to \mathbb{R}$ and $\ell : [a,b] \to \mathbb{R}$, $\ell(t) = t$ such that $f - m\ell$, $M\ell - f$ are monotonic nondecreasing, where m, M are given real numbers. If $(Tf)(a,b,\cdot)$ exists in every point $t \in (a,b)$, then we have the inequality*

$$\frac{(b-a)m}{\pi} \leq (Tf)(a,b;t) - \frac{1}{\pi} f(t) \ln\left(\frac{b-t}{t-a}\right) \leq \frac{(b-a)M}{\pi} \tag{2}$$

for all $t \in (a,b)$.

Remark 1 If the mapping is differentiable on (a, b), the condition that $f - m\ell$, $M\ell - f$ are monotonic nondecreasing is equivalent with the following more practical condition

$$m \le f'(t) \le M \text{ for all } t \in (a, b).$$

From (2) we may deduce the following approximation result

$$\left|(Tf)(a, b; t) - \frac{1}{\pi} f(t) \ln\left(\frac{b-t}{t-a}\right) - \frac{M+m}{2\pi}(b-a)\right| \le \frac{M-m}{2\pi}(b-a).$$

for all $t \in (a, b)$.

For several recent papers devoted to inequalities for the finite Hilbert transform (Tf), see [2–10, 12, 14–16] and [18, 19].

We can naturally generalize the concept of Hilbert transform as follows.

For a *continuous strictly increasing function* $g : [a, b] \to [g(a), g(b)]$ that is *differentiable* on (a, b) we define the following generalization of the finite Hilbert transform of a function $f : (a, b) \to \mathbb{C}$ by

$$\left(T_g f\right)(a, b; t) := \frac{1}{\pi} PV \int_a^b \frac{f(\tau) g'(\tau)}{g(\tau) - g(t)} d\tau \tag{3}$$

$$:= \lim_{\varepsilon \to 0+} \left[\int_a^{t-\varepsilon} + \int_{t+\varepsilon}^b\right] \frac{f(\tau) g'(\tau)}{\pi [g(\tau) - g(t)]} d\tau$$

$$:= \frac{1}{\pi} \lim_{\varepsilon \to 0+} \left[\int_a^{t-\varepsilon} \frac{f(\tau) g'(\tau)}{g(\tau) - g(t)} d\tau + \int_{t+\varepsilon}^b \frac{f(\tau) g'(\tau)}{g(\tau) - g(t)} d\tau\right]$$

for $t \in (a, b)$, provided the above PV exists.

For $[a, b] \subset (0, \infty)$ and $g(t) = \ln t$, $t \in [a, b]$ we have

$$(T_{\ln} f)(a, b; t) := \frac{1}{\pi} \lim_{\varepsilon \to 0+} \left[\int_a^{t-\varepsilon} \frac{f(\tau)}{\tau \ln\left(\frac{\tau}{t}\right)} d\tau + \int_{t+\varepsilon}^b \frac{f(\tau)}{\tau \ln\left(\frac{\tau}{t}\right)} d\tau\right] \tag{4}$$

where $t \in (a, b)$.

For $g(t) = \exp(\alpha t)$, $t \in [a, b] \subset \mathbb{R}$ with $\alpha > 0$ we have

$$\left(T_{\exp(\alpha)} f\right)(a, b; t) \tag{5}$$

$$:= \frac{1}{\pi} \lim_{\varepsilon \to 0+} \left[\int_a^{t-\varepsilon} \frac{f(\tau) \exp(\alpha\tau)}{\exp(\alpha\tau) - \exp(\alpha t)} d\tau + \int_{t+\varepsilon}^b \frac{f(\tau) \exp(\alpha\tau)}{\exp(\alpha\tau) - \exp(\alpha t)} d\tau\right]$$

where $t \in (a, b)$.

For $[a, b] \subset (0, \infty)$ and $g(t) = t^r$, $t \in [a, b]$, $r > 0$, we have

$$(T_r f)(a, b; t) := \frac{r}{\pi} \lim_{\varepsilon \to 0+} \left[\int_a^{t-\varepsilon} \frac{f(\tau) \tau^{r-1}}{\tau^r - t^r} d\tau + \int_{t+\varepsilon}^b \frac{f(\tau) \tau^{r-1}}{\tau^r - t^r} d\tau \right], \qquad (6)$$

where $t \in (a, b)$.

Similarly, we can consider the function $g(t) = -t^{-p}$, $t \in [a, b] \subset (0, \infty)$, $p > 0$, and then we have

$$\left(T_{-p} f \right)(a, b; t) := \frac{p}{\pi} \lim_{\varepsilon \to 0+} \left[\int_a^{t-\varepsilon} \frac{f(\tau) \tau^{-p-1}}{t^{-p} - \tau^{-p}} d\tau + \int_{t+\varepsilon}^b \frac{f(\tau) \tau^{-p-1}}{t^{-p} - \tau^{-p}} d\tau \right]$$

$$(7)$$

$$= \frac{p t^p}{\pi} \lim_{\varepsilon \to 0+} \left[\int_a^{t-\varepsilon} \frac{f(\tau)}{\tau (\tau^p - t^p)} d\tau + \int_{t+\varepsilon}^b \frac{f(\tau)}{\tau (\tau^p - t^p)} d\tau \right],$$

where $t \in (a, b)$.

For $[a, b] \subset \left[-\frac{\pi}{2\rho}, \frac{\pi}{2\rho} \right]$ and $g(t) = \sin(\rho t)$, $t \in [a, b]$ where $\rho > 0$, we have

$$\left(T_{\sin(\rho)} f \right)(a, b; t) \qquad (8)$$

$$:= \frac{\rho}{\pi} \lim_{\varepsilon \to 0+} \left[\int_a^{t-\varepsilon} \frac{f(\tau) \cos(\rho \tau)}{\sin(\rho \tau) - \sin(\rho t)} d\tau + \int_{t+\varepsilon}^b \frac{f(\tau) \cos(\rho \tau)}{\sin(\rho \tau) - \sin(\rho t)} d\tau \right]$$

where $t \in (a, b)$.

For $g(t) = \sinh(\sigma t)$, $t \in [a, b] \subset \mathbb{R}$ with $\sigma > 0$ we have

$$\left(T_{\sinh(\sigma)} f \right)(a, b; t) \qquad (9)$$

$$:= \frac{\sigma}{\pi} \lim_{\varepsilon \to 0+} \left[\int_a^{t-\varepsilon} \frac{f(\tau) \cosh(\sigma \tau)}{\sinh(\sigma \tau) - \sinh(\sigma t)} d\tau + \int_{t+\varepsilon}^b \frac{f(\tau) \cosh(\sigma \tau)}{\sinh(\sigma \tau) - \sinh(\sigma t)} d\tau \right]$$

where $t \in (a, b)$.

Similar transforms can be associated to the following functions as well:

$$g(t) = \tan(\rho t), \ t \in [a, b] \subset \left[-\frac{\pi}{2\rho}, \frac{\pi}{2\rho} \right] \text{ where } \rho > 0,$$

and

$$g(t) = \tanh(\sigma t), \ t \in [a, b] \subset \mathbb{R} \text{ with } \sigma > 0.$$

Motivated by the above facts, in this paper we consider the generalized finite Hilbert transform $\left(T_g f \right)(a, b; t)$ of complex valued functions f and establish some basic inequalities for several particular classes of interest. Applications for some particular instances of finite Hilbert transforms as the one presented in (4)–(9) are given as well.

2 Main Results

Consider the function $\mathbf{1}(t) = 1$, $t \in (a, b)$. We can state the following basic result:

Lemma 1 *For a continuous strictly increasing function $g : [a, b] \to [g(a), g(b)]$ that is differentiable on (a, b) we have*

$$\left(T_g \mathbf{1}\right)(a, b; t) = \frac{1}{\pi} \ln\left(\frac{g(b) - g(t)}{g(t) - g(a)}\right), \quad t \in (a, b). \tag{10}$$

We also have for $f : (a, b) \to \mathbb{C}$ that

$$\left(T_g f\right)(a, b; t) = \frac{1}{\pi} f(t) \ln\left(\frac{g(b) - g(t)}{g(t) - g(a)}\right) + \frac{1}{\pi} PV \int_a^b \frac{f(\tau) - f(t)}{g(\tau) - g(t)} g'(\tau) \, d\tau \tag{11}$$

for $t \in (a, b)$, provided that the PV from the right-hand side of the equality (11) exists.

Proof We have

$$\left(T_g \mathbf{1}\right)(a, b; t) = \frac{1}{\pi} \lim_{\varepsilon \to 0+} \left[\int_a^{t-\varepsilon} \frac{g'(\tau)}{g(\tau) - g(t)} d\tau + \int_{t+\varepsilon}^b \frac{g'(\tau)}{g(\tau) - g(t)} d\tau\right] \tag{12}$$

$$= \frac{1}{\pi} \lim_{\varepsilon \to 0+} \left[\ln|g(\tau) - g(t)|\big|_a^{t-\varepsilon} + \ln(g(\tau) - g(t))\big|_{t+\varepsilon}^b\right]$$

$$= \frac{1}{\pi} \lim_{\varepsilon \to 0+} \left[\ln(g(t) - g(t - \varepsilon)) - \ln(g(t) - g(a))\right.$$

$$\left. + \ln(g(b) - g(t)) - \ln(g(t + \varepsilon) - g(t))\right]$$

$$= \frac{1}{\pi} \ln\left(\frac{g(b) - g(t)}{g(t) - g(a)}\right) + \frac{1}{\pi} \lim_{\varepsilon \to 0+} \ln\left(\frac{g(t) - g(t - \varepsilon)}{g(t + \varepsilon) - g(t)}\right)$$

for $t \in (a, b)$.

Since g is differentiable, we have

$$\lim_{\varepsilon \to 0+} \frac{g(t) - g(t - \varepsilon)}{g(t + \varepsilon) - g(t)} = \lim_{\varepsilon \to 0+} \frac{\frac{g(t) - g(t-\varepsilon)}{\varepsilon}}{\frac{g(t+\varepsilon) - g(t)}{\varepsilon}} = \frac{g'(t)}{g'(t)} = 1$$

for $t \in (a, b)$, and by (12) we get (10).

From the definition (3) we have

$$
\left(T_g f\right)(a, b; t) := \frac{1}{\pi} PV \int_a^b \frac{\left(f(\tau) - f(t) + f(t)\right) g'(\tau)}{g(\tau) - g(t)} d\tau
$$

$$
= \frac{1}{\pi} PV \int_a^b \frac{\left(f(\tau) - f(t)\right) g'(\tau) d\tau}{g(\tau) - g(t)} + \frac{1}{\pi} PV \int_a^b \frac{f(t) g'(\tau) d\tau}{g(\tau) - g(t)}
$$

$$
= \frac{1}{\pi} PV \int_a^b \frac{\left(f(\tau) - f(t)\right) g'(\tau) d\tau}{g(\tau) - g(t)} + \frac{1}{\pi} f(t) PV \int_a^b \frac{g'(\tau) d\tau}{g(\tau) - g(t)}
$$

$$
= \frac{1}{\pi} f(t) \ln \left(\frac{g(b) - g(t)}{g(t) - g(a)}\right) + \frac{1}{\pi} PV \int_a^b \frac{\left(f(\tau) - f(t)\right) g'(\tau) d\tau}{g(\tau) - g(t)}
$$

for $t \in (a, b)$, which proves the identity (11).

The following result holds:

Theorem 3 *Assume that g is as in Lemma 1 and $f : [a, b] \to \mathbb{R}$ is continuous on $[a, b]$ and differentiable on (a, b). If*

$$
\left\| \frac{f'}{g'} \right\|_{(a,b),\infty} := \sup_{s \in (a,b)} \left| \frac{f'(s)}{g'(s)} \right| < \infty,
$$

then $\left(T_g f\right)(a, b; t)$ exists for all $t \in (a, b)$ and

$$
\left| \left(T_g f\right)(a, b; t) - \frac{1}{\pi} f(t) \ln \left(\frac{g(b) - g(t)}{g(t) - g(a)}\right) \right| \le \frac{1}{\pi} \left\| \frac{f'}{g'} \right\|_{(a,b),\infty} [g(b) - g(a)]
$$

$$
\tag{13}
$$

for all $t \in (a, b)$.

Proof By *Cauchy's mean value theorem*, for any $t, \tau \in (a, b)$ with $t \ne \tau$ there exists an s between t and τ such that

$$
\frac{f(\tau) - f(t)}{g(\tau) - g(t)} = \frac{f'(s)}{g'(s)},
$$

therefore for any $t, \tau \in (a, b)$ with $t \ne \tau$ we have

$$
\left| \frac{f(\tau) - f(t)}{g(\tau) - g(t)} \right| \le \left\| \frac{f'}{g'} \right\|_{(a,b),\infty}.
$$

This implies that

$$
\int_a^{t-\varepsilon} \left| \frac{f(\tau) - f(t)}{g(\tau) - g(t)} \right| g'(\tau) d\tau \le \left\| \frac{f'}{g'} \right\|_{(a,b),\infty} [g(t - \varepsilon) - g(a)]
$$

and

$$\int_{t+\varepsilon}^{b} \left| \frac{f(\tau) - f(t)}{g(\tau) - g(t)} \right| g'(\tau) \, d\tau \leq \left\| \frac{f'}{g'} \right\|_{(a,b),\infty} [g(b) - g(t+\varepsilon)]$$

for $t \in (a, b)$ and $\min\{t - a, b - t\} > \varepsilon > 0$.

By the triangle inequality for the modulus and the fact that $g'(\tau) > 0$ for $t \in (a, b)$, we have

$$\left| \int_{a}^{t-\varepsilon} \frac{f(\tau) - f(t)}{g(\tau) - g(t)} g'(\tau) \, d\tau + \int_{t+\varepsilon}^{b} \frac{f(\tau) - f(t)}{g(\tau) - g(t)} g'(\tau) \, d\tau \right| \quad (14)$$

$$\leq \int_{a}^{t-\varepsilon} \left| \frac{f(\tau) - f(t)}{g(\tau) - g(t)} \right| g'(\tau) \, d\tau + \int_{t+\varepsilon}^{b} \left| \frac{f(\tau) - f(t)}{g(\tau) - g(t)} \right| g'(\tau) \, d\tau$$

$$\leq \left\| \frac{f'}{g'} \right\|_{(a,b),\infty} [g(b) - g(t+\varepsilon) + g(t-\varepsilon) - g(a)]$$

for $t \in (a, b)$ and $\min\{t - a, b - t\} > \varepsilon > 0$.

By taking the limit over $\varepsilon \to 0+$ in (14) we get

$$\left| PV \int_{a}^{b} \frac{f(\tau) - f(t)}{g(\tau) - g(t)} g'(\tau) \, d\tau \right| \leq \left\| \frac{f'}{g'} \right\|_{(a,b),\infty} [g(b) - g(a)] \quad (15)$$

for $t \in (a, b)$.

By utilizing the equality (11) we obtain from (15) the desired result (13).

If g is a function which maps an interval I of the real line to the real numbers, and is both continuous and injective, then we can define the *g-mean of two numbers* $a, b \in I$ as

$$M_g(a, b) := g^{-1} \left(\frac{g(a) + g(b)}{2} \right).$$

If $I = \mathbb{R}$ and $g(t) = t$ is the *identity function*, then $M_g(a, b) = A(a, b) := \frac{a+b}{2}$, the *arithmetic mean*. If $I = (0, \infty)$ and $g(t) = \ln t$, then $M_g(a, b) = G(a, b) := \sqrt{ab}$, the *geometric mean*. If $I = (0, \infty)$ and $g(t) = \frac{1}{t}$, then $M_g(a, b) = H(a, b) := \frac{2ab}{a+b}$, the *harmonic mean*. If $I = (0, \infty)$ and $g(t) = t^p$, $p \neq 0$, then $M_g(a, b) = M_p(a, b) := \left(\frac{a^p + b^p}{2} \right)^{1/p}$, the *power mean with exponent p*. Finally, if $I = \mathbb{R}$ and $g(t) = \exp t$, then

$$M_g(a, b) = LME(a, b) := \ln \left(\frac{\exp a + \exp b}{2} \right),$$

the *LogMeanExp function*.

Corollary 4 *With the assumptions of Theorem 3, we have*

$$\left|\left(T_g f\right)(a, b; M_g(a, b))\right| \leq \frac{1}{\pi} \left\|\frac{f'}{g'}\right\|_{(a,b),\infty} [g(b) - g(a)]. \tag{16}$$

We also have

Theorem 4 *Let $g : [a, b] \to [g(a), g(b)]$ be a strictly increasing function that is differentiable on (a, b) and $f : (a, b) \to \mathbb{C}$ such that $f \circ g^{-1}$ is of H-r-Hölder type on $(g(a), g(b))$, where $H > 0$, $r \in (0, 1]$, then*

$$\left|\left(T_g f\right)(a, b; t) - \frac{1}{\pi} f(t) \ln\left(\frac{g(b) - g(t)}{g(t) - g(a)}\right)\right| \tag{17}$$

$$\leq \frac{H}{\pi r} \left[(g(b) - g(t))^r + (g(t) - g(a))^r\right]$$

for $t \in (a, b)$.

In particular, in the Lipschitz case, we have for $H = L$ that

$$\left|\left(T_g f\right)(a, b; t) - \frac{1}{\pi} f(t) \ln\left(\frac{g(b) - g(t)}{g(t) - g(a)}\right)\right| \leq \frac{L}{\pi r} [g(b) - g(a)] \tag{18}$$

for $t \in (a, b)$.

Proof For $t \in (a, b)$ and $\min\{t - a, b - t\} > \varepsilon > 0$ we have

$$\int_a^{t-\varepsilon} \left|\frac{f(\tau) - f(t)}{g(\tau) - g(t)}\right| g'(\tau) d\tau = \int_a^{t-\varepsilon} \left|\frac{f \circ g^{-1}(g(\tau)) - f \circ g^{-1}(g(t))}{g(\tau) - g(t)}\right| g'(\tau) d\tau$$

$$\leq H \int_a^{t-\varepsilon} \frac{|g(\tau) - g(t)|^r}{|g(\tau) - g(t)|} g'(\tau) d\tau$$

$$= H \int_a^{t-\varepsilon} |g(\tau) - g(t)|^{r-1} g'(\tau) d\tau$$

$$= H \int_a^{t-\varepsilon} (g(t) - g(\tau))^{r-1} g'(\tau) d\tau$$

$$= \frac{H}{r} \left[(g(t) - g(a))^r - (g(t) - g(t - \varepsilon))^r\right]$$

and

$$\int_{t+\varepsilon}^b \left|\frac{f(\tau) - f(t)}{g(\tau) - g(t)}\right| g'(\tau) d\tau = \int_{t+\varepsilon}^b \left|\frac{f \circ g^{-1}(g(\tau)) - f \circ g^{-1}(g(t))}{g(\tau) - g(t)}\right| g'(\tau) d\tau$$

$$\leq H \int_{t+\varepsilon}^b \frac{|g(\tau) - g(t)|^r}{|g(\tau) - g(t)|} g'(\tau) d\tau$$

$$= H \int_{t+\varepsilon}^{b} |g(\tau) - g(t)|^{r-1} g'(\tau) d\tau$$

$$= H \int_{t+\varepsilon}^{b} (g(\tau) - g(t))^{r-1} g'(\tau) d\tau$$

$$= \frac{H}{r} \left[(g(b) - g(t))^r - (g(t+\varepsilon) - g(t))^r \right].$$

By adding these two inequalities, we get

$$\int_{a}^{t-\varepsilon} \left| \frac{f(\tau) - f(t)}{g(\tau) - g(t)} \right| g'(\tau) d\tau + \int_{t+\varepsilon}^{b} \left| \frac{f(\tau) - f(t)}{g(\tau) - g(t)} \right| g'(\tau) d\tau \qquad (19)$$

$$\leq \frac{H}{r} \left[(g(b) - g(t))^r + (g(t) - g(a))^r \right.$$

$$\left. - (g(t+\varepsilon) - g(t))^r - (g(t) - g(t-\varepsilon))^r \right]$$

for $t \in (a, b)$ and $\min\{t - a, b - t\} > \varepsilon > 0$.

By using the triangle inequality and taking the limit over $\varepsilon \to 0+$, we get

$$\left| PV \int_{a}^{b} \frac{f(\tau) - f(t)}{g(\tau) - g(t)} g'(\tau) d\tau \right| \leq \frac{H}{r} \left[(g(b) - g(t))^r + (g(t) - g(a))^r \right]$$

for $t \in (a, b)$.

Finally, by making use of the equality (11) we deduce the desired result (17).

Corollary 5 *With the assumptions of Theorem 4, we have*

$$\left| (T_g f)(a, b; M_g(a, b)) \right| \leq \frac{H}{2^{r-1} \pi r} (g(b) - g(a))^r. \qquad (20)$$

In particular, for $r = 1$, we get

$$\left| (T_g f)(a, b; M_g(a, b)) \right| \leq \frac{L}{\pi} (g(b) - g(a)). \qquad (21)$$

For a function $f : (a, b) \to \mathbb{C}$ and an injective function $g : (a, b) \to \mathbb{C}$ we define the *divided difference*

$$[f, g; t, s] := \frac{f(t) - f(s)}{g(t) - g(s)} \quad \text{for } t, s \in (a, b), \ t \neq s.$$

Now, for $\gamma, \Gamma \in \mathbb{C}, \gamma \neq \Gamma$, an injective function $g : (a, b) \to \mathbb{C}$ and (a, b) a finite interval of real numbers, define the sets of complex-valued functions (see also [11] for a similar definition):

$$\bar{U}_{(a,b),g,d}\,(\gamma,\Gamma)$$

$$:= \left\{ f : (a,b) \to \mathbb{C} \mid \mathrm{Re}\left[(\Gamma - [f,g;t,s]) \left(\overline{[f,g;t,s]} - \bar{\gamma} \right) \right] \geq 0, \right.$$

$$\left. \text{for all } t,\, s \in (a,b),\, t \neq s \right\} \qquad (22)$$

and

$$\bar{\Delta}_{(a,b),g,d}\,(\gamma,\Gamma) := \left\{ f : (a,b) \to \mathbb{C} \mid \left| [f,g;t,s] - \frac{\gamma + \Gamma}{2} \right| \leq \frac{1}{2}|\Gamma - \gamma| \right.$$

$$\left. \text{for all } t,\, s \in (a,b),\, t \neq s \right\}. \qquad (23)$$

The following representation result may be stated.

Proposition 1 *For any* $\gamma,\, \Gamma \in \mathbb{C},\, \gamma \neq \Gamma$, *we have that* $\bar{U}_{(a,b),g,d}\,(\gamma,\Gamma)$ *and* $\bar{\Delta}_{(a,b),g,d}\,(\gamma,\Gamma)$ *are nonempty, convex, and closed sets and*

$$\bar{U}_{(a,b),d}\,(\gamma,\Gamma) = \bar{\Delta}_{(a,b),d}\,(\gamma,\Gamma). \qquad (24)$$

Proof We observe that for any $z \in \mathbb{C}$ we have the equivalence

$$\left| z - \frac{\gamma + \Gamma}{2} \right| \leq \frac{1}{2}|\Gamma - \gamma|$$

if and only if

$$\mathrm{Re}\left[(\Gamma - z)\,(\bar{z} - \bar{\gamma}) \right] \geq 0.$$

This follows by the equality

$$\frac{1}{4}|\Gamma - \gamma|^2 - \left| z - \frac{\gamma + \Gamma}{2} \right|^2 = \mathrm{Re}\left[(\Gamma - z)\,(\bar{z} - \bar{\gamma}) \right]$$

that holds for any $z \in \mathbb{C}$.

The equality (24) is thus a simple consequence of this fact.

On making use of the complex numbers field properties we can also state that:

Corollary 6 *For any* $\gamma,\, \Gamma \in \mathbb{C},\, \gamma \neq \Gamma$, *we have that*

$$\bar{U}_{(a,b),g,d}\,(\gamma,\Gamma) = \{ f : (a,b) \to \mathbb{C} \mid (\mathrm{Re}\,\Gamma - \mathrm{Re}\,[f,g;t,s])\,(\mathrm{Re}\,[f,g;t,s] - \mathrm{Re}\,\gamma)$$

$$+ (\mathrm{Im}\,\Gamma - \mathrm{Im}\,[f,g;t,s])\,(\mathrm{Im}\,[f,g;t,s] - \mathrm{Im}\,\gamma) \geq 0 \textit{ for all } t,\, s \in (a,b),\, t \neq s \}.$$

Now, if we assume that $\mathrm{Re}\,(\Gamma) \geq \mathrm{Re}\,(\gamma)$ and $\mathrm{Im}\,(\Gamma) \geq \mathrm{Im}\,(\gamma)$, then we can define the following set of functions as well:

$$\bar{S}_{(a,b),g,d}\,(\gamma,\Gamma) := \{f : (a,b) \to \mathbb{C} \mid \mathrm{Re}\,(\Gamma) \geq \mathrm{Re}\,[f,g;t,s] \geq \mathrm{Re}\,(\gamma)$$

$$\text{and } \mathrm{Im}\,(\Gamma) \geq \mathrm{Im}\,[f,g;t,s] \geq \mathrm{Im}\,(\gamma) \text{ for all } t,\ s \in (a,b),\ t \neq s\}. \qquad (25)$$

One can easily observe that $\bar{S}_{(a,b)g,d}\,(\gamma,\Gamma)$ is closed, convex, and

$$\emptyset \neq \bar{S}_{(a,b),g,d}\,(\gamma,\Gamma) \subseteq \bar{U}_{(a,b),g,d}\,(\gamma,\Gamma). \qquad (26)$$

We have

Theorem 5 *Let $g : [a,b] \to [g\,(a),g\,(b)]$ be a strictly increasing function that is differentiable on (a,b) and $f : (a,b) \to \mathbb{C}$ such that $f \in \bar{\Delta}_{(a,b),g,d}\,(\gamma,\Gamma)$ for some $\gamma,\ \Gamma \in \mathbb{C}, \gamma \neq \Gamma$. Then we have*

$$\left| (T_g f)\,(a,b;t) - \frac{1}{\pi}f\,(t)\ln\left(\frac{g\,(b)-g\,(t)}{g\,(t)-g\,(a)}\right) - \frac{\gamma+\Gamma}{2\pi}\,(g\,(b)-g\,(a)) \right| \qquad (27)$$

$$\leq \frac{1}{2\pi}|\Gamma - \gamma|\,(g\,(b)-g\,(a))$$

for all $t \in (a,b)$.

Proof Since $f \in \bar{\Delta}_{(a,b),d}\,(\gamma,\Gamma)$ it follows that

$$\left| f\,(t) - f\,(s) - \frac{\gamma+\Gamma}{2}\,(g\,(t)-g\,(s)) \right| \leq \frac{1}{2}|\Gamma - \gamma|\,|g\,(t)-g\,(s)|$$

for any $t,\ s \in (a,b)$.

By the continuity of the modulus property, we have

$$|f\,(t)-f\,(s)| - \left|\frac{\gamma+\Gamma}{2}\right|\,|g\,(t)-g\,(s)| \leq \left| f\,(t)-f\,(s) - \frac{\gamma+\Gamma}{2}\,(g\,(t)-g\,(s)) \right|$$

$$\leq \frac{1}{2}|\Gamma - \gamma|\,|g\,(t)-g\,(s)|,$$

for any $t,\ s \in (a,b)$, which implies that

$$|f\,(t)-f\,(s)| \leq \frac{1}{2}\,(|\gamma+\Gamma|+|\Gamma-\gamma|)\,|g\,(t)-g\,(s)|$$

for any $t,\ s \in (a,b)$. This can be written as

$$\left| f \circ g^{-1}\left(g\left(t\right)\right) - f \circ g^{-1}\left(g\left(s\right)\right) \right| \le \frac{1}{2}\left(\left|\gamma + \Gamma\right| + \left|\Gamma - \gamma\right|\right)\left|g\left(t\right) - g\left(s\right)\right|$$

for any $t,\ s\ \in\ (a,b)$, namely $f \circ g^{-1}$ is Lipschitzian with the constant $\frac{1}{2}\left(\left|\gamma + \Gamma\right| + \left|\Gamma - \gamma\right|\right)$ on $(g\left(a\right), g\left(b\right))$.

Therefore the Cauchy Principal value

$$PV \int_a^b \frac{f\left(\tau\right) - f\left(t\right)}{g\left(\tau\right) - g\left(t\right)} g'\left(\tau\right) d\tau$$

exists (see [13, Section 3.2] or [17, Lemma II.1.1]) and we have

$$\frac{1}{\pi} PV \int_a^b \left(\frac{f\left(\tau\right) - f\left(t\right)}{g\left(\tau\right) - g\left(t\right)} - \frac{\gamma + \Gamma}{2} \right) g'\left(\tau\right) d\tau \tag{28}$$

$$= \frac{1}{\pi} PV \int_a^b \frac{f\left(\tau\right) - f\left(t\right)}{g\left(\tau\right) - g\left(t\right)} g'\left(\tau\right) d\tau - \frac{\gamma + \Gamma}{2\pi}\left(g\left(b\right) - g\left(a\right)\right)$$

$$= \left(T_g f\right)\left(a, b; t\right) - \frac{1}{\pi} f\left(t\right) \ln\left(\frac{g\left(b\right) - g\left(t\right)}{g\left(t\right) - g\left(a\right)}\right) - \frac{\gamma + \Gamma}{2\pi}\left(g\left(b\right) - g\left(a\right)\right)$$

for any $t \in (a,b)$.

The following property of the Cauchy Principal Value follows by the properties of integral, modulus, and limit,

$$\left| PV \int_a^b A\left(t,s\right) ds \right| \le PV \int_a^b \left| A\left(t,s\right) \right| ds, \tag{29}$$

assuming that the PVs involved exist for all $t \in (a,b)$.

Using the equality (28) and the property (29) we get

$$\left| \left(T_g f\right)\left(a, b; t\right) - \frac{1}{\pi} f\left(t\right) \ln\left(\frac{g\left(b\right) - g\left(t\right)}{g\left(t\right) - g\left(a\right)}\right) - \frac{\gamma + \Gamma}{2\pi}\left(g\left(b\right) - g\left(a\right)\right) \right|$$

$$\le \frac{1}{\pi} PV \int_a^b \left| \frac{f\left(\tau\right) - f\left(t\right)}{g\left(\tau\right) - g\left(t\right)} - \frac{\gamma + \Gamma}{2} \right| g'\left(\tau\right) d\tau \le \frac{1}{2\pi}\left|\Gamma - \gamma\right| \int_a^b g'\left(\tau\right) d\tau$$

$$= \frac{1}{2\pi}\left|\Gamma - \gamma\right|\left(g\left(b\right) - g\left(a\right)\right)$$

for all $t \in (a,b)$ and the inequality (27) is obtained.

Corollary 7 *With the assumptions of Theorem 5 we have*

$$\left| \left(T_g f\right)\left(a, b; M_g\left(a, b\right)\right) - \frac{\gamma + \Gamma}{2\pi}\left(g\left(b\right) - g\left(a\right)\right) \right| \le \frac{1}{2\pi}\left|\Gamma - \gamma\right|\left(g\left(b\right) - g\left(a\right)\right). \tag{30}$$

The case of monotonic functions $f : (a, b) \to \mathbb{R}$ provides the following simple result:

Proposition 2 *Let $g : [a, b] \to [g(a), g(b)]$ be a strictly increasing function that is differentiable on (a, b) and $f : (a, b) \to \mathbb{R}$ a monotonic nondecreasing (non-increasing) function so that the generalized finite Hilbert transform $(T_g f)(a, b; t)$ exists, then*

$$(T_g f)(a, b; t) \geq (\leq) \frac{1}{\pi} f(t) \ln \left(\frac{g(b) - g(t)}{g(t) - g(a)} \right) \tag{31}$$

for any $t \in (a, b)$.

Proof The proof follows by the representation (11) on observing that if $f : (a, b) \to \mathbb{R}$ is a monotonic nondecreasing (nonincreasing) function on (a, b), then for any $t, \tau \in (a, b)$ we have

$$\frac{f(\tau) - f(t)}{g(\tau) - g(t)} \geq (\leq) 0,$$

which implies that

$$PV \int_a^b \frac{f(\tau) - f(t)}{g(\tau) - g(t)} g'(\tau) d\tau \geq (\leq) 0$$

for any $t \in (a, b)$. $\blacksquare$

Corollary 8 *Let $g : [a, b] \to [g(a), g(b)]$ be a strictly increasing function that is differentiable on (a, b) and $f : (a, b) \to \mathbb{R}$ a function such that for some real numbers $m < M$ we have that $f - mg$ and $Mg - f$ are monotonic nondecreasing on (a, b). If the generalized finite Hilbert transform $(T_g f)(a, b; t)$ exists, then we have*

$$\frac{m}{\pi}(g(b) - g(a)) \leq (T_g f)(a, b; t) - \frac{1}{\pi} f(t) \ln \left(\frac{g(b) - g(t)}{g(t) - g(a)} \right) \tag{32}$$

$$\leq \frac{M}{\pi}(g(b) - g(a))$$

for any $t \in (a, b)$.

This can be also written as

$$\left| (T_g f)(a, b; t) - \frac{1}{\pi} f(t) \ln \left(\frac{g(b) - g(t)}{g(t) - g(a)} \right) - \frac{1}{2\pi}(M + m)(g(b) - g(a)) \right| \tag{33}$$

$$\leq \frac{1}{2\pi}(M - m)(g(b) - g(a))$$

for any $t \in (a, b)$.

Proof Applying proposition (31) for the monotonic nondecreasing function $f - mg$ we have

$$\left(T_g \left(f - mg\right)\right)(a, b; t) \tag{34}$$

$$\geq \frac{1}{\pi} \left(f(t) - mg(t)\right) \ln \left(\frac{g(b) - g(t)}{g(t) - g(a)}\right)$$

$$= \frac{1}{\pi} f(t) \ln \left(\frac{g(b) - g(t)}{g(t) - g(a)}\right) - m \frac{1}{\pi} g(t) \ln \left(\frac{g(b) - g(t)}{g(t) - g(a)}\right)$$

for all $t \in (a, b)$.

By the linearity of the generalized Hilbert transform we also have

$$\left(T_g \left(f - mg\right)\right)(a, b; t) = \left(T_g f\right)(a, b; t) - m \left(T_g g\right)(a, b; t)$$

and by the identity (11) for $f = g$ we get

$$\left(T_g g\right)(a, b; t) = \frac{1}{\pi} g(t) \ln \left(\frac{g(b) - g(t)}{g(t) - g(a)}\right) + \frac{1}{\pi} PV \int_a^b \frac{g(\tau) - g(t)}{g(\tau) - g(t)} g'(\tau) d\tau$$

$$= \frac{1}{\pi} g(t) \ln \left(\frac{g(b) - g(t)}{g(t) - g(a)}\right) + \frac{1}{\pi} \left(g(b) - g(a)\right),$$

which gives that

$$\left(T_g \left(f - mg\right)\right)(a, b; t) \tag{35}$$

$$= \left(T_g f\right)(a, b; t) - \frac{m}{\pi} g(t) \ln \left(\frac{g(b) - g(t)}{g(t) - g(a)}\right) - \frac{m}{\pi} \left(g(b) - g(a)\right)$$

for all $t \in (a, b)$.

On making use of (34) and (35) we get

$$\left(T_g f\right)(a, b; t) - \frac{m}{\pi} g(t) \ln \left(\frac{g(b) - g(t)}{g(t) - g(a)}\right) - \frac{m}{\pi} \left(g(b) - g(a)\right)$$

$$\geq \frac{1}{\pi} f(t) \ln \left(\frac{g(b) - g(t)}{g(t) - g(a)}\right) - \frac{m}{\pi} g(t) \ln \left(\frac{g(b) - g(t)}{g(t) - g(a)}\right)$$

for all $t \in (a, b)$, which proves the first inequality in (32).

The second part follows in a similar way by considering the monotonic nondecreasing function $Mg - f$.

Remark 2 From (32) we get for $t = \frac{a+b}{2}$ that

$$\frac{m}{\pi} \left(g\left(b\right) - g\left(a\right)\right) \leq \left(T_g f\right) \left(a, b; \frac{a+b}{2}\right) \leq \frac{M}{\pi} \left(g\left(b\right) - g\left(a\right)\right), \tag{36}$$

where f and g are as in Corollary 8.

Remark 3 If f and g are as in Corollary 8, then we observe that

$$m \leq [f, g; t, s] = \frac{f\left(t\right) - f\left(s\right)}{g\left(t\right) - g\left(s\right)} \leq M$$

for all t, $s \in (a, b)$ with $t \neq s$, then by (27) for $\Gamma = M$ and $\gamma = m$ we recapture the inequality (33) as well.

Remark 4 We also observe that if $f : (a, b) \rightarrow \mathbb{R}$ is differentiable on (a, b) and

$$mg'\left(t\right) \leq f'\left(t\right) \leq Mg'\left(t\right) \text{ for all } t \in (a, b),$$

then the inequality (32) holds.

3 Related Results

The following identity is of interest as well:

Lemma 2 *Let* $g : [a, b] \rightarrow [g\left(a\right), g\left(b\right)]$ *be a strictly increasing function that is differentiable on* (a, b) *and* $f : (a, b) \rightarrow \mathbb{R}$ *a locally absolutely continuous function on* (a, b), *then*

$$\left(T_g f\right)\left(a, b; t\right) = \frac{1}{\pi} f\left(t\right) \ln\left(\frac{g\left(b\right) - g\left(t\right)}{g\left(t\right) - g\left(a\right)}\right)$$
$$+ \frac{1}{\pi} PV \int_a^b \left(\int_0^1 \frac{\left(f' \circ g^{-1}\right)\left(\left(1-s\right)g\left(\tau\right) + sg\left(t\right)\right)}{\left(g' \circ g^{-1}\right)\left(\left(1-s\right)g\left(\tau\right) + sg\left(t\right)\right)} ds\right) g'\left(\tau\right) d\tau \tag{37}$$

for any $t \in (a, b)$.

Proof For an absolutely continuous function $h : [c, d] \rightarrow \mathbb{C}$ and for x, $y \in [c, d]$ with $x \neq y$ we have

$$\frac{h\left(y\right) - h\left(x\right)}{y - x} = \frac{\int_x^y h'\left(u\right) du}{y - x}.$$

If we use the change of variable $u = (1 - s) x + sy$, $s \in [0, 1]$ we have $du = (y - x) ds$ and then

$$\frac{\int_x^y h'(u)\,du}{y-x} = \frac{(y-x)\int_0^1 h'((1-s)x+sy)\,ds}{y-x} = \int_0^1 h'((1-s)x+sy)\,ds.$$

For $t, \tau \in (a, b)$ with $t \neq \tau$ we then have

$$\frac{f(\tau)-f(t)}{g(\tau)-g(t)} = \frac{f\circ g^{-1}(g(\tau)) - f\circ g^{-1}(g(t))}{g(\tau)-g(t)}$$

$$= \int_0^1 \left(f\circ g^{-1}\right)'((1-s)g(\tau)+sg(t))\,ds.$$

For $z \in (g(a), g(b))$ we have

$$\left(f\circ g^{-1}\right)'(z) = \left(f'\circ g^{-1}\right)(z)\left(g^{-1}\right)'(z) = \frac{\left(f'\circ g^{-1}\right)(z)}{\left(g'\circ g^{-1}\right)(z)}$$

and therefore

$$\int_0^1 \left(f\circ g^{-1}\right)'((1-s)g(\tau)+sg(t))\,ds = \int_0^1 \frac{\left(f'\circ g^{-1}\right)((1-s)g(\tau)+sg(t))}{\left(g'\circ g^{-1}\right)((1-s)g(\tau)+sg(t))}\,ds$$

for $t, \tau \in (a, b)$ with $t \neq \tau$.

This implies that

$$PV\int_a^b \frac{f(\tau)-f(t)}{g(\tau)-g(t)}\,g'(\tau)\,d\tau$$

$$= PV\int_a^b \left(\int_0^1 \frac{\left(f'\circ g^{-1}\right)((1-s)g(\tau)+sg(t))}{\left(g'\circ g^{-1}\right)((1-s)g(\tau)+sg(t))}\,ds\right) g'(\tau)\,d\tau$$

for $t \in (a, b)$ and by the equality (11) we deduce (37).

Now, for $\varphi, \Phi \in \mathbb{C}$ and $[a, b]$ an interval of real numbers, define the sets of complex-valued functions

$$\bar{U}_{[a,b]}(\varphi, \Phi) := \left\{g : [a, b] \to \mathbb{C} \mid \operatorname{Re}\left[(\Phi - g(t))\left(\overline{g(t)} - \bar{\varphi}\right)\right] \geq 0 \text{ for a.e. } t \in [a, b]\right\}$$

and

$$\bar{\Delta}_{[a,b]}(\varphi, \Phi) := \left\{g : [a, b] \to \mathbb{C} \mid \left|g(t) - \frac{\varphi + \Phi}{2}\right| \leq \frac{1}{2}|\Phi - \varphi| \text{ for a.e. } t \in [a, b]\right\}.$$

The following representation result may be stated.

Proposition 3 *For any φ, $\Phi \in \mathbb{C}$, $\varphi \neq \Phi$, we have that $\bar{U}_{[a,b]}(\varphi, \Phi)$ and $\bar{\Delta}_{[a,b]}(\varphi, \Phi)$ are nonempty, convex, and closed sets and*

$$\bar{U}_{[a,b]}(\varphi, \Phi) = \bar{\Delta}_{[a,b]}(\varphi, \Phi). \tag{38}$$

The proof is as in Proposition 1.

Corollary 9 *For any φ, $\Phi \in \mathbb{C}$, $\varphi \neq \Phi$, we have that*

$$\bar{U}_{[a,b]}(\varphi, \Phi) = \{g : [a, b] \to \mathbb{C} \mid (\text{Re}\,\Phi - \text{Re}\,g(t))(\text{Re}\,g(t) - \text{Re}\,\varphi) \tag{39}$$
$$+ (\text{Im}\,\Phi - \text{Im}\,g(t))(\text{Im}\,g(t) - \text{Im}\,\varphi) \geq 0 \text{ for a.e. } t \in [a, b]\}.$$

Now, if we assume that $\text{Re}(\Phi) \geq \text{Re}(\varphi)$ and $\text{Im}(\Phi) \geq \text{Im}(\varphi)$, then we can define the following set of functions as well:

$$\bar{S}_{[a,b]}(\varphi, \Phi) := \{g : [a, b] \to \mathbb{C} \mid \text{Re}(\Phi) \geq \text{Re}\,g(t) \geq \text{Re}(\varphi) \tag{40}$$
$$\text{and } \text{Im}(\Phi) \geq \text{Im}\,g(t) \geq \text{Im}(\varphi) \text{ for a.e. } t \in [a, b]\}.$$

One can easily observe that $\bar{S}_{[a,b]}(\varphi, \Phi)$ is closed, convex, and

$$\emptyset \neq \bar{S}_{[a,b]}(\varphi, \Phi) \subseteq \bar{U}_{[a,b]}(\varphi, \Phi). \tag{41}$$

Theorem 6 *Let $g : [a, b] \to [g(a), g(b)]$ be a strictly increasing function that is differentiable on (a, b) and $f : (a, b) \to \mathbb{R}$ a locally absolutely continuous function on (a, b). Assume that there exists φ, $\Phi \in \mathbb{C}$, $\varphi \neq \Phi$, such that $\frac{f'}{g'} \in \bar{\Delta}_{[a,b]}(\varphi, \Phi)$, then we have*

$$\left| (T_g f)(a, b; t) - \frac{1}{\pi} f(t) \ln\left(\frac{g(b) - g(t)}{g(t) - g(a)} \right) - \frac{\varphi + \Phi}{2\pi}(g(b) - g(a)) \right| \tag{42}$$
$$\leq \frac{1}{2\pi} |\Phi - \varphi| (g(b) - g(a))$$

for all $t \in (a, b)$.

Proof Let t, $\tau \in (a, b)$ with $t \neq \tau$. Since $\frac{f'}{g'} \in \bar{\Delta}_{[a,b]}(\varphi, \Phi)$, hence

$$\left| \frac{(f' \circ g^{-1})((1-s)g(\tau) + sg(t))}{(g' \circ g^{-1})((1-s)g(\tau) + sg(t))} - \frac{\varphi + \Phi}{2} \right| \leq \frac{1}{2}|\Phi - \varphi|$$

for a.e. $s \in [0, 1]$.

Taking the integral over s in this inequality, we get

$$\left| \int_0^1 \frac{(f' \circ g^{-1})((1-s)g(\tau) + sg(t))}{(g' \circ g^{-1})((1-s)g(\tau) + sg(t))} ds - \frac{\varphi + \Phi}{2} \right|$$

$$\leq \int_0^1 \left| \frac{(f' \circ g^{-1})((1-s)g(\tau) + sg(t))}{(g' \circ g^{-1})((1-s)g(\tau) + sg(t))} - \frac{\varphi + \Phi}{2} \right| ds \leq \frac{1}{2} |\Phi - \varphi|$$

for $t, \tau \in (a, b)$ with $t \neq \tau$.

Using the property (29) we get

$$\left| PV \int_a^b \left(\int_0^1 \frac{(f' \circ g^{-1})((1-s)g(\tau) + sg(t))}{(g' \circ g^{-1})((1-s)g(\tau) + sg(t))} ds \right) g'(\tau) d\tau \right.$$

$$\left. - \frac{\varphi + \Phi}{2} (g(b) - g(a)) \right|$$

$$\leq PV \int_a^b \left| \int_0^1 \frac{(f' \circ g^{-1})((1-s)g(\tau) + sg(t))}{(g' \circ g^{-1})((1-s)g(\tau) + sg(t))} ds - \frac{\varphi + \Phi}{2} \right| g'(\tau) d\tau$$

$$\leq \frac{1}{2} |\Phi - \varphi| (g(b) - g(a))$$

for $t \in (a, b)$, and by the equality (37) we deduce the desired result (42).

4 Examples

Consider the following *logarithmic finite Hilbert transform*

$$(T_{\ln} f)(a, b; t) := \frac{1}{\pi} \lim_{\varepsilon \to 0+} \left[\int_a^{t-\varepsilon} \frac{f(\tau)}{\tau \ln\left(\frac{\tau}{t}\right)} d\tau + \int_{t+\varepsilon}^b \frac{f(\tau)}{\tau \ln\left(\frac{\tau}{t}\right)} d\tau \right] \tag{43}$$

where $t \in (a, b) \subset (0, \infty)$.

If we assume that if $f : (a, b) \to \mathbb{R}$ is differentiable on (a, b) and

$$\frac{m}{t} \leq f'(t) \leq \frac{M}{t} \text{ for all } t \in (a, b), \tag{44}$$

then by Remark 4 we have

$$\frac{m}{\pi} \ln\left(\frac{b}{a}\right) \leq (T_{\ln} f)(a, b; t) - \frac{1}{\pi} f(t) \ln\left(\frac{\ln\left(\frac{b}{t}\right)}{\ln\left(\frac{t}{a}\right)}\right) \leq \frac{M}{\pi} \ln\left(\frac{b}{a}\right) \tag{45}$$

for all $t \in (a, b)$.

In particular, we have

$$\frac{m}{\pi} \ln \left(\frac{b}{a} \right) \le (T_{\ln} f) (a, b; G (a, b)) \le \frac{M}{\pi} \ln \left(\frac{b}{a} \right), \tag{46}$$

where $G (a, b) := \sqrt{ab}$ is the *geometric mean* of $a, b > 0$.

This inequality can be extended for complex functions as follows: if $f : (a, b) \to \mathbb{C}$ is locally absolutely continuous on (a, b) and there exist the complex numbers φ, $\Phi \in \mathbb{C}, \varphi \ne \Phi$ such that

$$\left| t f' (t) - \frac{\varphi + \Phi}{2} \right| \le \frac{1}{2} |\Phi - \varphi| \text{ for a.e. } t \in [a, b], \tag{47}$$

then

$$\left| (T_{\ln} f) (a, b; t) - \frac{1}{\pi} f (t) \ln \left(\frac{\ln \left(\frac{b}{t} \right)}{\ln \left(\frac{t}{a} \right)} \right) - \frac{\varphi + \Phi}{2\pi} \ln \left(\frac{b}{a} \right) \right| \tag{48}$$

$$\le \frac{1}{2\pi} |\Phi - \varphi| \ln \left(\frac{b}{a} \right)$$

for all $t \in (a, b)$.

In particular, we have

$$\left| (T_{\ln} f) (a, b; G (a, b)) - \frac{\varphi + \Phi}{2\pi} \ln \left(\frac{b}{a} \right) \right| \le \frac{1}{2\pi} |\Phi - \varphi| \ln \left(\frac{b}{a} \right). \tag{49}$$

Now, observe that the fact that $f \circ \exp$ is of H-r-Hölder type on $(\ln a, \ln b)$, where $H > 0$, $r \in (0, 1]$ and $(a, b) \subset (0, \infty)$, is equivalent to the inequality

$$|f (t) - f (s)| \le H |\ln t - \ln s|^r \text{ for all } t, s \in (a, b),$$

then by (17) we get

$$\left| (T_{\ln} f) (a, b; t) - \frac{1}{\pi} f (t) \ln \left(\frac{\ln \left(\frac{b}{t} \right)}{\ln \left(\frac{t}{a} \right)} \right) \right| \le \frac{H}{\pi r} \left[\left(\ln \left(\frac{b}{t} \right) \right)^r + \left(\ln \left(\frac{t}{a} \right) \right)^r \right] \tag{50}$$

for all $t \in (a, b)$.

In particular, we have

$$\left| (T_g f) (a, b; G (a, b)) \right| \le \frac{H}{2^{r-1} \pi r} \left(\ln \left(\frac{b}{a} \right) \right)^r. \tag{51}$$

Consider the *exponential finite Hilbert transform*

$$\left(T_{\exp(\alpha)} f\right)(a, b; t) \tag{52}$$

$$:= \frac{1}{\pi} \lim_{\varepsilon \to 0+} \left[\int_a^{t-\varepsilon} \frac{f(\tau) \exp(\alpha \tau)}{\exp(\alpha \tau) - \exp(\alpha t)} d\tau + \int_{t+\varepsilon}^b \frac{f(\tau) \exp(\alpha \tau)}{\exp(\alpha \tau) - \exp(\alpha t)} d\tau \right]$$

$$= \frac{1}{\pi} \exp(-\alpha t)$$

$$\times \lim_{\varepsilon \to 0+} \left[\int_a^{t-\varepsilon} \frac{f(\tau) \exp(\alpha(\tau - t))}{\exp(\alpha(\tau - t)) - 1} d\tau + \int_{t+\varepsilon}^b \frac{f(\tau) \exp(\alpha(\tau - t))}{\exp(\alpha(\tau - t)) - 1} d\tau \right]$$

where $t \in (a, b) \subset \mathbb{R}$.

If we assume that if $f : (a, b) \to \mathbb{R}$ is differentiable on (a, b) and

$$n \exp(\alpha t) \le f'(t) \le N \exp(\alpha t) \text{ for all } t \in (a, b),$$

then by applying Remark 4 for $m = \frac{n}{\alpha}$, $M = \frac{N}{\alpha}$ we have

$$\frac{n}{\pi \alpha} \left(\exp(\alpha b) - \exp(\alpha a)\right) \tag{53}$$

$$\le \left(T_{\exp(\alpha)} f\right)(a, b; t) - \frac{1}{\pi} f(t) \ln \left(\frac{\exp(\alpha b) - \exp(\alpha t)}{\exp(\alpha t) - \exp(\alpha a)} \right)$$

$$\le \frac{N}{\pi \alpha} \left(\exp(\alpha b) - \exp(\alpha a)\right)$$

for any $t \in (a, b)$.

If we take in (53)

$$t = LME_\alpha(a, b) := \ln \left(\frac{\exp(\alpha a) + \exp(\alpha b)}{2} \right)^{1/\alpha},$$

then we get

$$\frac{n}{\pi \alpha} \left(\exp(\alpha b) - \exp(\alpha a)\right) \le \left(T_{\exp(\alpha)} f\right)(a, b; LME_\alpha(a, b)) \tag{54}$$

$$\le \frac{N}{\pi \alpha} \left(\exp(\alpha b) - \exp(\alpha a)\right).$$

This inequality can be extended for complex functions as follows: if $f : (a, b) \to \mathbb{C}$ is locally absolutely continuous on (a, b) and there exist the complex numbers φ, $\Phi \in \mathbb{C}$, $\varphi \ne \Phi$ such that

$$\left| \frac{f'(t)}{\exp(\alpha t)} - \frac{\varphi + \Phi}{2} \right| \le \frac{1}{2} |\Phi - \varphi| \text{ for a.e. } t \in [a, b],$$

then

$$\left| \left(T_{\exp(\alpha)} f \right)(a, b; t) - \frac{1}{\pi} f(t) \ln \left(\frac{\exp(\alpha b) - \exp(\alpha t)}{\exp(\alpha t) - \exp(\alpha a)} \right) \right.$$

$$\left. - \frac{\varphi + \Phi}{2\pi\alpha} \left(\exp(\alpha b) - \exp(\alpha a) \right) \right|$$

$$\leq \frac{1}{2\pi\alpha} |\Phi - \varphi| \left(\exp(\alpha b) - \exp(\alpha a) \right) \quad (55)$$

for any $t \in (a, b)$.

In particular, we get

$$\left| \left(T_{\exp(\alpha)} f \right)(a, b; LME_\alpha(a, b)) - \frac{\varphi + \Phi}{2\pi\alpha} \left(\exp(\alpha b) - \exp(\alpha a) \right) \right| \quad (56)$$

$$\leq \frac{1}{2\pi\alpha} |\Phi - \varphi| \left(\exp(\alpha b) - \exp(\alpha a) \right).$$

Now, observe that the fact that $f \circ \left(\frac{1}{\alpha} \ln \right)$ is of *H-r-Hölder type* on $(\exp(\alpha a), \exp(\alpha b))$, where $H > 0$, $r \in (0, 1]$ and $(a, b) \subset \mathbb{R}$, is equivalent to the inequality

$$|f(t) - f(s)| \leq H |\exp(\alpha t) - \exp(\alpha s)|^r \text{ for all } t, s \in (a, b), \quad (57)$$

then by the inequality (17) we get

$$\left| \left(T_{\exp(\alpha)} f \right)(a, b; t) - \frac{1}{\pi} f(t) \ln \left(\frac{\exp(\alpha b) - \exp(\alpha t)}{\exp(\alpha t) - \exp(\alpha a)} \right) \right| \quad (58)$$

$$\leq \frac{H}{\pi r} \left[\left(\exp(\alpha b) - \exp(\alpha t) \right)^r + \left(\exp(\alpha t) - \exp(\alpha a) \right)^r \right]$$

for any $t \in (a, b)$.

In particular, we have

$$\left| \left(T_{\exp(\alpha)} f \right)(a, b; LME_\alpha(a, b)) \right| \leq \frac{H}{2^{r-1}\pi r} \left(\exp(\alpha b) - \exp(\alpha a) \right)^r. \quad (59)$$

For $[a, b] \subset (0, \infty)$ and $g(t) = t^r$, $t \in [a, b]$, $r > 0$, we consider the *positive r-power Hilbert transform*

$$(T_r f)(a, b; t) := \frac{r}{\pi} \lim_{\varepsilon \to 0+} \left[\int_a^{t-\varepsilon} \frac{f(\tau) \tau^{r-1}}{\tau^r - t^r} d\tau + \int_{t+\varepsilon}^b \frac{f(\tau) \tau^{r-1}}{\tau^r - t^r} d\tau \right], \quad (60)$$

where $t \in (a, b)$.

If $f : (a, b) \to \mathbb{R}$ is differentiable on (a, b) and

$$mt^{r-1} \leq f'(t) \leq Mt^{r-1} \text{ for all } t \in (a, b),\qquad (61)$$

then by 32

$$\frac{m}{\pi r}\left(b^r - a^r\right) \leq (T_r f)(a, b; t) - \frac{1}{\pi} f(t) \ln\left(\frac{b^r - t^r}{t^r - a^r}\right) \leq \frac{M}{\pi r}\left(b^r - a^r\right)\qquad (62)$$

for all $t \in (a, b) \subset (0, \infty)$.

In particular, we have

$$\frac{m}{\pi r}\left(b^r - a^r\right) \leq (T_r f)(a, b; M_r(a, b)) \leq \frac{M}{\pi r}\left(b^r - a^r\right)\qquad (63)$$

where $M_r(a, b) := \left(\frac{a^r + b^r}{2}\right)^{1/r}$

Also, if $f : (a, b) \to \mathbb{C}$ is locally absolutely continuous on (a, b) and there exist the complex numbers $\varphi, \Phi \in \mathbb{C}, \varphi \neq \Phi$ such that

$$\left|\frac{f'(t)}{t^{r-1}} - \frac{\varphi + \Phi}{2}\right| \leq \frac{1}{2}|\Phi - \varphi| \text{ for a.e. } t \in [a, b],\qquad (64)$$

then

$$\left|(T_r f)(a, b; t) - \frac{1}{\pi} f(t) \ln\left(\frac{b^r - t^r}{t^r - a^r}\right) - \frac{\varphi + \Phi}{2\pi r}\left(b^r - a^r\right)\right|\qquad (65)$$

$$\leq \frac{1}{2\pi r}|\Phi - \varphi|\left(b^r - a^r\right)$$

for all $t \in (a, b) \subset (0, \infty)$.

In particular, we have

$$\left|(T_r f)(a, b; M_r(a, b)) - \frac{\varphi + \Phi}{2\pi r}\left(b^r - a^r\right)\right| \leq \frac{1}{2\pi r}|\Phi - \varphi|\left(b^r - a^r\right).\qquad (66)$$

The function $f \circ (\cdot)^{1/r}$ is of H-s-Hölder type on (a^r, b^r), where $H > 0$, $s \in (0, 1]$, is equivalent to

$$|f(t) - f(u)| \leq H\left|t^r - u^r\right|^s \text{ for all } t, u \in (a, b),$$

then by (17) we have

$$\left|(T_r f)(a, b; t) - \frac{1}{\pi} f(t) \ln\left(\frac{b^r - t^r}{t^r - a^r}\right)\right| \leq \frac{H}{\pi s}\left[\left(b^r - t^r\right)^s + \left(t^r - a^r\right)^s\right]\qquad (67)$$

for all $t \in (a, b) \subset (0, \infty)$.

In particular,

$$|(T_r f)(a, b; M_r(a, b))| \leq \frac{H}{2^{s-1}\pi s}(b^r - a^r)^s. \tag{68}$$

The case $r = 1$ provides the corresponding results for the regular Hilbert transform, see [1].

Similarly, we can consider the *negative p-power Hilbert transform*

$$\left(T_{-p} f\right)(a, b; t) := \frac{pt^p}{\pi} \lim_{\varepsilon \to 0+} \left[\int_a^{t-\varepsilon} \frac{f(\tau)}{\tau(\tau^p - t^p)} d\tau + \int_{t+\varepsilon}^b \frac{f(\tau)}{\tau(\tau^p - t^p)} d\tau \right], \tag{69}$$

for $[a, b] \subset (0, \infty)$ and $p > 0$.

If $f : (a, b) \to \mathbb{R}$ is differentiable on (a, b) and

$$m \leq t^{p+1} f'(t) \leq M \text{ for all } t \in (a, b), \tag{70}$$

then by (32) we have

$$\frac{m}{\pi p}\left(\frac{b^p - a^p}{a^p b^p}\right) \leq \left(T_{-p} f\right)(a, b; t) - \frac{1}{\pi} f(t) \ln\left(\frac{(b^p - t^p) a^p}{b^p (t^p - a^p)}\right) \tag{71}$$

$$\leq \frac{M}{\pi p}\left(\frac{b^p - a^p}{a^p b^p}\right)$$

for all $t \in (a, b)$.

In particular,

$$\frac{m}{\pi p}\left(\frac{b^p - a^p}{a^p b^p}\right) \leq \left(T_{-p} f\right)(a, b; M_{-p}(a, b)) - \frac{1}{\pi} f(t) \ln\left(\frac{(b^p - t^p) a^p}{b^p (t^p - a^p)}\right) \tag{72}$$

$$\leq \frac{M}{\pi p}\left(\frac{b^p - a^p}{a^p b^p}\right)$$

where $M_{-p}(a, b) := \left(\frac{a^{-p}+b^{-p}}{2}\right)^{-1/p}$.

The case $p = 1$ is of interest, since in this case

$$(T_{-1} f)(a, b; t) := \frac{t}{\pi} \lim_{\varepsilon \to 0+} \left[\int_a^{t-\varepsilon} \frac{f(\tau)}{\tau(\tau - t)} d\tau + \int_{t+\varepsilon}^b \frac{f(\tau)}{\tau(\tau - t)} d\tau \right], \tag{73}$$

and if

$$m \leq t^2 f'(t) \leq M \text{ for all } t \in (a, b), \tag{74}$$

then

$$\frac{m}{\pi}\left(\frac{b-a}{ab}\right) \le (T_{-1}f)(a,b;t) - \frac{1}{\pi}f(t)\ln\left(\frac{(b-t)a}{b(t-a)}\right) \le \frac{M}{\pi}\left(\frac{b-a}{ab}\right) \quad (75)$$

for all $t \in (a,b)$.

In particular, we have

$$\frac{m}{\pi}\left(\frac{b-a}{ab}\right) \le (T_{-1}f)(a,b;H(a,b)) \le \frac{M}{\pi}\left(\frac{b-a}{ab}\right) \quad (76)$$

where $H(a,b) := \frac{2ab}{a+b}$ is the *harmonic mean* of $a, b > 0$.

Also, if $f : (a,b) \to \mathbb{C}$ is locally absolutely continuous on (a,b) and there exist the complex numbers $\varphi, \Phi \in \mathbb{C}, \varphi \ne \Phi$ such that

$$\left|t^2 f'(t) - \frac{\varphi+\Phi}{2}\right| \le \frac{1}{2}|\Phi - \varphi| \text{ for a.e. } t \in [a,b], \quad (77)$$

then

$$\left|(T_{-1}f)(a,b;t) - \frac{1}{\pi}f(t)\ln\left(\frac{(b-t)a}{b(t-a)}\right) - \frac{\gamma+\Gamma}{2\pi}\left(\frac{b-a}{ab}\right)\right| \quad (78)$$

$$\le \frac{1}{2\pi}|\Gamma - \gamma|\left(\frac{b-a}{ab}\right)$$

for all $t \in (a,b)$.

In particular, we have

$$\left|(T_{-1}f)(a,b;H(a,b)) - \frac{\gamma+\Gamma}{2\pi}\left(\frac{b-a}{ab}\right)\right| \le \frac{1}{2\pi}|\Gamma - \gamma|\left(\frac{b-a}{ab}\right). \quad (79)$$

The fact that $f \circ \left[-(\cdot)^{-1}\right]$ is of *H-s-Hölder type* on $\left(-\frac{1}{a}, -\frac{1}{b}\right)$, where $K > 0$, $s \in (0,1]$, is equivalent to

$$|f(t) - f(u)| \le K\left|\frac{t-u}{tu}\right|^s \text{ for all } t, u \in (a,b),$$

then by (17) we have

$$\left|(T_{-1}f)(a,b;t) - \frac{1}{\pi}f(t)\ln\left(\frac{(b-t)a}{b(t-a)}\right)\right| \le \frac{K}{\pi s}\left[\left(\frac{b-t}{bt}\right)^s + \left(\frac{t-a}{ta}\right)^s\right]$$
$$(80)$$

for all $t \in (a,b)$.

In particular, we have

$$|(T_{-1}f)(a, b; H(a, b))| \leq \frac{K}{2^{s-1}\pi s}\left(\frac{b-a}{ba}\right)^s. \qquad (81)$$

References

1. N.M. Dragomir, S.S. Dragomir, P.M. Farrell, Some inequalities for the finite Hilbert transform, in *Inequality Theory and Applications*, Nova Science Publishers, vol. I (Huntington, NY, 2001), pp. 113–122
2. N.M. Dragomir, S.S. Dragomir, P.M. Farrell, Approximating the finite Hilbert transform via trapezoid type inequalities. Comput. Math. Appl. **43**(10–11), 1359–1369 (2002)
3. N.M. Dragomir, S.S. Dragomir, P.M. Farrell, G.W. Baxter, On some new estimates of the finite Hilbert transform. Libertas Math. **22**, 65–75 (2002)
4. N.M. Dragomir, S.S. Dragomir, P.M. Farrell, G.W. Baxter, A quadrature rule for the finite Hilbert transform via trapezoid type inequalities. J. Appl. Math. Comput. **13**(1–2), 67–84 (2003)
5. N.M. Dragomir, S.S. Dragomir, P.M. Farrell, G.W. Baxter, A quadrature rule for the finite Hilbert transform via midpoint type inequalities, in *Fixed Point Theory and Applications*, Nova Science Publishers, vol. 5, (Hauppauge, NY, 2004), pp. 11–22
6. S.S. Dragomir, Approximating the finite Hilbert transform via an Ostrowski type inequality for functions of bounded variation. J. Inequal. Pure Appl. Math. **3**(4), Article 51, 19 (2002)
7. S.S. Dragomir, Approximating the finite Hilbert transform via Ostrowski type inequalities for absolutely continuous functions. Bull. Korean Math. Soc. **39**(4), 543–559 (2002)
8. S.S. Dragomir, Inequalities for the Hilbert transform of functions whose derivatives are convex. J. Korean Math. Soc. **39**(5), 709–729 (2002)
9. S.S. Dragomir, Some inequalities for the finite Hilbert transform of a product. Commun. Korean Math. Soc. **18**(1), 39–57 (2003)
10. S.S. Dragomir, Sharp error bounds of a quadrature rule with one multiple node for the finite Hilbert transform in some classes of continuous differentiable functions. Taiwanese J. Math. **9**(1), 95–109 (2005)
11. S.S. Dragomir, The perturbed median principle for integral inequalities with applications, in *Nonlinear Analysis and Variational Problems*, Springer Optimization and Its Applications, vol. 35 (Springer, New York, 2010), pp. 53–63
12. S.S. Dragomir, Inequalities and approximations for the Finite Hilbert transform: a survey of recent results, Preprint *RGMIA Res. Rep. Coll.* **21**, Article 30, 560 (2018). http://rgmia.org/papers/v21/v21a30.pdf
13. F.D. Gakhov, *Boundary Value Problems* (English translation) (Pergamon Press, Oxford, 1966)
14. W. Liu, X. Gao, Approximating the finite Hilbert transform via a companion of Ostrowski's inequality for function of bounded variation and applications. Appl. Math. Comput. **247**, 373–385 (2014)
15. W. Liu, X. Gao, Y. Wen, Approximating the finite Hilberttransform via some companions of Ostrowski's inequalities. Bull. Malays. Math. Sci. Soc. **39**(4), 1499–1513 (2016)
16. W. Liu, N. Lu, Approximating the finite Hilbert transform via Simpson type inequalities and applications. Politehn. Univ. Bucharest Sci. Bull. Ser. A Appl. Math. Phys. **77**(3), 107–122 (2015)

17. S.G. Mikhlin, S. Prössdorf, *Singular Integral Operators* (English translation) (Springer Verlag, Berlin, 1986)
18. S. Wang, X. Gao, N. Lu, A quadrature formula in approximating the finite Hilbert transform via perturbed trapezoid type inequalities. J. Comput. Anal. Appl. **22**(2), 239–246 (2017)
19. S. Wang, N. Lu, X. Gao, A quadrature rule for the finite Hilberttransform via Simpson type inequalities and applications. J. Comput. Anal. Appl. **22**(2), 229–238 (2017)

Inequalities of Hermite–Hadamard Type for Composite Convex Functions

Silvestru Sever Dragomir

Abstract In this paper we obtain some inequalities of Hermite–Hadamard type for composite convex functions. Applications for AG, AH-convex functions, GA, GG, GH-convex functions, and HA, HG, HH-convex function are given. Applications for p, r-convex, and $LogExp$ convex functions are presented as well.

1 Introduction

The following inequality holds for any convex function f defined on $\mathbb{R}$

$$f\left(\frac{a+b}{2}\right) \leq \frac{1}{b-a} \int_a^b f(x)dx \leq \frac{f(a)+f(b)}{2}, \quad a,\, b \in \mathbb{R}, a < b. \tag{1}$$

It was firstly discovered by Ch. Hermite in 1881 in the journal *Mathesis* (see [18]). But this result was nowhere mentioned in the mathematical literature and was not widely known as Hermite's result.

E. F. Beckenbach, a leading expert on the history and the theory of convex functions, wrote that this inequality was proven by J. Hadamard in 1893 [3]. In 1974, D. S. Mitrinović found Hermite's note in *Mathesis* [18]. Since (1) was known as Hadamard's inequality, the inequality is now commonly referred to as the *Hermite–Hadamard inequality*.

In order to extend this result for other classes of functions, we need the following preparations.

S. S. Dragomir (✉)
Mathematics, College of Engineering & Science, Victoria University,
Melbourne City, MC, Australia

DST-NRF Centre of Excellence in the Mathematical and Statistical Sciences,
School of Computer Science & Applied Mathematics, University of the Witwatersrand,
Johannesburg, South Africa
e-mail: sever.dragomir@vu.edu.au

© Springer Nature Switzerland AG 2019
G. A. Anastassiou and J. M. Rassias (eds.), *Frontiers in Functional Equations and Analytic Inequalities*, https://doi.org/10.1007/978-3-030-28950-8_28

Let $g : [a, b] \to [g(a), g(b)]$ be a *continuous strictly increasing function* that is *differentiable* on (a, b).

Definition 1 A function $f : [a, b] \to \mathbb{R}$ will be called *composite-g^{-1} convex (concave) on* $[a, b]$ if the composite function $f \circ g^{-1} : [g(a), g(b)] \to \mathbb{R}$ is convex (concave) in the usual sense on $[g(a), g(b)]$.

In this way, any concept of convexity (log-convexity, harmonic convexity, trigonometric convexity, hyperbolic convexity, h-convexity, quasi-convexity, s-convexity, s-Godunova–Levin convexity, etc.) can be extended to the corresponding *composite-g^{-1} convexity*. The details however will not be presented here.

If $f : [a, b] \to \mathbb{R}$ is composite-g^{-1} convex on $[a, b]$, then we have the inequality

$$f \circ g^{-1}((1 - \lambda) u + \lambda v) \le (1 - \lambda) f \circ g^{-1}(u) + \lambda f \circ g^{-1}(v) \tag{2}$$

for any $u, v \in [g(a), g(b)]$ and $\lambda \in [0, 1]$.

This is equivalent to the condition

$$f \circ g^{-1}((1 - \lambda) g(t) + \lambda g(s)) \le (1 - \lambda) f(t) + \lambda f(s) \tag{3}$$

for any $t, s \in [a, b]$ and $\lambda \in [0, 1]$.

If we take $g(t) = \ln t, t \in [a, b] \subset (0, \infty)$, then the condition (3) becomes

$$f\left(t^{1-\lambda} s^{\lambda}\right) \le (1 - \lambda) f(t) + \lambda f(s) \tag{4}$$

for any $t, s \in [a, b]$ and $\lambda \in [0, 1]$, which is the concept of GA-*convexity* as considered in [1].

If we take $g(t) = -\frac{1}{t}, t \in [a, b] \subset (0, \infty)$, then (3) becomes

$$f\left(\frac{ts}{(1 - \lambda) s + \lambda t}\right) \le (1 - \lambda) f(t) + \lambda f(s) \tag{5}$$

for any $t, s \in [a, b]$ and $\lambda \in [0, 1]$, which is the concept of HA-*convexity* as considered in [1].

If $p > 0$ and we consider $g(t) = t^p, t \in [a, b] \subset (0, \infty)$, then the condition (3) becomes

$$f\left[((1 - \lambda) t^p + \lambda s^p)^{1/p}\right] \le (1 - \lambda) f(t) + \lambda f(s) \tag{6}$$

for any $t, s \in [a, b]$ and $\lambda \in [0, 1]$, which is the concept of p-*convexity* as considered in [22].

If we take $g(t) = \exp t, t \in [a, b]$, then the condition (3) becomes

$$f\left[\ln((1 - \lambda) \exp(t) + \exp g(s))\right] \le (1 - \lambda) f(t) + \lambda f(s) \tag{7}$$

which is the concept of *LogExp convex function* on $[a, b]$ as considered in [14].

Further, assume that $f : [a, b] \to J$, J an interval of real numbers and $k : J \to \mathbb{R}$ a continuous function on J that is *strictly increasing (decreasing)* on J.

Definition 2 We say that the function $f : [a, b] \to J$ is k-composite convex (concave) on $[a, b]$, if $k \circ f$ is convex (concave) on $[a, b]$.

In this way, any concept of convexity as mentioned above can be extended to the corresponding *k-composite convexity*. The details however will not be presented here.

With $g : [a, b] \to [g(a), g(b)]$ a *continuous strictly increasing function* that is *differentiable* on (a, b), $f : [a, b] \to J$, J an interval of real numbers and $k : J \to \mathbb{R}$ a continuous function on J that is *strictly increasing (decreasing)* on J, we can also consider the following concept:

Definition 3 We say that the function $f : [a, b] \to J$ is k-composite-g^{-1} convex (concave) on $[a, b]$, if $k \circ f \circ g^{-1}$ is convex (concave) on $[g(a), g(b)]$.

This definition is equivalent to the condition

$$k \circ f \circ g^{-1} ((1 - \lambda) g(t) + \lambda g(s)) \le (1 - \lambda)(k \circ f)(t) + \lambda (k \circ f)(s) \qquad (8)$$

for any $t, s \in [a, b]$ and $\lambda \in [0, 1]$.

If $k : J \to \mathbb{R}$ is *strictly increasing (decreasing)* on J, then the condition (8) is equivalent to:

$$f \circ g^{-1}((1 - \lambda) g(t) + \lambda g(s)) \le (\ge) k^{-1} [(1 - \lambda)(k \circ f)(t) + \lambda (k \circ f)(s)] \qquad (9)$$

for any $t, s \in [a, b]$ and $\lambda \in [0, 1]$.

If $k(t) = \ln t, t > 0$ and $f : [a, b] \to (0, \infty)$, then the fact that f is k-composite convex on $[a, b]$ is equivalent to the fact that f is *log-convex* or *multiplicatively convex* or *AG*-convex, namely, for all $x, y \in I$ and $t \in [0, 1]$ one has the inequality:

$$f(tx + (1 - t) y) \le [f(x)]^t [f(y)]^{1-t}. \qquad (10)$$

A function $f : I \to \mathbb{R} \setminus \{0\}$ is called *AH-convex (concave)* on the interval I if the following inequality holds [1]

$$f((1 - \lambda) x + \lambda y) \le (\ge) \frac{1}{(1 - \lambda) \frac{1}{f(x)} + \lambda \frac{1}{f(y)}} = \frac{f(x) f(y)}{(1 - \lambda) f(y) + \lambda f(x)} \qquad (11)$$

for any $x, y \in I$ and $\lambda \in [0, 1]$.

An important case that provides many examples is that one in which the function is assumed to be positive for any $x \in I$. In that situation the inequality (11) is equivalent to

$$(1 - \lambda) \frac{1}{f(x)} + \lambda \frac{1}{f(y)} \leq (\geq) \frac{1}{f((1 - \lambda)x + \lambda y)}$$

for any $x, y \in I$ and $\lambda \in [0, 1]$.

Taking into account this fact, we can conclude that the function $f : I \to (0, \infty)$ is AH-convex (concave) on I if and only if f is k-composite concave (convex) on I with $k : (0, \infty) \to (0, \infty)$, $k(t) = \frac{1}{t}$.

Following [1], we can introduce the concept of GH-convex (concave) function $f : I \subset (0, \infty) \to \mathbb{R}$ on an interval of positive numbers I as satisfying the condition

$$f\left(x^{1-\lambda}y^{\lambda}\right) \leq (\geq) \frac{1}{(1 - \lambda)\frac{1}{f(x)} + \lambda\frac{1}{f(y)}} = \frac{f(x) f(y)}{(1 - \lambda) f(y) + \lambda f(x)}. \tag{12}$$

Since

$$f\left(x^{1-\lambda}y^{\lambda}\right) = f \circ \exp\left[(1 - \lambda)\ln x + \lambda \ln y\right]$$

and

$$\frac{f(x) f(y)}{(1 - \lambda) f(y) + \lambda f(x)} = \frac{f \circ \exp(\ln x) f \circ \exp(\ln y)}{(1 - \lambda) f \circ \exp(y) + \lambda f \circ \exp(x)}$$

then $f : I \subset (0, \infty) \to \mathbb{R}$ is GH-convex (concave) on I if and only if $f \circ \exp$ is AH-convex (concave) on $\ln I := \{x \mid x = \ln t, \ t \in I\}$. This is equivalent to the fact that f is k-composite-g^{-1} concave (convex) on I with $k : (0, \infty) \to (0, \infty)$, $k(t) = \frac{1}{t}$ and $g(t) = \ln t$, $t \in I$.

Following [1], we say that the function $f : I \subset \mathbb{R} \setminus \{0\} \to (0, \infty)$ is HH-convex if

$$f\left(\frac{xy}{tx + (1 - t)y}\right) \leq \frac{f(x) f(y)}{(1 - t) f(y) + tf(x)} \tag{13}$$

for all $x, y \in I$ and $t \in [0, 1]$. If the inequality in (13) is reversed, then f is said to be HH-concave.

We observe that the inequality (13) is equivalent to

$$(1 - t) \frac{1}{f(x)} + t \frac{1}{f(y)} \leq \frac{1}{f\left(\frac{xy}{tx + (1 - t)y}\right)} \tag{14}$$

for all $x, y \in I$ and $t \in [0, 1]$.

This is equivalent to the fact that f is k-composite-g^{-1} concave on $[a, b]$ with $k : (0, \infty) \to (0, \infty)$, $k(t) = \frac{1}{t}$ and $g(t) = -\frac{1}{t}$, $t \in [a, b]$.

The function $f : I \subset (0, \infty) \to (0, \infty)$ is called GG-convex on the interval I of real umbers $\mathbb{R}$ if [1]

$$f\left(x^{1-\lambda}y^{\lambda}\right) \leq [f(x)]^{1-\lambda}[f(y)]^{\lambda} \tag{15}$$

for any $x, y \in I$ and $\lambda \in [0, 1]$. If the inequality is reversed in (15), then the function is called GG-concave.

This concept was introduced in 1928 by Montel [19], however, the roots of the research in this area can be traced long before him [20]. It is easy to see that [20], the function $f : [a, b] \subset (0, \infty) \to (0, \infty)$ is GG-convex if and only if the function $g : [\ln a, \ln b] \to \mathbb{R}$, $g = \ln \circ f \circ \exp$ is convex on $[\ln a, \ln b]$. This is equivalent to the fact that f is k-composite-g^{-1} convex on $[a, b]$ with $k : (0, \infty) \to \mathbb{R}$, $k(t) = \ln t$ and $g(t) = \ln t$, $t \in [a, b]$.

Following [1] we say that the function $f : I \subset \mathbb{R} \setminus \{0\} \to (0, \infty)$ is HG-convex if

$$f\left(\frac{xy}{tx + (1-t)y}\right) \leq [f(x)]^{1-t}[f(y)]^{t} \tag{16}$$

for all $x, y \in I$ and $t \in [0, 1]$. If the inequality in (3) is reversed, then f is said to be HG-concave.

Let $f : [a, b] \subset (0, \infty) \to (0, \infty)$ and define the associated functions $G_f : \left[\frac{1}{b}, \frac{1}{a}\right] \to \mathbb{R}$ defined by $G_f(t) = \ln f\left(\frac{1}{t}\right)$. Then f is HG-convex on $[a, b]$ iff G_f is convex on $\left[\frac{1}{b}, \frac{1}{a}\right]$. This is equivalent to the fact that f is k-composite-g^{-1} convex on $[a, b]$ with $k : (0, \infty) \to \mathbb{R}$, $k(t) = \ln t$ and $g(t) = -\frac{1}{t}$, $t \in [a, b]$.

Following [21], we say that the function $f : [a, b] \to (0, \infty)$ is r-convex, for $r \neq 0$, if

$$f((1-\lambda)x + \lambda y) \leq \left[(1-\lambda)f^r(y) + \lambda f^r(x)\right]^{1/r} \tag{17}$$

for any $x, y \in [a, b]$ and $\lambda \in [0, 1]$.

If $r > 0$, then the condition (17) is equivalent to

$$f^r((1-\lambda)x + \lambda y) \leq (1-\lambda)f^r(y) + \lambda f^r(x)$$

namely f is k-composite convex on $[a, b]$ where $k(t) = t^r$, $t \geq 0$.

If $r < 0$, then the condition (17) is equivalent to

$$f^r((1-\lambda)x + \lambda y) \geq (1-\lambda)f^r(y) + \lambda f^r(x)$$

namely f is k-composite concave on $[a, b]$ where $k(t) = t^r$, $t > 0$.

In this paper we obtain some inequalities of Hermite–Hadamard type for *composite convex functions*. Applications for various classes of convexity as provided above are given as well.

2 Some Refinements

We need the following refinement of Hermite–Hadamard inequality. This result was obtained for the first time by Barnett, Cerone, and Dragomir in 2002 in the paper [2, p. 10, Eq. (2.2)] where various applications for the Hermite–Hadamard divergence measure in Information Theory were also given. The same result was also rediscovered by El Farissi in 2010 with a similar proof, see [16].

Lemma 1 *Assume that $h : [c, d] \to \mathbb{R}$ is convex on $[c, d]$. Then for any $\lambda \in [0, 1]$ we have*

$$h\left(\frac{c+d}{2}\right) \leq \lambda h\left(\frac{\lambda d + (2-\lambda)c}{2}\right) + (1-\lambda)h\left(\frac{(1+\lambda)d + (1-\lambda)c}{2}\right)$$

$$\text{(18)}$$

$$\leq \frac{1}{d-c}\int_c^d h(u)\, du$$

$$\leq \frac{1}{2}\left[h((1-\lambda)c + \lambda d) + \lambda h(c) + (1-\lambda)h(d)\right] \leq \frac{h(c)+h(d)}{2}.$$

Proof For the sake of completeness, we give here a simple proof as in [2]. Applying the Hermite–Hadamard inequality on each subinterval $[c, (1-\lambda)c + \lambda d]$, $[(1-\lambda)c + \lambda d, d]$, where $\lambda \in (0, 1)$, then we have

$$h\left(\frac{c + (1-\lambda)c + \lambda d}{2}\right) \times [(1-\lambda)c + \lambda d - c]$$

$$\leq \int_c^{(1-\lambda)c+\lambda d} h(u)\, du$$

$$\leq \frac{h((1-\lambda)c + \lambda d) + h(c)}{2} \times [(1-\lambda)c + \lambda d - c]$$

and

$$h\left(\frac{(1-\lambda)c + \lambda d + d}{2}\right) \times [d - (1-\lambda)c - \lambda d]$$

$$\leq \int_{(1-\lambda)c+\lambda d}^d h(u)\, du$$

$$\leq \frac{h(d) + h((1-\lambda)c + \lambda d)}{2} \times [d - (1-\lambda)c - \lambda d],$$

which are clearly equivalent to

$$\lambda h \left(\frac{\lambda d + (2 - \lambda) c}{2} \right) \le \frac{1}{d - c} \int_c^{(1-\lambda)c+\lambda d} h (u) \, du \tag{19}$$

$$\le \frac{\lambda h ((1 - \lambda) c + \lambda d) + \lambda h (c)}{2}$$

and

$$(1 - \lambda) h \left(\frac{(1 + \lambda) d + (1 - \lambda) c}{2} \right) \le \frac{1}{d - c} \int_{(1-\lambda)c+\lambda d}^d h (u) \, du \tag{20}$$

$$\le \frac{(1 - \lambda) h (d) + (1 - \lambda) h ((1 - \lambda) c + \lambda d)}{2},$$

respectively.

Summing (19) and (20), we obtain the second and first inequality in (18).

By the convexity property, we obtain

$$\lambda h \left(\frac{\lambda d + (2 - \lambda) c}{2} \right) + (1 - \lambda) h \left(\frac{(1 + \lambda) d + (1 - \lambda) c}{2} \right)$$

$$\ge h \left[\lambda \left(\frac{\lambda d + (2 - \lambda) c}{2} \right) + (1 - \lambda) \left(\frac{(1 + \lambda) d + (1 - \lambda) c}{2} \right) \right]$$

$$= h \left(\frac{c + d}{2} \right)$$

and the first inequality in (18) is proved.

For various inequalities of Hermite–Hadamard type, see the monograph online [15] and the more recent survey paper [12].

If g is a function which maps an interval I of the real line to the real numbers, and is both continuous and injective, then we can define the g-mean of two numbers $a, b \in I$ as

$$M_g (a, b) := g^{-1} \left(\frac{g (a) + g (b)}{2} \right). \tag{21}$$

If $I = \mathbb{R}$ and $g (t) = t$ is the *identity function*, then $M_g (a, b) = A (a, b) := \frac{a+b}{2}$, the *arithmetic mean*. If $I = (0, \infty)$ and $g (t) = \ln t$, then $M_g (a, b) = G (a, b) := \sqrt{ab}$, the *geometric mean*. If $I = (0, \infty)$ and $g (t) = \frac{1}{t}$, then $M_g (a, b) = H (a, b) := \frac{2ab}{a+b}$, the *harmonic mean*. If $I = (0, \infty)$ and $g (t) = t^p$, $p \ne 0$, then $M_g (a, b) = M_p (a, b) := \left(\frac{a^p + b^p}{2} \right)^{1/p}$, the *power mean with exponent p*. Finally, if $I = \mathbb{R}$ and $g (t) = \exp t$, then

$$M_g(a, b) = LME(a, b) := \ln\left(\frac{\exp a + \exp b}{2}\right),\tag{22}$$

the *LogMeanExp function.*

Theorem 1 *Let* $g : [a, b] \to [g(a), g(b)]$ *be a continuous strictly increasing function that is differentiable on* (a, b). *If* $f : [a, b] \to \mathbb{R}$ *is composite-*g^{-1} *convex on* $[a, b]$, *then*

$$f\left(M_g(a, b)\right) \leq \lambda f \circ g^{-1}\left(\frac{\lambda g(b) + (2 - \lambda) g(a)}{2}\right)\tag{23}$$

$$+ (1 - \lambda) f \circ g^{-1}\left(\frac{(1 + \lambda) g(b) + (1 - \lambda) g(a)}{2}\right)$$

$$\leq \frac{1}{g(b) - g(a)} \int_a^b f(t) g'(t) \, dt$$

$$\leq \frac{1}{2}\left[f \circ g^{-1}((1 - \lambda) g(a) + \lambda g(b)) + \lambda f(a) + (1 - \lambda) f(b)\right]$$

$$\leq \frac{f(a) + f(b)}{2}$$

for any $\lambda \in [0, 1]$.

Proof From the inequality (18) we have for the convex function $f \circ g^{-1}$ and c, $d \in [g(a), g(b)]$ that

$$f \circ g^{-1}\left(\frac{c + d}{2}\right)\tag{24}$$

$$\leq \lambda f \circ g^{-1}\left(\frac{\lambda d + (2 - \lambda) c}{2}\right) + (1 - \lambda) f \circ g^{-1}\left(\frac{(1 + \lambda) d + (1 - \lambda) c}{2}\right)$$

$$\leq \frac{1}{d - c} \int_c^d f \circ g^{-1}(u) \, du$$

$$\leq \frac{1}{2}\left[f \circ g^{-1}((1 - \lambda) c + \lambda d) + \lambda f \circ g^{-1}(c) + (1 - \lambda) f \circ g^{-1}(d)\right]$$

$$\leq \frac{f \circ g^{-1}(c) + f \circ g^{-1}(d)}{2}$$

for any $\lambda \in [0, 1]$.

If we take $c = g(a)$ and $d = g(b)$, then we get

$$f \circ g^{-1}\left(\frac{g(a) + g(b)}{2}\right)\tag{25}$$

$$\leq \lambda f \circ g^{-1}\left(\frac{\lambda g(b) + (2 - \lambda) g(a)}{2}\right)$$

$$+ (1 - \lambda) f \circ g^{-1} \left(\frac{(1 + \lambda) g (b) + (1 - \lambda) g (a)}{2} \right)$$

$$\leq \frac{1}{g (b) - g (a)} \int_{g(a)}^{g(b)} f \circ g^{-1} (u) \, du$$

$$\leq \frac{1}{2} \left[f \circ g^{-1} ((1 - \lambda) g (a) + \lambda g (b)) + \lambda f (a) + (1 - \lambda) f (b) \right]$$

$$\leq \frac{f (a) + f (b)}{2}$$

for any $\lambda \in [0, 1]$.

Using the change of variable $g^{-1} (u) = t, t \in [a, b]$ we have $u = g (t)$, $du = g' (t) \, dt$ and

$$\int_{g(a)}^{g(b)} f \circ g^{-1} (u) \, du = \int_a^b f (t) g' (t) \, dt$$

and by (25) we get the desired result (23).

Corollary 1 *With the assumptions of Theorem 1 we have*

$$f \left(M_g (a, b) \right) \leq \frac{1}{2} \left[f \circ g^{-1} \left(\frac{g (b) + 3g (a)}{4} \right) + f \circ g^{-1} \left(\frac{g (a) + 3g (b)}{4} \right) \right] \tag{26}$$

$$\leq \frac{1}{g (b) - g (a)} \int_a^b f (t) g' (t) \, dt$$

$$\leq \frac{1}{2} \left[f \left(M_g (a, b) \right) + \frac{f (a) + f (b)}{2} \right] \leq \frac{f (a) + f (b)}{2}.$$

Remark 1 Using the change of variable $u = (1 - s) c + sd$, $s \in [0, 1]$, then we have $du = (d - c) \, ds$, which gives that

$$\frac{1}{d - c} \int_c^d h (u) \, du = \int_0^1 h ((1 - s) c + sd) \, ds.$$

Using this fact, we have from Theorem 1 the following inequality

$$f \left(M_g (a, b) \right) \leq \lambda f \circ g^{-1} \left(\frac{\lambda g (b) + (2 - \lambda) g (a)}{2} \right) \tag{27}$$

$$+ (1 - \lambda) f \circ g^{-1} \left(\frac{(1 + \lambda) g (b) + (1 - \lambda) g (a)}{2} \right)$$

$$\leq \frac{b-a}{g(b)-g(a)} \int_0^1 f((1-s)a+sb)\, g'((1-s)a+sb)\, ds$$

$$= \int_0^1 f \circ g^{-1}((1-\tau)g(a)+\tau g(b))\, d\tau$$

$$\leq \frac{1}{2}\left[f \circ g^{-1}((1-\lambda)g(a)+\lambda g(b))+\lambda f(a)+(1-\lambda)f(b)\right]$$

$$\leq \frac{f(a)+f(b)}{2}$$

for all $\lambda \in [0, 1]$.

Corollary 2 *Let* $g : [a, b] \rightarrow [g(a), g(b)]$ *be a continuous strictly increasing function that is differentiable on* (a, b), $f : [a, b] \rightarrow J$, *J an interval of real numbers, and* $k : J \rightarrow \mathbb{R}$ *a continuous function on J that is strictly increasing (decreasing) on J. If the function* $f : [a, b] \rightarrow J$ *is k-composite-g^{-1} convex on* $[a, b]$, *then*

$$f\left(M_g(a, b)\right)$$

$$\leq (\geq) k^{-1} \left\{ \lambda k \circ f \circ g^{-1}\left(\frac{\lambda g(b)+(2-\lambda)g(a)}{2}\right) \right.$$

$$\left. + (1-\lambda)k \circ f \circ g^{-1}\left(\frac{(1+\lambda)g(b)+(1-\lambda)g(a)}{2}\right)\right\}$$

$$\leq (\geq) k^{-1}\left(\frac{1}{g(b)-g(a)}\int_a^b k \circ f(t)\, g'(t)\, dt\right)$$

$$\leq (\geq) k^{-1}\left\{\frac{1}{2}\left[k \circ f \circ g^{-1}((1-\lambda)g(a)+\lambda g(b))+\lambda k \circ f(a)+(1-\lambda)k \circ f(b)\right]\right\}$$

$$\leq (\geq) k^{-1}\left(\frac{k \circ f(a)+k \circ f(b)}{2}\right) \qquad (28)$$

for any $\lambda \in [0, 1]$.

Proof From (23) we have

$$k \circ f\left(M_g(a, b)\right) \qquad (29)$$

$$\leq \lambda k \circ f \circ g^{-1}\left(\frac{\lambda g(b)+(2-\lambda)g(a)}{2}\right)$$

$$+ (1-\lambda)k \circ f \circ g^{-1}\left(\frac{(1+\lambda)g(b)+(1-\lambda)g(a)}{2}\right)$$

$$\leq \frac{1}{g(b) - g(a)} \int_a^b k \circ f(t)\, g'(t)\, dt$$

$$\leq \frac{1}{2}\left[k \circ f \circ g^{-1}\left((1 - \lambda)\, g(a) + \lambda g(b)\right) + \lambda k \circ f(a) + (1 - \lambda)\, k \circ f(b)\right]$$

$$\leq \frac{k \circ f(a) + k \circ f(b)}{2}$$

for any $\lambda \in [0, 1]$.

Taking k^{-1} in (29) we obtain the desired result (28).

In 1906, Fejér [17], while studying trigonometric polynomials, obtained the following inequalities which generalize that of Hermite and Hadamard:

Theorem 2 (Fejér's Inequality) *Consider the integral $\int_a^b h(x)\, w(x)\, dx$, where h is a convex function in the interval (a, b) and w is a positive function in the same interval such that*

$$w(x) = w(a + b - x), \text{ for any } x \in [a, b]$$

i.e., $y = w(x)$ is a symmetric curve with respect to the straight line which contains the point $\left(\frac{1}{2}(a + b), 0\right)$ and is normal to the x-axis. Under those conditions the following inequalities are valid:

$$h\left(\frac{a + b}{2}\right) \int_a^b w(x)\, dx \leq \int_a^b h(x)\, w(x)\, dx \leq \frac{h(a) + h(b)}{2} \int_a^b w(x)\, dx.$$

$$(30)$$

If h is concave on (a, b), then the inequalities reverse in (30).

If $w : [a, b] \to \mathbb{R}$ is continuous and positive on the interval $[a, b]$, then the function $W : [a, b] \to [0, \infty)$, $W(x) := \int_a^x w(s)\, ds$ is strictly increasing and differentiable on (a, b) and the inverse $W^{-1} : \left[a, \int_a^b w(s)\, ds\right] \to [a, b]$ exists.

Corollary 3 *Assume that $w : [a, b] \to \mathbb{R}$ is continuous and positive on the interval $[a, b]$ and $f : [a, b] \to \mathbb{R}$ is composite-W^{-1} convex on $[a, b]$, then we have the following Fejér's type inequality*

$$f\left[W^{-1}\left(\frac{1}{2}\int_a^b w(s)\, ds\right)\right]$$

$$\leq \lambda f\left[W^{-1}\left(\frac{1}{2}\lambda \int_a^b w(s)\, ds\right)\right] + (1 - \lambda)\, f\left[W^{-1}\left(\frac{1}{2}(1 + \lambda)\int_a^b w(s)\, ds\right)\right]$$

$$\leq \frac{1}{\int_a^b w(s)} \int_a^b f(t)\, w(t)\, dt$$

$$\leq \frac{1}{2} \left[f\left[W^{-1}\left(\lambda \int_a^b w\left(s\right)ds \right) \right] + \lambda f\left(a\right) + \left(1-\lambda\right) f\left(b\right) \right] \leq \frac{f\left(a\right)+f\left(b\right)}{2} \tag{31}$$

for all $\lambda \in [0, 1]$.

In particular, we have

$$f\left[W^{-1}\left(\frac{1}{2}\int_a^b w\left(s\right)ds \right) \right]$$

$$\leq \frac{1}{2} f\left[W^{-1}\left(\frac{1}{4}\int_a^b w\left(s\right)ds \right) \right] + \frac{1}{2} f\left[W^{-1}\left(\frac{3}{4}\int_a^b w\left(s\right)ds \right) \right]$$

$$\leq \frac{1}{\int_a^b w\left(s\right)} \int_a^b f\left(t\right) w\left(t\right)dt$$

$$\leq \frac{1}{2} \left[f\left[W^{-1}\left(\frac{1}{2}\int_a^b w\left(s\right)ds \right) \right] + \frac{f\left(a\right)+f\left(b\right)}{2} \right] \leq \frac{f\left(a\right)+f\left(b\right)}{2}. \tag{32}$$

Remark 2 Assume that $w : [a, b] \to \mathbb{R}$ is continuous and positive on the interval $[a, b]$, $f : [a, b] \to J$, J an interval of real numbers, and $k : J \to \mathbb{R}$ a continuous function on J that is *strictly increasing (decreasing)* on J. If the function $f : [a, b] \to J$ is k-composite-W^{-1} convex on $[a, b]$, then

$$f\left[W^{-1}\left(\frac{1}{2}\int_a^b w\left(s\right)ds \right) \right]$$

$$\leq (\geq) k^{-1} \left\{ \lambda k \circ f\left[W^{-1}\left(\frac{1}{2}\lambda \int_a^b w\left(s\right)ds \right) \right] \right.$$

$$\left. + \left(1-\lambda\right) k \circ f\left[W^{-1}\left(\frac{1}{2}\left(1+\lambda\right) \int_a^b w\left(s\right)ds \right) \right] \right\}$$

$$\leq (\geq) k^{-1} \left(\frac{1}{\int_a^b w\left(s\right)} \int_a^b k \circ f\left(t\right) w\left(t\right)dt \right)$$

$$\leq (\geq) k^{-1} \left\{ \frac{1}{2}\left[k \circ f\left[W^{-1}\left(\lambda \int_a^b w\left(s\right)ds \right) \right] + \lambda k \circ f\left(a\right) + \left(1-\lambda\right) k \circ f\left(b\right) \right] \right\}$$

$$\leq (\geq) k^{-1} \left(\frac{k \circ f\left(a\right)+k \circ f\left(b\right)}{2} \right) \tag{33}$$

for all $\lambda \in [0, 1]$.

In particular, we have

$$f\left[W^{-1}\left(\frac{1}{2}\int_a^b w(s)\,ds\right)\right]$$

$$\leq(\geq)k^{-1}\left\{\frac{1}{2}k\circ f\left[W^{-1}\left(\frac{1}{4}\int_a^b w(s)\,ds\right)\right]+\frac{1}{2}k\circ f\left[W^{-1}\left(\frac{3}{4}\int_a^b w(s)\,ds\right)\right]\right\}$$

$$\leq(\geq)k^{-1}\left(\frac{1}{\int_a^b w(s)}\int_a^b k\circ f(t)\,w(t)\,dt\right)$$

$$\leq(\geq)k^{-1}\left\{\frac{1}{2}\left[k\circ f\left[W^{-1}\left(\frac{1}{2}\int_a^b w(s)\,ds\right)\right]+\frac{1}{2}k\circ f(a)+\frac{1}{2}k\circ f(b)\right]\right\}$$

$$\leq(\geq)k^{-1}\left(\frac{k\circ f(a)+k\circ f(b)}{2}\right). \qquad (34)$$

3 Reverse Inequalities

The following reverse inequalities may be stated:

Theorem 3 *Let* $g:[a,b]\to[g(a),g(b)]$ *be a continuous strictly increasing function that is differentiable on* (a,b). *If* $f:[a,b]\to\mathbb{R}$ *is composite-g^{-1} convex on* $[a,b]$, *then*

$$0\leq\frac{1}{g(b)-g(a)}\int_a^b f(t)\,g'(t)\,dt-f\left(M_g(a,b)\right) \qquad (35)$$

$$\leq\frac{1}{8}(g(b)-g(a))\left[\frac{f'_-(b)}{g'_-(b)}-\frac{f'_+(a)}{g'_+(a)}\right]$$

and

$$0\leq\frac{f(a)+f(b)}{2}-\frac{1}{g(b)-g(a)}\int_a^b f(t)\,g'(t)\,dt \qquad (36)$$

$$\leq\frac{1}{8}(g(b)-g(a))\left[\frac{f'_-(b)}{g'_-(b)}-\frac{f'_+(a)}{g'_+(a)}\right],$$

provided that the lateral derivatives $f'_+(a)$, $g'_+(a)$, $f'_-(b)$ *and* $g'_-(b)$ *are finite.*

Proof Let $h:[c,d]\to\mathbb{R}$ be a convex function on $[c,d]$. We use the inequality that has been established in [4]

$$0\leq\frac{1}{d-c}\int_c^d h(u)\,du-h\left(\frac{c+d}{2}\right)\leq\frac{1}{8}(d-c)\left[h'_-(d)-h'_+(c)\right] \qquad (37)$$

and the inequality obtained in [5]

$$0 \leq \frac{h(c) + h(d)}{2} - \frac{1}{d-c} \int_c^d h(u)\, du \leq \frac{1}{8}(d-c)\left[h'_-(d) - h'_+(c)\right]. \quad (38)$$

The constant $\frac{1}{8}$ is best possible in both (37) and (38).

From the inequalities (37) and (38) we have for the convex function $h = f \circ g^{-1}$ and $c, d \in [g(a), g(b)]$ that

$$0 \leq \frac{1}{d-c} \int_c^d \left(f \circ g^{-1}\right)(u)\, du - \left(f \circ g^{-1}\right)\left(\frac{c+d}{2}\right) \quad (39)$$

$$\leq \frac{1}{8}(d-c)\left[\left(f \circ g^{-1}\right)'_-(d) - \left(f \circ g^{-1}\right)'_+(c)\right]$$

and

$$0 \leq \frac{\left(f \circ g^{-1}\right)(c) + \left(f \circ g^{-1}\right)(d)}{2} - \frac{1}{d-c} \int_c^d \left(f \circ g^{-1}\right)(u)\, du \quad (40)$$

$$\leq \frac{1}{8}(d-c)\left[\left(f \circ g^{-1}\right)'_-(d) - \left(f \circ g^{-1}\right)'_+(c)\right].$$

Since $f \circ g^{-1}$ has lateral derivatives for $z \in (g(a), g(b))$ it follows f has lateral derivatives in each point of (a, b) and by the chain rule and the derivative of the inverse function, we have

$$\left(f \circ g^{-1}\right)'_\pm(z) = \left(f'_\pm \circ g^{-1}\right)(z)\left(g^{-1}\right)'(z) = \frac{\left(f'_\pm \circ g^{-1}\right)(z)}{\left(g' \circ g^{-1}\right)(z)}. \quad (41)$$

Therefore, by (39) and (40) we get

$$0 \leq \frac{1}{d-c} \int_c^d \left(f \circ g^{-1}\right)(u)\, du - \left(f \circ g^{-1}\right)\left(\frac{c+d}{2}\right) \quad (42)$$

$$\leq \frac{1}{8}(d-c)\left[\frac{\left(f'_- \circ g^{-1}\right)(d)}{\left(g' \circ g^{-1}\right)(d)} - \frac{\left(f'_+ \circ g^{-1}\right)(c)}{\left(g' \circ g^{-1}\right)(c)}\right]$$

and

$$0 \leq \frac{\left(f \circ g^{-1}\right)(c) + \left(f \circ g^{-1}\right)(d)}{2} - \frac{1}{d-c} \int_c^d \left(f \circ g^{-1}\right)(u)\, du \quad (43)$$

$$\leq \frac{1}{8}(d-c)\left[\frac{\left(f'_- \circ g^{-1}\right)(d)}{\left(g' \circ g^{-1}\right)(d)} - \frac{\left(f'_+ \circ g^{-1}\right)(c)}{\left(g' \circ g^{-1}\right)(c)}\right]$$

and by taking $c = g(a)$ and $d = g(b)$ in (42) and (43), then we get the desired results (35) and (36).

Corollary 4 *Assume that $w : [a, b] \to \mathbb{R}$ is continuous and positive on the interval $[a, b]$. If $f : [a, b] \to \mathbb{R}$ is composite-W^{-1} convex on $[a, b]$, then we have the following weighted reverse integral inequalities*

$$0 \leq \frac{1}{\int_a^b w(s)} \int_a^b f(t) w(t) \, dt - f \left[W^{-1} \left(\frac{1}{2} \int_a^b w(s) \, ds \right) \right] \tag{44}$$

$$\leq \frac{1}{8} \left[\frac{f'_-(b)}{w(b)} - \frac{f'_+(a)}{w(a)} \right] \int_a^b w(s) \, ds$$

and

$$0 \leq \frac{f(a) + f(b)}{2} - \frac{1}{\int_a^b w(s)} \int_a^b f(t) w(t) \, dt \tag{45}$$

$$\leq \frac{1}{8} \left[\frac{f'_-(b)}{w(b)} - \frac{f'_+(a)}{w(a)} \right] \int_a^b w(s) \, ds,$$

provided that $f'_-(b)$ and $f'_+(a)$ are finite.

Remark 3 Let $g : [a, b] \to [g(a), g(b)]$ be a *continuous strictly increasing function* that is *differentiable* on (a, b), $f : [a, b] \to J$, J an interval of real numbers, and $k : J \to \mathbb{R}$ a continuous function on J that is *strictly increasing* on J and differentiable on the interior of J. If the function $f : [a, b] \to J$ is k-composite-g^{-1} convex on $[a, b]$ and $f'_+(a)$, $g'_+(a)$, $f'_-(b), g'_-(b)$, $k'(f(a))$ and $k'(f(b))$ are finite, then by Theorem 3 we have

$$0 \leq \frac{1}{g(b) - g(a)} \int_a^b (k \circ f)(t) g'(t) \, dt - k \circ f \left(M_g(a, b) \right) \tag{46}$$

$$\leq \frac{1}{8} (g(b) - g(a)) \left[\frac{k'(f(b)) f'_-(b)}{g'_-(b)} - \frac{k'(f(a)) f'_+(a)}{g'_+(a)} \right]$$

and

$$0 \leq \frac{k \circ f(a) + k \circ f(b)}{2} - \frac{1}{g(b) - g(a)} \int_a^b (k \circ f)(t) g'(t) \, dt \tag{47}$$

$$\leq \frac{1}{8} (g(b) - g(a)) \left[\frac{k'(f(b)) f'_-(b)}{g'_-(b)} - \frac{k'(f(a)) f'_+(a)}{g'_+(a)} \right].$$

Assume that $w : [a, b] \to \mathbb{R}$ is continuous and positive on the interval $[a, b]$, $f : [a, b] \to J$, J an interval of real numbers, and $k : J \to \mathbb{R}$ a continuous function on J that is *strictly increasing* on J and differentiable on the interior of J.

If the function $f : [a, b] \rightarrow J$ is k-composite-W^{-1} convex on $[a, b]$ and $f'_+(a)$, $f'_-(b)$, $k'(f(a))$ and $k'(f(b))$ are finite, then we have the weighted inequalities

$$0 \leq \frac{1}{g(b) - g(a)} \int_a^b (k \circ f)(t) w(t) dt - k \circ f\left(W^{-1}\left(\frac{1}{2} \int_a^b w(s) ds\right)\right)$$

(48)

$$\leq \frac{1}{8}(g(b) - g(a))\left[\frac{k'(f(b)) f'_-(b)}{w(b)} - \frac{k'(f(a)) f'_+(a)}{w(a)}\right]$$

and

$$0 \leq \frac{k \circ f(a) + k \circ f(b)}{2} - \frac{1}{g(b) - g(a)} \int_a^b (k \circ f)(t) w(t) dt \qquad (49)$$

$$\leq \frac{1}{8}(g(b) - g(a))\left[\frac{k'(f(b)) f'_-(b)}{w(b)} - \frac{k'(f(a)) f'_+(a)}{w(a)}\right].$$

4   Applications for AG and AH-Convex Functions

The function $f : [a, b] \rightarrow (0, \infty)$ is AG-convex means that f is k-composite convex on $[a, b]$ with $k(t) = \ln t$, $t > 0$. By making use of Corollary 2 for $g(t) = t$, we get

$$f\left(\frac{a + b}{2}\right) \leq f^\lambda\left(\frac{\lambda b + (2 - \lambda) a}{2}\right) f^{1-\lambda}\left(\frac{(1 + \lambda) b + (1 - \lambda) a}{2}\right)$$

$$\leq \exp\left(\frac{1}{b - a} \int_a^b \ln f(t) dt\right)$$

$$\leq \sqrt{f((1 - \lambda) a + \lambda b) f^\lambda(a) f^{1-\lambda}(b)} \leq \sqrt{f(a) f(b)} \qquad (50)$$

for any $\lambda \in [0, 1]$, see also [11].

If we use Remark 3 for $g(t) = t$, then we get

$$0 \leq \frac{1}{b - a} \int_a^b \ln f(t) dt - \ln f\left(\frac{a + b}{2}\right) \leq \frac{1}{8}(b - a)\left[\frac{f'_-(b)}{f(b)} - \frac{f'_+(a)}{f(a)}\right]$$

(51)

and

$$0 \leq \frac{\ln f(a) + \ln f(b)}{2} - \frac{1}{b - a} \int_a^b \ln f(t) dt \leq \frac{1}{8}(b - a)\left[\frac{f'_-(b)}{f(b)} - \frac{f'_+(a)}{f(a)}\right].$$

(52)

By taking the exponential in (51) and (52) we get the equivalent inequalities

$$1 \le \frac{\exp\left(\frac{1}{b-a} \int_a^b \ln f(t)\, dt\right)}{f\left(\frac{a+b}{2}\right)} \le \exp\left\{\frac{1}{8}(b-a)\left[\frac{f'_-(b)}{f(b)} - \frac{f'_+(a)}{f(a)}\right]\right\} \qquad (53)$$

and

$$1 \le \frac{\sqrt{f(a) f(b)}}{\exp\left(\frac{1}{b-a} \int_a^b \ln f(t)\, dt\right)} \le \exp\left\{\frac{1}{8}(b-a)\left[\frac{f'_-(b)}{f(b)} - \frac{f'_+(a)}{f(a)}\right]\right\} \qquad (54)$$

that was obtained in [11].

The function $f : [a, b] \to (0, \infty)$ is AH-convex on $[a, b]$ means that f is k-composite concave on $[a, b]$ with $k : (0, \infty) \to (0, \infty)$, $k(t) = \frac{1}{t}$. By making use of Corollary 2 for $g(t) = t$, we get

$$f\left(\frac{a+b}{2}\right)$$

$$\le \left\{\lambda f^{-1}\left(\frac{\lambda b + (2 - \lambda) a}{2}\right) + (1 - \lambda) f^{-1}\left(\frac{(1 + \lambda) b + (1 - \lambda) a}{2}\right)\right\}^{-1}$$

$$\le \left(\frac{1}{b-a} \int_a^b f^{-1}(t)\, dt\right)^{-1}$$

$$\le \left\{\frac{1}{2}\left[f^{-1}((1 - \lambda) a + \lambda b) + \lambda f^{-1}(a) + (1 - \lambda) f^{-1}(b)\right]\right\}^{-1}$$

$$\le \left(\frac{f^{-1}(a) + f^{-1}(b)}{2}\right)^{-1} \qquad (55)$$

for any $\lambda \in [0, 1]$.

By taking the power -1, this inequality is equivalent to

$$f^{-1}\left(\frac{a+b}{2}\right)$$

$$\ge \lambda f^{-1}\left(\frac{\lambda b + (2 - \lambda) a}{2}\right) + (1 - \lambda) f^{-1}\left(\frac{(1 + \lambda) b + (1 - \lambda) a}{2}\right)$$

$$\ge \frac{1}{b-a} \int_a^b f^{-1}(t)\, dt$$

$$\ge \frac{1}{2}\left[f^{-1}((1 - \lambda) a + \lambda b) + \lambda f^{-1}(a) + (1 - \lambda) f^{-1}(b)\right] \ge \frac{f^{-1}(a) + f^{-1}(b)}{2}$$

$$(56)$$

for any $\lambda \in [0, 1]$.

If we use Remark 3 for $g(t) = t$, then we get

$$0 \le f^{-1}\left(\frac{a+b}{2}\right) - \frac{1}{b-a}\int_a^b f^{-1}(t)\,dt \le \frac{1}{8}(b-a)\left[\frac{f'_-(b)}{f^2(b)} - \frac{f'_+(a)}{f^2(a)}\right]$$

(57)

and

$$0 \le \frac{1}{b-a}\int_a^b f^{-1}(t)\,dt - \frac{f^{-1}(a) + f^{-1}(b)}{2} \le \frac{1}{8}(b-a)\left[\frac{f'_-(b)}{f^2(b)} - \frac{f'_+(a)}{f^2(a)}\right].$$

(58)

5 Applications for GA, GG, and GH-Convex Functions

If we take $g(t) = \ln t$, $t \in [a, b] \subset (0, \infty)$, then $f : [a, b] \to \mathbb{R}$ is GA-convex on $[a, b]$ means that that $f : [a, b] \to \mathbb{R}$ composite-g^{-1} convex on $[a, b]$. By making use of Corollary 2 for $k(t) = t$, we get

$$f\left(\sqrt{ab}\right) \le (1-\lambda)f\left(a^{\frac{1-\lambda}{2}}b^{\frac{\lambda+1}{2}}\right) + \lambda f\left(a^{\frac{2-\lambda}{2}}b^{\frac{\lambda}{2}}\right)$$

(59)

$$\le \frac{1}{\ln\left(\frac{b}{a}\right)}\int_a^b \frac{f(t)}{t}\,dt$$

$$\le \frac{1}{2}\left[f\left(a^{1-\lambda}b^\lambda\right) + (1-\lambda)f(b) + \lambda f(a)\right] \le \frac{f(a) + f(b)}{2}$$

for any $\lambda \in [0, 1]$. This result was obtained in [6].

If we use Remark 3 for $k(t) = t$, then we get

$$0 \le \frac{1}{\ln\left(\frac{b}{a}\right)}\int_a^b \frac{f(t)}{t}\,dt - f\left(\sqrt{ab}\right) \le \frac{1}{8}\ln\left(\frac{b}{a}\right)\left[bf'_-(b) - af'_+(a)\right]$$

(60)

and

$$0 \le \frac{f(a) + f(b)}{2} - \frac{1}{\ln\left(\frac{b}{a}\right)}\int_a^b \frac{f(t)}{t}\,dt \le \frac{1}{8}\ln\left(\frac{b}{a}\right)\left[bf'_-(b) - af'_+(a)\right].$$

(61)

These results were also obtained in [6].

The function $f : I \subset (0, \infty) \to (0, \infty)$ is GG-convex means that f is k-composite-g^{-1} convex on $[a, b]$ with $k : (0, \infty) \to \mathbb{R}$, $k(t) = \ln t$ and $g(t) = \ln t$, $t \in [a, b]$. By making use of Corollary 2 we get

$$f\left(\sqrt{ab}\right) \le f^\lambda\left(a^{\frac{2-\lambda}{2}}b^{\frac{\lambda}{2}}\right)f^{1-\lambda}\left(a^{\frac{1-\lambda}{2}}b^{\frac{\lambda+1}{2}}\right)$$

$$\leq \exp\left(\frac{1}{\ln\left(\frac{b}{a}\right)} \int_a^b \frac{\ln f(t)}{t} dt\right)$$

$$\leq \sqrt{f\left(a^{1-\lambda}b^\lambda\right) f^\lambda(a) f^{1-\lambda}(b)} \leq \sqrt{f(a) f(b)} \qquad (62)$$

for any $\lambda \in [0, 1]$. This result was obtained in [7], see also [13].

If we use Remark 3, then we have the inequalities

$$1 \leq \frac{\sqrt{f(a) f(b)}}{\exp\left(\frac{1}{\ln b - \ln a} \int_a^b \frac{\ln f(s)}{s} ds\right)} \leq \left(\frac{b}{a}\right)^{\frac{1}{8}\left[\frac{f'_-(b)b}{f(b)} - \frac{f'_+(a)a}{f(a)}\right]} \qquad (63)$$

and

$$1 \leq \frac{\exp\left(\frac{1}{\ln b - \ln a} \int_a^b \frac{\ln f(s)}{s} ds\right)}{f\left(\sqrt{ab}\right)} \leq \left(\frac{b}{a}\right)^{\frac{1}{8}\left[\frac{f'_-(b)b}{f(b)} - \frac{f'_+(a)a}{f(a)}\right]}. \qquad (64)$$

These results were obtained in [7], see also [13].

We also have that $f : [a, b] \subset (0, \infty) \to \mathbb{R}$ is GH-convex on $[a, b]$ is equivalent to the fact that f is k-composite-g^{-1} concave on $[a, b]$ with $k : (0, \infty) \to (0, \infty)$, $k(t) = \frac{1}{t}$ and $g(t) = \ln t$, $t \in I$. By making use of Corollary 2 we get

$$f\left(\sqrt{ab}\right) \leq \left[\lambda f^{-1}\left(a^{\frac{2-\lambda}{2}}b^{\frac{\lambda}{2}}\right) + (1-\lambda) f^{-1}\left(a^{\frac{1-\lambda}{2}}b^{\frac{\lambda+1}{2}}\right)\right]^{-1}$$

$$\leq \left(\frac{1}{\ln\left(\frac{b}{a}\right)} \int_a^b \frac{f^{-1}(t)}{t} dt\right)^{-1}$$

$$\leq \left\{\frac{1}{2}\left[f^{-1}\left(a^{1-\lambda}b^\lambda\right) + \lambda f^{-1}(a) + (1-\lambda) f^{-1}(b)\right]\right\}^{-1}$$

$$\leq \left(\frac{f^{-1}(a) + f^{-1}(b)}{2}\right)^{-1} \qquad (65)$$

for any $\lambda \in [0, 1]$.

This is equivalent to

$$f^{-1}\left(\sqrt{ab}\right) \geq \lambda f^{-1}\left(a^{\frac{2-\lambda}{2}}b^{\frac{\lambda}{2}}\right) + (1-\lambda) f^{-1}\left(a^{\frac{1-\lambda}{2}}b^{\frac{\lambda+1}{2}}\right)$$

$$\geq \frac{1}{\ln\left(\frac{b}{a}\right)} \int_a^b \frac{f^{-1}(t)}{t} dt$$

$$\geq \frac{1}{2}\left[f^{-1}\left(a^{1-\lambda}b^{\lambda}\right) + \lambda f^{-1}(a) + (1-\lambda) f^{-1}(b)\right]$$

$$\geq \frac{f^{-1}(a) + f^{-1}(b)}{2}. \qquad (66)$$

If we use Remark 3, then we get

$$0 \leq f^{-1}\left(\sqrt{ab}\right) - \frac{1}{\ln\left(\frac{b}{a}\right)} \int_a^b \frac{f^{-1}(t)}{t} dt \leq \frac{1}{8} \ln\left(\frac{b}{a}\right)\left[\frac{bf'_-(b)}{f^2(b)} - \frac{af'_+(a)}{f^2(a)}\right] \qquad (67)$$

and

$$0 \leq \frac{1}{\ln\left(\frac{b}{a}\right)} \int_a^b \frac{f^{-1}(t)}{t} dt - \frac{f^{-1}(a) + f^{-1}(b)}{2}$$

$$\leq \frac{1}{8} \ln\left(\frac{b}{a}\right)\left[\frac{bf'_-(b)}{f^2(b)} - \frac{af'_+(a)}{f^2(a)}\right]. \qquad (68)$$

6 Applications for HA, HG, and HH-Convex Functions

Let $f : [a, b] \subset (0, \infty) \to \mathbb{R}$ be an HA-convex function on the interval $[a, b]$. This is equivalent to the fact that f is composite-g^{-1} convex on $[a, b]$ with the increasing function $g(t) = -\frac{1}{t}$. Then by applying Corollary 2 for $k(t) = t$, we have the inequalities

$$f\left(\frac{2ab}{a+b}\right) \leq (1-\lambda) f\left(\frac{2ab}{(1-\lambda) a + (\lambda+1) b}\right) + \lambda f\left(\frac{2ab}{(2-\lambda) a + \lambda b}\right) \qquad (69)$$

$$\leq \frac{ab}{b-a} \int_a^b \frac{f(t)}{t^2} dt$$

$$\leq \frac{1}{2}\left[f\left(\frac{ab}{(1-\lambda) a + \lambda b}\right) + (1-\lambda) f(a) + \lambda f(b)\right]$$

$$\leq \frac{f(a) + f(b)}{2}$$

for any $\lambda \in [0, 1]$. This result was obtained in [9].

If we use Remark 3, then we get

$$0 \leq \frac{ab}{b-a} \int_a^b \frac{f(t)}{t^2} dt - f\left(\frac{2ab}{a+b}\right) \leq \frac{1}{8}\left[\frac{f'_-(b) b^2 - f'_+(a) a^2}{ab}\right] (b-a) \qquad (70)$$

and

$$0 \leq \frac{f(a) + f(b)}{2} - \frac{ab}{b-a} \int_a^b \frac{f(t)}{t^2} dt \leq \frac{1}{8} \left[\frac{f'_-(b) b^2 - f'_+(a) a^2}{ab} \right] (b-a).$$

(71)

These results were obtained in [9].

Let $f : [a, b] \subset (0, \infty) \to (0, \infty)$ be an HG-convex function on the interval $[a, b]$. This is equivalent to the fact that f is k-composite-g^{-1} convex on $[a, b]$ with $k : (0, \infty) \to \mathbb{R}$, $k(t) = \ln t$ and $g(t) = -\frac{1}{t}$, $t \in [a, b]$. Then by applying Corollary 2, we have the inequalities

$$f\left(\frac{2ab}{a+b}\right) \leq f^{1-\lambda}\left(\frac{2ab}{(1-\lambda)a + (\lambda+1)b}\right) f^\lambda\left(\frac{2ab}{(2-\lambda)a + \lambda b}\right) \quad (72)$$

$$\leq \exp\left(\frac{ab}{b-a} \int_a^b \frac{\ln f(t)}{t^2} dt\right)$$

$$\leq \sqrt{f\left(\frac{ab}{(1-\lambda)a + \lambda b}\right) [f(a)]^{1-\lambda} [f(b)]^\lambda} \leq \sqrt{f(a) f(b)}$$

for any $\lambda \in [0, 1]$. This result was obtained in [10].

If we use Remark 3, then we get

$$1 \leq \frac{\exp\left(\frac{ab}{b-a} \int_a^b \frac{\ln f(t)}{t^2} dt\right)}{f\left(\frac{2ab}{a+b}\right)} \leq \exp\left(\frac{1}{8}\left[\frac{f'_-(b) b^2}{f(b)} - \frac{f'_+(a) a^2}{f(a)}\right]\frac{b-a}{ab}\right) \quad (73)$$

and

$$1 \leq \frac{\sqrt{f(a) f(b)}}{\exp\left(\frac{ab}{b-a} \int_a^b \frac{\ln f(t)}{t^2} dt\right)} \leq \exp\left(\frac{1}{8}\left[\frac{f'_-(b) b^2}{f(b)} - \frac{f'_+(a) a^2}{f(a)}\right]\frac{b-a}{ab}\right).$$

(74)

These results were obtained in [10].

Let $f : [a, b] \subset (0, \infty) \to (0, \infty)$ be an HH-convex function on the interval $[a, b]$. This is equivalent to the fact that f is k-composite-g^{-1} concave on $[a, b]$ with $k : (0, \infty) \to (0, \infty)$, $k(t) = \frac{1}{t}$ and $g(t) = -\frac{1}{t}$, $t \in [a, b]$. Then by applying Corollary 2, we have the inequalities

$$f\left(\frac{2ab}{a+b}\right)$$

$$\leq \left\{\lambda f^{-1}\left(\frac{2ab}{(2-\lambda)a + \lambda b}\right) + (1-\lambda) f^{-1}\left(\frac{2ab}{(1-\lambda)a + (\lambda+1)b}\right)\right\}^{-1}$$

$$\leq \left(\frac{ab}{b-a} \int_a^b \frac{f^{-1}(t)}{t^2} dt \right)^{-1}$$

$$\leq \left\{ \frac{1}{2} \left[f^{-1} \left(\frac{ab}{(1-\lambda)a + \lambda b} \right) + \lambda f^{-1}(a) + (1-\lambda) f^{-1}(b) \right] \right\}^{-1}$$

$$\leq \left(\frac{f^{-1}(a) + f^{-1}(b)}{2} \right)^{-1} \quad (75)$$

for any $\lambda \in [0, 1]$.

By taking the power -1 in (75), then we get

$$f^{-1} \left(\frac{2ab}{a+b} \right)$$

$$\geq \lambda f^{-1} \left(\frac{2ab}{(2-\lambda)a + \lambda b} \right) + (1-\lambda) f^{-1} \left(\frac{2ab}{(1-\lambda)a + (\lambda+1)b} \right)$$

$$\geq \frac{ab}{b-a} \int_a^b \frac{f^{-1}(t)}{t^2} dt$$

$$\geq \frac{1}{2} \left[f^{-1} \left(\frac{ab}{(1-\lambda)a + \lambda b} \right) + \lambda f^{-1}(a) + (1-\lambda) f^{-1}(b) \right] \geq \frac{f^{-1}(a) + f^{-1}(b)}{2} \quad (76)$$

for any $\lambda \in [0, 1]$.

If we use Remark 3, then we get

$$0 \leq f^{-1} \left(\frac{2ab}{a+b} \right) - \frac{ab}{b-a} \int_a^b \frac{f^{-1}(t)}{t^2} dt$$

$$\leq \frac{1}{8} \left[\frac{b^2 f'_-(b)}{f^2(b)} - \frac{a^2 f'_+(a)}{f^2(a)} \right] \frac{ab}{b-a} \quad (77)$$

and

$$0 \leq \frac{ab}{b-a} \int_a^b \frac{f^{-1}(t)}{t^2} dt - \frac{f^{-1}(a) + f^{-1}(b)}{2}$$

$$\leq \frac{1}{8} \left[\frac{b^2 f'_-(b)}{f^2(b)} - \frac{a^2 f'_+(a)}{f^2(a)} \right] \frac{ab}{b-a}. \quad (78)$$

For related results, see [8].

7 Applications for p, r-Convex, and *LogExp* Convex Functions

If $p > 0$ and we consider $g(t) = t^p$, $t \in [a, b] \subset (0, \infty)$, then $f : [a, b] \subset (0, \infty) \to (0, \infty)$ is p-convex on $[a, b]$ is equivalent to the fact that f is composite-g^{-1} convex on $[a, b]$. Using Corollary 2 for $k(t) = t$ we get

$$f\left(M_p(a, b)\right)$$

$$\leq \lambda f\left[\left(\frac{\lambda b^p + (2 - \lambda) a^p}{2}\right)^{1/p}\right] + (1 - \lambda) f\left[\left(\frac{(1 + \lambda) b^p + (1 - \lambda) a^p}{2}\right)^{1/p}\right]$$

$$\leq \frac{p}{b^p - a^p} \int_a^b f(t) t^{p-1} dt$$

$$\leq \frac{1}{2}\left\{f\left[((1 - \lambda) a^p + \lambda b^p)^{1/p}\right] + \lambda f(a) + (1 - \lambda) f(b)\right\} \leq \frac{f(a) + f(b)}{2}$$

$$(79)$$

for any $\lambda \in [0, 1]$, where $M_p(a, b) := \left(\frac{a^p + b^p}{2}\right)^{1/p}$. This improves the corresponding result from [22].

If we use Remark 3, then we get

$$0 \leq \frac{p}{b^p - a^p} \int_a^b f(t) t^{p-1} dt - f\left(M_p(a, b)\right) \leq \frac{1}{8p}\left(b^p - a^p\right)\left[\frac{f'_-(b)}{b^{p-1}} - \frac{f'_+(a)}{a^{p-1}}\right]$$

$$(80)$$

and

$$0 \leq \frac{a^p + b^p}{2} - \frac{p}{b^p - a^p} \int_a^b f(t) t^{p-1} dt \leq \frac{1}{8p}\left(b^p - a^p\right)\left[\frac{f'_-(b)}{b^{p-1}} - \frac{f'_+(a)}{a^{p-1}}\right].$$

$$(81)$$

Assume that the function $f : [a, b] \to (0, \infty)$ is r-convex, for $r > 0$. This is equivalent to the fact that f is k-composite convex with $k(t) = t^r$, $t > 0$, and by Corollary 2 for $g(t) = t$ we get

$$f\left(\frac{a + b}{2}\right)$$

$$\leq \left\{\lambda f^r\left(\frac{\lambda a + (2 - \lambda) b}{2}\right) + (1 - \lambda) f^r\left(\frac{(1 + \lambda) b + (1 - \lambda) a}{2}\right)\right\}^{1/r}$$

$$\leq \left(\frac{1}{b - a} \int_a^b f^r(t) dt\right)^{1/r}$$

$$\leq \left\{ \frac{1}{2} \left[f^r \left((1 - \lambda) a + \lambda b \right) + \lambda f^r (a) + (1 - \lambda) f^r (b) \right] \right\}^{1/r} \leq \left(\frac{f^r (a) + f^r (b)}{2} \right)^{1/r}$$

(82)

for any $\lambda \in [0, 1]$.

By taking the power $r > 0$, we get the equivalent inequality

$$f^r \left(\frac{a + b}{2} \right)$$

$$\leq \lambda f^r \left(\frac{\lambda a + (2 - \lambda) b}{2} \right) + (1 - \lambda) f^r \left(\frac{(1 + \lambda) b + (1 - \lambda) a}{2} \right)$$

$$\leq \frac{1}{b - a} \int_a^b f^r (t) \, dt$$

$$\leq \frac{1}{2} \left[f^r \left((1 - \lambda) a + \lambda b \right) + \lambda f^r (a) + (1 - \lambda) f^r (b) \right] \leq \frac{f^r (a) + f^r (b)}{2}$$

(83)

for any $\lambda \in [0, 1]$.

From Remark 3, we get for $g(t) = t$ that

$$0 \leq \frac{1}{b - a} \int_a^b f^r (t) \, dt - f^r \left(\frac{a + b}{2} \right)$$

$$\leq \frac{r}{8} (b - a) \left[f^{r-1} (b) f_-' (b) - f^{r-1} (a) f_+' (a) \right] \quad (84)$$

and

$$0 \leq \frac{f^r (a) + f^r (b)}{2} - \frac{1}{b - a} \int_a^b f^r (t) \, dt$$

$$\leq \frac{r}{8} (b - a) \left[f^{r-1} (b) f_-' (b) - f^{r-1} (a) f_+' (a) \right]. \quad (85)$$

Assume that $f : [a, b] \to \mathbb{R}$ is *LogExp convex function* on $[a, b]$ as considered in [14]. This is equivalent to the fact that f is composite-g^{-1} convex with $g(t) = \exp t$. By utilizing Corollary 2 for $k(t) = t$ we get

$$f (LME (a, b))$$

$$\leq \lambda f \left[\ln \left(\frac{\lambda \exp b + (2 - \lambda) \exp a}{2} \right) \right] + (1 - \lambda) f \left[\ln \left(\frac{(1 + \lambda) \exp b + (1 - \lambda) \exp a}{2} \right) \right]$$

$$\leq \frac{1}{\exp b - \exp a} \int_a^b f(t) \exp t \, dt$$

$$\leq \frac{1}{2} \left[f \left[\ln \left((1 - \lambda) \exp(a) + \lambda \exp(b) \right) \right] + \lambda f(a) + (1 - \lambda) f(b) \right] \leq \frac{f(a) + f(b)}{2} \tag{86}$$

for $\lambda \in [a, b]$, where $LME(a, b) := \ln \left(\frac{\exp a + \exp b}{2} \right)$.

If we use Remark 3, then we get

$$0 \leq \frac{1}{\exp b - \exp a} \int_a^b f(t) \exp t \, dt - f(LME(a, b)) \tag{87}$$

$$\leq \frac{1}{8} (\exp b - \exp a) \left[\exp(-b) f'_-(b) - \exp(-a) f'_+(a) \right]$$

and

$$0 \leq \frac{f(a) + f(b)}{2} - \frac{1}{\exp b - \exp a} \int_a^b f(t) \exp t \, dt \tag{88}$$

$$\leq \frac{1}{8} (\exp b - \exp a) \left[\exp(-b) f'_-(b) - \exp(-a) f'_+(a) \right].$$

References

1. G.D. Anderson, M.K. Vamanamurthy, M. Vuorinen, Generalized convexity and inequalities, J. Math. Anal. Appl. **335**, 1294–1308 (2007)
2. N.S. Barnett, P. Cerone, S.S. Dragomir, Some new inequalities for Hermite-Hadamard divergence in information theory, in *Stochastic Analysis & Applications*, ed. by Y.J. Cho, J.K. Kim, Y.K. Choi. Nova Science Publishers, vol. 3 (2003), pp. 7–19, ISBN 1-59033-860-X. Preprint *RGMIA Res. Rep. Coll.* **5**(4), Art. 8 (2002). http://rgmia.org/papers/v5n4/NIHHDIT. pdf
3. E.F. Beckenbach, Convex functions. Bull. Amer. Math. Soc. **54**, 439–460 (1948)
4. S.S. Dragomir, An inequality improving the first Hermite-Hadamard inequality for convex functions defined on linear spaces and applications for semi-inner products. J. Inequal. Pure Appl. Math. **3**(2), 31 (2002). https://www.emis.de/journals/JIPAM/article183.html?sid=183
5. S.S. Dragomir, An inequality improving the second Hermite-Hadamard inequality for convex functions defined on linear spaces and applications for semi-inner products. J. Inequal. Pure Appl. Math. **3**(3), 35 (2002). https://www.emis.de/journals/JIPAM/article187.html?sid=187
6. S.S. Dragomir, Inequalities of Hermite-Hadamard type for GA-convex functions, in *Annales Mathematicae Silesianae*, Preprint *RGMIA Res. Rep. Coll.* **18**, 30 (2015). http://rgmia.org/papers/v18/v18a30.pdf
7. S.S. Dragomir, Inequalities of Hermite-Hadamard type for GG-convex functions, Preprint RGMIA, Res. Rep. Coll. **18**, Article 71, 15 (2015). http://rgmia.org/papers/v18/v18a71.pdf
8. S.S. Dragomir, Inequalities of Hermite-Hadamard type for HH-convex functions, in Acta et Commentationes Universitatis Tartuensis de Mathematica, Preprint, RGMIA Res. Rep. Coll. **18**, Art. 80 (2015). http://rgmia.org/papers/v18/v18a80.pdf

9. S.S. Dragomir, Inequalities of Hermite-Hadamard type for HA-convex functions. Maroccan J. Pure & Appl. Anal. **3**(1), 83–101 (2017). Preprint, RGMIA Res. Rep. Coll. **18**, Art. 38 (2015). http://rgmia.org/papers/v18/v18a38.pdf

10. S.S. Dragomir, Inequalities of Hermite-Hadamard type for HG-convex functions. Probl. Anal. Issues Anal. **6** (24), No. (2), 1–17 (2017). Preprint, RGMIA Res. Rep. Coll. **18**, Art. 79 (2015). http://rgmia.org/papers/v18/v18a79.pdf

11. S.S. Dragomir, New inequalities of Hermite-Hadamard type for log-convex functions. Khayyam J. Math. **3**(2), 98–115 (2017)

12. S.S. Dragomir, Ostrowski type inequalities for Lebesgue integral: a survey of recent results. Aust. J. Math. Anal. Appl. **14**(1), Article 1, 283 (2017). http://ajmaa.org/cgi-bin/paper.pl?string=v14n1/V14I1P1.tex

13. S.S. Dragomir, Some integral inequalities of Hermite-Hadamard type for GG-convex functions. Mathematica (Cluj), **59**(82), No 1–2, 47–64 (2017). Preprint RGMIA, Res. Rep. Coll., **18**, Art. 74 (2015). http://rgmia.org/papers/v18/v18a74.pdf

14. S.S. Dragomir, Inequalities for a generalized finite Hilbert transform of convex functions, Preprint RGMIA Res. Rep. Coll. **21** (2018)

15. S.S. Dragomir, C.E.M. Pearce, *Selected Topics on Hermite-Hadamard Inequalities and Applications*, RGMIA Monographs, (2000). http://rgmia.org/monographs/hermite_hadamard.html

16. A. El Farissi, Simple proof and refinement of Hermite-Hadamard inequality. J. Math. Ineq. **4**(3), 365–369 (2010)

17. L. Fejér, Über die Fourierreihen, II, (In Hungarian). Math. Naturwiss, Anz. Ungar. Akad. Wiss. **24**, 369–390 (1906)

18. D.S. Mitrinović, I.B. Lacković, Hermite and convexity. Aequationes Math. **28**, 229–232 (1985)

19. P. Montel, Sur les functions convexes et les fonctions sousharmoniques. Journal de Math. **9**(7), 29–60 (1928)

20. C.P. Niculescu, Convexity according to the geometric mean. Math. Inequal. Appl. **3**(2), 155–167 (2000)

21. C.E.M. Pearce, J. Pečarić, V. Šimić, Stolarsky means and Hadamard's inequality. J. Math. Anal. Appl. **220**, 99–109 (1998)

22. K.S. Zhang, J.P. Wan, p-convex functions and their properties. Pure Appl. Math. **23**(1), 130–133 (2007)

Error Estimation for Approximate Solutions of Delay Volterra Integral Equations

Oktay Duman

Abstract This work is related to inequalities in the approximation theory. Mainly, we study numerical solutions of delay Volterra integral equations by using a collocation method based on sigmoidal function approximation. Error estimation and convergence analysis are provided. At the end of the paper we display numerical simulations verifying our results.

1 Introduction

The theory of integral equations is an important subject in pure and applied mathematics. There are many applications of this theory to problems in the physical and biological sciences. In general, integral equations are used as mathematical models for many and varied physical situations. In order to find numerical solutions of integral equations, we use various techniques, such as degenerate kernel methods, projection methods including collocation and Galerkin methods, and the Nystrom method, iteration methods. For details about the topic we suggest the books by Atkinson [5] and Brunner [7]. To model more realistic and complex structures, primarily taken from the biological sciences literature, we usually need delay dynamics, such as delay integral equations and delay differential equations. In the present paper we mainly focus on the numerical solutions of an integral equation with constant delay. In this study, for numerical solution of a delayed Volterra equation, we use a collocation method based on sigmoidal function approximation, which was first studied by Costarelli and Spigler (see [10–12]). For the neural network operators based on sigmoidal functions and their approximation properties, we refer the papers and cited therein: [1–4, 9, 13]. For other variants of collocation methods used in numerical solution of Volterra-type integral equations see the recent papers: [5–7, 15–18].

O. Duman (✉)
TOBB University of Economics and Technology, Department of Mathematics, Söğütözü, Ankara, Turkey
e-mail: oduman@etu.edu.tr

© Springer Nature Switzerland AG 2019
G. A. Anastassiou and J. M. Rassias (eds.), *Frontiers in Functional Equations and Analytic Inequalities*, https://doi.org/10.1007/978-3-030-28950-8_29

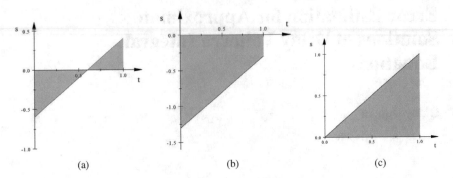

Fig. 1 Domains D_τ and D. (**a**) $0 < \tau < 1$. (**b**) $\tau > 1$. (**c**) No delay

We consider the following delay Volterra integral equation

$$f(t) = \begin{cases} g(t) + \int_0^{t-\tau} K_1(t,s)f(s)ds + \int_0^t K_2(t,s)f(s)ds, & \text{if } t \in [0,1] \\ \phi(t), & \text{if } t \in [-\tau,0), \end{cases}$$
(1)

where $\tau > 0$ is a delay term. We should note that Eq. (1) and its special cases have frequently been encountered in physical and biological modeling processes (see, for instance, [8]). For example, in the case of $K_2 = 0$, the equation turns out to the theory of a circulating fuel nuclear reactor (see [14]). Also, the case of $K_2 = -K_1$ reduces to the process of population dynamics. Recently, some numerical solutions of Eq. (1) have been obtained in different approaches, such as iterated collocation method, Runge–Kutta method, a multilevel correction method.

In (1) the kernel functions K_1 and K_2 are defined on the following domains D_τ and D, respectively (see Fig. 1):

$$\begin{aligned} D_\tau = \{(t,s) : 0 \le s \le t-\tau, \ \tau \le t \le 1\} \\ \cup \{(t,s) : t-\tau \le s \le 0, \ 0 \le t \le \tau\} \end{aligned} \quad \text{for } 0 < \tau < 1,$$
(2)

$$D_\tau = \{(t,s) : t-\tau \le s \le 0, \ 0 \le t \le 1\} \text{ for } \tau \ge 1,$$
(3)

and

$$D = \{(t,s) : 0 \le s \le t, \ 0 \le t \le 1\}.$$
(4)

We assume that $K_1 : D_\tau \to \mathbb{R}$, $K_2 : D \to \mathbb{R}$, $g : [0,1] \to \mathbb{R}$, and $\phi : [-\tau,0] \to \mathbb{R}$ are given sufficiently smooth functions on their domains such that (5) possesses a unique solution $f \in C[0,1]$, For the continuity ϕ at 0, we need the following condition

$$\phi(0) = g(0) - \int_{-\tau}^0 K_1(0,s)\phi(s)ds.$$
(5)

Note that f is already determined with the function ϕ on $[-\tau, 0]$. Hence, throughout the paper we assume that $\phi \in C[-\tau, 0]$ having the condition (5). About the kernels, at first, we assume that, for each fixed $t \in [0, 1]$, the kernel function $K_1(t, s)$ is integrable with respect to s such that $(t, s) \in D_\tau$, and $K_2(t, s)$ is integrable with respect to s such that $(t, s) \in D$.

In this paper we seek numerical solutions of (1) with respect to the given delay term $\tau > 0$.

We should note that if we take $\tau \geq 1$, then (1) reduces to the standard Volterra integral equation:

$$f(t) = h(t) + \int_0^t K(t, s) f(s) ds, \tag{6}$$

where

$$h(t) := g(t) - \int_{t-\tau}^0 K_1(t, s) \phi(s) ds \quad \text{and} \quad K(t, s) := K_2(t, s).$$

Since the numerical solutions of (6) via the collocation method based on neural network approximation were systematically investigated by Costarelli and Spigler (see [11]), in the present paper we just focus on the case of $0 < \tau < 1$.

It is not hard to see that, for $0 < \tau < 1$, Eq. (1) can be written in the following form:

$$f(t) = \begin{cases} g(t) + \int_0^{t-\tau} K_1(t, s) f(s) ds + \int_0^t K_2(t, s) f(s) ds, & \text{if } t \in [\tau, 1] \\ g(t) - \int_{t-\tau}^0 K_1(t, s) \phi(s) ds + \int_0^t K_2(t, s) f(s) ds, & \text{if } t \in [0, \tau) \\ \phi(t), & \text{if } t \in [-\tau, 0). \end{cases} \tag{7}$$

Our strategy in this paper is as follows:

- In Sect. 2, we prove the existence and uniqueness of the approximate solution of (1) for $0 < \tau < 1$ via the collocation method based on neural network approximation.
- In Sect. 3, under suitable smoothness conditions on the functions K_1, K_2, g, and ϕ, we obtain the uniform convergence of this approximate solution to the exact solution of (1) on the interval $[0, 1]$.
- In Sect. 4, we display some numerical simulations verifying our results.
- The last section is devoted to the concluding remarks.

2 Collocation Method Based on Unit Step Functions

Now, for $k = 0, 1, 2, \ldots, N$, define the functions $H_k : [0, 1] \to \{0, 1\}$ by

$$H_k(t) = H(t - t_k) \text{ for } k = 1, 2, \ldots, N \quad \text{and} \quad H_0(t) = H(t - t_{-1}) = 1,$$

where $t_k = \frac{k}{N}$ and H is the unit step function, $H(t) := 1$ for $t \geq 0$, and $H(t) := 0$ for $t < 0$. Then, Costarelli and Spigler [10] proved the following neural network approximation theorem: for every $f \in C[0, 1]$,

$$\lim_{N \to \infty} \|G_N(f) - f\|_{[0,1]} = 0,$$

where $\|\cdot\|_{[0,1]}$ denotes the usual supremum norm on the unit interval $[0, 1]$, and

$$G_N(f; t) = \sum_{k=1}^{N} (f(t_k) - f(t_{k-1})) H_k(t) + f(0). \tag{8}$$

Later they applied the above approximation to the numerical solutions of Volterra integral equations (see [11, 12]). The main idea is first to take unknown coefficients in (8) and then to determine them by forcing a collocation method on the finite dimensional space $\Sigma_N = span\{H_k : k = 0, 1, \ldots, N\}$ and the collocation points $C_N = \{t_0, t_1, \ldots, t_N\}$. We should note that such a collocation method and further improvements have been investigated by Brunner (see, for instance, [6]).

For numerical solutions of (1) we first assume that

$$G_N^*(t) = \sum_{k=0}^{N} \alpha_k H_k(t) \quad \text{for } t \in [0, 1] \tag{9}$$

denotes the approximation to the exact solution f on $[0, 1]$. We also remark that there is no need to approximate to f on $[-\tau, 0)$ since f is already determined on $[-\tau, 0)$ as a given function ϕ.

Now, as stated before, we just consider the term τ belonging to $(0, 1)$. Since Eq. (7) holds for this delay τ, the corresponding residual on $[0, 1]$ is given by

$$r_N(t) = \begin{cases} G_N^*(t) - g(t) - \int_0^{t-\tau} K_1(t, s) G_N^*(s) ds & \text{if } t \in [\tau, 1] \\ - \int_0^t K_2(t, s) G_N^*(s) ds, & \\ G_N^*(t) - g(t) + \int_{t-\tau}^0 K_1(t, s) \phi(s) ds & \text{if } t \in [0, \tau). \\ - \int_0^t K_2(t, s) G_N^*(s) ds, & \end{cases} \tag{10}$$

Intuitively, we try to make $r_N(t)$ as small as possible. For that, the unknown coefficients α_k in (9) are determined by forcing

$$r_N(t_j) = 0, \quad j = 0, 1, \ldots, N, \tag{11}$$

at the collocation points $t_j = \frac{j}{N}$. Then, from (7) and (9)–(11) we may write that

$$\sum_{k=0}^{N} \left(H_k(t_j) - \int_0^{t_j - \tau} K_1(t_j, s) H_k(s) ds - \int_0^{t_j} K_2(t_j, s) H_k(s) ds \right) \alpha_k = g(t_j)$$

$$\text{for } t_j \in [\tau, 1] \ (j = 0, 1, \ldots, N),$$

$$\tag{12}$$

and

$$\sum_{k=0}^{N} \left(H_k(t_j) - \int_0^{t_j} K_2(t_j, s) H_k(s) ds \right) \alpha_k = g(t_j) - \int_{t_j-\tau}^{0} K_1(t_j, s) \phi(s) ds$$

$$\text{for } t_j \in [0, \tau) \ \ (j = 0, 1, \dots, N).$$

(13)

The last linear system of Eqs. (12)–(13) can be written in the matrix form as follows:

$$A_N X_N = B_N,$$

where $X_N = \begin{bmatrix} \alpha_0 \ \alpha_1 \ \cdots \ \alpha_N \end{bmatrix}^T$ is the unknown matrix and the matrices $A_N = [a_{jk}]_{j,k=0,\dots,N}$ and $B_N = \begin{bmatrix} b_0 \ b_1 \ \cdots \ b_N \end{bmatrix}^T$ are given as follows:

$$a_{jk} = \left[\left(H_k(t_j) - \int_0^{t_j-\tau} K_1(t_j, s) H_k(s) ds - \int_0^{t_j} K_2(t_j, s) H_k(s) ds \right) \right] \text{ for } t_j \in [\tau, 1]$$

$$a_{jk} = \left[\left(H_k(t_j) - \int_0^{t_j} K_2(t_j, s) H_k(s) ds \right) \right] \text{ for } t_j \in [0, \tau), \quad (j, k = 0, 1, \dots, N)$$

and

$$b_j = g(t_j) \text{ for } t_j \in [\tau, 1]$$

$$b_j = g(t_j) - \int_{t_j-\tau}^{0} K_1(t_j, s) \phi(s) ds \text{ for } t_j \in [0, \tau) \ \ (j = 0, 1, \dots, N).$$

Since

$$H_k(t_j) = H(t_j - t_k) = \begin{cases} 1, \text{ if } j \geq k \\ 0, \text{ if } j < k, \end{cases}$$

the entries of the matrix $A_N = [a_{jk}]$ becomes

$$a_{jk} := \begin{cases} 0, & \text{if } j < k \\ 1, & \text{if } j = k \\ 1 - \int_{t_k}^{t_j} K_2(t_j, s) ds, & \text{if } j > k \text{ and } t_j \leq \tau + t_k \\ 1 - \int_{t_k}^{t_j-\tau} K_1(t_j, s) ds - \int_{t_k}^{t_j} K_2(t_j, s) ds, & \text{if } j > k \text{ and } t_j > \tau + t_k \end{cases}$$

(14)

for $j, k = 0, 1, \dots, N$. Then, $A_N = [a_{jk}]$ is a lower triangular matrix with $\det A = 1$. Therefore, for the unknown matrix X_N, we have a unique solution

$$X_N = A_N^{-1} B_N,$$

which immediately implies that the Volterra integral Eq. (7) with delay $0 < \tau < 1$ admits a unique approximate solution in the form of (9).

As a result, we get the next theorem.

Theorem 1 *For a given delay Volterra integral equation as in* (1) *with a delay term* $0 < \tau < 1$, *let* $\phi \in C[-\tau, 0]$ *satisfy the condition* (5), *and let, for each fixed* $t \in [0, 1]$, *the kernel function* $K_1(t, s)$ *be integrable with respect to s such that* $(t, s) \in D_\tau$, *and* $K_2(t, s)$ *be integrable with respect to s such that* $(t, s) \in D$. *Then, the collocation method for solving* (1) *admits a unique solution, which means that the approximate solution* G_N^* *having the form* (9) *is unique, and the corresponding matrix to the linear system is lower triangular, whose terms are given by* (14).

3   Error Estimation and Convergence Analysis

Now we show that, under some smoothness conditions, the approximate solution obtained in the previous section is uniformly convergent to the exact solution f on $[0, 1]$.

Theorem 2 *For a given delay Volterra integral Eq.* (7) *with a delay* $0 < \tau < 1$, *in addition to the conditions of Theorem 1, assume that the followings hold:*

(a) *for each* $i = 1, 2$, *there exist* $M_i > 0$ *and* $\beta_i \in (0, 1]$ *such that*

$$|K_i(t, s) - K_i(u, s)| \leq M_i |t - u|^{\beta_i} \text{ for all } (t, s), (u, s) \in D_\tau \text{ and } D, \text{ resp.,}$$

(b) *for each* $i = 1, 2$, *there exist positive constants* C_i *such that*

$$|K_i(t, s)| \leq C_i \text{ for all } (t, s) \in D_\tau \text{ and } D, \text{ resp.,}$$

(c) *there exist* $M_3 > 0$ *and* $\beta_3 \in (0, 1]$ *such that*

$$|g(t) - g(u)| \leq M_3 |t - u|^{\beta_3} \text{ for all } t, u \in [0, 1].$$

Then, for every $t \in [0, 1]$ *and for all* $N \in \mathbb{N}$, *we have the error estimation between the approximate solution* G_N^* *having the form* (9) *and the exact solution* $f \in C[0, 1]$ *as follows:*

$$|e_N(t)| := \left| f(t) - G_N^*(t) \right| \leq \frac{M_\tau e^C}{N^\beta}, \tag{15}$$

where $\beta =: \min\{\beta_1, \beta_2, \beta_3\}$ *and* M_τ, C *are certain positive constants independent of t and N.*

Proof Let $\tau \in (0, 1)$ be given. We first assume that $t \in [0, \tau]$. For a given $N \in \mathbb{N}$, putting $j = j(N, t) := \lfloor Nt \rfloor$, we get

$$|e_N(t)| = \left| f(t) - G_N^*(t) \right| \leq \left| f(t) - f(t_j) \right| + \left| f(t_j) - G_N^*(t) \right|.$$

Also since $G_N^*(t) = G_N^*(t_j)$, we obtain that

$$|e_N(t)| \leq |f(t) - f(t_j)| + |f(t_j) - G_N^*(t_j)|. \tag{16}$$

Then, using (7) and (11), we see that

$$|e_N(t)| \leq |g(t) - g(t_j)|$$

$$+ \left| \int_{t_j - \tau}^{0} K_1(t_j, s)\phi(s)ds - \int_{t-\tau}^{0} K_1(t, s)\phi(s)ds \right|$$

$$+ \left| \int_{0}^{t} K_2(t, s)f(s)ds - \int_{0}^{t_j} K_2(t_j, s)f(s)ds \right|$$

$$+ \int_{0}^{t_j} |K_2(t_j, s)| |f(s) - G_N^*(s)| ds.$$

The last inequality implies that

$$|e_N(t)| \leq |g(t) - g(t_j)| + \|\phi\|_{[-\tau, 0]} \int_{t-\tau}^{0} |K_1(t, s) - K_1(t_j, s)| ds$$

$$+ \|\phi\|_{[-\tau, 0]} \int_{t_j - \tau}^{t-\tau} |K_1(t_j, s)| ds$$

$$+ \|f\|_{[0, 1]} \int_{0}^{t_j} |K_2(t, s) - K_2(t_j, s)| ds$$

$$+ \|f\|_{[0, 1]} \int_{t_j}^{t} |K_2(t, s)| ds + \int_{0}^{t_j} |K_2(t_j, s)| |e_N(s)| ds,$$

where, as usual, $\|\cdot\|_{[a, b]}$ denotes the usual supremum norm on $[a, b]$. From the hypotheses $(a) - (c)$, we conclude that

$$|e_N(t)| \leq M_3 (t - t_j)^{\beta_3} + \tau M_1 \|\phi\|_{[-\tau, 0]} (t - t_j)^{\beta_1}$$

$$+ C_1 \|\phi\|_{[-\tau, 0]} (t - t_j) + M_2 \|f\|_{[0, 1]} (t - t_j)^{\beta_2}$$

$$+ C_2 \|f\|_{[0, 1]} (t - t_j) + C_2 \int_{0}^{t_j} |e_N(s)| ds.$$

By the fact that $j = \lfloor Nt \rfloor$ we observe $0 \leq t - t_j \leq \frac{1}{N} \leq 1$. Then, taking $\beta := \min\{\beta_1, \beta_2, \beta_3\}$ and using the fact that $t_j \leq t$, we may write that

$$|e_N(t)| \leq \frac{M_3 + (\tau M_1 + C_1) \|\phi\|_{[-\tau, 0]} + (C_2 + M_2) \|f\|_{[0, 1]}}{N^\beta}$$

$$+ C_2 \int_{0}^{t} |e_N(s)| ds.$$

Then, it follows from Gronwall's lemma that

$$|e_N(t)| \leq \frac{\left(M_3 + (\tau M_1 + C_1)\, \|\phi\|_{[-\tau,0]} + (C_2 + M_2)\, \|f\|_{[0,1]}\right) e^{C_2}}{N^\beta}, \tag{17}$$

Assume now that $t \in (\tau, 1]$. For a given $N \in \mathbb{N}$, take again $j := \lfloor Nt \rfloor$. Then, there are two possible cases: $t_j \leq \tau$ or $t_j > \tau$. If $t_j \leq \tau$, then we may write from (7), (10), (11), and (16) that

$$
\begin{aligned}
|e_N(t)| \leq\ & \left|g(t) - g\left(t_j\right)\right| \\
&+ \int_0^{t-\tau} |K_1(t,s)|\, |f(s)|\, ds + \int_{t_j-\tau}^0 \left|K_1(t_j,s)\right|\, |\phi(s)|\, ds \\
&+ \left|\int_0^t K_2(t,s) f(s)\, ds - \int_0^{t_j} K_2(t_j,s) f(s)\, ds\right| \\
&+ \left|\int_0^{t_j} K_2(t_j,s) f(s)\, ds - \int_0^{t_j} K_2(t_j,s) G_N^*(s)\, ds\right|.
\end{aligned}
$$

Hence, from $(a) - (c)$, we get

$$
\begin{aligned}
|e_N(t)| \leq\ & M_3(t - t_j)^{\beta_3} + C_1\, \|f\|_{[0,1]}\, (t - \tau) \\
&+ C_1\, \|\phi\|_{[-\tau,0]}\, \left(\tau - t_j\right) + M_2\, \|f\|_{[0,1]}\, (t - t_j)^{\beta_2} \\
&+ C_2\, \|f\|_{[0,1]}\, (t - t_j) + C_2 \int_0^t |e_N(s)|\, ds.
\end{aligned}
$$

Since $t - \tau \leq t - t_j$ and $t_j \leq t$, we arrive

$$|e_N(t)| \leq \frac{\left(M_3 + C_1\, \|\phi\|_{[-\tau,0]} + (C_1 + C_2 + M_2)\, \|f\|_{[0,1]}\right) e^{C_2}}{N^\beta}. \tag{18}$$

Finally, if $t_j > \tau$, then

$$
\begin{aligned}
|e_N(t)| \leq\ & \left|g(t) - g\left(t_j\right)\right| \\
&+ \left|\int_0^{t-\tau} K_1(t,s) f(s)\, ds - \int_0^{t_j-\tau} K_1(t_j,s) f(s)\, ds\right| \\
&+ \left|\int_0^t K_2(t,s) f(s)\, ds - \int_0^{t_j} K_2(t_j,s) f(s)\, ds\right| \\
&+ \left|\int_0^{t_j-\tau} K_2(t_j,s) f(s)\, ds - \int_0^{t_j-\tau} K_2(t_j,s) G_N^*(s)\, ds\right| \\
&+ \left|\int_0^{t_j} K_2(t_j,s) f(s)\, ds - \int_0^{t_j} K_2(t_j,s) G_N^*(s)\, ds\right|,
\end{aligned}
$$

which implies

$$|e_N(t)| \leq M_3(t - t_j)^{\beta_3}$$
$$+ M_1 \|f\|_{[0,1]} (t - t_j)^{\beta_1} + C_1 \|f\|_{[0,1]} (t - t_j)$$
$$+ M_2 \|f\|_{[0,1]} (t - t_j)^{\beta_2} + C_2 \|f\|_{[0,1]} (t - t_j)$$
$$+ 2C_2 \int_0^t |e_N(s)| \, ds.$$

Hence

$$|e_N(t)| \leq \frac{\left(M_3 + (C_1 + C_2 + M_1 + M_2) \|f\|_{[0,1]}\right) e^{2C_2}}{N^\beta} \tag{19}$$

Then, taking $C := 2C_2$ and

$$M_\tau := M_3 + (C_1 + \tau M_1) \|\phi\|_{[-\tau,0]} + (C_1 + C_2 + M_1 + M_2) \|f\|_{[0,1]}$$

and also combining (17)–(19), for every $t \in [0, 1]$ and for all $N \in \mathbb{N}$ we arrive the inequality (15).

Corollary 1 *Under the conditions of Theorem 2, the approximate solution G_N^* having the form (9) is uniformly convergent to the exact solution $f \in C[0, 1]$.*

Proof From (15), observe that $\|e_N\|_{[0,1]} \leq \dfrac{M_\tau e^C}{N^\beta}$ for every $N \in \mathbb{N}$.

4 Numerical Simulations

In this section we display two numerical applications verifying our results obtained in the previous sections.

Example 1 Consider the Volterra integral Eq. (1) with the delay $\tau = \frac{1}{2}$ and the following given functions:

$$K_1(t, s) = t + s, \quad K_2(t, s) = ts, \quad \phi(t) = 6t^3 - 9t^2 + 4t$$

and

$$g(t) = \frac{t}{960} \left(1152t^5 - 2448t^4 + 3440t^3 - 9160t^2 + 11\,040t - 4171\right).$$

Then, the exact solution of (1) will be

$$f(t) = 6t^3 - 9t^2 + 4t.$$

Fig. 2 Graphs of $G_{10}^*(t)$ and $G_{20}^*(t)$, which are numerical solutions of (1), approximate to the exact solution $f(t) = 6t^3 - 9t^2 + 4t$

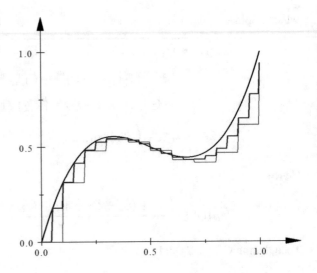

Table 1 Maximum errors for numerical solutions in Example 1

N	$\|e_N\|_{[0,1]}$
5	5.8195×10^{-1}
10	3.8642×10^{-1}
20	2.2434×10^{-1}

We understand from Fig. 2 and Table 1 that the corresponding numerical solutions G_N^* defined by (9) approximate to f when N is sufficiently large. We know from Theorem 2 that the upper bounds of the errors depend on $\beta_1, \beta_2, \beta_3, C_1, C_2, M_1, M_2, M_3$, and τ.

Example 2 Consider the following delay differential equation

$$
\begin{cases}
f''(t) + \left(2t - \frac{1}{3}\right) f'\left(t - \frac{1}{3}\right) - 2tf'(t) + 3f\left(t - \frac{1}{3}\right) - 3f(t) = g''(t), \; t \in \left[\frac{1}{3}, 1\right] \\[2mm]
f''(t) - 2tf'(t) - 3f(t) = g''(t) - \frac{5t}{3} + \frac{4}{9}, \; t \in \left[0, \frac{1}{3}\right)
\end{cases}
\tag{20}
$$

with the initial and delay conditions

$$
f(0) = 0, \; f'(0) = \frac{1}{3} \text{ and } f(t) = \phi(t) = \frac{t}{3} \text{ for } t \in \left[-\frac{1}{3}, 0\right).
$$

Then, by taking $\tau = \frac{1}{3}$, $K(t,s) := K_1(t,s) = -K_2(t,s) = -(t+s)$, $\phi(t) = \frac{t}{3}$ and

$$
g(t) =
\begin{cases}
\frac{1}{972}(3t-1)^2\left(21t^2 + 20t - 3\right) - \frac{1}{36}t^3(7t+10) + \frac{1}{3}t(t+1) & \text{if } \tau \le t \le 1 \\[2mm]
\frac{1}{486}(3t-1)^2(15t-2) - \frac{1}{36}t^3(7t+10) + \frac{1}{3}t(t+1) & \text{if } 0 \le t < \tau,
\end{cases}
$$

the Eq. (20) turns out to be the following delay integral equation:

$$f(t) = \begin{cases} g(t) + \int_{t-\tau}^{t} K(t,s) f(s) ds & \text{if } 0 \leq t \leq 1 \\ \phi(t) & \text{if } -\tau \leq t < 0, \end{cases} \qquad (21)$$

which is a special case of (1). Then, observe that the exact solution of (20), or equivalently (21), becomes

$$f(t) = \frac{t(t+1)}{3} \text{ for } t \in [0,1].$$

Table 2 and Fig. 3 verify Theorems 1 and 2.

N	$\|e_N\|_{[0,1]}$
7	1.7792×10^{-1}
15	8.9126×10^{-2}
21	6.5134×10^{-2}

Table 2 Maximum errors for numerical solutions in Example 2

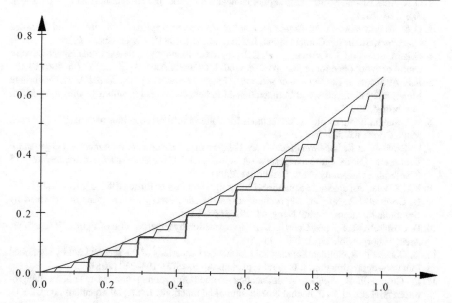

Fig. 3 Graphs of $G_7^*(t)$ and $G_{21}^*(t)$, which are numerical solutions of (20), approximate to the exact solution $f(t) = \frac{t(t+1)}{3}$

5 Concluding Remarks

In this paper we obtain numerical solutions of delay Volterra integral equation having linear structure by using a collocation method based on neural network approximation introduced by Costarelli and Spigler [11]. For a future study, it would be interesting to investigate nonlinear delay Volterra equations and also delay Volterra integro-equations from this point of view. We should note that although we mainly use the unit interval [0, 1], all results can easily be moved any compact interval by a simple translation. We understand from this paper that the collocation method based on the process of unit step functions is quite an effective and easy way for obtaining the approximate solutions of Volterra integral equations. Another interesting situation is the multivariate case of this process.

References

1. G.A. Anastassiou, in *Intelligent Systems: Approximation by Artificial Neural Networks*, Intelligent Systems Reference Library, vol. 19 (Springer-Verlag, Berlin, 2011)
2. G.A. Anastassiou, Multivariate sigmoidal neural network approximation. Neural Netw. **24**(4), 378–386 (2011)
3. G.A. Anastassiou, in *Intelligent Systems II: Complete Approximation by Neural Network Operators*, Studies in Computational Intelligence, vol. 608 (Springer, Cham, 2016)
4. G.A. Anastassiou, L. Coroianu, S.G. Gal, Approximation by a nonlinear Cardaliaguet-Euvrard neural network operator of max-product kind. J. Comput. Anal. Appl. **12**(2), 396–406 (2010)
5. K.E. Atkinson, in *The Numerical Solution of Integral Equations of the Second Kind*, Cambridge Monographs on Applied and Computational Mathematics, vol. 4 (Cambridge University Press, Cambridge, 1997)
6. H. Brunner, Iterated collocation methods for Volterra integral equations with delay arguments. Math. Comp. **62**, 581–599 (1994)
7. H. Brunner, in *Collocation Methods for Volterra Integral and Related Functional Differential Equations*, Cambridge Monographs on Applied and Computational Mathematics, vol. 15 (Cambridge University Press, Cambridge 2004)
8. K.L. Cooke, An epidemic equation with immigration. Math. Biosci. **29**(1–2), 135–158 (1976)
9. D. Costarelli, R. Spigler, Approximation results for neural network operators activated by sigmoidal functions. Neural Netw. **44**, 101–106 (2013)
10. D. Costarelli, R. Spigler, Constructive approximation by superposition of sigmoidal functions. Anal. Theory Appl. **29**(2), 169–196 (2013)
11. D. Costarelli, R. Spigler, Solving Volterra integral equations of the second kind by sigmoidal functions approximation. J. Integral Equations Appl. **25**(2), 193–222 (2013)
12. D. Costarelli, R. Spigler, A collocation method for solving nonlinear Volterra integro-differential equations of neutral type by sigmoidal functions. J. Integral Equations Appl. **26**(1), 15–52 (2014)
13. D. Costarelli, G. Vinti, Pointwise and uniform approximation by multivariate neural network operators of the max-product type. Neural Netw. **81**, 81–90 (2016)
14. J.K. Hale, S.M.V. Lunel, in *Introduction to Functional-Differential Equations*, Applied Mathematical Sciences, vol. 99 (Springer-Verlag, New York, 1993)
15. W. Ming, C. Huang, Collocation methods for Volterra functional integral equations with non-vanishing delays. Appl. Math. Comput. **296**, 198–214 (2017)

16. W. Ming, C. Huang, M. Li, Superconvergence in collocation methods for Volterra integral equations with vanishing delays. J. Comput. Appl. Math. **308**, 361–378 (2016)
17. S. Nemati, Numerical solution of Volterra-Fredholm integral equations using Legendre collocation method. J. Comput. Appl. Math. **278**, 29–36 (2015)
18. K. Yang, Zhang, R.: Analysis of continuous collocation solutions for a kind of Volterra functional integral equations with proportional delay. J. Comput. Appl. Math. **236**(5), 743–752 (2011)

Harmonic and Trace Inequalities
in Lipschitz Domains

Soumia Touhami, Abdellatif Chaira, and Delfim F. M. Torres

Abstract We prove boundary inequalities in arbitrary bounded Lipschitz domains
on the trace space of Sobolev spaces. For that, we make use of the trace operator,
its Moore–Penrose inverse, and of a special inner product. We show that our trace
inequalities are particularly useful to prove harmonic inequalities, which serve as
powerful tools to characterize the harmonic functions on Sobolev spaces of non-
integer order.

2010 Mathematics Subject Classification 47A30, 47J20

1 Introduction

In this article we establish some new and important operator inequalities connected
with traces on Hilbert spaces. Trace inequalities find several interesting applications,
e.g., to problems from quantum statistical mechanics and information theory
[2, 5, 11]. Here we establish new trace inequalities in Lipschitz domains, that is,
in a domain of the Euclidean space whose boundary is "sufficiently regular," in the
sense that it can be thought of as, locally, being the graph of a Lipschitz continuous
function [14]. The study of Lipschitz domains is an important research area per
se, since many of the Sobolev embedding theorems require them as the natural
domain of study [19]. Consequently, many partial differential equations found in

S. Touhami · A. Chaira
Moulay Ismail University, Faculté des Sciences, Laboratoire de Mathématiques et leures
Applications, Equipe EDP et Calcul Scientifique, Meknes, Morocco
e-mail: s.touhami@edu.umi.ac.ma; a.chaira@fs.umi.ac.ma

D. F. M. Torres (✉)
Center for Research and Development in Mathematics and Applications (CIDMA),
Department of Mathematics, University of Aveiro, Aveiro, Portugal
e-mail: delfim@ua.pt

© Springer Nature Switzerland AG 2019
G. A. Anastassiou and J. M. Rassias (eds.), *Frontiers in Functional Equations
and Analytic Inequalities*, https://doi.org/10.1007/978-3-030-28950-8_30

applications and variational problems are defined on Lipschitz domains [3, 6, 12]. In our case, we investigate the application of the obtained trace inequalities in Lipschitz domains to harmonic functions [8], which is a subject of strong current research [10, 15, 18, 22].

The paper is organized as follows. In Sect. 2, we fix notations and recall necessary definitions and results, needed in the sequel. Our contribution is then given in Sect. 3: we prove a Moore–Penrose inverse equality (Theorem 1), trace inequalities (Theorem 2), and harmonic inequalities (Theorems 3 and 4). As an application of Theorem 4, we obtain a functional characterization of the harmonic Hilbert spaces for the range of values $0 \leq s \leq 1$ (Corollary 1).

2 Preliminaries

Let $\mathcal{H}_1$ and $\mathcal{H}_2$ be two Hilbert spaces with inner products $(\cdot, \cdot)_{\mathcal{H}_1}$ and $(\cdot, \cdot)_{\mathcal{H}_2}$ and associated norms $\| \cdot \|_{\mathcal{H}_1}$ and $\| \cdot \|_{\mathcal{H}_2}$, respectively. We denote by $\mathcal{L}(\mathcal{H}_1, \mathcal{H}_2)$ the space of all linear operators from $\mathcal{H}_1$ into $\mathcal{H}_2$ and $\mathcal{L}(\mathcal{H}_1, \mathcal{H}_1)$ is briefly denoted by $\mathcal{L}(\mathcal{H}_1)$. For an operator $A \in \mathcal{L}(\mathcal{H}_1, \mathcal{H}_2)$, $\mathcal{D}(A)$, $\mathcal{R}(A)$, and $\mathcal{N}(A)$ denote its domain, its range, and its null space, respectively. The set of all bounded operators from $\mathcal{H}_1$ into $\mathcal{H}_2$ is denoted by $\mathcal{B}(\mathcal{H}_1, \mathcal{H}_2)$, while $\mathcal{B}(\mathcal{H}_1, \mathcal{H}_1)$ is briefly denoted by $\mathcal{B}(\mathcal{H}_1)$. The set of all closed densely defined operators from $\mathcal{H}_1$ into $\mathcal{H}_2$ is denoted by $C(\mathcal{H}_1, \mathcal{H}_2)$ and, analogously as before, $C(\mathcal{H}_1, \mathcal{H}_1)$ is denoted by $C(\mathcal{H}_1)$. For $A \in C(\mathcal{H}_1, \mathcal{H}_2)$, its adjoint operator is denoted by $A^* \in C(\mathcal{H}_2, \mathcal{H}_1)$.

The Moore–Penrose inverse of a closed densely defined operator $A \in C(\mathcal{H}_1, \mathcal{H}_2)$, denoted by $A^\dagger$, is defined as the unique linear operator in $C(\mathcal{H}_2, \mathcal{H}_1)$ such that

$$\mathcal{D}(A^\dagger) = \mathcal{R}(A) \oplus \mathcal{N}(A^*), \quad \mathcal{N}(A^\dagger) = \mathcal{N}(A^*),$$

and

$$\begin{cases} AA^\dagger A = A, \\ A^\dagger AA^\dagger = A^\dagger, \end{cases} \quad \begin{cases} AA^\dagger \subset P_{\overline{\mathcal{R}(A)}}, \\ A^\dagger A \subset P_{\overline{\mathcal{R}(A^\dagger)}}, \end{cases}$$

where $\overline{\mathcal{E}}$ denotes the closure of $\mathcal{E}$, $\mathcal{E} \in \{\mathcal{R}(A), \mathcal{R}(A^\dagger)\}$, and $P_{\overline{\mathcal{E}}}$ the orthogonal projection on the closed subspace $\overline{\mathcal{E}}$. The following lemma is used in the proof of our Moore–Penrose inverse equality (Theorem 1).

Lemma 1 (See Lemma 2.5 and Corollary 2.6 of [13]) *Let* $A \in C(\mathcal{H}_1, \mathcal{H}_2)$ *and* $B \in C(\mathcal{H}_2, \mathcal{H}_1)$ *be such that* $B = A^\dagger$. *Then,*

1. $A(I + A^*A)^{-1} = B^*(I + BB^*)^{-1}$;
2. $(I + A^*A)^{-1} + (I + BB^*)^{-1} = I + P_{\mathcal{N}(B^*)}$;
3. $A^*(I + AA^*)^{-1} = B(I + B^*B)^{-1}$;

4. $(I + AA^*)^{-1} + (I + B^*B)^{-1} = I + P_{N(A^*)}$;
5. $(I + AA^*)^{-1} + (I + B^*B)^{-1} = I$ (if A^* is injective);
6. $N(A^*(I + AA^*)^{-1/2}) = N(A^*) = N(B)$.

Lemma 2 (See Theorem 3.5 of [21]) *Let $\mathcal{H}_1$ and $\mathcal{H}_2$ be two Hilbert spaces, $A \in \mathcal{B}(\mathcal{H}_1, \mathcal{H}_2)$, and B be its Moore–Penrose inverse. Then, the operator $B^*(I + BB^*)^{-1/2}$ is bounded with closed range and has a bounded Moore–Penrose inverse given by*

$$T_B = B(I + B^*B)^{-1/2} + A^*(I + B^*B)^{-1/2}.$$

Moreover, the adjoint operator of T_B is T_{B^*}, where

$$T_{B^*} = B^*(I + BB^*)^{-1/2} + A(I + BB^*)^{-1/2}.$$

Lemma 3 (See Theorem 3.8 of [21]) *Let $\mathcal{H}_1$ and $\mathcal{H}_2$ be two Hilbert spaces, $A \in \mathcal{B}(\mathcal{H}_1, \mathcal{H}_2)$, and B be its Moore–Penrose inverse. Then, the decomposition*

$$A = (I + B^*B)^{-1/2} T_{B^*}$$

holds, where $T_{B^*} = B^*(I + BB^*)^{-1/2} + A(I + BB^*)^{-1/2}$.

Lemma 4 (See Corollary 3.7 of [21]) *Let $A \in \mathcal{B}(\mathcal{H}_1, \mathcal{H}_2)$ and B be its Moore–Penrose inverse. Then, T_B is an isomorphism from $\mathcal{R}(B^*)$ to $N(B^*)^{\perp}$, where $N(B^*)^{\perp}$ denotes the orthogonal complement of $N(B^*)$.*

Let Ω be an open subset of $\mathbb{R}^d$ with boundary $\partial\Omega$ and closure $\overline{\Omega}$. We say that $\partial\Omega$ is Lipschitz continuous if for every $x \in \partial\Omega$ there exists a coordinate system $(\widehat{y}, y_d) \in \mathbb{R}^{d-1} \times \mathbb{R}$, a neighborhood $Q_{\delta,\delta'}(x)$ of x, and a Lipschitz function $\gamma_x : \widehat{Q}_\delta \to \mathbb{R}$, with the following properties:

1. $\Omega \cap Q_{\delta,\delta'}(x) = \{(\widehat{y}, y_d) \in Q_{\delta,\delta'}(x) \,/\, \gamma_x(\widehat{x}) < y_d\}$;
2. $\partial\Omega \cap Q_{\delta,\delta'}(x) = \{(\widehat{y}, y_d) \in Q_{\delta,\delta'}(x) \,/\, \gamma_x(\widehat{x}) = y_d\}$;

where $Q_{\delta,\delta'}(x) = \{(\widehat{y}, y_d) \in \mathbb{R}^d \,/\, \|\widehat{y} - \widehat{x}\|_{\mathbb{R}^{d-1}} < \delta \text{ and } |y_d - x_d| < \delta'\}$ and

$$\widehat{Q}_\delta(x) = \left\{\widehat{y} \in \mathbb{R}^{d-1} \,/\, \|\widehat{y} - \widehat{x}\|_{\mathbb{R}^{d-1}} < \delta\right\}$$

for $\delta, \delta' > 0$. An open connected subset $\Omega \subset \mathbb{R}^d$, whose boundary is Lipschitz continuous, is called a Lipschitz domain. In the rest of this paper, Ω denotes a bounded Lipschitz domain in $\mathbb{R}^d$, $d \geq 2$. We denote by $C^k(\Omega)$, $k \in \mathbb{N}$ or $k = \infty$, the space of real k times continuously differentiable functions on Ω. The space C^∞ of all real functions on Ω with a compact support in Ω is denoted by $C_c^\infty(\Omega)$. We say that a sequence $(\varphi_n)_{n \geq 1} \in C_c^\infty(\Omega)$ converges to $\varphi \in C_c^\infty(\Omega)$, if there exists a compact $Q \subset \Omega$ such that $\text{supp}(\varphi_n) \subset Q$ for all $n \geq 1$ and, for all multi-index $\alpha \in \mathbb{N}^d$, the sequence $(\partial^\alpha \varphi_n)_{n \geq 1}$ converges uniformly to $\partial^\alpha \varphi$, where ∂^α denotes the partial derivative of order α. The space $C_c^\infty(\Omega)$, induced by this convergence, is denoted by $\mathscr{D}(\Omega)$, while $\mathscr{D}'(\Omega)$ is the space of distributions on Ω. For $k \in \mathbb{N}$, $H^k(\Omega)$ is the space of all distributions u defined on Ω such that all partial derivatives of order at most k lie in $L^2(\Omega)$, i.e., $\partial^\alpha u \in L^2(\Omega)$ $\forall |\alpha| \leq k$. This is a Hilbert space with the scalar product

$$(u, v)_{k,\Omega} = \sum_{|\alpha| \leq k} \int_\Omega \partial^\alpha u \, \partial^\alpha v \, dx,$$

where dx is the Lebesgue measure and $u, v \in H^k(\Omega)$. The corresponding norm, denoted by $\| \cdot \|_{k,\Omega}$, is given by

$$\|u\|_{k,\Omega} = \left(\sum_{|\alpha| \leq k} \int_\Omega |\partial^\alpha u|^2 \, dx \right)^{1/2}.$$

Sobolev spaces $H^s(\Omega)$, for non-integers s, are defined by the real interpolation method [1, 16, 20]. The trace spaces $H^s(\partial\Omega)$ can be defined by using charts on $\partial\Omega$ and partitions of unity subordinated to the covering of $\partial\Omega$. If Ω is a Lipschitz hypograph, then there exists a Lipschitz function $\gamma : \mathbb{R}^{d-1} \to \mathbb{R}$ such that $\Omega = \{x \in \mathbb{R}^{d-1} \, / \, x_d < \gamma(\widehat{x}) \text{ for all } \widehat{x} \in \mathbb{R}^{d-1}\}$. This allows to construct Sobolev spaces on the boundary $\partial\Omega$, in terms of Sobolev spaces on $\mathbb{R}^{d-1}$ [16]. This is done as follows. For $g \in L^2(\partial\Omega)$, we define $g_\gamma(\widehat{x}) = g(\widehat{x}, \gamma(\widehat{x}))$ for $\widehat{x} \in \mathbb{R}^{d-1}$, we let

$$H^s(\partial\Omega) = \left\{ g \in L^2(\partial\Omega) \mid g_\gamma \in H^s(\mathbb{R}^{d-1}) \text{ for } 0 \leq s \leq 1 \right\},$$

and equip this space with the inner product $(g, y)_{H^s(\partial\Omega)} = (g_\gamma, y_\gamma)_{s,\mathbb{R}^{d-1}}$, where

$$(u, v)_{s,\mathbb{R}^{d-1}} = \int_{\mathbb{R}^{d-1}} (1 + |\xi|^2)^s \widehat{u}(\xi) \widehat{v}(\xi) \, d\xi$$

and $\widehat{u}$ denotes the Fourier transform of u. Recalling that any Lipschitz function is almost everywhere differentiable, we know that any Lipschitz hypograph Ω has a surface measure σ and an outward unit normal v that exists σ-almost everywhere on $\partial\Omega$. If Ω is a Lipschitz hypograph, then $d\sigma(x) = \sqrt{1 + \|\nabla\gamma(\widehat{x})\|_{\mathbb{R}^{d-1}}^2} \, d\widehat{x}$ and

$$v(x) = \frac{(-\nabla\gamma(\widehat{x}), 1)}{\sqrt{1 + \|\nabla\gamma(\widehat{x})\|_{\mathbb{R}^{d-1}}^2}}$$

for almost every $x \in \partial\Omega$. Suppose now that Ω is a Lipschitz domain. Because $\partial\Omega \subset \bigcup_{x \in \partial\Omega} Q_{\delta,\delta'}(x)$ and $\partial\Omega$ is compact, there exist $x^1, x^2, \ldots, x^n \in \partial\Omega$ such that

$$\partial\Omega \subset \bigcup_{j=1}^{n} Q_{\delta,\delta'}(x^j).$$

It follows that the family $(W_j) = (Q_{\delta,\delta'}(x^j))$ is a finite open cover of $\partial\Omega$, i.e., each W_j is an open subset of $\mathbb{R}^d$ and $\partial\Omega \subseteq \bigcup_j W_j$. Let (φ_j) be a partition of unity subordinate to the open cover (W_j) of $\partial\Omega$, i.e., $\varphi_j \in \mathscr{D}(W_j)$ and $\sum_j \varphi_j(x) = 1$ for all $x \in \partial\Omega$. The inner product in $H^s(\partial\Omega)$ is then defined by

$$(u, v)_{s,\partial\Omega} = \sum_j (\varphi_j u, \varphi_j v)_{H^s(\partial\Omega_j)},$$

where Ω_j can be transformed to a Lipschitz hypograph by a rigid motion, i.e., by a rotation plus a translation, and satisfies $W_j \cap \Omega = W_j \cap \Omega_j$ for each j. The associated norm will be denoted by $\| \cdot \|_{s,\partial\Omega}$. It is interesting to mention that a different choice of $(W_j), (\Omega_j)$ and (φ_j) would yield the same space $H^s(\partial\Omega)$ with an equivalent norm, for $0 \le s \le 1$. For more on the subject, we refer the interested reader to [1, 7, 16].

The trace operator maps each continuous function u on $\overline{\Omega}$ to its restriction onto $\partial\Omega$ and may be extended to be a bounded surjective operator, denoted by Γ_s, from $H^s(\Omega)$ to $H^{s-\frac{1}{2}}(\partial\Omega)$ for $1/2 < s < 3/2$ [4, 16]. The range and null space of Γ_s are given by $\mathcal{R}(\Gamma_s) = H^{s-1/2}(\partial\Omega)$ and $N(\Gamma_s) = H_0^s(\Omega)$, respectively, where $H_0^s(\Omega)$ is defined to be the closure in $H^s(\Omega)$ of infinitely differentiable functions compactly supported in Ω. For $s = 3/2$, this is no longer valid. For $s > 3/2$, the trace operator from $H^s(\Omega)$ to $H^1(\partial\Omega)$ is bounded [4].

Let us set $\Gamma = T_1 \Gamma_1$, where Γ_1 is the trace operator from $H^1(\Omega)$ to $H^{1/2}(\partial\Omega)$ and T_1 is the embedding operator from $H^{1/2}(\partial\Omega)$ into $L^2(\partial\Omega)$. According to a classical result of Gagliardo [9], we know that $\mathcal{R}(\Gamma) = H^{1/2}(\partial\Omega)$. Since Γ_1 is bounded and T_1 is compact [17], the trace operator Γ from $H^1(\Omega)$ to $L^2(\partial\Omega)$ is also compact.

Now, let us induce $H^1(\Omega)$ with the following inner product:

$$(u, v)_{\partial,\Omega} = \int_\Omega \nabla u \nabla v \, dx + \int_{\partial\Omega} \Gamma u \Gamma v \, d\sigma \quad \forall u, v \in H^1(\Omega).$$

The associated norm $\| \cdot \|_{\partial,\Omega}$ is given by

$$\|u\|_{\partial,\Omega} = \left(\|\nabla u\|_{0,\Omega}^2 + \|\Gamma u\|_{0,\partial\Omega}^2 \right)^{1/2}$$

and $H^1(\Omega)$, induced with the inner product $(\cdot, \cdot)_{\partial,\Omega}$, is denoted by $H_\partial^1(\Omega)$. A well-known result of Nečas [17], asserts that under the condition that Ω is a bounded Lipschitz domain, the norms $\| \cdot \|_{\partial,\Omega}$ and $\| \cdot \|_{1,\Omega}$ are equivalent.

The following characterization is useful to prove our trace inequalities in Sect. 3.

Lemma 5 (See Corollary 6.9 of [21]) *Let Γ be the trace operator from $H^1_\partial(\Omega)$ to $L^2(\partial\Omega)$, Λ its Moore–Penrose inverse, and Λ^* be its adjoint operator. Then, for $0 \le s \le 1$, we have that $H^s(\partial\Omega) = \mathcal{H}^s(\partial\Omega)$ with equivalence of norms, where $\mathcal{H}^s(\partial\Omega) = \{(I + \Lambda^*\Lambda)^{-s}g \mid g \in L^2(\partial\Omega)\}$.*

3 Main Results

We begin by proving an important equality that, together with the trace inequalities of Theorem 2, will be useful to prove our harmonic inequality of Theorem 3.

Theorem 1 (The Moore–Penrose Inverse Equality) *Let $\Gamma \in \mathcal{B}(H^1_\partial(\Omega), L^2(\partial\Omega))$ be the trace operator and $\Lambda \in C(L^2(\partial\Omega), H^1_\partial(\Omega))$ its Moore–Penrose inverse. Then, for a real s, the following equality holds:*

$$T_{\Lambda^*}(I + \Lambda\Lambda^*)^{-s} = (I + \Lambda^*\Lambda)^{-s}T_{\Lambda^*},$$

where $T_{\Lambda^*} = \Lambda^*(I + \Lambda\Lambda^*)^{-1/2} + \Gamma(I + \Lambda\Lambda^*)^{-1/2}$.

Proof From Lemma 1,

$$\begin{aligned}
\mathcal{N}(T_{\Lambda^*}(I + \Lambda\Lambda^*)^{-s}) &= \mathcal{N}(T_{\Lambda^*}) \\
&= \mathcal{N}(\Lambda^*(I + \Lambda\Lambda^*)^{-1/2}) \\
&= \mathcal{N}(\Lambda^*) \\
&= H^1_0(\Omega) \\
&= \mathcal{N}((I + \Lambda^*\Lambda)^{-s}T_{\Lambda^*}),
\end{aligned}$$

and we have

$$T_{\Lambda^*}(I + \Lambda\Lambda^*)^{-s}v = 0 = (I + \Lambda^*\Lambda)^{-s}T_{\Lambda^*}v$$

for all $v \in \mathcal{N}(\Lambda^*)$. Now let us consider the operator $\Gamma^*\Gamma : H^1_\partial(\Omega) \longrightarrow H^1_\partial(\Omega)$, where Γ^* is the adjoint of the trace operator Γ. Given the compactness of Γ, the operator $\Gamma^*\Gamma$ is compact and self-adjoint. Then there exists a sequence of pairs $(s_k, v_k)_{k \ge 1}$ associated to $\Gamma^*\Gamma$ such that

$$\Gamma^*\Gamma v_k = s_k^2 v_k.$$

To prove the equality on $\mathcal{N}(\Lambda^*)^\perp$, we show that

$$T_{\Lambda^*}(I + \Lambda\Lambda^*)^{-s}v_k = (I + \Lambda^*\Lambda)^{-s}T_{\Lambda^*}v_k.$$

To this end, let $\Gamma v_k = s_k z_k$. It follows that $\Gamma^* z_k = s_k v_k$ and, from Lemma 3, $\Gamma v_k = (I + \Lambda^* \Lambda)^{-1/2} T_{\Lambda^*} v_k$. On the other hand,

$$T_{\Lambda^*} v_k = \Lambda^* (I + \Lambda \Lambda^*)^{-1/2} v_k + \Gamma (I + \Lambda \Lambda^*)^{-1/2} v_k.$$

By putting $(I + \Lambda \Lambda^*)^{-1} v_k = w_k$, we have $v_k = w_k + \Lambda \Lambda^* w_k$ and

$$\Gamma^* \Gamma v_k = s_k^2 v_k = \Gamma^* \Gamma w_k + w_k,$$

which implies that

$$
\begin{aligned}
(I + \Gamma^* \Gamma)^{-1} \Gamma^* \Gamma v_k &= s_k^2 (I + \Gamma^* \Gamma)^{-1} v_k \\
&= w_k \\
&= \Gamma^* \Gamma (I + \Gamma^* \Gamma)^{-1} v_k.
\end{aligned}
\tag{1}
$$

Using Lemma 1, it follows from (1) that

$$(I + \Gamma^* \Gamma)^{-1} \Gamma^* \Gamma v_k = \Gamma^* \Lambda^* (I + \Lambda \Lambda^*)^{-1} v_k.$$

This leads, again from Lemma 1, to

$$
\begin{aligned}
(I + \Lambda \Lambda^*)^{-1} v_k &= s_k^2 (I + \Gamma^* \Gamma)^{-1} v_k \\
&= s_k^2 (v_k - (I + \Lambda \Lambda^*)^{-1} v_k),
\end{aligned}
$$

so that

$$(1 + s_k^2)(I + \Lambda \Lambda^*)^{-1} v_k = s_k^2 v_k.$$

Thus,

$$
\begin{aligned}
(I + \Lambda \Lambda^*)^{-1} v_k &= \frac{s_k^2}{1 + s_k^2} v_k \\
&= (1 + s_k^2)^{-1} s_k^2 v_k,
\end{aligned}
$$

which implies that

$$\left(I + \Lambda \Lambda^* \right)^{-s} v_k = \left(s_k^2 \left(1 + s_k^2 \right)^{-1} \right)^s v_k.$$

In particular,

$$(I + \Lambda \Lambda^*)^{-1/2} v_k = s_k (1 + s_k^2)^{-1/2} v_k.$$

Consequently,

$$T_{\Lambda^*} v_k = \Lambda^* (I + \Lambda\Lambda^*)^{-1/2} v_k + \Gamma (I + \Lambda\Lambda^*)^{-1/2} v_k$$

$$= \Lambda^* \left(\frac{s_k}{\sqrt{1 + s_k^2}} v_k \right) + \Gamma \left(\frac{s_k}{\sqrt{1 + s_k^2}} v_k \right)$$

$$= \frac{1}{\sqrt{1 + s_k^2}} z_k + \frac{s_k^2}{\sqrt{1 + s_k^2}} z_k$$

$$= \sqrt{1 + s_k^2} z_k.$$

Therefore,

$$T_{\Lambda^*} (I + \Lambda\Lambda^*)^{-s} v_k = \left(\frac{s_k^2}{1 + s_k^2} \right)^s \left(\frac{1 + s_k^2}{\sqrt{1 + s_k^2}} \right) z_k$$

$$= \left(\frac{s_k^2}{1 + s_k^2} \right)^s \sqrt{1 + s_k^2} z_k.$$

On the other hand, $\Gamma\Gamma^* z_k = \Gamma s_k v_k = s_k^2 z_k$. By putting $(I + \Lambda^*\Lambda)^{-1} z_k = e_k$, we have

$$z_k = e_k + \Lambda^* \Lambda e_k,$$

which implies that

$$\Gamma\Gamma^* z_k = s_k^2 z_k = \Gamma\Gamma^* e_k + e_k$$

and

$$(I + \Gamma\Gamma^*)^{-1} \Gamma\Gamma^* z_k = s_k^2 (I + \Gamma\Gamma^*)^{-1} z_k$$

$$= e_k$$

$$= \Gamma\Gamma^* (I + \Gamma\Gamma^*)^{-1} z_k.$$

Using Lemma 1, it follows that

$$(I + \Gamma\Gamma^*)^{-1} \Gamma\Gamma^* z_k = \Gamma\Lambda (I + \Lambda^*\Lambda)^{-1} z_k.$$

This leads, again from Lemma 1, to

$$(I + \Lambda^*\Lambda)^{-1}z_k = s_k^2(I + \Gamma\Gamma^*)^{-1}z_k = s_k^2\left(z_k - (I + \Lambda^*\Lambda)^{-1}z_k\right),$$

so that

$$(1 + s_k^2)(I + \Lambda^*\Lambda)^{-1}z_k = s_k^2 z_k$$

and

$$(I + \Lambda^*\Lambda)^{-1}z_k = s_k^2(1 + s_k^2)^{-1}z_k.$$

Consequently,

$$\left(I + \Lambda^*\Lambda\right)^{-s}z_k = \left(s_k^2\left(1 + s_k^2\right)^{-1}\right)^s z_k,$$

which implies that

$$\left(I + \Lambda^*\Lambda\right)^{-s} T_{\Lambda^*}v_k = \left(I + \Lambda^*\Lambda\right)^{-s}\sqrt{1 + s_k^2}\,z_k$$

$$= \sqrt{1 + s_k^2}\left(\frac{s_k^2}{1 + s_k^2}\right)^s z_k.$$

Hence, one has $(I + \Lambda^*\Lambda)^{-s} T_{\Lambda^*}v_k = T_{\Lambda^*}(I + \Lambda\Lambda^*)^{-s}v_k$ for all $k \geq 1$ and the proof is complete. $\qquad\square$

Let us now consider the family of Hilbert spaces

$$\mathcal{H}^s(\Omega) = \{v \in H^s(\Omega) \,/\, \Delta v = 0 \text{ in } \mathscr{D}'(\Omega)\}, \quad s \geq 0,$$

that consist of real harmonic functions on the usual Sobolev space $H^s(\Omega)$. For $1 < s < 3/2$, we equip $\mathcal{H}^s(\Omega)$ with the following norm:

$$\|u\|_{\mathcal{H}^s(\Omega)} = \|\Gamma_s u\|_{s-1/2,\partial\Omega}.$$

Theorem 2 (The Trace Inequalities) *Let $\Omega \subset \mathbb{R}^d$, $d \geq 2$, be a bounded Lipschitz domain with boundary $\partial\Omega$. Consider, for $1 < s < 3/2$, the trace operators Γ_s and Γ from $H^s(\Omega)$ to $H^{s-1/2}(\partial\Omega)$ and from $H_{\partial}^1(\Omega)$ to $L^2(\partial\Omega)$, respectively, and let Λ be the Moore–Penrose inverse of Γ. Then there exist two positive constants c_1 and c_2 such that the inequalities*

$$c_1\|\Gamma_s v\|_{s-1/2,\partial\Omega} \leq \|(I + \Lambda^*\Lambda)^{s-1/2}\Gamma\tilde{v}\|_{0,\partial\Omega} \leq c_2\|\Gamma_s v\|_{s-1/2,\partial\Omega} \qquad (2)$$

hold for all $v \in \mathcal{H}^s(\Omega)$, where $\tilde{v}$ is the embedding of v in $\mathcal{H}^1(\Omega)$.

Proof Assume $1 < s < 3/2$ and let $v \in \mathcal{H}^s(\Omega)$ and $\tilde{v}$ be its embedding in $\mathcal{H}^1(\Omega)$. Clearly, $\Gamma_s v \in H^{s-1/2}(\partial\Omega)$ and $\Gamma\tilde{v} \in L^2(\partial\Omega)$. From Lemma 5, it follows that

$$\mathcal{H}^{s-1/2}(\partial\Omega) = H^{s-1/2}(\partial\Omega)$$

with equivalence between the norm $\| \cdot \|_{s-1/2,\partial\Omega}$ and the graph norm defined for $g \in L^2(\partial\Omega)$ by

$$g \longmapsto \|(I + \Lambda^*\Lambda)^{s-1/2}g\|_{0,\partial\Omega}.$$

Equivalently, there exist two positive constants c_1 and c_2 such that (2) holds for all $v \in \mathcal{H}^s(\Omega)$. □

As an application of Theorems 1 and 2, we prove the harmonic inequality (3).

Theorem 3 (The Harmonic Inequality for $1 < s < 3/2$) *Assume $1 < s < 3/2$. Then, for all $v \in \mathcal{H}^s(\Omega)$, the following inequality holds:*

$$\|v\|_{\mathcal{H}^s(\Omega)} \leq \|T_{\Lambda^*}\| \, \|(I + \Lambda\Lambda^*)^{s-1}\tilde{v}\|_{\partial,\Omega}, \tag{3}$$

where

$$T_{\Lambda^*} = \Lambda^*(I + \Lambda\Lambda^*)^{-1/2} + \Gamma(I + \Lambda\Lambda^*)^{-1/2}$$

and $\tilde{v}$ is the embedding of v in $\mathcal{H}^1(\Omega)$.

Proof Consider $v \in \mathcal{H}^s(\Omega)$ and $\tilde{v}$ its embedding in $\mathcal{H}^1(\Omega)$. It follows from Theorem 2 that there exists a positive constant c_3 such that

$$\|v\|_{\mathcal{H}^s(\Omega)} = \|\Gamma_s v\|_{s-1/2,\partial\Omega} \leq c_3 \, \|(I + \Lambda^*\Lambda)^{s-1/2}\Gamma\tilde{v}\|_{0,\partial\Omega}.$$

Moreover, from Lemma 3, we have

$$\Gamma = (I + \Lambda^*\Lambda)^{-1/2}T_{\Lambda^*},$$

where

$$T_{\Lambda^*} = \Lambda^*(I + \Lambda\Lambda^*)^{-1/2} + \Gamma(I + \Lambda\Lambda^*)^{-1/2}.$$

It follows from Theorem 1 that

$$\|v\|_{\mathcal{H}^s(\Omega)} = \|(I + \Lambda^*\Lambda)^{s-1}T_{\Lambda^*}\tilde{v}\|_{0,\partial\Omega} = \|T_{\Lambda^*}(I + \Lambda\Lambda^*)^{s-1}\tilde{v}\|_{0,\partial\Omega}.$$

Lemma 2 asserts that T_{Λ^*} is bounded and the intended inequality (3) follows. □

Now consider the embedding operator E from $H^1(\Omega)$ into $L^2(\Omega)$ and its adjoint E^*, which is the solution operator of the following Robin problem for the Poisson equation:

$$\begin{cases} -\Delta u = f & \text{in } \Omega, \\ \partial_\nu u + \Gamma u = 0 & \text{on } \partial\Omega, \end{cases}$$

where $f \in L^2(\Omega)$ and ∂_ν denotes the normal derivative operator with exterior normal ν. Let E_0^* be the solution operator of the Dirichlet problem for the following Poisson equation:

$$\begin{cases} -\Delta u^0 = f & \text{in } \Omega, \\ \Gamma u^0 = 0 & \text{on } \partial\Omega. \end{cases}$$

By setting $E_1^* = E^* - E_0^*$ and $u^1 = E_1^* f$, it follows that u^1 is the solution of the Dirichlet problem for the following Laplace equation:

$$\begin{cases} -\Delta u^1 = 0 & \text{in } \Omega, \\ \Gamma u^1 = \Gamma u & \text{on } \partial\Omega. \end{cases}$$

Let $0 \le s \le 1$, F_1 be the Moore–Penrose inverse of E_1, F_1^* be its adjoint operator, and denote

$$X^s(\Omega) = \left\{ (I + F_1^* F_1)^{-s/2} v \mid v \in \mathcal{H}(\Omega) \right\},$$

where $\mathcal{H}(\Omega)$ is the Bergman space. Next we prove the harmonic inequalities for the case $s = 1$.

Theorem 4 (The Harmonic Inequalities for s = 1) *Let $\Omega \subset \mathbb{R}^d$, $d \ge 2$, be a bounded Lipschitz domain. Then, for all $v \in \mathcal{H}^1(\Omega)$, there exist two positive constants c_1' and c_2', not depending on v, such that*

$$c_1' \|v\|_{\partial,\Omega} \le \|(I + F_1^* F_1)^{1/2} E_1 v\|_{0,\Omega} \le c_2' \|v\|_{\partial,\Omega}. \tag{4}$$

Proof From Lemma 3, the decomposition

$$E_1 = (I + F_1^* F_1)^{-1/2} T_{F_1^*}$$

holds, where

$$T_{F_1^*} = F_1^*(I + F_1 F_1^*)^{-1/2} + E_1(I + F_1 F_1^*)^{-1/2}.$$

Moreover, since $\mathcal{R}(E_1) = \mathcal{H}^1(\Omega)$, it follows that $\mathcal{H}^1(\Omega) = \mathcal{X}^1(\Omega)$. Now consider the graph norm

$$\|u\|_{\mathcal{X}^1(\Omega)} = \|(I + F_1^* F_1)^{1/2} u\|_{0,\Omega}$$

for $u \in \mathcal{X}^1(\Omega)$. It follows that $E_1 v \in \mathcal{X}^1(\Omega)$ for $v \in \mathcal{H}^1(\Omega)$ and

$$\|(I + F_1^* F_1)^{1/2} E_1 v\|_{0,\Omega} = \|T_{F_1^*} v\|_{0,\Omega}.$$

In agreement with Lemma 4, we can view $T_{F_1^*}$ as an isomorphism from $\mathcal{H}^1(\Omega)$ into $\mathcal{H}(\Omega)$, and there exist two positive constants c_1' and c_2', not depending on v, such that (4) holds.   □

The harmonic inequalities of Theorem 4 are a useful tool to provide a functional characterization of the harmonic Hilbert spaces for the range of values $0 \leq s \leq 1$.

Corollary 1 *Assume* $0 \leq s \leq 1$. *Then* $\mathcal{H}^s(\Omega)$ *form an interpolatory family. Moreover,*

$$\mathcal{H}^s(\Omega) = \mathcal{X}^s(\Omega)$$

with equivalence of norms.

Proof For $s = 0$, one has the equality $\mathcal{X}(\Omega) = \mathcal{H}(\Omega)$ by definition. For $s = 1$, Theorem 4 asserts that $\mathcal{X}^1(\Omega) = \mathcal{H}^1(\Omega)$ with the equivalence of the norm $\|\cdot\|_{\mathcal{X}^1(\Omega)}$ with the norm on $\mathcal{H}^1(\Omega)$, which is the same as the one on $H_\partial^1(\Omega)$. The intended equality $\mathcal{H}^s(\Omega) = \mathcal{X}^s(\Omega)$ with equivalence of norms, $0 < s < 1$, follows from classical results on the theory of positive self-adjoint operators, which assert that both $\mathcal{X}^s(\Omega)$ and $\mathcal{H}^s(\Omega)$ form an interpolating family for $0 < s < 1$.   □

Acknowledgements This research is part of the first author's Ph.D. project, which is carried out at Moulay Ismail University, Meknes. It was essentially finished during a visit of Touhami to the Department of Mathematics of University of Aveiro, Portugal, November 2018. The hospitality of the host institution and the financial support of Moulay Ismail University, Morocco, and CIDMA, Portugal, are here gratefully acknowledged. Torres was partially supported by the Portuguese Foundation for Science and Technology (FCT) through CIDMA, project UID/MAT/04106/2019.

References

1. R.A. Adams, J.J.F. Fournier, *Sobolev Spaces* (Elsevier/Academic Press, Amsterdam 2003)
2. E. Carlen, Trace inequalities and quantum entropy: an introductory course, in *Entropy and the Quantum*. Contemporary Mathematics, vol. 529 (American Mathematical Society, Providence, RI, 2010), pp. 73–140

3. C.D. Collins, J.L. Taylor, Eigenvalue convergence on perturbed Lipschitz domains for elliptic systems with mixed general decompositions of the boundary, J. Differ. Equ. **265**(12), 6187–6209 (2018)
4. M. Costabel, Boundary integral operators on Lipschitz domains: elementary results. SIAM J. Math. Anal. **19**(3), 613–626 (1988)
5. X. Chen, R. Jiang, D. Yang, Hardy and Hardy-Sobolev spaces on strongly Lipschitz domains and some applications. Anal. Geom. Metr. Spaces **4**, 336–362 (2016)
6. B. Dacorogna, *Introduction to the Calculus of Variations*, 3rd edn. (Imperial College Press, London, 2015)
7. R. Dautray, J.-L. Lions, *Mathematical Analysis and Numerical Methods for Science and Technology*. Functional and Variational Methods, vol. 2 (Springer, Berlin, 1988)
8. L.C. Evans, *Partial Differential Equations*. Graduate Studies in Mathematics, vol. 19 (American Mathematical Society, Providence, 1998)
9. E. Gagliardo, Caratterizzazioni delle tracce sulla frontiera relative ad alcune classi di funzioni in *n* variabili. Rend. Sem. Mat. Univ. Padova **27**, 284–305 (1957)
10. V.K. Gupta, P. Sharma, Hypergeometric inequalities for certain unified classes of multivalent harmonic functions. Appl. Appl. Math. **13**(1), 315–332 (2018)
11. M. Hayajneh, S. Hayajneh, F. Kittaneh, On some classical trace inequalities and a new Hilbert-Schmidt norm inequality. Math. Inequal. Appl. **21**(4), 1175–1183 (2018)
12. M. Kohr, W.L. Wendland, Variational approach for the Stokes and Navier–Stokes systems with nonsmooth coefficients in Lipschitz domains on compact Riemannian manifolds. Calc. Var. Partial Differ. Equ. **57**(6), 165 (2018)
13. J.-Ph. Labrousse, Inverses généralisés d'opérateurs non bornés. Proc. Amer. Math. Soc. **115**(1), 125–129 (1992)
14. C.R. Loga, An extension theorem for matrix weighted Sobolev spaces on Lipschitz domains. Houston J. Math. **43**(4), 1209–1233 (2017)
15. G. Maze, U. Wagner, A note on the weighted harmonic-geometric-arithmetic means inequalities. Math. Inequal. Appl. **15**(1), 15–26 (2012)
16. W. McLean, *Strongly Elliptic Systems and Boundary Integral Equations* (Cambridge University Press, Cambridge, 2000)
17. J. Nečas, *Les méthodes directes en théorie des équations elliptiques* (Masson et Cie, Éditeurs, Paris, 1967)
18. S.M. Nikol'skiĭ, P.I. Lizorkin, Inequalities for harmonic, spherical and algebraic polynomials. Dokl. Akad. Nauk SSSR **289**(3), 541–545 (1986)
19. M. Prats, Sobolev regularity of the Beurling transform on planar domains. Publ. Mat. **61**(2), 291–336 (2017)
20. L. Tartar, *An Introduction to Sobolev Spaces and Interpolation Spaces*. Lecture Notes of the Unione Matematica Italiana, vol. 3 (Springer, Berlin, 2007)
21. S. Touhami, A. Chaira, D.F.M. Torres, Functional characterizations of trace spaces in Lipschitz domains. Banach J. Math. Anal. **13**, 407–426 (2019)
22. B. Wei, W. Wang, Some inequalities for general L_p-harmonic Blaschke bodies. J. Math. Inequal. **10**(1), 63–73 (2016)

Dirichlet Beta Function via Generalized Mathieu Series Family

P. Cerone

Abstract Integral representations for a generalized Mathieu series and its companions are used to obtain approximation and bounds for undertaking analysis leading to novel insights for the Dirichlet Beta function and its companions. The bounds are procured using a variety of approaches including utilizing integral representation and Čebyšev functional results. The relationship to Zeta type functions is also examined. It is demonstrated that the Dirichlet Beta function relations are particular cases of the generalized Mathieu companions.

1 Introduction

The series, known in the literature as the Mathieu series,

$$S(r) = \sum_{n=1}^{\infty} \frac{2n}{\left(n^2 + r^2\right)^2}, \quad r > 0, \tag{1}$$

has been extensively studied in the past since its introduction by Mathieu [24] in 1890, where it arose in connection with work on elasticity of solid bodies. The reader is directed to the references and the books [6], [4], and [29] for further illustration of various representations and bounds. The various applications areas involve the solution of the biharmonic equation in a rectangular two-dimensional domain using the so- called *superposition method* and the interested reader is referred to the work of Meleshko [25–27] for excellent coverage and further references. A literature search in MathScinet with 'Mathieu series' results in over 700 hits demonstrates that the area continues to attract many avenues of research and application. See also some of the recent activity such as in [4, 15, 17, 28, 31].

P. Cerone (✉)
Department of Mathematics and Statistics, La Trobe University, Melbourne, VIC, Australia
e-mail: p.cerone@latrobe.edu.au

© Springer Nature Switzerland AG 2019
G. A. Anastassiou and J. M. Rassias (eds.), *Frontiers in Functional Equations and Analytic Inequalities*, https://doi.org/10.1007/978-3-030-28950-8_31

One of the main questions addressed in relation to the series is obtaining sharp bounds.

Building on some results from [38], Alzer et al. [1] showed that the best constants a and b in

$$\frac{1}{x^2 + a} < S(x) < \frac{1}{x^2 + b}, \quad x \neq 0 \tag{2}$$

are $a = \frac{1}{2\zeta(3)}$ and $b = \frac{1}{6}$ where $\zeta(\cdot)$ denotes the Riemann zeta function defined by

$$\zeta(p) = \sum_{n=1}^{\infty} \frac{1}{n^p}. \tag{3}$$

An integral representation for $S(r)$ as given in (1) was presented in [16] and [18] as

$$S(r) = \frac{1}{r} \int_0^{\infty} \frac{x}{e^x - 1} \sin(rx) \, dx. \tag{4}$$

Guo [20] utilized (4) to obtain bounds on $S(r)$. Alternate bounds to (1) were obtained by Qi and coworkers in [33–35].

Guo in [20] posed the interesting problem as to whether there is an integral representation of the generalized Mathieu series

$$S_\mu(r) = \sum_{n=1}^{\infty} \frac{2n}{(n^2 + r^2)^{1+\mu}}, \quad r > 0, \ \mu > 0. \tag{5}$$

In [36] an integral representation was obtained for $S_m(r)$, where $m \in \mathbb{N}$, namely

$$S_m(r) = \frac{2}{(2r)^m \, m!} \int_0^{\infty} \frac{t^m}{e^t - 1} \cos\left(\frac{m\pi}{2} - rt\right) dt$$

$$- 2 \sum_{k=1}^{m} \left[\frac{(k-1)(2r)^{k-2m-1}}{k! \, (m-k+1)} \binom{-(m+1)}{m-k} \right.$$

$$\left. \times \int_0^{\infty} \frac{t^k \cos\left[\frac{\pi}{2}(2m-k+1) - rt\right]}{e^t - 1} dt \right]. \tag{6}$$

The challenge of Guo [20] to obtain an integral representation for $S_\mu(r)$ as defined in (5) was successfully answered by Cerone and Lenard [12] in which the following two theorems were proved.

Theorem 1 *The generalized Mathieu series $S_\mu(r)$ defined by (5) may be represented in the integral form*

$$S_\mu(r) = C_\mu(r) \int_0^\infty \frac{x^{\mu+\frac{1}{2}}}{e^x - 1} J_{\mu-\frac{1}{2}}(rx)\, dx, \qquad \mu > 0, \tag{7}$$

where

$$C_\mu(r) = \frac{\sqrt{\pi}}{(2r)^{\mu-\frac{1}{2}} \Gamma(\mu+1)} \tag{8}$$

and $J_\nu(z)$ is the νth order Bessel function of the first kind.

Theorem 2 *For m a positive integer we have*

$$S_m(r) = \frac{1}{2^{m-1}} \cdot \frac{1}{r^{2m-1}} \cdot \frac{1}{m} \sum_{k=0}^{m-1} \frac{(-1)^{\lfloor \frac{3k}{2} \rfloor}}{k!} r^k \left[\delta_{k\ even} A_k(r) + \delta_{k\ odd} B_k(r) \right], \tag{9}$$

where

$$A_k(r) = \int_0^\infty \frac{x^{k+1}}{e^x - 1} \sin(rx)\, dx, \quad B_k(r) = \int_0^\infty \frac{x^{k+1}}{e^x - 1} \cos(rx)\, dx, \tag{10}$$

with $\delta_{condition} = 1$ if condition holds and zero otherwise and $\lfloor x \rfloor$ is the largest integer not greater than x.

The emphasis as in [12] became the derivation of bounds for the generalized Mathieu series $S_\mu(r)$. The first approach utilized sharp bounds for the Bessel function $|J_\nu(z)|$. To this end, in an article by Landau [21], the best possible uniform bounds were obtained for Bessel functions using monotonicity arguments. Landau showed

$$|J_\nu(z)| < \frac{b_L}{\nu^{\frac{1}{3}}} \tag{11}$$

uniformly in the argument $z > 0$ and is best possible in the exponent $\frac{1}{3}$ and constant

$$b_L = 2^{\frac{1}{3}} \sup_z Ai(z) = 0.674885\cdots, \tag{12}$$

where $Ai(z)$ is the Airy function satisfying

$$w'' - zw = 0.$$

Landau also showed that for $z > 0$

$$|J_\nu (z)| \leq \frac{c_L}{z^{\frac{1}{3}}} \tag{13}$$

uniformly in the order $\nu > 0$ and the exponent $\frac{1}{3}$ is best possible with

$$c_L = \sup_z z^{\frac{1}{3}} J_0 (z) = 0.78574687\ldots. \tag{14}$$

The following theorem, based on the Landau bounds (11)–(14), was obtained in [12].

Theorem 3 *The generalized Mathieu series $S_\mu (r)$ are bounded above, for $\mu > \frac{1}{2}$ and $r > 0$, are given by,*

$$S_\mu (r) \leq b_L \frac{\sqrt{\pi}}{(2r)^{\mu - \frac{1}{2}}} \cdot \frac{1}{\left(\mu - \frac{1}{2}\right)^{\frac{1}{3}}} \cdot \frac{\Gamma\left(\mu + \frac{3}{2}\right)}{\Gamma(\mu + 1)} \zeta\left(\mu + \frac{3}{2}\right), \tag{15}$$

and

$$S_\mu (r) \leq c_L \cdot \frac{\sqrt{\pi}}{2^{\mu - \frac{1}{2}} r^{\mu - \frac{1}{6}}} \cdot \Gamma\left(\mu + \frac{7}{6}\right) \zeta\left(\mu + \frac{7}{6}\right), \tag{16}$$

where b_L and c_L are given by (12) and (14), respectively.

The following corollary was also obtained in [12] for $S(r) = S_1(r)$. The first of these results is corrected below.

Corollary 1 *The Mathieu series $S(r)$ satisfies the following bounds:*

$$S(r) \leq \frac{3\pi}{2^{\frac{13}{6}}} b_L \cdot \zeta\left(\frac{5}{2}\right) \cdot \frac{1}{\sqrt{r}} \tag{17}$$

and

$$S(r) \leq \frac{7c_L}{36} \cdot \sqrt{\frac{\pi}{2}} \cdot \Gamma\left(\frac{1}{6}\right) \zeta\left(\frac{13}{6}\right) \cdot r^{-\frac{5}{6}}, \tag{18}$$

where b_L and c_L are given by (12) and (14), respectively.

The following results were obtained in [12] using a weighted Čebyšev functional approach. See also [10] where the approach was utilized for a greater variety of special functions.

Theorem 4 *For $\mu > 0$ and $r > 0$ the generalized Mathieu series $S_\mu(r)$ satisfies*

$$\left| S_\mu(r) - \frac{\pi^2}{12\mu \left(r^2 + \frac{1}{4}\right)^\mu} \right| \tag{19}$$

$$\leq \kappa \left[\frac{1}{\sqrt{\pi}} \cdot \frac{\Gamma\left(2\mu - \frac{1}{2}\right)}{2^{2\mu-1}\Gamma^2(\mu+1)} \int_0^{\frac{\pi}{2}} \frac{\cos^{2\mu-1}\phi}{\left[\left(\frac{1}{4}\right)^2 + r^2\cos^2\phi\right]^{2\mu-\frac{1}{2}}} d\phi - \frac{1}{2\mu^2 \left(r^2 + \frac{1}{4}\right)^{2\mu}} \right]^{\frac{1}{2}}$$

$$\leq \kappa \left[\frac{\Gamma\left(2\mu - \frac{1}{2}\right)\Gamma\left(\mu + \frac{1}{2}\right)}{2^{2\mu}\Gamma^3(\mu+1)} \cdot \frac{1}{r^{4\mu-1}} - \frac{1}{2\mu^2\left(r^2 + \frac{1}{4}\right)^{2\mu}} \right]^{\frac{1}{2}},$$

where

$$\kappa = \left[\pi^2 \left(1 - \frac{\pi^2}{72}\right) - 7\zeta(3) \right]^{\frac{1}{2}} = 0.3198468959\ldots\ldots \tag{20}$$

Corollary 2 *The Mathieu series $S(r)$ satisfies the following bounds:*

$$\left| \sum_{n=1}^\infty \frac{2n}{(n^2+r^2)^2} - \frac{\pi^2}{12\left(r^2 + \frac{1}{4}\right)} \right| \leq 2\sqrt{2} \cdot \kappa \left\{ \frac{2}{1+(4r)^2} - \frac{1}{\left[1+(2r)^2\right]^2} \right\}^{\frac{1}{2}}$$

$$\tag{21}$$

where κ is as given by (20).

As explained in Milovanović and Pogany [28], motivated by [12], a family of *Mathieu a-series* were introduced by Pogany et al.[32] together with their integral representations, various approaches and results were used to procure bounds.

The *alternating generalized Mathieu series*, companion to $S_\mu(r)$, was introduced by Pogany and Tomovski [31] and is represented by

$$\tilde{S}_\mu(r) = \sum_{n=1}^\infty \frac{(-1)^{n-1} \cdot 2n}{(n^2 + r^2)^{1+\mu}}, \quad r > 0, \ \mu > 0. \tag{22}$$

which can be also expressed in the following integral form

$$\tilde{S}_\mu(r) = C_\mu(r) \int_0^\infty \frac{x^{\mu+\frac{1}{2}}}{e^x + 1} J_{\mu-\frac{1}{2}}(rx)\, dx, \quad \mu > 0, \tag{23}$$

where $C_\mu(r)$ is as given in (8).

In the paper [13], bounds were obtained for the alternating generalized Mathieu series $\tilde{S}_\mu(r)$ in Sect. 3; and, the odd $\phi_\mu(r)$ and even $\psi_\mu(r)$ generalized Mathieu series, as defined in Sect. 4. This was accomplished by using their integral representations via Čebyšev Functional bounds which is presented in Sect. 2. The methodology produces both the approximation and bounds for the companion series of the *generalized Mathieu series*. In Sect. 5 some properties of the generalized Mathieu series and its companions are given with an emphasis on the moments in terms of Beta and Zeta functions. The paper's emphasis was to analyze the odd and even counterparts for the generalized Mathieu series as has been accomplished. It is further demonstrated that the relationship between the Zeta function, the alternating Zeta function, and the odd Zeta function is recaptured by allowing $r- > 0$ in the relationship between the generalized Mathieu series, the alternating and odd Mathieu series in Theorem 8 and Remark 7.

The emphasis of the current article is to extend this methodology to the Dirichlet Beta L- function family through the generalized Mathieu series, whereas the work in [13] emphasized the extension to a generalized Mathieu series $S_\mu(r)$ by the Zeta function, $\zeta(\cdot)$ as generator. The current work is based on the generator as the sum of the reciprocal powers of odd positive numbers, $\lambda(s)$. The Dirichlet Beta function $\beta(s) = \sum_{n=1}^{\infty} \frac{(-1)^{n-1}}{(2n-1)^s}$ has the honor as the lead of this family though. The generalized Mathieu series are based on two parameters r and μ, as exemplified by (5) and (7), in addition to various generators. The Dirichlet L-series have played a great deal of attention in number theory. These are also relevant to lattice sums which may be represented by lower dimensional lattice sums. The classic example of this was first given by Lorenz [22] as [39], is given by,

$$\sum_{m,n\neq 0,0}^{\infty} \frac{1}{\left(m^2 + n^2\right)^s} = 4\zeta(s) \cdot \beta(s). \tag{24}$$

The reader is encouraged to refer to [5, 39] and [14] for interest and further references.

2 Some Results on Bounding the Čebyšev Functional

The current section presents a key methodology to procure approximation and bounds for the integral representations of the Zeta and Dirichlet Beta function companions as generators of generalized Mathieu series.

The weighted Čebyšev functional defined by

$$T(f, g; p) := M(fg; p) - M(f; p)M(g; p), \tag{25}$$

where M is the weighted integral mean

$$M(h; p) := \frac{\int_a^b p(x) h(x) dx}{\int_a^b p(x) dx}, \qquad (26)$$

has been extensively investigated in the literature with the view of determining its bounds. The unweighted Čebyšev functional $T(f, g; 1)$ was bounded by Grüss in [19] by the product of the difference of the functions and their function bounds.

There has been much activity in procuring bounds for $T(f, g; p)$ and the interested reader is referred to [7, 11]. The functional $T(f, g; p)$ is known to satisfy a number of identities. Included amongst these are identities of Sönin type, namely

$$P \cdot T(f, g; p) = \int_a^b p(t) [f(t) - K] [g(t) - M(g; p)] dt, \qquad (27)$$

where K is a constant and

$$P = \int_a^b p(x) dx. \qquad (28)$$

The constant $K \in \mathbb{R}$, but in the literature some of the more popular values have been taken as

$$0, \ \frac{\Delta + \delta}{2}, \ f\left(\frac{a+b}{2}\right) \text{ and } M(f; p),$$

where $-\infty < \delta \le f(t) \le \Delta < \infty$ and $t \in [a, b]$.

An identity attributed to Körkine, viz.,

$$P^2 \cdot T(f, g; p) = \frac{1}{2} \int_a^b \int_a^b p(x) p(y) (f(x) - f(y)) (g(x) - g(y)) dx dy \qquad (29)$$

may also easily be shown to hold.

Remark 1 For $-\infty < \delta \le f(t) \le \Delta < \infty$ for $t \in [a, b]$ Cerone and Dragomir [11] showed that

$$P \cdot |T(f, g; p)| \le \frac{1}{2} (\Delta - \delta) \int_a^b p(t) |g(t) - M(g; p)| dt \qquad (30)$$

$$\le \frac{1}{2} (\Delta - \delta) \left(\int_a^b p(t) |g(t) - M(g; p)|^\alpha dt\right)^{\frac{1}{\alpha}}, \ 1 \le \alpha < \infty$$

$$\le \frac{1}{2} (\Delta - \delta) ess \sup_{t \in [a,b]} |g(t) - M(g; p)|.$$

Specifically, if $-\infty < \phi \le g(t) \le \Phi < \infty$ for $t \in [a, b]$, then

$$|T(f, g; p)| \le \frac{1}{2} (\Delta - \delta) \int_a^b p(t) |g(t) - \mathcal{M}(g; p)| \, dt \tag{31}$$

$$\le \frac{1}{2} (\Delta - \delta) \left[\frac{1}{P} \int_a^b p(t) g^2(t) \, dt - \mathcal{M}^2(g; p) \right]^{\frac{1}{2}}$$

$$\le \frac{1}{4} (\Delta - \delta) (\Phi - \phi).$$

The results in (30) were obtained from the Sönin type identity (27) on taking $K = \frac{\Delta + \delta}{2}$.

It is instructive to show from (27) that the best K, in the sense of providing the sharpest bound for the Euclidean or two-norm, results when $K = \mathcal{M}(f; p)$.

Lemma 1 *The sharpest bound for the Čebyšev functional involving the Euclidean norm is given by*

$$P \cdot |T(f, g; p)| \tag{32}$$

$$\le \inf_K \left[\int_a^b p(t) (f(t) - K)^2 \, dt \right]^{\frac{1}{2}} \left[\int_a^b p(t) (g(t) - \mathcal{M}(g; p))^2 \, dt \right]^{\frac{1}{2}}$$

$$= \left[\int_a^b p(t) f^2(t) \, dt - \mathcal{M}^2(f; p) \right]^{\frac{1}{2}} \left[\int_a^b p(t) g^2(t) \, dt - \mathcal{M}^2(g; p) \right]^{\frac{1}{2}}.$$

Proof From (27) we have, on using the Cauchy-Buniakowsky-Schwartz, inequality

$$P \cdot |T(f, g; p)| \le \left(\int_a^b p(t) (f(t) - K)^2 \, dt \right)^{\frac{1}{2}} \left(\int_a^b p(t) (g(t) - \mathcal{M}(g; p))^2 \, dt \right)^{\frac{1}{2}}.$$

Now, the sharpest bound is obtained by taking the infimum over $K \in \mathbb{R}$. That is,

$$\inf_{K \in \mathbb{R}} \left(\int_a^b p(t) (f(t) - K)^2 \, dt \right)^{\frac{1}{2}} = \inf_{K \in \mathbb{R}} \left(\int_a^b p(t) \left(f^2(t) - 2Kf(t) + K^2 \right) dt \right)^{\frac{1}{2}}$$

$$= \inf_{K \in \mathbb{R}} \left[\int_a^b p(t) f^2(t) \, dt \right.$$

$$\left. + P \cdot K (K - 2\mathcal{M}(f; p)) \right]^{\frac{1}{2}}$$

$$= \left(\int_a^b p(t) f^2(t) \, dt - P \cdot \mathcal{M}^2(f; p) \right)^{\frac{1}{2}},$$

and the infimum occurs when $K = \mathcal{M}(f; p)$.

In the next section Lemma 1 is used to obtain bounds.
We note that the first inequality in (30) results from

$$|P \cdot T(f, g; p)| \leq \inf_K \|f(\cdot) - K\|_\infty \int_a^b p(t) |g(t) - M(g; p)| \, dt \qquad (33)$$

$$\leq \|f(\cdot) - K\|_\infty \int_a^b p(t) |g(t) - M(g; p)| \, dt,$$

which are tighter than those in Lemma 1.

However, (33) relies on knowing where the shifted functions are positive and where they are negative. This is not always an easy task.

The first result in (30) arises from (33) with $K = \frac{\Delta + \delta}{2}$ so that

$$\left\| f(\cdot) - \frac{\Delta + \delta}{2} \right\|_\infty = \sup_{t \in [a,b]} \left| f(t) - \frac{\Delta + \delta}{2} \right| = \frac{\Delta - \delta}{2},$$

where $-\infty < \delta \leq f(t) \leq \Delta < \infty$ for $t \in [a, b]$.

3 Bounds for $\tilde{S}_\mu(r)$ via the Čebyšev Functional

Bounds on the Čebyšev functional (25) may be looked upon as estimating the distance of the weighted mean of the product of two functions from the product of the weighted means of the two functions. This proves to be quite useful since the individual means are invariably easier to evaluate.

The following technical lemma involving the Euler beta function will $B(x, y)$ be required, which is represented in terms of the gamma function by

$$B(x, y) = \frac{\Gamma(x)\Gamma(y)}{\Gamma(x + y)}. \qquad (34)$$

Lemma 2 *The following result holds (see [13] for the proof)*

$$\frac{1}{2} \cdot \frac{B(\frac{1}{2}, \mu)}{[\alpha^2 + r^2]^{2\mu - \frac{1}{2}}} \leq \int_0^{\frac{\pi}{2}} \frac{\cos^{2\mu-1} \phi}{[\alpha^2 + r^2 \cos^2 \phi]^{2\mu - \frac{1}{2}}} d\phi \leq \frac{1}{2} \cdot \frac{B(\frac{1}{2}, \mu)}{\alpha^{4\mu-1}}, \qquad (35)$$

It is noted that equality follows in (35) when $r = 0$.

Theorem 5 (see [13] for the Proof) *For $\mu > 0$ and $r > 0$, the alternating generalized Mathieu series $\tilde{S}_\mu(r)$ satisfies the following bounds:*

$$\left| \tilde{S}_\mu\,(r) - \frac{\pi^2}{24\mu\left(r^2 + \frac{1}{4}\right)^\mu} \right| \tag{36}$$

$$\le \tilde{\kappa}\left[\frac{1}{\sqrt{\pi}} \cdot \frac{\Gamma\left(2\mu - \frac{1}{2}\right)}{2^{2\mu-1}\Gamma^2\,(\mu+1)} \int_0^{\frac{\pi}{2}} \frac{\cos^{2\mu-1}\phi}{\left[\left(\frac{1}{4}\right)^2 + r^2\cos^2\phi\right]^{2\mu-\frac{1}{2}}}d\phi - \frac{1}{2\mu^2\left(r^2 + \frac{1}{4}\right)^{2\mu}} \right]^{\frac{1}{2}}$$

$$\le \tilde{\kappa}\left[\frac{1}{2^{3\mu-1}\mu^2(\mu-\frac{1}{2})B(\mu,\mu-\frac{1}{2})} - \frac{1}{2\mu^2\left(r^2 + \frac{1}{4}\right)^{2\mu}} \right]^{\frac{1}{2}},$$

where $\tilde{\kappa}$ is as given by

$$\tilde{\kappa} = \left[\frac{\pi^3}{4} - 8\cdot G - 2\cdot\left(\frac{\pi^2}{24}\right)^2 \right]^{\frac{1}{2}} = 0.29260623049\ldots. \tag{37}$$

Corollary 3 *The alternating Mathieu series $\tilde{S}\,(r)$ satisfies the result*

$$\left| \sum_{n=1}^\infty \frac{(-1)^{n-1}\cdot 2n}{(n^2 + r^2)^2} - \frac{\pi^2}{24\left(r^2 + \frac{1}{4}\right)} \right| \le 2\sqrt{2}\cdot\tilde{\kappa}\left\{ \frac{2}{1 + (4r)^2} - \frac{1}{\left[1 + (2r)^2\right]^2} \right\}^{\frac{1}{2}}, \tag{38}$$

where $\tilde{\kappa}$ is as given by (37).

Proof Let $\mu = 1$ in (36) and using (1) and (5) gives the above result (38), on noting that

$$2^6 \int_0^{\frac{\pi}{2}} \frac{\cos\phi}{\left[1 + (4r\cos\phi)^2\right]^{\frac{3}{2}}}d\phi = \frac{64}{1 + (4r)^2}$$

and after some simplification.

Remark 2 The result of Theorem 5 holds for any $\mu > 0$ and $r > 0$ whereas those obtained in [12] were valid for $\mu > \frac{1}{2}$.

Remark 3 From (37) we may infer, since $\tilde{\kappa} > 0$,

$$G < \frac{\pi^3}{32}\left(1 - \frac{\pi}{72}\right) = 0.9266678949\ldots.$$

4 Odd and Even Generalized Mathieu Series

Using the generalized Mathieu series, $S_\mu(r)$ as given in (1) and (7)–(8) together with the *alternating* generalized Mathieu series $\tilde{S}_\mu(r)$ as given in (22)–(23) we introduce the *odd generalized Mathieu series*, $\phi_\mu(r)$ and the *even generalized Mathieu series*, $\psi_\mu(r)$. These are given by [13]

$$\phi_\mu(r) := \frac{S_\mu(r) + \tilde{S}_\mu(r)}{2} \tag{39}$$

$$= \sum_{n=1}^{\infty} \frac{2 \cdot (2n-1)}{\left((2n-1)^2 + r^2\right)^{1+\mu}}$$

$$= C_\mu(r) \cdot 2 \int_0^\infty \frac{x^{\mu+\frac{1}{2}}}{e^x - e^{-x}} J_{\mu-\frac{1}{2}}(rx)\,dx, \quad r, \mu > 0,$$

and

$$\psi_\mu(r) := \frac{S_\mu(r) - \tilde{S}_\mu(r)}{2} \tag{40}$$

$$= \sum_{n=1}^{\infty} \frac{2 \cdot (2n)}{\left((2n)^2 + r^2\right)^{1+\mu}}$$

$$= C_\mu(r) \cdot 2 \int_0^\infty \frac{x^{\mu+\frac{1}{2}}}{e^{2x} - 1} J_{\mu-\frac{1}{2}}(rx)\,dx, \quad r, \mu > 0,$$

where $C_\mu(r)$ is positive as defined in (8).

Remark 4 It may be noticed that if we have *identities* for *any two* of the generalized Mathieu type series $S_\mu(r)$, $\tilde{S}_\mu(r)$, $\phi_\mu(r)$, $\psi_\mu(r)$ then we may deduce the other two. In particular, $S_\mu(r) = \frac{\phi_\mu(r) + \psi_\mu(r)}{2}$ and $\tilde{S}_\mu(r) = \frac{\phi_\mu(r) - \psi_\mu(r)}{2}$. This, however, is **not** the case with regard to inequalities or bounds since recourse to the triangle inequality would result in a coarser bound. We may further notice that their integral representation may be given by

$$2C_\mu(r) \int_0^\infty H(x) \cdot x^{\mu-\frac{1}{2}} J_{\mu-\frac{1}{2}}(rx)\,dx, \quad r, \mu > 0 \tag{41}$$

where $C_\mu(r)$ is positive as defined in (8) and $H(x)$ is one of the following

$$H_M(x) = \frac{x}{e^x - 1}, \quad H_A(x) = \frac{x}{e^x + 1}, \quad H_O(x) = \frac{x}{e^x - e^{-x}}, \quad H_E(x) = \frac{x}{e^{2x} - 1}, \tag{42}$$

where the subscripts relate to the generalized Mathieu, alternating Mathieu, odd Mathieu, and even Mathieu series integral representations, respectively.

Remark 5 It should be emphasized that the $H.(\cdot)$ in (42) represent the weights associated with the integral representation of the generalized Mathieu and its companions. They satisfy the following conditions:

$$H_A(x) < H_E(x) < H_O(x) < H_M(x) \quad , x < \ln(2) \tag{43}$$

$$H_E(x) < H_A(x) < H_O(x) < H_M(x) \quad , x > \ln(2).$$

In [13] the odd and even generalized Mathieu series bounds were obtained via a Čebyšev functional approach. If we allow the subscripts of O and E to represent the cases related to $\phi_\mu(r)$ (odd) and $\psi_\mu(r)$ (even).

We note from (39) that

$$\frac{\phi_\mu(r)}{2C_\mu(r)} = \int_0^\infty H_O(x) \cdot x^{\mu-\frac{1}{2}} J_{\mu-\frac{1}{2}}(rx)\, dx, \quad r, \mu > 0 \tag{44}$$

where from (42)

$$H_O(x) = \frac{x}{e^x - e^{-x}}. \tag{45}$$

The following theorem is a correction of the result in [13]. The $(\frac{1}{2})^2$ within the integral was $(1)^2$.

Theorem 6 (see [13] for the Proof) *For $\mu > 0$ and $r > 0$ the **odd** generalized Mathieu series $\phi_\mu(r)$ satisfies the following relationship, namely*

$$\left| \phi_\mu(r) - \frac{\pi^2}{4\mu \left(r^2 + 1\right)^\mu} \right| \tag{46}$$

$$\leq \kappa_O \left[\frac{4\Gamma\left(2\mu - \frac{1}{2}\right)}{2^{2\mu-1}\sqrt{\pi}\,\Gamma^2(\mu+1)} \int_0^{\frac{\pi}{2}} \frac{\cos^{2\mu-1}\phi}{\left[(\frac{1}{2})^2 + r^2\cos^2\phi \right]^{2\mu-\frac{1}{2}}} d\phi - \frac{4}{\mu^2\left(1^2+r^2\right)^{2\mu}} \right]^{\frac{1}{2}}$$

$$\leq \kappa_O \left[\frac{2^{2\mu+3}}{\mu^2(\mu - \frac{1}{2}) \cdot B\left(\mu, \mu - \frac{1}{2}\right)} - \frac{4}{\mu^2\left(1^2+r^2\right)^{2\mu}} \right]^{\frac{1}{2}},$$

where

$$\kappa_O = \left[\frac{\pi^2}{8} (1 - \frac{\pi^2}{8}) + \frac{7}{8} \zeta (3) \right]^{\frac{1}{2}}$$

and $B(x, y)$ is the Euler beta function given by (34).

We note from (39) that

$$\frac{\psi_\mu(r)}{2C_\mu(r)} = \int_0^\infty H_E(x) \cdot x^{\mu-\frac{1}{2}} J_{\mu-\frac{1}{2}}(rx) \, dx, \quad r, \mu > 0 \tag{47}$$

where from (42)

$$H_E(x) = \frac{x}{e^{2x} - 1}. \tag{48}$$

Theorem 7 (see [13] for the Proof) *For $\mu > 0$ and $r > 0$ the **even** generalized Mathieu series $\psi_\mu(r)$ satisfies the following relationship:*

$$\left| \psi_\mu(r) - \frac{\pi^2}{6\mu \left(r^2 + 2^2 \right)^\mu} \right| \tag{49}$$

$$\leq \kappa_E \left[\frac{4\Gamma \left(2\mu - \frac{1}{2} \right)}{2^{2\mu-1} \sqrt{\pi} \Gamma^2 (\mu + 1)} \int_0^{\frac{\pi}{2}} \frac{\cos^{2\mu-1} \phi}{\left[(1)^2 + r^2 \cos^2 \phi \right]^{2\mu-\frac{1}{2}}} d\phi - \frac{8}{\mu^2 \left(1^2 + r^2 \right)^{2\mu}} \right]^{\frac{1}{2}}$$

$$\leq \kappa_E \left[\frac{1}{4^{\mu-1} \mu^2 (\mu - \frac{1}{2}) \cdot B \left(\mu, \mu - \frac{1}{2} \right)} - \frac{8}{\mu^2 \left(1^2 + r^2 \right)^{2\mu}} \right]^{\frac{1}{2}},$$

where

$$\kappa_E = \left[\frac{\pi^2}{24} (1 - \frac{\pi^2}{12}) \right]^{\frac{1}{2}} \tag{50}$$

and $B(x, y)$ is the Euler beta function given by (34).

The remainder of the results in this section were developed in [13] to complete the interplay between the generators as depicted by the Zeta family and the generalized Mathieu series expressions.

The following lemma demonstrates the relationship for the generalized Mathieu series and its companions.

Lemma 3 *The companion generalized Mathieu series may be expressed in terms of the generalized Mathieu series, namely*

$$\tilde{S}_\mu (r) = S_\mu (r) - 4^{-\mu} S_\mu \left(\frac{r}{2}\right)$$

$$\phi_\mu(r) = S_\mu (r) - 2^{-2\mu-1} S_\mu \left(\frac{r}{2}\right) \tag{51}$$

$$\psi_\mu(r) = 2^{-2\mu-1} S_\mu \left(\frac{r}{2}\right).$$

Proof From the generalized Mathieu series (5) it may be shown that

$$S_\mu \left(\frac{r}{2}\right) = 2^{2\mu+1} \sum_{n=1}^{\infty} \frac{2 \cdot (2n)}{\left((2n)^2 + r^2\right)^{1+\mu}} \tag{52}$$

and so from (40) and (52) gives $\psi_\mu(r) = 2^{-2\mu-1} S_\mu \left(\frac{r}{2}\right)$, the third result. Further, the first result of (51) readily follows on noting that $2 \cdot \psi_\mu(r) = S_\mu (r) - \tilde{S}_\mu (r)$. The second result is procured from (39), $2 \cdot \phi_\mu(r) = S_\mu (r) + \tilde{S}_\mu (r)$ and substituting the first result for $\tilde{S}_\mu (r)$.

Remark 6 It is important to emphasize, as mentioned earlier, that obtaining bounds for the companions in terms of those of the generalized Mathieu series would produce inferior bounds *from using* the triangle inequality required for the first two results in (51).

Theorem 8 *The following relationship holds:*

$$S_\mu (r) = 2\phi_\mu(r) - \tilde{S}_\mu (r). \tag{53}$$

Proof The relationship (53) follows easily from (51) by subtracting the first equation from twice the second.

Remark 7 Equation (53) recaptures, on allowing $r - > 0$, the well-known result involving the Zeta function $\zeta(x)$

$$\zeta(x) = 2\lambda(x) - \eta(x) \tag{54}$$

where $\lambda(x)$ is the odd zeta, $\eta(x)$ is the alternating zeta, and $x = 2\mu + 1$. This demonstrates that (53) is an extension of the Zeta expression (54) through the variable r of Mathieu type functions.

Remark 8 The Čebyšev Functional bounds have been used to procure bounds for the Mathieu family of special functions. Much effort has been expended in the literature as to various ways of bounding the Mathieu series (1). The accuracy of bounds over particular regions of parameters cannot be determined a priori (see also, for example, [3, 28]). A comparison of the bounds using (2) and (38) demonstrates that the upper bound for the Mathieu series is better for $0 < r < 0.855662$ and for the lower bound, better over $0 < r < 1.206377$ for (38) and better for the remainder of r for the bounds (2). It must be remembered however that (38) is valid for the more general result involving parameters r and μ.

5 Dirichlet Beta Function Generalized Mathieu Series Bounds

The Dirichlet beta function or Dirichlet L-function is given by [17]

$$\beta(x) = \sum_{n=0}^{\infty} \frac{(-1)^n}{(2n+1)^x}, \quad x > 0 \tag{55}$$

where $\beta(2) = G$, Catalan's constant. See [9] and [8] in which sharp double bounds were obtained.

It is readily observed that $\beta(x)$ is the alternating version of $\lambda(x)$, however, it cannot be directly related to $\zeta(x)$. It is also related to $\eta(x)$ in that only the odd terms are summed.

The beta function may be evaluated explicitly at positive odd integer values of x, namely

$$\beta(2n+1) = (-1)^n \frac{E_{2n}}{2(2n)!} \left(\frac{\pi}{2}\right)^{2n+1}, \tag{56}$$

where E_n are the Euler numbers generated by

$$\operatorname{sech}(x) = \frac{2e^x}{e^{2x}+1} = \sum_{n=0}^{\infty} E_n \frac{x^n}{n!}.$$

The Dirichlet beta function may be analytically continued over the whole complex plane by the functional equation

$$\beta(1-z) = \left(\frac{2}{\pi}\right)^z \sin\left(\frac{\pi z}{2}\right) \Gamma(z) \beta(z).$$

The function $\beta(z)$ is defined everywhere in the complex plane and has no singularities, unlike the Riemann zeta function, $\zeta(s) = \sum_{n=1}^{\infty} \frac{1}{n^s}$, which has a simple pole at $s = 1$.

The Dirichlet beta function and the zeta function have important applications in a number of branches of mathematics, and in particular in Analytic number theory. See, for example, [2, 14].

Further, $\beta(x)$ has an alternative integral representation [17, p. 56], namely

$$\beta(x) = \frac{1}{2\Gamma(x)} \int_0^{\infty} \frac{t^{x-1}}{\cosh(t)} dt, \quad x > 0.$$

That is,

$$\beta(x) = \frac{1}{\Gamma(x)} \int_0^{\infty} \frac{t^{x-1}}{e^t + e^{-t}} dt, \quad x > 0. \tag{57}$$

The function $\beta(x)$ is also connected to prime number theory [17] which may perhaps be best summarized by

$$\beta(x) = \prod_{\substack{p \text{ prime} \\ p \equiv 1 \bmod 4}} \left(1 - p^{-x}\right)^{-1} \cdot \prod_{\substack{p \text{ prime} \\ p \equiv 3 \bmod 4}} \left(1 + p^{-x}\right)^{-1} = \prod_{\substack{p \text{ odd} \\ \text{prime}}} \left(1 - (-1)^{\frac{p-1}{2}} p^{-x}\right)^{-1},$$

where the rearrangement of factors is permitted because of absolute convergence.

The main thrust of the article is to investigate the Dirichlet Beta function via generalized Mathieu series approach, which may be looked upon as an alternating odd generalized Mathieu series, $\tilde{\phi}_\mu(r)$, namely

$$\tilde{\phi}_\mu(r) = \sum_{n=1}^\infty \frac{(-1)^{n-1} 2(2n-1)}{\left((2n-1)^2 + r^2\right)^{1+\mu}} \tag{58}$$

$$= C_\mu(r) \cdot 2 \int_0^\infty \frac{x^{\mu+\frac{1}{2}}}{e^x + e^{-x}} J_{\mu-\frac{1}{2}}(rx)\, dx, \quad r, \mu > 0.$$

This is, in part, inspired by the alternating odd zeta function, $\beta(s) = \sum_{n=1}^\infty \frac{(-1)^{n-1}}{(2n-1)^s}$ which has explicit closed form solution in terms of Euler polynomials for $s = 2m + 1$ whereas $\zeta(2m)$ for $m \in \mathbb{N}$ is explicitly given in terms of Bernoulli polynomials. This is so, since using a limiting argument $\tilde{\phi}_\mu(0) = 2\beta(2\mu + 1)$.

Theorem 9 *For $\mu > 0$ and $r > 0$ the alternating odd generalized Mathieu series $\tilde{\phi}_\mu(r)$ satisfies the following relationship, namely*

$$\left| \tilde{\phi}_\mu(r) - \frac{2 \cdot G}{\mu \left(r^2 + 1\right)^\mu} \right| \tag{59}$$

$$\leq \kappa_{\tilde{o}} \left[\frac{4\Gamma\left(2\mu - \frac{1}{2}\right)}{2^{2\mu-1}\sqrt{\pi}\,\Gamma^2(\mu+1)} \int_0^{\frac{\pi}{2}} \frac{\cos^{2\mu-1}\phi}{\left[\left(\frac{1}{2}\right)^2 + r^2 \cos^2\phi\right]^{2\mu-\frac{1}{2}}} d\phi - \frac{4}{\mu^2\left(1^2 + r^2\right)^{2\mu}} \right]^{\frac{1}{2}}$$

$$\leq \kappa_{\tilde{o}} \left[\frac{2^{2\mu+3}}{\mu^2(\mu-\frac{1}{2}) \cdot B\left(\mu, \mu - \frac{1}{2}\right)} - \frac{4}{\mu^2\left(1^2 + r^2\right)^{2\mu}} \right]^{\frac{1}{2}},$$

where

$$\kappa_{\tilde{o}} = \left[G(1 - G) + \frac{\pi^3}{32} \right]^{\frac{1}{2}} \tag{60}$$

and $B(x, y)$ is the Euler beta function given by (34)

Proof We notice that $\frac{\tilde{\phi}_\mu(r)}{2C_\mu(r)}$ from (58) may be written in the form

$$\frac{\tilde{\phi}_\mu(r)}{2C_\mu(r)} = \int_0^\infty e^{-x} \cdot \frac{x}{1+e^{-2x}} \cdot x^{\mu-\frac{1}{2}} J_{\mu-\frac{1}{2}}(rx)\,dx, \quad r,\mu > 0 \qquad (61)$$

If we now let

$$p_{\tilde{O}}(x) = e^{-x}, \quad f_{\tilde{O}}(x) = \frac{x}{1+e^{-2x}}, \quad g(x) = x^{\mu-\frac{1}{2}} J_{\mu-\frac{1}{2}}(rx) \qquad (62)$$

then from (26)

$$P_{\tilde{O}} = \int_0^\infty p_{\tilde{O}}(x)\,dx = \int_0^\infty e^{-x}dx = 1, \qquad (63)$$

$$P_{\tilde{O}} \cdot M\left(f_{\tilde{O}}; p\right) = \int_0^\infty e^{-x} \cdot \frac{x}{1+e^{-2x}}dx = G \qquad (64)$$

and

$$P_{\tilde{O}} \cdot M(g; p) = \int_0^\infty e^{-x} \cdot x^{\mu-\frac{1}{2}} J_{\mu-\frac{1}{2}}(rx)\,dx = \frac{(2r)^{\mu-\frac{1}{2}} \Gamma(\mu)}{\sqrt{\pi}\left(1^2+r^2\right)^\mu}, \qquad (65)$$

where we have used the fact that G is Catalan's constant (see [30, p. 610]) to procure (64), and from Watson [37, p. 386],

$$\int_0^\infty e^{-\alpha x} \cdot x^\nu J_\nu(\beta x)\,dx = \frac{(2\beta)^\nu}{\sqrt{\pi}} \cdot \frac{\Gamma\left(\nu+\frac{1}{2}\right)}{\left(\alpha^2+\beta^2\right)^{\nu+\frac{1}{2}}}, \quad \text{Re}(\nu) > \frac{1}{2}, \quad \text{Re}(\alpha) > |\text{Im}(\beta)|,$$

with $\alpha = 1$, $\nu = \mu - \frac{1}{2}$, $\beta = r$ to obtain (65).

Now, from (25)–(28) we have on using (61)–(65)

$$\frac{\tilde{\phi}_\mu(r)}{2C_\mu(r)} - \frac{(2r)^{\mu-\frac{1}{2}} \Gamma(\mu)}{\sqrt{\pi}\left(1^2+r^2\right)^\mu} \cdot G$$

$$= \int_0^\infty e^{-x} \left(x^{\mu-\frac{1}{2}} J_{\mu-\frac{1}{2}}(rx) - K\right)\left(\frac{x}{1+e^{-2x}} - G\right)dx.$$

Further, using the Cauchy-Buniakowsky-Schwartz inequality, we have from (32) to give

$$\left| \frac{\tilde{\phi}_\mu (r)}{2C_\mu (r)} - \frac{(2r)^{\mu - \frac{1}{2}} \Gamma (\mu)}{\sqrt{\pi} (1^2 + r^2)^\mu} \cdot G \right|$$

$$= \left(\int_0^\infty e^{-x} \left(x^{\mu - \frac{1}{2}} J_{\mu - \frac{1}{2}} (rx) - K \right)^2 dx \right)^{\frac{1}{2}} \left(\int_0^\infty e^{-x} \left(\frac{x}{1 + e^{-2x}} - G \right)^2 dx \right)^{\frac{1}{2}}.$$

$$(66)$$

As mentioned in Sect. 2, the appropriate choice of K is the weighted integral mean as given from (65), namely

$$K = K_* = \frac{(2r)^{\mu - \frac{1}{2}} \Gamma (\mu)}{\sqrt{\pi} (1^2 + r^2)^\mu}. \tag{67}$$

Now using the result

$$\int_a^b p(t) [h(t) - M(h; p)]^2 dt = \int_a^b p(t) h^2(t) dt - P \cdot M^2(h; p). \tag{68}$$

to evaluate the two expressions on the right-hand side of (66) produces; firstly,

$$\int_0^\infty e^{-x} \left(\frac{x}{1 + e^{-2x}} - G \right)^2 dx = \int_0^\infty e^{-x} \left(\frac{x}{1 + e^{-2x}} \right)^2 dx - 1 \cdot G^2. \tag{69}$$

and secondly, allowing for the permissible interchange of integration and summation, we have

$$\int_0^\infty e^{-x} \left(\frac{x}{1 + e^{-2x}} \right)^2 dx = \int_0^\infty e^{-x} x^2 \cdot \left(\sum_{n=1}^\infty (-1)^{(n-1)} n e^{-2(n-1)x} \right) dx$$

$$(70)$$

$$= \sum_{n=0}^\infty (-1)^n (n+1) \cdot \int_0^\infty e^{-(2n+1)x} x^2 dx$$

$$= \sum_{n=0}^\infty \frac{(-1)^n (n+1) \Gamma(3)}{(2n+1)^3} = \sum_{n=0}^\infty \frac{(-1)^n \cdot (2n+2)}{(2n+1)^3}$$

$$= \sum_{n=0}^\infty (-1)^n \left(\frac{1}{(2n+1)^2} + \frac{1}{(2n+1)^3} \right)$$

$$= G + \beta(3)$$

where $\sum_{n=0}^{\infty} \frac{(-1)^n}{(2n+1)^x} = \beta(x)$ [30, p. 602] and, in (70) we have used the fact that

$$\int_0^\infty e^{-\alpha x} x^p dx = \frac{\Gamma(p+1)}{\alpha^{p+1}}.$$

Hence, from (69) and (70) we have

$$\left[\int_0^\infty e^{-x} \left(\frac{x}{1-e^{-2x}} - G\right)^2 dx\right]^{\frac{1}{2}} = \left[G(1-G) + \frac{\pi^3}{32}\right]^{\frac{1}{2}}. \tag{71}$$

where we have used the fact that $\beta(3) = \frac{\pi^3}{32}$.

Now, for the first expression on the right-hand side of (66), we have, on using (68) and (65)

$$\int_0^\infty e^{-x} \left(x^{\mu-\frac{1}{2}} J_{\mu-\frac{1}{2}}(rx) - K_*\right)^2 dx = \int_0^\infty e^{-x} x^{2\mu-1} J_{\mu-\frac{1}{2}}^2(rx)\, dx - 1 \cdot K_*^2. \tag{72}$$

A result in Watson [37, p. 290] states that

$$\int_0^\infty e^{-2at} J_\alpha(\gamma t) J_\beta(\gamma t) t^{\alpha+\beta} dt$$

$$= \frac{\Gamma\left(\alpha+\beta+\frac{1}{2}\right)}{\pi^{\frac{3}{2}}} \gamma^{\alpha+\beta} \int_0^{\frac{\pi}{2}} \frac{\cos^{\alpha+\beta}\phi \cos(\alpha-\beta)\phi}{(a^2+\gamma^2 \cos^2\phi)^{\alpha+\beta+\frac{1}{2}}} d\phi \tag{73}$$

and so taking $a = \frac{1}{2}$, $\alpha = \beta = \mu - \frac{1}{2}$ and $\gamma = r$ in (73) gives

$$\int_0^\infty e^{-x} x^{2\mu-1} J_{\mu-\frac{1}{2}}^2(rx)\, dx$$

$$= \frac{\Gamma\left(2\mu-\frac{1}{2}\right) r^{2\mu-1}}{\pi^{\frac{3}{2}}} \int_0^{\frac{\pi}{2}} \frac{\cos^{2\mu-1}\phi}{\left(\left(\frac{1}{2}\right)^2 + r^2 \cos^2\phi\right)^{2\mu-\frac{1}{2}}} d\phi \tag{74}$$

That is, from (72) and (72) we have

$$\left[\int_0^\infty e^{-x} \left(x^{\mu-\frac{1}{2}} J_{\mu-\frac{1}{2}} (rx) - K_* \right)^2 dx \right]^{\frac{1}{2}}$$

$$= \left[\frac{\Gamma\left(2\mu - \frac{1}{2}\right)}{\pi^{\frac{3}{2}}} r^{2\mu-1} \int_0^{\frac{\pi}{2}} \frac{\cos^{2\mu-1}\phi}{\left[\left(\frac{1}{2}\right)^2 + r^2 \cos^2 \phi \right]^{2\mu-\frac{1}{2}}} d\phi - 1 \cdot K_*^2 \right]^{\frac{1}{2}}. \qquad (75)$$

Placing (75) and (71) into (66) produces the stated result (59) upon multiplication by $2C_\mu(r)$ and using (8).

For the coarser bound in (59) we have from (35) of Lemma 2

$$\int_0^{\frac{\pi}{2}} \frac{\cos^{2\mu-1}\phi}{\left[(\frac{1}{2})^2 + r^2 \cos^2 \phi \right]^{2\mu-\frac{1}{2}}} d\phi \leq 2^{4\mu} B(\frac{1}{2}, \mu) = 2^{4\mu} \frac{\sqrt{\pi}\Gamma(\mu)}{\Gamma\left(\mu + \frac{1}{2}\right)}$$

and so, on substitution into the first result in (59) produces the second, upon some simplification.

5.1 Dirichlet L-Function Generalized Mathieu Series

The previous work investigating the generalized Mathieu series was extended to the alternating, odd, and even generalized Mathieu series. The odd ($\phi_\mu(r)$) and alternating ($\tilde{\phi}_\mu(r)$) generalized Mathieu series will be used to obtain other results concerning $L_{(4,1)}(\cdot)$ and $L_{(4,3)}(\cdot)$ as generators.

Let

$$\Phi_\mu^+(r) := \frac{\phi_\mu(r) + \tilde{\phi}_\mu(r)}{2} \qquad (76)$$

$$= \sum_{n=0}^\infty \frac{2 \cdot (4n+1)}{((4n+1)^2 + r^2)^{1+\mu}}$$

$$= C_\mu(r) \cdot 2 \int_0^\infty \frac{xe^x}{e^{2x} - e^{-2x}} x^{\mu-\frac{1}{2}} J_{\mu-\frac{1}{2}} (rx) dx, \quad r, \mu > 0,$$

and

$$\Phi_\mu^-(r) := \frac{\phi_\mu(r) - \tilde{\phi}_\mu(r)}{2} \tag{77}$$

$$= \sum_{n=0}^{\infty} \frac{2 \cdot (4n+3)}{\left((4n+3)^2 + r^2\right)^{1+\mu}}$$

$$= C_\mu(r) \cdot 2 \int_0^\infty \frac{xe^{-x}}{e^{2x} - e^{-2x}} x^{\mu - \frac{1}{2}} J_{\mu - \frac{1}{2}}(rx)\, dx, \quad r, \mu > 0,$$

where

$$C_\mu(r) = \frac{\sqrt{\pi}}{(2r)^{\mu - \frac{1}{2}} \Gamma(\mu + 1)}. \tag{78}$$

Theorem 10 *For $\mu > 0$ and $r > 0$ the alternating odd generalized Mathieu series $\tilde{\phi}_\mu(r)$ satisfies the following relationship, namely*

$$\left| \frac{\Phi_\mu^+(r)}{2C_\mu(r)} - \frac{(2r)^{\mu - \frac{1}{2}} \Gamma(\mu)}{\sqrt{\pi} \left(1^2 + r^2\right)^\mu} \cdot L_{(4,1)}(2) \right| \tag{79}$$

$$\leq \kappa_{\phi^+} \left[\frac{4\Gamma\left(2\mu - \frac{1}{2}\right)}{2^{2\mu - 1} \sqrt{\pi} \Gamma^2(\mu + 1)} \int_0^{\frac{\pi}{2}} \frac{\cos^{2\mu - 1} \phi}{\left[\left(\frac{1}{2}\right)^2 + r^2 \cos^2 \phi\right]^{2\mu - \frac{1}{2}}} d\phi - \frac{4}{\mu^2 \left(1^2 + r^2\right)^{2\mu}} \right]^{\frac{1}{2}}$$

$$\leq \kappa_{\phi^+} \left[\frac{2^{2\mu + 3}}{\mu^2(\mu - \frac{1}{2}) \cdot B\left(\mu, \mu - \frac{1}{2}\right)} - \frac{4}{\mu^2 \left(1^2 + r^2\right)^{2\mu}} \right]^{\frac{1}{2}},$$

where

$$\kappa_{\phi^+} = \left[L_{(4,1)}(2)\left(\frac{1}{2} - L_{(4,1)}(2)\right) + \frac{3}{2} L_{(4,1)}(3) \right]^{\frac{1}{2}} \tag{80}$$

and $B(x, y)$ is the Euler beta function given by (34)

Proof We notice that $\frac{\Phi_\mu^+(r)}{2C_\mu(r)}$ from (76) may be written in the form

$$\frac{\Phi_\mu^+(r)}{2C_\mu(r)} = \int_0^\infty e^{-x} \cdot \frac{x}{1 - e^{-4x}} \cdot x^{\mu - \frac{1}{2}} J_{\mu - \frac{1}{2}}(rx)\, dx, \quad r, \mu > 0. \tag{81}$$

If we now let

$$p_{\phi^+}(x) = e^{-x}, \quad f_{\phi^+}(x) = \frac{x}{1 - e^{-4x}}, \quad g(x) = x^{\mu - \frac{1}{2}} J_{\mu - \frac{1}{2}}(rx) \tag{82}$$

then from (26)

$$P_{\phi^+} = \int_0^\infty p_{\phi^+}(x)\, dx = \int_0^\infty e^{-x} dx = 1, \tag{83}$$

$$P_{\phi^+} \cdot \mathcal{M}(f_{\tilde{o}}; p) = \int_0^\infty e^{-x} \cdot \frac{x}{1 - e^{-4x}} dx = L_{(4,1)}(2) \tag{84}$$

and

$$P_{\phi^+} \cdot \mathcal{M}(g; p) = \int_0^\infty e^{-x} \cdot x^{\mu - \frac{1}{2}} J_{\mu - \frac{1}{2}}(rx)\, dx = \frac{(2r)^{\mu - \frac{1}{2}} \Gamma(\mu)}{\sqrt{\pi} \left(1^2 + r^2\right)^\mu}, \tag{85}$$

where we have used the fact that $L_{(4,1)}(x) = \sum_{n=0}^\infty \frac{1}{(4n+1)^x}$ to procure (84), and from Watson [37, p. 386]

$$\int_0^\infty e^{-\alpha x} \cdot x^\nu J_\nu(\beta x)\, dx = \frac{(2\beta)^\nu}{\sqrt{\pi}} \cdot \frac{\Gamma\left(\nu + \frac{1}{2}\right)}{\left(\alpha^2 + \beta^2\right)^{\nu + \frac{1}{2}}}, \quad \text{Re}(\nu) > \frac{1}{2}, \ \text{Re}(\alpha) > |\text{Im}(\beta)|,$$

with $\alpha = 1$, $\nu = \mu - \frac{1}{2}$, $\beta = r$ to obtain (85).

Now, from (25)–(28) we have on using (81)–(85)

$$\frac{\Phi_\mu^+(r)}{2C_\mu(r)} - \frac{(2r)^{\mu - \frac{1}{2}} \Gamma(\mu)}{\sqrt{\pi} \left(1^2 + r^2\right)^\mu} \cdot L_{(4,1)}(2)$$

$$= \int_0^\infty e^{-x} \left(x^{\mu - \frac{1}{2}} J_{\mu - \frac{1}{2}}(rx) - K\right) \left(\frac{x}{1 - e^{-4x}} - L_{(4,1)}(2)\right) dx. \tag{86}$$

Further, using the Cauchy-Buniakowsky-Schwartz inequality, we have from (32). to give

$$\left| \frac{\Phi_\mu^+(r)}{2C_\mu(r)} - \frac{(2r)^{\mu - \frac{1}{2}} \Gamma(\mu)}{\sqrt{\pi} \left(1^2 + r^2\right)^\mu} \cdot L_{(4,1)}(2) \right|$$

$$= \left(\int_0^\infty e^{-x} \left(x^{\mu - \frac{1}{2}} J_{\mu - \frac{1}{2}}(rx) - K\right)^2 dx \right)^{\frac{1}{2}}$$

$$\times \left(\int_0^\infty e^{-x} \left(\frac{x}{1 - e^{-4x}} - L_{(4,1)}(2)\right)^2 dx \right)^{\frac{1}{2}}. \tag{87}$$

As mentioned in Sect. 2, the appropriate choice of K is the weighted integral mean as given from (85), namely

$$K = K_* = \frac{(2r)^{\mu-\frac{1}{2}} \Gamma(\mu)}{\sqrt{\pi}(1^2 + r^2)^\mu}. \tag{88}$$

Now using the result

$$\int_a^b p(t)[h(t) - M(h; p)]^2 dt = \int_a^b p(t) h^2(t) dt - P \cdot M^2(h; p). \tag{89}$$

to evaluate the two expressions on the right-hand side of (87) produces; firstly,

$$\int_0^\infty e^{-x} \left(\frac{x}{1 + e^{-2x}} - L_{(4,1)}(2) \right)^2 dx = \int_0^\infty e^{-x} \left(\frac{x}{1 - e^{-4x}} \right)^2 dx - 1 \cdot (L_{(4,1)}(2))^2. \tag{90}$$

and secondly, allowing for the permissible interchange of integration and summation, we have

$$\int_0^\infty e^{-x} \left(\frac{x}{1 - e^{-4x}} \right)^2 dx = \int_0^\infty e^{-x} x^2 \cdot \left(\sum_{n=1}^\infty n e^{-4(n-1)x} \right) dx \tag{91}$$

$$= \sum_{n=0}^\infty (n+1) \cdot \int_0^\infty x^2 e^{-(4n+1)x} x^2 dx$$

$$= \sum_{n=0}^\infty (n+1) \cdot \int_0^\infty e^{-(4n+1)x} x^2 dx$$

$$= \sum_{n=0}^\infty \frac{(n+1)\Gamma(3)}{(4n+1)^3} = \sum_{n=0}^\infty \frac{(2n+2)}{(4n+1)^3}$$

$$= \frac{1}{2} \sum_{n=0}^\infty \left(\frac{1}{(4n+1)^2} + \frac{3}{(4n+1)^3} \right)$$

$$= \frac{1}{2} L_{(4,1)}(2) + \frac{3}{2} L_{(4,1)}(3)$$

where we have used the fact that

$$\int_0^\infty e^{-\alpha x} x^p dx = \frac{\Gamma(p+1)}{\alpha^{p+1}}.$$

Hence, from (90) and (91) we have

$$\left[\int_0^\infty e^{-x} \left(\frac{x}{1 - e^{-2x}} - L_{(4,1)}(2)\right)^2 dx\right]^{\frac{1}{2}}$$

$$= \left[L_{(4,1)}(2)\left(\frac{1}{2} - L_{(4,1)}(2)\right) + \frac{3}{2}L_{(4,1)}(3)\right]^{\frac{1}{2}}. \tag{92}$$

Now, for the first expression on the right-hand side of (87), we have, on using (89) and (85)

$$\int_0^\infty e^{-x} \left(x^{\mu - \frac{1}{2}} J_{\mu - \frac{1}{2}}(rx) - K_*\right)^2 dx = \int_0^\infty e^{-x} x^{2\mu - 1} J_{\mu - \frac{1}{2}}^2(rx)\, dx - 1 \cdot K_*^2. \tag{93}$$

A result in Watson [37, p. 290] states that

$$\int_0^\infty e^{-2at} J_\alpha(\gamma t) J_\beta(\gamma t)\, t^{\alpha + \beta}\, dt$$

$$= \frac{\Gamma\left(\alpha + \beta + \frac{1}{2}\right)}{\pi^{\frac{3}{2}}} \gamma^{\alpha + \beta} \int_0^{\frac{\pi}{2}} \frac{\cos^{\alpha + \beta} \phi \cos(\alpha - \beta)\phi}{\left(a^2 + \gamma^2 \cos^2 \phi\right)^{\alpha + \beta + \frac{1}{2}}} d\phi \tag{94}$$

and so taking $a = \frac{1}{2}$, $\alpha = \beta = \mu - \frac{1}{2}$ and $\gamma = r$ in (94) gives

$$\int_0^\infty e^{-x} x^{2\mu - 1} J_{\mu - \frac{1}{2}}^2(rx)\, dx$$

$$= \frac{\Gamma\left(2\mu - \frac{1}{2}\right) r^{2\mu - 1}}{\pi^{\frac{3}{2}}} \int_0^{\frac{\pi}{2}} \frac{\cos^{2\mu - 1}\phi}{\left(\left(\frac{1}{2}\right)^2 + r^2 \cos^2 \phi\right)^{2\mu - \frac{1}{2}}} d\phi \tag{95}$$

That is, from (93) and (95) we have

$$\left[\int_0^\infty e^{-x} \left(x^{\mu - \frac{1}{2}} J_{\mu - \frac{1}{2}}(rx) - K_*\right)^2 dx\right]^{\frac{1}{2}}$$

$$= \left[\frac{\Gamma\left(2\mu - \frac{1}{2}\right)}{\pi^{\frac{3}{2}}} r^{2\mu - 1} \int_0^{\frac{\pi}{2}} \frac{\cos^{2\mu - 1}\phi}{\left[\left(\frac{1}{2}\right)^2 + r^2 \cos^2 \phi\right]^{2\mu - \frac{1}{2}}} d\phi - 1 \cdot K_*^2\right]^{\frac{1}{2}}. \tag{96}$$

Placing (96) and (92) into (87) produces the stated result (79) upon multiplication by $2C_\mu(r)$ and using (8).

For the coarser bound in (79) we have from (35) of Lemma 2

$$\int_0^{\frac{\pi}{2}} \frac{\cos^{2\mu-1}\phi}{\left[(\frac{1}{2})^2 + r^2\cos^2\phi\right]^{2\mu-\frac{1}{2}}} d\phi \le 2^{4\mu} B(\frac{1}{2}, \mu) = 2^{4\mu} \frac{\sqrt{\pi}\,\Gamma(\mu)}{\Gamma(\mu+\frac{1}{2})}$$

and so on substitution into the first result in (79) produces the second, upon some simplification.

Theorem 11 *For $\mu > 0$ and $r > 0$ the alternating odd generalized Mathieu series $\tilde{\phi}_\mu(r)$ satisfies the following relationship, namely*

$$\left| \Phi_\mu^-(r) - \frac{(2r)^{\mu-\frac{1}{2}}\Gamma(\mu)}{\sqrt{\pi}\,(3^2+r^2)^\mu} \cdot L_{(4,3)}(2) \right| \tag{97}$$

$$\le \kappa_{\phi^-} \left[\frac{4\Gamma(2\mu-\frac{1}{2})}{2^{2\mu-1}\sqrt{\pi}\,\Gamma^2(\mu+1)} \int_0^{\frac{\pi}{2}} \frac{\cos^{2\mu-1}\phi}{\left[(\frac{3}{2})^2 + r^2\cos^2\phi\right]^{2\mu-\frac{1}{2}}} d\phi - \frac{1}{3}\frac{(2r)^{\mu-\frac{1}{2}}\Gamma(\mu)}{\sqrt{\pi}\,(3^2+r^2)^\mu} \right]^{\frac{1}{2}}$$

$$\le \kappa_{\phi^-} \left[\frac{2^{2\mu+3}}{\mu^2(\mu-\frac{1}{2})\cdot B(\mu,\mu-\frac{1}{2})} - \frac{1}{3}\frac{(2r)^{\mu-\frac{1}{2}}\Gamma(\mu)}{\sqrt{\pi}\,(3^2+r^2)^\mu} \right]^{\frac{1}{2}},$$

where

$$\kappa_{\phi^-} = \left[\frac{1}{2}\left(L_{(4,3)}(2) + L_{(4,3)}(3)\right) - \frac{1}{3}\left(L_{(4,3)}(2)\right)^2 \right]^{\frac{1}{2}} \tag{98}$$

and $B(x, y)$ is the Euler beta function given by (34)

Proof We notice that $\frac{\Phi_\mu^-(r)}{2C_\mu(r)}$ from (77) may be written in the form

$$\frac{\Phi_\mu^-(r)}{2C_\mu(r)} = \int_0^\infty e^{-3x} \cdot \frac{x}{1-e^{-4x}} \cdot x^{\mu-\frac{1}{2}} J_{\mu-\frac{1}{2}}(rx)\,dx, \quad r, \mu > 0. \tag{99}$$

If we now let

$$P_{\phi^-}(x) = e^{-3x}, \quad f_{\phi^-}(x) = \frac{x}{1-e^{-4x}}, \quad g(x) = x^{\mu-\frac{1}{2}} J_{\mu-\frac{1}{2}}(rx) \tag{100}$$

then from (26)

$$P_{\phi^-} = \int_0^\infty P_{\phi^-}(x)\,dx = \int_0^\infty e^{-3x}\,dx = \frac{1}{3}, \tag{101}$$

$$P_{\phi^-} \cdot M\left(f_{\tilde{O}}; p\right) = \int_0^\infty e^{-3x} \cdot \frac{x}{1 - e^{-4x}}\,dx = L_{(4,3)}(2) \tag{102}$$

and

$$P_{\phi^-} \cdot M(g; p) = \int_0^\infty e^{-3x} \cdot x^{\mu - \frac{1}{2}} J_{\mu - \frac{1}{2}}(rx)\,dx = \frac{(2r)^{\mu - \frac{1}{2}}\,\Gamma(\mu)}{\sqrt{\pi}\,\left(3^2 + r^2\right)^\mu}, \tag{103}$$

where we have used the fact that $L_{(4,3)}(x) = \sum_{n=0}^\infty \frac{1}{(4n+3)^x}$ to procure (102), and from Watson [37, p. 386],

$$\int_0^\infty e^{-\alpha x} \cdot x^\nu J_\nu(\beta x)\,dx = \frac{(2\beta)^\nu}{\sqrt{\pi}} \cdot \frac{\Gamma\left(\nu + \frac{1}{2}\right)}{\left(\alpha^2 + \beta^2\right)^{\nu + \frac{1}{2}}}, \quad \mathrm{Re}\,(\nu) > \frac{1}{2}, \quad \mathrm{Re}\,(\alpha) > |\mathrm{Im}\,(\beta)|,$$

with $\alpha = 1$, $\nu = \mu - \frac{1}{2}$, $\beta = r$ to obtain (103).

Now, from (25)–(28) we have on using (99)–(103)

$$\frac{\Phi_\mu^-(r)}{2C_\mu(r)} - \frac{(2r)^{\mu - \frac{1}{2}}\,\Gamma(\mu)}{\sqrt{\pi}\,\left(3^2 + r^2\right)^\mu} \cdot L_{(4,3)}(2)$$

$$= \int_0^\infty e^{-3x} \left(x^{\mu - \frac{1}{2}} J_{\mu - \frac{1}{2}}(rx) - K\right)\left(\frac{x}{1 - e^{-4x}} - L_{(4,3)}(2)\right)dx. \tag{104}$$

Further, using the Cauchy-Buniakowsky-Schwartz inequality, we have from (32), to give

$$\left|\frac{\Phi_\mu^-(r)}{2C_\mu(r)} - \frac{(2r)^{\mu - \frac{1}{2}}\,\Gamma(\mu)}{\sqrt{\pi}\,\left(3^2 + r^2\right)^\mu} \cdot L_{(4,3)}(2)\right| \tag{105}$$

$$= \left(\int_0^\infty e^{-3x} \left(x^{\mu - \frac{1}{2}} J_{\mu - \frac{1}{2}}(rx) - K\right)^2 dx\right)^{\frac{1}{2}}$$

$$\times \left(\int_0^\infty e^{-3x} \left(\frac{x}{1 - e^{-4x}} - L_{(4,3)}(2)\right)^2 dx\right)^{\frac{1}{2}}.$$

As mentioned in Sect. 2, the appropriate choice of K is the weighted integral mean as given from (103), namely

$$K = K_* = \frac{(2r)^{\mu - \frac{1}{2}}\,\Gamma(\mu)}{\sqrt{\pi}\,\left(3^2 + r^2\right)^\mu}, \tag{106}$$

Now using the result

$$\int_a^b p(t)[h(t) - M(h; p)]^2 dt = \int_a^b p(t) h^2(t) dt - P \cdot M^2(h; p). \quad (107)$$

to evaluate the two expressions on the right-hand side of (105) produces; firstly,

$$\int_0^\infty e^{-x}\left(\frac{x}{1+e^{-2x}} - L_{(4,3)}(2)\right)^2 dx = \int_0^\infty e^{-x}\left(\frac{x}{1-e^{-4x}}\right)^2 dx - \frac{1}{3}\cdot(L_{(4,3)}(2))^2. \quad (108)$$

and secondly, allowing for the permissible interchange of integration and summation, we have

$$\int_0^\infty e^{-3x}\left(\frac{x}{1-e^{-4x}}\right)^2 dx = \int_0^\infty e^{-3x}x^2\cdot\left(\sum_{n=1}^\infty ne^{-4(n-1)x}\right)dx \quad (109)$$

$$= \sum_{n=0}^\infty (n+1)\cdot\int_0^\infty x^2 e^{-(4n+1)x}x^2 dx$$

$$= \sum_{n=0}^\infty (n+1)\cdot\int_0^\infty e^{-(4n+3)x}x^2 dx$$

$$= \sum_{n=0}^\infty \frac{(n+1)\Gamma(3)}{(4n+3)^3} = \sum_{n=0}^\infty \frac{(2n+2)}{(4n+3)^3}$$

$$= \frac{1}{2}\sum_{n=0}^\infty\left(\frac{1}{(4n+3)^2} + \frac{1}{(4n+3)^3}\right)$$

$$= \frac{1}{2}\left(L_{(4,3)}(2) + L_{(4,3)}(3)\right).$$

where we have used the fact that

$$\int_0^\infty e^{-\alpha x}x^p dx = \frac{\Gamma(p+1)}{\alpha^{p+1}}.$$

Hence, from (107) and (102) we have

$$\left[\int_0^\infty e^{-3x}\left(\frac{x}{1-e^{-2x}} - L_{(4,3)}(2)\right)^2 dx\right]^{\frac{1}{2}}$$

$$= \left[\frac{1}{2}\left(L_{(4,3)}(2) + L_{(4,3)}(3)\right) - \frac{1}{3}\left(L_{(4,3)}(2)\right)^2\right]^{\frac{1}{2}}. \quad (110)$$

Now, for the first expression on the right-hand side of (105), we have, on using (107) and (103)

$$\int_0^\infty e^{-x} \left(x^{\mu-\frac{1}{2}} J_{\mu-\frac{1}{2}}(rx) - K_* \right)^2 dx = \int_0^\infty e^{-x} x^{2\mu-1} J_{\mu-\frac{1}{2}}^2(rx)\, dx - \frac{1}{3} \cdot K_*^2.$$

(111)

A result in Watson [37, p. 290] states that

$$\int_0^\infty e^{-2at} J_\alpha(\gamma t) J_\beta(\gamma t)\, t^{\alpha+\beta}\, dt$$

$$= \frac{\Gamma\left(\alpha+\beta+\frac{1}{2}\right)}{\pi^{\frac{3}{2}}} \gamma^{\alpha+\beta} \int_0^{\frac{\pi}{2}} \frac{\cos^{\alpha+\beta}\phi \cos(\alpha-\beta)\phi}{\left(a^2+\gamma^2\cos^2\phi\right)^{\alpha+\beta+\frac{1}{2}}}\, d\phi$$

(112)

and so taking $a = \frac{3}{2}$, $\alpha = \beta = \mu - \frac{1}{2}$ and $\gamma = r$ in (112) gives

$$\int_0^\infty e^{-x} x^{2\mu-1} J_{\mu-\frac{1}{2}}^2(rx)\, dx$$

$$= \frac{\Gamma\left(2\mu-\frac{1}{2}\right) r^{2\mu-1}}{\pi^{\frac{3}{2}}} \int_0^{\frac{\pi}{2}} \frac{\cos^{2\mu-1}\phi}{\left(\left(\frac{3}{2}\right)^2 + r^2\cos^2\phi\right)^{2\mu-\frac{1}{2}}}\, d\phi$$

(113)

That is, from (111) and (113) we have

$$\left[\int_0^\infty e^{-x} \left(x^{\mu-\frac{1}{2}} J_{\mu-\frac{1}{2}}(rx) - K_* \right)^2 dx \right]^{\frac{1}{2}}$$

$$= \left[\frac{\Gamma\left(2\mu-\frac{1}{2}\right)}{\pi^{\frac{3}{2}}} r^{2\mu-1} \int_0^{\frac{\pi}{2}} \frac{\cos^{2\mu-1}\phi}{\left[\left(\frac{3}{2}\right)^2 + r^2\cos^2\phi\right]^{2\mu-\frac{1}{2}}}\, d\phi - \frac{1}{3} \cdot K_*^2 \right]^{\frac{1}{2}}.$$

(114)

Placing (114) and (110) into (105) produces the stated result (97) upon multiplication by $2C_\mu(r)$ and using (8).

For the coarser bound in (97) we have from (35) of Lemma 2

$$\int_0^{\frac{\pi}{2}} \frac{\cos^{2\mu-1}\phi}{\left[(\frac{3}{2})^2 + r^2\cos^2\phi\right]^{2\mu-\frac{1}{2}}}\, d\phi \le \frac{1}{2}\left(\frac{2}{3}\right)^{4\mu-1} B(\frac{1}{2}, \mu) = \left(\frac{3}{4}\right)\left(\frac{2}{3}\right)^{4\mu} \cdot \frac{\sqrt{\pi}\,\Gamma(\mu)}{\Gamma\left(\mu+\frac{1}{2}\right)}$$

and so on substitution into the first result in (97) produces the second, upon some simplification.

The following theorem demonstrates the relationship for the generalized Mathieu series related to the Beta L- function family. This can be compared with the end of Sect. 4 where the Zeta function family results were discussed.

Theorem 12 *The following relationships hold, namely*

$$\phi_\mu (r) = \begin{cases} 2 \cdot \Phi_\mu^+(r) - \tilde{\phi}_\mu (r) \\ 2 \cdot \Phi_\mu^-(r) + \tilde{\phi}_\mu (r) \end{cases} \tag{115}$$

$$\tilde{\phi}_\mu (r) = \begin{cases} 2 \cdot \Phi_\mu^+(r) - \phi_\mu (r) \\ \phi_\mu (r) - 2 \cdot \Phi_\mu^-(r) \end{cases}$$

where $\phi_\mu (r)$ is defined in (39), $\tilde{\phi}_\mu (r)$ is given in (58), and $\Phi_\mu^+(r)$ and $\Phi_\mu^-(r)$ are defined in (76) and (77), respectively. These entities represent the generalized Mathieu series propagated by series of reciprocal powers of odd numbers, alternating odd numbers, $L_{(4,1)}(\cdot)$ and $L_{(4,3)}(\cdot)$.

Proof This is trivial since $\Phi_\mu^+(r) := \frac{\phi_\mu(r)+\tilde{\phi}_\mu(r)}{2}$ and $\Phi_\mu^-(r) := \frac{\phi_\mu(r)-\tilde{\phi}_\mu(r)}{2}$ are defined (76) and (77).

Remark 9 If r is allowed to tend to zero for the first result at (115), namely $\phi_\mu (r) = 2 \cdot \Phi_\mu^+(r) - \tilde{\phi}_\mu (r)$, then the relationship $\lambda(x) = 2L_{(4,1)}(x) - \beta(x)$ where $x = 2\mu + 1$ results. The similar process relating the Zeta function $\zeta(x)$ produced $\zeta(x) = 2\lambda(x) - \eta(x)$, where $\lambda(x)$ is the odd zeta, $\eta(x)$ is the alternating zeta, and $x = 2\mu + 1$ at (54).

6 Some Properties of the Generalized Mathieu Series and Its Companions Including Beta Related L-Functions

Let

$$G_\mu(r; H) = \gamma_\mu \int_0^\infty H(x)x^{\mu-\frac{1}{2}} \cdot \frac{J_{\mu-\frac{1}{2}}(rx)}{r^{\mu-\frac{1}{2}}} dx, \quad r, \mu > 0 \tag{116}$$

where from (42). Let

$$\gamma_\mu = \begin{cases} C_\mu = \dfrac{\sqrt{\pi}}{2^{\mu-\frac{1}{2}}\Gamma(\mu+1)}, & \text{for } H_M(\cdot) \text{ and } H_A(\cdot) \\ 2C_\mu, & \text{for } H_O(\cdot) \text{ and } H_E(\cdot) \end{cases} \tag{117}$$

The following proposition determines the moments of (116). See also [15] where a Mellin transform approach has been used only for the generalized Mathieu series.

Proposition 1 (see [13]) *The moments of $G_\mu(r; H)$ from (116) and (117) given by*

$$M^{(k)} = \int_0^\infty r^k G_\mu(r; H) dr \tag{118}$$

$$= [1 or 2] \frac{B(\frac{k}{2} + \frac{1}{2}, \mu - \frac{k}{2} + \frac{1}{2})}{\Gamma(2\mu - k)} \int_0^\infty x^{2\mu - k - 2} H(x) dx$$

where $B(x, y)$ is the Euler beta function given by (34), and

$$[1 or 2] = \begin{cases} 1, & for\ H_M(\cdot)\ and\ H_A(\cdot) \\ 2, & for\ H_O(\cdot)\ and\ H_E(\cdot) \end{cases} . \tag{119}$$

Proof From (116), (117) and (118) we have

$$M^{(k)} = \gamma_\mu \int_0^\infty x^{\mu - \frac{1}{2}} H(x) \int_0^\infty r^k \cdot \frac{J_{\mu - \frac{1}{2}}(rx)}{r^{\mu - \frac{1}{2}}} dr dx \tag{120}$$

and so the substitution of $\omega = rx$ produces

$$\int_0^\infty r^k \cdot \frac{J_{\mu - \frac{1}{2}}(rx)}{r^{\mu - \frac{1}{2}}} dr = x^{\mu - k - \frac{3}{2}} \int_0^\infty \omega^{k - (\mu - \frac{1}{2})} \cdot J_{\mu - \frac{1}{2}}(\omega) d\omega \tag{121}$$

$$= x^{\mu - k - \frac{3}{2}} \cdot 2^{k - \mu + \frac{1}{2}} \frac{\Gamma\left(\frac{k}{2} + \frac{1}{2}\right)}{\Gamma\left(\mu - \frac{k}{2}\right)}$$

where we have used the result

$$\int_0^\infty \omega^{\lambda - \nu} \cdot J_\nu(\omega) d\omega = 2^{\lambda - \nu} \frac{\Gamma\left(\frac{\lambda}{2} + \frac{1}{2}\right)}{\Gamma(\nu - \frac{\lambda}{2} + \frac{1}{2})}$$

with $\nu = \mu - \frac{1}{2}$ and $\lambda = k$. Thus, substituting (121) into (120) gives

$$M^{(k)} = \gamma_\mu \delta_{\mu,k} \int_0^\infty x^{2\mu - k - 2} H(x) dx \tag{122}$$

where

$$\delta_{\mu,k} = 2^{k - \mu + \frac{1}{2}} \frac{\Gamma\left(\frac{k}{2} + \frac{1}{2}\right)}{\Gamma\left(\mu - \frac{k}{2}\right)}. \tag{123}$$

Further, using the duplication formula for the gamma function,

$$\sqrt{\pi} \Gamma(2z) = 2^{2z - 1} \Gamma(z) \Gamma\left(z + \frac{1}{2}\right) \text{ with } z = \mu - \frac{k}{2}$$

gives $\Gamma\left(\mu - \frac{k}{2}\right) = \frac{\sqrt{\pi}\Gamma(2\mu-k)}{2^{2\mu-k-1}\Gamma\left(\mu-\frac{k}{2}+\frac{1}{2}\right)}$ and so, from (117) and (123), we have

$$\gamma_\mu \, \delta_{\mu,k} = [1 or 2] \, \frac{\Gamma\left(\frac{k}{2}+\frac{1}{2}\right)\Gamma\left(\mu-\frac{k}{2}+\frac{1}{2}\right)}{\Gamma(\mu+1)\,\Gamma(2\mu-k)}. \tag{124}$$

Substitution of (124) into (122) produces the statement of the proposition.

The following corollary gives the moments for the generalized Mathieu series and its companions.

Corollary 4 (see [13]) *Let the subscript of M, A, O, E indicate the generalized: Mathieu, Alternating, Odd and Even series moments, respectively (see [13] for the proof). These are then given by*

$$M_M^{(k)} = B(\frac{k}{2}+\frac{1}{2}, \mu-\frac{k}{2}+\frac{1}{2})\zeta(2\mu-k), \qquad 2\mu-k > 1 \tag{125}$$

$$M_A^{(k)} = B(\frac{k}{2}+\frac{1}{2}, \mu-\frac{k}{2}+\frac{1}{2})(1-2^{-(2\mu-k-1)})\zeta(2\mu-k), \quad 2\mu-k > 0$$

$$M_O^{(k)} = 2B(\frac{k}{2}+\frac{1}{2}, \mu-\frac{k}{2}+\frac{1}{2})(1-2^{-(2\mu-k)})\zeta(2\mu-k), \quad 2\mu-k > 1$$

$$M_E^{(k)} = 2B(\frac{k}{2}+\frac{1}{2}, \mu-\frac{k}{2}+\frac{1}{2})2^{-(2\mu-k)}\zeta(2\mu-k), \quad 2\mu-k > 1$$

where $B(x, y)$ is the Euler beta function given by (34).

We now analyze the moments associated with the Dirichlet Beta-L function results via Mathieu type series.

Let

$$G_\mu(r; H) = \gamma_\mu \int_0^\infty H(x)x^{\mu-\frac{1}{2}} \cdot \frac{J_{\mu-\frac{1}{2}}(rx)}{r^{\mu-\frac{1}{2}}}dx, \qquad r, \mu > 0 \tag{126}$$

where $C_\mu = \dfrac{\sqrt{\pi}}{2^{\mu-\frac{1}{2}}\Gamma(\mu+1)}$, and

$$\gamma_\mu = 2C_\mu \text{ for } H_\phi(x) := \frac{x}{e^x - e^{-x}}, \; H_{\tilde{\phi}}(x) := \frac{x}{e^x + e^{-x}}, \tag{127}$$

$$H_{L(4,1)}(x) := \frac{xe^x}{e^{2x} - e^{-2x}}, \; H_{L(4,3)}(x) := \frac{xe^{-x}}{e^{2x} - e^{-2x}}.$$

Proposition 2 *The moments of $G_\mu(r; H)$ from (126) and (127) are given by*

$$M^{(k)} = \int_0^\infty r^k G_\mu(r; H) dr \tag{128}$$

$$= \frac{2B(\frac{k}{2} + \frac{1}{2}, \mu - \frac{k}{2} + \frac{1}{2})}{\Gamma(2\mu - k)} \int_0^\infty x^{2\mu - k - 2} H(x) dx$$

Proof Similar to the previous proposition.

Corollary 5 *Let the subscript of O, $\tilde{O}$, $L_{(4,1)}$, $L_{(4,3)}$ indicate the generalized: Odd, Alternating Odd, Dirichlet $L_{(4,1)}$ and $L_{(4,3)}$, series moments, respectively. These are then given by*

$$M_O^{(k)} = 2B(\frac{k}{2} + \frac{1}{2}, \mu - \frac{k}{2} + \frac{1}{2})(1 - 2^{-(2\mu - k)})\zeta(2\mu - k), \ 2\mu - k > 1 \tag{129}$$

$$M_{\tilde{O}}^{(k)} = 2B(\frac{k}{2} + \frac{1}{2}, \mu - \frac{k}{2} + \frac{1}{2})\Gamma(2\mu - k)\beta(2\mu - k), \qquad 2\mu - k > 0$$

$$M_{L_{(4,1)}}^{(k)} = 2B(\frac{k}{2} + \frac{1}{2}, \mu - \frac{k}{2} + \frac{1}{2})\Gamma(2\mu - k)L_{(4,1)}(2\mu - k), \ 2\mu - k > 1$$

$$M_{L_{(4,3)}}^{(k)} = 2B(\frac{k}{2} + \frac{1}{2}, \mu - \frac{k}{2} + \frac{1}{2})\Gamma(2\mu - k)L_{(4,3)}(2\mu - k), \ 2\mu - k > 1$$

where $B(x, y)$ is the Euler beta function given by (34)

Proof From (128) we require to evaluate the integral for the various $H(x)$ representing each of the generalized Mathieu functions as given in (127).

That is, we require to evaluate

$$M(q; H) = \int_0^\infty x^{q-1} H(x) dx \tag{130}$$

where $q = 2\mu - k - 1$ and for each of $H(x)$ as given in (127).

For the **odd generalized Mathieu series** we have from (127)

$$M(q; H_O) = \int_0^\infty \frac{x^q}{e^x - e^{-x}} dx, \ q = 2\mu - k - 1 \tag{131}$$

$$= \Gamma(2\mu - k)(1 - 2^{-(2\mu - k)})\zeta(2\mu - k)$$

where we have used the result

$$\int_0^\infty e^{-x} \frac{x^q}{1 - e^{-2x}} dx = \int_0^\infty e^{-x} \frac{x^q}{1 - e^{-2x}} dx$$

$$= \int_0^\infty e^{-x} x^q \left(\sum_{n=0}^\infty e^{-2nx} \right) dx$$

$$= \sum_{n=0}^{\infty} \int_0^{\infty} e^{-(2n+1)x} x^q dx$$

$$= \sum_{n=0}^{\infty} \frac{\Gamma(q+1)}{(2n+1)^{q+1}}$$

$$= \Gamma(q+1)(1 - 2^{-(q+1)})\varsigma(q+1)$$

where $\sum_{n=0}^{\infty} \frac{1}{(2n+1)^s} = (1-2^{-s})\varsigma(s)$, we have allowed the permissible interchange of integration and summation and we have used the result

$$\int_0^{\infty} e^{-\alpha x} x^p dx = \frac{\Gamma(p+1)}{\alpha^{p+1}}. \tag{132}$$

Substituting (131) into (128) and noting (132) gives the first result.
For the **alternating odd generalized Mathieu series** we have from (127)

$$M(q; H_{\hat{o}}) = \int_0^{\infty} \frac{x^q}{e^x + e^{-x}} dx, \quad q = 2\mu - k - 1 \tag{133}$$

$$= \Gamma(2\mu - k)\beta(2\mu - k)$$

where we have used the result,

$$\int_0^{\infty} e^{-x} \frac{x^q}{1 + e^{-2x}} dx = \int_0^{\infty} e^{-x} \frac{x^q}{1 + e^{-2x}} dx$$

$$= \int_0^{\infty} e^{-x} x^q \left(\sum_{n=0}^{\infty} (-1)^n e^{-2nx} \right) dx$$

$$= \sum_{n=0}^{\infty} (-1)^n \int_0^{\infty} e^{-(2n+1)x} x^q dx$$

$$= \sum_{n=0}^{\infty} \frac{(-1)^n \Gamma(q+1)}{(2n+1)^{q+1}}$$

$$= \Gamma(q+1)\beta(q+1)$$

where $\sum_{n=0}^{\infty} \frac{(-1)^n}{(2n+1)^s} = \beta(s)$, $\mathrm{Re}\, s > 1$ [30, p. 602, 25.2.2], we have allowed the permissible interchange of integration and summation and we have used the result (132).

Substituting (133) into (128) and noting (132) gives the second result.

Thirdly, for the **Dirichlet** $L_{(4,1)}$ generalized Mathieu series we have from (127)

$$M(q; H_{L_{(4,1)}}) = \int_0^\infty \frac{x^q \cdot e^x}{e^{2x} - e^{-2x}} dx, \ q = 2\mu - k - 1 \tag{134}$$

$$= \Gamma(2\mu - k) L_{(4,1)}(2\mu - k)$$

where we have

$$\int_0^\infty \frac{x^q e^{-x}}{1 - e^{-4x}} dx = \int_0^\infty e^{-x} x^q \left(\sum_{n=0}^\infty e^{-4nx} \right) dx$$

$$= \sum_{n=0}^\infty \int_0^\infty e^{-(4n+1)x} x^q dx$$

$$= \Gamma(q + 1) L_{(4,1)}(q + 1)$$

giving (134).

Substituting (134) into (128) and noting (132) gives the third result.

Finally, for the **Dirichlet** $L_{(4,3)}$ generalized Mathieu series we have from (127)

$$M(q; H_{L_{(4,3)}}) = \int_0^\infty \frac{x^q \cdot e^{-x}}{e^{2x} - e^{-2x}} dx, \ q = 2\mu - k - 1 \tag{135}$$

$$= \Gamma(2\mu - k) L_{(4,3)}(2\mu - k)$$

where we have

$$\int_0^\infty \frac{x^q \cdot e^{-3x}}{1 - e^{-4x}} dx = \int_0^\infty e^{-3x} x^q \left(\sum_{n=0}^\infty e^{-4nx} \right) dx$$

$$= \sum_{n=0}^\infty \int_0^\infty e^{-(4n+3)x} x^q dx$$

$$= \Gamma(q + 1) L_{(4,3)}(q + 1)$$

giving (135).

Substituting (135) into (128) and noting (132) gives the last result and thus completing the proof.

The generalized Mathieu series $S_\mu(r)$ is a positive, decreasing function of both μ and r for $\mu > 0, r > 0$.

The following interesting results hold (see also [12])

Corollary 6 *The generalized Mathieu series as defined in (5) satisfies the identity*

$$\int_0^\infty S_\mu(r)\, dr = \sqrt{\pi} \cdot \frac{\Gamma\left(\mu + \frac{1}{2}\right)}{\mu \Gamma(\mu)} \zeta(2\mu), \quad \mu > 0. \tag{136}$$

For m a positive integer, then

$$\int_0^\infty S_m(r)\, dr = \frac{(-1)^{m-1} 2^{2m-1} \pi^{2m+\frac{1}{2}}}{m!\,(2m)!} \Gamma\left(m + \frac{1}{2}\right) B_{2m}, \tag{137}$$

where we have used a 1748 result of Euler states that for $m \in \mathbb{N}$

$$\zeta(2m) = (-1)^{m-1} \frac{2^{2m-1} \pi^{2m}}{(2m)!} B_{2m}. \tag{138}$$

and, B_k are the Bernoulli numbers defined by

$$\frac{x}{e^x - 1} = \sum_{k=0}^\infty \frac{x^k}{k!} B_k, \quad |x| < 2\pi.$$

Remark 10 An alternative representation to (137) is given in a 1999 paper by Lin in Chinese (see [23]), namely

$$\zeta(2m) = A_m \pi^{2m}, \tag{139}$$

where A_m satisfies the recurrence relation

$$A_m = (-1)^{m-1} \cdot \frac{m}{(2m+1)!} + \sum_{j=1}^{m-1} \frac{(-1)^{j-1}}{(2j+1)!} A_{m-j} \tag{140}$$

and by convention the sum is neglected for $m = 1$ so that $A_1 = \frac{1}{3!}$. Thus an equivalent result to (137) may be obtained as, from (139),

$$\int_0^\infty S_m(r)\, dr = \frac{\Gamma\left(m + \frac{1}{2}\right)}{m!} \pi^{2m+\frac{1}{2}} \cdot A_m$$

with A_m being given by (140).

Remark 11 Similar results to the above corollary may be obtained for the companion zeroth moments by taking in (125). These are obviously related, for example,

$$M_A^{(0)} = M_M^{(0)} - M_E^{(0)} \quad \text{and} \quad M_O^{(0)} = 2M_M^{(0)} - M_E^{(0)}.$$

Corollary 7 *From (115) the Moments associated with the Dirichlet beta-L series are given by the following relationship, namely*

$$M_O^{(k)} = \left\{ \begin{array}{c} 2 \cdot M_{L_{(4,1)}}^{(k)} - M_{\mathring{O}}^{(k)} \\ 2 \cdot M_{L_{(4,3)}}^{(k)} + M_{\mathring{O}}^{(k)} \end{array} \right\} \tag{141}$$

$$M_{\mathring{O}}^{(k)} = \left\{ \begin{array}{c} 2 \cdot M_{L_{(4,1)}}^{(k)} - M_O^{(k)} \\ M_O^{(k)} - 2 \cdot M_{L_{(4,3)}}^{(k)} \end{array} \right\}$$

Remark 12 The moments may be used to approximate the class of generalized Mathieu series and obtain bounds for the remainders. Further, the current paper has aimed at investigating odd and even members of generalized Mathieu series, which it is believed not to have been treated in the literature. Their relationship to the Zeta function has also been highlighted throughout the paper and in particular in Theorem 8 and Remark 7. The Dirichlet Beta-L function and its companion generators have been explored via the generalized Mathieu series in terms of procuring approximation and bounds. The moments have been determined and identities have been recaptured for the generators by allowing r->0 for the generalized Mathieu representations in terms of both series and integrals.

References

1. H. Alzer, J.L. Brenner, O.G. Ruehr, On Mathieu's inequality. J. Math. Anal. Appl. **218**, 607–610 (1998)
2. T.M. Apostol, *Analytic Number Theory* (Springer, New York, 1976)
3. A. Bagdasaryan, A new lower bound for Mathieu's series. N. Z. J. Math. **44**, 75–81 (2014)
4. Á. Baricz, P.L. Butzer, T.K. Pogány, Alternating Mathieu series, Hilbert-Eisenstein series and their generalized Omega functions, in *Analytic Number Theory, Approximation Theory, and Special Functions* (Springer, New York, 2014), pp. 775–808
5. J.M. Borwein, I.J. Zucker, J. Boersma, The evaluation of character Euler double sums. Ramanujan J. **15** , 377–405 (2008)
6. P.S. Bullen, *A Dictionary of Inequalities* (Addison Wesley Longman, Boston, 1998)
7. P. Cerone, On an identity of the Chebychev functional and some ramifications. J. Inequal. Pure Appl. Math. **3**(1), Art. 4 (2002). http://jipam.vu.edu.au/v3n1/
8. P. Cerone, Bounds for Zeta and related functions. J. Inequal. Pure Appl. Math. **6**(5), Art. 134 (2005). http://jipam.vu.edu.au/v6n5/
9. P. Cerone, On a double inequality for the Dirichlet Beta function. Tamsui Oxf. J. Math. Sci. **24**(3), 269–276 (2008)
10. P. Cerone, Special functions approximation and bounds via integral representation, in *Advances in Inequalities for Special Functions*, ed. by P. Cerone, S. Dragomir (Nova Science Publishers, New York, 2008), pp. 1–36
11. P. Cerone, S.S. Dragomir, New upper and lower bounds for the Chebyshev functional. J. Inequal. Pure Appl. Math. **3**(5), Art. 77 (2002). http://jipam.vu.edu.au/v3n5/

12. P. Cerone, C. Lenard, On integral forms of generalised Mathieu series. JIPAM J. Inequal. Pure Appl. Math. **4**(5), 11 pp. (2003). Art 100 (electronic). http://jipam.vu.edu.au/v4n5/
13. P. Cerone, C. Lenard, On generalised Mathieu series and its companions. Hacet. J. Math. Stat. (2018, accepted for publication)
14. J.B. Conrey, The Riemann hypothesis. Notices Am. Math. Soc. **50**, 341–353 (2003)
15. N. Elezović, H.M. Srivastava, Ž. Tomovski, Integral representations and integral transforms of some families of Mathieu type series. Integral Transforms Spec. Funct. **19**(7), 481–495 (2008)
16. O.E. Emersleben, Über die Reihe $\sum_{k=1}^{\infty} k(k^2 + c^2)^{-2}$. Math. Ann. **125**, 165–171 (1952)
17. S.R. Finch, *Mathematical Constants* (Cambridge University Press, Cambridge, 2003)
18. I. Gavrea, Some remarks on Mathieu's series, in *Mathematical Analysis and Approximation Theory* (Burg Verlag, Rehau, 2002), pp. 113–117
19. G. Grüss, Über das maximum des absoluten Betrages von $\frac{1}{b-a} \int_a^b f(x) g(x) dx - \frac{1}{b-a} \int_a^b f(x) dx \cdot \frac{1}{b-a} \int_a^b g(x) dx$. Math. Z. **39**, 215–226 (1934)
20. B.-N. Guo, Note on Mathieu's inequality. RGMIA Res. Rep. Coll. **3**(3), Article 5 (2000). Available online at http://rgmia.vu.edu.au/v3n3.html
21. L. Landau, Monotonicity and bounds on Bessel functions. Electron. J. Differ. Equ. **4**, 147–154 (2002)
22. L. Lorenz, Bidrag tiltalienes theori. Tidsskrift Math. **1**, 97–114 (1871)
23. Q.-M. Luo, B.-N. Guo, F. Qi, On evaluation of Riemann zeta function $\zeta(s)$. RGMIA Res. Rep. Coll. **6**(1), Article 8 (2003). http://rgmia.vu.edu.au/v6n1.html
24. E. Mathieu, *Traité de physique mathématique, VI–VII: Théorie de l'élasticité des corps solides* (Gauthier-Villars, Paris, 1890)
25. V.V. Meleshko, Equilibrium of elastic rectangle: Mathieu-Inglis-Pickett solution revisited. J. Elast. **40**(3), 207–238 (1995)
26. V.V. Meleshko, Biharmonic problem in a rectangle. Appl. Sci. Res. **58**(1–4), 217–249 (1997)
27. V.V. Meleshko, Bending of an elastic rectangular clamped plate: exact versus "engineering" solutions. J. Elast. **48**(1), 1–50 (1997)
28. G.V. Milovanović, T.K. Pogány, New integral forms of generalized Mathieu series and related applications. Appl. Anal. Discrete Math. **7**(1), 180–192 (2013)
29. D.S. Mitrinović, J.E. Pečarić, A.M. Fink, *Classical and New Inequalities in Analysis* (Kluwer Academic Publishers, Dordrecht, 1993)
30. F.W.J. Olver, D.W. Lozier, R.F. Boisvert, C.W. Clark, *NIST Handbook of Mathematical Functions* (National Bureau of Standards and Technology/Cambridge University Press, New York, 2010)
31. T.K. Pogány, Ž. Tomovski, Bounds improvement for alternating Mathieu type series. J. Math. Inequal. **4**(3), 315–324 (2010)
32. T.K. Pogány, H.M. Srivastava, Ž. Tomovski, Some families of Mathieu a-series and alternating Mathieu a-series. Appl. Math. Comput. **173**, 69–108 (2006)
33. F. Qi, Inequalities for Mathieu's series. Int. J. Pure Appl. Math. (2003, in press)
34. F. Qi, Integral expressions and inequalities of Mathieu type series. RGMIA Res. Rep. Coll. **6**(2), Article 5 (2003). Available online at http://rgmia.vu.edu.au/v6n2.html
35. F. Qi, Ch.-P. Chen, B.-N. Guo, Notes on double inequalities of Mathieu's series. Int. J. Math. Math. Sci. **2005**(16), 2547–2554 (2005)
36. Ž. Tomovski, K. Trenčevski, On an open problem of Bai-Ni Guo and Feng Qi. J. Inequal. Pure Appl. Math. **4**(2), 1–7 (2003). http://jipam.vu.edu.au/v4n2/
37. G.N. Watson, *A Treatise on the THEORY of Bessel Functions*, 2nd edn. (Cambridge University Press, Cambridge, 1966)
38. J.E. Wilkins Jr., Solution of problem 97–1. SIAM Rev.
39. I.J. Zucker, R.C. McPhedran, Dirichlet L-series with real and complex characters and their application to solving double sums. Proc. R. Soc. A **464**, 1405–1422 (2008)

Recent Research on Levinson's Inequality

Jadranka Mićić and Marjan Praljak

Abstract In this paper we review the results on Levinson's inequality and present some of its generalizations using several new approaches. We provide a probabilistic version for the family of 3-convex functions at a point. We also show that this is the largest family of continuous functions for which the inequality holds. From the obtained inequality, we derive new families of exponentially convex functions and related results. We also give a monotonic refinement of the probabilistic version of Levinson's inequality. Levinson's type inequality of Hilbert space operators is discussed as well for unital fields of positive linear mappings and a large class of functions. Order among quasi-arithmetic means is considered under similar conditions.

1 Introduction

In this section we will review the history of research of Levison's inequality.

A well-known inequality due to Levinson [11] is given in the following theorem.

Theorem 1 *If $f : (0, 2c) \to \mathbb{R}$ satisfies $f''' \geq 0$ and $p_i, x_i, y_i, i = 1, 2, \ldots, n$, are such that $p_i > 0$, $\sum_{i=1}^{n} p_i = 1$, $0 \leqslant x_i \leqslant c$ and*

$$x_1 + y_1 = x_2 + y_2 = \ldots = x_n + y_n = 2c, \tag{1}$$

then the inequality

J. Mićić (✉)
Faculty of Mechanical Engineering and Naval Architecture, University of Zagreb, Zagreb, Croatia
e-mail: jmicic@fsb.hr

M. Praljak
Faculty of Food Technology and Biotechnology, University of Zagreb, Zagreb, Croatia
e-mail: mpraljak@pbf.hr

© Springer Nature Switzerland AG 2019
G. A. Anastassiou and J. M. Rassias (eds.), *Frontiers in Functional Equations and Analytic Inequalities*, https://doi.org/10.1007/978-3-030-28950-8_32

$$\sum_{i=1}^{n} p_i f(x_i) - f(\overline{x}) \leqslant \sum_{i=1}^{n} p_i f(y_i) - f(\overline{y}) \tag{2}$$

holds, where $\overline{x} = \sum_{i=1}^{n} p_i x_i$ and $\overline{y} = \sum_{i=1}^{n} p_i y_i$ denote the weighted arithmetic means.

A function $f : I \rightarrow \mathbb{R}$ is called k-convex if kth order divided difference satisfies $[x_0, \ldots, x_k] f \geq 0$ for all choices of $k + 1$ distinct points $x_0, \ldots, x_k \in I$. If the kth derivative $f^{(k)}$ of a k-convex function exists, then $f^{(k)} \geq 0$, but $f^{(k)}$ may not exist (for properties of divided differences and k-convex functions, see [22]).

Popoviciu [23] showed that in Theorem 1 it is enough to assume that f is 3-convex. Bullen [4] gave another proof of Popoviciu's result, as well as a converse of Levinson's inequality (rescaled to a general interval $[a, b]$). Bullen's result is the following:

Theorem 2

(a) If $f : [a, b] \rightarrow \mathbb{R}$ is 3-convex and $p_i, x_i, y_i, i = 1, 2, \ldots, n$, are such that $p_i > 0$, $\sum_{i=1}^{n} p_i = 1$, $a \leqslant x_i, y_i \leqslant b$, (1) holds (for some $c \in [a, b]$) and

$$\max(x_1, \ldots, x_n) \leq \min(y_1, \ldots, y_n), \tag{3}$$

then (2) holds.

(b) If for a continuous function f inequality (2) holds for all n, all $c \in [a, b]$, all $2n$ distinct points satisfying (1) and (3) and all weights $p_i > 0$ such that $\sum_{i=1}^{n} p_i = 1$, then f is 3-convex.

Pečarić [20] proved that one can weaken the assumption (3) and still guarantee that inequality (2) holds, i.e. the following result holds:

Theorem 3 If $f : [a, b] \rightarrow \mathbb{R}$ is 3-convex and $p_i, x_i, y_i, i = 1, 2, \ldots, n$, are such that $p_i > 0$, $\sum_{i=1}^{n} p_i = 1$, $a \leqslant x_i, y_i \leqslant b$, (1) holds (for some $c \in [a, b]$) and

$$x_i + x_{n-i+1} \leq 2c, \quad \frac{p_i x_i + p_{n-i+1} x_{n-i+1}}{p_i + p_{n-i+1}} \leq c, \quad i = 1, 2, \ldots, n, \tag{4}$$

then (2) holds.

The inequality from Theorem 3 for uniform weights $p_i = \frac{1}{n}$ was proven by Lawrence and Segalman [10]. A shorter proof of Lawrence and Segalman's result for a wider class of functions was obtained by Pečarić [21]. More recently, Hussain et al. [9] gave a refinement of the inequality from Theorem 3.

All of the generalizations of Levinson's inequality mentioned so far assume that (1) holds, i.e. that the distribution of the points x_i is equal to the distribution of the points y_i reflected around the point $c \in [a, b]$. Recently, Mercer [12] made a significant improvement by replacing this condition of symmetric distribution with the weaker one that the variances of the two sequences are equal.

Theorem 4 *If* $f : [a, b] \to \mathbb{R}$ *satisfies* $f''' \geq 0$ *and* $p_i, x_i, y_i, i = 1, 2, \ldots, n$, *are such that* $p_i > 0$, $\sum_{i=1}^n p_i = 1$, $a \leqslant x_i, y_i \leqslant b$, (3) *holds and*

$$\sum_{i=1}^n p_i (x_i - \overline{x})^2 = \sum_{i=1}^n p_i (y_i - \overline{y})^2, \tag{5}$$

then (2) holds.

Witkowski [24] showed that, similarly as before, the assumptions on differentiability of f can be weakened and for Theorem 4 to hold it is enough to assume that f is 3-convex. Furthermore, Witkowski weakened the assumption (5) as well and showed that equality of variances can be replaced by inequality in certain direction.

Theorem 5 *If* $f : (a, b) \to \mathbb{R}$ *is 3-convex,* $p_i > 0$ *for* $i = 1, 2, \ldots, n$, $\sum_{i=1}^n p_i = 1$, $a \leqslant x_i, y_i \leqslant b$ *are such that (3) holds and*

(a) $f''_-(\max x_i) > 0$ *and* $\sum_{i=1}^n p_i (x_i - \overline{x})^2 \leq \sum_{i=1}^n p_i (y_i - \overline{y})^2$, *or*
(b) $f''_+(\min y_i) < 0$ *and* $\sum_{i=1}^n p_i (x_i - \overline{x})^2 \geq \sum_{i=1}^n p_i (y_i - \overline{y})^2$, *or*
(c) $f''_-(\max x_i) \leq 0 \leq f''_+(\min y_i)$,

then (2) holds.

Witkowski [24] extended this result in several ways. Firstly, he showed that Levinson's inequality can be stated in a more general setting with random variables. Furthermore, he showed that it is enough to assume that f is 3-convex and that the assumption (5) of equality of the variances can be weakened to inequality in a certain direction. In the following, $\mathbb{E}(Z)$ and $\mathrm{Var}(Z)$ denote the expectation and variance, respectively, of a random variable Z.

Theorem 6 *Let I be an open interval of R (bounded or unbounded), $f : I \to R$ be a 3-convex function, and $X, Y : (\Omega, \mu) \to I$ be two random variables satisfying*

(i) $\mathbb{E}(X^2)$, $\mathbb{E}(Y^2)$, $\mathbb{E}(f(X))$, $\mathbb{E}(f(Y))$, $\mathbb{E}(f'(X))$, $\mathbb{E}(f'(Y))$, $\mathbb{E}(Xf'(X))$, $\mathbb{E}(Yf'(Y))$ *are finite,*
(ii) *ess sup* $X \leq$ *ess inf* Y,
(iii) $f''_+(\text{ess sup } X) > 0$ *and* $\mathrm{Var}(X) \leq \mathrm{Var}(Y)$, *or* $f''_-(\text{ess inf } Y) < 0$ *and* $\mathrm{Var}(X) \geq \mathrm{Var}(Y)$, *or* $f''_+(\text{ess sup } X) < 0 < f''_-(\text{ess inf } Y)$,

then

$$\mathbb{E}(f(X)) - f(\mathbb{E}(X)) \leq \mathbb{E}(f(Y)) - f(\mathbb{E}(Y)). \tag{6}$$

2 Generalization of Levinson's Inequality

In this section we will build on and extend the methods of Witkowski [24]. We will introduce a new class of functions $\mathcal{K}_1^c(a, b)$ that extends 3-convex functions and can be interpreted as functions that are "3-convex at point c".

Definition 1 Let $f : I \to \mathbb{R}$ be a real valued function on an arbitrary interval I in $\mathbb{R}$ and $c \in I^\circ$, where I° is the interior of I.

We say that $f \in \mathcal{K}_1^c(I)$ (resp. $f \in \mathcal{K}_2^c(I)$) if there exists a constant α such that the function $F(t) = f(t) - \frac{\alpha}{2}t^2$ is concave (resp. convex) on $I \bigcap (-\infty, c]$ and convex (resp. concave) on $I \bigcap [c, \infty)$.

Before stating our main results, we will introduce a new class of functions and show some of its properties.

Remark 1

(1) If $f \in \mathcal{K}_i^c(a, b)$, $i = 1, 2$, and $f''(c)$ exists, then $f''(c) = A$. If $f \in \mathcal{K}_1^c(a, b)$, we have

$$[x_1, x_2, x_3]F = [x_1, x_2, x_3]f - \frac{A}{2} \le 0 \le [y_1, y_2, y_3]f - \frac{A}{2} = [y_1, y_2, y_3]F.$$

Therefore, if $f''_-(c)$ and $f''_+(c)$ exist, letting $x_j \nearrow c$ and $y_j \searrow c$, we get $f''_-(c) \le A \le f''_+(c)$.

(2) If $f : (a, b) \to \mathbb{R}$ is 3-convex (3-concave), then $f \in \mathcal{K}_1^c(a, b)$ ($f \in \mathcal{K}_2^c(a, b)$) for every $c \in (a, b)$. Indeed, if f is 3-convex, then f', f''_- and f''_+ exist and f' is convex (see [22]). Hence, for every $\alpha_1, \alpha_2 \in (a, c]$, $\beta_1, \beta_2 \in [c, b)$ and $A \in [f''_-(c), f''_+(c)]$ the function $F(x) = f(x) - \frac{A}{2}x^2$ satisfies

$$\frac{F'(\alpha_2) - F'(\alpha_1)}{\alpha_2 - \alpha_1} \le 0 \le \frac{F'(\beta_2) - F'(\beta_1)}{\beta_2 - \beta_1},$$

so F' is nonincreasing on $(a, c]$ and nondecreasing on $[c, b)$. The next theorem shows that this property characterizes 3-convex (3-concave) functions.
    On the other hand, $f(x) = x^4$ is an example of a function that belongs to $\mathcal{K}_1^2(-1, 3)$, but is not 3-convex on $(-1, 3)$. Furthermore, $f(x) = |x|$ is an example of a function that belongs to $\mathcal{K}_1^0(-1, 1)$, but f is not differentiable at zero, a point in the interval $(-1, 1)$.

(3) If $f \in \mathcal{K}_1^c(a, b)$ ($f \in \mathcal{K}_2^c(a, b)$) for every $c \in (a, b)$, then f is 3-convex (3-concave), see [3, Theorem 2.4].

Taking into account (2) and (3), we can describe the property from the definition of $\mathcal{K}_1^c(a, b)$ as "3-convexity at point c". Therefore, we have shown that a function f is 3-convex on (a, b) if and only if it is 3-convex at every $c \in (a, b)$.

We will generalize Theorem 4 by weakening the assumptions on the function f.

Theorem 7 *Let $a < x_i \leq c \leq y_i < b$, $p_i > 0$ for $i = 1, 2, \ldots, n$, $\sum_{i=1}^{n} p_i = 1$ and (5) holds. If $f \in \mathcal{K}_1^c(a, b)$, then inequality (2) holds and if $f \in \mathcal{K}_2^c(a, b)$, then (2) holds with reverse sign of inequality.*

Proof For $0 \leq t \leq 1$, let $x_i(t) = \bar{x} + t(x_i - \bar{x})$ and $y_i(t) = \bar{y} + t(y_i - \bar{y})$. If $f \in \mathcal{K}_1^c(a, b)$, then the function

$$U(t) := \sum_{i=1}^{n} p_i f(y_i(t)) - f(\bar{y}) - \sum_{i=1}^{n} p_i f(x_i(t)) + f(\bar{x})$$

is convex and $U'_+(0) \geq 0$ (see [3, Theorem 2.46]). So $U(0) \leq U(1)$, which is inequality (2). $\qquad \blacksquare$

The following theorem represents a probabilistic version of Levinson's inequality under the assumption of equal variances.

Theorem 8 *Let $X, Y : \Omega \rightarrow I$ be two random variables such that*

$$\text{Var}(X) = \text{Var}(Y) < \infty \tag{7}$$

and that there exists $c \in I^{\circ}$ such that

$$\text{ess sup } X \leq c \leq \text{ess inf } Y. \tag{8}$$

Then for every $f \in \mathcal{K}_1^c(I)$ such that $\mathbb{E}(f(X))$ and $\mathbb{E}(f(Y))$ are finite inequality (6) holds.

Proof Let $F(x) = f(x) - \frac{A}{2}x^2$, where A is the constant from Definition 1. Since $F : I \cap (-\infty, c] \rightarrow \mathbb{R}$ is concave, Jensen's inequality implies

$$0 \leq F(\mathbb{E}(X)) - \mathbb{E}(F(X)) = f(\mathbb{E}(X)) - \mathbb{E}(f(X)) + \frac{A}{2}\text{Var}(X). \tag{9}$$

Similarly, $F : I \cap [c, \infty) \rightarrow \mathbb{R}$ is convex, so

$$0 \leq \mathbb{E}(f(Y)) - f(\mathbb{E}(Y)) - \frac{A}{2}\text{Var}(Y). \tag{10}$$

Adding up (9) and (10) we obtain that (6) holds. $\qquad \blacksquare$

2.1 Mean Value Theorems

Notice that Levinson's inequality (6) is linear in f. This motivates us to define the following linear functional: for fixed random variables $X, Y : \Omega \rightarrow I$ and $c \in I^{\circ}$ such that (7) and (8) hold, we define

$$\Lambda(f) = \mathbb{E}(f(Y)) - f(\mathbb{E}(Y)) - \mathbb{E}(f(X)) + f(\mathbb{E}(X)) \qquad (11)$$

for functions $f : I \to \mathbb{R}$ such that $\mathbb{E}(f(X))$ and $\mathbb{E}(f(Y))$ are finite. Notice that Theorem 8 guarantees that $\Lambda(f) \geq 0$ for $f \in \mathcal{K}_1^c(I)$.

We will give two mean value results.

Theorem 9 *Let* $-\infty < a < c < b < \infty$, $I = [a, b]$, $X, Y : \Omega \to I$ *be two random variables such that (7) and (8) hold, and let* Λ *be given by (11). Then for* $f \in C^3([a, b])$ *there exists* $\xi \in [a, b]$ *such that*

$$\Lambda(f) = \frac{f'''(\xi)}{6} \left[\mathbb{E}(Y^3 - X^3) - \mathbb{E}^3(Y) + \mathbb{E}^3(X) \right]. \qquad (12)$$

Proof Since f is bounded, $\mathbb{E}(f(X))$ and $\mathbb{E}(f(Y))$ are finite and $\Lambda(f)$ is well defined. Furthermore, since $f \in C^3([a, b])$, there exist $m = \min_{x \in [a,b]} f'''(x)$ and $M = \max_{x \in [a,b]} f'''(x)$. The functions $f_1(x) = f(x) - \frac{m}{6}x^3$ and $f_2(x) = \frac{M}{6}x^3 - f(x)$ are 3-convex. Hence, by Theorem 8 we have $\Lambda(f_i) \geq 0$, $i = 1, 2$, and we get

$$\frac{m}{6}\Lambda(id^3) \leq \Lambda(f) \leq \frac{M}{6}\Lambda(id^3), \qquad (13)$$

where $id(x) = x$. Since id^3 is 3-convex, by Theorem 8 we have

$$0 \leq \Lambda(id^3) = \mathbb{E}(Y^3 - X^3) - \mathbb{E}^3(Y) + \mathbb{E}^3(X).$$

If $\Lambda(id^3) = 0$, then (13) implies $\Lambda(f) = 0$ and (12) holds for every $\xi \in [a, b]$. Otherwise, dividing (13) by $0 < \Lambda(id^3)/6$ we get

$$m \leq \frac{6\Lambda(f)}{\Lambda(id^3)} \leq M,$$

so continuity of f''' insures existence of $\xi \in [a, b]$ satisfying (12).

Theorem 10 *Let* I, c, X, Y, *and* Λ *be as in Theorem 9 and let* $f, g \in C^3([a, b])$. *If* $\Lambda(g) \neq 0$, *then there exists* $\xi \in [a, b]$ *such that either*

$$\frac{\Lambda(f)}{\Lambda(g)} = \frac{f'''(\xi)}{g'''(\xi)},$$

or $f'''(\xi) = g'''(\xi) = 0$.

Proof Define $h \in C^3([a, b])$ by $h(x) = \alpha f(x) - \beta g(x)$, where $\alpha = \Lambda(g)$, $\beta = \Lambda(f)$. Due to the linearity of Λ we have $\Lambda(h) = 0$. Now, by Theorem 9 there exist $\xi, \xi_1 \in [a, b]$ such that

$$0 = \Lambda(h) = \frac{h'''(\xi)}{6}\Lambda(id^3) \quad \text{and} \quad 0 \neq \Lambda(g) = \frac{g'''(\xi_1)}{6}\Lambda(id^3).$$

Therefore, $\Lambda(id^3) \neq 0$ and $0 = h'''(\xi) = \alpha f'''(\xi) - \beta g'''(\xi)$, which gives the claim of the theorem.

Remark 2 Theorems 9 and 10 are generalizations of mean value results from [1]. Indeed, let $I = [0, 2a]$, $c = a$ be the midpoint of the segment and X be the discrete random variable taking values $x_i \in [0, c]$ with probabilities p_i, $i = 1, \ldots, n$. The random variables $Y_1 = 2a - X$ and $Y_2 = X + a$ satisfy $\text{Var}(Y_1) = \text{Var}(Y_2) = \text{Var}(X)$. The results from [1] can be recovered by applying Theorems 9 and 10 with the pair of random variables X and Y_1 or X and Y_2.

2.2 Exponential Convexity

In this subsection we will give refinements of the results obtained in the above subsection by constructing certain exponentially convex functions.

We will first give some basic definitions and results on exponential convexity that we will use in this subsection.

Definition 2 A function $g : I \rightarrow \mathbb{R}$, where I is an interval in $\mathbb{R}$, is *n*-exponentially convex in the Jensen sense on I if

$$\sum_{i,j=1}^{n} \xi_i \xi_j g \left(\frac{x_i + x_j}{2} \right) \geq 0$$

holds for all choices $\xi_i \in \mathbb{R}$ and $x_i \in I$, $i = 1, \ldots, n$.

A function $g : I \rightarrow \mathbb{R}$ is *n*-exponentially convex on I if it is *n*-exponentially convex in the Jensen sense and continuous on I.

Definition 3 A function $g : I \rightarrow \mathbb{R}$ is exponentially convex in the Jensen sense on I if it is *n*-exponentially convex in the Jensen sense on I for every $n \in \mathbb{N}$. A function $g : I \rightarrow \mathbb{R}$ is exponentially convex on I if it is exponentially convex in the Jensen sense and continuous on I.

Remark 3 A function $g : I \rightarrow \mathbb{R}$ is log-convex in the Jensen sense, i.e.

$$g \left(\frac{x_1 + x_2}{2} \right)^2 \leq g(x_1) g(x_2), \quad \text{for all } x_1, x_2 \in I, \tag{14}$$

if and only if

$$\xi_1^2 g(x_1) + 2 \xi_1 \xi_2 g \left(\frac{x_1 + x_2}{2} \right) + \xi_2^2 g(x_2) \geq 0$$

holds for every $\xi_1, \xi_2 \in \mathbb{R}$ and $x_1, x_2 \in I$, i.e., if and only if g is 2-exponentially convex in the Jensen sense. If $g(x_1) = 0$ for some x_1 and $[a, b] \subset I$ is an arbitrary interval containing x_1, then it follows from (14) and non-negativity of g that g vanishes on $[a_1, b_1]$, where $a_1 = (a + x_1)/2$ and $b_1 = (x_1 + b)/2$. Applying the same reasoning to intervals $[a, a_1]$ and $[b_1, b]$ we obtain sequences $a_n \searrow a$ and $b_n \nearrow b$ with g vanishing on $[a_n, b_n]$. Thus g is zero on (a, b) and a function that is 2-exponentially convex in the Jensen sense is either identically equal to zero or it is strictly positive and log-convex in the Jensen sense.

The following results will enable us to construct exponentially convex functions.

Theorem 11 *Let $X, Y : \Omega \to I$ be two random variables and $c \in I^\circ$ such that (7) and (8) hold and let Λ be given by (11). Furthermore, let $\Upsilon = \{f_t : I \to \mathbb{R} \mid t \in J\}$, where J is an interval in $\mathbb{R}$, be a family of functions such that, for every $t \in J$, $\mathbb{E}(f_t(X))$ and $\mathbb{E}(f_t(Y))$ are finite and for every four mutually different points $u_0, u_1, u_2, u_3 \in I$ the mapping $t \mapsto [u_0, u_1, u_2, u_3]f_t$ is n-exponentially convex. Then the mapping $t \mapsto \Lambda(f_t)$ is n-exponentially convex in the Jensen sense on J. If the mapping $t \mapsto \Lambda(f_t)$ is continuous on J, then it is n-exponentially convex on J.*

Proof For $\xi_i \in \mathbb{R}$ and $t_i \in J, i = 1, \ldots, n$, we define the function

$$f(x) = \sum_{i,j=1}^{n} \xi_i \xi_j f_{\frac{t_i + t_j}{2}}(x).$$

Due to linearity of the divided differences and the assumption that the function $t \mapsto [u_0, u_1, u_2, u_3]f_t$ is n-exponentially convex in the Jensen sense we have

$$[u_0, u_1, u_2, u_3]f = \sum_{i,j=1}^{n} \xi_i \xi_j [u_0, u_1, u_2, u_3]f_{\frac{t_i + t_j}{2}} \geq 0.$$

This implies that f is 3-convex, so $f \in \mathcal{K}_1^c(I)$. Due to linearity of the expectation, $\mathbb{E}(f(X))$ and $\mathbb{E}(f(Y))$ are finite, so by Theorem 8

$$0 \leq \Lambda(f) = \sum_{i,j=1}^{n} \xi_i \xi_j \Lambda(f_{\frac{t_i + t_j}{2}}).$$

Therefore, the mapping $t \mapsto \Lambda(f_t)$ is n-exponentially convex. If it is also continuous, it is n-exponentially convex by definition. $\square$

If the assumptions of Theorem 11 hold for all $n \in \mathbb{N}$, then we immediately get the following corollary.

Corollary 1 *Let $X, Y, c,$ and Λ be as in Theorem 11. Furthermore, let $\Upsilon = \{f_t : I \to \mathbb{R} \mid t \in J\}$, where J is an interval in $\mathbb{R}$, be a family of functions such that, for every $t \in J$, $\mathbb{E}(f_t(X))$ and $\mathbb{E}(f_t(Y))$ are finite and for every four mutually different points $u_0, u_1, u_2, u_3 \in I$ the mapping $t \mapsto [u_0, u_1, u_2, u_3]f_t$ is exponentially convex. Then the mapping $t \mapsto \Lambda(f_t)$ is exponentially convex in the Jensen sense on J. If the mapping $t \mapsto \Lambda(f_t)$ is continuous on J, then it is exponentially convex on J.*

Corollary 2 *Let X, Y, c, and Λ be as in Theorem 11. Furthermore, let $\Upsilon = \{f_t : I \to \mathbb{R} \mid t \in J\}$, where J is an interval in $\mathbb{R}$, be a family of functions such that, for every $t \in J$, $\mathbb{E}(f_t(X))$ and $\mathbb{E}(f_t(Y))$ are finite and for every four mutually different points $u_0, u_1, u_2, u_3 \in I$ the mapping $t \mapsto [u_0, u_1, u_2, u_3] f_t$ is 2-exponentially convex in the Jensen sense. Then the following statements hold:*

1. *If the mapping $t \mapsto \Lambda(f_t)$ is continuous on J, then for $r, s, t \in J$ such that $r < s < t$, we have*

$$\Lambda(f_s)^{t-r} \le \Lambda(f_r)^{t-s} \Lambda(f_t)^{s-r}. \tag{15}$$

2. *If the mapping $t \mapsto \Lambda(f_t)$ is strictly positive and differentiable on J, then for all $s, t, u, v \in J$ such that $s \le u$ and $t \le v$ we have*

$$\mu_{s,t}(\Upsilon) \le \mu_{u,v}(\Upsilon),$$

where

$$\mu_{s,t}(\Upsilon) = \begin{cases} \left(\frac{\Lambda(f_s)}{\Lambda(f_t)} \right)^{\frac{1}{s-t}}, & s \ne t, \\ \exp \left(\frac{\frac{d}{ds}(\Lambda(f_s))}{\Lambda(f_s)} \right), & s = t. \end{cases} \tag{16}$$

Consider now the family of functions

$$\Upsilon_1 = \{f_t : I \to \mathbb{R} \mid t \in \mathbb{R}\}, \qquad I \subset (0, \infty),$$

defined by

$$f_t(x) = \begin{cases} \frac{x^t - \frac{t(t-1)}{2} x^2 + t(t-2)x - \frac{(t-1)(t-2)}{2}}{t(t-1)(t-2)}, & t \ne 0, 1, 2 \\ \frac{1}{2} \ln x, & t = 0, \\ -x \ln x, & t = 1, \\ \frac{1}{2} x^2 \ln x, & t = 2. \end{cases} \tag{17}$$

The functions f_t are 3-convex since $f_t'''(x) = x^{t-3} \ge 0$. Moreover, the function

$$f(x) = \sum_{i,j=1}^{n} \xi_i \xi_j f_{\frac{t_i+t_j}{2}}(x)$$

satisfies

$$f'''(x) = \sum_{i,j=1}^{n} \xi_i \xi_j f_{\frac{t_i+t_j}{2}}'''(x) = \left(\sum_{i=1}^{n} \xi_i e^{\frac{t_i-3}{2} \ln x} \right)^2 \ge 0,$$

so f is 3-convex. Therefore

$$0 \le [u_0, u_1, u_2, u_3]f = \sum_{i,j=1}^{n} \xi_i \xi_j [u_0, u_1, u_2, u_3] f_{\frac{t_i + t_j}{2}}$$

so the mapping $t \mapsto [u_0, u_1, u_2, u_3]f_t$ is n-exponentially convex in the Jensen sense. As this holds for all $n \in \mathbb{N}$, we see that the family Υ_1 satisfies the assumptions of Corollary 1. For the remainder of this section we assume that $\mathbb{E}(f_t(X))$ and $\mathbb{E}(f_t(Y))$ are finite for all f_t given by (17). Hence, by Corollary 1, the mapping $t \mapsto \Lambda(f_t)$ is exponentially convex in the Jensen sense. It is straightforward to check that it is also continuous, so the mapping $t \mapsto \Lambda(f_t)$ is exponentially convex. An immediate consequence of Corollary 2 (i) is the following result.

Corollary 3 *Let* $I \subset (0, \infty)$, $c \in I°$, *and let* $X, Y : \Omega \to I$ *be two random variables such that (7) and (8) hold. If* $\mathbb{E}(Y^t - X^t) - \mathbb{E}^t(Y) + \mathbb{E}^t(X) \ne 0$ *for some* $t \in \mathbb{R}\backslash\{0, 1, 2\}$, *then for all* $r, s, t \in \mathbb{R}\backslash\{0, 1, 2\}$ *such that* $r < s < t$ *we have*

$$\frac{\mathbb{E}(Y^t - X^t) - \mathbb{E}^t(Y) + \mathbb{E}^t(X)}{t(t-1)(t-2)} \ge \left(\frac{\mathbb{E}(Y^s - X^s) - \mathbb{E}^s(Y) + \mathbb{E}^s(X)}{s(s-1)(s-2)} \right)^{\frac{t-r}{s-r}} \cdot$$

$$\cdot \left(\frac{\mathbb{E}(Y^r - X^r) - \mathbb{E}^r(Y) + \mathbb{E}^r(X)}{r(r-1)(r-2)} \right)^{\frac{s-t}{s-r}} > 0. \qquad (18)$$

Applying Theorem 10 for the functions $f = f_t$ and $g = f_s$ given by (17) and defined on a segment $I = [a, b] \subset (0, \infty)$, we conclude that there exist $\xi \in I$ such that

$$\xi = \left(\frac{f_s'''}{f_t'''} \right)^{-1} \left(\frac{\Lambda(f_s)}{\Lambda(f_t)} \right) = \left(\frac{\Lambda(f_s)}{\Lambda(f_t)} \right)^{\frac{1}{s-t}}, \quad s \ne t.$$

Moreover, $\mu_{s,t}(\Upsilon_1)$ given by (16) for the family Υ_1 can be calculated in the limiting cases $s \to t$ as well and equal

$$\mu_{s,t}(\Upsilon_1) = \begin{cases} \left(\frac{\Lambda(f_s)}{\Lambda(f_t)} \right)^{\frac{1}{s-t}}, & s \ne t, \\ \exp\left(\frac{2\Lambda(f_s f_0)}{\Lambda(f_0)} - \frac{3s^2 - 6s + 2}{s(s-1)(s-2)} \right), & s = t \ne 0, 1, 2, \\ \exp\left(\frac{\Lambda(f_0^2)}{\Lambda(f_0)} + \frac{3}{2} \right), & s = t = 0, \\ \exp\left(\frac{\Lambda(f_0 f_1)}{\Lambda(f_1)} \right), & s = t = 1, \\ \exp\left(\frac{\Lambda(f_0 f_2)}{\Lambda(f_2)} - \frac{3}{2} \right), & s = t = 2. \end{cases}$$

By Corollary 2 (ii), $\mu_{s,t}(\Upsilon_1)$ are monotone in parameters s and t.

Remark 4 By applying Corollary 1 to the family of functions Υ_1 given by (17) and the pair of discrete random variables X and Y_1 or X and Y_2 from Remark 2 we conclude that the mapping $t \mapsto [u_0, u_1, u_2, u_3] f_t$ is exponentially convex which generalizes the result from [2] where the log-convexity of the mapping was proven. Also, the inequalities obtained in [2] can be recovered from Corollary 2 (*i*). Furthermore, $\mu_{s,t}(\Upsilon_1)$ applied for the same family of functions and random variables yield the Cauchy means obtained in [1].

2.3   *A Monotonic Refinement of Levinson's Inequality*

In this subsection we will construct the corresponding two mappings in connection with Levinson's inequality and show their monotonicity and convexity properties.

Theorem 12 *Let $f : [a, b] \to \mathbb{R}$ be 3-convex at point c, $x : \Omega \to [a, c]$ and $y : \Omega \to [c, b]$ such that $\mathrm{Var}(x) = \mathrm{Var}(y)$ and $H, V : [0, 1] \to \mathbb{R}$ the mappings*

$$H(t) = \frac{1}{\mu(\Omega)} \int_\Omega [f(ty(s) + (1 - t)\mathbb{E}[y]) - f(tx(s) + (1 - t)\mathbb{E}[x])] \, d\mu(s)$$

and

$$V(t) = \frac{1}{\mu(\Omega)^2} \int_\Omega \int_\Omega [f(ty(s) + (1 - t)y(u)) - f(tx(s) + (1 - t)x(u))] \, d\mu(s)d\mu(u).$$

Then:

(a) *the mappings H and V are convex on $[0, 1]$,*
(b) *the mapping H is nondecreasing on $[0, 1]$, while the mapping V is nonincreasing on $[0, \frac{1}{2}]$ and nondecreasing on $[\frac{1}{2}, 1]$,*
(c) *the following equalities hold:*

$$\inf_{t \in [0,1]} H(t) = H(0) = f(\mathbb{E}[y]) - f(\mathbb{E}[x]),$$

$$\sup_{t \in [0,1]} H(t) = H(1) = \mathbb{E}[f(y)] - \mathbb{E}[f(x)]),$$

$$\inf_{t \in [0,1]} V(t) = V(\frac{1}{2}) = \frac{1}{\mu(\Omega)^2} \iint_{\Omega\Omega} \left[f\left(\frac{y(s) + y(u)}{2}\right) - f\left(\frac{x(s) + x(u)}{2}\right) \right]$$

$$d\mu(s)d\mu(u),$$

$$\sup_{t \in [0,1]} V(t) = V(0) = V(1) = \mathbb{E}[f(y)] - \mathbb{E}[f(x)]),$$

(d) *$V(t) \geq \max\{H(t), H(1 - t)\}$ holds for all $t \in [0, 1]$.*

Proof Since the function y takes values in $[c, b]$, so does the function $y_{(t)} = ty + (1 - t)\mathbb{E}[y]$ for every $t \in [0, 1]$. Furthermore, since the function F is convex on $[c, b]$, by [6] the mapping

$$H_1(t) = \frac{1}{\mu(\Omega)} \int_\Omega F(ty(s) + (1 - t)\mathbb{E}[y]) \, d\mu(s)$$

is convex an nondecreasing on $[0, 1]$, and we have

$$H_1(t) = \mathbb{E}[f(y_{(t)})] - \frac{A}{2}t^2 \mathrm{Var}(y) - \frac{A}{2}\mathbb{E}^2[y].$$

Similarly, the function $x_{(t)} = tx + (1 - t)\mathbb{E}[x]$ takes values in $[a, c]$ for every $t \in [0, 1]$ and $-F$ is convex on $[a, c]$, so by [6] the mapping

$$H_2(t) = -\frac{1}{\mu(\Omega)} \int_\Omega F(tx(s) + (1 - t)\mathbb{E}[x]) \, d\mu(s)$$

is convex and nondecreasing on $[0, 1]$, and we have

$$H_2(t) = -\mathbb{E}[f(x_{(t)})] + \frac{A}{2}t^2 \mathrm{Var}(x) + \frac{A}{2}\mathbb{E}^2[x].$$

Let us also denote the (constant) mapping $H_3(t) = \frac{A}{2}\left(\mathbb{E}^2[y] - \mathbb{E}^2[x]\right)$. All three of the mappings H_i, $i = 1, 2, 3$, are convex and nondecreasing and, therefore, so is their sum. Since $\mathrm{Var}(x) = \mathrm{Var}(y)$ we have $H = H_1 + H_2 + H_3$ and this proves the convexity and monotonicity properties of H from parts (a) and (b), while the first two equalities in (c) follow by simple calculation.

As for the mapping V, first of all, it is easy to see that $V(t) = V(1 - t)$ for all $t \in [0, 1]$, that is, V is symmetric with respect to $t = \frac{1}{2}$. Next, since y takes values in $[c, b]$ and F is convex on that interval, then the mapping

$$V_1(t) = \frac{1}{\mu(\Omega)^2} \int_\Omega \int_\Omega F(ty(s) + (1 - t)y(u)) \, d\mu(s)d\mu(u)$$

is convex on $[0, 1]$, nondecreasing on $[\frac{1}{2}, 1]$ and we have

$$V_1(t) = \frac{1}{\mu(\Omega)^2} \iint_{\Omega\Omega} f(ty(s)+(1-t)y(u)) \, d\mu(s)d\mu(u)+At(1-t)\mathrm{Var}(y)-\frac{A}{2}\mathbb{E}[y^2].$$

Similarly, since x takes values in $[a, c]$ and $-F$ is convex on that interval, then the mapping

$$V_2(t) = -\frac{1}{\mu(\Omega)^2} \int_\Omega \int_\Omega F(tx(s) + (1 - t)x(u)) \, d\mu(s)d\mu(u)$$

is convex on $[0, 1]$, nondecreasing on $[\frac{1}{2}, 1]$ and we have

$$V_2(t) = -\frac{1}{\mu(\Omega)^2} \iint_{\Omega\Omega} f(tx(s)+(1-t)x(u))\, d\mu(s)d\mu(u) - At(1-t)\mathrm{Var}(x) + \frac{A}{2}\mathbb{E}[x^2].$$

Let us also denote the (constant) mapping $V_3(t) = \frac{A}{2}\left(\mathbb{E}[y^2] - \mathbb{E}[x^2]\right)$. All three of the mappings $V_i, i = 1, 2, 3$, are convex and nondecreasing on $[\frac{1}{2}, 1]$ and, therefore, so is their sum. Since $\mathrm{Var}(x) = \mathrm{Var}(y)$ we have $V = V_1 + V_2 + V_3$. Furthermore, since V is symmetric around $t = \frac{1}{2}$, it follows that it is nonincreasing on $[0, \frac{1}{2}]$, its minimum is attained at $t = \frac{1}{2}$ and its maximum is attained at $t = 0$ and $t = 1$. This proves the convexity and monotonicity properties of V.

Finally, as for part (d), since V is symmetric around $t = \frac{1}{2}$ and H is nondecreasing, it is enough to prove that $V(t) \geq H(t)$ for $t \in [\frac{1}{2}, 1]$. This inequality holds since $V_1(t) \geq H_1(t)$ and $V_2(t) \geq H_2(t)$ by Theorem by [6] (d) and $V_3(t) = H_3(t)$ since $\mathrm{Var}(x) = \mathrm{Var}(y)$ and this finishes the proof.

A monotonic refinement of Levinson's inequality (6) based on Theorem 12 is the following: if $x_{(t)}$ and $y_{(t)}$ for $t \in [0, 1]$ are as in the proof of Theorem 12, then $H(t) = \mathbb{E}[f(y_{(t)})] - \mathbb{E}[f(x_{(t)})])$ and for $0 \leq s \leq t \leq 1$ it holds

$$f(\mathbb{E}[y]) - f(\mathbb{E}[x]) = H(0) \leq \mathbb{E}[f(y_{(s)})] - \mathbb{E}[f(x_{(s)})])$$
$$\leq \mathbb{E}[f(y_{(t)})] - \mathbb{E}[f(x_{(t)})]) \leq H(1) = \mathbb{E}[f(y)] - \mathbb{E}[f(x)]).$$

3 Levinson's Operator Inequality

In this section we give Levinson's operator inequality and Levinson's mappings.

Let $\mathcal{B}(H)$ be the algebra of all bounded linear operators on a complex Hilbert space H. We denote by $\mathcal{B}_h(H)$ the real subspace of all self-adjoint operators on H.

We assume that the reader is familiar with basic notions about operator theory.

If a function f is operator convex, then the so-called *Choi-Davis-Jensen's inequality* (or in short *Jensen's operator inequality*) $f(\Phi(X)) \leq \Phi(f(X))$ holds for any unital positive linear mapping Φ on $\mathcal{B}(H)$ and any $X \in \mathcal{B}_h(H)$ with spectrum contained in I. Many other versions of Jensen's operator inequality can be found in [7, 8].

Now we give the definition of classes of functions for which we observe Levinson's operator inequality.

Definition 4 Let $f \in C(I)$ be a real valued function on an arbitrary interval I in $\mathbb{R}$ and $c \in I^\circ$, where I° is the interior of I.

We say that $f \in \mathcal{K}_1^c(I)$ (resp. $f \in \mathcal{K}_2^c(I)$) if F is operator concave (resp. operator convex) on $I \cap (-\infty, c]$ and operator convex (resp. operator concave) on $I \cap [c, \infty)$.

3.1 Levinson's Inequality with Operator Convexity and Concavity

First, we observe Levinson's operator inequality for $f \in \overset{\bullet}{\mathcal{K}}_1^c(I)$.

Theorem 13 *Let $(X_1, \ldots, X_n)$ be an n-tuple and $(Y_1, \ldots, Y_k)$ be a k-tuple of self-adjoint operators $X_i, Y_j \in \mathcal{B}_h(H)$ with spectra contained in $[m_x, M_x]$ and $[m_y, M_y]$, respectively, such that $a < m_x \leq M_x \leq c \leq m_y \leq M_y < b$ for some $a, b, c \in \mathbb{R}$. Let $(\Phi_1, \ldots, \Phi_n)$ be a unital n-tuple and $(\Psi_1, \ldots, \Psi_k)$ be a unital k-tuple of positive linear mappings $\Phi_i, \Psi_j : \mathcal{B}(H) \to \mathcal{B}(K)$.*

If $f \in \overset{\bullet}{\mathcal{K}}_1^c((a, b))$ and $\alpha X \leq \alpha Y$, then

$$\sum_{i=1}^{n} \Phi_i\big(f(X_i)\big) - f\Big(\sum_{i=1}^{n} \Phi_i(X_i)\Big) \leq \frac{\alpha}{2} X \leq \frac{\alpha}{2} Y \leq \sum_{j=1}^{k} \Psi_j\big(f(Y_j)\big) - f\Big(\sum_{j=1}^{k} \Psi_j(Y_j)\Big),$$
(19)

where $X := \sum_{i=1}^{n} \Phi_i\big(X_i^2\big) - \Big(\sum_{i=1}^{n} \Phi_i(X_i)\Big)^2$, $Y := \sum_{j=1}^{k} \Psi_j\big(Y_j^2\big) - \Big(\sum_{j=1}^{k} \Psi_j(Y_j)\Big)^2$.

If $f \in \overset{\bullet}{\mathcal{K}}_2^c((a, b))$ and $\alpha X \geq \alpha Y$, then the reverse inequalities are valid in (19).

Proof If $f \in \overset{\bullet}{\mathcal{K}}_1^c((a, b))$, then there is a constant α such that $F(t) = f(t) - \frac{\alpha}{2} t^2$ is operator concave on $[m_x, c] \subset (a, c]$. Then the reverse of Jensen's operator inequality gives

$$0 \leq f\Big(\sum_{i=1}^{n} \Phi_i(X_i)\Big) - \frac{\alpha}{2}\Big(\sum_{i=1}^{n} \Phi_i(X_i)\Big)^2 - \sum_{i=1}^{n} \Phi_i\big(f(X_i)\big) + \frac{\alpha}{2} \sum_{i=1}^{n} \Phi_i\big(X_i^2\big).$$

It follows

$$\sum_{i=1}^{n} \Phi_i\big(f(X_i)\big) - f\Big(\sum_{i=1}^{n} \Phi_i(X_i)\Big) \leq \frac{\alpha}{2} X.$$
(20)

Also, since F is operator convex on $[c, M_y] \subset [c, b)$, Jensen's operator inequality gives

$$\frac{\alpha}{2} Y \leq \sum_{j=1}^{k} \Psi_j\big(f(Y_j)\big) - f\Big(\sum_{j=1}^{k} \Psi_j(Y_j)\Big).$$
(21)

Combining (20) and (21) and taking into account that $\alpha X \leq \alpha Y$, we obtain (19).

Next, we give the following obvious corollary to Theorem 13 with convex combinations of operators X_i, $i = 1, \ldots, n$ and Y_j, $j = 1, \ldots, k$. This is a generalization of [3, Theorem 2.6].

Corollary 4 *Let operators and a, b, c be as in Theorem 13 and $(p_1, \ldots, p_n)$ be an n-tuple and $(q_1, \ldots, q_k)$ be a k-tuple of positive scalars such that $\sum_{i=1}^{n} p_i = 1$ and $\sum_{j=1}^{k} q_j = 1$.*

If $f \in \overset{\bullet}{\mathcal{K}_1^c}((a, b))$ and $\alpha P \le \alpha Q$, then

$$\sum_{i=1}^{n} p_i f(X_i) - f(\bar{X}) \le \frac{\alpha}{2} P \le \frac{\alpha}{2} Q \le \sum_{j=1}^{k} q_j f(Y_j) - f(\bar{Y}), \tag{22}$$

where $P := \sum_{i=1}^{n} p_i (X_i - \bar{X})^2$, $Q := \sum_{j=1}^{k} q_j (Y_j - \bar{Y})^2$, $\bar{X} := \sum_{i=1}^{n} p_i X_i$, $\bar{Y} := \sum_{j=1}^{k} q_j Y_j$ denote the weighted arithmetic means of operators.

If $f \in \overset{\bullet}{\mathcal{K}_2^c}([m_x, M_y])$ and $\frac{\alpha}{2} P \ge \frac{\alpha}{2} Q$, then reverse inequalities are valid in (22).

Remark 5

(a) If f is convex, then $f''_-(c) \le \alpha \le f''_+(c)$ (see [3]). So, condition $\alpha X \le \alpha Y$ (resp. $\alpha X \ge \alpha Y$) in Theorem 13 can be weakened to $X \le Y$ (resp. $Y \le X$).

(b) Setting $n = k$, $p_i = q_i$ and $A := \sum_{i=1}^{n} p_i (X_i - \bar{X})^2 = \sum_{i=1}^{n} p_i (Y_i - \bar{Y})^2$ in Corollary 4, we get the operators version of (2) and a generalization of Mercer's result given in Theorem 4.

3.2 Levinson's Inequality Without Operator Concavity and Convexity

In this subsection we give Levinson's operator inequality for $f \in \mathcal{K}_i^c(I)$, $i = 1, 2$.

Operator convexity plays an essential role in (19). This inequality will be false if we replace an operator convex function by a general convex function (see [18, Counterexample 1]). Now, we give a general result when Levinson's operator inequality holds for $f \in \mathcal{K}_1^c([m_x, M_y])$ with conditions on the spectra of operators. There have been many interesting works devoted to obtain operator inequalities under spectra conditions. The reader is referred to [16, 19] and the references therein.

Theorem 14 *Let mappings and operators be as in Theorem 13, m_{X_i}, M_{X_i} be bounds of X_i and m_{Y_j}, M_{Y_j} be bounds of Y_j, such that $a < m_{X_i} \le M_{X_i} \le c \le m_{Y_j} \le M_{Y_j} < b$ for some $a, b, c \in \mathbb{R}$, $i = 1, \ldots, n$, $j = 1, \ldots, k$, and let m_X, M_X and m_Y, M_Y be bounds of $\bar{X} = \sum_{i=1}^{n} \Phi_i(X_i)$ and $\bar{Y} = \sum_{j=1}^{k} \Psi_j(Y_j)$, respectively, such that*

$$(m_X, M_X) \cap [m_{X_i}, M_{X_i}] = \varnothing, \quad i = 1, \ldots, n,$$
$$(m_Y, M_Y) \cap [m_{Y_j}, M_{Y_j}] = \varnothing, \quad j = 1, \ldots, k. \tag{23}$$

If $f \in \mathcal{K}_1^c((a, b))$ and $\alpha X \leq \alpha Y$ hold, then (19) is valid, where

$$X := \sum_{i=1}^{n} \Phi_i(X_i^2) - \bar{X}^2 \quad \text{and} \quad Y := \sum_{j=1}^{k} \Psi_j(Y_j^2) - \bar{Y}^2.$$

If $f \in \mathcal{K}_2^c((a, b))$ and $\alpha X \geq \alpha Y$ hold, then reverse inequalities are valid in (19).

Proof The proof is similar to the one for Theorem 13, when we apply Jensen's operator inequality without operator convexity and concavity (see [16, Theorem 1.]).

As an application of Theorem 14, we obtain many interesting inequalities. For example, we obtain the following inequalities for some power functions.

Example 1 The function $f(t) = t^n, n = 3, 4, \ldots$ is an element of $\mathcal{K}_1^c((-c, \infty))$ for $\alpha = n(n-1)c^{n-2}$ and $c \in \mathbb{R}^+$. Let mappings Φ_i, Ψ_j and operators X_i, Y_j, X, Y, $\bar{X}, \bar{Y}$ be as in Theorem 14. If spectra conditions (23) hold and $X \leq Y$, then

$$\sum_{i=1}^{n} \Phi_i(X_i^n) - \bar{X}^n \leq \binom{n}{2} c^{n-2} X \leq \binom{n}{2} c^{n-2} Y \leq \sum_{j=1}^{k} \Psi_j(Y_j^n) - \bar{Y}^n.$$

Next, we give a version of Levison's operator inequality with the scalar product. We omit the proof.

Theorem 15 *Let operators and a, b, c be as in Theorem 13. Let $(z_1, \ldots, z_n)$ be an n-tuple and $(w_1, \ldots, w_k)$ be a k-tuple of vectors $z_i, w_j \in H$, such that $\sum_{i=1}^{n} \| z_i \|^2 = 1$ and $\sum_{i=1}^{k} \| w_i \|^2 = 1$. If $f \in \mathcal{K}_1^c((a, b))$ and $\alpha \mathrm{x} \leq \alpha \mathrm{y}$, then*

$$\sum_{i=1}^{n} \langle f(X_i) z_i, z_i \rangle - f(\bar{\mathrm{x}}) \leq \frac{\alpha}{2} \mathrm{x} \leq \frac{\alpha}{2} \mathrm{y} \leq \sum_{j=1}^{k} \langle f(Y_j) w_j, w_j \rangle - f(\bar{\mathrm{y}}), \tag{24}$$

where $\mathrm{x} := \sum_{i=1}^{n} \langle (X_i - \bar{\mathrm{x}} 1_H)^2 z_i, z_i \rangle$, $\bar{\mathrm{x}} := \sum_{i=1}^{n} \langle X_i z_i, z_i \rangle$, $\mathrm{y} := \sum_{j=1}^{k} \langle (Y_j - \bar{\mathrm{y}} 1_H)^2$ $w_j, w_j \rangle$, $\bar{\mathrm{y}} := \sum_{j=1}^{k} \langle Y_j w_j, w_j \rangle$. But, if $f \in \mathcal{K}_2^c((a, b))$ and $\alpha \mathrm{x} \geq \alpha \mathrm{y}$, then reverse inequalities are valid in (24).

3.3 Converse of Levinson's Operator Inequality

In this subsection we give converse of the inequality (19) for $f \in \mathcal{K}_1^c(I)$. First, for convenience we introduce some abbreviations:

Let $f : [m, M] \to \mathbb{R}$, $m < M$ and $\alpha \in \mathbb{R}$. We denote a linear function through the points $(m, F(m))$ and $(M, F(M))$ by $f^{line}_{\alpha, [m,M]}$, i.e.

$$f^{line}_{\alpha, [m,M]}(t) = \frac{M - t}{M - m} f(m) + \frac{t - m}{M - m} f(M) - \frac{\alpha}{2} \left((M + m)t - mM \right), \quad t \in \mathbb{R},$$

and the slope of the line through $(m, F(m))$ and $(M, F(M))$ by $k_{\alpha, f[m,M]}$, i.e.

$$k_{\alpha, f[m,M]} = \frac{f(M) - f(m)}{M - m} - \frac{\alpha}{2}(M + m).$$

Next, we give converse of Levinson's operator inequality for two operators.

Theorem 16 *Let $X, Y \in \mathcal{B}_h(H)$ be self-adjoint operators with spectra contained in $[m, M]$ and $[n, N]$, respectively, such that $a < m \le M \le c \le n \le N < b$. Let Φ, Ψ be normalized positive linear mappings $\Phi, \Psi : \mathcal{B}(H) \to \mathcal{B}(K)$ and m_x, M_x, $(m_x \le M_x)$ and n_y, N_y, $(n_y \le N_y)$ be bounds of operators $\Phi(X)$ and $\Psi(Y)$, respectively (see Fig. 1). If $f \in \mathcal{K}^c_1((a, b))$ and $C_1 \ge C_2$, then*

$$\Phi\big(f(X)\big) - f\big(\Phi(X)\big) + \beta_1 1_K \ge C_1 \ge C_2 \ge \Psi\big(f(Y)\big) - f\big(\Psi(Y)\big) + \beta_2 1_K, \quad (25)$$

where

$$C_1 := \frac{\alpha}{2} \left[\Phi\big(X^2\big) - \Phi(X)^2 \right], \quad C_2 := \frac{\alpha}{2} \left[\Psi\big(Y^2\big) - \Psi(Y)^2 \right], \quad (26)$$

and

$$\begin{aligned}
\beta_1 &= \max_{m_x \le t \le M_x} \left\{ f(t) - \tfrac{\alpha}{2} t^2 - f^{line}_{\alpha, [m,M]}(t) \right\} \ge 0, \\
\beta_2 &= \min_{n_y \le t \le N_y} \left\{ f(t) - \tfrac{\alpha}{2} t^2 - f^{line}_{\alpha, [n,N]}(t) \right\} \le 0.
\end{aligned} \quad (27)$$

In the dual case, if $f \in \mathcal{K}^c_2((a, b))$ and $C_1 \le C_2$ hold, then reverse inequalities are valid in (25), where $\beta_1 \le 0$ with min *instead of* max *and $\beta_2 \ge 0$ with* max *instead of* min *in (27).*

Fig. 1 Bounds of operators for converse of Levinson's inequality

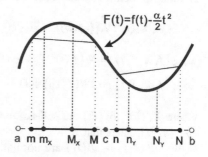

$F(t) = f(t) - \frac{\alpha}{2} t^2$

a m m_x M_x M c n n_y N_y N b

Proof If $f \in \mathcal{K}_1^c((a, b))$, then there is a constant α such that $F(t) = f(t) - \frac{\alpha}{2}t^2$ is concave on $[m, M] \subset (a, c]$. The converse of Jensen's operator inequality gives (see [17, Theorem 3.4])

$$\Phi\big(f(X)\big) - \frac{\alpha}{2}\Phi\big(X^2\big) - f\big(\Phi(X)\big) + \frac{\alpha}{2}\Phi(X)^2 + \beta_1 1_K \geq 0$$

$$\Rightarrow \Phi\big(f(X)\big) - f\big(\Phi(X)\big) + \beta_1 1_K \geq C_1. \tag{28}$$

Similarly, since F is operator convex on $[n, N] \subset [c, b)$, then Jensen's operator inequality gives

$$C_2 \geq \Psi\big(f(Y)\big) - f\big(\Psi(Y)\big) + \beta_2 1_K. \tag{29}$$

Combining inequalities (28) and (29) and taking into account $C_1 \geq C_2$ we obtain (25).

Remark 6 Applying Theorem 16 we obtain a version of the converse of Levinson's inequality (19) with more operators. We omit the details. Applying this result we can obtain the following converse of (22).

Let X_i, Y_j, p_i, q_j, P, Q, a, b, c be as in Corollary 4 and β_1, β_2 be defined by (27).

If $f \in \mathcal{K}_1^c((a, b))$ and $\alpha P \geq \alpha Q$, then

$$\sum_{i=1}^{k_1} p_i f(X_i) - f\big(\bar{X}\big) + \beta_1 1_K \geq \frac{\alpha}{2}P \geq \frac{\alpha}{2}Q \geq \sum_{j=1}^{k_2} q_j f(Y_j) - f\big(\bar{Y}\big) + \beta_2 1_K.$$

3.4 Levinson's Mapping and Its Properties

In this subsection we observe some Levinson's mappings for two operators. Analogously, we can observe Levinson's mapping for more operators. We omit the details.

1. First, we define two Levinson's mappings

$$\mathfrak{L}_{\Phi, \Psi}, \bar{\mathfrak{L}}_{\Phi, \Psi} : \overset{\bullet}{\mathcal{K}_1^c}((a, b)) \times \mathcal{B}_h(\mathcal{H}) \times \mathcal{B}_h(\mathcal{H}) \times [0, 1] \to \mathcal{B}_h(\mathcal{H}) \text{ as}$$

$$\mathfrak{L}_{\Phi, \Psi}(f, X, Y, t) := \Psi\big(f\,(tY + (1-t)\Psi(Y))\big) - \Phi\big(f\,(tX + (1-t)\Phi(X))\big), \tag{30}$$

$$\bar{\mathfrak{L}}_{\Phi, \Psi}(f, X, Y, t) := t\Psi\big(f(Y)\big) + (1-t)f(\Psi(Y)) - \big[t\Phi\big(f(X)\big) + (1-t)f(\Phi(X))\big],$$

where $X, Y \in \mathcal{B}_h(H)$ are self-adjoint operators with spectra contained in $[m, M]$ and $[n, N]$, such that $a < m \leq M \leq c \leq n \leq N < b$, $t \in [0, 1]$ and $\Phi, \Psi : \mathcal{B}(H) \to \mathcal{B}(H)$ are normalized positive linear mappings. Moreover, let

Φ preserve the operator $\bar{X} := \Phi(X)$ and the product of operators X and $\bar{X}$, and let Ψ, analogous, preserve $\bar{Y} := \Psi(Y)$ and the product of Y and $\bar{Y}$ in the mapping (30).

Theorem 17 *If $C_1 \leq C_2$ holds (see (26)), then $\mathfrak{L}_{\Phi,\Psi}(f, X, Y, \cdot)$ and $\bar{\mathfrak{L}}_{\Phi,\Psi}(f, X, Y, \cdot)$ is convex and monotone increasing on $[0, 1]$. So,*

$$\inf_{t\in[0,1]} \mathfrak{L}_{\Phi,\Psi}(f, X, Y, t) = \inf_{t\in[0,1]} \bar{\mathfrak{L}}_{\Phi,\Psi}(f, X, Y, t) = f(\Psi(Y)) - f(\Phi(X)),$$

$$\sup_{t\in[0,1]} \mathfrak{L}_{\Phi,\Psi}(f, X, Y, t) = \sup_{t\in[0,1]} \bar{\mathfrak{L}}_{\Phi,\Psi}(f, X, Y, t) = \Psi(f(Y)) - \Phi(f(X)).$$

These properties are proven in [15, Theorems 3.1 and 3.2].

2. Next, we define the operator-valued functional

$$\Delta_{\Phi,\Psi} : \overset{\bullet}{\mathcal{K}_1^c}((a, b)) \times \mathcal{B}_h(\mathcal{H}) \times \mathcal{B}_h(\mathcal{H}) \times \mathcal{B}_h(\mathcal{H}) \times \mathcal{B}_h(\mathcal{H}) \times [0, 1] \to \mathcal{B}_h(\mathcal{H})$$

related to Levinson's inequality as a difference between respective mappings $\bar{\mathfrak{L}}_{\Phi,\Psi}$ and $\mathfrak{L}_{\Phi,\Psi}$, i.e.

$$\begin{aligned}\Delta_{\Phi,\Psi}(f, A, B; C, D, t) &= (1 - t)\Psi\big(f(C)\big) + t\Psi\big(f(D)\big) - \Psi\big(f((1 - t)C + tD)\big) \\ &\quad - \big[(1 - t)\Phi\big(f(A)\big) + t\Phi\big(f(B)\big) - \Phi\big(f((1 - t)A + tB)\big)\big]\end{aligned}$$

where $\Phi, \Psi : \mathcal{B}(H) \to \mathcal{B}(H)$ are normalized positive linear mappings, $f \in \overset{\bullet}{\mathcal{K}_1^c}((a, b))$, $A, B, C, D \in \mathcal{B}_h(H)$ are self-adjoint operators with spectra of A, C contained in $[m, M]$ and spectra of B, D contained in $[n, N]$, such that $a < m \leq M \leq c \leq n \leq N < b$ and $t \in [0, 1]$.

For the sake of convenience let us define operator functions:

$$\delta_\Phi(A, B) = \Phi\big((A - B)^2\big) \tag{31}$$

and

$$\Delta_\Phi(f, A, B, t) = (1 - t)\Phi\big(f(A)\big) + t\Phi\big(f(B)\big) - \Phi\big(f((1 - t)A + tB)\big).$$

So, we can read

$$\Delta_{\Phi,\Psi}(f, A, B; C, D, t) = \Delta_\Psi(f, C, D, t) - \Delta_\Phi(f, A, B, t). \tag{32}$$

We can also consider the following functional

$$\begin{aligned}\Theta_{\Phi,\Psi}(f, A, B; C, D) &= \Psi\big(f(C)\big) + \Psi\big(f(D)\big) - \Psi\Big(\int_0^1 f((1 - t)C + tD)\, dt\Big) \\ &\quad - \Phi\big(f(A)\big) - \Phi\big(f(B)\big) + \Phi\Big(\int_0^1 f((1 - t)A + tB)\, dt\Big).\end{aligned}$$

We observe that

$$\Theta_{\Phi,\Psi}(f, A, B; C, D)$$
$$= \int_0^1 \Delta_{\Phi,\Psi}(f, A, B; C, D, t)dt = \int_0^1 \Delta_{\Phi,\Psi}(f, A, B; C, D, 1-t)dt \geq 0.$$
$$(33)$$

Now, we can show an operator quasi-linearity property for the functional (32), as operator superadditive and operator monotone as a function of intervals.

Theorem 18 *If* $\alpha \, \delta_\Psi(C, D) \geq \alpha \, \delta_\Phi(A, B)$, *then for every* $A_1 = (1-s)A + sB \in$ $[A, B]$ *and* $C_1 = (1-s)C + sD \in [C, D]$, *we have*

$$0 \leq \Delta_{\Phi,\Psi}(f, A, A_1; C, C_1, t) + \Delta_{\Phi,\Psi}(f, A_1, B; C_1, D, t)$$

$$\leq \Delta_{\Phi,\Psi}(f, A, B; C, D, t).$$

Moreover, if $B_1 = (1-r)A + rB$ *and* $D_1 = (1-r)C + rD$ *so that* $[A_1, B_1] \subset [A, B]$ *and* $[C_1, D_1] \subset [C, D]$, *then*

$$0 \leq \Delta_{\Phi,\Psi}(f, A_1, B_1; C_1, D_1, t) \leq \Delta_{\Phi,\Psi}(f, A, B; C, D, t).$$

These properties are proven in [15, Theorem 3.4].

Applying the above two inequalities we are able to state the following bounds. For the details of the proof, see in [15, Corollary 3.6].

Corollary 5 *Let* $\Phi, \Psi : \mathcal{B}(H) \to \mathcal{B}(H)$ *be normalized positive linear mappings,* $A, C \in \mathcal{B}_h(H)$ *be self-adjoint operators with spectra of* A *and* C *contained in* $[m, M]$ *and* $[n, N]$, *respectively, such that* $a < m \leq M \leq c \leq n \leq N < b$, $\bar{A} = \Phi(A)$ *and* $\bar{C} = \Psi(C)$.
If $\alpha \, \delta_\Psi(C, \bar{C}) \geq \alpha \, \delta_\Phi(A, \bar{A})$, $B = (1-s)A + s\bar{A}$ *and* $D = (1-s)C + s\bar{C}$, *then*

$$\inf_{\substack{B \in [A, \bar{A}] \\ D \in [C, \bar{C}]}} \left\{ \Psi\big(f\big((1-t)C + tD\big)\big) + \Psi\big(f\big((1-t)D + t\bar{C}\big)\big) - \Psi\big(f(D)\big) \right.$$

$$\left. -\Phi\big(f\big((1-t)A + tB\big)\big) - \Phi\big(f\big((1-t)B + t\bar{A}\big)\big) + \Phi\big(f(B)\big) \right\}$$

$$= \Psi\big(f\big((1-t)C + t\bar{C}\big)\big) - \Phi\big(f\big((1-t)A + t\bar{A}\big)\big)$$

holds for every $f \in \overset{\bullet}{\mathcal{K}}_1^c((a, b))$ *and* $t \in [0, 1]$.
Moreover, if $B_1 = (1-s)A + s\bar{A}$, $B_2 = (1-r)A + rB$, $D_1 = (1-s)C + s\bar{C}$ *and* $D_2 = (1-r)C + rD$ *for* $r, s \in [0, 1]$, *then*

$$\sup_{\substack{B_1, B_2 \in [A, \bar{A}] \\ D_1, D_2 \in [C, \bar{C}]}} \left\{ (1-t)\Psi\big(f(D_1)\big) + t\Psi\big(f(D_2)\big) - \Psi\big(f((1-t)D_1 + tD_2)\big) \right.$$

$$-(1-t)\Phi\big(f(B_1)\big) - t\Phi\big(f(B_2)\big) + \Phi\big(f((1-t)B_1 + tB_2)\big) \Big\}$$

$$= (1-t)\Psi\big(f(C)\big) + t\Psi\big(f(\bar{C})\big) - \Psi\big(f((1-t)C + t\bar{C})\big)$$

$$- (1-t)\Phi\big(f(A)\big) - t\Phi\big(f(\bar{A})\big) + \Phi\big(f((1-t)A + t\bar{A})\big).$$

(34)

If Φ and Ψ preserve the operator $f(\bar{A})$ and $f(\bar{C})$, respectively, then supremum in (34) is equal to $\bar{\mathfrak{L}}_{\Phi,\Psi}(f, A, C, t) - \mathfrak{L}_{\Phi,\Psi}(f, A, C, t)$.

Similarly, utilizing the representation (33), we can obtain bounds of

$$\Psi\big(\textstyle\int_0^1 f((1-t)C + tD)\, dt\big) + \Psi\big(\textstyle\int_0^1 f((1-t)D + t\bar{C})\, dt\big) - \Psi\big(f(D)\big)$$
$$- \Phi\big(\textstyle\int_0^1 f((1-t)A + tB)\, dt\big) - \Phi\big(\textstyle\int_0^1 f((1-t)B + t\bar{A})\, dt\big) + \Phi\big(f(B)\big).$$

We omit the details.

4 Quasi-Arithmetic Means of Operators

In this section we give order among quasi-arithmetic means of operators.

We define the quasi-arithmetic mean as follows:

$$\mathfrak{M}_\varphi(\mathbf{X}, \mathbf{\Phi}, n) := \varphi^{-1}\left(\sum_{i=1}^n \Phi_i\big(\varphi(X_i)\big)\right), \qquad (35)$$

where $(X_1, \ldots, X_n)$ is an n-tuple of self-adjoint operators in $\mathcal{B}_h(H)$ with spectra in I, $(\Phi_1, \ldots, \Phi_n)$ is a unital n-tuple of positive linear mappings $\Phi_i : \mathcal{B}(H) \to \mathcal{B}(K)$, and $\varphi : I \to \mathbb{R}$ is a strictly monotone function. There have been many works devoted to observing the order among these means, see, e.g., [7, 8, 13, 14].

The power mean is a special case of the quasi-arithmetic mean:

$$\mathfrak{M}_r(\mathbf{X}, \mathbf{\Phi}, n) := \begin{cases} \left(\sum_{i=1}^n \Phi_i\big(X_i^r\big)\right)^{1/r}, & r \in \mathbb{R}\backslash\{0\}, \\ \exp\left(\sum_{i=1}^n \Phi_i\big(\ln(X_i)\big)\right), & r = 0, \end{cases} \qquad (36)$$

where $X_1, \ldots, X_n$ are positive operators.

4.1 Results with Operator Convexity and Concavity

As a generalization of [5, Corollary] on operators and quasi-arithmetic means, we obtain the following results.

Theorem 19 *Let mappings, operators and a, b, c_1 be as in Theorem 13. Let $\psi, \varphi :$ $(a, b) \to \mathbb{R}$ be strictly monotone functions, $c = \varphi(c_1)$ and I is the open interval between $\varphi(a)$ and $\varphi(b)$.*

If $\psi \circ \varphi^{-1} \in \overset{\bullet}{\mathcal{K}}^c_1(I)$ and $\alpha X_\varphi \leq \alpha Y_\varphi$, then

$$
\begin{aligned}
&\psi\big(\mathfrak{M}_\psi(\mathbf{X}, \mathbf{\Phi}, n)\big) - \psi\big(\mathfrak{M}_\varphi(\mathbf{X}, \mathbf{\Phi}, n)\big) \\
&\leq \tfrac{\alpha}{2} X_\varphi \leq \tfrac{\alpha}{2} Y_\varphi \leq \psi\big(\mathfrak{M}_\psi(\mathbf{Y}, \mathbf{\Psi}, k)\big) - \psi\big(\mathfrak{M}_\varphi(\mathbf{Y}, \mathbf{\Psi}, k)\big),
\end{aligned}
\tag{37}
$$

where

$$
\begin{aligned}
X_\varphi &:= \sum_{i=1}^n \Phi_i\big(\varphi(X_i)^2\big) - \Big(\sum_{i=1}^n \Phi_i\left(\varphi(X_i)\right)\Big)^2, \\
Y_\varphi &:= \sum_{j=1}^k \Psi_j\big(\varphi(Y_j)^2\big) - \Big(\sum_{j=1}^k \Psi_j\left(\varphi(Y_j)\right)\Big)^2.
\end{aligned}
\tag{38}
$$

If $\psi \circ \varphi^{-1} \in \overset{\bullet}{\mathcal{K}}^c_2(I)$ and $\alpha X_\varphi \geq \alpha Y_\varphi$ hold, then reverse inequalities are valid in (37).

Proof Suppose that φ is a strictly increasing function in (a, b). For a function $f \in \overset{\bullet}{\mathcal{K}}^c_1((\varphi(a), \varphi(b)))$ there is a constant α such that $F(t) = f(t) - \frac{\alpha}{2} t^2$ is operator concave on $[\varphi(m_x), c] \subset (\varphi(a), c]$. Then the converse of Jensen's operator inequality gives

$$
\sum_{i=1}^n \Phi_i\big(f(\varphi(X_i))\big) - f\Big(\sum_{i=1}^n \Phi_i(\varphi(X_i))\Big) \leq \frac{\alpha}{2}\Big[\sum_{i=1}^n \Phi_i\big(\varphi(X_i)^2\big) - \Big(\sum_{i=1}^n \Phi_i(\varphi(X_i))\Big)^2\Big].
$$

Also, since F is operator convex on $[c, \varphi(M_y)] \subset [c, \varphi(b))$, Jensen's inequality gives

$$
\frac{\alpha}{2}\Big[\sum_{j=1}^k \Psi_j\big(\varphi(Y_j)^2\big) - \Big(\sum_{j=1}^k \Psi_j(\varphi(Y_j))\Big)^2\Big] \leq \sum_{j=1}^k \Psi_j\big(f(\varphi(Y_j))\big) - f\Big(\sum_{j=1}^k \Psi_j(\varphi(Y_j))\Big).
$$

Setting $f = \psi \circ \varphi^{-1}$ in the above two inequalities and taking into account that $\alpha X_\varphi \leq \alpha Y_\varphi$ holds, we obtain the desired inequality (37).

Setting ψ equal to the identity function in Theorem 19, we can obtain inequality (39). We give this result with weakened assumptions.

Theorem 20 *Let mappings, operators and a, b, c_1 be as in Theorem 13. Let f : $(a, b) \to \mathbb{R}$ such that $\varphi := f|_{(a,c_1]}$, $\psi := f|_{[c_1,b)}$ be strictly monotone functions, $c = \varphi(c_1)$ and I is the open interval between $f(a)$ and $f(b)$.*

If $f^{-1} \in \overset{\bullet}{\mathcal{K}_1^c}(I)$ and $\alpha Y_\psi \le \alpha X_\varphi$, then

$$\mathfrak{M}_\psi(Y, \Psi, k) - \mathfrak{M}_1(Y, \Psi, k) \le \frac{\alpha}{2} Y_\psi \le \frac{\alpha}{2} X_\varphi \le \mathfrak{M}_\varphi(X, \Phi, n) - \mathfrak{M}_1(X, \Phi, n),$$
(39)

where

$$Y_\psi := \left(\sum_{j=1}^{k} \Psi_j \left(\psi(Y_j)\right)\right)^2 - \sum_{j=1}^{k} \Psi_j \left(\psi(Y_j)^2\right), \quad X_\varphi := \left(\sum_{i=1}^{n} \Phi_i \left(\varphi(X_i)\right)\right)^2 - \sum_{i=1}^{n} \Phi_i \left(\varphi(X_i)^2\right).$$

If $f^{-1} \in \overset{\bullet}{\mathcal{K}_2^c}(I)$ and $\alpha Y_\psi \ge \alpha X_\varphi$ holds, then reverse inequalities are valid in (39).

Remark 7 Let Φ_i, Ψ_j be mappings, X_i, Y_j be positive operators as in Theorem 19 and $0 < m_x \le M_x \le c \le m_y \le M_y < b$. Setting $\alpha = 0$, $f(t) = t^s$, $s \ge 1$ for $t \in (0, c]$, $f(t) = c^{s/r} t^r$, $r \le -1$ or $\frac{1}{2} \le r \le 1$ for $t \in [c, \infty)$ in Theorem 20, we obtain

$$\mathfrak{M}_r(Y, \Psi, k) - \mathfrak{M}_1(Y, \Psi, k) \le 0 \le \mathfrak{M}_s(X, \Phi, n) - \mathfrak{M}_1(X, \Phi, n).$$
(40)

The inequality (40) holds for all positive operators X_i, Y_j without condition $M_x \le c \le m_y$. Really, LHS (resp. RHS) of (40) holds since $t \mapsto t^s$ (resp. $t \mapsto t^r$) is operator concave (resp. operator convex) on $(0, \infty)$, see [8, 14].

Setting $\alpha \ne 0$ in Theorem 20, we can obtain a refinement of (40) for some r, s:

Corollary 6 *Let Φ_i, Ψ_j be mappings, X_i, Y_j be positive operators as in Theorem 19, $0 < m_x \le M_x \le c \le m_y \le M_y < b$ and*

$$C_{1/2} := \left(\sum_{j=1}^{k} \Psi_i(\sqrt{Y_j})\right)^2 - \sum_{j=1}^{k} \Psi_j(Y_j), \quad C_s := \left(\sum_{i=1}^{n} \Phi_i(X_i^s)\right)^2 - \sum_{i=1}^{n} \Phi_i(X_i^{2s}),$$
$$C_{\exp} := \left(\sum_{i=1}^{n} \Phi_i(\exp X_i)\right)^2 - \sum_{i=1}^{n} \Phi_i\left((\exp X_i)^2\right).$$

1. If $s \ge 1$ and $C_{1/2} \le C_s$, then for every $\alpha \in (0, 2c^{1-2s})$

$$\mathfrak{M}_{1/2}(Y, \Psi, k) - \mathfrak{M}_1(Y, \Psi, k) \le \alpha C_{1/2} \le \alpha C_s \le \mathfrak{M}_s(X, \Phi, n) - \mathfrak{M}_1(X, \Phi, n).$$
(41)

2. If $C_{1/2} \le C_{\exp}$, then for every $\alpha \in (0, 2c^{1-2s})$

$$\mathfrak{M}_{1/2}(Y, \Psi, k) - \mathfrak{M}_1(Y, \Psi, k) \le \alpha C_{1/2} \le \alpha C_{\exp} \le \mathfrak{M}_{\exp}(X, \Phi, n) - \mathfrak{M}_1(X, \Phi, n).$$
(42)

4.2 Results Without Operator Convexity and Concavity

For wider application it is interesting to consider inequalities involving quasi-arithmetic means under similar conditions as in Sect. 3.2. Thus, if spectra conditions hold, then (37) is valid for all strictly monotone functions $\varphi, \psi : (a, b) \to \mathbb{R}$ such that $\psi \circ \varphi^{-1} \in \mathcal{K}_1^c(I)$:

Theorem 21 *Let Φ_i, Ψ_j be mappings, X_i, Y_j be operators as in Theorem 14, $a < m_x \le M_x \le c \le m_y \le M_y < b$, and $m_{\varphi X}, M_{\varphi X}$ and $m_{\varphi Y}, M_{\varphi Y}$ be bounds of $\mathfrak{M}_\varphi(\mathbf{X}, \mathbf{\Phi}, n)$ and $\mathfrak{M}_\varphi(\mathbf{Y}, \mathbf{\Psi}, k)$, respectively, such that*

$$
\begin{aligned}
(m_{\varphi X}, M_{\varphi X}) \cap [m_{X_i}, M_{X_i}] = \varnothing, \quad i = 1, \ldots, n, \\
(m_{\varphi Y}, M_{\varphi Y}) \cap [m_{Y_j}, M_{Y_j}] = \varnothing, \quad j = 1, \ldots, k.
\end{aligned}
\tag{43}
$$

Let $\psi, \varphi : (a, b) \to \mathbb{R}$ be strictly monotone functions, $c = \varphi(c_1)$ and I is the open interval between $\varphi(a)$ and $\varphi(b)$.

If $\psi \circ \varphi^{-1} \in \overset{\bullet}{\mathcal{K}}_1^c(I)$ and $\alpha X_\varphi \le \alpha Y_\varphi$, then (37) is valid, where X_φ, Y_φ are defined by (38).

If $\psi \circ \varphi^{-1} \in \overset{\bullet}{\mathcal{K}}_2^c(I)$ and $\alpha X_\varphi \ge \alpha Y_\varphi$ hold, then reverse inequalities are valid in (37).

Proof We use the same technique as in the proof of Theorem 19, taking into account spectra conditions (43). We omit the details.

Applying Theorem 21, we obtain a generalization and refining of Bullen's result [5, Corollary] for power means.

Corollary 7 *Let the assumptions of Theorem 21 hold with spectra conditions (43). Let*

$$
C_{sX} := \begin{cases} \left(\sum_{i=1}^n \Phi_i(X_i^s)\right)^2 - \sum_{i=1}^n \Phi_i(X_i^{2s}), & s \ne 0 \\ \left(\sum_{i=1}^n \Phi_i(\ln X_i)\right)^2 - \sum_{i=1}^n \Phi_i(\ln^2(X_i)), & s = 0, \end{cases}
$$

and C_{sY}, C_{0Y} be analogous notations for operators $Y_1, \ldots, Y_k$.

Let $\alpha = \frac{r}{s}\left(\frac{r}{s} - 1\right) c^{\frac{r}{s}-2}$ for $rs \ne 0$, $\alpha = r^2 \exp(cr)$ for $s = 0$ and $\alpha = -\frac{1}{s}c^{-2}$ for $r = 0$.

(i) *If $r < 0 < s$, $2s \le r \le s < 0$, $s < 0 < r$ or $0 < s \le r \le 2s$, $C_{sX} \le C_{sY}$, then*

$$
\mathfrak{M}_s(\mathbf{X}, \mathbf{\Phi}, n)^r - \mathfrak{M}_r(\mathbf{X}, \mathbf{\Phi}, n)^r \le \alpha C_{sX} \le \alpha C_{sY} \le \mathfrak{M}_s(\mathbf{Y}, \mathbf{\Psi}, k)^r - \mathfrak{M}_r(\mathbf{Y}, \mathbf{\Psi}, k)^r.
\tag{44}
$$

(ii) *If $r \le 2s < 0$, $0 < 2s \le r$, $C_{sX} \ge C_{sY}$ or $s \le r < 0$ or $0 < r \le s$, $C_{sX} \le C_{sY}$, then reverse inequalities are valid in (44).*

(iii) If $r < 0$, $C_{0X} \leq C_{0Y}$, then (44) is valid for $s = 0$.

But, if $r > 0$, $C_{0X} \geq C_{0Y}$, then reverse inequalities are valid in (44) for $s = 0$.

(iv) If $s > 0$, $C_{sX} \leq C_{sY}$, then

$$\ln\left(\frac{\mathfrak{M}_0(\mathbf{X}, \mathbf{\Phi}, n)}{\mathfrak{M}_s(\mathbf{X}, \mathbf{\Phi}, n)}\right) \leq \alpha C_{sX} \leq \alpha C_{sY} \leq \ln\left(\frac{\mathfrak{M}_0(\mathbf{Y}, \mathbf{\Psi}, k)}{\mathfrak{M}_s(\mathbf{Y}, \mathbf{\Psi}, k)}\right). \tag{45}$$

But, if $s < 0$, $C_{sX} \geq C_{sY}$ and spectra conditions:

$$(m_{\varphi_Y}, M_{\varphi_Y}) \cap [m_{Y_i}, M_{Y_i}] = \varnothing, \quad i = 1, \ldots, k, \tag{46}$$

hold, then reverse inequalities are valid in (45).

Proof We prove only (i). We set $\varphi(t) = t^s$, $\psi(t) = t^r$ and $f(t) = t^{\frac{r}{s}}$, $r, s \neq 0$. Let us consider a function $F(t) = t^{\frac{r}{s}} - \frac{\alpha}{2}t^2$ for $\alpha = \frac{r}{s}(\frac{r}{s} - 1)c^{\frac{r}{s}-2}$. Since $F''(t) = \frac{r}{s}(\frac{r}{s} - 1)(t^{\frac{r}{s}-2} - c^{\frac{r}{s}-2})$, then c is inflection point of F. If $\frac{r}{s} < 0$ or $1 \leq \frac{r}{s} \leq 2$, then $f \in \mathcal{K}_2^c((0, \infty))$ and $\alpha > 0$. So, applying Theorem 21 we obtain (44) in the case (i).

Finally, we give version of Theorem 20 without operator convexity or concavity. The proof is similar to the one for Theorem 21 and we omit it.

Theorem 22 *Let the assumptions of Theorem 21 hold with spectra conditions* (43). *Let* $f : (a, b) \to \mathbb{R}$ *such that* $\varphi := f|_{(a,c_1]}$, $\psi := f|_{[c_1,b)}$ *be strictly monotone functions,* $c = \varphi(c_1)$ *and* I *is the open interval between* $f(a)$ *and* $f(b)$.

If $f^{-1} \in \mathcal{K}_1^c(I)$ *and* $\alpha Y_\psi \leq \alpha X_\varphi$, *then* (39) *is valid, where* X_φ, Y_φ *are defined by* (38).

If $f^{-1} \in \mathcal{K}_2^c(I)$ *and* $\alpha Y_\psi \geq \alpha X_\varphi$ *holds, then reverse inequalities are valid in* (39).

Remark 8 By setting $r = 1$ in Corollary 7, we obtain order between $\mathfrak{M}_s(\mathbf{Y}, \mathbf{\Psi}, k)$ and $\mathfrak{M}_s(\mathbf{X}, \mathbf{\Phi}, n)$. Moreover, setting $\alpha = 0$ and $f(t) = t^s$, $s \geq 1$ for $t \in (0, c]$, $f(t) = c^{s/r} t^r$, $r \leq 1$ for $t \in [c, \infty)$ in Theorem 22, we obtain obvious inequality

$$\mathfrak{M}_r(\mathbf{Y}, \mathbf{\Psi}, k) - \mathfrak{M}_1(\mathbf{Y}, \mathbf{\Psi}, k) \leq 0 \leq \mathfrak{M}_s(\mathbf{X}, \mathbf{\Phi}, n) - \mathfrak{M}_1(\mathbf{X}, \mathbf{\Phi}, n)$$

under spectra conditions (43).

But, for some $\alpha > 0$, we can obtain refining of the above inequalities: *Let* C_{sX}, C_{rY}, *and* C_{0Y} *as in Corollary 20. If* $s \geq 1$, $0 \leq r \leq 1$, $C_{rY} \leq C_{sX}$, *and spectra conditions* (46) *hold, then*

$$\mathfrak{M}_r(\mathbf{Y}, \mathbf{\Psi}, k) - \mathfrak{M}_1(\mathbf{Y}, \mathbf{\Psi}, k) \leq \alpha C_{sY} \leq \alpha C_{sX} \leq \mathfrak{M}_s(\mathbf{X}, \mathbf{\Phi}, n) - \mathfrak{M}_1(\mathbf{X}, \mathbf{\Phi}, n)$$

is valid for every $\alpha \in (0, c^{-2+(1-r)/r^2} (1 - r)/r^2)$.

References

1. M. Anwar, J. Pečarić, Cauchy's means of Levinson type. J. Inequal. Pure Appl. Math. **9**(4), 8 pp. (2008). Article 120
2. M. Anwar, J. Pečarić, On logarithmic convexity for Ky-Fan inequality. J. Inequal. Appl. **2008**, 4 pp. (2008). Art. ID 870950
3. I.A. Baloch, J. Pečarić, M. Praljak, Generalization of Levinson's inequality. J. Math. Inequal. **9**(2), 571–586 (2015)
4. P.S. Bullen, An inequality of N. Levinson. Univ. Beograd Publ. Elektrotehm. Fak. Ser. Mat. Fiz. **421–460**, 109–112 (1973)
5. P.S. Bullen, An inequality of N. Levinson. Univ. Beograd. Publ. Elektrotehn. Fak. Ser. Mat. Fiz. **412–460**, 109–112 (1973)
6. Y.J. Cho, M. Matić, J. Pečarić, Two mappings in connection to Jensen's inequality. Panamer. Math. J. **12**, 43–50 (2002)
7. T. Fujii, J. Mićić Hot, J. Pečarić, Y. Seo, *Recent Developments of Mond-Pečarić Method in Operator Inequalities*. Monographs in Inequalities, vol. 4 (Element, Zagreb, 2012)
8. T. Furuta, J. Mićić Hot, J. Pečarić, Y. Seo, *Mond-Pečarić Method in Operator Inequalities*. Monographs in Inequalities, vol. 1 (Element, Zagreb, 2005)
9. S. Hussain, J. Pečarić, I. Perić, Jensen's inequality for convex-concave antisymmetric functions and applications. J. Inequal. Appl. **2008**, 6 pp. (2008). Article ID 185089
10. S. Lawrence, D. Segalman, A generalization of two inequalities involving means. Proc. Amer. Math. Soc. **35**, 96–100 (1972)
11. N. Levinson, Generalisation of an inequality of Ky Fan. J. Math. Anal. Appl. **8**, 133–134 (1964)
12. A.McD. Mercer, Short proof of Jensen's and Levinson' inequalities. Math. Gazette **94**, 492–495 (2010)
13. J. Mićić, Refinements of quasi-arithmetic means inequalities for Hilbert space operators. Banach J. Math. Anal. **9**, 111–126 (2015)
14. J. Mićić, K. Hot, Inequalities among quasi-arithmetic means for continuous field of operators. Filomat **26**(5), 977–991 (2012)
15. J. Mićić, J. Pečarić, Some mappings related to Levinson's inequality for Hilbert space operators. Filomat **31**(7), 1995–2009 (2017)
16. J. Mićić, Z. Pavić, J. Pečarić, Jensen's inequality for operators without operator convexity. Linear Algebra Appl. **434**, 1228–1237 (2011)
17. J. Mićić, Z. Pavić, J. Pečarić, Some better bounds in converses of the Jensen operator inequality. Oper. Matrices **6**, 589–605 (2012)
18. J. Mićić, J. Pečarić, M. Praljak, Levinson's inequality for Hilbert space operators. J. Math. Inequal. **2015**(127), 1–15 (2015)
19. M.S. Moslehian, J. Mićić, M. Kian, An operator inequality and its consequences. Linear Algebra Appl. **439**, 584–591 (2013)
20. J. Pečarić, On an inequality of N. Levinson. Univ. Beograd Publ. Elektrotehn. Fak. Ser. Mat. Fiz. **678–715**, 71–74 (1980)
21. J. Pečarić, On Levinson's inequality. Real Anal. Exchange **15**, 710–712 (1989/1990)
22. J. Pečarić, F. Proschan, Y.L. Tong, *Convex Functions, Partial Orderings, and Statistical Applications* (Academic, Boston, 1992)
23. T. Popoviciu, Sur une inegalite de N. Levinson. Mathematica (Cluj) **6**, 301–306 (1964)
24. A. Witkowski, On Levinson's inequality. Ann. Univ. Paedagog. Crac. Stud. Math. **12**, 59–67 (2013)

Integral Norm Inequalities for Various Operators on Differential Forms

Shusen Ding, Dylan Helliwell, Gavin Pandya, and Arthur Yae

Abstract We obtain integral norm estimates for the homotopy operator, the potential operator, and their composition applied to differential forms. Initial results are established for all differential forms, while stronger results are shown to hold for solutions to the A-harmonic equation.

1 Introduction

The objective of this paper is to develop the upper bound estimates for the homotopy operator T, the potential operator P, and their composition $T \circ P$ acting on differential forms. It is well known that differential forms have been used in many fields of science and engineering such as PDEs, analysis, theoretical physics, and general relativity, see [1, 2, 4–6, 8, 14, 15]. For example, they were used to define various systems of partial differential equations and geometric structures on manifolds, see [1, 9, 14].

Additionally, both the homotopy operator T and the potential operator P are key operators which are widely studied and well used in several areas of mathematics, such as partial differential equations, potential analysis, and harmonic analysis, see [1, 3, 7, 10, 12, 15].

In many situations, we need to study the upper bound of the operators and their composition. In so doing, we encounter expressions such as $P\omega = d(TP\omega) + T(dP\omega)$, and thus we develop L^p estimates for $TP\omega$ and $dP\omega$ in terms of the L^p-norm of the differential form ω.

The homotopy operator T brings subtlety to these arguments because of the, sometimes implicit, dependence on the so-called weight function that is necessary to establish the algebraic properties of T. As such, we have attempted to more explicitly show how the choice of the weight function affects various results.

S. Ding (✉) · D. Helliwell · G. Pandya · A. Yae
Department of Mathematics, Seattle University, Seattle, WA, USA
e-mail: sding@seattleu.edu; helliwed@seattleu.edu; pandyag@seattleu.edu; yaea@seattleu.edu

© Springer Nature Switzerland AG 2019
G. A. Anastassiou and J. M. Rassias (eds.), *Frontiers in Functional Equations and Analytic Inequalities*, https://doi.org/10.1007/978-3-030-28950-8_33

677

This paper is organized as follows: we first introduce basic notation, definitions, and lemmas in Sect. 2. This includes a detailed discussion of the homotopy operator, the potential operator, and the A-Harmonic equation on differential forms. Then in Sect. 3 we present and prove our main results. Finally, in Sect. 4, we demonstrate applications of our main results by applying our main theorems to some particular differential forms to develop norm estimates for some operators.

This work was carried out mainly over the summer of 2018, and we gratefully acknowledge the Seattle University's College of Science and Engineering Summer Undergraduate Research Program, and Seattle University's Department of Mathematics for their generous support.

2 Background

This section provides the necessary background material for our main results. First, we establish our conventions and notation for differential forms and their norms. Then we discuss the homotopy operator and include versions of the Poincaré and Sobolev inequalities for differential forms. We next provide details about the potential operator, including an explicit bound on the operator norm. Finally, we discuss the A-harmonic equation and its solutions.

2.1 Differential Forms

In what follows, Ω is always an open, bounded, convex subset of $\mathbb{R}^n$, while D denotes an arbitrary bounded open set in $\mathbb{R}^n$. The Lebesgue measure of D is written $|D|$. We denote by B an open ball and, given a positive constant σ, σB is the ball centered at the same point and with radius equal to σ times that of B.

Let $\Lambda^k = \Lambda^k(D)$ be the set of differential k-forms on D of the form

$$\omega = \sum_I \omega_I dx_{i_1} \wedge dx_{i_2} \wedge \cdots \wedge dx_{i_k} = \sum_I \omega_I dx_I,$$

where the sum is over all ordered multi-indices $I = (i_1, \ldots, i_k)$ with $i_p < i_q$ for $p < q$, and the ω_I are measurable functions on D. To indicate the value of ω at a point $x \in D$ and vectors $v_1, v_2, \ldots, v_k$, we write

$$\omega(x; v_1, v_2, \ldots, v_k).$$

If v is a vector field on D, we define the interior product $v \lrcorner \omega$ to be the $(k-1)$-form on D such that

$$(v \lrcorner \omega)(x; v_1, v_2, \ldots, v_{k-1}) = \omega(x; v, v_1, v_2, \ldots, v_k).$$

We define $L^p(D, \Lambda^k)$ to be the set of k-forms whose component functions are each in $L^p(D)$ and we define the norm on $L^p(D, \Lambda^k)$ to be

$$\|\omega\|_{p,D} = \left(\int_D \sum_I |\omega_I(x)|^p \, dx \right)^{\frac{1}{p}}.$$

Similarly, we define $W^{1,p}(D, \Lambda^k)$ to be the set of k-forms whose component functions are each in the Sobolev space $W^{1,p}(D)$, and we define the norm on $W^{1,p}(D, \Lambda^k)$ to be

$$\|\omega\|_{1,p,D} = \left(\int_D \sum_I |\omega_I(x)|^p + \sum_{I,j} \left| \frac{\partial \omega_I}{\partial x_j}(x) \right|^p \, dx \right)^{\frac{1}{p}}.$$

For simplicity of notation, the domain may not appear if it is unambiguous.

We will make use of local versions of these spaces as well and write $\omega \in L^p_{loc}(D, \Lambda^k)$ if for all open sets G such that $\overline{G} \subset D$, we have $\omega \in L^p(G, \Lambda^k)$, with an analogous definition for $W^{1,p}_{loc}(D, \Lambda^k)$

From the monotonic property of the L^p spaces, we have the following lemma.

Lemma 1 *Suppose* $1 < p < n$, $1 < q < np/(n-p)$ *and* $\omega \in L^{np/(n-p)}(D, \Lambda^k)$. *Then for all balls* $B \subset D$

$$\|\omega\|_{q,B} \le |B|^{\frac{1}{n}+\frac{1}{q}-\frac{1}{p}} \|\omega\|_{np/(n-p),B}.$$

2.2 The Homotopy Operator

Given a point y in a convex domain Ω for a k-form ω, we define the $(k-1)$-form $K_y\omega$ as follows: given a point $x \in \Omega$ and a set of vectors $v_1, \ldots, v_{k-1}$ at x, extend the vectors to be constant on the segment connecting x and y. Then

$$K_y\omega(x; v_1, \ldots v_{k-1}) = \int_0^1 t^{k-1}\left((x-y) \lrcorner \omega\right)(tx + (1-t)y; v_1, \ldots, v_{k-1}) \, dt.$$

Since the vectors v_i do not play a significant role in the analysis, they are often omitted and we write

$$K_y\omega(x) = \int_0^1 t^{k-1}\left((x-y) \lrcorner \omega\right)(tx + (1-t)y) \, dt. \tag{1}$$

Next, let $\varphi \in L^\infty(\Omega)$ with the property that $\int_\Omega \varphi = 1$, and define the Homotopy operator

$$T_\varphi \omega(x) = \int_\Omega \varphi(y) K_y \omega(x) \, dy. \tag{2}$$

In mathematical physics the homotopy operator comes up in the context of Lagrangian mechanics. Suppose that a given mechanical system has a smooth manifold X of dimension n as its space of independent variables. If the configuration space of the system is a fiber bundle Q over X, then a Lagrangian is an n-form η on the jet bundle $J^k Q$, which essentially encodes the possible kth order behaviors of a smooth section of Q. Given a section σ of Q, the k-jet $j^k \sigma$ is a section of $J^k Q$ where $j^k \sigma|_x$ encodes the kth order behavior of σ at x. Then the action of the Lagrangian η is the map

$$S : \Gamma(Q) \longrightarrow \mathbb{R}$$

$$\sigma \longmapsto \int_X (j^k \sigma)^* \eta.$$

It can be shown that two Lagrangians η and ζ have the same extremals if and only if $\eta - \zeta = d\omega$ for some $n - 1$ form ω.

Motivated by this theorem, we introduce the notion of a homotopy operator to address the general question "when can we invert an exterior derivative?" Let $f, g : C \to D$ be cochain maps. Recall that f and g are called cochain homotopic if there exists a map $\psi : C \to D$ such that $d_D \psi + \psi d_C = g - f$, in which case we call ψ a homotopy operator. Pictorially, we have

where ψ is a homotopy operator if

$$d(\psi_{n-1} a) + \psi_n (da) = g_n a - f_n a$$

for all $a \in C_n$.

In the case that C and D are the DeRham complexes of manifolds M and N, an explicit homotopy operator can be constructed. Specifically, suppose $f, g : M \to N$ are smooth maps which are connected by a homotopy

$$h : I \times M \to N, \ h(0, x) = f(x), \ h(1, x) = g(x).$$

Then the pullback maps f^* and g^* are cochain homotopic, with explicit homotopy operator ψ given by

$$\psi\omega(x; v_1, \ldots, v_{k-1}) = \int_I \left(\frac{\partial}{\partial t} \lrcorner h^*\omega\right)\left((t, x); \tilde{v}_1, \ldots, \tilde{v}_{k-1}\right) dt,$$

where $\frac{\partial}{\partial t}$ is the unit vector field tangent to I and $\tilde{v}_i$ are vector fields on $I \times M$ that project down to v_i.

In a contractible manifold, we may take f to be the constant map and g to be the identity, in which case

$$d(\psi\omega) + \psi(d\omega) = g^*\omega - f^*\omega = \omega.$$

Thus if ω is closed, ψ inverts the exterior derivative, i.e.

$$\omega = d(\psi\omega).$$

Therefore if we are investigating two Lagrangians η and ζ, we may find whether or not they are equivalent by simply checking whether or not $\eta - \zeta$ is closed.

With a particular choice of the constant y, the subsequent homotopy operator is denoted by K_y. Moreover, any linear combination of the K_y normalized to unity satisfies the same algebraic requirement. That is, if $d\mu$ is a volume form on M and $\varphi \in L^\infty(\mu)$ is such that $\int_M \varphi d\mu = 1$, then the quantity

$$T_\varphi\omega(x) = \int_M \varphi(y)K_y\omega(x)d\mu(y)$$

satisfies

$$\omega = T_\varphi(d\omega) + d(T_\varphi\omega).$$

We call φ the weight function for T_φ.

On a bounded, convex domain $\Omega \subset \mathbb{R}^n$, we can use the linear homotopy $h_y(x, t) = tx + (1 - t)y$. In this case K_y takes the form given by (1) and the homotopy operator takes the form of Eq. (2).

Note that if ω is a closed form, then $T(d\omega)$ vanishes and $d(T\omega) = \omega$, so we have successfully inverted the exterior derivative. The homotopy operator also arises in other contexts where applications of differential forms are found, as the inversion of the exterior derivative is broadly useful, see [1, 5, 7].

It was established in [9] that the homotopy operator is bounded on $L^p(\Omega, \Lambda^k)$. Because of the estimates made there, it is easy to lose track of the dependence on the weight function, so we restate the result here and include a proof which provides an explicit bound on the operator norm:

Lemma 2 *Let $1 \leq k \leq n$ and let $\varphi \in L^\infty(\Omega)$. Then for all $1 < p < \infty$ the homotopy operator*

$$T_\varphi : L^p(\Omega, \Lambda^k) \longrightarrow L^p(\Omega, \Lambda^{k-1})$$

is a bounded operator. That is, there exists a constant $C > 0$ such that

$$\|T_\varphi \omega\|_p \le C \|\omega\|_p$$

for all $\omega \in L^p(\Omega, \Lambda^k)$. In particular, we show that

$$C \le \|\varphi\|_\infty \frac{n\pi^{\frac{n}{2}}(\operatorname{diam}\Omega)^{n+1}2^{n-1}}{k\Gamma(\frac{n}{2}+1)}.$$

Proof This proof follows in three steps. We make a change of variables to isolate ω in the integrand. Then, we use the boundedness of Ω to restrict the domain of integration and formulate a bound on what remains. And finally, we apply Young's convolution inequality to complete the proof. Let ω be a k-form on Ω, $k \ge 1$. Applying Fubini's theorem to the integral defining T_φ yields

$$T_\varphi \omega(x) = \int_0^1 \int_\Omega \varphi(y) t^{k-1}(x-y)\lrcorner\omega(tx+(1-t)y)dydt.$$

Using the change of variables $(y,t) \mapsto (z,s) : z = tx + (1-t)y, s = \frac{t}{1-t}$, we obtain

$$T_\varphi \omega(x) = \int_0^\infty \int_{\Omega_s} s^{k-1}(1+s)^{n-k}\varphi(z-s(x-z))(x-z)\lrcorner\omega(z)dzds,$$

where $\Omega_s = \{z : z - s(x-z)\}$ is obtained by shrinking Ω by a factor of $\frac{1}{1+s}$ about x. The convexity of Ω then guarantees that $\Omega_s \subset \Omega$.

By extending φ to be zero outside Ω we have $\varphi(z - s(x-z)) = 0$ for $z \notin \Omega_s$. This allows us to expand the domain of integration above, and then apply Fubini's theorem to get

$$T_\varphi \omega(x) = \int_0^\infty \int_\Omega s^{k-1}(1+s)^{n-k}\varphi(z-s(x-z))(x-z)\lrcorner\omega(z) \, dz \, ds$$

$$= \int_\Omega \int_0^\infty s^{k-1}(1+s)^{n-k}\varphi(z-s(x-z))(x-z)\lrcorner\omega(z) \, ds \, dz$$

$$= \int_\Omega \left(\int_0^\infty s^{k-1}(1+s)^{n-k}\varphi(z-s(x-z)) \, ds\right)(x-z)\lrcorner\omega(z) \, dz.$$

Define a function $F : \Omega \times \mathbb{R}^n \to \mathbb{R}$ by

$$F(u,v) = \int_0^\infty s^{k-1}(1+s)^{n-k}|\varphi(u-sv)|ds.$$

Since φ is supported in Ω, the integrand vanishes whenever

$$s > a_v := (\operatorname{diam}\Omega)/|v|$$

and hence

$$F(u, v) = \int_0^{a_v} s^{k-1}(1+s)^{n-k}|\varphi(u - sv)|\,ds.$$

Now we estimate:

$$|F(u, v)| = \left| \int_0^{a_v} s^{k-1}(1+s)^{n-k}|\varphi(u - sv)|\,ds \right|$$

$$\leq \|\varphi\|_\infty \int_0^{a_v} s^{k-1} \sum_{j=0}^{n-k} \binom{n-k}{j} s^j\,ds$$

$$= \|\varphi\|_\infty \sum_{j=0}^{n-k} \binom{n-k}{j} \int_0^{a_v} s^{j+k-1}\,ds$$

$$= \|\varphi\|_\infty \sum_{j=0}^{n-k} \binom{n-k}{j} \frac{(a_v)^{j+k}}{j+k}$$

$$= \|\varphi\|_\infty \sum_{j=0}^{n-k} \binom{n-k}{j} \frac{(\text{diam } \Omega)^{j+k}}{(j+k)|v|^{j+k}}.$$

Using this estimate, along with the fact that $|v \lrcorner \eta| \leq |v||\eta|$ for any vector field v and form η, we have

$$|T_\varphi \omega(x)| \leq \int_\Omega \|\varphi\|_\infty \sum_{j=0}^{n-k} \binom{n-k}{j} \frac{(\text{diam } \Omega)^{j+k}}{(j+k)|x-z|^{j+k-1}}|\omega(z)|\,dz$$

$$= \|\varphi\|_\infty \sum_{j=0}^{n-k} \binom{n-k}{j} \frac{(\text{diam } \Omega)^{j+k}}{j+k} \int_\Omega \frac{|\omega(z)|}{|x-z|^{j+k-1}}\,dz.$$

Let $S = \{x - y : x, y \in \Omega\}$. For $0 \leq j \leq n-k$, set $f_j(u) = \chi_S(u)\frac{y}{|y|^{j+k}}$, where χ_S is the characteristic function. Since these functions have poles of order at most $n-1$ and are compactly supported in $\mathbb{R}^n$, each $f_j \in L^1(\mathbb{R}^n)$. Thus we have

$$|T\omega(x)| \leq \|\varphi\|_\infty \sum_{j=0}^{n-k} \binom{n-k}{j} \frac{(\text{diam } \Omega)^{j+k}}{j+k} \int_\Omega |f_j(x-z)||\omega(z)|\,dz$$

$$= \|\varphi\|_\infty \sum_{j=0}^{n-k} \binom{n-k}{j} \frac{(\text{diam } \Omega)^{j+k}}{j+k}(|f_j| * |\omega|)(x)$$

(where we set ω identically zero outside of Ω). Finally, by Young's convolution inequality,

$$\|T\omega\|_p \leq \left[\|\varphi\|_\infty \sum_{j=0}^{n-k} \binom{n-k}{j} \frac{(\operatorname{diam} \Omega)^{j+k}}{j+k} \|f_j\|_1 \right] \|\omega\|_p.$$

Note that

$$\|f_j\|_1 \leq \frac{\sigma_n (\operatorname{diam} \Omega)^{n-j-k+1}}{2^{n-j-k+1}(n-j-k+1)},$$

where $\sigma_n = n\pi^{\frac{n}{2}}/\Gamma(\frac{n}{2}+1)$ is the surface area of the unit sphere in $\mathbb{R}^n$. Thus we find

$$\|T\omega\|_p \leq \|\varphi\|_\infty \frac{n\pi^{\frac{n}{2}}(\operatorname{diam} \Omega)^{n+1} 2^{n-1}}{k\Gamma(\frac{n}{2}+1)} \|\omega\|_p.$$

$\square$

In [9], it is demonstrated that, given Ω, a choice of φ can be made such that

$$\|\varphi\|_\infty \leq \kappa(\Omega)(\operatorname{diam} \Omega)^{-n}$$

where $\kappa(\Omega)$ is a scale invariant quantity that captures the shape of Ω. As such, the estimate above can be simplified to

$$\|T\omega\|_p \leq C(\operatorname{diam} \Omega)\|\omega\|_p$$

where C depends on the dimension and the shape of Ω but not its size. Implicit in this is the choice of φ.

For our purposes, the domains of interest will be balls, and we make the estimates above a bit more explicit. Let φ_0 be a weight function on the unit ball centered at the origin. Then, given a ball B with radius r centered at x_0, the function

$$\varphi_B(x) = r^{-n}\varphi_0\left(r^{-1}(x-x_0)\right)$$

is a weight function for B and

$$\|\varphi_B\|_\infty = \left(\frac{\operatorname{diam} B}{2}\right)^{-n} \|\varphi_0\|_\infty$$

so that, for $T = T_{\varphi_B}$, we have

$$\|T\omega\|_{p,B} \leq \frac{C}{k}\|\varphi_0\|_\infty(\operatorname{diam} B)\|\omega\|_{p,B}$$

where

$$C = \frac{n\pi^{\frac{n}{2}}2^{2n-1}}{\Gamma(\frac{n}{2}+1)}$$

is a purely dimensional constant. We encode this as follows to use later.

Lemma 3 *Let φ_0 be a weight function for the unit ball and let $T_B = T_{\varphi_B}$ be the associated homotopy operator on a ball B. On a bounded domain D, let $\omega \in L^p_{loc}(D, \Lambda^k)$, $1 < p < \infty$. Then there is a constant C depending only on n, k and φ_0, such that for all balls $B \subset D$*

$$\|T_B\omega\|_{p,B} \le C|B|^{1/n}\|\omega\|_{p,B}.$$

Note that since we have switched from diam B to $|B|^{\frac{1}{n}}$, the dimensional contribution to the constant is not quite the same as that written above.

For $\omega \in L^p(\Omega, \Lambda^k)$ and for any weight function φ supported on Ω, we define

$$\omega_\Omega = \begin{cases} \dfrac{1}{|\Omega|}\displaystyle\int_\Omega \omega(x)dx & \text{if } k = 0 \\[2mm] d(T_\varphi\omega) & \text{if } 1 \le k \le n. \end{cases}$$

This serves as an average of sorts for forms in that there is a constant C such that for all closed forms $\eta \in L^p(\Omega, \Lambda^k)$,

$$\|\omega - \omega_\Omega\|_p \le C\|\omega - \eta\|_p.$$

In this sense, ω_Ω minimizes the L^p-distance between ω and the subspace of exact forms to within a factor of C. This notation is meant to parallel the fact that the average $u_\Omega = \frac{1}{|\Omega|}\int_\Omega u$ of a real-valued function u uniquely minimizes the L^1-distance between u and the subspace of closed 0-forms (constants):

$$\|u - u_\Omega\|_{1,\Omega} < \|u - c\|_{1,\Omega} \text{ for all } c \ne u_\Omega,$$

and approximately minimizes the L^p-distance to within a factor of 2:

$$\|u - u_\Omega\|_{p,\Omega} \le 2\|u - c\|_{p,\Omega} \text{ for all } c \in \mathbb{R}.$$

We state here two lemmas that further illustrate the averaging nature of ω_B and will also prove to be useful later. Here and later, for a given ball B, the weight function φ_B being used is determined as above in terms of a given weight function φ_0 on the unit ball.

Lemma 4 (Poincaré Inequality) *Let ω and $d\omega$ be elements of $L^p(D, \Lambda^k)$, $1 < p < \infty$, and let φ_0 be a weight function for the unit ball. Then there is a constant C depending only on n, p, and φ_0, such that for all balls $B \subset D$*

$$\|\omega - \omega_B\|_{p,B} \le C|B|^{1/n}\|d\omega\|_{p,B}.$$

Lemma 5 (Sobolev Inequality) *Let ω and $d\omega$ be elements of $L^p(D, \Lambda^k)$, $1 < p < n$. Then there is a constant C depending only on n, p, and φ_0, such that for all balls $B \subset D$*

$$\|\omega - \omega_B\|_{\frac{np}{n-p}, B} \le C\|d\omega\|_{p,B}.$$

2.3 The Potential Operator

The name "potential operator" may refer to several different constructions. The one we investigate here arises from harmonic analysis and the theory of singular integrals. For $\alpha > 0$, the *Riesz Potential Operator* on $D \subset \mathbb{R}^n$ is defined as

$$P_\alpha : L^p(D) \longrightarrow L^p(D)$$

$$P_\alpha f(x) = \int_D \frac{f(y)dy}{|x - y|^{n-\alpha}}.$$

To see that P_α is a well-defined (and indeed bounded) operator $L^p(D) \to L^p(D)$, note that $g(x) = 1/|x|^{n-\alpha}$ is integrable on any bounded domain in $\mathbb{R}^n$, and that $P_\alpha f = g * f$. By Young's convolution inequality, $\|P_\alpha f\|_p \le \|g\|_1 \|f\|_p$.

The desirable properties of the Riesz potential operators can be captured by a standard integral operator

$$Pf(x) = \int_D K(x, y)f(y)dy$$

for a measurable kernel satisfying

$$\mathcal{K} = \max \left\{ \operatorname*{ess\,sup}_{x \in D} \left(\int_D |K(x, y)| \, dy \right), \operatorname*{ess\,sup}_{y \in D} \left(\int_D |K(x, y)| \, dx \right) \right\} < \infty.$$

We generalize this operator to differential forms as follows: given a point $x \in D$ and a set of vectors $v_1, \ldots, v_k$ at x, extend the vectors to constant vector fields on D. Then

$$(P\omega)(x; v_1, \ldots, v_k) = \int_D K(x, y)\omega(y; v_1, \ldots, v_k)dy.$$

We note that, by evaluating on standard basis vectors $\{e_i : 1 \le i \le n\}$, this is the same as applying the operator component-wise. Thus we define the potential operator as

$$P\omega(x) = \int_D K(x, y)\omega(y)dy$$

$$= \sum_I \left(\int_D K(x, y)\omega_I(y)dy \right) dx_I$$

where again K satisfies the standard estimates.

With the potential operator defined, we have the following lemma, the proof of which is similar to the proof for Young's convolution inequality.

Lemma 6 *For* $1 < q < \infty$, $P : L^q(D, \Lambda^k) \to L^q(D, \Lambda^k)$ *is a bounded operator with*

$$\|P\omega\|_q \leq \mathcal{K} \|\omega\|_q.$$

The operator P is non-local so some care is necessary when developing local arguments. To handle this, we restrict the domain of definition for K. More specifically, if P is defined on a domain D with kernel K, and $U \subset D$ is open, we define the restricted potential operator $P_U : L^q(U, \Lambda^k) \longrightarrow L^q(U, \Lambda^k)$ as follows:

$$P_U\omega(x) = \int_U K(x, y)\omega(y)\, dy.$$

This is equivalent to using the kernel $K_U(x, y) = \chi_U(y)K(x, y)$ on the original domain D and as such, if K satisfies the necessary estimates, so does K_U, and we may conclude that P_U is bounded as long as P is and the operator norm for P_U is bounded by that of P. To use later, we state this formally as follows:

Lemma 7 *Let* P *be the potential operator on domain* D, *and let* $U \subset D$ *be open. Then for* $1 < q < \infty$, *the restricted potential operator* $P_U : L^q(U, \Lambda^k) \longrightarrow L^q(U, \Lambda^k)$ *is bounded with*

$$\|P_U\omega\|_{q,U} \leq \mathcal{K} \|\omega\|_{q,U}.$$

2.4 The A-Harmonic Equation

The A-harmonic equation, as its name suggests, generalizes Laplace's equation. Its solutions, called A-harmonic tensors, are differential forms $\omega \in \Lambda^k(D)$ such that

$$d^\star A(x, d\omega) = 0$$

where $d^\star$ is the Hodge codifferential operator and $A : D \times \Lambda^l(D) \to \Lambda^l(D)$ satisfies the conditions:

$$|A(x, \xi)| \le a|\xi|^{p-1} \quad \text{and} \quad \langle A(x, \xi), \xi \rangle \ge |\xi|^p$$

for almost every $x \in D$ and all $\xi \in \Lambda^l(D)$. Here $a, b > 0$ are constants and $1 < p < \infty$ is a fixed exponent associated with the A-harmonic equation. Here, the inner product and point-wise norm are defined in terms of the Hodge star operator as follows: $\langle \alpha, \beta \rangle = \star(\alpha \wedge \star\beta)$ and $|\alpha| = \langle \alpha, \alpha \rangle^{\frac{1}{2}}$

A weak solution to the A-harmonic equation is an element of the Sobolev space $W_{loc}^{1,p}(D, \Lambda^{l-1})$ such that

$$\int_D \langle A(x, d\omega), d\varphi \rangle \, dx = 0$$

for all $\varphi \in W_{loc}^{1,p}(D, \Lambda^{l-1})$ with compact support.

This is defined in analogy to the scalar A-harmonic equation:

$$\text{div } A(x, \nabla f(x)) = 0,$$

where $f(x)$ is a function in D.

If a k-form ω solves the A-harmonic equation in a domain D, we write $\omega \in \mathscr{A}(D, \Lambda^k)$.

Some of the properties that A-harmonic tensors enjoy include the following Weak Reverse Hölder Inequality [11].

Lemma 8 (Weak Reverse Hölder Inequality) *Suppose $\omega \in \mathscr{A}(D, \Lambda^k)$. Then for all $1 < p < \infty$, $\omega \in L_{loc}^p(D, \Lambda^k)$. Moreover for $s, t, \sigma > 1$ and for all balls B with $\sigma B \subset D$*

$$\|\omega\|_{s,B} \le C|B|^{(t-s)/st} \|\omega\|_{t,\sigma B}.$$

where the constant C is independent of B and ω.

See [1, 11, 15] for more results about A-harmonic tensors.

3   Main results

This section contains our main results. First, we establish an estimate for $T \circ P$ acting on A-harmonic tensors. Then we show that $d \circ P$ is bounded. Finally, we develop the upper bound for the norm $\|TP(\omega) - (TP(\omega))_B\|_{p,B}$ in terms of the norm of ω.

Theorem 1 *Let φ_0 be a weight function for the unit ball and let $T_B = T_{\varphi_B}$ be the associated homotopy operator on a ball B. Let P be the potential operator on a domain D. Let $\omega \in \mathscr{A}(D, \Lambda^k)$ and let $p, q \in (1, \infty)$. Then for all balls B with $\sigma B \subset D$ where $\sigma > 1$,*

$$\|T_B P_B \omega\|_{q,B} \le C |B|^{\frac{1}{n}+\frac{1}{p}-\frac{1}{q}} \|\omega\|_{p,\sigma B}$$

where the constant C is independent of B and ω.

Proof By Lemma 8, for all B with $\sigma B \subset D$, $\omega \in L^q(B, \Lambda^k)$. Then, combining Lemmas 3, 7, and 8, we find

$$\|T_B P_B \omega\|_{q,B} \le C_1 |B|^{\frac{1}{n}} \|P_B \omega\|_{q,B}$$

$$\le C_2 |B|^{\frac{1}{n}} \|\omega\|_{q,B}$$

$$\le C |B|^{\frac{1}{n}+\frac{1}{p}-\frac{1}{q}} \|\omega\|_{p,\sigma B}.$$

$\square$

Next we show that not only is the potential operator bounded on L^p spaces, but also as a map from L^p spaces to Sobolev spaces.

Theorem 2 *Suppose P is the potential operator in a domain $D \subset R^n$ with kernel $K \in W^{1,p}(D \times D)$ for some $p \ge 2$ and let $U \subset D$ be open. Then $P_U : L^q(U, \Lambda^k) \to W^{1,p}(U, \Lambda^k)$ is bounded, where $q = \frac{p}{p-1}$ is the Hölder conjugate of p. More specifically*

$$\|P_U \omega\|_{1,p,U} \le C \|\omega\|_{\frac{p}{p-1},U}$$

where C is independent of ω and U.

Proof For any multi-index I and index j,

$$\frac{\partial (P_U \omega)_I}{\partial x_j} = \int_U \frac{\partial K}{\partial x_j}(x, y) \omega_I(y) dy.$$

Thus

$$\left\| \frac{\partial (P_U \omega)_I}{\partial x_j} \right\|_{p,U}^p \le \int_U \left(\int_U \left| \frac{\partial K}{\partial x_j}(x, y) \omega_I(y) \right| dy \right)^p dx$$

$$\le \int_U \left[\left(\int_U \left| \frac{\partial K}{\partial x_j}(x, y) \right|^p dy \right)^{\frac{1}{p}} \left(\int_U |\omega_I(y)|^{\frac{p}{p-1}} dy \right)^{\frac{p-1}{p}} \right]^p dx$$

$$= \|\omega_I\|_{\frac{p}{p-1},U}^p \int_U \int_U \left| \frac{\partial K}{\partial x_j}(x, y) \right|^p dy dx$$

$$\le \|\omega\|_{\frac{p}{p-1},U}^p \|K\|_{1,p,U \times U}^p$$

where in the second inequality we have used Hölder's inequality. Thus,

$$\|P_U\omega)\|_{1,p,U}^p = \int_U \sum_I |(P_U\omega)_I(x)|^p + \sum_{I,j} \left|\frac{\partial(P_U\omega)_I}{\partial x_j}(x)\right|^p dx$$

$$= \|P_U\omega\|_{p,U}^p + \sum_{I,j} \left\|\frac{\partial(P_U\omega)_I}{\partial x_i}\right\|_{p,U}^p$$

$$\leq \mathscr{K}\|\omega\|_{p,U}^p + N\|\omega\|_{\frac{p}{p-1},B}^p \|K\|_{1,p,U\times U}^p$$

$$\leq C^p\|\omega\|_{\frac{p}{p-1},U}^p$$

where $N = n\binom{n}{k}$. The first inequality arises from Lemma 7 for the first term and the previous estimate for the second. The second inequality follows from the inclusion of $L^p(U, \Lambda^k)$ into $L^q(U < \Lambda^k)$ since $q = \frac{p}{p-1} \leq p$. Finally, note that the constant C depends on the kernel for the potential operator, and hence implicitly on D, but not on the subset U.   □

As a corollary, we find that the map $\omega \mapsto dP_U\omega$ is bounded on L^p spaces:

Corollary 1 *Suppose P is the potential operator in a domain $D \subset R^n$ with kernel $K \in W^{1,p}(D \times D)$ for some $p \geq 2$ and let $U \subset D$ be open. Then $dP_U : L^p(U, \Lambda^k) \to L^p(U, \Lambda^k)$ is bounded. Specifically, there is a constant C, independent of ω and U, such that*

$$\|dP_U\omega\|_{p,U} \leq C|U|^{\frac{p-2}{p}} \|\omega\|_{p,U}.$$

Proof Note that

$$\|dP_U\omega\|_{p,U} \leq \|P_U\omega\|_{1,p,U}$$

so by Theorem 2 and Hölder's inequality we have

$$\|d(P_U\omega)\|_{p,U} \leq C\|\omega\|_{\frac{p}{p-1},U}$$

$$\leq C|U|^{\frac{p-2}{p}} \|\omega\|_{p,U}.$$

                                                                                            □

It is worth noting that Theorem 2 and Corollary 1 are both valid for any open subsets of D, including D itself.

Combining Lemma 4 and Corollary 1 proves the following:

Corollary 2 *Suppose P is the potential operator in a domain $D \subset R^n$ with kernel $K \in W^{1,p}(D \times D)$ for some $p \geq 2$, and let φ_0 be a weight function for the unit ball. Then for all balls $B \subset D$ and for all $\omega \in L^p(D, \Lambda^k)$,*

$$\|P_B\omega - (P_B\omega)_B\|_{p,B} \le C|B|^{\frac{1}{n}+\frac{p-2}{p}}\|\omega\|_{p,B}$$

where the constant C is independent of B and ω.

Theorem 3 *Suppose P is the potential operator in a domain D with kernel $K \in W^{1,p}(D \times D)$ for some $p \ge 2$. Let φ_0 be a weight function for the unit ball and let $T_B = T_{\varphi_B}$ be the associated homotopy operator on a ball B. Let $1 < q < np/(n-p)$. Then for all balls $B \subset D$ and for all $\omega \in L^p(D, \Lambda^k)$,*

$$\|T_B P_B\omega - (T_B P_B\omega)_B\|_{q,B} \le C|B|^{\frac{1}{n}+\frac{1}{q}-\frac{1}{p}}\|\omega\|_{p,B}$$

where the constant C is independent of B and ω.

Proof By combining a number of our earlier results, we have

$\|T_B P_B\omega-(T_B P_B\omega)_B\|_{q,B}$

$$\le |B|^{\frac{1}{n}+\frac{1}{q}-\frac{1}{p}}\|T_B P_B\omega - (T_B P_B\omega)_B\|_{np/(n-p),B} \qquad \text{(Lemma 1)}$$

$$\le C_1|B|^{\frac{1}{n}+\frac{1}{q}-\frac{1}{p}}\|d(T_B P_B\omega)\|_{p,B} \qquad \text{(Lemma 5)}$$

$$= C_1|B|^{\frac{1}{n}+\frac{1}{q}-\frac{1}{p}}\|(P_B\omega)_B\|_{p,B}$$

$$\le C_2|B|^{\frac{1}{n}+\frac{1}{q}-\frac{1}{p}}\left[\|P_B\omega - (P_B\omega)_B\|_{p,B} + \|P_B\omega\|_{p,B}\right]$$

$$\le C_2|B|^{\frac{1}{n}+\frac{1}{q}-\frac{1}{p}}\left[C_3|B|^{\frac{1}{n}+\frac{p-2}{p}}\|\omega\|_{p,B} + \|P_B\omega\|_{p,B}\right] \qquad \text{(Corollary 2)}$$

$$\le C_2|B|^{\frac{1}{n}+\frac{1}{q}-\frac{1}{p}}\left[C_4\|\omega\|_{p,B} + C_5\|\omega\|_{p,B}\right] \qquad \text{(Lemma 6)}$$

$$= C|B|^{\frac{1}{n}+\frac{1}{q}-\frac{1}{p}}\|\omega\|_{p,B}.$$

$\square$

4 Applications

It can be quite hard to obtain the upper bounds of the composite operator. The following example shows that the norm inequalities proved in this paper provide us easy ways to estimate the upper bounds for the norms of the composite operator.

Example 1 Let $\Omega = \{(x, y, z) : x^2 + y^2 + z^2 < 1\} \subset \mathbb{R}^3$ and ω be the following 1-form defined by

$$\omega(x, y, z) = \frac{xdx + ydy + zdz}{\sqrt{1 + x^2 + y^2 + z^2}}.$$

Then, ω is smooth in Ω and it is a solution of the A-harmonic equation for any operators A satisfying the required conditions mentioned in Sect. 2.4. Note that

$$|\omega(x, y, z)| = \frac{\sqrt{x^2 + y^2 + z^2}}{\sqrt{1 + x^2 + y^2 + z^2}} < 1.$$

Therefore,

$$\|\omega\|_{p,\Omega} = \left(\int_{\Omega} |\omega|^p dv\right)^{1/p} \leq |\Omega| = 4\pi/3.$$

Using Lemma 3 and Corollary 1, we obtain the following upper bound estimates for $T\omega$ and $dP\omega$

$$\|T\omega\|_{p,\Omega} \leq C_1 \|\omega\|_{p,\Omega} \leq C_2$$

and

$$\|dP\omega\|_{p,\Omega} \leq C_3 \|\omega\|_{p,\Omega} \leq C_4,$$

respectively.

Example 2 Similar to the case in $\mathbb{R}^3$, we can easily check that the following differential form

$$\omega(x_1, \cdots x_n) = \sum_{i=1}^{n} \frac{x_i}{\sqrt{1 + x_1^2 + \cdots + x_n^2}} dx_i$$

is also a solution of the A-harmonic equation for any operators A. Using Lemma 3 and Corollary 1 proved above as we did in the last example, we obtain the upper bound estimates for the operators T and dP applied to the 1-form ω defined above.

We conclude with two remarks: First, the inequalities proved in Theorem 3 can also be extended into the global cases, such as the L^p-averaging domains; see [1] or [13] for more properties of L^p-averaging domains. Second, most of the L^p norm inequalities obtained in this paper can be extended to the L^ψ estimates, where ψ is a convex function.

References

1. R.P. Agarwal, S. Ding, C. Nolder, *Inequalities for Differential Forms* (Springer, New York, 2009). MR 2552910
2. H. Bi, Weighted inequalities for potential operators on differential forms. J. Inequal. Appl. **2010**, Art. ID 713625 (2010). MR 2600211

3. S.M. Buckley, P. Koskela, Orlicz-Hardy inequalities. Ill. J. Math. **48**(3), 787–802 (2004). MR 2114252
4. H. Bi, Y. Xing, Poincaré-type inequalities with $L^p(\log L)^\alpha$-norms for Green's operator. Comput. Math. Appl. **60**(10), 2764–2770 (2010). MR 2734317
5. H. Cartan, *Differential Forms* (Translated from the French, Houghton). (Mifflin Co., Boston, MA, 1970). MR 0267477
6. S. Ding, B. Liu, Dirac-harmonic equations for differential forms. Nonlinear Anal. **122**, 43–57 (2015). MR 3348064
7. S. Ding, G. Shi, Y. Xing, Higher integrability of iterated operators on differential forms. Nonlinear Anal. **145**, 83–96 (2016). MR 3547675
8. V. Gol'dshtein, M. Troyanov, Sobolev inequalities for differential forms and $L_{q,p}$-cohomology. J. Geom. Anal. **16**(4), 597–631 (2006). MR 2271946
9. T. Iwaniec, A. Lutoborski, Integral estimates for null Lagrangians. Arch. Ration. Mech. Anal. **125**(1), 25–79 (1993). MR 1241286
10. J.M. Martell, Fractional integrals, potential operators and two-weight, weak type norm inequalities on spaces of homogeneous type. J. Math. Anal. Appl. **294**(1), 223–236 (2004). MR 2059882
11. C.A. Nolder, Hardy-Littlewood theorems for A-harmonic tensors. Ill. J. Math. **43**(4), 613–632 (1999). MR 1712513
12. C. Scott, L^p theory of differential forms on manifolds. Trans. Am. Math. Soc. **347**(6), 2075–2096 (1995). MR 1297538
13. S.G. Staples, L^p-averaging domains and the Poincaré inequality. Ann. Acad. Sci. Fenn. Ser. A I Math. **14**(1), 103–127 (1989). MR 997974
14. F.W. Warner, *Foundations of Differentiable Manifolds and Lie Groups*. Graduate Texts in Mathematics, vol. 94 (Springer, New York/Berlin, 1983). Corrected reprint of the 1971 edition. MR 722297
15. Y. Wang, C. Wu, Global Poincaré inequalities for Green's operator applied to the solutions of the nonhomogeneous A-harmonic equation. Comput. Math. Appl. **47**(10–11), 1545–1554 (2004). MR 2079864

Hadamard Integral Inequality for the Class of Harmonically (γ, η)-Convex Functions

Hamid Vosoughian, Sadegh Abbaszadeh, and Maryam Oraki

Abstract In this paper, harmonically (γ, η)-convex inequality is introduced as

$$f\left(\frac{1}{\gamma_{\frac{1}{y}, \frac{1}{x}}(t)}\right) \leq \frac{1}{\eta_{\frac{1}{f(y)}, \frac{1}{f(x)}}(t)},$$

in which γ and η are two geodesic arcs. Then, some refinements of Hadamard integral inequality for harmonically (γ, η)-convex functions in the case of Lebesgue and Sugeno integral are studied.

1 Introduction

Harmonic mean is a kind of averaging functions which is a generalization of arithmetic mean and has applications in many branches of science. In the theory of aggregation functions, harmonic mean is a quasi-arithmetic and anti-Lagrangian mean generated by the function $\frac{1}{x^2}$. Based on the definition of harmonic mean, the concept of harmonically convexity was introduced by Das [7]. A function $f : I \subset \mathbb{R}/\{0\} \to \mathbb{R}$ is said to be harmonically convex, if

$$f\left(\frac{xy}{(1-t)x + ty}\right) \leq (1-t)f(x) + tf(y) \tag{1}$$

H. Vosoughian
Department of Mathematics, Faculty of Mathematics, Statistics and Computer Sciences, Semnan University, Semnan, Iran

S. Abbaszadeh (✉)
Department of Computer Science, Paderborn University, Paderborn, Germany

M. Oraki
Department of Mathematics, Payame Noor University, Tehran, Iran

© Springer Nature Switzerland AG 2019
G. A. Anastassiou and J. M. Rassias (eds.), *Frontiers in Functional Equations and Analytic Inequalities*, https://doi.org/10.1007/978-3-030-28950-8_34

for all $x, y \in I$ and $t \in [0, 1]$. Like other generalization of convexity, harmonically convexity may be exploited in multi-objective nonlinear programming problems [8, 11], where the global optimum is the point in which the direction of the monotonicity is changed.

The Hadamard inequality provides an upper bound for the mean value of a convex function $f : [a, b] \longrightarrow \mathbb{R}$,

$$\int_0^1 f\big((1 - t)a + tb\big)dt \leq \frac{f(a) + f(b)}{2}. \tag{2}$$

The above inequality would be reversed in the case that f is concave. Several refinements of Hadamard inequality for the class of harmonically convex functions have been studied by different authors; Işcan [9, 10] obtained new estimates on generalization of Hadamard inequality for harmonically quasi-convex functions, via the Riemann–Liouville fractional integral. Chen [6] gave some extensions of the Hermite–Hadamard inequality for harmonically convex functions via fractional integrals. The aim of this paper is to consider the inequality (2) for a new class of convex functions, which is called harmonically (γ, η)-convex functions.

This paper follows this organization: Some preliminaries and specifically, the definition of Sugeno integral and its properties are presented in Sect. 2. In Sect. 3, harmonically (γ, η)-convex functions are introduced, and some refinements of Hadamard inequality for this new class are considered. In Sect. 4, some applications for harmonically (γ, η)-convexity in the case of new geodesic arcs are dealt with. Eventually, a conclusion is given in Sect. 5.

2   Preliminaries

The geodesic path is defined as follows.

Definition 1 ([15, 16]) A geodesic is a $\mathbb{C}^\infty$ smooth path γ whose tangent is parallel along the path γ. Let M be a complete n-dimensional Riemannian manifold. For all $x, y \in M$, the mapping $\gamma_{x,y} : [0, 1] \to M$ is a geodesic joining the points x and y if $\gamma_{x,y}(0) = y$ and $\gamma_{x,y}(1) = x$.

Let X be a non-empty set and Σ be a σ-algebra of subsets of X.

Definition 2 ([14]) Let $\mu : \Sigma \longrightarrow [0, \infty)$ be a set function. μ is called a fuzzy measure if

1. $\mu(\emptyset) = 0$.
2. $E, F \in \Sigma$ and $E \subset F$ imply $\mu(E) \leq \mu(F)$.
3. $E_n \in \Sigma$ $(n \in \mathbb{N})$, $E_1 \subset E_2 \subset \ldots$, imply
   $\lim_{n\to\infty} \mu(E_n) = \mu(\bigcup_{n=1}^\infty E_n)$ (continuity from below).
4. $E_n \in \Sigma$ $(n \in \mathbb{N})$, $E_1 \supset E_2 \supset \ldots$, $\mu(E_1) < \infty$, imply
   $\lim_{n\to\infty} \mu(E_n) = \mu(\bigcap_{n=1}^\infty E_n)$ (continuity from above).

The triple (X, Σ, μ) is called a fuzzy measure space.

Let (X, Σ, μ) be a fuzzy measure space. The set of all μ-measurable functions, $\mathcal{F}_\mu(X)$, is defined as

$$\mathcal{F}_\mu(X) = \{f : X \longrightarrow [0, \infty) : f \text{ is measurable with respect to } \Sigma\}.$$

For $f \in \mathcal{F}_\mu(X)$ and $\alpha > 0$, we denote by F_α and $F_{\tilde{\alpha}}$ the following sets

$$F_\alpha = \{x \in X : f(x) \geq \alpha\} \quad \text{and} \quad F_{\tilde{\alpha}} = \{x \in X : f(x) > \alpha\}.$$

Note that if $\alpha \leq \beta$, then $F_\beta \subset F_\alpha$ and $F_{\tilde{\beta}} \subset F_{\tilde{\alpha}}$.

Definition 3 ([12, 17, 19]) Let (X, Σ, μ) be a fuzzy measure space, $f \in \mathcal{F}_\mu(X)$ and $A \in \Sigma$, then the Sugeno integral of f on A with respect to the fuzzy measure μ is defined by

$$(S) \int_A f d\mu = \bigvee_{\alpha \geq 0} \left(\alpha \wedge \mu(A \cap F_\alpha)\right),$$

where $\wedge$ is just the prototypical t-norm minimum and $\vee$ the prototypical t-conorm maximum. If $A = X$, then

$$(S) \int_A f d\mu = \bigvee_{\alpha \geq 0} \left(\alpha \wedge \mu(F_\alpha)\right).$$

The following properties of Sugeno integral are well known and can be found in [12, 19].

Theorem 1 *Let (X, Σ, μ) be a fuzzy measure space, $A, B \in \Sigma$ and $f, g \in \mathcal{F}_\mu(X)$; then*

(F_1) $(S) \int_A f d\mu \leq \mu(A)$.
(F_2) $(S) \int_A k d\mu = k \wedge \mu(A)$, k *non-negative constant.*
(F_3)   *If $f \leq g$ on A then $(S) \int_A f d\mu \leq (S) \int_A g d\mu$.*
(F_4)   *If $A \subset B$ then $(S) \int_A f d\mu \leq (S) \int_B f d\mu$.*

Most well-known integral inequalities have been proved for Sugeno integral, see [1–5, 13, 18].

3   Main Results

From now on, let (X, Σ, μ) be a fuzzy measure space. If $f \in \mathcal{F}^\mu(X)$ and $A \in \Sigma$, we set

$$\Gamma = \left\{\alpha \mid \alpha \geq 0, \mu(A \cap F_\alpha) > \mu(A \cap F_\beta) \text{ for any } \beta > \alpha\right\}.$$

Then

$$(S) \int_A f \mathrm{d}\mu = \bigvee_{\alpha \in \Gamma} \left(\alpha \wedge \mu(A \cap F_\alpha) \right).$$

If $X = \mathbb{R}$ is the set of real numbers, Σ is the Borel field and μ is the Lebesgue measure, (X, Σ, μ) is a fuzzy measure space; but it is important to know that the Sugeno integral is not an extension of the Lebesgue integral.

The main contribution of this paper is to study a new concept of convexity in which both sides of convex inequality are based on harmonic mean. This generalized form is defined by

$$f\left(\frac{xy}{(1-t)x + ty}\right) \leq \frac{f(x)f(y)}{(1-t)f(x) + tf(y)} \tag{3}$$

for all $x, y \in I$ and $t \in [0, 1]$.

Theorem 2 *Let $I \subseteq \mathbb{R}^+$ and $a, b \in I$ with $a < b$. If $f : I \to I$ is function satisfying (3) with $af(b) \neq bf(a)$, then*

$$\frac{1}{b-a} \int_a^b f(x)\mathrm{d}x$$

$$\leq \frac{f(a)f(b)(b-a)}{bf(a) - af(b)} + \frac{ab(f(b) - f(a))f(a)f(b)}{(bf(a) - af(b))^2} \left(\ln\left(\frac{b}{a}\right) + \ln\left(\frac{f(a)}{f(b)}\right) \right).$$

Proof One can easily see that the function $f : [a, b] \to [a, b]$, where $a, b \in I \subset \mathbb{R}^+$ and $a < b$, satisfies the inequality

$$f(x) \leq \frac{f(a)f(b)(b-a)x}{(bf(a) - af(b))x + ab(f(b) - f(a))}.$$

Integrating both sides of the above inequality over $[a, b]$, we obtain the assertion of theorem. $\square$

Consequently, the concept of harmonically (γ, η)-convex functions is introduced as follows:

Definition 4 For two closed subintervals I and J of $(0, +\infty)$, let $\gamma_{x,y} : [0, 1] \to I$ be a geodesic arc joining the points $x, y \in I$ and $\eta_{u,v} : [0, 1] \to J$ be a geodesic arc joining the points $u, v \in J$. A real valued function $f : I \to J$ is said to be harmonically (γ, η)-convex if

$$f\left(\frac{1}{\gamma_{\frac{1}{y}, \frac{1}{x}}(t)}\right) \leq \frac{1}{\eta_{\frac{1}{f(y)}, \frac{1}{f(x)}}(t)}$$

for all $t \in [0, 1]$.

Remark 1 By the following method, one can distinguish different cases of harmonically (γ, η)-convex functions from each other: If $f : [a, b] \rightarrow [c, d]$ satisfies the inequality

$$f(x) = f\left(\frac{1}{\gamma_{\frac{1}{b}, \frac{1}{a}}\left(\gamma_{\frac{1}{b}, \frac{1}{a}}^{-1}\left(\frac{1}{x}\right)\right)}\right) \leq \left(\frac{1}{\eta_{\frac{1}{f(b)}, \frac{1}{f(a)}}\left(\gamma_{\frac{1}{b}, \frac{1}{a}}^{-1}\left(\frac{1}{x}\right)\right)}\right) \tag{4}$$

for all $x \in [a, b]$, then f is a harmonically (γ, η)-convex function. For the geodesic arcs $\gamma : [0, 1] \rightarrow I$ and $\eta : [0, 1] \rightarrow J$ defined by $\gamma_{x,y}(t) = (1 - t)x + ty$ and $\eta_{u,v}(t) = u^{1-t}v^t$, respectively, by (4) we have the following inequalities:

- Harmonically (γ, γ)-convex functions satisfy the inequality

$$f(x) \leq \left(\frac{1}{f(a)} + \left(\frac{1}{f(b)} - \frac{1}{f(a)}\right)\left(\frac{ab - bx}{ax - bx}\right)\right)^{-1} \tag{5}$$

for all $x \in [a, b]$.
- Harmonically (γ, η)-convex functions satisfy the inequality

$$f(x) \leq f(a)\left(\frac{f(b)}{f(a)}\right)^{\frac{ab-bx}{ax-bx}} \tag{6}$$

for all $x \in [a, b]$.
- Harmonically (η, γ)-convex functions satisfy the inequality

$$f(x) \leq \left(\frac{1}{f(a)} + \left(\frac{1}{f(b)} - \frac{1}{f(a)}\right)\left(\frac{\ln(a) - \ln(x)}{\ln(a) - \ln(b)}\right)\right)^{-1} \tag{7}$$

for all $x \in [a, b]$.
- Harmonically (η, η)-convex functions satisfy the inequality

$$f(x) \leq f(a)\left(\frac{f(b)}{f(a)}\right)^{\frac{\ln(a)-\ln(x)}{\ln(a)-\ln(b)}} \tag{8}$$

for all $x \in [a, b]$.

In the next theorem, we consider the refinement of Hadamard inequality for harmonically (γ, η)-convex functions in the context of nonlinear integrals.

Theorem 3 *Consider the fuzzy measure space* $(\mathbb{R}, \Sigma, \mu)$. *Let* $\gamma : [0, 1] \rightarrow [a, b]$ *and* $\eta : [0, 1] \rightarrow [c, d]$ *be two invertible geodesic arcs. If* $f : [a, b] \rightarrow [c, d]$ *is a harmonically* (γ, η)-*convex function, then*

$$(S)\int_a^b f\mathrm{d}\mu \le \begin{cases} \bigvee\limits_{\alpha\in[f(a),f(b))}\left(\alpha\wedge\mu\left(\left[\dfrac{1}{\gamma_{\frac{1}{b},\frac{1}{a}}\left(\eta^{-1}_{\frac{1}{f(b)},\frac{1}{f(a)}}(\frac{1}{\alpha})\right)},b\right]\right)\right), \\[2pt] \gamma,\eta \ are \ comonotone, \\[18pt] \bigvee\limits_{\alpha\in[f(b),f(a))}\left(\alpha\wedge\mu\left(\left[a,\dfrac{1}{\gamma_{\frac{1}{b},\frac{1}{a}}\left(\eta^{-1}_{\frac{1}{f(b)},\frac{1}{f(a)}}(\frac{1}{\alpha})\right)}\right]\right)\right), \\[2pt] \gamma,\eta \ are \ countermonotone. \end{cases}$$

Proof It follows from the (γ,η)-convexity of f and the property (F3) of fuzzy measures that

$$(S)\int_a^b f(x)\mathrm{d}\mu = (S)\int_a^b f\left(\frac{1}{\gamma_{\frac{1}{b},\frac{1}{a}}\left(\gamma^{-1}_{\frac{1}{b},\frac{1}{a}}(\frac{1}{x})\right)}\right)\mathrm{d}\mu$$

$$\le (S)\int_a^b \left(\frac{1}{\eta_{\frac{1}{f(b)},\frac{1}{f(a)}}\left(\gamma^{-1}_{\frac{1}{b},\frac{1}{a}}(\frac{1}{x})\right)}\right)\mathrm{d}\mu. \tag{9}$$

Firstly, we assume that γ and η are comonotone, then $\eta\circ\gamma^{-1}$ is an increasing function. So, according to Definition 3, we have

$$(S)\int_a^b \left(\frac{1}{\eta_{\frac{1}{f(b)},\frac{1}{f(a)}}\left(\gamma^{-1}_{\frac{1}{b},\frac{1}{a}}(\frac{1}{x})\right)}\right)\mathrm{d}\mu$$

$$= \bigvee_{\alpha>0}\left(\alpha\wedge\mu\left([a,b]\cap\frac{1}{\eta_{\frac{1}{f(b)},\frac{1}{f(a)}}\left(\gamma^{-1}_{\frac{1}{b},\frac{1}{a}}(\frac{1}{x})\right)}\ge\alpha\right)\right)$$

$$= \bigvee_{\alpha>0}\left(\alpha\wedge\mu\left([a,b]\cap\eta_{\frac{1}{f(b)},\frac{1}{f(a)}}\left(\gamma^{-1}_{\frac{1}{b},\frac{1}{a}}(\frac{1}{x})\right)\le(\frac{1}{\alpha})\right)\right)$$

$$= \bigvee_{\alpha>0}\left(\alpha\wedge\mu\left([a,b]\cap\{\frac{1}{x}\le\gamma_{\frac{1}{b},\frac{1}{a}}\left(\eta^{-1}_{\frac{1}{f(b)},\frac{1}{f(a)}}(\frac{1}{\alpha})\right)\}\right)\right) \tag{10}$$

$$= \bigvee_{\alpha>0}\left(\alpha\wedge\mu\left([a,b]\cap\{x\ge\frac{1}{\gamma_{\frac{1}{b},\frac{1}{a}}\left(\eta^{-1}_{\frac{1}{f(b)},\frac{1}{f(a)}}(\frac{1}{\alpha})\right)}\}\right)\right)$$

$$= \bigvee_{\alpha>0}\left(\alpha\wedge\mu\left(\left[\frac{1}{\gamma_{\frac{1}{b},\frac{1}{a}}\left(\eta^{-1}_{\frac{1}{f(b)},\frac{1}{f(a)}}(\frac{1}{\alpha})\right)},b\right]\right)\right).$$

Since $\eta \circ \gamma^{-1}$ is increasing, we have

$$a \leq \frac{1}{\gamma_{\frac{1}{b},\frac{1}{a}}\left(\eta^{-1}_{\frac{1}{f(b)},\frac{1}{f(a)}}\left(\frac{1}{\alpha}\right)\right)} < b$$

$$\Rightarrow \frac{1}{b} < \gamma_{\frac{1}{b},\frac{1}{a}}\left(\eta^{-1}_{\frac{1}{f(b)},\frac{1}{f(a)}}\left(\frac{1}{\alpha}\right)\right) \leq \frac{1}{a}$$

$$\Rightarrow \eta_{\frac{1}{f(b)},\frac{1}{f(a)}}\left(\gamma^{-1}_{\frac{1}{b},\frac{1}{a}}\left(\frac{1}{b}\right)\right) < \frac{1}{\alpha} \leq \eta_{\frac{1}{f(b)},\frac{1}{f(a)}}\left(\gamma^{-1}_{\frac{1}{b},\frac{1}{a}}\left(\frac{1}{a}\right)\right)$$

$$\Rightarrow \frac{1}{\eta_{\frac{1}{f(b)},\frac{1}{f(a)}}\left(\gamma^{-1}_{\frac{1}{b},\frac{1}{a}}\left(\frac{1}{a}\right)\right)} \leq \alpha < \frac{1}{\eta_{\frac{1}{f(b)},\frac{1}{f(a)}}\left(\gamma^{-1}_{\frac{1}{b},\frac{1}{a}}\left(\frac{1}{b}\right)\right)} \qquad (11)$$

$$\Rightarrow \frac{1}{\eta_{\frac{1}{f(b)},\frac{1}{f(a)}}(1)} \leq \alpha < \frac{1}{\eta_{\frac{1}{f(b)},\frac{1}{f(a)}}(0)}$$

$$\Rightarrow f(a) \leq \alpha < f(b).$$

Thus, $\Gamma = \left[f(a), f(b)\right)$ and we only need to consider $\alpha \in \left[f(a), f(b)\right)$. It follows from (9), (10), and (11) that

$$(S)\int_a^b \left(\frac{1}{\eta_{\frac{1}{f(b)},\frac{1}{f(a)}}\left(\gamma^{-1}_{\frac{1}{b},\frac{1}{a}}\left(\frac{1}{x}\right)\right)}\right)d\mu$$

$$\leq \bigvee_{\alpha \in \left[f(b),f(a)\right]} \left(\alpha \wedge \mu\left(\left[\frac{1}{\gamma_{\frac{1}{b},\frac{1}{a}}\left(\eta^{-1}_{\frac{1}{f(b)},\frac{1}{f(a)}}\left(\frac{1}{\alpha}\right)\right)}, b\right]\right)\right).$$

Secondly, we assume that γ and η are countermonotone, then $\eta \circ \gamma^{-1}$ is a decreasing function. So, by Definition 3 we have

$$(S)\int_a^b \left(\frac{1}{\eta_{\frac{1}{f(b)},\frac{1}{f(a)}}\left(\gamma^{-1}_{\frac{1}{b},\frac{1}{a}}\left(\frac{1}{x}\right)\right)}\right)d\mu$$

$$= \bigvee_{\alpha > 0} \left(\alpha \wedge \mu\left([a,b] \cap \frac{1}{\eta_{\frac{1}{f(b)},\frac{1}{f(a)}}\left(\gamma^{-1}_{\frac{1}{b},\frac{1}{a}}\left(\frac{1}{x}\right)\right)} \geq \alpha\right)\right)$$

$$= \bigvee_{\alpha > 0} \left(\alpha \wedge \mu\left([a,b] \cap \eta_{\frac{1}{f(b)},\frac{1}{f(a)}}\left(\gamma^{-1}_{\frac{1}{b},\frac{1}{a}}\left(\frac{1}{x}\right)\right) \leq \left(\frac{1}{\alpha}\right)\right)\right) \qquad (12)$$

$$= \bigvee_{\alpha > 0} \left(\alpha \wedge \mu\left([a,b] \cap \{\frac{1}{x} \geq \gamma_{\frac{1}{b},\frac{1}{a}}\left(\eta^{-1}_{\frac{1}{f(b)},\frac{1}{f(a)}}\left(\frac{1}{\alpha}\right)\right)\}\right)\right)$$

$$= \bigvee_{\alpha > 0} \left(\alpha \wedge \mu \left([a, b] \cap \{ x \le \frac{1}{\gamma_{\frac{1}{b}, \frac{1}{a}} \left(\eta^{-1}_{\frac{1}{f(b)}, \frac{1}{f(a)}} (\frac{1}{\alpha}) \right)} \} \right) \right)$$

$$= \bigvee_{\alpha > 0} \left(\alpha \wedge \mu \left(\left[a, \frac{1}{\gamma_{\frac{1}{b}, \frac{1}{a}} \left(\eta^{-1}_{\frac{1}{f(b)}, \frac{1}{f(a)}} (\frac{1}{\alpha}) \right)} \right] \right) \right).$$

Since $\eta \circ \gamma^{-1}$ is decreasing, we have

$$a < \frac{1}{\gamma_{\frac{1}{b}, \frac{1}{a}} \left(\eta^{-1}_{\frac{1}{f(b)}, \frac{1}{f(a)}} (\frac{1}{\alpha}) \right)} \le b$$

$$\Rightarrow \frac{1}{b} \le \gamma_{\frac{1}{b}, \frac{1}{a}} \left(\eta^{-1}_{\frac{1}{f(b)}, \frac{1}{f(a)}} (\frac{1}{\alpha}) \right) < \frac{1}{a}$$

$$\Rightarrow \eta_{\frac{1}{f(b)}, \frac{1}{f(a)}} \left(\gamma^{-1}_{\frac{1}{b}, \frac{1}{a}} (\frac{1}{a}) \right) < \frac{1}{\alpha} \le \eta_{\frac{1}{f(b)}, \frac{1}{f(a)}} \left(\gamma^{-1}_{\frac{1}{b}, \frac{1}{a}} (\frac{1}{b}) \right) \qquad (13)$$

$$\Rightarrow \frac{1}{\eta_{\frac{1}{f(b)}, \frac{1}{f(a)}} \left(\gamma^{-1}_{\frac{1}{b}, \frac{1}{a}} (\frac{1}{b}) \right)} \le \alpha < \frac{1}{\eta_{\frac{1}{f(b)}, \frac{1}{f(a)}} \left(\gamma^{-1}_{\frac{1}{b}, \frac{1}{a}} (\frac{1}{a}) \right)}$$

$$\Rightarrow \frac{1}{\eta_{\frac{1}{f(b)}, \frac{1}{f(a)}} (1)} \le \alpha < \frac{1}{\eta_{\frac{1}{f(b)}, \frac{1}{f(a)}} (0)}$$

$$\Rightarrow f(b) \le \alpha < f(a).$$

Thus, $\Gamma = [f(b), f(a))$ and we only need to consider $\alpha \in [f(b), f(a))$. It follows from (9), (12), and (13) that

$$(S) \int_a^b \left(\frac{1}{\eta_{\frac{1}{f(b)}, \frac{1}{f(a)}} \left(\gamma^{-1}_{\frac{1}{b}, \frac{1}{a}} (\frac{1}{x}) \right)} \right) d\mu$$

$$\le \bigvee_{\alpha \in [f(b), f(a))} \left(\alpha \wedge \mu \left(\left[a, \frac{1}{\gamma_{\frac{1}{b}, \frac{1}{a}} \left(\eta^{-1}_{\frac{1}{f(b)}, \frac{1}{f(a)}} (\frac{1}{\alpha}) \right)} \right] \right) \right).$$

$\square$

Corollary 1 *Let $f : [a, b] \to [c, d]$ be a harmonically (γ, η)-convex function, Σ be the Borel field, and μ be the Lebesgue measure on $\mathbb{R}$. Then*

$$(S) \int_a^b f \, d\mu \leq \begin{cases} \displaystyle\bigvee_{\alpha \in [f(a), f(b))} \left(\alpha \wedge \left(b - \frac{1}{\gamma_{\frac{1}{b}, \frac{1}{a}} \left(\eta^{-1}_{\frac{1}{f(b)}, \frac{1}{f(a)}} (\frac{1}{\alpha}) \right)} \right) \right), \\ \quad \gamma, \eta \text{ are comonotone,} \\[2em] \displaystyle\bigvee_{\alpha \in [f(b), f(a))} \left(\alpha \wedge \left(\frac{1}{\gamma_{\frac{1}{b}, \frac{1}{a}} \left(\eta^{-1}_{\frac{1}{f(b)}, \frac{1}{f(a)}} (\frac{1}{\alpha}) \right)} - a \right) \right), \\ \quad \gamma, \eta \text{ are countermonotone.} \end{cases}$$

The rest of the results of this section are concerned about investigating particular geodesic arcs $\gamma : [0, 1] \to I$ and $\eta : [0, 1] \to J$ defined by $\gamma_{x,y}(t) = (1 - t)x + ty$ and $\eta_{u,v}(t) = u^{1-t} v^t$.

Theorem 4 *Let* $(\mathbb{R}, \Sigma, \mu)$ *be a fuzzy measure space and* $f : [a, b] \to [a, b]$ *be a harmonically* (γ, γ)*-convex function. Then*

$$(S) \int_a^b f \, d\mu \leq \begin{cases} \displaystyle\bigvee_{\alpha \in [f(a), f(b))} \left(\alpha \wedge \mu \left(\left[\frac{ab\alpha(f(b) - f(a))}{f(a)(f(b) - \alpha)(b - a) + a\alpha(f(b) - f(a))}, b \right] \right) \right), \\ \quad f(a) < f(b), \\[1.5em] f(a) \wedge \mu([a, b]), \quad f(a) = f(b), \\[1.5em] \displaystyle\bigvee_{\alpha \in [f(b), f(a))} \left(\alpha \wedge \mu \left(\left[a, \frac{ab\alpha(f(b) - f(a))}{f(a)(f(b) - \alpha)(b - a) + a\alpha(f(b) - f(a))} \right] \right) \right), \\ \quad f(a) > f(b). \end{cases}$$

Proof It is easy to obtain that

$$\frac{1}{\gamma_{\frac{1}{b}, \frac{1}{a}} \left(\gamma^{-1}_{\frac{1}{f(b)}, \frac{1}{f(a)}} (\frac{1}{\alpha}) \right)} = \frac{ab\alpha(f(b) - f(a))}{f(a)(f(b) - \alpha)(b - a) + a\alpha(f(b) - f(a))}$$

and the assertion of the theorem comes from the assertion of Theorem 3 with particular geodesic arcs. $\qquad \square$

Theorem 5 *Let* $(\mathbb{R}, \Sigma, \mu)$ *be a fuzzy measure space and* $f : [a, b] \to [c, d]$ *be a harmonically* (γ, η)*-convex function. Then*

$$(S) \int_a^b f \, d\mu \le \begin{cases} \displaystyle\bigvee_{\alpha \in [f(a), f(b)]} \left(\alpha \wedge \mu \left(\left[\frac{ab}{(b-a)\left(\log_{\frac{f(b)}{f(a)}} \frac{f(b)}{\alpha} \right) + a}, b \right] \right) \right), \\ \qquad f(a) < f(b), \\[2ex] f(a) \wedge \mu([a, b]), \qquad f(a) = f(b), \\[2ex] \displaystyle\bigvee_{\alpha \in [f(b), f(a)]} \left(\alpha \wedge \mu \left(\left[a, \frac{ab}{(b-a)\left(\log_{\frac{f(b)}{f(a)}} \frac{f(b)}{\alpha} \right) + a} \right] \right) \right), \\ \qquad f(a) > f(b). \end{cases}$$

Proof One can easily see that

$$\frac{1}{\gamma_{\frac{1}{b}, \frac{1}{a}} \left(\gamma^{-1}_{\frac{1}{f(b)}, \frac{1}{f(a)}} \left(\frac{1}{\alpha} \right) \right)} = \frac{ab}{(b-a)\left(\log_{\frac{f(b)}{f(a)}} \frac{f(b)}{\alpha} \right) + a}$$

and the assertion of the theorem comes from the assertion of Theorem 3 with particular geodesic arcs. □

Theorem 6 *Let* $(\mathbb{R}, \Sigma, \mu)$ *be a fuzzy measure space and* $f : [c, d] \to [a, b]$ *be a harmonically* (η, γ)-*convex function. Then*

$$(S) \int_a^b f \, d\mu \le \begin{cases} \displaystyle\bigvee_{\alpha \in [f(a), f(b)]} \left(\alpha \wedge \mu \left(\left[b \left(\frac{a}{b} \right)^{\frac{f(a)(f(b)-\alpha)}{\alpha(f(b)-f(a))}}, b \right] \right) \right), \\ \qquad f(a) < f(b), \\[2ex] f(a) \wedge \mu([a, b]), \qquad f(a) = f(b), \\[2ex] \displaystyle\bigvee_{\alpha \in [f(b), f(a)]} \left(\alpha \wedge \mu \left(\left[a, b \left(\frac{a}{b} \right)^{\frac{f(a)(f(b)-\alpha)}{\alpha(f(b)-f(a))}} \right] \right) \right), \\ \qquad f(a) > f(b). \end{cases}$$

Proof Obviously,

$$\frac{1}{\gamma_{\frac{1}{b}, \frac{1}{a}} \left(\gamma^{-1}_{\frac{1}{f(b)}, \frac{1}{f(a)}} \left(\frac{1}{\alpha} \right) \right)} = b \left(\frac{a}{b} \right)^{\frac{f(a)(f(b)-\alpha)}{\alpha(f(b)-f(a))}}$$

and the assertion of the theorem concluded by the assertion of Theorem 3 with particular geodesic arcs. □

Theorem 7 *Let* $(\mathbb{R}, \Sigma, \mu)$ *be a fuzzy measure space and* $f : [c, d] \to [c, d]$ *be a harmonically* (η, η)-*convex function. Then*

$$(S) \int_a^b f \mathrm{d}\mu \leq \begin{cases} \bigvee_{\alpha \in \left[f(a), f(b)\right]} \left(\alpha \wedge \mu \left(\left[b \left(\frac{b}{a}\right)^{\log \frac{f(b)}{f(a)} \frac{\alpha}{f(b)}}, b \right] \right) \right), \\ \quad f(a) < f(b), \\[4pt] f(a) \wedge \mu \left([a, b] \right), \quad f(a) = f(b), \\[4pt] \bigvee_{\alpha \in \left[f(b), f(a)\right]} \left(\alpha \wedge \mu \left(\left[a, b \left(\frac{b}{a}\right)^{\log \frac{f(b)}{f(a)} \frac{\alpha}{f(b)}} \right] \right) \right), \\ \quad f(a) > f(b). \end{cases}$$

Proof Clearly,

$$\frac{1}{\eta_{\frac{1}{b}, \frac{1}{a}} \left(\eta^{-1}_{\frac{1}{f(b)}, \frac{1}{f(a)}} \left(\frac{1}{\alpha} \right) \right)} = b \left(\frac{b}{a} \right)^{\log \frac{f(b)}{f(a)} \frac{\alpha}{f(b)}}$$

and the assertion of the theorem concluded by the assertion of Theorem 3 with particular geodesic arcs. □

Example 1 Suppose that the geodesic arcs $\gamma : [0, 1] \to I$ and $\eta : [0, 1] \to J$ are defined by $\gamma_{x,y}(t) = (1 - t)x + ty$ and $\eta_{u,v}(t) = u^{1-t}v^t$, respectively. Denoting the right-hand side functions of the inequalities (5)–(8) by $g(x)$, we have

- The function $f : [1, 2] \to (0, +\infty)$ defined by $f(x) = \sqrt{x}$ is harmonically (γ, γ)-convex (see (5)), and satisfies the assertion of Theorem 4, see Fig. 1 left.

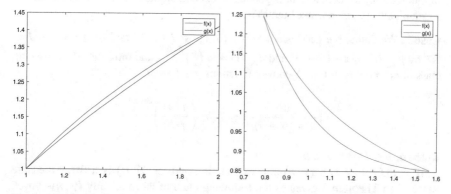

Fig. 1 A harmonically (γ, γ)-convex function (left) and a harmonically (γ, η)-convex function (right) dominated by the corresponding $g(x)$

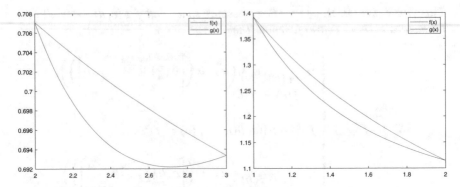

Fig. 2 A harmonically (η, γ)-convex function (left) and a harmonically (η, η)-convex function (right) dominated by the corresponding $g(x)$

- The function $f : \left[\frac{\pi}{4}, \frac{\pi}{2}\right] \to (0, +\infty)$ defined by $f(x) = \frac{1}{x}^{\sin^2(\frac{1}{x})}$ is harmonically (γ, η)-convex (see (6)), and satisfies the assertion of Theorem 5, see Fig. 2 right.

- The function $f : [2, 3] \to (0, +\infty)$ defined by $f(x) = \frac{1}{x}^{\frac{1}{x}}$ is harmonically (η, γ)-convex (see (7)), and satisfies the assertion of Theorem 6, see Fig. 1 left.

- The function $f : [1, 2] \to (0, +\infty]$ defined by $f(x) = \sqrt[4]{\cosh(\frac{2}{x})}$ is harmonically (η, η)-convex (see (8)), and satisfies the assertion of Theorem 7, see Fig. 2 right.

4 Application

In this section, we consider new geodesic arcs and applied them in the structure of harmonically (γ, η)-convex functions.

Case 1 We define the geodesic arcs $\gamma : [0, 1] \to I \subset \mathbb{R}^+$ and $\eta : [0, 1] \to J \subset \mathbb{R}^+$ by $\gamma_{x,y}(t) = x + (y - x)t$ and $\eta_{u,v}(t) = u\left(\frac{v}{u}\right)^{\sin\frac{\pi}{2}t}$, and introduce the following class of harmonically (γ, η)-convex functions $f : I \to J$:

$$f\left(\frac{ab}{a + (b - a)t}\right) \le f(b)\left(\frac{f(a)}{f(b)}\right)^{\sin\frac{\pi}{2}t},$$

where $a, b \in I$ with $a < b$.

If $f : I \to J$ is a (γ, η)-convex function and $(\mathbb{R}, \Sigma, \mu)$ is a fuzzy measure space, then Theorem 3 gives us the following class of harmonically (γ, η)-convex functions:

$$(S) \int_a^b f \, d\mu \leq \begin{cases} \bigvee_{\alpha \in [f(a), f(b)]} \left(\alpha \wedge \mu \left(\left[\dfrac{\pi ab}{\pi a + 2(b-a) \arcsin\left(\log_{\frac{f(b)}{f(a)}} \left(\frac{f(b)}{\alpha} \right) \right)}, b \right] \right) \right), \\ \qquad f(b) > f(a), \\[2em] \bigvee_{\alpha \in [f(b), f(a)]} \left(\alpha \wedge \mu \left(\left[a, \dfrac{\pi ab}{\pi a + 2(b-a) \arcsin\left(\log_{\frac{f(b)}{f(a)}} \left(\frac{f(b)}{\alpha} \right) \right)} \right] \right) \right), \\ \qquad f(a) > f(b). \end{cases}$$

By replacing $\eta_{u,v}(t) = u \left(\frac{v}{u} \right)^{\sin \frac{\pi}{2} t}$ with $\eta_{u,v}(t) = u \left(\frac{v}{u} \right)^{\cos \frac{\pi}{2} t}$ we obtain

$$f \left(\frac{ab}{a + (b-a)t} \right) \leq f(b) \left(\frac{f(a)}{f(b)} \right)^{\cos \frac{\pi}{2} t},$$

Thus, we obtain

$$(S) \int_a^b f \, d\mu \leq \begin{cases} \bigvee_{\alpha \in [f(a), f(b)]} \left(\alpha \wedge \mu \left(\left[\dfrac{\pi ab}{\pi a + 2(b-a) \arccos\left(\log_{\frac{f(b)}{f(a)}} \left(\frac{f(b)}{\alpha} \right) \right)}, b \right] \right) \right), \\ \qquad f(b) > f(a), \\[2em] \bigvee_{\alpha \in [f(b), f(a)]} \left(\alpha \wedge \mu \left(\left[a, \dfrac{\pi ab}{\pi a + 2(b-a) \arccos\left(\log_{\frac{f(b)}{f(a)}} \left(\frac{f(b)}{\alpha} \right) \right)} \right] \right) \right), \\ \qquad f(a) > f(b). \end{cases}$$

Case 2 We define the geodesic arcs $\gamma : [0, 1] \to I \subset \mathbb{R}^+$ and $\eta : [0, 1] \to J \subset \mathbb{R}^+$ by $\gamma_{x,y}(t) = x \left(\frac{y}{x} \right)^{\sin \frac{\pi}{2} t}$ and $\eta_{u,v}(t) = u + (v - u)t$, the following class of harmonically (γ, η)-convex functions $f : I \to J$ is introduced:

$$f \left(b \left(\frac{a}{b} \right)^{\sin \frac{\pi}{2} t} \right) \leq \frac{f(a) f(b)}{f(a) + (f(b) - f(a))t},$$

where $a, b \in I$ with $a < b$.

If $f : I \to J$ is a (γ, η)-convex function and $(\mathbb{R}, \Sigma, \mu)$ is a fuzzy measure space, then by Theorem 3:

$$(S) \int_a^b f \mathrm{d}\mu \leq \begin{cases} \bigvee_{\alpha \in [f(a), f(b)]} \left(\alpha \wedge \mu \left(\left[b \left(\frac{a}{b} \right)^{\sin\left(\frac{\pi f(a)(f(b)-\alpha)}{2\alpha(f(b)-f(a))} \right)}, b \right] \right) \right), \\ \quad f(b) > f(a), \\[2ex] \bigvee_{\alpha \in [f(b), f(a)]} \left(\alpha \wedge \mu \left(\left[a, b \left(\frac{a}{b} \right)^{\sin\left(\frac{\pi f(a)(f(b)-\alpha)}{2\alpha(f(b)-f(a))} \right)} \right] \right) \right), \\ \quad f(a) > f(b). \end{cases}$$

Also replacing $\gamma_{x,y}(t) = x \left(\frac{y}{x} \right)^{\sin \frac{\pi}{2} t}$ with $\gamma_{x,y}(t) = x \left(\frac{y}{x} \right)^{\cos \frac{\pi}{2} t}$ we have

$$f \left(b \left(\frac{a}{b} \right)^{\cos \frac{\pi}{2} t} \right) \leq \frac{f(a) f(b)}{f(a) + (f(b) - f(a))t},$$

So,

$$(S) \int_a^b f \mathrm{d}\mu \leq \begin{cases} \bigvee_{\alpha \in [f(a), f(b)]} \left(\alpha \wedge \mu \left(\left[b \left(\frac{a}{b} \right)^{\cos\left(\frac{\pi f(a)(f(b)-\alpha)}{2\alpha(f(b)-f(a))} \right)}, b \right] \right) \right), \\ \quad f(b) > f(a), \\[2ex] \bigvee_{\alpha \in [f(b), f(a)]} \left(\alpha \wedge \mu \left(\left[a, b \left(\frac{a}{b} \right)^{\cos\left(\frac{\pi f(a)(f(b)-\alpha)}{2\alpha(f(b)-f(a))} \right)} \right] \right) \right), \\ \quad f(a) > f(b). \end{cases}$$

Case 3 We define the geodesic arcs $\gamma : [0, 1] \to I \subset \mathbb{R}^+$ and $\eta : [0, 1] \to J \subset \mathbb{R}^+$ by $\gamma_{x,y}(t) = x \left(\frac{y}{x} \right)^t$ and $\eta_{u,v}(t) = u \left(\frac{v}{u} \right)^{\sin \frac{\pi}{2} t}$, the following class of harmonically (γ, η)-convex functions $f : I \to J$ is introduced:

$$f \left(b \left(\frac{a}{b} \right)^t \right) \leq f(b) \left(\frac{f(a)}{f(b)} \right)^{\sin \frac{\pi}{2} t},$$

where $a, b \in I$ with $a < b$.

If $f : I \to J$ is a (γ, η)-convex function and $(\mathbb{R}, \Sigma, \mu)$ is a fuzzy measure space, then according to Theorem 3 we have the following result:

$$(S)\int_a^b f d\mu \le \begin{cases} \bigvee\limits_{\alpha \in [f(a), f(b))} \left(\alpha \wedge \mu \left(\left[b\left(\frac{a}{b}\right)^{\frac{2}{\pi}\arcsin\left(\log_{\frac{f(b)}{f(a)}}\frac{f(b)}{\alpha}\right)}, b \right] \right) \right), \\ \qquad f(b) > f(a), \\ \\ \bigvee\limits_{\alpha \in [f(b), f(a))} \left(\alpha \wedge \mu \left(\left[a, b\left(\frac{a}{b}\right)^{\frac{2}{\pi}\arcsin\left(\log_{\frac{f(b)}{f(a)}}\frac{f(b)}{\alpha}\right)} \right] \right) \right), \\ \qquad f(a) > f(b). \end{cases}$$

By replacing $\eta_{u,v}(t) = u\left(\frac{v}{u}\right)^{\sin\frac{\pi}{2}t}$ with $\eta_{u,v}(t) = u\left(\frac{v}{u}\right)^{\cos\frac{\pi}{2}t}$ we get

$$f\left(b\left(\frac{a}{b}\right)^t\right) \le f(b)\left(\frac{f(a)}{f(b)}\right)^{\cos\frac{\pi}{2}t},$$

and so

$$(S)\int_a^b f d\mu \le \begin{cases} \bigvee\limits_{\alpha \in [f(a), f(b))} \left(\alpha \wedge \mu \left(\left[b\left(\frac{a}{b}\right)^{\frac{2}{\pi}\arccos\left(\log_{\frac{f(b)}{f(a)}}\frac{f(b)}{\alpha}\right)}, b \right] \right) \right), \\ \qquad f(b) > f(a), \\ \\ \bigvee\limits_{\alpha \in [f(b), f(a))} \left(\alpha \wedge \mu \left(\left[a, b\left(\frac{a}{b}\right)^{\frac{2}{\pi}\arccos\left(\log_{\frac{f(b)}{f(a)}}\frac{f(b)}{\alpha}\right)} \right] \right) \right), \\ \qquad f(a) > f(b). \end{cases}$$

Case 4 We define the geodesic arcs $\gamma : [0, 1] \to I \subset \mathbb{R}^+$ and $\eta : [0, 1] \to J \subset \mathbb{R}^+$ by $\gamma_{x,y}(t) = x\left(\frac{y}{x}\right)^{\sin\frac{\pi}{2}t}$ and $\eta_{u,v}(t) = u\left(\frac{v}{u}\right)^t$, the following class of geodesic convex functions $f : I \to J$ is introduced:

$$f\left(b\left(\frac{a}{b}\right)^{\sin\frac{\pi}{2}t}\right) \le f(b)\left(\frac{f(a)}{f(b)}\right)^t,$$

where $a, b \in I$ with $a < b$.

If $f : I \to J$ is a (γ, η)-convex function and $(\mathbb{R}, \Sigma, \mu)$ is a fuzzy measure space, then according to Theorem 3 we obtain the following result:

$$(S) \int_a^b f \, d\mu \leq \begin{cases} \bigvee_{\alpha \in [f(a), f(b)]} \left(\alpha \wedge \mu \left(\left[b \left(\frac{a}{b} \right)^{\sin \left(\frac{\pi}{2} \log_{\frac{f(b)}{f(a)}} \frac{f(b)}{\alpha} \right)}, b \right] \right) \right), \\ \qquad f(b) > f(a), \\ \bigvee_{\alpha \in [f(b), f(a)]} \left(\alpha \wedge \mu \left(\left[a, b \left(\frac{a}{b} \right)^{\sin \left(\frac{\pi}{2} \log_{\frac{f(b)}{f(a)}} \frac{f(b)}{\alpha} \right)} \right] \right) \right), \\ \qquad f(a) > f(b). \end{cases}$$

In addition, by replacing $\gamma_{x,y}(t) = x \left(\frac{y}{x} \right)^{\sin \frac{\pi}{2} t}$ with $\gamma_{x,y}(t) = x \left(\frac{y}{x} \right)^{\cos \frac{\pi}{2} t}$, we have

$$f \left(b \left(\frac{a}{b} \right)^{\cos \frac{\pi}{2} t} \right) \leq f(b) \left(\frac{f(a)}{f(b)} \right)^t,$$

and then

$$(S) \int_a^b f \, d\mu \leq \begin{cases} \bigvee_{\alpha \in [f(a), f(b)]} \left(\alpha \wedge \mu \left(\left[b \left(\frac{a}{b} \right)^{\cos \left(\frac{\pi}{2} \log_{\frac{f(b)}{f(a)}} \frac{f(b)}{\alpha} \right)}, b \right] \right) \right), \\ \qquad f(b) > f(a), \\ \bigvee_{\alpha \in [f(b), f(a)]} \left(\alpha \wedge \mu \left(\left[a, b \left(\frac{a}{b} \right)^{\cos \left(\frac{\pi}{2} \log_{\frac{f(b)}{f(a)}} \frac{f(b)}{\alpha} \right)} \right] \right) \right), \\ \qquad f(a) > f(b). \end{cases}$$

5 Conclusion

Any generalization of the concept of convexity may cause a modification in nonlinear optimization, e.g., in necessary and sufficient optimality criteria and in the connectedness of the solution set in linear and nonlinear complementarity systems. This paper deals with the concept of harmonically convexity, and introduces some generalizations for this concept. Moreover, some refinements of Hadamard inequality for harmonically convex functions are considered, in the context of linear and nonlinear integrals.

References

1. S. Abbaszadeh, A. Ebadian, Nonlinear integrals and Hadamard-type inequalities. Soft Comput. **22**, 2843–2849 (2018)
2. S. Abbaszadeh, M. Eshaghi, A Hadamard type inequality for fuzzy integrals based on r-convex functions. Soft Comput. **20**, 3117–3124 (2016)
3. S. Abbaszadeh, M. Eshaghi, M. de la Sen, The Sugeno fuzzy integral of log-convex functions. J. Inequal. Appl. **2015**, 362 (2015)
4. S. Abbaszadeh, M.E. Gordji, E. Pap, A. Szakál, Jensen-type inequalities for Sugeno integral. Inf. Sci. **376**, 148–157 (2017)
5. S. Abbaszadeh, A. Ebadian, M. Jaddi, Hölder type integral inequalities with different pseudo-operations. Asian Eur. J. Math. **12**, 1950032 (2018)
6. F. Chen, Extensions of the Hermite-Hadamard inequality for harmonic convex functions via fractional integrals. Appl. Math. Comput. **268**, 121–128 (2015)
7. C. Das, *Mathematical Programming in Complex Space*, Ph.D. Dissertation, Sambalpur University (1975)
8. C. Das, K.L. Roy, K.N. Jena, Harmonic convexity and application to optimization problems. Math. Educ. India **37**, 58–64 (2003)
9. I. Işcan, On generalization of different type inequalities for harmonically quasi-convex functions via fractional integrals. Appl. Math. Comput. **275** 287–298 (2016)
10. I. Işcan, S. Wu, Hermite-Hadamard type inequalities for harmonically convex functions via fractional integrals. Appl. Math. Comput. **238**, 237–244 (2014)
11. L. Pang, W. Wang, Z. Xia, Duality theorems on multi-objective programming of generalized functions. Acta Math. Appl. Sin. **22**, 49–58 (2006)
12. E. Pap, *Null-Additive Set Functions* (Kluwer, Dordrecht, 1995)
13. E. Pap, M. Štrboja, Generalizations of integral inequalities for integrals based on nonadditive measures, in *Intelligent Systems: Models and Applications*, ed. by E. Pap. Topics in Intelligent Engineering and Informatics, vol. 3 (Springer, Berlin, 2013), pp. 3–22
14. D. Ralescu, G. Adams, The fuzzy integral. J. Math. Anal. Appl. **75**, 562–570 (1980)
15. T. Rapcsák, *Geodesic Convexity on* $\mathbb{R}^n$, *Generalized Convexity* (Springer, Berlin/Heidelberg, 1994), pp. 91–103
16. T. Rapcsák, *Smooth Nonlinear Optimization in* $\mathbb{R}^n$, vol. 19 (Springer, New York, 2013)
17. M. Sugeno, *Theory of Fuzzy Integrals and Its Applications*, Ph.D. Dissertation, Tokyo Institute of Technology (1974)
18. A. Szakál, E. Pap, S. Abbaszadeh, M.E. Gordji, Jensen inequality with subdifferential for Sugeno integral, in *Mexican International Conference on Artificial Intelligence* (Springer, Cham, 2016), pp. 193–199
19. Z. Wang, G. Klir, *Fuzzy Measure Theory* (Plenum, New York, 1992)

Norm Inequalities for Singular Integrals Related to Operators and Dirac-Harmonic Equations

Ravi P. Agarwal, Shusen Ding, and Yuming Xing

Abstract In this paper, we establish norm inequalities and imbedding inequalities for singular integrals of the solutions to the Dirac-harmonic equation and operators acting on these solutions. These inequalities can be used to study the integrability of the operators and their compositions.

1 Introduction

Let $\Omega \subset \mathbb{R}^n$, $n \geq 2$, be a domain with Lebesgue measure $|\Omega| < \infty$. Let B and σB be the balls with the same center and $\mathrm{diam}(\sigma B) = \sigma \mathrm{diam}(B)$. We do not distinguish the balls from cubes in this paper. We use $D'(\Omega, \wedge^l)$ to denote the space of all differential l-forms in Ω and $L^p(\Omega, \wedge^l)$ is the space of all l-forms $u(x) = \sum_I u_I(x)dx_I$ in Ω satisfying $\int_\Omega |u_I|^p < \infty$ for all ordered l-tuples I, $l = 1, 2, \cdots, n$. We always use $d : D'(\Omega, \wedge^l) \to D'(\Omega, \wedge^{l+1})$, $l = 0, 1, \cdots, n-1$, to denote the exterior derivative. The Hodge star operator $\star : \wedge^k \to \wedge^{n-k}$ is defined as follows. If

$$\omega = \omega_{i_1 i_2 \cdots i_k}(x_1, x_2, \cdots, x_n)dx_{i_1} \wedge dx_{i_2} \wedge \cdots \wedge dx_{i_k} = \omega_I dx_I, \quad i_1 < i_2 < \cdots < i_k,$$

is a differential k-form, then

$$\star\omega = \star(\omega_{i_1 i_2 \cdots i_k}dx_{i_1} \wedge dx_{i_2} \wedge \cdots \wedge dx_{i_k}) = (-1)^{\sum(I)}\omega_I dx_J,$$

R. P. Agarwal
Department of Mathematics, Texas A&M University-Kingsville, Kingsville, TX, USA
e-mail: Ravi.Agarwal@tamuk.edu

S. Ding
Department of Mathematics, Seattle University, Seattle, WA, USA
e-mail: sding@seattleu.edu

Y. Xing (✉)
Department of Mathematics, Harbin Institute of Technology, Harbin, People's Republic of China
e-mail: xyuming@hit.edu.cn

© Springer Nature Switzerland AG 2019
G. A. Anastassiou and J. M. Rassias (eds.), *Frontiers in Functional Equations and Analytic Inequalities*, https://doi.org/10.1007/978-3-030-28950-8_35

where $I = (i_1, i_2, \cdots, i_k)$, $J = \{1, 2, \cdots, n\} - I$, and $\sum(I) = \frac{k(k+1)}{2} + \sum_{j=1}^{k} i_j$.
The Hodge codifferential operator $d^\star : D'(\Omega, \wedge^{l+1}) \to D'(\Omega, \wedge^l)$ is defined by
$d^\star = (-1)^{nl+1} \star d\star$ on $D'(\Omega, \wedge^{l+1})$, $l = 0, 1, \cdots, n - 1$.

A differential k-form $u(x)$ is generated by $\{dx_{i_1} \wedge dx_{i_2} \wedge \cdots \wedge dx_{i_k}\}$, $k = 1, 2, \cdots, n$, that is,

$$u(x) = \sum_I \omega_I(x)dx_I = \sum \omega_{i_1 i_2 \cdots i_k}(x)dx_{i_1} \wedge dx_{i_2} \wedge \cdots \wedge dx_{i_k}, \tag{1}$$

where the coefficients $\omega_{i_1 i_2 \cdots i_k}(x)$ are differentiable functions defined in Ω and $I = (i_1, i_2, \cdots, i_k)$, $1 \le i_1 < i_2 < \cdots < i_k \le n$. A function $u(x_1, x_2, \cdots, x_n)$ is called a 0-form. Let $\wedge^l = \wedge^l(\mathbb{R}^n)$ be the set of all l-forms in $\mathbb{R}^n$ and $\wedge = \wedge(\mathbb{R}^n) = \oplus_{l=0}^{n} \wedge^l ((\mathbb{R}^n))$ be a graded algebra with respect to the exterior products. The Dirac-harmonic equation for differential forms was introduced in [3] in 2015, which includes the well-known A-harmonic equations as its special cases. Specifically, the following equation for differential forms

$$d^\star A(x, D\omega) = 0, \tag{2}$$

is called the Dirac-harmonic equation, where $D = d + d^*$ is the Dirac operator and the operator $A : \Omega \times \wedge(\mathbb{R}^n) \to \wedge(\mathbb{R}^n)$ satisfies the following conditions:

$$|A(x, \xi)| \le a|\xi|^{p-1} \quad \text{and} \quad < A(x, \xi), \xi > \ge |\xi|^p \tag{3}$$

for almost every $x \in \Omega$ and all $\xi \in \wedge(\mathbb{R}^n)$. Here $a > 0$ is a constant and $1 < p < \infty$ is a fixed exponent related to the condition of the operator A satisfies. Let $W_{p,loc}^1(\Omega, \wedge^{l-1}) = \cap W_p^1(\Omega', \wedge^{l-1})$, where the intersection is for all Ω' compactly contained in Ω. A solution to the Dirac-harmonic equation is an element of the Sobolev space $W_{p,loc}^1(\Omega, \wedge^{l-1})$ such that

$$\int_\Omega < A(x, D\omega), D\varphi >= 0 \tag{4}$$

for all $\varphi \in W_p^1(\Omega, \wedge^{l-1})$ with compact support.

We should notice that if ω is a function (0-form), both the A-harmonic equation $d^* A(x, d\omega) = 0$ and the Dirac-harmonic equation $d^* A(x, D\omega) = 0$ become the usual A-harmonic equation

$$\text{div} A(x, \nabla\omega) = 0 \tag{5}$$

for functions. See [1, 6, 8] for more results about different versions of the A-harmonic equation. Also, if $A : \Omega \times \wedge^l(\mathbb{R}^n) \to \wedge^l(\mathbb{R}^n)$ is defined by $A(x, \xi) = \xi|\xi|^{p-2}$ with $p > 1$, then, A satisfies the required conditions (3) and Eq. (2) becomes

$$d^\star(D\omega|D\omega|^{p-2}) = 0,$$

which is called the p-Dirac-harmonic equation for differential forms. Let $C^\infty(\Omega, \wedge^l)$ be the space of smooth l-forms on Ω and the Green's operator G be defined on $C^\infty(\Omega, \wedge^l)$ by assigning $G(u)$ to be a solution of the Poisson's equation $\Delta G(u) = u - H(u)$, where H is the harmonic projection operator, see [4, 7, 9] for more results about Green's operator. We use $W^{1,p}(E, \wedge^l)$ to denote the Sobolev space of l-forms which equals $L^p(E, \wedge^l) \cap L_1^p(E, \wedge^l)$ with norm

$$\|u\|_{W^{1,p}(E)} = \|u\|_{W^{1,p}(E,\wedge^l)} = diam(E)^{-1}\|u\|_{p,E} + \|\nabla u\|_{p,E} \tag{6}$$

for any subset $E \subset \mathbb{R}^n$ and $p > 0$.

The purpose of this paper is to establish the norm inequalities for solutions to the Dirac-harmonic equation and the composition $G \circ F$ of Green's operator G and a bounded operator F applied to these solutions defined in $\Omega \subset \mathbb{R}^n$. In Sect. 2, we prove the norm inequalities with singular factors for the composition $G \circ F$ of Green's operator and any bounded operator F. In Sect. 3, we establish the weighted Poincaré inequalities for solutions to the Dirac-harmonic equation of differential forms. In Sect. 4, we obtain the Sobolev imbedding inequalities with singular factors for the composition of operators and the solutions to the Dirac-harmonic equation. Many existing results are special cases of our new inequalities obtained in this paper. In real applications, we often face to estimate the integrals with singular factors. For example, let us assume that the object P_1 with mass m_1 is located at the origin and the object P_2 with mass m_2 is located at (x_1, x_2, x_3) in $\mathbb{R}^3$. Then, Newton's Law of Gravitation states that the magnitude of the gravitational force between two objects P_1 and P_2 is $|\mathbf{F}| = m_1 m_2 G/d^2(P_1, P_2)$, where $d(P_1, P_2) = \sqrt{x_1^2 + x_2^2 + x_3^2}$ is the distance between P_1 and P_2, and G is the gravitational constant. Thus, we have to evaluate a singular integral whenever the integrand contains $|\mathbf{F}|$ as a factor and the integral domain includes the origin.

2 Estimates for the Norms of Operators

In this section, we will prove some norm inequalities for the composition $G \circ F$ of Green's operator G and a bounded operator F applied to differential forms, see [4, 7–9] for more results about Green's operator and other operators on differential forms. We will need the following Caccioppoli inequality and Weak Reverse Hölder inequality obtained in [3] in 2015.

Lemma 1 *Suppose that u is a solution of the Dirac-harmonic equation in a domain Ω. Then there is a constant C, which is independent of u, such that*

$$\left(\int_B |Du|^p dx\right)^{1/p} \leq C|B|^{-1/n} \left(\int_{\sigma B} |u - c|^p dx\right)^{1/p} \tag{7}$$

for all balls or cubes B with $\sigma B \subset \Omega$, where c is any differential form with $Dc = 0$, $\sigma > 1$ and $p > 1$ are constants.

Lemma 2 *Let u be a solution of the Dirac-harmonic equation in a domain Ω and $0 < s, t < \infty$. Then, there exists a constant C, independent of u, such that*

$$\|u\|_{s,B} \leq C|B|^{(t-s)/st} \|u\|_{t,\sigma B} \tag{8}$$

for all balls or cubes B with $\sigma B \subset \Omega$ for some $\sigma > 1$.

We will also use the following norm inequality appeared in [7].

Lemma 3 *Let $u \in C^\infty(\wedge^l M)$ and $l = 1, 2, \cdots, n$, $1 < s < \infty$. Then, there exists a positive constant C, independent of u, such that*

$$\|dd^*G(u)\|_{s,M} + \|d^*dG(u)\|_{s,M} + \|dG(u)\|_{s,M} + \|d^*G(u)\|_{s,M} + \|G(u)\|_{s,M} \leq C\|u\|_{s,M}. \tag{9}$$

We first prove the following local Poincaré-type estimate for the operators $G \circ F$.

Theorem 1 *Let $u \in L^s_{loc}(\Omega, \wedge^l)$, $l = 1, 2, \cdots, n$, $1 < s < \infty$, G be Green's operator, and F be a bounded operator. Then, there exists a constant C, independent of u, such that*

$$\|G(F(u)) - (G(F(u)))_B\|_{s,B} \leq C|B|diam(B)\|u\|_{s,B} \tag{10}$$

for all balls B with $B \subset \Omega$.

Proof We know that

$$\|Tv\|_{s,B} \leq C|B|diam(B)\|v\|_{s,B}$$

and $v_B = d(Tv)$ for any l form v with $l \neq 0$, where T is the homotopy operator, see [2, 5] for more properties of the homotopy operator T, By decomposition of differential forms, we have

$$v = d(Tv) + T(dv) = v_B + T(dv).$$

Thus,

$$v - v_B = T(dv) \tag{11}$$

for any form v. Therefore, replacing the above v by $G(F(u))$ and using Lemma 3, we have

$$\|G(F(u)) - (G(F(u)))_B\|_{s,B}$$

$$= \|Td(G(F(u)))\|_{s,B}$$

$$\leq C_1|B|diam(B)\|d(G(F(u)))\|_{s,B}$$

$$\leq C_2|B|diam(B)\|F(u)\|_{s,B} \tag{12}$$

$$\leq C_3|B|diam(B)\|u\|_{s,B}$$

since the operator F is bounded. The proof of Theorem 1 has been completed. □

In applications, such as in calculating electric or magnetic fields, we often face the fact that the integrand contains a singular factor. So, we extend our result to the following singular weighted case.

Theorem 2 *Let $u \in L^s_{loc}(\Omega, \wedge^l)$, $l = 1, 2, \cdots, n$, $1 < s < \infty$, be a solution of the Dirac-harmonic equation in a bounded domain Ω, G be Green's operator, and F be a bounded operator. Then, there exists a constant C, independent of u, such that*

$$\left(\int_B |G(F(u)) - (G(F(u)))_B|^s \frac{1}{|x - x_B|^\alpha} dx\right)^{1/s} \leq C|B|^\gamma \left(\int_{\sigma B} |u|^s \frac{1}{|x - x_B|^\lambda} dx\right)^{1/s} \tag{13}$$

for all balls B with $\sigma B \subset \Omega$ and any real numbers α and λ with $\alpha > \lambda \geq 0$, where $\gamma = 1 + \frac{1}{n} - \frac{\alpha - \lambda}{ns}$ and x_B is the center of ball B and $\sigma > 1$ is a constant.

Proof Let $\varepsilon \in (0, 1)$ be a constant small enough such that $\varepsilon n < \alpha - \lambda$ and $B \subset \Omega$ be a ball with center x_B and radius r_B. Select $t = s/(1 - \varepsilon)$. Then, we have $t > s$. Choose $\beta = t/(t - s)$, By the Hölder inequality and Theorem 1, we find that

$$\left(\int_B \left(|G(F(u)) - (G(F(u)))_B|\right)^s \frac{1}{|x-x_B|^\alpha} dx\right)^{1/s}$$

$$= \left(\int_B \left(|G(F(u)) - (G(F(u)))_B| \frac{1}{|x-x_B|^{\alpha/s}}\right)^s dx\right)^{1/s}$$

$$\leq \|G(F(u)) - (G(F(u)))_B\|_{t,B} \left(\int_B \left(\frac{1}{|x-x_B|}\right)^{t\alpha/(t-s)} dx\right)^{(t-s)/st} \tag{14}$$

$$= \|G(F(u)) - (G(F(u)))_B\|_{t,B} \left(\int_B |x - x_B|^{-\alpha\beta} dx\right)^{1/\beta s}$$

$$\leq C_1|B|diam(B)\|u\|_{t,B}\||x - x_B|^{-\alpha}\|_{\beta,B}^{1/s}.$$

We can assume that $x_B = 0$. Otherwise, we can easily move the center to the origin by a simple transformation. Thus, for any $x \in B$, it follows that

$$|x - x_B| \geq |x| - |x_B| = |x|.$$

Using the polar coordinate substitution, we obtain that

$$\int_B |x - x_B|^{-\alpha\beta} dx \leq C \int_0^{r_B} \rho^{-\alpha\beta} \rho^{n-1} d\rho \leq \frac{C}{n - \alpha\beta} (r_B)^{n-\alpha\beta}. \tag{15}$$

Select $m = nst/(ns + \alpha t - \lambda t)$, then $0 < m < s$. Using Lemma 2, we find that

$$\|u\|_{t,B} \leq C_2 |B|^{\frac{m-t}{mt}} \|u\|_{m,\sigma B}, \tag{16}$$

where $\sigma > 1$ is a constant. By Lemma 2 again, we obtain

$$\|u\|_{m,\sigma B} = \left(\int_{\sigma B} \left(|u| |x - x_B|^{-\lambda/s} |x - x_B|^{\lambda/s} \right)^m dx \right)^{1/m}$$

$$\leq \left(\int_{\sigma B} \left(|u| |x - x_B|^{-\lambda/s} \right)^s dx \right)^{1/s} \left(\int_{\sigma B} \left(|x - x_B|^{\lambda/s} \right)^{\frac{ms}{s-m}} dx \right)^{\frac{s-m}{ms}}$$

$$\leq \left(\int_{\sigma B} |u|^s |x - x_B|^{-\lambda} dx \right)^{1/s} C_3 (\sigma r_B)^{\lambda/s + n(s-m)/ms}$$

$$\leq C_4 \left(\int_{\sigma B} |u|^s |x - x_B|^{-\lambda} dx \right)^{1/s} (r_B)^{\lambda/s + n(s-m)/ms}.$$

$$\tag{17}$$

It should be noticed that

$$diam(B) \cdot |B|^{1 + \frac{1}{t} - \frac{1}{m}} = |B|^{1 + \frac{1}{n} + \frac{1}{t} - \frac{ns + \alpha t - \lambda t}{nst}} = |B|^{1 + \frac{1}{n} - \frac{\alpha - \lambda}{ns}}. \tag{18}$$

Substituting (15), (16) and (17) in (14) and using (18), we have

$$\left(\int_B (|G(F(u)) - (G(F(u)))_B|)^s \frac{1}{|x - x_B|^\alpha} dx \right)^{1/s} \leq C_5 |B|^\gamma \left(\int_{\sigma B} |u|^s |x - x_B|^{-\lambda} dx \right)^{1/s}.$$

We have completed the proof of Theorem 2.   □

Replacing α by 2α and λ by α in Theorem 2, we have the following version of the symmetric inequality.

Corollary 1 *Assume that u, G, F satisfy the same conditions as mentioned in Theorem 2. Then,*

$$\left(\int_B |G(F(u)) - (G(F(u)))_B|^s \frac{1}{|x - x_B|^{2\alpha}} dx \right)^{1/s} \leq C |B|^{1 + \frac{1}{n} - \frac{\alpha}{ns}} \left(\int_{\sigma B} |u|^s \frac{1}{|x - x_B|^\alpha} dx \right)^{1/s}. \tag{19}$$

for all balls B with $\sigma B \subset \Omega$ and any real numbers $\alpha \geq 0$.

Sometimes, it is more convenient to use if the right-hand side does not contain any singular factor, that is, the integral on right-hand side has no weight. Thus, choosing $\lambda = 0$ in Theorem 2, inequality (13) reduces to the following useful version.

Corollary 2 *Assume that u, G, F satisfy the same conditions as mentioned in Theorem 2. Then,*

$$\left(\int_B |G(F(u)) - (G(F(u)))_B|^s \frac{1}{|x - x_B|^\alpha} dx \right)^{1/s} \leq C|B|^{1+\frac{1}{n}-\frac{\alpha}{ns}} \left(\int_{\sigma B} |u|^s dx \right)^{1/s} \tag{20}$$

for all balls B with $\sigma B \subset \Omega$ and any real numbers $\alpha \geq 0$.

Note that it does not contain a singular factor in the integral on the right-hand side of the above inequality.

In the proof of Theorem 2, if we substitute (15) into (14), then using the weak reverse Hölder inequality (Lemma 2),

$$\|u\|_{t,B} \leq C|B|^{1/t-1/s} \|u\|_{s,\sigma B},$$

where $\sigma > 1$ is a constant, and by a simple calculation, we will obtain the following inequality which enables us to estimate the norm of $G \circ F$ with a singular factor in terms of the simple norm of u.

Theorem 3 *Let $u \in L_{loc}^s(\Omega, \wedge^l), l = 1, 2, \cdots, n, 1 < s < \infty$, be a solution of the Dirac-harmonic equation in a bounded domain Ω, G be Green's operator, and F be a bounded operator. Then, there exists a constant C, independent of u, such that*

$$\left(\int_B |G(F(u)) - (G(F(u)))_B|^s \frac{1}{|x - x_B|^\alpha} dx \right)^{1/s} \leq C \left(\int_{\sigma B} |u|^s dx \right)^{1/s} \tag{21}$$

for all balls B with $\sigma B \subset \Omega$ and any real numbers α with $0 \leq \alpha \leq s(n+1)$, where x_B is the center of ball B and $\sigma > 1$ is a constant.

Note that we can also obtain the above Theorem 3 directly from Corollary 2 by selecting α such that $0 \leq \alpha \leq s(n+1)$.

The following definition of $L^s(\mu)$-averaging domains can be found in [1].

Definition 1 We call a proper subdomain $\Omega \subset \mathbb{R}^n$ an $L^s(\mu)$-averaging domain, $s \geq 1$, if $\mu(\Omega) < \infty$ and there exists a constant C such that

$$\left(\frac{1}{\mu(\Omega)} \int_\Omega |u - u_{B_0}|^s d\mu \right)^{1/s} \leq C \sup_{4B \subset \Omega} \left(\frac{1}{\mu(B)} \int_B |u - u_B|^s d\mu \right)^{1/s} \tag{22}$$

for some ball $B_0 \subset \Omega$ and all $u \in L_{loc}^s(\Omega; \mu)$. Here the supremum is over all balls $B \subset \Omega$ with $4B \subset \Omega$ and μ is a measure defined by $d\mu = w(x)dx$ for a weight $w(x)$ and $u_B = \frac{1}{\mu(B)} \int_B u(x)dx$.

Theorem 4 *Let $u \in D'(\Omega, \wedge^1)$ be a solution of the Dirac-harmonic equation, G be Green's operator, and F be a bounded operator. Assume that s is a fixed exponent associated with the Dirac-harmonic equation. Then, there exists a constant C, independent of u, such that*

$$\left(\int_\Omega |G(F(u)) - (G(F(u)))_{B_0}|^s) \frac{1}{d(x, \partial\Omega)^\alpha} dx \right)^{1/s} \le C \left(\int_\Omega |u|^s \frac{1}{d(x, \partial\Omega)^\lambda} dx \right)^{1/s}$$
(23)

for any bounded and convex $L^s(\mu)$-*averaging domain* $\Omega \subset \mathbb{R}^n$. *Here* $B_0 \subset \Omega$ *is a fixed ball and* α *and* λ *are constants with* $0 \le \lambda < \alpha < \min\{n, s + \lambda + n(s - 1)\}$.

Proof Let r_B be the radius of a ball $B \subset \Omega$. We may assume the center of B is 0. Then, $d(x, \partial\Omega) \ge r_B - |x|$ for any $x \in B$. Therefore, $d^{-1}(x, \partial\Omega) \le \frac{1}{r_B - |x|}$ for any $x \in B$. Similar to the proof of Theorem 2, we have

$$\left(\int_B |G(F(u)) - (G(F(u)))_B|^s \frac{1}{d(x, \partial\Omega)^\alpha} dx \right)^{1/s} \le C_1 |B|^\gamma \left(\int_{\sigma B} |u|^s \frac{1}{d(x, \partial\Omega)^\lambda} dx \right)^{1/s}$$
(24)

for all balls B with $\sigma B \subset \Omega, \sigma > 1$, and any real numbers α and λ with $\alpha > \lambda \ge 0$, where $\gamma = 1 + \frac{1}{n} - \frac{\alpha - \lambda}{ns}$. Write $d\mu = \frac{1}{d(x, \partial\Omega)^\alpha} dx$. Then,

$$\mu(B) = \int_B d\mu = \int_B \frac{1}{d(x, \partial\Omega)^\alpha} dx \ge \int_B \frac{1}{(diam(\Omega))^\alpha} dx = C_1 |B|,$$

and hence $\frac{1}{\mu(B)} \le \frac{C_2}{|B|}$. Since Ω is an $L^s(\mu)$-averaging domain, using (22) as well as Theorem 2, and noticing that $\gamma - 1/s = (1 - 1/s) + (s + \lambda - \alpha)/ns > 0$, we obtain

$$\left(\frac{1}{\mu(\Omega)} \int_\Omega |G(F(u)) - (G(F(u)))_{B_0}|^s) \frac{1}{d(x, \partial\Omega)^\alpha} dx \right)^{1/s}$$

$$= \left(\frac{1}{\mu(\Omega)} \int_\Omega |G(F(u)) - (G(F(u)))_{B_0}|^s) d\mu \right)^{1/s}$$

$$\le C_3 \sup_{4B \subset \Omega} \left(\frac{1}{\mu(B)} \int_B |G(F(u)) - (G(F(u)))_B|^s d\mu \right)^{1/s}$$

$$\le C_4 \sup_{4B \subset \Omega} \left(\frac{1}{|B|} \int_B |G(F(u)) - (G(F(u)))_B|^s d\mu \right)^{1/s}$$

$$\le C_5 \sup_{4B \subset \Omega} |B|^{\gamma - 1/s} \left(\int_{\sigma B} |u|^s \frac{1}{d(x, \partial\Omega)^\lambda} dx \right)^{1/s}$$

$$\le C_5 |\Omega|^{\gamma - 1/s} \left(\int_\Omega |u|^s \frac{1}{d(x, \partial\Omega)^\lambda} dx \right)^{1/s}$$

$$\le C_6 \left(\int_\Omega |u|^s \frac{1}{d(x, \partial\Omega)^\lambda} dx \right)^{1/s},$$

which is equivalent to

$$\left(\int_\Omega |G(F(u)) - (G(F(u)))_{B_0}|^s) \frac{1}{d(x,\partial\Omega)^\alpha} dx\right)^{1/s} \leq C \left(\int_\Omega |u|^s \frac{1}{d(x,\partial\Omega)^\lambda} dx\right)^{1/s}.$$

We have completed the proof of Theorem 4. $\square$

Similar to the case of the local inequality, choose $\lambda = 0$ in Theorem 4, we have the following estimate for the weighted norm of $G \circ F$ in terms of the simple norm of u.

Corollary 3 *Let* $u \in D'(\Omega, \wedge^1)$ *be a solution of the Dirac-harmonic equation, G be Green's operator, and F be a bounded operator. Assume that s is a fixed exponent associated with the Dirac-harmonic equation. Then, there exists a constant C, independent of u, such that*

$$\left(\int_\Omega |G(F(u)) - (G(F(u)))_{B_0}|^s) \frac{1}{d(x,\partial\Omega)^\alpha} dx\right)^{1/s} \leq C \left(\int_\Omega |u|^s dx\right)^{1/s}$$

(25)

for any bounded and convex $L^s(\mu)$*-averaging domain* $\Omega \subset \mathbb{R}^n$*. Here* $B_0 \subset \Omega$ *is a fixed ball and* α *is a constant with* $0 < \alpha < \min\{n, s + +n(s-1)\}$.

3 Poincaré Inequalities

When we deal with the integral of the vector field $\mathbf{F} = \nabla f$, we need to study the singular integral if the potential function f contains a singular factor, such as the potential energy in physics. It is obvious that the singular integrals are more interesting to us since they have wide applications in different fields of mathematics and physics. The basic integral estimates have been well developed in [3]. In this section, we will prove the global inequalities with singular factors for solutions to the Dirac-harmonic equation in the John domains. We will use the following Covering Lemma appearing in [6].

Lemma 4 *Each* Ω *has a modified Whitney cover of cubes* $\mathcal{V} = \{Q_i\}$ *such that*

$$\cup_i Q_i = \Omega, \quad \sum_{Q_i \in \mathcal{V}} \chi_{\sqrt{\frac{5}{4}}Q} \leq N\chi_\Omega$$

for all $x \in \mathbb{R}^n$ *and some* $N > 1$*, and if* $Q_i \cap Q_j \neq \emptyset$*, then there exists a cube R (this cube need not be a member of* $\mathcal{V}$*) in* $Q_i \cap Q_j$ *such that* $Q_i \cup Q_j \subset NR$*. Moreover, if* Ω *is* δ*-John, then there is a distinguished cube* $Q_0 \in \mathcal{V}$ *which can be connected with every cube* $Q \in \mathcal{V}$ *by a chain of cubes* $Q_0, Q_1, \cdots, Q_k = Q$ *from* $\mathcal{V}$ *and such that* $Q \subset \rho Q_i$*,* $i = 0, 1, 2, \cdots, k$*, for some* $\rho = \rho(n, \delta)$.

We will also need the following version of the weak reverse Hölder inequality proved in [3] in 2015.

Lemma 5 *Let ω be a solution to the Dirac-harmonic equation in Ω, $\sigma > 1$ be some constant, and $0 < r, s < \infty$ be any constants. Then, there exists a constant C, independent of ω, such that*

$$\|\omega - \omega_B\|_{s,B} \leq C|B|^{(r-s)/rs}\|\omega - \omega_B\|_{r,\sigma B} \qquad (26)$$

for all cubes or balls B with $\sigma B \subset \Omega$.

The following Poincaré inequality with the Dirac operator also appeared in [3] in 2015.

Lemma 6 *Let $u \in D'(Q, \wedge^l)$ be a differential form and $Du \in L^p(Q, \wedge)$, $p > 1$. Then, $u - u_Q$ is in $L^p(Q, \wedge)$ and*

$$\|u - u_Q\|_{p,Q} \leq C|Q|^{1/n}\|Du\|_{p,Q}. \qquad (27)$$

for any cube $Q \subset \mathbb{R}^n$, where C is a constant, independent of u.

Using Lemma 5, Lemma 6, and the same method developed in the proof of Theorem 2, we can easily prove the following weighted Poincaré inequality.

Theorem 5 *Let $u \in L_{loc}^s(\Omega, \wedge^l)$, $l = 1, 2, \cdots, n$, $1 < s < \infty$, be a solution of the Dirac-harmonic equation in a bounded domain Ω. Then, there exists a constant C, independent of u, such that*

$$\left(\int_U |u - u_U|^s) \frac{1}{d^\alpha(x, \partial\Omega)} dx\right)^{1/s} \leq C|U|^\gamma \left(\int_{\sigma U} |Du|^s \frac{1}{d^\lambda(x, \partial\Omega)} dx\right)^{1/s}, \qquad (28)$$

for all balls U with $\sigma U \subset \Omega$ and any real numbers α and λ with $\alpha > \lambda \geq 0$, where $\gamma = 1 + \frac{1}{n} - \frac{\alpha - \lambda}{ns}$ and x_U is the center of ball U and $\sigma > 1$ is a constant.

Definition 2 A proper subdomain $\Omega \subset \mathbb{R}^n$ is called a δ-John domain, $\delta > 0$, if there exists a point $x_0 \in \Omega$ which can be joined with any other point $x \in \Omega$ by a continuous curve $\gamma \subset \Omega$ so that

$$d(\xi, \partial\Omega) \geq \delta|x - \xi|$$

for each $\xi \in \gamma$. Here $d(\xi, \partial\Omega)$ is the Euclidean distance between ξ and $\partial\Omega$.

We are ready to prove the following weighted norm Poincaré inequality for solutions to the Dirac-harmonic equation.

Theorem 6 *Let $u \in D'(\Omega, \wedge^0)$, $1 < s < \infty$, be a solution of the Dirac-harmonic equation in δ-John domain $\Omega \subset \mathbb{R}^n$ and $Du \in L^s(\Omega, \wedge^1)$. Then, there exists a constant C, independent of u, such that*

$$\left(\int_\Omega |u - u_{U_0}|^s\right)\frac{1}{d^\alpha(x, \partial\Omega)}dx\right)^{1/s} \le C\left(\int_\Omega |Du|^s \frac{1}{d^\lambda(x, \partial\Omega)}dx\right)^{1/s},$$

where $U_0 \subset \Omega$ is a fixed cube and α and λ are constants with $0 \le \lambda < \alpha < min\{n, s + \lambda + n(s - 1)\}$.

Proof From Lemma 4, there is a modified Whitney cover of cubes $\mathscr{V} = \{U_i\}$ for Ω such that $\Omega = \cup U_i$, and $\sum_{U_i \in \mathscr{V}} \chi_{\sqrt{\frac{5}{4}}U_i} \le N\chi_\Omega$ for some $N > 1$. Since $\Omega = \cup U_i$, for any $x \in \Omega$, it follows that $x \in U_i$ for some i. Apply (28) to U_i, we find that

$$\left(\int_{U_i} |u - u_{U_i}|^s\right)\frac{1}{d^\alpha(x, \partial\Omega)}dx\right)^{1/s} \le C|U_i|^\gamma \left(\int_{\sigma U_i} |Du|^s \frac{1}{d^\lambda(x, \partial\Omega)}dx\right)^{1/s},$$
(29)

where $\sigma > 1$ is a constant. Let $\mu(x)$ and $\mu_1(x)$ be measures defined by $d\mu = \frac{1}{d^\alpha(x, \partial\Omega)}dx$ and $d\mu_1(x) = \frac{1}{d^\lambda(x, \partial\Omega)}dx$. Then,

$$\mu(U) = \int_U \frac{1}{d^\alpha(x, \partial\Omega)}dx \ge \int_U \frac{1}{(diam(\Omega))^\alpha}dx = M|U|,$$
(30)

where M is a positive constant. Then, by the elementary inequality $(a + b)^s \le 2^s(|a|^s + |b|^s)$, $s \ge 0$, we have

$$\left(\int_\Omega |u - u_{U_0}|^s \frac{1}{d^\alpha(x, \partial\Omega)}dx\right)^{1/s}$$

$$= \left(\int_{\cup U_i} |u - u_{U_0}|^s d\mu\right)^{1/s}$$

$$\le \left(\sum_{U \in \mathscr{V}} \left(2^s \int_U |u - u_U|^s d\mu\right.\right.$$

$$\left.\left. + 2^s \int_U |u_U - u_{U_0}|^s d\mu\right)\right)^{1/s}$$
(31)

$$\le C_1\left(\sum_{U \in \mathscr{V}} \int_U |u - u_U|^s d\mu\right)^{1/s}$$

$$+ \left(\sum_{U \in \mathscr{V}} \int_U |u_U - u_{U_0}|^s d\mu\right)^{1/s}$$

for a fixed $U_0 \subset \Omega$. The first sum in (31) can be estimated by using (29) and Covering Lemma,

$$\sum_{U \in \mathcal{V}} \int_U |u - u_U|^s d\mu$$

$$\leq C_2 \sum_{U \in \mathcal{V}} |U|^{\gamma s} \int_{\rho U} |Du|^s d\mu_1$$

$$\leq C_3 |\Omega|^{\gamma s} \sum_{U \in \mathcal{V}} \int_U |Du|^s d\mu_1) \leq C_4 |\Omega|^{\gamma s} \int_\Omega |Du|^s d\mu_1 \tag{32}$$

$$\leq C_5 \int_\Omega |Du|^s \frac{dx}{d^\lambda(x, \partial\Omega)}.$$

To estimate the second sum in (31), we need to use the property of δ-John domain. Fix a cube $U \in \mathcal{V}$ and let $U_0, U_1, \cdots, U_k = U$ be the chain in Lemma x.

$$|u_U - u_{U_0}| \leq \sum_{i=0}^{k-1} |u_{U_i} - u_{U_{i+1}}|. \tag{33}$$

The chain $\{U_i\}$ also has property that, for each i, $i = 0, 1, \cdots, k - 1$, with $U_i \cap U_{i+1} \neq \emptyset$, there exists a cube D_i such that $D_i \subset U_i \cap U_{i+1}$ and $U_i \cup U_{i+1} \subset N D_i$, $N > 1$.

$$\frac{max\{|U_i|, |U_{i+1}|\}}{|U_i \cap U_{i+1}|} \leq \frac{max\{|U_i|, |U_{i+1}|\}}{|D_i|} \leq C_6$$

For such D_j, $j = 0, 1, \cdots, k - 1$, Let $E = min\{D_0, D_1, \cdots, D_{k-1}\}$, then

$$\frac{max\{|U_i|, |U_{i+1}|\}}{|U_i \cap U_{i+1}|} \leq \frac{max\{|U_i|, |U_{i+1}|\}}{|E|} \leq C_7. \tag{34}$$

By (30), (34), and Lemma 6, we have

$$|u_{U_i} - u_{U_{i+1}}|^s$$

$$= \frac{1}{\mu(U_i \cap U_{i+1})} \int_{U_i \cap U_{i+1}} |u_{U_i} - u_{U_{i+1}}|^s \frac{dx}{d^\alpha(x, \partial\Omega)}$$

$$\leq C_8 \frac{1}{|U_i \cap U_{i+1}|} \int_{U_i \cap U_{i+1}} |u_{U_i} - u_{U_{i+1}}|^s \frac{dx}{d^\alpha(x, \partial\Omega)}$$

$$\leq C_8 \frac{C_7}{max\{|U_i|, |U_{i+1}|\}} \int_{U_i \cap U_{i+1}} |u_{U_i} - u_{U_{i+1}}|^s d\mu \tag{35}$$

$$\leq C_9 \sum_{j=i}^{i+1} \frac{1}{|U_j|} \int_{U_j} |u - u_U|^s d\mu$$

$$\leq C_{10} \sum_{j=i}^{i+1} \frac{|U_j|^{\gamma s}}{|U_j|} \int_{\rho U_j} |Du|^s d\mu_1$$

$$= C_{10} \sum_{j=i}^{i+1} |U_j|^{\gamma s - 1} \int_{\rho U_j} |Du|^s \frac{dx}{d^{\lambda}(x, \partial \Omega)}$$

Since $U \subset NU_j$ for $j = i, i+1, 0 \leq i \leq k-1$, From (35)

$$|u_{U_i} - u_{U_{i+1}}|^s \chi_U(x) \leq C_{11} \sum_{j=i}^{i+1} \chi_{NU_j}(x) |U_j|^{\gamma s - 1} \int_{\rho U_j} |Du|^s \frac{dx}{d^{\lambda}(x, \partial \Omega)}$$

$$\leq C_{12} \sum_{j=i}^{i+1} \chi_{NU_j}(x) |\Omega|^{\gamma s - 1} \int_{\rho U_j} |Du|^s d\mu_1. \tag{36}$$

We see that $|\Omega|^{\gamma - 1/s} < \infty$ since $\gamma - \frac{1}{s} = 1 + \frac{1}{n} + \frac{\lambda}{ns} - \frac{1}{s} - \frac{\alpha}{ns} > 0$ when $\alpha < s + \lambda + n(s-1)$. Thus, from $(a+b)^{1/s} \leq 2^{1/s}(|a|^{1/s} + |b|^{1/s})$, (33) and (36),

$$|u_U - u)_{U_0}|\chi_U(x) \leq C_{13} \sum_{E \in \mathcal{V}} \left(\int_{\rho E} |Du|^s d\mu_1 \right)^{1/s} \cdot \chi_{NE}(x)$$

for every $x \in \mathbb{R}^n$. Then,

$$\sum_{U \in \mathcal{V}} \int_U |u_U - u_{U_0}|^s d\mu \leq C_{13} \int_n \left| \sum_{E \in \mathcal{V}} \left(\int_{\rho E} |Du|^s d\mu_1 \right)^{1/s} \chi_{NE}(x) \right|^s d\mu.$$

Notice that

$$\sum_{E \in \mathcal{V}} \chi_{NE}(x) \leq \sum_{E \in \mathcal{V}} \chi_{\rho NE}(x) \leq N \chi_{\Omega}(x).$$

Using elementary inequality $|\sum_{i=1}^N t_i|^s \leq N^{s-1} \sum_{i=1}^N |t_i|^s$, we finally have

$$\sum_{U \in \mathcal{V}} \int_U |u_U - u_{U_0}|^s d\mu$$

$$\leq C_{14} \int_n \left(\sum_{E \in \mathcal{V}} (\int_{\rho E} |Du|^s d\mu_1) \chi_{NE}(x) \right) d\mu$$

$$= C_{14} \sum_{E \in \mathcal{V}} (\int_{\rho E} |Du|^s d\mu_1) \leq C_{15} \int_{\Omega} |Du|^s \frac{dx}{d^{\lambda}(x, \partial \Omega)}. \tag{37}$$

Substituting (32) and (37) in (31), we have proved Theorem 6. $\qquad \square$

4 Sobolev Imbedding Inequalities

We all know that for any $v \in L^s_{loc}(\Omega, \wedge^1)$, it follows that $\|\nabla v\|_{s,B} = \|dv\|_{s,B}$. Thus, by Lemma 3, we have

$$\|\nabla G(F(u))\|_{s,B} = \|dG(F(u))\|_{s,B} \leq C_1\|F(u)\|_{s,B} \leq C_2\|u\|_{s,B}. \tag{38}$$

Starting from (38), using the similar method developed in Sect. 2 and noticing that $\frac{1}{d(x,\partial\Omega)} \leq \frac{1}{r_B - |x|}$ for any $x \in B$ where r_B is the radius of ball B, we can prove the following lemma.

Lemma 7 *Let* $u \in L^s_{loc}(\Omega, \wedge^1)$, $1 < s < \infty$, *be a solution of Dirac-harmonic equation in a bounded domain* Ω, *G be Green's operator, and F be a bounded operator. There exists a constant C, independent of u, such that*

$$\left(\int_B |\nabla G(F(u))|^s \frac{1}{d(x,\partial\Omega)^\alpha} dx\right)^{1/s} \leq C|B|^\gamma \left(\int_{\rho B} |u|^s \frac{1}{d(x,\partial\Omega)^\lambda} dx\right)^{1/s}. \tag{39}$$

for all balls B with $\rho B \subset \Omega$ and any numbers α, λ with $\alpha > \lambda \geq 0$ and $\gamma = 1 + \frac{1}{n} - \frac{\alpha - \lambda}{ns}$. Here x_B is the center of ball B.

We should notice that (39) can also be written as

$$\|\nabla G F(u)\|_{s,B,w_1} \leq C|B|^\gamma \|u\|_{s,\rho B, w_2}, \tag{40}$$

where the weights are defined by $w_1(x) = \frac{1}{d^\alpha(x,\partial\Omega)}$ and $w_2(x) = \frac{1}{d^\lambda(x,\partial\Omega)}$. Next, we prove the imbedding inequality with a singular factor in John domain.

Theorem 7 *Let* $u \in D'(\Omega, \wedge^1)$ *be a solution of the Dirac-harmonic equation, G be Green's operator, and F be a bounded operator. Then, there exists a constant C, independent of u, such that*

$$\|G(F(u)) - (G(F(u)))_{B_0}\|_{W^{1,s}(\Omega), w_1} \leq C\|u\|_{s,\Omega,w_2} \tag{41}$$

for any $L^s(\mu)$-averaging domain $\Omega \subset \mathbb{R}^n$. Here the weights are defined by $w_1(x) = \frac{1}{d^\alpha(x,\partial\Omega)}$ and $w_2(x) = \frac{1}{d^\lambda(x,\partial\Omega)}$. $B_0 \subset \Omega$ is a fixed cube. α and λ are constants with $0 \leq \lambda < \alpha < \min\{n, \lambda + n(s - 1)\}$.

Proof Using Lemma 4, (40) and noticing that

$$\gamma - \frac{1}{s} - \frac{1}{n} = 1 - \frac{1}{s} - \frac{\alpha - \lambda}{ns} > 0$$

by the condition $\alpha < \min\{n, n(s - 1) + \lambda\}$. We obtain

$$\|\nabla GF(u)\|_{s,\Omega,w_1} \leq \sum_{B \in \mathscr{V}} \|\nabla GF(u)\|_{s,B,w_1}$$

$$\leq \sum_{B \in \mathscr{V}} \left(C_1 |B|^{\gamma} \|u\|_{s,\rho B,w_2} \right)$$

$$\leq \sum_{B \in \mathscr{V}} \left(C_1 |\Omega|^{\gamma} \|u\|_{s,\Omega,w_2} \right) \tag{42}$$

$$\leq C_2 N \|u\|_{s,\Omega,w_2}$$

$$\leq C_3 \|u\|_{s,\Omega,w_2}.$$

We know that $(G(F(u)))_{B_0}$ is a closed form, $\nabla((G(F(u)))_{B_0}) = d((G(F(u)))_{B_0}) = 0$. Thus, by using Theorem 4 and (42), and using the condition

$$\gamma - \frac{1}{n} - \frac{1}{s} = 1 - \frac{1}{s} - \frac{\alpha - \lambda}{ns} > 0.$$

We find that

$$\|G(F(u)) - (G(F(u)))_{B_0}\|_{W^{1,s}(\Omega),w_1}$$

$$= diam(\Omega)^{-1} \|G(F(u)) - (G(F(u)))_{B_0}\|_{s,\Omega,w_1} + \|\nabla(G(F(u)) - (G(F(u)))_{B_0})\|_{s,\Omega,w_1}$$

$$= diam(\Omega)^{-1} \|G(F(u)) - (G(F(u)))_{B_0}\|_{s,\Omega,w_1} + \|\nabla(G(F(u)))\|_{s,\Omega,w_1}$$

$$\leq C_1 |\Omega|^{-1/n} |\Omega|^{\gamma - 1/s} \|u\|_{s,\Omega,w_2} + C_2 \|u\|_{s,\Omega,w_2}$$

$$\leq C_3 \|u\|_{s,\Omega,w_2}$$

holds. We have completed the proof of the Theorem 7.   □

Using Theorem 6 and the similar method to the proof of Theorem 7 above, we have the following imbedding inequality with singular factors.

Theorem 8 *Let $u \in L^s(\Omega, \wedge^0)$, $1 < s < \infty$, be a solution of the Dirac-harmonic equation in δ-John domain $\Omega \subset \mathbb{R}^n$ and $Du \in L^s(\Omega, \wedge^1)$, Then, there exists a constant C, independent of u, such that*

$$\|u - u_{U_0}\|_{W^{1,s}(\Omega),w_1} \leq C \|Du\|_{s,\Omega,w_2} \tag{43}$$

where $U_0 \subset \Omega$ is a fixed cube; α and λ are constants with $0 \leq \lambda < \alpha < min\{n, s + \lambda + n(s - 1)\}$ and the weights are defined by $w_1(x) = \frac{1}{d^{\alpha}(x,\partial\Omega)}$ and $w_2(x) = \frac{1}{d^{\lambda}(x,\partial\Omega)}$.

Using the Poincaré inequality and Caccioppoli inequality, we have

$$\|u - u_B\|_{p,B} \le C_1 \|u - c\|_{p,\sigma B}$$

for any $B \subset \Omega$, any closed form c and some constant $\sigma > 1$. Choose $c = 0$, we obtain

$$\|u - u_B\|_{p,B} \le C_1 \|u\|_{p,\sigma B} \tag{44}$$

We should notice that in Theorem 8, the right-hand side is the weighted norm of Du. However, sometimes, it would be complicated to evaluate the norm $\|Du\|_{s,\Omega,w_2}$. Hence, we are motivated to develop inequalities with the simple norm $\|u\|_{s,\Omega,w_2}$ on the right-hand side. For this purpose, starting with (44) and using the same method developed in the proof of Theorem 6, we can easily obtain the following inequality with the simple weighted norm of u on the right-hand side.

Theorem 9 *Let $u \in D'(\Omega, \wedge^0)$, $1 < s < \infty$, be a solution of the Dirac-harmonic equation in δ-John domain $\Omega \subset \mathbb{R}^n$. Then, there exists a constant C, independent of u, such that*

$$\left(\int_\Omega |u - u_{U_0}|^s \right) \frac{1}{d^\alpha(x, \partial\Omega)} dx \right)^{1/s} \le C \left(\int_\Omega |u|^s \frac{1}{d^\lambda(x, \partial\Omega)} dx \right)^{1/s}, \tag{45}$$

where $U_0 \subset \Omega$ is a fixed cube and α and λ are constants with $0 \le \lambda < \alpha < min\{n, s + \lambda + n(s-1)\}$.

Since λ is any real number with $\lambda \ge 0$ in (45), choosing $\lambda = 0$ in (45), we have

$$\left(\int_\Omega |u - u_{U_0}|^s \right) \frac{1}{d^\alpha(x, \partial\Omega)} dx \right)^{1/s} \le C \left(\int_\Omega |u|^s dx \right)^{1/s}, \tag{46}$$

where $U_0 \subset \Omega$ is a fixed cube and α is a constant with $0 < \alpha < min\{n, s+n(s-1)\}$. From (46) and noticing the fact that $\nabla u = du$ for $u \in D'(\Omega, \wedge^0)$ and Caccioppoli inequality, we have the following simple version of imbedding inequality

$$\|u - u_{U_0}\|_{W^{1,s}(\Omega),w_1} \le C\|u\|_{s,\Omega},$$

where $U_0 \subset \Omega$ is a fixed cube, $w_1(x) = \frac{1}{d^\alpha(x,\partial\Omega)}$, and α is a constant with $0 < \alpha < min\{n, s + n(s - 1)\}$. This would be a very useful global inequality since the right-hand side only contains a simple norm of the solution u. Hence, we should pay enough attention to it and summarize it as the following theorem.

Theorem 10 *Let $u \in D'(\Omega, \wedge^0)$, $1 < s < \infty$, be a solution of the Dirac-harmonic equation in δ-John domain $\Omega \subset \mathbb{R}^n$. Then, there exists a constant C, independent of u, such that*

$$\|u - u_{U_0}\|_{W^{1,s}(\Omega),w_1} \leq C\|u\|_{s,\Omega},$$

where $U_0 \subset \Omega$ is a fixed cube, $w_1(x) = \frac{1}{d^\alpha(x,\partial\Omega)}$, and α is a constant with $0 < \alpha < min\{n, s + n(s - 1)\}$.

Remark 1. Since the Dirac-harmonic equation is an extension of the A-harmonic equations and p-harmonic equations, many existing results about harmonic equations are the special cases of our theorems.

2. The L^s-norm estimates developed in this paper can be extended into the L^φ-norm estimates. Considering the length of the paper, we do not include the L^φ-norm estimates here.

References

1. R.P. Agarwal, S. Ding, C. Nolder, *Inequalities for Differential Forms* (Springer, New York, 2009). MR 2552910
2. H. Cartan, *Differential Forms* (Houghton Mifflin, Boston, 1970). MR 0267477, Translated from the French
3. S. Ding, B. Liu, Dirac-harmonic equations for differential forms. Nonlinear Anal. **122**, 43–57 (2015). MR 3348064
4. S. Ding, G. Shi, Y. Xing, Higher integrability of iterated operators on differential forms. Nonlinear Anal. **145**, 83–96 (2016). MR 3547675
5. T. Iwaniec, A. Lutoborski, Integral estimates for null Lagrangians. Arch. Rational Mech. Anal. **125**(1), 25–79 (1993). MR 1241286
6. C.A. Nolder, Hardy-Littlewood theorems for A-harmonic tensors. Ill. J. Math. **43**(4), 613–632 (1999). MR 1712513
7. C. Scott, L^p theory of differential forms on manifolds. Trans. Amer. Math. Soc. **347**(6), 2075–2096 (1995). MR 1297538
8. Y. Wang, C. Wu, Global Poincaré inequalities for Green's operator applied to the solutions of the nonhomogeneous A-harmonic equation. Comput. Math. Appl. **47**(10–11), 1545–1554 (2004). MR 2079864
9. F.W. Warner, *Foundations of Differentiable Manifolds and Lie Groups*. Graduate Texts in Mathematics, vol. 94 (Springer, New York, 1983). Corrected reprint of the 1971 edition. MR 722297

Inequalities for Analytic Functions Defined by a Fractional Integral Operator

Alina Alb Lupaş

Abstract In this paper we have introduced and studied the subclass $\mathscr{DR}_{m,n}(\lambda, d, \alpha, \beta, \gamma)$ using the fractional integral associated with the convolution product of generalized Sălăgean operator and Ruscheweyh derivative. The main objective is to obtain some inequalities that give several properties such as coefficient estimates, distortion theorems, closure theorems, neighborhoods and the radii of starlikeness, convexity and close-to-convexity of functions belonging to the class $\mathscr{DR}_{m,n}(\lambda, d, \alpha, \beta, \gamma)$.

1 Introduction

Denote by U the unit disc of the complex plane, $U = \{z \in \mathbb{C} : |z| < 1\}$ and $\mathscr{H}(U)$ the space of holomorphic functions in U.

Let $\mathscr{A}(p, l) = \{f \in \mathscr{H}(U) : f(z) = z^p + \sum_{j=p+l}^{\infty} a_j z^j, \ z \in U\}$, with $\mathscr{A}(1, 1) = \mathscr{A}$ and $\mathscr{H}[a, l] = \{f \in \mathscr{H}(U) : f(z) = a + a_l z^l + a_{l+1} z^{l+1} + \dots, \ z \in U\}$, where $p, l \in \mathbb{N}, a \in \mathbb{C}$.

Definition 1 (Al Oboudi [1]) For $f \in \mathscr{A}, \alpha \geq 0$ and $m \in \mathbb{N}$, the operator D_α^m is defined by $D_\alpha^m : \mathscr{A} \to \mathscr{A}$ as follows:

$$D_\alpha^0 f(z) = f(z)$$

$$D_\alpha^1 f(z) = (1 - \alpha) f(z) + \alpha z f'(z) = D_\alpha f(z)$$

$$\dots$$

$$D_\alpha^m f(z) = (1 - \alpha) D_\alpha^{m-1} f(z) + \alpha z \left(D_\alpha^m f(z) \right)' = D_\alpha \left(D_\alpha^{m-1} f(z) \right), \ z \in U.$$

A. A. Lupaş (✉)

Department of Mathematics and Computer Science, University of Oradea, Oradea, Romania

e-mail: dalb@uoradea.ro

© Springer Nature Switzerland AG 2019

G. A. Anastassiou and J. M. Rassias (eds.), *Frontiers in Functional Equations and Analytic Inequalities*, https://doi.org/10.1007/978-3-030-28950-8_36

Remark 1 If $f \in \mathscr{A}$ and $f(z) = z + \sum_{j=2}^{\infty} a_j z^j$, then

$$D_\alpha^m f(z) = z + \sum_{j=2}^{\infty} [1 + (j-1)\alpha]^m a_j z^j, \ z \in U.$$

Remark 2 For $\alpha = 1$ in the above definition we obtain the Sălăgean differential operator [5].

Definition 2 (S.T. Ruscheweyh [4]) For $f \in \mathscr{A}$ and $n \in \mathbb{N}$, the operator R^n is defined by $R^n : \mathscr{A} \to \mathscr{A}$ as follows:

$$R^0 f(z) = f(z)$$
$$R^1 f(z) = zf'(z)$$
$$\dots$$
$$(n+1) R^{n+1} f(z) = z \left(R^n f(z) \right)' + n R^n f(z), \ z \in U.$$

Remark 3 If $f \in \mathscr{A}$, $f(z) = z + \sum_{j=2}^{\infty} a_j z^j$, then $R^n f(z) = z + \sum_{j=2}^{\infty} \frac{\Gamma(n+j)}{\Gamma(n+1)\Gamma(j)} a_j z^j$ for $z \in U$.

Definition 3 Let $n, m \in \mathbb{N}$. Denote by $DR_\alpha^{m,n} : \mathscr{A} \to \mathscr{A}$ the operator given by the convolution product of the generalized Sălăgean operator S^m and the Ruscheweyh derivative R^n:

$$DR_\alpha^{m,n} f(z) = \left(D_\alpha^m * R^n \right) f(z), \tag{1}$$

for any $z \in U$ and each nonnegative integers m, n.

Remark 4 If $f \in \mathscr{A}$ and $f(z) = z + \sum_{j=2}^{\infty} a_j z^j$, then

$$DR_\alpha^{m,n} f(z) = z + \sum_{j=2}^{\infty} [1 + (j-1)\alpha]^m \frac{\Gamma(n+j)}{\Gamma(n+1)\Gamma(j)} a_j^2 z^j, \ z \in U.$$

Definition 4 ([2]) The fractional integral of order λ ($\lambda > 0$) is defined for a function f by

$$D_z^{-\lambda} f(z) = \frac{1}{\Gamma(\lambda)} \int_0^z \frac{f(t)}{(z-t)^{1-\lambda}} dt, \tag{2}$$

where f is an analytic function in a simply-connected region of the z-plane containing the origin, and the multiplicity of $(z-t)^{\lambda-1}$ is removed by requiring $\log(z-t)$ to be real, when $(z-t) > 0$.

From Definition 3 and Definition 4 we get the fractional integral associated with the convolution product of generalized Sălăgean operator and Ruscheweyh derivative,

$$D_z^{-\lambda} D R_\alpha^{m,n} f(z) = \frac{1}{\Gamma(\lambda)} \int_0^z \frac{D R_\alpha^{m,n} f(t)}{(z-t)^{1-\lambda}} dt =$$

$$\frac{1}{\Gamma(\lambda)} \int_0^z \frac{t}{(z-t)^{1-\lambda}} dt + \sum_{j=2}^\infty [1+(j-1)\alpha]^m \frac{\Gamma(n+j)}{\Gamma(\lambda)\Gamma(n+1)\Gamma(j)} a_j^2 \int_0^z \frac{t^j}{(z-t)^{1-\lambda}} dt,$$

which has the following form, after a simple calculation,

$$D_z^{-\lambda} D R_\alpha^{m,n} f(z) = \frac{1}{\Gamma(\lambda+2)} z^{\lambda+1} + \sum_{j=2}^\infty \frac{[1+(j-1)\alpha]^m j \Gamma(n+j)}{\Gamma(n+1)\Gamma(j+\lambda+1)} a_j^2 z^{j+\lambda},$$

for the function $f(z) = z + \sum_{j=2}^\infty a_j z^j \in \mathscr{A}$. We note that $D_z^{-\lambda} D R_\alpha^{m,n} f(z) \in \mathscr{A}(\lambda+1, 1)$.

Remark 5 For $\alpha = 1$ we obtain the operator $D_z^{-\lambda} S R^{m,n}$ defined and studied in [3].

Definition 5 Let the function $f \in \mathscr{A}$. Then f is said to be in the class $\mathscr{D}\mathscr{R}_{m,n}(\lambda, d, \alpha, \beta, \gamma)$ if it satisfies the following criterion:

$$\left| \frac{1}{d} \left(\frac{z(D_z^{-\lambda} D R_\alpha^{m,n} f(z))' + \gamma z^2 (D_z^{-\lambda} D R_\alpha^{m,n} f(z))''}{(1-\gamma) D_z^{-\lambda} D R_\alpha^{m,n} f(z) + \gamma z (D_z^{-\lambda} D R_\alpha^{m,n} f(z))'} - 1 \right) \right| < \beta, \tag{3}$$

where $\lambda > 0$, $d \in \mathbb{C} - \{0\}$, $\alpha \geq 0$, $0 < \beta \leq 1$, $0 \leq \gamma \leq 1$, $m, n \in \mathbb{N}$, $z \in U$.

In this paper we shall first deduce a necessary and sufficient condition for a function f to be in the class $\mathscr{D}\mathscr{R}_{m,n}(\lambda, d, \alpha, \beta, \gamma)$. Then obtain the distortion and growth theorems, closure theorems, neighborhood and radii of univalent starlikeness, convexity and close-to-convexity of order δ, $0 \leq \delta < 1$, for these functions.

2 Coefficient Inequality

Theorem 1 Let the function $f \in \mathscr{A}$. Then f is said to be in the class $\mathscr{D}\mathscr{R}_{m,n}(\lambda, d, \alpha, \beta, \gamma)$ if and only if

$$\sum_{j=2}^\infty \frac{[1+(j-1)\alpha]^m j \Gamma(n+j)}{\Gamma(j+\lambda+1)} \cdot$$

$$\left\{ \gamma j^2 + [\gamma(2\lambda - 2 + \beta|d|) + 1] j + [\gamma(\lambda-1) + 1](\lambda - 1 + \beta|d|) \right\} a_j^2$$

$$\leq (\gamma\lambda + 1)(\beta|d| - \lambda) \frac{\Gamma(n+1)}{\Gamma(\lambda+2)}, \tag{4}$$

where $\lambda > 0$, $d \in \mathbb{C} - \{0\}$, $\alpha \geq 0$, $0 < \beta \leq 1$, $0 \leq \gamma \leq 1$, $m, n \in \mathbb{N}$, $z \in U$.

Proof Let $f \in \mathscr{DR}_{m,n}(\lambda, d, \alpha, \beta, \gamma)$. Assume that inequality (4) holds true. Then we find that

$$\left| \frac{z(D_z^{-\lambda} DR_\alpha^{m,n} f(z))' + \gamma z^2 (D_z^{-\lambda} DR_\alpha^{m,n} f(z))''}{(1-\gamma) D_z^{-\lambda} DR_\alpha^{m,n} f(z) + \gamma z (D_z^{-\lambda} DR_\alpha^{m,n} f(z))'} - 1 \right| =$$

$$\left| \frac{\frac{\lambda(\gamma\lambda+1)}{\Gamma(\lambda+2)} z^{\lambda+1} + \sum_{j=2}^{\infty} \frac{[1+(j-1)\alpha]^m j \Gamma(n+j)}{\Gamma(n+1)\Gamma(j+\lambda+1)} \left\{ \gamma j^2 + [2\gamma(\lambda-1)+1]j + (\lambda-1)[\gamma(\lambda-1)+1] \right\} a_j^2 z^{j+\lambda}}{\frac{\gamma\lambda+1}{\Gamma(\lambda+2)} z^{\lambda+1} + \sum_{j=2}^{\infty} \frac{[1+(j-1)\alpha]^m j \Gamma(n+j)}{\Gamma(n+1)\Gamma(j+\lambda+1)} [\gamma j + \gamma(\lambda-1)+1] a_j^2 z^{j+\lambda}} \right| \leq$$

$$\frac{\frac{\lambda(\alpha\lambda+1)}{\Gamma(\lambda+2)} + \sum_{j=2}^{\infty} \frac{[1+(j-1)\alpha]^m j \Gamma(n+j)}{\Gamma(n+1)\Gamma(j+\lambda+1)} \left\{ \gamma j^2 + [2\gamma(\lambda-1)+1]j + (\lambda-1)[\gamma(\lambda-1)+1] \right\} a_j^2 \left| z^{j-1} \right|}{\frac{\alpha\lambda+1}{\Gamma(\lambda+2)} - \sum_{j=2}^{\infty} \frac{[1+(j-1)\alpha]^m j \Gamma(n+j)}{\Gamma(n+1)\Gamma(j+\lambda+1)} [\gamma j + \gamma(\lambda-1)+1] a_j^2 \left| z^{j-1} \right|} \leq \beta|d|.$$

Choosing values of z on real axis and letting $z \to 1^-$, we have

$$\sum_{j=2}^{\infty} \frac{[1+(j-1)\alpha]^m j \Gamma(n+j)}{\Gamma(j+\lambda+1)} \cdot$$

$$\left\{ \gamma j^2 + [\gamma(2\lambda-2+\beta|d|)+1]j + [\gamma(\lambda-1)+1](\lambda-1+\beta|d|) \right\} a_j^2$$

$$\leq (\gamma\lambda+1)(\beta|d|-\lambda) \frac{\Gamma(n+1)}{\Gamma(\lambda+2)}.$$

Conversely, assume that $f \in \mathscr{DR}_{m,n}(\lambda, d, \alpha, \beta, \gamma)$, then we get the following inequality

$$Re\left\{ \frac{z(D_z^{-\lambda} DR_\alpha^{m,n} f(z))' + \gamma z^2 (D_z^{-\lambda} DR_\alpha^{m,n} f(z))''}{(1-\gamma) D_z^{-\lambda} DR_\alpha^{m,n} f(z) + \gamma z (D_z^{-\lambda} DR_\alpha^{m,n} f(z))'} - 1 \right\} > -\beta|d|$$

$$Re\left\{ \frac{\frac{\lambda(\gamma\lambda+1)}{\Gamma(\lambda+2)} z^{\lambda+1} + \sum_{j=2}^{\infty} \frac{[1+(j-1)\alpha]^m j \Gamma(n+j)}{\Gamma(n+1)\Gamma(j+\lambda+1)} \left\{ \gamma j^2 + [2\gamma(\lambda-1)+1]j + (\lambda-1)[\gamma(\lambda-1)+1] \right\} a_j^2 z^{j+\lambda}}{\frac{\gamma\lambda+1}{\Gamma(\lambda+2)} z^{\lambda+1} + \sum_{j=2}^{\infty} \frac{[1+(j-1)\alpha]^m j \Gamma(n+j)}{\Gamma(n+1)\Gamma(j+\lambda+1)} [\gamma j + \gamma(\lambda-1)+1] a_j^2 z^{j+\lambda}} - 1 + \beta|d| \right\} > 0$$

$$Re \frac{\frac{(\gamma\lambda+1)(\beta|d|-\lambda)}{\Gamma(\lambda+2)} z^{\lambda+1} + \sum_{j=2}^{\infty} \frac{[1+(j-1)\alpha]^m j \Gamma(n+j)}{\Gamma(n+1)\Gamma(j+\lambda+1)} \left\{ \gamma j^2 + [\gamma(2\lambda-2+\beta|d|)+1]j + [\gamma(\lambda-1)+1](\lambda-1+\beta|d|) \right\} a_j^2 z^{j+\lambda}}{\frac{\gamma\lambda+1}{\Gamma(\lambda+2)} z^{\lambda+1} + \sum_{j=2}^{\infty} \frac{[1+(j-1)\alpha]^m j \Gamma(n+j)}{\Gamma(n+1)\Gamma(j+\lambda+1)} [\gamma j + \gamma(\lambda-1)+1] a_j^2 z^{j+\lambda}} > 0.$$

Since $Re(-e^{i\theta}) \geq -|e^{i\theta}| = -1$, the above inequality reduces to

$$\frac{\frac{(\gamma\lambda+1)(\beta|d|-\lambda)}{\Gamma(\lambda+2)} r^{\lambda+1} - \sum_{j=2}^{\infty} \frac{[1+(j-1)\alpha]^m j \Gamma(n+j)}{\Gamma(n+1)\Gamma(j+\lambda+1)} \left\{ \gamma j^2 + [\gamma(2\lambda-2+\beta|d|)+1]j + [\gamma(\lambda-1)+1](\lambda-1+\beta|d|) \right\} a_j^2 r^{j+\lambda j}}{\frac{\gamma\lambda+1}{\Gamma(\lambda+2)} r^{\lambda+1} - \sum_{j=2}^{\infty} \frac{[1+(j-1)\alpha]^m j \Gamma(n+j)}{\Gamma(n+1)\Gamma(j+\lambda+1)} [\gamma j + \gamma(\lambda-1)+1] a_j^2 r^{j+\lambda}} > 0.$$

Letting $r \to 1^-$ and by the mean value theorem we have desired inequality (4). This completes the proof of Theorem 1

Corollary 1 *Let the function* $f \in \mathscr{A}$ *be in the class* $\mathscr{DR}_{m,n}(\lambda, d, \alpha, \beta, \gamma)$. *Then*

$$a_j \leq$$

$$\sqrt{\frac{(\gamma\lambda+1)\,(\beta\,|d|-\lambda)\,\frac{\Gamma(n+1)}{\Gamma(\lambda+2)}}{\frac{[1+(j-1)\alpha]^m\,j\,\Gamma(n+j)}{\Gamma(j+\lambda+1)}\left\{\gamma\,j^2+[\gamma\,(2\lambda-2+\beta\,|d|)+1]\,j+[\gamma\,(\lambda-1)+1]\,(\lambda-1+\beta\,|d|)\right\}}},$$

$$j \geq 2.$$

3 Distortion Theorems

Theorem 2 *Let the function* $f \in \mathscr{A}$ *be in the class* $\mathscr{DR}_{m,n}(\lambda, d, \alpha, \beta, \gamma)$. *Then for* $|z| = r < 1$, *we have*

$$r - \sqrt{\frac{(\gamma\lambda+1)\,(\lambda+2)\,(\beta\,|d|-\lambda)}{2\,(1+\alpha)^m\,(n+1)\,\{(\lambda+1)\,[\gamma\,(\lambda+1+\beta\,|d|)+1]+\beta\,|d|\}}}\,r^2 \leq |f(z)|$$

$$\leq r + \sqrt{\frac{(\gamma\lambda+1)\,(\lambda+2)\,(\beta\,|d|-\lambda)}{2\,(1+\alpha)^m\,(n+1)\,\{(\lambda+1)\,[\gamma\,(\lambda+1+\beta\,|d|)+1]+\beta\,|d|\}}}\,r^2.$$

The result is sharp for the function f *given by*

$$f(z) = z + \sqrt{\frac{(\gamma\lambda+1)\,(\lambda+2)\,(\beta\,|d|-\lambda)}{2\,(1+\alpha)^m\,(n+1)\,\{(\lambda+1)\,[\gamma\,(\lambda+1+\beta\,|d|)+1]+\beta\,|d|\}}}\,z^2, \quad z \in U.$$

Proof Given that $f \in \mathscr{DR}_{m,n}(\lambda, d, \alpha, \beta, \gamma)$, from the Eq. (4) and since

$$2\,(1+\alpha)^m\,(n+1)\,\{(\lambda+1)\,[\gamma\,(\lambda+1+\beta\,|d|)+1]+\beta\,|d|\}$$

is nondecreasing and positive for $j \geq 2$, then we have

$$\sqrt{2\,(1+\alpha)^m\,(n+1)\,\{(\lambda+1)\,[\gamma\,(\lambda+1+\beta\,|d|)+1]+\beta\,|d|\}}\sum_{j=2}^{\infty} a_j \leq$$

$$\sum_{j=2}^{\infty}\sqrt{\frac{[1+(j-1)\alpha]^m\,j\,\Gamma(n+j)}{\Gamma(j+\lambda+1)}\left\{\gamma\,j^2+[\gamma\,(2\lambda-2+\beta\,|d|)+1]\,j+[\gamma\,(\lambda-1)+1]\,(\lambda-1+\beta\,|d|)\right\}a_j}$$

$$\leq \sqrt{(\gamma\lambda+1)\,(\beta\,|d|-\lambda)\,\frac{\Gamma(n+1)}{\Gamma(\lambda+2)}},$$

which is equivalent to

$$\sum_{j=2}^{\infty} a_j \le \sqrt{\frac{(\gamma\lambda+1)\,(\lambda+2)\,(\beta\,|d|-\lambda)}{2\,(1+\alpha)^m\,(n+1)\,\{(\lambda+1)\,[\gamma\,(\lambda+1+\beta\,|d|)+1]+\beta\,|d|\}}}. \tag{5}$$

Using (5), we obtain

$$f(z) = z + \sum_{j=2}^{\infty} a_j z^j$$

$$|f(z)| \le |z| + \sum_{j=2}^{\infty} a_j |z|^j \le r + \sum_{j=2}^{\infty} a_j r^j \le r + r^2 \sum_{j=2}^{\infty} a_j$$

$$\le r + \sqrt{\frac{(\gamma\lambda+1)\,(\lambda+2)\,(\beta\,|d|-\lambda)}{2\,(1+\alpha)^m\,(n+1)\,\{(\lambda+1)\,[\gamma\,(\lambda+1+\beta\,|d|)+1]+\beta\,|d|\}}}\,r^2.$$

Similarly,

$$|f(z)| \ge r - \sqrt{\frac{(\gamma\lambda+1)\,(\lambda+2)\,(\beta\,|d|-\lambda)}{2\,(1+\alpha)^m\,(n+1)\,\{(\lambda+1)\,[\gamma\,(\lambda+1+\beta\,|d|)+1]+\beta\,|d|\}}}\,r^2.$$

This completes the proof of Theorem 2.

Theorem 3 *Let the function* $f \in \mathscr{A}$ *be in the class* $\mathscr{DR}_{m,n}(\lambda, d, \alpha, \beta, \gamma)$. *Then for* $|z| = r < 1$, *we have*

$$-\sqrt{\frac{2\,(\gamma\lambda+1)\,(\lambda+2)\,(\beta\,|d|-\lambda)}{(1+\alpha)^m\,(n+1)\,\{(\lambda+1)\,[\gamma\,(\lambda+1+\beta\,|d|)+1]+\beta\,|d|\}}}\,r \le |f'(z)|$$

$$\le \sqrt{\frac{2\,(\gamma\lambda+1)\,(\lambda+2)\,(\beta\,|d|-\lambda)}{(1+\alpha)^m\,(n+1)\,\{(\lambda+1)\,[\gamma\,(\lambda+1+\beta\,|d|)+1]+\beta\,|d|\}}}\,r.$$

The result is sharp for the function f *given by*

$$f(z) = z + \sqrt{\frac{(\gamma\lambda+1)\,(\lambda+2)\,(\beta\,|d|-\lambda)}{2\,(1+\alpha)^m\,(n+1)\,\{(\lambda+1)\,[\gamma\,(\lambda+1+\beta\,|d|)+1]+\beta\,|d|\}}}\,z^2, \quad z \in U.$$

Proof From (5)

$$f'(z) = 1 + \sum_{j=2}^{\infty} j a_j z^{j-1}$$

$$|f'(z)| \le 1 - \sum_{j=2}^{\infty} j a_j |z|^{j-1} \le 1 + \sum_{j=2}^{\infty} j a_j r^{j-1} \le$$

$$1 + \sqrt{\frac{2(\gamma\lambda + 1)(\lambda + 2)(\beta |d| - \lambda)}{(1+\alpha)^m (n+1)\{(\lambda + 1)[\gamma(\lambda + 1 + \beta |d|) + 1] + \beta |d|\}}} r.$$

Similarly,

$$|f'(z)| \ge 1 - \sqrt{\frac{2(\gamma\lambda + 1)(\lambda + 2)(\beta |d| - \lambda)}{(1+\alpha)^m (n+1)\{(\lambda + 1)[\gamma(\lambda + 1 + \beta |d|) + 1] + \beta |d|\}}} r.$$

This completes the proof of Theorem 3.

4 Closure Theorems

Theorem 4 *Let the functions f_k, $k = 1, 2, \ldots, l$, defined by*

$$f_k(z) = z + \sum_{j=2}^{\infty} a_{j,k} z^j, \quad a_{j,k} \ge 0, \ z \in U, \tag{6}$$

be in the class $\mathscr{DR}_{m,n}(\lambda, d, \alpha, \beta, \gamma)$. Then the function h defined by

$$h(z) = \sum_{k=1}^{l} \mu_k f_k(z), \quad \mu_k \ge 0, \ z \in U,$$

is also in the class $\mathscr{DR}_{m,n}(\lambda, d, \alpha, \beta, \gamma)$, where

$$\sum_{k=1}^{l} \mu_k = 1.$$

Proof We can write

$$h(z) = \sum_{k=1}^{l} \mu_k z + \sum_{k=1}^{l} \sum_{j=2}^{\infty} \mu_k a_{j,k} z^j = z + \sum_{j=2}^{\infty} \sum_{k=1}^{l} \mu_k a_{j,k} z^j.$$

Furthermore, since the functions f_k, $k = 1, 2, \ldots, l$, are in the class $\mathscr{DR}_{m,n}(\lambda, d, \alpha, \beta, \gamma)$, then from Corollary 1 we have

$$\sum_{j=2}^{\infty} \sqrt{\frac{[1+(j-1)\alpha]^m \, j\Gamma\,(n+j)}{\Gamma\,(j+\lambda+1)}} \left\{\gamma j^2 + [\gamma\,(2\lambda-2+\beta\,|d|)+1]\,j + [\gamma\,(\lambda-1)+1]\,(\lambda-1+\beta\,|d|)\right\}a_j$$

$$\leq \sqrt{(\gamma\lambda+1)\,(\beta\,|d|-\lambda)\,\frac{\Gamma\,(n+1)}{\Gamma\,(\lambda+2)}}$$

Thus it is enough to prove that

$$\sum_{j=2}^{\infty} \sqrt{\frac{[1+(j-1)\alpha]^m \, j\Gamma\,(n+j)}{\Gamma\,(j+\lambda+1)}} \left\{\gamma j^2 + [\gamma\,(2\lambda-2+\beta\,|d|)+1]\,j + [\gamma\,(\lambda-1)+1]\,(\lambda-1+\beta\,|d|)\right\}\left(\sum_{k=1}^{m}\mu_k a_{j,k}\right) =$$

$$\sum_{k=1}^{m}\mu_k \sum_{j=2}^{\infty} \sqrt{\frac{[1+(j-1)\alpha]^m \, j\Gamma\,(n+j)}{\Gamma\,(j+\lambda+1)}} \left\{\gamma j^2 + [\gamma\,(2\lambda-2+\beta\,|d|)+1]\,j + [\gamma\,(\lambda-1)+1]\,(\lambda-1+\beta\,|d|)\right\}a_{j,k}$$

$$\leq \sum_{k=1}^{m}\mu_k\sqrt{(\gamma\lambda+1)\,(\beta\,|d|-\lambda)\,\frac{\Gamma\,(n+1)}{\Gamma\,(\lambda+2)}} = \sqrt{(\gamma\lambda+1)\,(\beta\,|d|-\lambda)\,\frac{\Gamma\,(n+1)}{\Gamma\,(\lambda+2)}}.$$

Hence the proof is complete.

Corollary 2 *Let the functions* f_k, $k = 1, 2$, *defined by (6) be in the class* $\mathscr{DR}_{m,n}(\lambda, d, \alpha, \beta, \gamma)$. *Then the function h defined by*

$$h(z) = (1 - \zeta)f_1(z) + \zeta f_2(z), \quad 0 \leq \zeta \leq 1, \, z \in U,$$

is also in the class $\mathscr{DR}_{m,n}(\lambda, d, \alpha, \beta, \gamma)$.

Theorem 5 *Let*

$$f_1(z) = z,$$

and

$$f_j(z)=z+ \sqrt{\frac{(\gamma\lambda+1)\,(\beta\,|d|-\lambda)\,\frac{\Gamma(n+1)}{\Gamma(\lambda+2)}}{\frac{[1+(j-1)\alpha]^m j\Gamma(n+j)}{\Gamma(j+\lambda+1)}\left\{\gamma j^2 + [\gamma\,(2\lambda-2+\beta\,|d|)+1]\,j + [\gamma\,(\lambda-1)+1]\,(\lambda-1+\beta\,|d|)\right\}}}\,z^j,$$

$j \geq 2$, $z \in U$.

    Then the function f is in the class $\mathscr{DR}_{m,n}(\lambda, d, \alpha, \beta, \gamma)$ *if and only if it can be expressed in the form*

$$f(z) = \mu_1 f_1(z) + \sum_{j=2}^{\infty}\mu_j f_j(z), \quad z \in U,$$

where $\mu_1 \geq 0$, $\mu_j \geq 0$, $j \geq 2$ *and* $\mu_1 + \sum_{j=2}^{\infty} \mu_j = 1$.

Proof Assume that f can be expressed in the form

$$f(z) = \mu_1 f_1(z) + \sum_{j=2}^{\infty} \mu_j f_j(z) =$$

$$z + \sum_{j=2}^{\infty} \sqrt{\frac{(\gamma\lambda+1)(\beta|d|-\lambda)\frac{\Gamma(n+1)}{\Gamma(\lambda+2)}}{\frac{[1+(j-1)\alpha]^m j\Gamma(n+j)}{\Gamma(j+\lambda+1)}\{\gamma j^2+[\gamma(2\lambda-2+\beta|d|)+1]j+[\gamma(\lambda-1)+1](\lambda-1+\beta|d|)\}}}\,\mu_j z^j.$$

Thus

$$\sum_{j=2}^{\infty} \sqrt{\frac{\frac{[1+(j-1)\alpha]^m j\Gamma(n+j)}{\Gamma(j+\lambda+1)}\{\gamma j^2+[\gamma(2\lambda-2+\beta|d|)+1]j+[\gamma(\lambda-1)+1](\lambda-1+\beta|d|)\}}{(\gamma\lambda+1)(\beta|d|-\lambda)\frac{\Gamma(n+1)}{\Gamma(\lambda+2)}}}.$$

$$\sqrt{\frac{(\gamma\lambda+1)(\beta|d|-\lambda)\frac{\Gamma(n+1)}{\Gamma(\lambda+2)}}{\frac{[1+(j-1)\alpha]^m j\Gamma(n+j)}{\Gamma(j+\lambda+1)}\{\gamma j^2+[\gamma(2\lambda-2+\beta|d|)+1]j+[\gamma(\lambda-1)+1](\lambda-1+\beta|d|)\}}}\,\mu_j =$$

$$\sum_{j=2}^{\infty} \mu_j = 1-\mu_1 \leq 1.\text{Hence } f \in \mathcal{DR}_{m,n}(\lambda, d, \alpha, \beta, \gamma).$$

Conversely, assume that $f \in \mathcal{DR}_{m,n}(\lambda, d, \alpha, \beta, \gamma)$.
Setting

$$\mu_j = \sqrt{\frac{\frac{[1+(j-1)\alpha]^m j\Gamma(n+j)}{\Gamma(j+\lambda+1)}\{\gamma j^2+[\gamma(2\lambda-2+\beta|d|)+1]j+[\gamma(\lambda-1)+1](\lambda-1+\beta|d|)\}}{(\gamma\lambda+1)(\beta|d|-\lambda)\frac{\Gamma(n+1)}{\Gamma(\lambda+2)}}}\,a_j,$$

since

$$\mu_1 = 1 - \sum_{j=2}^{\infty} \mu_j.$$

Thus

$$f(z) = \mu_1 f_1(z) + \sum_{j=2}^{\infty} \mu_j f_j(z).$$

Hence the proof is complete.

Corollary 3 *The extreme points of the class* $\mathcal{DR}_{m,n}(\lambda, d, \alpha, \beta, \gamma)$ *are the functions*

$$f_1(z) = z,$$

and

$$f_j(z) = z +$$

$$\sqrt{\dfrac{(\gamma\lambda + 1)(\beta\,|d| - \lambda)\dfrac{\Gamma(n+1)}{\Gamma(\lambda+2)}}{\dfrac{[1+(j-1)\alpha]^m j\Gamma(n+j)}{\Gamma(j+\lambda+1)}\left\{\gamma j^2 + [\gamma(2\lambda - 2 + \beta\,|d|) + 1]\,j + [\gamma(\lambda - 1) + 1](\lambda - 1 + \beta\,|d|)\right\}}}\,z^j,$$

$j \geq 2,\ z \in U.$

5 Inclusion and Neighborhood Results

We define the δ- neighborhood of a function $f \in \mathscr{A}$ by

$$N_\delta(f) = \{g \in \mathscr{A} : g(z) = z + \sum_{j=2}^{\infty} b_j z^j \text{ and } \sum_{j=2}^{\infty} j|a_j - b_j| \leq \delta\}. \tag{7}$$

In particular, for $e(z) = z$

$$N_\delta(e) = \{g \in \mathscr{A} : g(z) = z + \sum_{j=2}^{\infty} b_j z^j \text{ and } \sum_{j=2}^{\infty} j|b_j| \leq \delta\}. \tag{8}$$

Furthermore, a function $f \in \mathscr{A}$ is said to be in the class $\mathscr{DR}_{m,n}^{\xi}(\lambda, d, \alpha, \beta, \gamma)$ if there exists a function $h \in \mathscr{DR}_{m,n}(\lambda, d, \alpha, \beta, \gamma)$ such that

$$\left|\frac{f(z)}{h(z)} - 1\right| < 1 - \xi, \quad z \in U, \quad 0 \leq \xi < 1. \tag{9}$$

Theorem 6 *If*

$$\delta = \sqrt{\frac{2(\gamma\lambda + 1)(\lambda + 2)(\beta\,|d| - \lambda)}{(1+\alpha)^m(n+1)\{(\lambda + 1)[\gamma(\lambda + 1 + \beta\,|d|) + 1] + \beta\,|d|\}}},$$

then

$$\mathscr{DR}_{m,n}(\lambda, d, \alpha, \beta, \gamma) \subset N_\delta(e).$$

Proof Let $f \in \mathscr{DR}_{m,n}(\lambda, d, \alpha, \beta, \gamma)$. Then in view of assertion of Corollary 1 and since

$$\frac{[1 + (j - 1)\alpha]^m \, j\Gamma(n + j)}{\Gamma(j + \lambda + 1)} \cdot$$

$$\left\{ \gamma j^2 + [\gamma(2\lambda - 2 + \beta |d|) + 1]j + [\gamma(\lambda - 1) + 1](\lambda - 1 + \beta |d|) \right\}$$

$$\geq \frac{(1 + \alpha)^m \, \Gamma(n + 2)}{4\Gamma(\lambda + 3)} \left\{ (\lambda + 1)[\gamma(\lambda + 1 + \beta |d|) + 1] + \beta |d| \right\}$$

for $j \geq 2$, we get

$$\sqrt{\frac{(1 + \alpha)^m \, \Gamma(n + 2)}{4\Gamma(\lambda + 3)} \left\{ (\lambda + 1)[\gamma(\lambda + 1 + \beta |d|) + 1] + \beta |d| \right\}} \sum_{j=2}^{\infty} a_j \leq$$

$$\sum_{j=2}^{\infty} \sqrt{\frac{[1 + (j - 1)\alpha]^m \, j\Gamma(n + j)}{\Gamma(j + \lambda + 1)} \left\{ \gamma j^2 + [\gamma(2\lambda - 2 + \beta |d|) + 1]j + [\gamma(\lambda - 1) + 1](\lambda - 1 + \beta |d|) \right\} a_j}$$

$$\leq \sqrt{(\gamma\lambda + 1)(\beta |d| - \lambda) \frac{\Gamma(n + 1)}{\Gamma(\lambda + 2)}},$$

which implies

$$\sum_{j=2}^{\infty} a_j \leq \sqrt{\frac{(\gamma\lambda + 1)(\lambda + 2)(\beta |d| - \lambda)}{2(1 + \alpha)^m (n + 1)\{(\lambda + 1)[\gamma(\lambda + 1 + \beta |d|) + 1] + \beta |d|\}}}. \tag{10}$$

Applying assertion of Corollary 1 in conjunction with (10), we obtain

$$\sum_{j=2}^{\infty} j a_j \leq \sqrt{\frac{2(\gamma\lambda + 1)(\lambda + 2)(\beta |d| - \lambda)}{(1 + \alpha)^m (n + 1)\{(\lambda + 1)[\gamma(\lambda + 1 + \beta |d|) + 1] + \beta |d|\}}} = \delta,$$

by virtue of (7), we have $f \in N_\delta(e)$.

This completes the proof of the Theorem 6.

Theorem 7 *If $h \in \mathscr{DR}_{m,n}(\lambda, d, \alpha, \beta, \gamma)$ and*

$$\xi = 1 + \frac{\delta}{2} \sqrt{\frac{(\gamma\lambda + 1)(\lambda + 2)(\beta |d| - \lambda)}{2(1 + \alpha)^m (n + 1)\{(\lambda + 1)[\gamma(\lambda + 1 + \beta |d|) + 1] + \beta |d|\}}}, \tag{11}$$

then

$$N_\delta(h) \subset \mathscr{DR}_{m,n}^\xi(\lambda, d, \alpha, \beta, \gamma).$$

Proof Suppose that $f \in N_\delta(h)$, we then find from (7) that

$$\sum_{j=2}^{\infty} j|a_j - b_j| \leq \delta,$$

which readily implies the following coefficient inequality

$$\sum_{j=2}^{\infty} |a_j - b_j| \leq \frac{\delta}{2}. \tag{12}$$

Next, since $h \in \mathscr{DR}_{m,n}(\lambda, d, \alpha, \beta, \gamma)$ in the view of (10), we have

$$\sum_{j=2}^{\infty} b_j \leq \sqrt{\frac{(\gamma\lambda + 1)(\lambda + 2)(\beta|d| - \lambda)}{2(1+\alpha)^m (n+1)\{(\lambda+1)[\gamma(\lambda+1+\beta|d|)+1] + \beta|d|\}}}. \tag{13}$$

Using (12) and (13), we get

$$\left|\frac{f(z)}{h(z)} - 1\right| \leq \frac{\sum_{j=2}^{\infty} |a_j - b_j|}{1 - \sum_{j=2}^{\infty} b_j} \leq$$

$$\frac{\delta}{2\left(1 - \sqrt{\frac{(\gamma\lambda+1)(\lambda+2)(\beta|d|-\lambda)}{2(1+\alpha)^m(n+1)\{(\lambda+1)[\gamma(\lambda+1+\beta|d|)+1]+\beta|d|\}}}\right)} = 1 - \xi,$$

provided that ξ is given by (11), thus by condition (9), $f \in \mathscr{DR}_{m,n}^{\xi}(\lambda, d, \alpha, \beta, \gamma)$, where ξ is given by (11).

6 Radii of Starlikeness, Convexity, and Close-to-Convexity

Theorem 8 *Let the function $f \in \mathscr{A}$ be in the class $\mathscr{DR}_{m,n}(\lambda, d, \alpha, \beta, \gamma)$. Then f is univalent starlike of order δ, $0 \leq \delta < 1$, in $|z| < r_1$, where*

$$r_1 = \inf_j$$

$$\left\{ \frac{(1-\delta)^2 \frac{[1+(j-1)\alpha]^m j \Gamma(n+j)}{\Gamma(j+\lambda+1)} \left\{ \gamma j^2 + [\gamma(2\lambda-2+\beta|d|)+1]j + [\gamma(\lambda-1)+1](\lambda-1+\beta|d|) \right\}}{(\gamma\lambda+1)(\beta|d|-\lambda)\frac{\Gamma(n+1)}{\Gamma(\lambda+2)}(j-\delta)^2} \right\}^{\frac{1}{2(j-1)}}$$

The result is sharp for the function f given by

$$f_j(z) = z +$$

$$\sqrt{\frac{(\gamma\lambda+1)(\beta|d|-\lambda)\frac{\Gamma(n+1)}{\Gamma(\lambda+2)}}{\frac{[1+(j-1)\alpha]^m j\Gamma(n+j)}{\Gamma(j+\lambda+1)}\left\{\gamma j^2 + [\gamma(2\lambda-2+\beta|d|)+1]j + [\gamma(\lambda-1)+1](\lambda-1+\beta|d|)\right\}}} z^j, \quad j \geq 2.$$

Proof It suffices to show that

$$\left|\frac{zf'(z)}{f(z)} - 1\right| \leq 1 - \delta, \quad |z| < r_1.$$

Since

$$\left|\frac{zf'(z)}{f(z)} - 1\right| = \left|\frac{\sum_{j=2}^{\infty}(j-1)a_j z^{j-1}}{1+\sum_{j=2}^{\infty}a_j z^{k-1}}\right| \leq \frac{\sum_{j=2}^{\infty}(j-1)a_j|z|^{j-1}}{1-\sum_{j=2}^{\infty}a_j|z|^{j-1}}.$$

To prove the theorem, we must show that

$$\frac{\sum_{j=2}^{\infty}(j-1)a_j|z|^{j-1}}{1-\sum_{j=2}^{\infty}a_j|z|^{j-1}} \leq 1 - \delta.$$

It is equivalent to

$$\sum_{j=2}^{\infty}(j-\delta)a_j|z|^{j-1} \leq 1 - \delta,$$

using Theorem 1, we obtain

$$|z| \leq \left\{\frac{(1-\delta)^2\frac{[1+(j-1)\alpha]^m j\Gamma(n+j)}{\Gamma(j+\lambda+1)}\left\{\gamma j^2 + [\gamma(2\lambda-2+\beta|d|)+1]j + [\gamma(\lambda-1)+1](\lambda-1+\beta|d|)\right\}}{(\gamma\lambda+1)(\beta|d|-\lambda)\frac{\Gamma(n+1)}{\Gamma(\lambda+2)}(j-\delta)^2}\right\}^{\frac{1}{2(j-1)}}.$$

Hence the proof is complete.

Theorem 9 *Let the function $f \in \mathscr{A}$ be in the class $\mathscr{DR}_{m,n}(\lambda, d, \alpha, \beta, \gamma)$. Then f is univalent convex of order δ, $0 \leq \delta \leq 1$, in $|z| < r_2$, where*

$$r_2 = \inf_j$$

$$\left\{\frac{(1-\delta)^2\frac{[1+(j-1)\alpha]^m j\Gamma(n+j)}{\Gamma(j+\lambda+1)}\left\{\gamma j^2 + [\gamma(2\lambda-2+\beta|d|)+1]j + [\gamma(\lambda-1)+1](\lambda-1+\beta|d|)\right\}}{(\gamma\lambda+1)(\beta|d|-\lambda)\frac{\Gamma(n+1)}{\Gamma(\lambda+2)}(j-\delta)^2}\right\}^{\frac{1}{2(j-1)}}.$$

The result is sharp for the function f given by

$$f_j(z) = z + \tag{14}$$

$$\sqrt{\frac{(\gamma\lambda + 1)(\beta|d| - \lambda)\frac{\Gamma(n+1)}{\Gamma(\lambda+2)}}{\frac{[1+(j-1)\alpha]^m j\Gamma(n+j)}{\Gamma(j+\lambda+1)}\left\{\gamma j^2 + [\gamma(2\lambda - 2 + \beta|d|) + 1]j + [\gamma(\lambda - 1) + 1](\lambda - 1 + \beta|d|)\right\}}}\, z^j,$$

$j \geq 2$.

Proof It suffices to show that

$$\left|\frac{zf''(z)}{f'(z))}\right| \leq 1 - \delta, \quad |z| < r_2.$$

Since

$$\left|\frac{zf''(z)}{f'(z)}\right| = \left|\frac{\sum_{j=2}^{\infty} j(j-1)a_j z^{j-1}}{1 + \sum_{j=2}^{\infty} ja_j z^{j-1}}\right| \leq \frac{\sum_{j=2}^{\infty} j(j-1)a_j|z|^{j-1}}{1 - \sum_{j=2}^{\infty} ja_j|z|^{j-1}}.$$

To prove the theorem, we must show that

$$\frac{\sum_{j=2}^{\infty} j(j-1)a_j|z|^{j-1}}{1 - \sum_{j=2}^{\infty} ja_j|z|^{j-1}} \leq 1 - \delta,$$

$$\sum_{j=2}^{\infty} j(j - \delta)a_j|z|^{j-1} \leq 1 - \delta,$$

using Theorem 1, we obtain

$$|z|^{j-1} \leq \frac{(1-\delta)}{j(j-\delta)}\sqrt{\frac{\frac{[1+(j-1)\alpha]^m j\Gamma(n+j)}{\Gamma(j+\lambda+1)}\left\{\gamma j^2 + [\gamma(2\lambda - 2 + \beta|d|) + 1]j + [\gamma(\lambda - 1) + 1](\lambda - 1 + \beta|d|)\right\}}{(\gamma\lambda + 1)(\beta|d| - \lambda)\frac{\Gamma(n+1)}{\Gamma(\lambda+2)}}},$$

or

$$|z| \leq \left\{\frac{(1-\delta)^2\frac{[1+(j-1)\alpha]^m j\Gamma(n+j)}{\Gamma(j+\lambda+1)}\left\{\gamma j^2 + [\gamma(2\lambda - 2 + \beta|d|) + 1]j + [\gamma(\lambda - 1) + 1](\lambda - 1 + \beta|d|)\right\}}{(\gamma\lambda + 1)(\beta|d| - \lambda)\frac{\Gamma(n+1)}{\Gamma(\lambda+2)}(j-\delta)^2}\right\}^{\frac{1}{2(j-1)}}.$$

Hence the proof is complete.

Theorem 10 *Let the function* $f \in \mathscr{A}$ *be in the class* $\mathscr{DR}_{m,n}(\lambda, d, \alpha, \beta, \gamma)$. *Then* f *is univalent close-to-convex of order* δ, $0 \leq \delta < 1$, *in* $|z| < r_3$, *where*

$$r_3 = \inf_j \left\{ \frac{(1-\delta)^2 \frac{[1+(j-1)\alpha]^m \Gamma(n+j)}{j^2 \Gamma(j+\lambda+1)} \left\{ \gamma j^2 + [\gamma (2\lambda - 2 + \beta |d|) + 1] j + [\gamma (\lambda - 1) + 1](\lambda - 1 + \beta |d|) \right\}}{(\gamma \lambda + 1)(\beta |d| - \lambda) \frac{\Gamma(n+1)}{\Gamma(\lambda+2)}} \right\}^{\frac{1}{2(j-1)}}.$$

The result is sharp for the function f *given by (14).*

Proof It suffices to show that

$$|f'(z) - 1| \leq 1 - \delta, \quad |z| < r_3.$$

Then

$$|f'(z) - 1| = \left| \sum_{j=2}^{\infty} j a_j z^{j-1} \right| \leq \sum_{j=2}^{\infty} j a_j |z|^{j-1}.$$

Thus $|f'(z) - 1| \leq 1 - \delta$ if $\sum_{j=2}^{\infty} \frac{j a_j}{1-\delta} |z|^{j-1} \leq 1$. Using Theorem 1, the above inequality holds true if

$$|z|^{j-1} \leq \frac{(1-\delta)}{j} \cdot$$

$$\sqrt{\frac{\frac{[1+(j-1)\alpha]^m j \Gamma(n+j)}{\Gamma(j+\lambda+1)} \left\{ \gamma j^2 + [\gamma (2\lambda - 2 + \beta |d|) + 1] j + [\gamma (\lambda - 1) + 1](\lambda - 1 + \beta |d|) \right\}}{(\gamma \lambda + 1)(\beta |d| - \lambda) \frac{\Gamma(n+1)}{\Gamma(\lambda+2)}}}$$

or

$$|z| \leq \left\{ \frac{(1-\delta)^2 \frac{[1+(j-1)\alpha]^m \Gamma(n+j)}{j^2 \Gamma(j+\lambda+1)} \left\{ \gamma j^2 + [\gamma (2\lambda - 2 + \beta |d|) + 1] j + [\gamma (\lambda - 1) + 1](\lambda - 1 + \beta |d|) \right\}}{(\gamma \lambda + 1)(\beta |d| - \lambda) \frac{\Gamma(n+1)}{\Gamma(\lambda+2)}} \right\}^{\frac{1}{2(j-1)}}.$$

Hence the proof is complete.

References

1. F.M. Al-Oboudi, On univalent functions defined by a generalized Sălăgean operator. Int. J. Math. Math. Sci. **27**, 1429–1436 (2004)
2. N.E. Cho, A.M.K. Aouf, Some applications of fractional calculus operators to a certain subclass of analytic functions with negative coefficients. Turk. J. Math. **20**, 553–562 (1996)
3. A.A. Lupas, Properties on a subclass of analytic functions defined by a fractional integral operator. J. Comput. Anal. Appl. **27**(3), 506–510 (2019)
4. S. Ruscheweyh, New criteria for univalent functions. Proc. Am. Math. Soc. **49**, 109–115 (1975)
5. G.S. Sălăgean, *Subclasses of Univalent Functions*. Lecture Notes in Mathematics, vol. 1013 (Springer, Berlin, 1983), pp. 362–372

Index

© Springer Nature Switzerland AG 2019
G. A. Anastassiou and J. M. Rassias (eds.), *Frontiers in Functional Equations and Analytic Inequalities*, https://doi.org/10.1007/978-3-030-28950-8

Printed in the United States
By Bookmasters